D1244064

# Algebraic and Geometric Topology

PROCEEDINGS OF SYMPOSIA IN PURE MATHEMATICS

VOLUME XXXII — Part 1

# ALGEBRAIC AND GEOMETRIC TOPOLOGY

PROCEEDINGS OF SYMPOSIA
IN PURE MATHEMATICS
Volume XXXII, Part 1

# ALGEBRAIC AND GEOMETRIC TOPOLOGY

AMERICAN MATHEMATICAL SOCIETY
PROVIDENCE, RHODE ISLAND
1978

PROCEEDINGS OF THE SYMPOSIUM IN PURE MATHEMATICS
OF THE AMERICAN MATHEMATICAL SOCIETY

HELD AT STANFORD UNIVERSITY

STANFORD, CALIFORNIA
AUGUST 2–21, 1976

EDITED BY

R. JAMES MILGRAM

Prepared by the American Mathematical Society
with partial support from National Science Foundation grant MCS 76-01696

**Library of Congress Cataloging in Publication Data**

Symposium in Pure Mathematics, Stanford University, 1976.
Algebraic and geometric topology.
(Proceedings of symposia in pure mathematics; v. 32)
"Proceedings of the Symposium in Pure Mathematics of the American Mathematical Society, held at Stanford University, Stanford, California, August 2–21, 1976."
Bibliography: v. 1, p. v. 2, p.
1. Algebraic topology–Congresses. 2. Manifolds (Mathematics)–Congresses. 3. Global analysis (Mathematics)–Congresses. I. Milgram, R. James. II. American Mathematical Society. III. Title. IV. Series.
QA612.S93 1976 514′.2 78–14304
ISBN 0-8218-1432-X (v. 1)
ISBN 0-8218-1433-8 (v. 2)

AMS (MOS) subject classifications (1970). Primary 53C15, 53C30, 53C35, 53C40, 55-XX, 57Axx, 57Bxx, 57Cxx, 57Dxx, 58Bxx, 58C25, 58Dxx, 58Gxx.

# TABLE OF CONTENTS

## Algebraic $K$- and $L$-Theory



## Surgery and its Applications



## Group Actions

TABLE OF CONTENTS

## PREFACE

The American Mathematical Society held its 24th Summer Research Institute at Stanford University from August 2–21, 1976. The topic of the meeting was Algebraic and Geometric Topology. Particular emphasis was placed on Algebraic *K*- and *L*-Theory, Surgery and Surgery Classifying Spaces, Group Actions on Manifolds, and 3 and 4 Manifold Theory.

The organizing committee consisted of Raoul Bott, William Browder (Co-chairman), Pierre Conner, Robion Kirby, Richard Lashof, R. James Milgram (Chairman), Daniel Quillen, and P. Emery Thomas (Co-chairman).

The main lecturers were Raoul Bott, Richard Lashof and Mel Rothenberg, Sylvain Cappell and Julius Shaneson, C. McA. Gordon, Robert Edwards, Allen Hatcher and Jack Wagoner, Wu. C. Hsiang, Max Karoubi, R. James Milgram, Daniel Quillen, Laurence Siebenmann, and C. T. C. Wall.

Special hour lectures were given by Gregory Brumfiel, William Jaco, Robion Kirby, Ronnie Lee, James Lin, Ib Madsen, John Morgan, Robert Oliver, Ted Petrie, P. Emery Thomas, F. Waldhausen, and James West.

There was also a series of problem sessions run by Raoul Bott (foliations), W. C. Hsiang (group actions and surgery), R. Kirby (3 and 4 manifolds), and C. T. C. Wall (algebraic *K*- and *L*-theory).

Among the more active seminars were 3 and 4 manifolds (Kirby), Kervaire invariant (Browder—E. Brown), Homotopy theory (E. Thomas—M. Mahowald), Algebraic *K*- and *L*- theory (J. Wagoner), Group actions (W. C. Hsiang), Foliations (Bott), Geometry of manifolds (L. Siebenmann), and Surgery and surgery classifying spaces (R. Lashof).

These proceedings include write ups of most of the main lectures, as well as selected seminar talks, and most of the problem sessions.

The institute was sponsored by the National Science Foundation under contract number MCS 76–01696.

The organizing committee wishes to thank Dorothy Smith of the American Mathematical Society for her help in running and organizing the conference.

R. JAMES MILGRAM
STANFORD UNIVERSITY

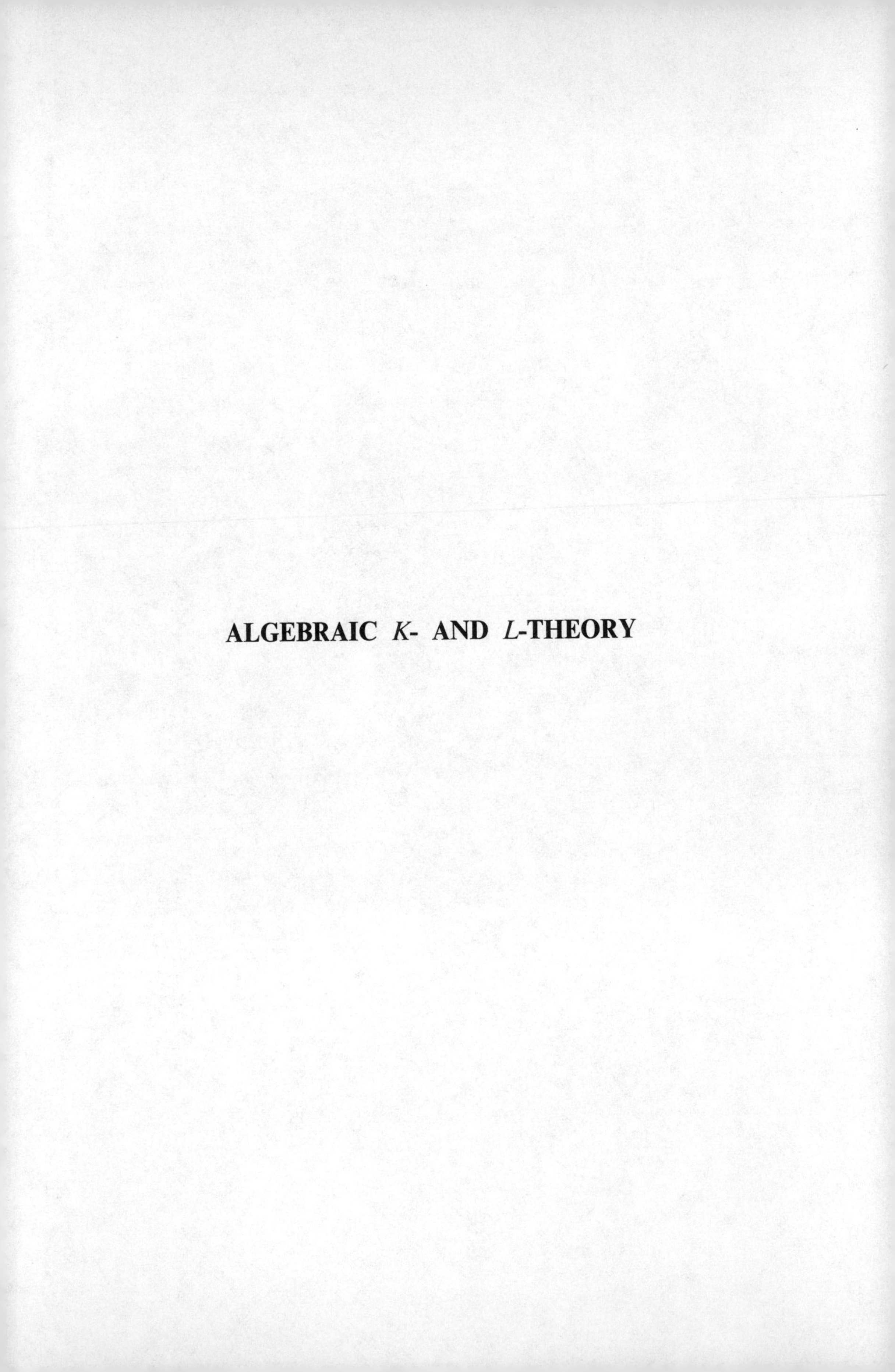

# ALGEBRAIC *K*- AND *L*-THEORY

Proceedings of Symposia in Pure Mathematics
Volume 32, 1978

# CONCORDANCE SPACES, HIGHER SIMPLE-HOMOTOPY THEORY, AND APPLICATIONS

A. E. HATCHER

While much is now known, through surgery theory, about the classification problem for manifolds of dimension at least five, information about the automorphism groups of such manifolds is as yet rather sparse. In fact, it seems that there is not a single closed manifold $M$ of dimension greater than three for which the homotopy type of the automorphism space Diff($M$), PL($M$), or TOP($M$) in the smooth, PL, or topological category, respectively, is in any sense known. (As usual, Diff($M$) is given the $C^\infty$ topology, PL($M$) is a simplicial group, and TOP($M$) is the singular complex of the homeomorphism group with the compact-open topology.) Besides surgery theory, the principal tool in studying homotopy properties of these automorphism spaces is the *concordance space* functor $C(M) = \{$automorphisms of $M \times I$ fixed on $M \times 0\}$. This paper is a survey of some of the main results to date on concordance spaces.

Here is an outline of the contents. In §1 we describe how, in a certain stable dimension range, $C(M)$ is a homotopy functor of $M$, which we denote by $\mathscr{C}(M)$. The application to automorphism spaces is outlined in §2. In §3 we recall the explicit calculations which have been made for $\pi_0\mathscr{C}(M)$ and $\pi_1\mathscr{C}(M)$, along the lines pioneered by Cerf, and apply them in §4 to compute the group of isotopy classes of automorphisms of the $n$-torus, $n \geqq 5$. §5 is concerned with a stabilized version of $\mathscr{C}(M)$, defined roughly as $\Omega^\infty\mathscr{C}(S^\infty M)$, together with the curious equivalence of $\Omega^\infty\mathscr{C}_{\mathrm{PL}}(S^\infty M)$ with $\mathscr{C}_{\mathrm{PL}}(M)/\mathscr{C}_{\mathrm{Diff}}(M)$, due to Burghelea-Lashof (based on earlier fundamental work of Morlet). In §6, $\mathscr{C}_{\mathrm{PL}}(M)$ is "reduced" to higher simple-homotopy theory. This has some interest in its own right, e.g., it provides a fibered form of Wall's obstruction to finiteness. The important new work of Waldhausen relating $\mathscr{C}_{\mathrm{PL}}(M)$ to algebraic $K$-theory is outlined, very briefly and imperfectly, in §7. This seems to be the most promising area for future developments in the sub-

---

*AMS (MOS) subject classifications* (1970). Primary 57C10, 57D50.

ject. §8 describes how the expected extension of Waldhausen's work to $\mathscr{C}_{\text{Diff}}(M)$ leads to an apparent contradiction with the known calculation of $K_3(\boldsymbol{Z})$, using Igusa's work on $\pi_1\mathscr{C}_{\text{Diff}}(M)$. The only way out of this dilemma seems to be the rather unlikely prospect that Waldhausen's and Igusa's definitions of a certain $K_3$-type invariant, though both quite natural, do not agree.

Finally, two short appendices provide a product formula and iterated deloopings for concordance spaces.

**1. The concordance functor $\mathscr{C}$.** Let $M$ be a compact manifold, and let $C(M)$ be the space of concordances of $M$, i.e., automorphisms of $M \times I$ fixed on $M \times 0$. According to the category, these will be diffeomorphisms, PL homeomorphisms, or topological homeomorphisms, and we write $C_{\text{Diff}}(M)$, $C_{\text{PL}}(M)$, or $C_{\text{TOP}}(M)$ when we wish to specify the category. It is known that when $M$ is a PL manifold of dimension $\geqq 5$, the natural map $C_{\text{PL}}(M) \to C_{\text{TOP}}(M)$ is a homotopy equivalence **[3]**, **[18]** (essentially because $\text{Top}(n)/\text{PL}(n) \simeq \text{TOP/PL}$ for $n \geqq 5$). So we shall restrict our attention primarily to $C_{\text{Diff}}$ and $C_{\text{PL}}$.

It is sometimes useful to replace $C(M)$ by the subspace $C(M \text{ rel } \partial M)$ consisting of concordances fixed on $\partial M \times I$. For example if $M \to N$ is a codimension-zero embedding, then there is induced $C(M \text{ rel } \partial M) \to C(N \text{ rel } \partial N)$ by extending concordances via the identity on $(N - M) \times I$. Of course, $C(M)$ can be identified with $C(M \text{ rel } \partial M)$ since $(M \times I, M \times 0)$ is isomorphic to $(M \times I, M \times 0 \cup \partial M \times I)$ by "bending around the corners". (This involves the usual smoothing of corners in Diff.) We will usually not distinguish between $C(M)$ and $C(M \text{ rel } \partial M)$, leaving the reader to determine by context which is meant.

Concordance spaces satisfy an important stability property:

THEOREM 1.1. *The inclusion* $C(M) \to C(M \times I), f \mapsto f \times \text{id}_I$, *is k-connected provided* $\dim M \gg k$.

This is proved in **[12]** for $C_{\text{PL}}$, in the range $\dim M \geqq 3k + 10$. With more care the same methods could probably be improved to yield $\dim M \geqq 2k + 8$. Burghelea-Lashof **[4]** reduced the theorem for $C_{\text{Diff}}$ to the PL case, but with the dimension estimate doubled. Quite probably these stable dimension ranges can be considerably improved.

COROLLARY 1.2. $C(M) \to C(M \times I) \to C(M \times I^2) \to \cdots$ *is eventually an isomorphism on any* $\pi_i$.

DEFINITION. $\mathscr{C}(M) = \bigcup_n C(M \times I^n)$.

In the remainder of this section we shall show:

PROPOSITION 1.3. $\mathscr{C}(—)$ *is a homotopy functor.*

The proof utilizes a transfer map for concordance spaces, which we now define. Let $p: E \to M$ be a locally trivial bundle in the category of compact (smooth or PL) manifolds. The transfer map will be $p^*: C(M) \to C(E)$. Observe first that a concordance $F \in C(M)$ determines (1) a function $f = (\text{proj}) \circ F: M \times (I, 0, 1) \to (I, 0, 1)$ and (2) a one-dimensional foliation $\mathscr{F} = F^{-1}$ (product foliation on $M \times I$) such that $f$ restricts to a homeomorphism from each leaf of $\mathscr{F}$ to $I$. And conversely, a function $f$ and a foliation $\mathscr{F}$ related in this way determine a concordance $F \in C(M)$.

So to define $p^*$ we set $p^*(f) = f \circ (p \times id_I)$, and $p^*(\mathscr{F})$ we define via a local trivialization $M \times I \times$ (fiber of $p$) of $p \times \mathrm{id}_I$, setting $p^*(\mathscr{F}) = \mathscr{F} \times$ (point foliation). Any other local trivialization is related to this one by a transformation of the form $(m, t, x) \to (m, t, \phi_m(x))$, preserving $p^*(\mathscr{F})$, so $p^*(\mathscr{F})$ is well defined. This defines $p^*: C(M) \to C(E)$. It clearly commutes with the stabilization $M \to M \times I$, $E \to E \times I$, and so defines also $p^*: \mathscr{C}(M) \to \mathscr{C}(E)$.

PROOF OF 1.3. Let $f: M \to N$ be a map. Replacing $N$ by $N \times I^k$, $k$ large, we may assume $f$ is an embedding, with a neighborhood of its image in $N$ a disc bundle $p: E \to M$. Then $f_*: \mathscr{C}(M) \to \mathscr{C}(N)$ is defined as the composition $i_* p^*$, where $i_*: \mathscr{C}(E) \to \mathscr{C}(N)$ is induced by the inclusion $i: E \to N$. With this definition of $f_*$, it is clear that $\mathscr{C}(—)$ becomes a homotopy functor on the category of compact manifolds (and continuous maps), or equivalently, on the category of finite complexes. One can trivially extend the domain of $\mathscr{C}$ to infinite (but locally finite) complexes by simply taking the direct limit over finite subcomplexes. (On a noncompact manifold this would amount to taking compactly supported concordances.)

**2. Relation with automorphism groups.** We will let $A(M)$ stand for one of the automorphism spaces Diff($M$), PL($M$), TOP($M$) of diffeomorphisms, PL, or topological homeomorphisms of $M$. For convenience we assume $M$ closed, though the results in this section hold also for compact $M$ provided everything is taken rel $\partial M$.

The idea in trying to say something about the homotopy type of $A(M)$ is to compare it with $G(M)$, the $H$-space of self-homotopy equivalences of $M$, about which much more is currently known. For example, if $M$ is a $K(\pi, 1)$, then as an easy application of obstruction theory, $G(M) \simeq \mathrm{Out}(\pi) \times K(\mathrm{Center}(\pi), 1)$, where Out $(\pi)$ is the outer automorphism group of $\pi = \pi_1 M$, i.e., automorphisms modulo inner automorphisms.

One can interpolate between $A(M)$ and $G(M)$ the space $\tilde{A}(M)$ of block automorphisms of $M$. This is the simplicial group whose $k$-simplices are automorphisms of $M \times \Delta^k$ which leave invariant each $M \times$ (face of $\Delta^k$). $\tilde{A}(M)$ contains $A(M)$ as the automorphisms of $M \times \Delta^k$ preserving projection to $\Delta^k$. Similarly, one can define $\tilde{G}(M)$, but clearly $\tilde{G}(M) \simeq G(M)$, and we shall regard $\tilde{A}(M)$ as contained in $G(M)$.

According to surgery theory, there is a fibration (see §17A of [29]) $G(M)/\tilde{A}(M) \to (G/A)^M \to L^s(M)$ where $L^s(M)$ is Quinn's surgery space, $G/A$ is $G/O$, $G/\mathrm{PL}$, or $G/\mathrm{TOP}$, and dim $M \geqq 5$. This fits into a braid of fibrations

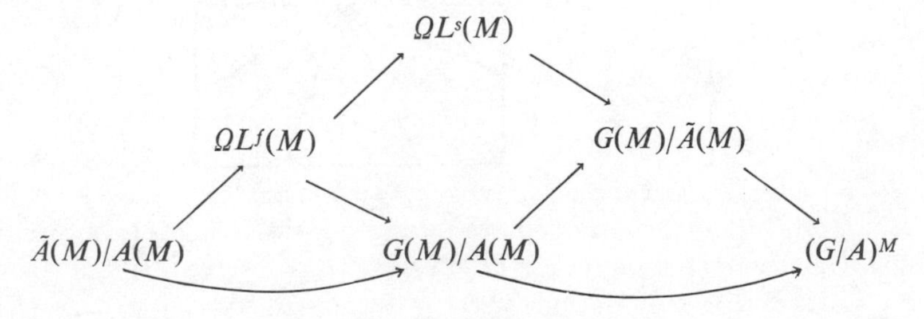


where $\Omega L^f(M)$, the homotopy fiber of $G(M)/A(M) \to (G/A)^M$, can be regarded as a fibered-surgery form of $\Omega L^s(M)$. (See [15]. In this paper we will make no use of $\Omega L^f(M)$.)

So information about $G(M)/A(M)$ can be derived from surgery theory and information about $\tilde{A}(M)/A(M)$. The latter is intimately related to concordance spaces, as the following shows:

PROPOSITION 2.1. *There is a spectral sequence with* $E^1_{pq} = \pi_q C(M \times I^p)$ *converging to* $\pi_{p+q+1}(\tilde{A}(M)/A(M))$.

We outline the construction of this spectral sequence. An element of $\pi_k\tilde{A}(M)/A(M) = \pi_k(\tilde{A}(M), A(M))$ is represented by an automorphism of $M \times I^k$ which preserves projection to $I^k$ over $\partial I^k$. Let $\bar{A}(M \times I^k)$ be the group of all such automorphisms, modulo the subgroup of those which preserve projection to $I^k$ over all of $I^k$. Let $\bar{C}(M \times I^k)$ be the group of all automorphisms of $M \times I^k \times I$ which preserve projection to $I^k \times I$ over $I^k \times 0 \cup \partial I^k \times I$, modulo the subgroup of those preserving projection over all of $I^k \times I$. It is easy to verify that the natural map $C(M \times I^k) \to \bar{C}(M \times I^k)$ is a homotopy equivalence. There are fibrations $\bar{A}(M \times I^{k+1}) \to \bar{C}(M \times I^k) \to \bar{A}(M \times I^k)$ which give an exact couple

$$\begin{array}{ccc} \sum_{j,k} \pi_j \bar{A}(M \times I^k) & \xrightarrow{\partial} & \sum_{j,k} \pi_j \bar{A}(M \times I^{k+1}) \\ & \nwarrow \quad \swarrow & \\ & \sum_{j,k} \pi_j \bar{C}(M \times I^k) & \end{array}$$

and hence a spectral sequence. The chain of homomorphisms

$$0 = \pi_k \bar{A}(M \times I^0) \xrightarrow{\partial} \pi_{k-1} \bar{A}(M \times I^1) \xrightarrow{\partial} \cdots \xrightarrow{\partial} \pi_0 \bar{A}(M \times I^k) \longrightarrow \pi_k(\tilde{A}(M), A(M))$$

gives a filtration of $\pi_k(\tilde{A}(M), A(M))$, according to how far back a given element can be pulled; successive obstructions to pulling back lie in $\pi_i\bar{C}(M \times I^{k-i-1})$, $i = 0, 1, \cdots, k-1$. The $E^\infty$ term of the spectral sequence is associated to this filtration of $\pi_k(\tilde{A}(M), A(M))$.

The first differential is induced from $\delta\colon C(M \times I^p) \to C(M \times I^{p-1})$, where $\delta(f) = f \mid M \times I^p \times 1$, regarded as lying in $C(M \times I^{p-1})$. ($C(X)$ means $C(X \text{ rel } \partial X)$ here.) Now suppose $f = \Sigma g$, where $\Sigma\colon C(M \times I^{p-1}) \to C(M \times I^p)$ is stabilization. Note that in $C(X \text{ rel } \partial X)$, stabilization looks like:

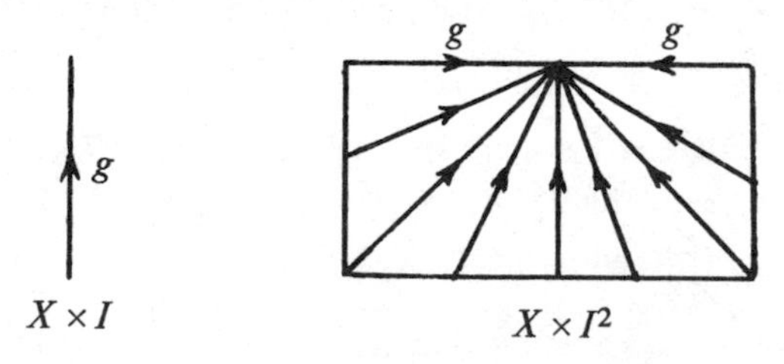


FIGURE 1

Thus $\delta(\Sigma g) = g + \bar{g}$, where the duality involution "—" on $C(X)$ is induced by reflecting $I$ through its midpoint. We shall show in Appendix I that "—" anticommutes with $\Sigma$ (up to homotopy), so we may define a $Z_2$-action on $\pi_*\mathscr{C}(M)$,

$[f] \to [\bar{f}]$, by letting it be $[f] \to (-1)^k [\bar{f}]$ on $\pi_* C(M \times I^k)$. (The sign of this involution on $\pi_* \mathscr{C}(M)$ depends on the parity of dim $M$.) These arguments yield:

PROPOSITION 2.2. *In the stable range* $p + \dim M \gg q$, $E^2_{pq} = H_p(\mathbf{Z}_2; \pi_q \mathscr{C}(M))$.

**3. The brute force calculations of $\pi_0 \mathscr{C}(M)$ and $\pi_1 \mathscr{C}(M)$.** The first major result about concordance spaces was Cerf's theorem that $\pi_0 C_{\mathrm{Diff}}(M) = 0$ if $\pi_1 M = 0$ and dim $M \geqq 5$ [6]. (In the PL category this is a much easier theorem, due to Rourke [24].) A refinement of Cerf's techniques yielded:

THEOREM 3.1 (Diff, PL, OR TOP). *For* dim $M \geqq 5$ *there is a natural exact sequence*

$$H_0(\pi_1 M; (\pi_2 M)[\pi_1 M]/(\pi_2 M)[1]) \longrightarrow \pi_0 C(M)$$
$$\longrightarrow Wh_2(\pi_1 M) \oplus H_0(\pi_1 M; \mathbf{Z}_2[\pi_1 M]/\mathbf{Z}_2[1]) \longrightarrow 0.$$

*If the first k-invariant of M (in $H^3(\pi_1 M; \pi_2 M)$) vanishes, this is a split short exact sequence (but the splitting is not natural).*

For a $\pi$-module $A$, $H_0(\pi; A)$ is just $A$ modulo the $\pi$-action. In the present case, $\pi_1 M$ acts in the usual way on $\pi_i M$ (hence by conjugation on itself) and trivially on $\mathbf{Z}_2$. $Wh_2(\pi_1 M)$ is a certain quotient of $K_2 \mathbf{Z}[\pi_1 M]$.

In the smooth category this theorem is proved in [14] for dim $M \geqq 6$. (The case dim $M = 5$ is due to K. Igusa.) However, when I wrote Part II of [14], I was not aware that I was using the vanishing of the $k$-invariant in Lemma 3.7, p. 262. (Igusa pointed out the error.) The lemma is actually false without the $k$-invariant hypothesis, but as far as I know the theorem may not require it. Volodin [27] announced the result without any restriction on $k$-invariants.

For the PL and TOP categories, one can show (see [23]) that $\pi_0 C(M)$ depends only on a neighborhood of the 3-skeleton of $M$, which can be smoothed, and then appeal to the equivalence $C_{\mathrm{PL}}(M) \simeq C_{\mathrm{TOP}}(M)$ for $M$ PL, mentioned in §1, and to the results on $C_{\mathrm{PL}}(M)/C_{\mathrm{Diff}}(M)$ in §5 below. (Alternatively, the methods of [12] allow the Diff proof to be translated into PL.)

Igusa has gone much deeper with Cerf theory to obtain:

THEOREM 3.2 [16]. *Suppose the first two k-invariants of M vanish, and* dim $M$ *is sufficiently large* ($\geqq$ 10 *certainly suffices*). *Then there is an exact sequence*

$$0 \longrightarrow H_1(\pi_1 M; (\mathbf{Z}_2 \times \pi_2 M)[\pi_1 M]/(\mathbf{Z}_2 \times \pi_2 M)[1]) \oplus H_0(\pi_1 M; \Omega_2^{fr}(\Omega M)/\pi_3 M)$$
$$\longrightarrow \pi_1 C_{\mathrm{Diff}}(M) \longrightarrow Wh_3(\pi_1 M) \longrightarrow 0.$$

Here $Wh_3(\pi)$ is defined as a certain quotient of $K_3 \mathbf{Z}[\pi]$. In particular, $Wh_3(0)$ is the cokernel of $\pi_3^S \to K_3 \mathbf{Z}$ which is $\mathbf{Z}_{24} \subsetneq \mathbf{Z}_{48}$ according to [20]. Thus $Wh_3(0) \approx \mathbf{Z}_2$.

COROLLARY 3.3. $\pi_1 C_{\mathrm{Diff}}(D^n)$ *has order* 4 (*n large*).

R. Lee has shown independently that $\pi_1 C_{\mathrm{Diff}}(D^n)$maps onto $Wh_3(0)$ for large enough $n$.

The mere fact that $\pi_1 C_{\mathrm{Diff}}(D^n)$ is nonzero is in many respects a striking result. It gives a new kind of difference between the smooth and PL categories, not traceable to exotic spheres. (Note that $C_{\mathrm{PL}}(D^n) \simeq *$ by the Alexander trick.) It follows that there are really two kinds of higher Whitehead groups, $Wh_i^{\mathrm{Diff}}(\pi)$ and $Wh_i^{\mathrm{PL}}(\pi)$, which coincide for $i \leqq 2$ only "by accident". For $i = 3$, $Wh_3^{PL}(0) = 0$, but $Wh_3^{Diff}(0)$ is the $Wh_3(0)$ above, which is $Z_2$. More generally, it follows from 5.5 and 5.6 below that $\pi_1 C_{PL}(M^n) \approx \pi_1 C_{Diff}(M^n)/\pi_1 C_{Diff}(D^n)$, $n$ large, so one would expect that $Wh_3^{PL}(\pi) \approx Wh_3^{Diff}(\pi)/Wh_3^{Diff}(0)$ for any $\pi$.

The calculations of $\pi_0 C(M)$ and $\pi_1 C(M)$ suggest that the $n$-type of $\mathscr{C}(M)$ depends only on the $(n+2)$-type of $M$. This is indeed true; see 5.2 below. Igusa **[17]** has shown this is best possible, in general: If in the Postnikov tower $\{M_k\}$ of $M$, the fibration $K(\pi_{n+2}M, n+2) \to M_{n+2} \to M_{n+1}$ has a homotopy-section, then $H_0(\pi_1 M; (\pi_{n+2}M)[\pi_1 M]/(\pi_{n+2}M)[1])$ is a direct summand of $\pi_n \mathscr{C}(M)$.

**4. The isotopy classification of automorphisms of the *n*-torus.** A good example for the preceding machinery is the calculation of $\pi_0 A(T^n)$, $n \geqq 5$, for $A =$ Diff, PL, or TOP. The steps go as follows.

(1) $G(T^n) \simeq T^n \times \mathrm{GL}(n, Z)$, and the map $A(T^n) \to G(T^n)$ has a section up to homotopy. Hence there is a split exact sequence

$$0 \longrightarrow \pi_1 G(T^n)/A(T^n) \longrightarrow \pi_0 A(T^n) \longrightarrow \mathrm{GL}(n, Z) \longrightarrow 0.$$

(2) $G(T^n)/\widetilde{\mathrm{TOP}}(T^n) \simeq *$. For $n \geqq 5$ this is a result in surgery theory, see, e.g., **[25]**. However there is an elementary proof, using only the local contractibility of $\mathrm{TOP}(M)$, which works for all $n$ **[19]**. Hence

$$\pi_1 G(T^n)/\mathrm{TOP}(T^n) \approx \pi_1 \widetilde{\mathrm{TOP}}(T^n)/\mathrm{TOP}(T^n).$$

(3) Using 2.2 (this is almost overkill), $\pi_1 \tilde{A}(T^n)/A(T^n) \approx \pi_0 C(T^n)/\{x \pm \bar{x}\}$, $n \geqq 5$. Then by 3.1, since $Wh_2(\pi_1 T^n) = 0$, we have

$$\pi_0 C(T^n) \approx Z_2[t_1, t_1^{-1}, \cdots, t_n, t_n^{-1}]/Z_2[1]$$

where the $t_i$'s generate $\pi_1 T^n$. According to **[14]**, the involution "—" on $\pi_0 C(T^n)$ is induced by $t_i \mapsto t_i^{-1}$. Hence $\pi_1 \tilde{A}(T^n)/A(T^n) \approx Z_2[t_1, \cdots, t_n]/Z_2[1]$, $n \geqq 5$, independent of $A$.

(4) The preceding steps give the case $A =$ TOP. For $A =$ PL or Diff we consider the diagram

$$\begin{array}{ccccccc}
\longrightarrow \pi_1 \tilde{A}(T^n)/A(T^n) & \longrightarrow & \pi_1 G(T^n)/A(T^n) & \longrightarrow & \pi_1 G(T^n)/\tilde{A}(T^n) & \longrightarrow \\
\downarrow \approx & & \downarrow & & \downarrow & \\
0 \longrightarrow \pi_1 \widetilde{\mathrm{TOP}}(T^n)/\mathrm{TOP}(T^n) & \xrightarrow{\approx} & \pi_1 G(T^n)/\mathrm{TOP}(T^n) & \longrightarrow & 0 &
\end{array}$$

This shows $\pi_1 G(T^n)/A(T^n) \approx \pi_1 \tilde{A}(T^n)/A(T^n) \oplus \pi_1 G(T^n)/\tilde{A}(T^n)$.

(5) Again from surgery theory **[25]**,

$$\pi_1 G(T^n)/\mathrm{Diff}(T^n) \approx hS(T^n \times I \text{ rel } \partial) \approx [\Sigma T^n, \mathrm{TOP}/O]$$

$$\approx \sum_{i=0}^{n} H_{n-i}(T^n; \pi_{i+1} \mathrm{TOP}/O) \approx \sum_{i=0}^{n} \binom{n}{i} \Gamma_{i+1} \oplus \binom{n}{2} Z_2.$$

Similarly,

$$\pi_1 G(T^n)/\mathrm{PL}(T^n) \approx \binom{n}{2} Z_2.$$

Thus we have:

THEOREM 4.1. *If $n \geqq 5$ there are split exact sequences*

$$0 \longrightarrow Z_2^\infty \longrightarrow \pi_0\,\mathrm{TOP}(T^n) \longrightarrow \mathrm{GL}(n, Z) \longrightarrow 0,$$

$$0 \longrightarrow Z_2^\infty \oplus \binom{n}{2} Z_2 \longrightarrow \pi_0\,\mathrm{PL}(T^n) \longrightarrow \mathrm{GL}(n, Z) \longrightarrow 0,$$

$$0 \longrightarrow Z_2^\infty \oplus \binom{n}{2} Z_2 \oplus \sum_{i=0}^{n} \binom{n}{i} \Gamma_{i+1} \longrightarrow \pi_0\,\mathrm{Diff}(T^n) \longrightarrow \mathrm{GL}(n, Z) \longrightarrow 0.$$

REMARKS. (1) This result was obtained also by Hsiang-Sharpe **[15]**.

(2) The same analysis allows one to compute $\pi_0\, A(T^n \times D^k \text{ rel } \partial)$, $n + k \geqq 5$, with no extra work. We leave this to the reader.

(3) The split extensions in 4.1 are nontrivial. The conjugation action of $\mathrm{GL}(n, Z)$ on

$$Z_2^\infty \approx Z_2[t_1, t_1^{-1}, \cdots, t_n, t_n^{-1}]/Z_2[t_i + t_1^{-1}, \cdots, t_n + t_n^{-1}]$$

is induced by the usual action of $\mathrm{GL}(n, Z)$ on $Z^n$, the monomials. The action on $\binom{n}{2}Z_2 \approx H_{n-2}(T^n; Z_2)$ and $\Sigma\binom{n}{i}\Gamma_{i+1} \approx \Sigma H_{n-i}(T^n; \Gamma_{i+1})$ just comes from the action on $T^n$.

(4) The automorphisms in the subgroup $Z_2^\infty \subset \pi_0\, A(T^n)$ are diffeomorphisms which are concordant (smoothly) but not isotopic (even topologically) to the identity. These are rather delicate creatures. (a) They are annihilated by lifting to 2-fold covers (an observation of Laudenbach), hence by covers of any even order; also by certain odd order covers (depending on the diffeomorphism). (b) The product map $\pi_0\, A(T^n) \to \pi_0\, A(T^{n+1})$, $[f] \to [f \times \mathrm{id}_{S^1}]$, kills $Z_2^\infty$ as we shall show in Appendix I. (c) On $T^n \# S^i \times S^{n-i}$ $(3 \leqq i \leqq n-3)$ any automorphism concordant to the identity is isotopic to the identity **[13]**. Thus $Z_2^\infty \subset \pi_0\, A(T^n)$ dies in $\pi_0\, A(T^n \# S^i \times S^{n-i})$.

(5) The subgroups $\Gamma_{i+1} \subset \pi_0\, A(T^n)$ are represented by diffeomorphisms of $D^i$ rel $\partial D^i$ cross the identity on a factor $T^{n-i}$ of $T^n$. The elements of $\binom{n}{2}Z_2$ are represented by diffeomorphisms whose mapping tori are fake tori (homeomorphic but not PL homeomorphic to $T^{n+1}$).

To finish this section we will describe an explicit construction due to Farrell (unpublished), of a diffeomorphism $f\colon S^1 \times D^{n-1} \to S^1 \times D^{n-1}$ rel $\partial$ ($n$ large) which is concordant to the identity but not obviously isotopic to the identity (everything rel $\partial$ here). To show that $f$ is in fact not isotopic to the identity seems to require most of the machinery of **[14]**. A simpler proof of this would be quite welcome.

Embedding $S^1 \times D^{n-1}$ in $T^n$ to represent an element $\alpha \in \pi_1 T^n$, one extends $f$ on $S^1 \times D^{n-1}$ to a diffeomorphism $f_\alpha\colon T^n \to T^n$ via the identity outside $S^1 \times D^{n-1}$. We leave it as an exercise to check that, in the subgroup $Z_2^\infty \approx Z_2[t_1, \cdots, t_n]/Z_2[1]$ of $\pi_0\, A(T^n)$, $f_\alpha$ represents the monomial generator $\alpha = t_1^{p_1} \cdots t_n^{p_n}$.

To construct $f$ we will perform two embedded surgeries on the interior of the codimension one slice $D_0^{n-1} = * \times D^{n-1}$, producing a new disc $D_1^{n-1} \subset S^1 \times D^{n-1}$ with $\partial D_1^{n-1} = \partial D_0^{n-1}$ and $S^1 \times D^{n-1} - D_1^{n-1}$ still an $n$-ball, so that $D_1^{n-1} = f(D_0^{n-1})$ for some homeomorphism (in fact, diffeomorphism) $f$ of $S^1 \times D^{n-1}$ rel $\partial$.

In a neighborhood of $D_0^{n-1}$ label the two sides of $D_0^{n-1}$ as $+$ and $-$. In the $+$ side, attach an embedded $i$-handle $D^i \times D^{n-i}$ to $D_0^{n-1}$ in the trivial way. This effects a surgery on $D_0^{n-1}$ to $\chi(D_0^{n-1})$, say. We could undo the effect of this surgery by now attaching an embedded $(i + 1)$-handle $D^{i+1} \times D^{n-i-1}$ on the $+$ side of $\chi(D_0^{n-1})$, in the trivial way so that the surgered $\chi(D_0^{n-1})$ would be an $(n-1)$-disc isotopic to $D_0^{n-1}$. (All of this would occur near $D_0^{n-1}$.) The Farrell construction is to take instead a new embedding of the $(i+1)$-handle, but attached to $\chi(D_0^{n-1})$ in the same way so that $\chi(D_0^{n-1})$ is again surgered to a disc, this time the desired $D_1^{n-1}$. The new $(i+1)$-handle is obtained from the old by replacing the old core $D^{i+1}$ by $D^{i+1} \# S^{i+1}$, the (interior) connected sum with a certain $S^{i+1} \subset (S^1 \times D^{n-1}) - \chi(D_0^{n-1})$. This $S^{i+1}$ is constructed as follows. The core $D^i$ of the $i$-handle can be completed to a sphere $S^i$ by adding another $i$-disc on the $-$ side of $D_0^{n-1}$. In a neighborhood of this $S^i$, embed $S^{i+1}$ so as to represent the Hopf map $S^{i+1} \to S^i$. For this we must assume $i \geqq 2$ and $n$ large enough to get $S^{i+1}$ actually embedded. Finally, to form the connected sum of $D^{i+1}$, which is on the $+$ side of $\chi(D_0^{n-1})$, with $S^{i+1}$, which is on the $-$ side, we must connect $D^{i+1}$ to $S^{i+1}$ by an arc which circles around the $S^1$ factor of $S^1 \times D^{n-1}$.

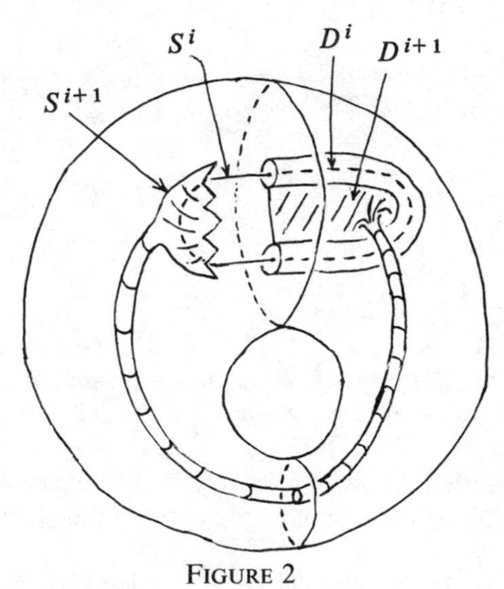


FIGURE 2

The construction actually gives a concordance from $D_0^{n-1}$ to $D_1^{n-1}$, namely the trace of the two surgeries, that is,

$$(D_0^{n-1} \times [0, \tfrac{1}{3}]) \cup (h^i \times \tfrac{1}{3}) \cup (\chi(D_0^{n-1}) \times [\tfrac{1}{3}, \tfrac{2}{3}]) \cup (h^{i+1} \times \tfrac{2}{3}) \cup (D_1^{n-1} \times [\tfrac{2}{3}, 1])$$

in $S^1 \times D^{n-1} \times I$, where $h^i$ and $h^{i+1}$ are the $i$- and $(i+1)$-handles, respectively.

**5. The functor $\mathscr{C}$ stabilized.** Besides the stability Theorem 1.1, the other fundamental general result about concordance spaces is Morlet's Lemma of Disjunction. This is stated in terms of spaces of concordances of embeddings, defined as follows:

DEFINITION. Let $N \subset M$ be a proper submanifold ($N \cap \partial M = \partial N$). Then $CE(N, M)$ is the space of (proper) concordances $F: N \times I \hookrightarrow M \times I$ rel $N \times 0 \cup \partial N \times I$ with $F(N \times 1) \subset M \times 1$.

LEMMA OF DISJUNCTION (Diff OR PL). *Let $D^p \subset V^n$, $D^q \subset V^n$ be disjoint properly embedded discs, $n - p \geqq 3$, $n - q \geqq 3$. Then*

$$\pi_i(CE(D^p, V), CE(D^p, V - D^q)) = 0 \quad \textit{for } i \leqq 2n - p - q - 5.$$

For published proofs see [5] or [22].

The lemma of disjunction can be reformulated in terms of *relative* concordance spaces $C(M, N) = C(M)/C(N)$, where $N \subset M$ is a codimension zero submanifold. (The relative $\mathscr{C}$ is defined similarly.) These give fibrations $C(N) \to C(M) \to C(M, N)$.

PROPOSITION 5.1 (EXCISION). *The lemma of disjunction is equivalent to*: $C(A, A \cap B) \to C(A \cup B, B)$ *is* $(k + l - 3)$*-connected if* $(A, A \cap B)$ *is* $k$*-connected* $(k > 1)$ *and* $(B, A \cap B)$ *is* $l$*-connected* $(l > 1)$.

PROOF. It suffices to consider the case that $A = M^n \cup (D^{k+1} \times D^{n-k-1})$ and $B = M^n \cup (D^{l+1} \times D^{n-l-1})$, i.e., disjoint $(k + 1)$- and $(l + 1)$-handles are attached to $M = A \cap B$. Then $C(A, A \cap B)$ can be regarded as the space $CE(D^{k+1} \times D^{n-k-1}, A)$ of concordances of $D^{k+1} \times D^{n-k-1}$ in $A$ rel $\partial D^{k+1} \times D^{n-k-1}$. Similarly, $C(A \cup B, B) = CE(D^{k+1} \times D^{n-k-1}, A \cup B)$. Consider the fibrations

$$\begin{array}{ccccc} CE(D^{k+1} \times D^{n-k-1}, A \text{ rel } 0 \times D^{n-k-1}) & \to & CE(D^{k+1} \times D^{n-k-1}, A) & \to & CE(0 \times D^{n-k-1}, A) \\ \downarrow & & \downarrow & & \downarrow \\ CE(D^{k+1} \times D^{n-k-1}, A \cup B \text{ rel } 0 \times D^{n-k-1} & \to & CE(D^{k+1} \times D^{n-k-1}, A \cup B) & \to & CE(0 \times D^{n-k-1}, A \cup B) \end{array}$$

The two fibers are homotopy equivalent, essentially because one can shrink concordances of $D^{k+1} \times D^{n-k-1}$ rel $0 \times D^{n-k-1}$ to their germ near $0 \times D^{n-k-1}$. Choosing $D^p = 0 \times D^{n-k-1}$ and $D^q = 0 \times D^{n-l-1}$, the result now follows.

COROLLARY 5.2. *If* $X \to Y$ *is* $k$*-connected, the induced map* $\mathscr{C}(X) \to \mathscr{C}(Y)$ *is* $(k - 2)$*-connected.*

PROOF. We may take $Y = X \cup e^{k+1}$ with $k \geqq 2$. Splitting $e^{k+1}$ down the middle, we get $e^{k+1} = e^k \cup e_1^{k+1} \cup e_2^{k+1}$; hence a fibration

$$\mathscr{C}(X \cup e^k, X) \longrightarrow \mathscr{C}(X \cup e^k \cup e_1^{k+1}, X) \longrightarrow \mathscr{C}(X \cup e^k \cup e_1^{k+1}, X \cup e^k)$$

with contractible total space. By induction on $k$, $\mathscr{C}(X \cup e^k, X)$ is $(k-3)$-connected, and therefore $\mathscr{C}(X \cup e^k \cup e_1^{k+1}, X \cup e^k)$ is $(k-2)$-connected. By excision (5.1)

$$\mathscr{C}(X \cup e^k \cup e_1^{k+1}, X \cup e^k) \longrightarrow \mathscr{C}(X \cup e^{k+1}, X \cup e^k \cup e_2^{k+1}) \simeq \mathscr{C}(X \cup e^{k+1}, X)$$

is $(2k - 3)$-connected, and so $\mathscr{C}(X \cup e^{k+1}, X)$ is $(k - 2)$-connected, as desired. (This argument is lifted from [17].)

A purely formal consequence of 5.1, obtained by choosing $A$ and $B$ to be cones on $X = A \cap B$, is:

COROLLARY 5.3. *The natural suspension map* $\mathscr{C}(X, *) \to \Omega\,\mathscr{C}(SX, *)$ *is* $(2n - 2)$*-connected if* $X$ *is* $n$*-connected.*

This allows us to make the following:

DEFINITION. $\mathscr{C}^S(X, *) = \lim_n \Omega^n\, \mathscr{C}(S^n X, *)$.

LEMMA 5.4. $\mathscr{C}^S_{\mathrm{Diff}}(X, *)$ *is contractible for all* $X$.

PROOF. $\pi_*\mathscr{C}^S(X, *)$ is a homology theory, since by 5.1 it satisfies the excision axiom. So it suffices to prove 5.4 when $X = S^n$, $* = D^n$. We claim: $C_{\text{Diff}}(S^n \times D^k, D^n \times D^k)$ is $(2n-4)$-connected, for any $k$. For, an application of 5.1 gives

$$\pi_i C(S^n \times D^k, D^n \times D^k) \approx \pi_{i+1} C(S^{n+1} \times D^{k-1}, D^{n+1} \times D^{k-1}), \qquad i \leqq 2n - 4.$$

Iterating, we eventually get to $\pi_{i+k}C(S^{n+k}, D^{n+k})$. But in the smooth category, $C_{\text{Diff}}(S^{n+k}, D^{n+k}) \simeq CE_{\text{Diff}}(D^0, S^{n+k})$; and the latter space is contractible, by an amusing elementary argument which we leave to the reader. (*Hint.* Cap off $S^{n+k} \times I$ with an $(n + k + 1)$-ball attached to $S^{n+k} \times 1$.)

The main result of this section is due to Burghelea-Lashof [**4**]:

THEOREM 5.5. $\mathscr{C}_{\text{Diff}}(X, *) \to \mathscr{C}_{\text{PL}}(X, *) \to \mathscr{C}^S_{\text{PL}}(X, *)$ *is a fibration, up to homotopy.*

This is quite an amazing result. In the older approach to smooth concordance spaces, begun by Cerf, one studies $k$-parameter families of $C^\infty$ functions $M \times I \to I$, and the first problem one encounters is the local one of understanding the singularities of codimension $\leqq k$. For example, when $k \leqq 4$ one encounters Thom's seven "elementary" catastrophes (these are all actually used in Igusa's work on $\pi_1 C(M)$ mentioned in §3). The complexity of these singularities increases rapidly with $k$, and they have only been completely classified (by Arnold) for relatively small values of $k$. So as an approach to smooth concordance spaces, this seems hopeless in general. Fortunately, the theorem gives an alternative approach in terms of PL concordance spaces, which are considerably more tractable as we shall see in §§6 and 7 below.

Theorem 5.5 is proved by considering the diagram

$$\begin{array}{ccccc}
\mathscr{C}_{\text{Diff}}(X, *) & \longrightarrow & \mathscr{C}_{\text{PL}}(X, *) & \longrightarrow & \mathscr{C}_{\text{PL}}(X, *)/\mathscr{C}_{\text{Diff}}(X, *) \\
\downarrow & & \downarrow & & \downarrow \simeq \\
\mathscr{C}^S_{\text{Diff}}(X, *) & \longrightarrow & \mathscr{C}^S_{\text{PL}}(X, *) & \xrightarrow{\simeq} & \mathscr{C}^S_{\text{PL}}(X, *)/\mathscr{C}^S_{\text{Diff}}(X, *).
\end{array}$$

The horizontal arrow labelled a homotopy equivalence is such because $\mathscr{C}^S_{\text{Diff}}(X, *) \simeq *$ by 5.4. The other homotopy equivalence follows from the fact that $\pi_* \mathscr{C}_{\text{PL}}(X, *)/\mathscr{C}_{\text{Diff}}(X, *)$ is a homology theory (hence already stable), which comes from fibered smoothing theory—see [**4**] for details.

COROLLARY 5.6. *The homology theory* $\pi_*\mathscr{C}^S_{\text{PL}}(X, *)$ *has coefficients* $\pi_{*-1}\mathscr{C}_{\text{Diff}}(*)$.

This follows by choosing $X = S^0$, since $\mathscr{C}_{\text{PL}}(S^0, *) = \mathscr{C}_{\text{PL}}(*) = \lim_n C_{\text{PL}}(D^n)$ is contractible by the Alexander trick.

Recall from §3 that $\pi_0\mathscr{C}_{\text{Diff}}(*) = 0$, but $\pi_1\mathscr{C}_{\text{Diff}}(*)$ is a group of order four. Nothing is yet known about $\pi_i\mathscr{C}_{\text{Diff}}(*)$ for $i > 1$. A very interesting question is whether or not $\pi_* \mathscr{C}_{\text{Diff}}(*)$ is all torsion.[1]

REMARK. According to [**4**], $\pi_*\mathscr{C}^S_{\text{PL}}(X, *)$ can also be described as the homology theory associated to the spectrum

[1]See note added in proof, below.

$$\frac{\mathrm{Top}(n+1)}{O(n+1)} \Big/ \frac{\mathrm{Top}(n)}{O(n)}.$$

**6. Higher simple-homotopy theory.** According to the usual pattern, one takes a geometric problem, reduces it to a homotopy problem, then tries to attack the homotopy problem by the big algebraic topology machine. In this section we describe part of the reduction of $\mathscr{C}_{\mathrm{PL}}$ to homotopy theory, though the homotopy theory which arises is not of the usual sort: It is a higher simple-homotopy theory, generalizing J. H. C. Whitehead **[30]**. In the following section (§7) this higher simple-homotopy theory is then related to more usual constructions in homotopy theory. One can anticipate that within a few years the algebraic topologists will have done something with this homotopy theory to make the whole program worthwhile.

As a motivation for higher simple-homotopy theory, we pose the following:

*Problem* 6.1 (*fibered obstruction to finiteness*). Let $\pi: E \to B$ be a fibration, with $B$ a finite polyhedron and with fibers homotopy equivalent to a finite polyhedron $X$. Is $\pi$ fiber-homotopy equivalent to a fibration $\pi': E' \to B$ such that

(a) $E'$ is a finite polyhedron and $\pi'$ is PL?

(b) $\pi'$ is the projection of a locally trivial bundle with fiber a compact PL manifold $M$ and structure group PL($M$)?

(c) $E'$ is a compact ANR and $\pi'$ is a proper map?

It can be shown that (a), (b), (c) are equivalent (see **[7]**, **[8]** for (c)); we will focus on (a).

By a polyhedral version of the path space construction, one could easily construct a PL fibration $\pi': E' \to B$ fiber-homotopy equivalent to the given $\pi$, with $E'$ an infinite polyhedron. On the other hand, if $\pi'$ is required only to be a quasi-fibration, then $E'$ can be taken to be a finite polyhedron (and in fact, one can take all fibers to be PL homeomorphic to $X$). The problem is to have both $E'$ finite and $\pi'$ satisfying the covering homotopy property.

Problem 6.1 can be reformulated as a lifting problem. As is well known, $\pi$ is classified by a map $B \to BG(X)$, where $G(X)$ is the $H$-space of self-homotopy equivalences of $X$. One can construct a universal space $B(X)$ for PL fibrations of finite polyhedra, as follows. $B(X)$ is the simplicial set whose $k$-simplices are the PL maps $\pi: E \to \Delta^k$ satisfying the covering homotopy property, with $E$ a finite polyhedron and fibers $\simeq X$. There is a natural forgetful map $B(X) \to BG(X)$, and Problem 6.1 becomes the lifting problem:

(6.2)

$B(X)$ has an amusing heuristic interpretation, as "the space of all finite polyhedra of the homotopy type of $X$", or more precisely as the singular complex of this "space". For the fibers of a $k$-simplex $\pi: E \to \Delta^k$ in $B(X)$ form a "continuous" $k$-parameter family of finite polyhedra, the "continuity" being expressed in the covering homotopy property for $\pi$.

The space $B(X)$ can be related to Whitehead's simple-homotopy theory. Recall the basic notion of an elementary collapse, sometimes written $L_0 \searrow^e L_1$, where $L_0$ is $L_1$ with a ball attached along one of its faces. Collapsing (projecting) the ball to this face induces the map $L_0 \to L_1$. The equivalence relation on finite complexes generated by elementary collapses is, by definition, simple homotopy equivalence.

A nice generalization of elementary collapse is given in the following definition (in the polyhedral category, for convenience):

DEFINITION. A PL map of finite polyhedra $f: L_0 \to L_1$ is a *simple map* if $f^{-1}(x) \simeq x$ for all $x \in L_1$.

The definition is due to M. M. Cohen [9], who used the term "contractible mapping". (In a somewhat more general setting the terminology "*CE* map" or "cell-like map" is also used.) Cohen proved that a simple map is a simple homotopy equivalence.

Let $S$ be the category of finite polyhedra, with morphisms the simple maps. (It is easy to verify that simple maps are closed under composition.) The classifying space $BS$ is then defined; it is the simplicial set whose $k$-simplices are the chains $L_0 \to L_1 \to \cdots \to L_k$ in $S$. Note that $\pi_0 BS$ is just the set of simple homotopy types of finite complexes, since simple maps generate the relation of simple homotopy equivalence.

Let $BS_X$ denote those components of $BS$ containing polyhedra of the homotopy type of the given $X$.

THEOREM 6.3 [12]. $BS_X \simeq B(X)$.

The map $BS_X \to B(X)$ can be defined as follows. A $k$-simplex of $BS_X$ is a chain $L_0 \xrightarrow{f_1} L_1 \to \cdots \xrightarrow{f_k} L_k$. One forms its iterated mapping cylinder $M(f_1, \cdots, f_k)$, which is defined inductively as the ordinary mapping cylinder of the composition $M(f_1, \cdots, f_{k-1}) \to L_{k-1} \xrightarrow{f_k} L_k$. ($M(f_1)$ is the usual mapping cylinder.) Then one proves that the natural projection $M(f_1, \cdots, f_k) \to \Delta^k$ is a fibration if and only if the maps $f_i$ are simple maps. The map $BS_X \to B(X)$ sends $L_0 \xrightarrow{f_1} L_1 \to \cdots \xrightarrow{f_k} L_k$ to $M(f_1, \cdots, f_k) \to \Delta^k$.

In view of the lifting problem (6.2), one is interested in the homotopy-theoretic fiber of $B(X) \to BG(X)$. This will be described using the following:

FUNDAMENTAL DEFINITION. $S(X)$ is the category whose objects are finite polyhedra containing the given finite subpolyhedron $X$ as a deformation retract, and whose morphisms are the simple maps restricting to the identity on $X$.

THEOREM 6.4 [12]. *$BS(X) \to B(X) \to BG(X)$ is a fibration, up to homotopy.*

Thus obstructions to solving Problem 6.1 come from $\pi_* BS(X)$. Whitehead's fundamental theorem can be reformulated (much along the lines of [10]) as the calculation $\pi_0 BS(X) \approx Wh_1(\pi_1 X)$, the algebraic Whitehead torsion group, a quotient of $K_1 \mathbf{Z}[\pi_1 X]$. As an example, let $f: X \to X$ be a homotopy equivalence with nonzero torsion, inducing the identity on $\pi_1 X$. Let $T(f)$ be the mapping torus, and $\pi: E \to S^1$ the path space construction applied to the obvious projection $T(f) \to S^1$, so that $\pi$ is a fibration with fibers $\simeq X$. Then the answer to 6.1 is negative; the torsion of $f$ is the obstruction (see [7] for more details).

Now to relate $BS(X)$ with concordance spaces, let $X$ be a compact PL manifold.

Then there is a natural map of $BC_{\mathrm{PL}}(X)$ to the homotopy-fiber of $B(X) \to BG(X)$, since $BC_{\mathrm{PL}}(X)$ classifies PL bundles with fiber $X \times I$, trivialized on the subfiber $X \times 0$ (so, projection $X \times I \to X \times 0$ induces a fiber-homotopy trivialization). Stabilizing, one obtains a map of $B\mathscr{C}_{\mathrm{PL}}(X)$ to the homotopy-fiber of $B(X) \to BG(X)$.

THEOREM 6.5 [12]. *The natural map of $B\mathscr{C}_{\mathrm{PL}}(X)$ to the homotopy-fiber $BS(X)$ of $B(X) \to BG(X)$ is a homotopy equivalence onto the identity component of $BS(X)$.*

The other components of $BS(X)$ correspond to nontrivial $h$-cobordisms on $X$. Indeed, 6.5 can be regarded as a parametrized $h$-cobordism theorem, in the PL category.

A natural question is, does $B\mathscr{C}_{\mathrm{Diff}}(X)$ also have a categorical description? One obvious candidate is the category $E(X)$ whose objects are the same as those of $S(X)$, but whose morphisms are the finite compositions of elementary collapses. Then is $B\mathscr{C}_{\mathrm{Diff}}(X)$ homotopy equivalent to the identity component of $BE(X)$?

**7. Waldhausen's "Quillenization" of $\mathscr{C}_{\mathrm{PL}}$.** Waldhausen's basic idea is to imitate the exact sequence defining $Wh_1(\pi)$,

$$0 \to K_1(\mathbf{Z}) \oplus H_1(\pi) \to K_1\mathbf{Z}[\pi] \to Wh_1(\pi) \to 0$$

by constructing a diagram of fibrations

$$(7.1)\qquad \begin{array}{ccccc} h(X; K(*)) & \longrightarrow & K(X) & \longrightarrow & Wh(X) \\ \downarrow & & \downarrow & & \downarrow \\ h(B\pi; K(\mathbf{Z})) & \longrightarrow & K(\mathbf{Z}[\pi]) & \longrightarrow & Wh(\pi) \end{array} \qquad \pi = \pi_1 X$$

where:

(a) $Wh(X)$ is a delooping of $BS(X)$, hence a double delooping of $\mathscr{C}_{\mathrm{PL}}(X)$. (This differs from the notation of [12], where "$Wh(X)$" was equal to $BS(X)$.)

(b) $K(X)$ is "the algebraic $K$-theory of the topological space $X$." Waldhausen defines this (he calls it $A(X)$, but we have already used the letter $A$ for automorphism spaces) using a very nice generalization of the Quillen $Q$-construction, but it seems that a plus-construction is also possible, and we give this definition. Let $GL(X_n)$ be the $H$-space of homotopy equivalences $X_n \bigvee_{j=1}^k S_j^i \to X_n \bigvee_{j=1}^k S_j^i$ rel $X_n$, stabilized over $k$ and $i$, where $\{X_n\}$ is the Postnikov tower of $X$. (One needs $i \gg n$ in order to get canonical retractions $X_n \bigvee_{j=1}^k S_j^i \to X_n$ by means of which the stabilization with respect to $i$ is defined.) We would like now to let $GL(X) = \lim_n GL(X_n)$, though strictly speaking, inverse limits are not generally defined. Nonetheless, there is an $H$-space $GL(X)$ which is the inverse limit of the system $\{GL(X_n)\}$ in the same sense that a space is the inverse limit of its Postnikov tower. For all practical purposes one can just choose large finite values $i \gg n \gg 0$, and only let $k \to \infty$, which is no problem. As an important special case, $GL(*)$ is (exactly) the $H$-space of base pointed homotopy equivalences of $\bigvee_{j=1}^k S_j^i$, stabilized over $k$ and $i$ in the obvious way.

PROPOSITION 7.2. *$\pi_0\, GL(X) \approx GL(\mathbf{Z}\pi_1 X)$, and for $i > 0$,*

$$\pi_i GL(X) \approx M(\Omega_i^{fr}(\Omega X)) = \lim_n M_n(\Omega_i^{fr}(\Omega X)),$$

*where $M_n$ is the additive group of $n \times n$ matrices and $M_n \subset M_{n+1}$ by adjoining zeros.*

DEFINITION. $K(X) = BGL(X)^{+} \times K_0(\mathbf{Z}\pi_1 X)$, where "+" is with respect to the commutator subgroup of $\pi_1 BGL(X)$.

(c) $\pi_* h(—; K(*))$ is the homology theory associated to $K(*)$ (all the spaces in 7.1 are infinite loopspaces).

(d) $K(\mathbf{Z}\pi) = \mathrm{BGL}(\mathbf{Z}\pi)^{+} \times K_0(\mathbf{Z}\pi)$.

(e) $\pi_* h(—; K(\mathbf{Z}))$ is the homology theory associated to $K(\mathbf{Z})$.

(f) $Wh(\pi)$ is by definition a delooping of the homotopy-fiber of a natural map $h(B\pi; K(\mathbf{Z})) \to K(\mathbf{Z}\pi)$. See **[28]**. Waldhausen defines the higher Whitehead group $Wh_i(\pi)$ as $\pi_i Wh(\pi)$. For $i \leqq 2$ this agrees with earlier definitions. For $i = 3$ it is the $Wh_3^{\mathrm{PL}}(\pi)$ of §3. According to the main result of **[28]**, $Wh(\pi)$ is contractible for a large class of groups, e.g., free abelian groups.

(g) The map $K(X) \to K(\mathbf{Z}\pi_1 X)$ is induced from $BGL(X) \to BGL(\mathbf{Z}\pi_1 X)$, the first stage in the Postnikov tower for $BGL(X)$.

REMARK. The map $\Sigma_\infty \to GL(\mathbf{Z})$ factors through $GL(*)$, as permutations of the spheres in $\bigvee_{j=1}^{\infty} S_j^i$. Hence $\pi_*^S \to K_*(\mathbf{Z})$ factors through $\pi_* K(*)$.

An immediate consequence of the definitions and 7.2 is:

COROLLARY 7.3. *$K(B\pi) \to K(\mathbf{Z}\pi)$ is a $\mathbf{Q}$-equivalence. In particular, $K(*) \to K(\mathbf{Z})$ is a $\mathbf{Q}$-equivalence; hence also $Wh(B\pi) \to Wh(\pi)$.*

Thus if $\pi$ is in Waldhausen's class of groups for which $Wh(\pi) \simeq *$, $Wh(B\pi)$ is a torsion space, in both senses! Going back to §§2 and 4, we can conclude from 7.3 as a very special case:

COROLLARY 7.4. *$\pi_i\mathrm{TOP}(T^n) \to \pi_i G(T^n)$ is an isomorphism* mod *torsion, for $i \ll n$. (And likewise for* PL.)

One good potential application of Waldhausen's work depends on the following:

CONJECTURE. *For any simply-connected $X$, $H_i(BGL(X))$ is finitely generated (f.g.) for all $i$.*

More generally, one might hope this is true if $\pi_1 X$ is finite.

PROPOSITION 7.5. *If the conjecture is true, then*

(a) *$\pi_i\mathrm{PL}(M^n)$ and $\pi_i\mathrm{TOP}(M^n)$ are f.g. for $i \ll n$ and $M$ a simply-connected* PL *manifold.*

(b) *$\pi_i\mathrm{Diff}(M^n)$ is f.g. for $i \ll n$ and $\pi_1 M = 0$.*

PROOF. $H_*(BGL(M)) = H_* BGL(M)^{+}$ f.g. $\Rightarrow \pi_* K(M)$ f.g. $\Rightarrow \pi_* Wh(M)$ f.g. $\Rightarrow \pi_i C_{\mathrm{PL}}(M)$ f.g., $i \ll n$ (by 6.2) $\Rightarrow \pi_i \widetilde{\mathrm{PL}}(M)/\mathrm{PL}(M)$ f.g. (by 2.1) $\Rightarrow \pi_i G(M)/\mathrm{PL}(M)$ f.g. (since $\pi_* G(M)/\widetilde{\mathrm{PL}}(M)$ f.g. by surgery theory) $\Rightarrow \pi_i\mathrm{PL}(M)$ f.g. (since $\pi_* G(M)$ f.g.). And similarly for TOP. To go from PL to Diff, one uses 5.5.

One approach to studying $K(X)$ might be to take the Postnikov tower $\{BGL(X)_n\}$ for $BGL(X)$ and apply the plus construction. Thus one would have fibrations

$$\begin{array}{ccccc} K(M(\Omega_n^{fr}(\Omega X)), n+1) & \longrightarrow & BGL(X)_{n+1} & \longrightarrow & BGL(X)_n \\ \downarrow & & \downarrow & & \downarrow \\ F_{n+1}(X) & \longrightarrow & (BGL(X)_{n+1})^{+} & \longrightarrow & BGL(X)_n)^{+} \end{array}$$

and the question would be, what is the relation of the new fiber $F_{n+1}(X)$ to the old one $K(M(\Omega_n^{fr}(\Omega X)), n+1)$? It is not hard to see that $F_{n+1}(X)$ is $n$-connected, and that $\pi_{n+1}F_{n+1}(X)$ is just $M(\Omega_n^{fr}(\Omega X))$ modulo the conjugation action of $\mathrm{GL}(\boldsymbol{Z}\pi_1 X)$. This in turn is computable as $H_0(\pi_1 X; \Omega_n^{fr}(\Omega X))$, that is, $\Omega_n^{fr}(\Omega X)$ modulo the conjugation action of $\pi_1 X$ (see **[26]**). In view of 3.2, one might guess that $\pi_{n+2}F_{n+1}(X) \approx H_1(\pi_1 X; \Omega_n^{fr}(\Omega X))$. More generally, does $F_{n+1}(X)$ depend only on $M(\Omega_n^{fr}(\Omega X))$ and the conjugation action by $\mathrm{GL}(\boldsymbol{Z}\pi_1 X)$, at least in some stable range (below dimension $2n$, say)?

**8. A hypothetical splitting.** The following would seem to be a reasonable thing to hope for, and Waldhausen asserts in **[31]** that it is true:

*Hypothesis.* The map $K(X) \to Wh(X)$ of 7.1 factors through $Wh_{\mathrm{Diff}}(X)$, a double delooping of $\mathscr{C}_{\mathrm{Diff}}(X)$. (Recall that $Wh(X)$ is a double delooping of $\mathscr{C}_{\mathrm{PL}}(X)$.)

The hypothesis implies that there is a diagram of fibrations

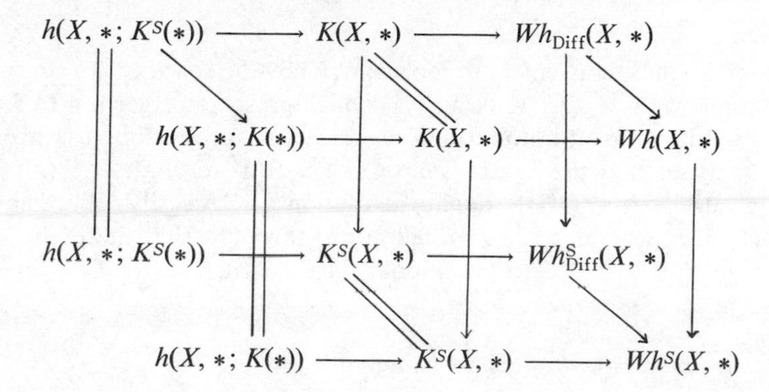


The fibration containing $Wh^S(X, *)$ is obtained by stabilizing the one containing $Wh(X, *)$. The fiber $h(X, *; K(*))$ is the same for both since $\pi_* h(-; K(*))$ is a homology theory, and so is already stable. Similarly, $\pi_*$ of the fiber of $K(X, *) \to Wh_{\mathrm{Diff}}(X, *)$ is a homology theory (since this is true for the PL analog $K(X, *) \to Wh(X, *)$ and $\pi_*\mathscr{C}_{\mathrm{PL}}(X, *)/\mathscr{C}_{\mathrm{Diff}}(X, *)$ is a homology theory, as mentioned after 5.5), so the fiber of $K(X, *) \to Wh_{\mathrm{Diff}}(X, *)$ is the same as the fiber of $K^S(X, *) \to Wh^S_{\mathrm{Diff}}(X, *) \simeq *$. Hence this fiber is $K^S(X, *)$, for which we are also using the notation $h(X, *; K^S(*))$.

The diagram yields:

COROLLARY OF THE HYPOTHESIS. $K(X, *) \simeq K^S(X, *) \times Wh_{\mathrm{Diff}}(X, *)$.

However, this seems to lead to a contradiction. Let $X = S^0$, and consider the diagram

$$\begin{array}{ccc} \pi_3 K(*) & \longrightarrow & \pi_1\mathscr{C}_{\mathrm{Diff}}(*) \\ \downarrow & & \downarrow \\ K_3(\boldsymbol{Z}) & \longrightarrow & Wh_3^{\mathrm{Diff}}(0) \end{array}$$

where the maps to $Wh_3^{\mathrm{Diff}}(0)$ are those defined by Igusa (see §3). It seems reasonable to *suppose that the diagram commutes.* By the preceding corollary, Corollary 3.3, and **[20]**, we then get a commutative diagram

$$\begin{array}{ccccccccc}
0 & \longrightarrow & \boldsymbol{Z}_2 & \longrightarrow & \pi_1\mathscr{C}_{\mathrm{Diff}}(*) & \longrightarrow & Wh_3^{\mathrm{Diff}}(0) & \longrightarrow & 0 \\
 & & & & \downarrow & & \Vert & & \\
 & & & & K_3(\boldsymbol{Z}) & \longrightarrow & Wh_3^{\mathrm{Diff}}(0) & \longrightarrow & 0 \\
 & & & & \wr\wr & & \wr\wr & & \\
 & & & & \boldsymbol{Z}_{48} & & \boldsymbol{Z}_2 & &
\end{array}$$

which is impossible. Can it be that the trouble is in the commutativity of the diagram?

**Appendix I. A product formula for concordances.** Let $M$ and $N$ be compact manifolds. The product map $p: C(M) \to C(M \times N)$ is given by $p(F) = F \times id_N$. For $D^n \subset N^n$ there is also the inclusion-induced map $i: C(M) \xrightarrow{\Sigma^n} C(M \times D^n) \to C(M \times N)$.

PROPOSITION (Diff OR PL). *$p \simeq \chi(N) \cdot i$, where $\chi(N)$ is the Euler characteristic of $N$.*

For the proof it will be convenient to replace $C(X)$ by the space $C'(X)$ consisting of automorphisms of $(X \times I, X \times 0, X \times 1)$ which preserve projection to $I$ over $\partial X$, modulo the subgroup of automorphisms preserving projection to $I$ over all of $X$. It is easy to see that the natural map $C(X) \approx C(X \text{ rel } \partial X) \to C'(X)$ is a homotopy equivalence. In $C'(X)$ the duality involution "—" is easily defined as conjugation by $id_X \times r$, where $r: I \to I$ is reflection through the midpoint. The stabilization map $\Sigma: C'(X) \to C'(X \times I)$ looks just like it does in $C(X \text{ rel } \partial X)$; see Figure 1 in §2.

LEMMA. *$\Sigma$ anticommutes with —, up to homotopy.*

A proof is suggested by the following pictures:

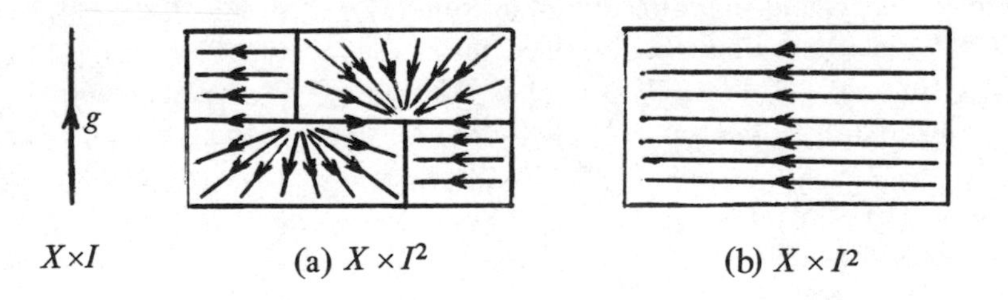


FIGURE 3

The two large rectangles in (a) represent $\Sigma\bar{g}$ and $\overline{\Sigma g}$; the two smaller squares are isotopies (level preserving), hence trivial in $C'(X \times I)$. The whole of (a) is clearly isotopic to (b), itself an isotopy, hence trivial in $C'(X \times I)$. Thus $\Sigma\bar{g} + \overline{\Sigma g} \simeq 0$.

PROOF OF THE PROPOSITION. For convenience we choose the smooth category and assume $N$ is closed. Let $g \in C'(M)$. As in Figure 4 below, first deform $g \times \mathrm{id}_N$ to $G$ in $C'(M \times N)$ by shrinking vertically the support of $g$ on each slice $M \times \{x\} \times$

$[\phi(x) - \varepsilon, \phi(x) + \varepsilon]$, where $\phi: N \to (0, 1)$ is a Morse function. Then deform $G$ to $G'$ by tilting the slices $M \times \{x\} \times [\phi(x) - \varepsilon, \phi(x) + \varepsilon]$ so that they are horizontal away from the critical points of $\phi$, and so that near a critical point of index $p$, $G'$ is just $(\overline{\Sigma^-})^p (\Sigma)^{n-p} (g)$. By the lemma, this equals $(-1)^p\, i\, (g)$. Summing over all critical points of $\phi$ gives the result.

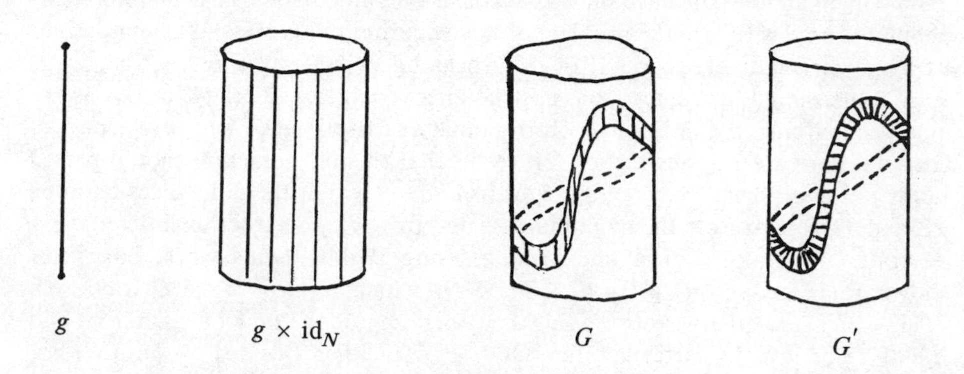


FIGURE 4

**Appendix II. An infinite delooping of $\mathscr{C}(M)$.** First, we mention the purely formal infinite delooping of $\mathscr{C}(M)$ coming from the obvious $k$-fold little cubes structure on $C(M \times I^k)$. This seems however not to have much geometric significance. More interesting is a delooping in terms of the spaces $C^b(M \times \mathbf{R}^k)$ of concordances $M \times \mathbf{R}^k \times I \to M \times \mathbf{R}^k \times I$ of bounded distance from the identity, given by the following:

PROPOSITION. *There are natural equivalences $C^b(M \times \mathbf{R}^k \times I) \simeq \Omega C^b(M \times \mathbf{R}^{k+1})$ compatible with stabilization and hence inducing $\mathscr{C}^b(M \times \mathbf{R}^k) \simeq \Omega\mathscr{C}^b(M \times \mathbf{R}^{k+1})$. In particular, $\mathscr{C}(M) \simeq \Omega^k\mathscr{C}^b(M \times \mathbf{R}^k)$.*

PROOF. Let $N = M \times \mathbf{R}^k$. A map $\lambda: C^b(N \times I) \to \Omega C^b(N \times \mathbf{R})$ can be defined as follows. Let $i: C^b(N \times I) \to C^b(N \times \mathbf{R})$ be induced by $I \subset \mathbf{R}$. Two null-homotopies of $i$ are obtained by translating $I$ to $+\infty$ and $-\infty$ in $\mathbf{R}$. This defines for each $f \in C^b(N \times I)$ the loop $\lambda(f)$ of concordances in $C^b(N \times \mathbf{R})$. To see that $\lambda$ is a homotopy equivalence we consider the fibrations

$$\begin{array}{ccccc} C^b(N \times \mathbf{R} \text{ rel } N \times 0) & \longrightarrow & C^b(N \times \mathbf{R}) & \xrightarrow{r} & CE^b(N, N \times \mathbf{R}) \\ & & & & \| \\ C^b(N \times [-\infty, 0] \text{ rel } \partial) & \longrightarrow & \mathscr{E} & \xrightarrow{r} & CE^b(N, N \times (-\infty, \infty)) \end{array}$$

where $CE$ denotes concordances of embeddings,

$$\mathscr{E} = CE^b(N \times [-\infty, 0], N \times [-\infty, \infty] \text{ rel } N \times -\infty),$$

and the two maps $r$ are restriction to $N \times 0$. (The boundedness condition assures the covering homotopy property for $r$.) Clearly $C^b(N \times \boldsymbol{R} \text{ rel } N \times 0)$ and $\mathscr{E}$ are contractible, and $C^b(N \times [-\infty, 0] \text{ rel } \partial) \simeq C^b(N \times I)$. It is not hard to check that the resulting equivalence $C^b(N \times I) \simeq \Omega C^b(N \times \boldsymbol{R})$ is given by $\lambda$.

THEOREM [2]. *The first $k$ homotopy groups of $C^b(M^n \times \boldsymbol{R}^k)$, for $n + k \geqq 5$, are* $K_{-k+2}(\boldsymbol{Z}\pi_1 M), \cdots, K_{-1}(\boldsymbol{Z}\pi_1 M), \tilde{K}_0(\boldsymbol{Z}\pi_1 M), Wh_1(\pi_1 M)$.

Anderson and Hsiang have shown in [1] that the functors $K_{-i}$ have an interesting geometric application to the problem of existence and uniqueness of triangulations of locally triangulable spaces. Roughly speaking, what they show is that, away from dimension 4 and apart from obstructions which arise already in the case of closed manifolds, the only other obstructions to the existence and uniqueness of triangulations are $K_{-i}$ obstructions. It seems that this phenomenon should persist in the automorphism spaces of a polyhedron, namely that the differences between PL and TOP stem from the manifold case and from $K_{-i}$ obstructions.

ADDED IN PROOF. Farrell and Hsiang, using Waldhausen's work, have now shown that $\pi_i \mathscr{C}_{\text{Diff}}(*) \otimes \boldsymbol{Q} \approx K_{i+2}(\boldsymbol{Z}) \otimes \boldsymbol{Q}$, which is now to be $\boldsymbol{Q}$ for $i = 3, 7, 11, \cdots$ and zero otherwise. From this they can compute the rank of $\pi_i \text{Diff}(S^n)$ and $\pi_i \text{Diff}(T^n)$, $i \ll n$. In particular, for odd $n \geqq 37$, neither $O(n+1) \to \text{Diff}(S^n)$ nor $\text{Diff}(T^n) \to G(T^n)$ is a $\boldsymbol{Q}$-equivalence!

## REFERENCES


**1.** D. R. Anderson and W. C. Hsiang, *Extending combinatorial* PL *structures on stratified spaces*, Invent. Math. **32** (1976), 179–204.

**2.** ———, *The functors $K_{-i}$ and pseudo-isotopies of polyhedra*, Ann. of Math. (2) **105** (1977), 201–223.

**3.** D. Burghelea and R. Lashof, *The homotopy type of the space of diffeomorphisms*. I, II, Trans. Amer. Math. Soc. **196** (1974), 1–50. MR **50** #8574.

**4.** ———, *Stability of concordances and the suspension homomorphism*, Ann. of Math. (2) **105** (1977), 449–472.

**5.** D. Burghelea, R. Lashof and M. Rothenberg, *Groups of automorphisms of manifolds*, Lecture Notes in Math., vol. 473, Springer-Verlag, Berlin and New York, 1975. MR **52** #1738.

**6.** J. Cerf, *La stratification naturelle des espaces de fonctions différentiables réeles et le théorème de la pseudo-isotopie*, Inst. Hautes Études Sci. Publ. Math. No. **39** (1970), 5–173. MR **45** #1176.

**7.** T. A. Chapman and S. Ferry, *Hurewicz fiberings of ANRs* (to appear).

**8.** ———, *Fibering Hilbert cube manifolds over ANRs* (to appear).

**9.** M. M. Cohen, *Simplicial structures and transverse cellularity*, Ann. of Math. (2) **85** (1967), 218–245. MR **35** #1037.

**10.** ———, *A course in simple-homotopy theory*, Springer-Verlag, 1973. MR **50** #14762.

**11.** A. Hatcher, *Pseudo-isotopy and $K_2$*, Lecture Notes in Math., vol. 342, Springer-Verlag, Berlin and New York, 1973. MR **51** #4278.

**12.** ———, *Higher simple homotopy theory*, Ann. of Math. (2) **102** (1975), 101–137. MR **52** #4305.

**13.** A. Hatcher and T. Lawson, *Stability theorems for "concordance implies isotopy"*, Duke Math. J. **43** (1976), 555–560.

**14.** A. Hatcher and J. Wagoner, *Pseudo-isotopies of compact manifolds*, Astérisque **6** (1973). MR **50** #5821.

**15.** W. C. Hsiang and R. W. Sharpe, *Parametrized surgery and isotopy*, Pacific J. Math. **67** (1976), 401–459.

**16.** K. Igusa (to appear).

**17.** ———, *Postnikov invariants and pseudo-isotopy* (preprint).

**18.** R. Kirby and L. Siebenmann, *Foundational essays on topological manifolds, smoothing, and triangulations*, Ann. of Math. Studies, no. **88**, Princeton Univ. Press, Princeton, N.J., 1977.

**19.** T. Lawson, *Homeomorphisms of* $B^k \times T^n$, Proc. Amer. Math. Soc. **56** (1976), 349–350.

**20.** R. Lee and R. Szczarba, $K_3(\mathbf{Z})$ *is cyclic of order* 48, Ann. of Math. (2) **104** (1976), 31–60.

**21.** J.-L. Loday, *Higher Whitehead groups and stable homotopy*, Bull. Amer. Math. Soc. **82** (1976), 134–136.

**22.** K. Millett, *Piecewise linear concordances and isotopies*, Mem. Amer. Math. Soc. No. **153** (1974). MR **51** #1837.

**23.** E. K. Pedersen, *Topological concordances*, Bull. Amer. Math. Soc. **80** (1974), 658–660. MR **49** #3950.

**24.** C. Rourke, *Embedded handle theory, concordance and isotopy*, Topology of Manifolds, (Proc. Inst., Univ. of Georgia, Athens, Ga., 1969), edited by J. Cantrell and C. Edwards, Markham, Chicago, Ill., 1970, pp. 431–438. MR **43** #5537.

**25.** L. Siebenmann, *Disruption of low-dimensional handlebody theory by Rohlin's theorem*, (Proc. Inst., Univ. of Georgia, Athens, Ga., 1969), edited by J. Cantrell and C. Edwards, Markham, Chicago, Ill., pp. 57–76. MR **42** #6836.

**26.** J. Stallings, *Centerless groups—an algebraic formulation of Gottlieb's theorem*, Topology **4** (1965), 129–134. MR **34** #2666.

**27.** I. A. Volodin, *Generalized Whitehead groups, and pseudoisotopies*, Uspehi Mat. Nauk **27** (1972), 229–230. (Russian) MR **52** #15501.

**28.** F. Waldhausen, *Algebraic K-theory of generalized free products*, Ann. of Math. (to appear).

**29.** C. T. C. Wall, *Surgery on compact manifolds*, Academic Press, New York, 1970.

**30.** J. H. C. Whitehead, *Simple homotopy types*, Amer. J. Math. **72** (1950), 1–57. MR **11**, 735.

**31.** F. Waldhausen, *Algebraic K-theory of topological spaces*. I, these PROCEEDINGS, Part I, pp. 35–60.

PRINCETON UNIVERSITY

Proceedings of Symposia in Pure Mathematics
Volume 32, 1978

# DIFFEOMORPHISMS, $K_2$, AND ANALYTIC TORSION

J. B. WAGONER*

This expository article and those of A. Hatcher and F. Waldhausen on higher $K$-theory in these PROCEEDINGS treat current developments in the theory of concordances and higher simple homotopy theory in relation to algebraic $K$-theory. We shall concentrate here on "lower $K$-theory"; that is, the geometry of $K_1$ and $K_2$ via the space of smooth functions on a manifold, a $K_2$-type invariant for diffeomorphisms of closed manifolds, and combinatorial and analytic torsion for manifolds. We shall also briefly speculate about the problem of higher analytic torsion for detecting elements of $\pi_i \operatorname{Diff}(D^n)$ and of $\pi_i \mathscr{P}(S^{n-1})$ where $\mathscr{P}(S^{n-1})$ is the space of smooth concordances of $S^{n-1}$. Consideration of these homotopy groups is a theme relating the present article to those of Hatcher and of Waldhausen which, among other things, discuss the role of $\pi_i \mathscr{P}(S^{n-1})$ in higher simple homotopy theory and in homotopy groups of spaces of concordances.

Consider the problem of computing $\pi_i \operatorname{Diff}(D^n)$ where $i$ is fixed and $n$ is very large. Standard techniques in differential topology show that in this situation it suffices to consider the $i$th homotopy group of the space of diffeomorphisms keeping the origin fixed. Taking the derivative at the origin then leads to the split exact sequence

$$0 \to \pi_i \mathscr{P}(S^{n-1}) \to \pi_i \operatorname{Diff}(D^n) \to \pi_i \operatorname{GL}_n(R) \to 0.$$

Here for any compact manifold $M$, possibly with boundary $\partial M$, $\mathscr{P}(M)$ denotes the space of *pseudoisotopies* or *concordances* of $M$; that is, the space of diffeomorphisms of $M \times I$ to itself which are the identity on $(M \times 0) \cup (\partial M \times I)$. The sequence is split, of course, because $\pi_i \operatorname{GL}_n(R) \cong \pi_i O(n)$. The groups $\pi_i \operatorname{GL}_n(R)$ are well known from Bott periodicity, so the problem reduces to computing $\pi_i \mathscr{P}(S^{n-1})$. Two main questions in the subject are:

(a) Is $\pi_i \mathscr{P}(S^{n-1})$ finitely generated for $i \ll n$?

---

*AMS (MOS) subject classifications* (1970). Primary 18F25, 57D50, 57D80, 58G99.

*Partially supported by the A. P. Sloan Foundation and NSF Grant MCS 74–03423.

(b) Is $\pi_i \mathscr{P}(S^{n-1}) \otimes Q \cong K_{i+2}(Z) \otimes Q$ for $i \ll n$?

One motive for thinking (b) might be reasonable comes from the connection of the $h$-cobordism and pseudoisotopy theorems with $K_1$ and $K_2$ as reviewed briefly below. The situation for larger $i$ is unclear, but the papers of Hatcher and Waldhausen in this volume discuss a theory which offers a program for deciding the question. One highly conjectural hope of this article is that an affirmative answer to (b) would fit in well with the problem of higher analytic torsion discussed later on.

**1. Torsion and $h$-cobordisms.** Let $W$ be a compact connected smooth manifold of dimension $n$ with boundary $\partial W$ the disjoint union $\partial W = N \cup N'$ where $N$ and $N'$ may be both empty and are not necessarily connected. Let $F$ be a field with an automorphism ("conjugation") $a \to \bar{a}$ of order two. Let $A^*$ denote the conjugate transpose of an $r \times r$ matrix $A$ with entries in $F$. Let $U(r, F)$ be the *unitary* group of those $r \times r$ matrices $A$ for which $A \cdot A^* = 1$. Let $\pi = \pi_1 W$ and suppose we are given an *acyclic* representation $\rho\colon \pi \to U(r, F)$ in the sense that the homology groups $H_*(W, N; \rho)$ of the complex $F^r \otimes_\rho C_*(\tilde{W}, \tilde{N})$ vanish. Here $\tilde{W}$ denotes the universal cover of $W$ and for any subset $X$ of $W$ we let $\tilde{X}$ be the part of $W$ lying over $X$. Then as in **[11]**, **[12]**, **[6]** it is possible to define the Reidemeister-Franz-de Rham *torsion*

$$\tau_\rho(W, N) \in F^*/\pm \det \rho(\pi)$$

where $F^*$ is the multiplicative group of nonzero elements of $F$ and $\pm\det \rho(\pi)$ is the subgroup of elements of the form $\pm\det \rho(g)$ for $g \in \pi$.

Recall briefly the definition of torsion following **[6]**. Let $f\colon (W; N, N') \to (I; 0, 1)$ be a nice Morse function as in **[11]**. This gives rise to a chain complex

$$(1) \qquad 0 \leftarrow \tilde{C}_0 \leftarrow \tilde{C}_1 \leftarrow \cdots \leftarrow \tilde{C}_n \leftarrow 0$$

of free $Z[\pi]$ modules with each $\tilde{C}_i$ having one basis element corresponding to each critical point of index $i$ of $f$. Let $\tilde{\partial}_i\colon \tilde{C}_i \to \tilde{C}_{i-1}$ denote the boundary operator. Let $C_* = F^r \otimes_\rho \tilde{C}_*$ with boundary operator $\partial_i\colon C_i \to C_{i-1}$ given by $\partial_i = 1 \otimes \tilde{\partial}_i$. By hypothesis $C_*$ is acyclic so it is possible to choose a *contraction operator* $d_i\colon C_i \to C_{i+1}$ satisfying $d_i d_{i+1} = 0$ and $d_i \partial_{i+1} + \partial_i d_{i-1} = 1$. Here as in the remainder of the paper we read composition from left to right. Let $C_{\text{ev}} = \bigoplus_k C_{2k}$ and $C_{\text{odd}} = \bigoplus_k C_{2k+1}$. Let $\partial_{\text{ev}}\colon C_{\text{ev}} \to C_{\text{odd}}$ and $d_{\text{ev}}\colon C_{\text{ev}} \to C_{\text{odd}}$ be the corresponding sums of the boundary and contraction operators. Acyclicity of $C_*$ implies $\partial_{\text{ev}} + d_{\text{ev}}\colon C_{\text{ev}} \to C_{\text{odd}}$ is an isomorphism. The basis for $\tilde{C}_i$ is well defined up to ordering and multiplication by $\pm g$ for $g$ in $\pi$. Hence $\tilde{C}_i$ has a preferred basis determined up to transformation by a *monomial* matrix; that is, by a square matrix which has exactly one nonzero entry of the form $\pm g$ in each row and column. The isomorphism $\partial_{\text{ev}} + d_{\text{ev}}$ can therefore be viewed as a $q \times q$ invertible matrix over $F$ where $q = \dim_F C_{\text{ev}} = \dim_F C_{\text{odd}}$. The determinant $\det(\partial_{\text{ev}} + d_{\text{ev}}) \in F^*$ is well defined up to elements of the form $\pm\det \rho(g)$ for $g \in \pi$ and one defines

$$\tau_\rho(W, N) = \det(\partial_{\text{ev}} + d_{\text{ev}}) \in F^*/\pm\det \rho(\pi).$$

For any group $G$ let $M(G)$ denote the group of monomial matrices over $G$ of infinite size which are eventually the identity. Let $\mathrm{GL}(F) = \lim_r \mathrm{GL}(r, F)$ and $U(F) = \lim_r U(r, F)$. For $i \geqq 1$ let $K_i(F) = \pi_i\, B\,\mathrm{GL}(F)^+$ be the algebraic $K$-theory groups defined by Quillen (see **[8]**, **[15]** for example), let $KU_i(F) = \pi_i BU(F)^+$, and let $K_i'(F)$ denote the quotient of $K_i(F)$ by the image of the homomorphism $KU_i(F) \to K_i(F)$. In the special case where $\rho: \pi \to F^*$ is a unitary character we let $K_i^\rho(F)$ denote the quotient of $K_i(F)$ by the image of the homomorphism $\pi_i B(M(\pm\rho(\pi)))^+ \to \pi_i B\,\mathrm{GL}(F)^+$. There is a natural map $K_i^\rho(F) \to K_i'(F)$. Clearly $K_1^\rho(F) = F^*/\pm\rho(\pi)$ so that $\tau_\rho(W, N) \in K_1^\rho(F)$. In the second section we discuss an invariant of diffeomorphisms of $(W; N, N')$ taking values in $K_2^\rho(F)$ and $K_2'(F)$.

If $F = \boldsymbol{C}$ and $\tau_\rho(W, N)$ is considered as lying in $K_1'(\boldsymbol{C}) \cong$ positive reals, it is customary to call $\tau_\rho$ the *R-torsion*. This was used classically to study manifolds of the form $W = S^n/\pi$ where $\pi$ is a finite group acting freely and orthogonally on the $n$-sphere $S^n$. Such manifolds are called Clifford-Klein manifolds. The following theorem was proved by Franz **[7]** for lens spaces and by de Rham **[5]** in the general case. See **[11]** and **[3]** also.

PROPOSITION 1.1. *A Clifford-Klein manifold W is determined up to isometry by its fundamental group* $\pi$ *together with the collection of R-torsions* $\tau_\rho(W)$.

Torsion satisfies product and duality formulae **[3**, §23], **[12]**.

PRODUCT FORMULA 1.2. Let $Q$ denote a closed, compact, simply connected manifold with Euler-Poincaré characteristic $\chi(Q)$. Then

$$\tau_\rho(W \times Q, N \times Q) = \tau_\rho(W, N)^{\chi(Q)}.$$

DUALITY FORMULA 1.3. Let $z \to \bar{z}$ denote the involution on $K_i^\rho(F)$ or $K_i'(F)$ induced by taking a matrix to its conjugate transpose inverse. Let $\varepsilon(n) = (-1)^n$ where $n = \dim W$. Assume $W$ is orientable. Then

$$\tau_\rho(W, N') = [\bar{\tau}_\rho(W, N)]^{\varepsilon(n)}.$$

Note the difference in parity from **[12]**. This is accounted for by using "conjugate transpose inverse" instead of just "conjugate transpose" to get the involution.

The formula (1.3) shows that $R$-torsion vanishes on a closed, orientable, even dimensional manifold.

A second example of torsion is in knot theory. See **[12]**. Let $F = Q(t)$ be the field of rational functions over $Q$ with involution $\bar{p}(t) = p(t^{-1})$. Let $S^n \to S^{n+2}$ be a smooth knot and let $W$ denote the complement of an open tubular neighborhood of $S^n$. $W$ is a homology circle. Let $\rho: \pi_1(W) \to T$ be the usual projection onto the infinite cyclic group $T \cong H_1(W)$. Then $\tau_\rho(W)$ is defined in $K_1^\rho(Q(t))$ and for $n = 1$ is given by

$$\tau_\rho(W) = \frac{A(t)}{1-t} \bmod (\pm t^k)$$

where $A(t)$ is the Alexander polynomial.

The concept of torsion subsequently developed into a key tool for classification of manifolds via Whitehead's theory of simple homotopy types and the $s$-cobordism theorem. See **[11]** and **[3]**. An *h-cobordism* $W$ between closed manifolds $N$ and $N'$ where $\partial W = N \cup N'$ satisfies the condition that the inclusions $N \to W$ and $N' \to W$ are homotopy equivalences. Usually $W$ is also taken to be connected. When $\partial N$ is not empty we require that $\partial W = N \cup U \cup N'$ where $N \cap U = \partial N$, $N' \cap U = \partial N'$, and there exists some product structure

(2) $$g: \partial N \times I \to U$$

with $g|\partial N \times 0 = \mathrm{id}$ and $g(\partial N \times 1) = \partial N'$. The complex (1) above is acyclic and $\partial_{\mathrm{ev}} + d_{\mathrm{ev}}$ is a large invertible matrix over the group ring $Z[\pi]$. The torsion $\tau(W, N)$ is defined to be the class of $\partial_{\mathrm{ev}} + d_{\mathrm{ev}}$ in

$$Wh(\pi) = K_1(Z[\pi])/[\pm g].$$

PROPOSITION 1.4 ($s$-COBORDISM). *Let $W$ be an $h$-cobordism with a product structure $g: \partial N \times I \to U$ as above. Let* $\dim W \geqslant 6$. *Then there is an extension to a diffeomorphism $G: N \times I \to W$ with $G|N \times 0 = \mathrm{id}$ and $G(N \times 1) = N'$ iff $\tau(W, N) = 1$. Moreover every element of $Wh(\pi)$ can be realized as the torsion of some $h$-cobordism on $N$.*

The Whitehead torsion $\tau(W, N)$ satisfies product and duality formulae similar to those for torsion **[3]**, **[11]**.

**2. Pseudoisotopies and $K_2$-type torsion.** The $s$-cobordism theorem gives necessary and sufficient conditions for the existence of a product structure on an $h$-cobordism. The pseudoisotopy theorem **[10]**, **[9**, §10**]**, **[19]** measures the uniqueness of such product structures and is connected with the group $K_2$.

PROPOSITION 2.1. *Let* $\dim N \geqq 6$ *and suppose that the first $k$-invariant lying in $H^3(\pi_1 N; \pi_2 N)$ vanishes. Then*

$$\pi_0 \mathscr{P}(N) \cong Wh_2(\pi) \oplus Wh_1^+(\pi_1; Z_2 \times \pi_2).$$

*Product and duality formulae also hold.*

This result was proved in the early 1970's following naturally the development of the $s$-cobordism theorem, Cerf's result that $\pi_0\mathscr{P}(N) = 0$ whenever $N$ is simply connected and $\dim N \geqslant 5$, and Milnor's definition **[13]** of the functor $K_2$. But let us go "backwards historically" and ask whether there is a $K_2$-type invariant for diffeomorphisms of a manifold $(W; N, N')$ with an acyclic representation as in the situation where torsion is defined. The answer is yes and the theorem below is intended to suggest the pattern for the problem of higher analytic torsion discussed later on.

Suppose $W$ is a manifold with an acyclic representation $\rho: \pi \to U(r, F)$ as in §1. Here $\pi = \pi_1(W, x)$ where $x$ is a fixed base point in $W$. Let $\mathscr{D}$ denote the group of diffeomorphisms of $W$ fixing the base point and inducing the identity on $\pi/\kappa$ where $\kappa$ is the kernel of $\rho$. The following is proved in **[20]**.

PROPOSITION 2.2. *There is a homomorphism*

$$\Sigma_\rho: \pi_0 \mathscr{D} \to K_2'(F)$$

*which satisfies product and duality formulae. Moreover if $\rho: \pi \to F^*$ is a unitary character* (*i.e.*, $r = 1$), *then $\Sigma_\rho$ takes values in $K_2^0(F)$.*

Examples below show $\Sigma_\rho$ can be nontrivial on diffeomorphisms of closed manifolds.

The construction of $\Sigma_\rho$ uses one parameter Cerf-Morse theory as in **[10]**, but the algebra is somewhat more complicated and, curiously, uses an idea developed later on in **[21]** in connection with higher algebraic $K$-theory.

PRODUCT FORMULA 2.3. Let $Q$, $\chi(Q)$, etc., be as in (1.2). Then

$$\Sigma_\rho(G \times 1_Q) = \Sigma_\rho(G)^{\chi(Q)}.$$

DUALITY FORMULA 2.4. Assume $W$ is orientable and let $G \in D$ be a diffeomorphism of the triple $(W; N, N')$ which preserves orientation. Let $\bar{G}$ denote $G$ considered as a diffeomorphism of the triple $(W; N', N)$. Let $z \to \bar{z}$ be the involution of $K_2'(F)$ and $K_2^\varrho(F)$ as in (1.3). Let $\varepsilon(n) = (-1)^n$. Then

$$\Sigma_\rho(\bar{G}) = [\bar{\Sigma}_\rho(G)]^{\varepsilon(n)}.$$

For example, suppose $F = \boldsymbol{C}$ and let us consider $\Sigma_\rho(G)$ as lying in $K_2'(\boldsymbol{C})$. In [**1**, Corollary 1.3] and [**18**] it is shown that $K_2'(\boldsymbol{C})$ is a rational vector space of uncountable dimension and is the negative eigenspace of $K_2(\boldsymbol{C})$ under the involution $-: K_2(\boldsymbol{C}) \to K_2(\boldsymbol{C})$ of (1.3) which, incidentally, coincides with the involution of $K_2(\boldsymbol{C})$ induced by complex conjugation on $\boldsymbol{C}$. If $W$ is a closed manifold, then $\bar{G} = G$; so the duality formula (2.4) implies

COROLLARY 2.5. *If* dim $W$ *is even,* $W$ *is closed and orientable, and* $G$ *preserves the orientation, then* $\Sigma_\rho(G) = 1$ *in* $K_2'(\boldsymbol{C})$.

Here $K_2'(\boldsymbol{C})$ is written multiplicatively.

REMARK 2.6. Let $W$ be closed. Then $\Sigma_\rho(G)$ vanishes in $K_2'(\boldsymbol{C})$ whenever $G$ is isotopic keeping the base point $x$ fixed to an isometry in some Riemannian metric on $W$. This is because the group of isometries of $W$ is a compact Lie group with finitely many components. Therefore $G$ has finite order in $\pi_0\mathscr{D}$ and $\Sigma_\rho(G)$ is trivial because $\Sigma_\rho$ is a homomorphism into a rational vector space. Also $\Sigma_\rho(G)$ vanishes in $K_2'(\boldsymbol{C})$ whenever $\pi_1 W$ is finite. For then the representation $\rho: \pi \to U(r, \boldsymbol{C})$ is unitarily equivalent to a representation into $U(r, \bar{Q})$ where $\bar{Q}$ is the algebraic closure of $Q$, and hence $\Sigma_\rho$ factors through the torsion group $K_2(\bar{Q})$. In fact $K_2(\bar{Q})$ is zero. See [**4**, §§14, 15] and [**1**, Corollary 1.3].

Here is an example where $\Sigma_\rho(G)$ is nontrivial in $K_2'(\boldsymbol{C})$ when $W = L_5 \times S^3 \times S^1$: Let $L_5$ denote the three dimensional lens space obtained by identifying $(z_0, z_1)$ in $S^3$ with $(\alpha z_0, \alpha z_1)$ where $\alpha = e^{2\pi i/5}$. Let $\pi$ be the cyclic group of order five and $T$ be the infinite cyclic group. Then $\pi_1(L_5 \times S^3 \times S^1) = \pi \times T$ and we know from [**22**] that $Wh(\pi)$ is contained as a direct factor of $Wh_2(\pi \times T)$. A construction due to Siebenmann [**23**] gives a pseudoisotopy $G: L_5 \times S^2 \times S^1 \times I \to L_5 \times S^2 \times S^1 \times I$ so that in the notation of [**10**] we have $\Sigma(G) \in Wh(\pi) \subset Wh_2(\pi \times T)$. Any element of $Wh(\pi)$ can be so realized. Let $2G$ denote the "double" of $G$. Thus $2G$ is a diffeomorphism of $L_5 \times S^2 \times S^1 \times [0, 2]$ which is the identity on $L_5 \times S^2 \times S^1 \times 0$ and $L_5 \times S^2 \times S^1 \times 2$. Moreover by the duality formula [**10**] we have $\Sigma(2G) = \Sigma(G)^2$ in $Wh_2(\pi \times T)$. Consider $L_5 \times S^3 \times S^1$ as the union

$$(L_5 \times D^3 \times S^1) \cup (L_5 \times S^2 \times S^1 \times [0, 2]) \cup (L_5 \times D^3 \times S^1)$$

and extend $2G$ to a diffeomorphism of $L_5 \times S^3 \times S^1$ by letting it be the identity on both of the $L_5 \times D^3 \times S^1$ parts. Let $\rho: \pi \times T \to S^1$ be the representation taking the generator $u$ of $\pi$ to $\alpha$ and the generator $t$ of $T$ to any element $z$ of $S^1$ which is part of a transcendence basis for $\boldsymbol{C}$ over $Q$. Now choose $G$ so that $\Sigma(G) \in Wh(\pi) \subset Wh_2(\pi \times \tau)$ is the Milnor symbol $\{u + u^{-1} - 1, t\}$ where $u + u^{-1} - 1$ is the infinite cyclic generator of $Wh(\pi)$. Then from [**20**] we have

(2.7) $$\Sigma_\rho(2G) = \{(\alpha + \alpha^{-1} - 1)^2, z\} \in K_2'(C)$$

which is nontrivial. See **[23]**, **[18]**, and **[13**, §11].

There is also a potential example of $\Sigma_\rho$ in knot theory. Let $F = Q(t)$. Let $S^n \to S^{n+2}$ be a codimension two knot and let $\text{Diff}(S^{n+2}, S^n)$ denote the group of diffeomorphisms of $S^{n+2}$ leaving $S^n$ setwise invariant and preserving the orientations of $S^{n+2}$ and $S^n$. Let $W$ denote the complement of an open tubular neighborhood of $S^n$ and let $\rho\colon \pi_1 W \to T$ be the usual abelianization homomorphism to the infinite cycle group $T$. Then methods of Proposition 2.2 can be used to construct a homomorphism

(2.8) $$\Sigma_\rho\colon \pi_0 \,\text{Diff}(S^{n+2}, S^n) \to K_2^\varrho(Q(t)) \bmod \{\tau_\rho(W), t\}.$$

It is not hard to construct nontrivial examples of (2.8). See **[20]**. Similarly, these ideas should also apply to get $K_2$-type invariants for $\text{Diff}(S_\varepsilon^{2n+1}, K)$ where $K = f^{-1}(0) \cap S_\varepsilon^{2n+1}$ is the link of an isolated singularity of a polynomial $f\colon C^{n+1} \to C$.

The group $K_2^\varrho(Q(t))$ is isomorphic to $K_2(Q(t))$ divided by the subgroup of order four $K_2(Z[t, t^{-1}]) = K_2(Z) \oplus K_1(Z)$ which is generated by the symbols $\{t, -1\}$ and $\{-1, -1\}$. See **[15**, Theorem 8], **[22]**, **[18**, §2]. It is therefore quite large **[13**, §11].

Finally, here is a very simple example. Let $W = S^n \times S^1$ and let $f\colon S^n \times S^1 \to S^n \times S^1$ and $(x, y)$ to $(r(x), y)$ where $r(x_1, \cdots, x_{n+1}) = (-x_1, \cdots, x_{n+1})$ for $(x_1, \cdots, x_{n+1}) \in S^n$. Let $\rho\colon \pi_1 W \to Q(t)^*$ send the generator to $t$. From **[20]** we know

(2.9) $$\Sigma_\rho(t) = \{-1, (t-1)^{\varepsilon(n)}\}$$

where $\varepsilon(n) = (-1)^n$. To see (2.9) is nontrivial let $K_2(Q(t)) \to Q^*$ be the symbol homomorphism obtained as in **[13**, Lemma 11.5] from the discrete valuation on $Q(t)$ corresponding to the prime element $t - 1$ in $Q[t]$. Then $\{-1, -1\}$ and $\{t, -1\}$ go to 1 under this map so there is an induced homomorphism $K_2^\varrho(Q(t)) \to Q^*$. According to **[13**, Lemma 11.5], the element $\{-1, (t-1)^{\varepsilon(n)}\}$ goes to $-1$ under this map.

**3. Analytic *R*-torsion.** The $R$-torsion invariant $\tau_\rho(W, N)$ is obtained from the finite dimensional, acyclic complex $C_*$ arising as in §1 from Morse theory or alternatively from a combinatorial triangulation compatible with the differentiable structure on $W$. Thus $\tau_\rho$ is often called the "combinatorial" torsion. For convenience assume $W$ is compact and closed. Let $0 \to D^0 \to D^1 \to \cdots \to D^n \to 0$ denote the de Rham complex with coboundary $d\colon D^q \to D^{q+1}$ where $D^q$ is the space of smooth sections of $\Lambda^q T^* W \otimes E_\rho$. $E_\rho$ is the flat bundle associated to the unitary representation $\rho$. $D^*$ is infinite dimensional but acyclic because $C_*$ is acyclic. The problem arises of trying to obtain $\tau_\rho$ from the analytic data contained in $D^*$. If $W$ has a Riemannian metric, the Laplacian $\Delta_q\colon D^q \to D^q$ is defined and A. Shapiro suggested that its eigenvalues could be used to get $\tau_\rho$. Ray and Singer in **[16]** have proposed such a definition of the "analytic" torsion $T_\rho$ and showed that it satisfied certain formal properties analogous to $\tau_\rho$. Recently J. Cheeger proved that $T_\rho = \tau_\rho$. In this section we very briefly review this work.

Let $C_* = C^r \otimes \tilde{C}_*$ be the acyclic complex as in §1 obtained from a nice Morse function on $W$. Since $\rho$ is unitary each chain group $C_q$ has a natural hermitian metric so the adjoint $\partial_{q+1}^*: C_q \to C_{q+1}$ of the boundary operator $\partial_{q+1}: C_{q+1} \to C_q$ is defined. The combinatorial Laplacian $\Delta_q^{(c)}: C_q \to C_q$ is given by

$$\Delta_q^{(c)} = -(\partial_q \partial_q^* + \partial_{q+1}^* \partial_{q+1}).$$

Remember the convention is to read composition from left to right. Since $C_*$ is acyclic, $\Delta_q^{(c)}$ has strictly negative eigenvalues $\lambda_1^q, \cdots, \lambda_{n_q}^q$. Ray and Singer first observe that

$$\log \tau_\rho(W) = \frac{1}{2} \sum_{q=0}^{n} (-1)^{q+1} q \log \det(-\Delta_q^{(c)}). \tag{3.1}$$

Then they point out that log $\det(-\Delta_q^{(c)})$ can be expressed in terms of the combinatorial zeta function $\zeta_q^{(c)}(s) = \sum_{1 \leqslant i \leqslant n_q} (-\lambda_i^q)^{-s}$. This is because $(d/ds)(-\lambda)^{-s} = -\log(-\lambda)(-\lambda)^{-s}$ and hence $-\log\det(-\Delta_q^{(c)})$ is just the value of the derivative of $\zeta_q^{(c)}(s)$ evaluated at $s = 0$. Thus (3.1) becomes

$$\log \tau_\rho(W) = \frac{1}{2} \sum_{q=0}^{n} (-1)^q q \zeta_q^{(c)\prime}(0). \tag{3.2}$$

Suppose as above that $W$ has a fixed Riemannian metric. Then there is the duality map $*: D^q \to D^{n-q}$ and an inner product on $D^q$. The formal adjoint of $d$ is $\delta: D^q \to D^{q+1}$ where $\delta = (-1)^{nq+n+1} * d *$ and the Laplacian is $\Delta = -(d\delta + \delta d)$. Let $\Delta_q: D^q \to D^q$ denote the restriction of $\Delta$ to the $q$-forms $D^q$. Then $\Delta_q$ is selfadjoint and negative definite with strictly negative eigenvalues $\lambda_i^q \to -\infty$ with each $\lambda_i^q$ of finite multiplicity. The zeta function $\zeta_{q,\rho}(s) = \sum_i (-\lambda_i^q)^{-s}$ converges for Re$(s)$ large and can be continued to a meromorphic function on the whole complex plane which is holomorphic at $s = 0$. Then using (3. 2) as motivation Ray and Singer proposed the definition of the *analytic torsion* $T_\rho(W)$ to be the positive real root of

$$\log T_\rho(W) = \frac{1}{2} \sum_{q=0}^{n} (-1)^q q \zeta_{q,\rho}'(0). \tag{3.3}$$

The following is shown in **[16]**.

PROPOSITION 3.4. (a) *$T_\rho(W)$ is independent of the choice of Riemannian metric when $\rho: \pi \to U(r, \mathbf{C})$ is acyclic.*

(b) $\log T_\rho(W \times Q) = \chi(Q) \log T_\rho(W)$ *whenever $Q$ is simply connected.*

(c) $\log T_\rho(W) = 0$ *if* dim $W$ *is even.*

In **[16]** it was suggested that the analytic and combinatorial definitions of torsion agree. This has in fact been proved recently by J. Cheeger:

PROPOSITION 3.5. *If $W$ is a closed, compact, Riemannian manifold with an acyclic unitary representation $\rho$, then $T_\rho(W) = \tau_\rho(W)$.*

Now consider the case of a closed, compact, complex manifold $W$ of complex dimension $n$ equipped with a hermitian metric. Here the analytic and combinatorial torsions vanish because the real dimension of $W$ is $2n$. However in **[17]** Ray and Singer use the $\bar{\partial}$ operator in the Dolbeault complex to define a sequence of real numbers $T_0(W, \rho), \cdots, T_n(W, \rho)$. These *$\bar{\partial}$-torsions* are in general nontrivial and, for example, their computation on a Riemann surface involves the Selberg zeta function.

Combinatorial and/or Morse theory methods do not seem fine enough to get nontrivial torsion invariants in the complex setting. One must use the bigrading of the de Rham complex and then carry over the analytic methods from the real case to the vertical columns $\bar{\partial}: D^{p,q} \to D^{p,q+1}$. A rough analogy is that one must use analytic methods to show the odd Betti numbers of a compact Kähler manifold are even. There seems to be no way to prove this topologically.

**4. Higher analytic torsion.** Let $W$ be a compact, closed $C^\infty$ manifold with base point $x$ and let $\rho: \pi \to U(r, \boldsymbol{C})$ be an acyclic representation as in §1. Let $\pi_{-1}$ Diff$(W, x)$ denote the set of compact, closed $C^\infty$ manifolds homotopy equivalent to $W$. More precisely an element of $\pi_{-1}$ Diff$(W, x)$ is a manifold $W'$ with base point $x'$ together with an acyclic representation $\rho'$ into $U(r, \boldsymbol{C})$ and a homotopy equivalence $h: (W', x') \to (W, x)$ compatible with $\rho$ and $\rho'$. We allow $h$ to vary by a base point preserving homotopy. Then $R$-torsion is defined and gives a map

$$\tau_\rho: \pi_{-1} \operatorname{Diff}(W, x) \to K_1'(\boldsymbol{C}).$$

The homomorphism log | det |: GL$(r, \boldsymbol{C}) \to \boldsymbol{R}$ is a continuous cohomology class and of course induces an isomorphism between $K_1'(\boldsymbol{C})$ and $\boldsymbol{R}$. While these $R$-torsions are not completely effective in general (as is the Whitehead torsion in the $s$-cobordism theorem), nevertheless for manifolds such as lens spaces they do capture the geometry as the Franz-de Rham theorem of §1 illustrates.

For $i \geqq 0$ let $\pi_i$ Diff$(W, x)$ denote the $i$th homotopy group of the topological group Diff$(W, x)$ of diffeomorphisms fixing the base point and inducing the identity on $\pi/\kappa$ where $\kappa$ is the kernel of $\rho: \pi \to U(r, \boldsymbol{C})$. In §2 we discussed the homomorphism $\Sigma_\rho: \pi_0$ Diff$(W, x) \to K_2'(\boldsymbol{C})$. This suggests that more generally there *might* be a homomorphism

$$\pi_i \operatorname{Diff}(W, x) \to K_{i+2}'(\boldsymbol{C}). \tag{4.1}$$

But how to construct such a map? The definition of $\Sigma_\rho$ involves two-parameter families of functions on $W$ and the argument is long. Consideration of two- and three-parameter families of functions should yield the desired homomorphism $\pi_1$ Diff$(W, x) \to K_3'(\boldsymbol{C})$. This probably could in fact be done, but the proof would be extremely long. For $i \geqq 2$ the construction (4.1) by consideration of $(i + 1)$- and $(i + 2)$-parameter families of functions on $W$ approaches the impossible from a practical viewpoint. Another way to get $K_{i+2}$ -type invariants for $\pi_i$ Diff$(W, x)$ might be through the analytic data of the de Rham complex. This is the program of "higher analytic torsion".

First consider the case of $\Sigma_\rho$ where $i = 0$. Choose a nice Morse function $f: W \to R$, let $H \in$ Diff$(W, x)$, and let $f' = f \circ H^{-1}$. Let $f_t$, $0 \leqq t \leqq 1$, be a nice one-parameter family of functions from $f$ to $f'$. Then the invertible matrices $\partial_{\text{ev}} + d_{\text{ev}}$ and $\partial_{\text{ev}}' + d_{\text{ev}}'$ obtained as in §1 from $f$ and $f'$ respectively are essentially the same, and the one-parameter family $f_t$ gives rise to a sequence of elementary operations from $\partial_{\text{ev}} + d_{\text{ev}}$ to $\partial_{\text{ev}}' + d_{\text{ev}}'$ defining an element of $K_2$. Suppose now that $W$ is given a Riemannian metric $\mathfrak{g}$. This determines the formal adjoint $\delta: D^q \to D^{q-1}$ to $d: D^q \to D^{q+1}$ in the de Rham complex and then one has the Laplacian $\Delta: D^q \to D^q$ depending on $\mathfrak{g}$. Let $\mathfrak{g}'$ denote the metric induced by $H$ and $\Delta'$ denote the corresponding Laplacian. $\Delta$ and $\Delta'$ have the same eigenvalues and multiplicities, although perhaps different eigenspaces. Thus the analytic torsions are the same. Now let $\mathfrak{g}_t$ be a path from $\mathfrak{g}$ to $\mathfrak{g}'$ in the contractible space of all Riemannian metrics on $W$ and let $\Delta_t$ be the corresponding one-parameter family of Laplacians with changing eigenvalues, multiplicities, and eigenspaces. Can an element of $K_2'(\boldsymbol{C})$ or some other $K_2$-type invariant be produced from this data? This analytic approach should be independent of the path chosen between $\Delta$ and $\Delta'$ and hence should clearly vanish whenever $H$ preserves the metric $\mathfrak{g}$. This is consistent with (2.6) where it was seen that $\Sigma_\rho$ vanishes on isometries. A potential problem here is that one might expect an "analytic" invariant to be continuous in some sense but that there are no continuous symbols on $K_2(\boldsymbol{C})$. However it is not really clear why in this situation the invariant would have to be continuous in a classical sense. Perhaps also it is too restrictive to look at the Laplacians $\Delta_t$; maybe one should consider the one-parameter family of operators of the form $d + \delta$ obtained from $\mathfrak{g}_t$.

Where there are continuous cohomology classes detecting algebraic $K$-theory classes the task of constructing "analytic" invariants may be easier. This is the situation for $i$ = odd in (4.1). Indeed for $i = -1$ the determinant gives $R$-torsion using combinatorial methods where things are finite dimensional, and the derivative of the zeta function at the origin serves as some sort of "log det" for the Laplacian in the infinite dimensional de Rham setting.

Let $H_c^*(\mathrm{GL}(n, \boldsymbol{C}); \boldsymbol{R})$ denote the continuous, or equivalently, the differentiable cohomology of $\mathrm{GL}(n, \boldsymbol{C})$ formed from those real valued Eilenberg-Mac Lane cochains $C(g_1, \cdots, g_k)$ which are continuous (or differentiable) in the variables $g_j$. The Van Est theorem implies this continuous cohomology is isomorphic to the exterior algebra with generators $U_1, U_3, \cdots, U_{2n-1}$ where $U_{2s+1}$ has degree $2s + 1$.

Now fix $i = 2s + 1$ and let $n$ be very large. Then one has

$$K_{2s+1}(\boldsymbol{C}) \cong \mathrm{Prim}\, H_{2s+1}(\mathrm{GL}(n, \boldsymbol{C}); \boldsymbol{R})$$

where "Prim" means the primitive elements and the homology groups are the Eilenberg-Mac Lane homology of the discrete group $\mathrm{GL}(n, \boldsymbol{C})$. See **[2]**. Thus the cohomology class $U_{2s+1}$ induces a homomorphism $U_{2s+1}\colon K_{2s+1}(\boldsymbol{C}) \to \boldsymbol{R}$. Since a continuous cohomology class vanishes on compact groups, we see that $U_{2s+1}$ induces a homomorphism $U'_{2s+1}\colon K'_{2s+1}(\boldsymbol{C}) \to \boldsymbol{R}$. If the mythical homomorphism (4.1) existed, composition with $U'_{2s+1}$ would be one natural way to obtain a map

$$\pi_{2s-1}\, \mathrm{Diff}(W, x) \to \boldsymbol{R}. \tag{4.2}$$

The question is whether there is a direct way to obtain such a homomorphism from the de Rham complex. As usual one would choose a Riemannian metric $\mathfrak{g}$ on $W$ and consider the $(2s-1)$-parameter family of metrics $\mathfrak{g}_t$ induced from $\mathfrak{g}$ by a $(2s-1)$-parameter family of diffeomorphisms $H_t$ representing an element of $\pi_{2s-1}\, \mathrm{Diff}(W, x)$. The family $\mathfrak{g}_t$, $t \in S^{2s-1}$, would be extended to a family $\mathfrak{g}_t$, $t \in D^{2s}$, by contractibility of the space of metrics, and then one would try to obtain a homomorphism as in (4.2) from the corresponding $2s$-parameter family of Laplacians.

Here is an example when $i = 3$. Take $W$ to be the lens space $L$ obtained by dividing $S^{4n+3}$ by the action of $\pi = Z/3$ which multiplies all the coordinates by the primitive cube root of unity $\xi = e^{2\pi i/3}$. Let $\rho\colon \pi \to Q(\xi)^*$ take the generator to $\xi$. It is likely that laborious work using two- and three-parameter families of functions on $L$ could produce a homomorphism $\gamma: \pi_1 \operatorname{Diff}(L, x) \to K_3^{\varrho}(Q(\xi))$. Suppose this has been done. Using the fact that $\pi_3 BM(\pm\pi)^+$ is finite for $\pi = Z/3$ together with the localization exact sequence from algebraic $K$-theory we obtain an isomorphism $K_3(Z[\xi]) \cong K_3^{\varrho}(Q(\xi))$ modulo torsion. Borel's work [2] shows that $K_3(Z[\xi])$ has rank one with an infinite cycle generator which is detected by the continuous cohomology class $U_3$. Thus we would like to find elements of $\pi_1 \operatorname{Diff}(L, x)$ which go to something nonzero under the composition $U_3' \circ \gamma$ and would like to have an analytic formula for the invariant. B. Jahren and W. -C. Hsiang have shown recently that for $n$ large enough

$$L^0_{4n+4}(Z/3) \otimes Q \subset \pi_1 \operatorname{Diff}(L, x) \otimes Q$$

where $L^0_{4n+4}(\pi)$ denotes the kernel of the homomorphism between Wall surgery groups $L_{4n+4}(\pi) \to L_{4n+4}(1)$ induced by the map of $\pi$ to the trivial group. Work of Petrie [14] and Wall [24] shows that $L^0_{4n+4}(Z/3)$ has rank one. This raises several questions. What is the dimension of $\pi_1 \operatorname{Diff}(L, x) \otimes Q$? Is $L^0_{4n+4}(Z/3) \otimes Q$ detectable by the hypothetical homomorphism $U_3' \circ \gamma$? If so, what explains the connection between the $G$-signature invariants used to compute surgery groups and the hoped for $K_3$-type invariant?

One could also pose the program of higher analytic torsion for detecting elements of $\pi_i \operatorname{Diff}(W; N, N')$ where $W$ is an acyclic manifold with $\partial W = N \cup N'$ and the de Rham complex consists of differential forms satisfying boundary conditions as in [16]. The simplest example is where $W = N \times I$ and one considers $\pi_i \mathscr{P}(N)$. Note that when $N = S^{n-1}$ as in question (b) of the introduction, Borel's work [2] shows all of $K_*(Z) \otimes Q$ is detected by continuous cohomology classes.

## References


1. H. Bass and J. Tate, *The Milnor ring of a global field*, Algebraic $K$-theory. II, Lecture Notes in Math., vol. 342, Springer-Verlag, Berlin and New York, 1973, pp. 349–428.

2. A. Borel, *Stable real cohomology of arithmetic groups*, Ann. Sci. École Norm. Sup. (4) **7** (1974), 235–272. MR **52** #8338.

3. M. M. Cohen, *A course in simple homotopy theory*, Graduate Texts in Math., vol. 10, Springer-Verlag, 1973. MR **50** #14762.

4. R. K. Dennis and M. R. Stein, *The functor $K_2$: a survey of computations and problems*, Algebraic $K$-theory. II, Lecture Notes in Math., vol. 342, Springer-Verlag, Berlin and New York, 1973, pp. 243–280. MR **50** #7292.

5. G. de Rham, *Complexes à automorphismes et homéomorphie différentiaple*, Ann. Inst. Fourier (Grenoble) **2** (1950), 51–67. MR **13**, 268.

6. G. de Rham, M. Kervaire and S. Maumary, *Torsion et type simple d'homotopie*, Lecture Notes in Math., vol. 48, Springer-Verlag, Berlin and New York, 1967. MR **36** #5943.

7. W. Franz, *Über die Torsion einer Überdeckung*, J. Reine Angew. Math. **173** (1935), 245–254.

8. S. Gersten, *Higher K-theory of rings*, Algebraic $K$-theory, I, Lecture Notes in Math., vol. 341, Springer-Verlag, Berlin and New York, 1973, pp. 3–42.

9. A. Hatcher, *Higher simple homotopy theory*, Ann. of Math. (2) **102** (1975), 101–137. MR **52** #4305.

10. A. Hatcher and J. Wagoner, *Pseudo-isotopies of compact manifolds*, Astérisque, No. 6, Société Mathématique de France, Paris, 1973. MR **50** #5821.

11. J. Milnor, *Whitehead torsion*, Bull. Amer. Math. Soc. **72** (1966), 358–426. MR **33** #4922.

12. ———, *A duality theorem for Reidemeister torsion*, Ann. of Math. (2) **76** (1962), 137–147. MR **25** #4526.

**13.** J. Milnor, *Introduction to algebraic K-theory*, Ann. of Math. Studies No. 72, Princeton Univ. Press, Princeton, N.J., 1971. MR **50** #2304.

**14.** T. Petrie, *The Atiyah-Singer invariant, the Wall groups* $L_n(\pi, 1)$, *and the function* $(te^x + 1)/(te^x - 1)$, Ann. of Math. (2) **92** (1970), 179–187. MR **47** #7761.

**15.** D. Quillen, *Higher algebraic K-theory*. I, Algebric *K*-theory, I, Lecture Notes In Math., vol. 341, Springer-Verlag, Berlin and New York, 1973, pp. 85–147. MR **49** #2895.

**16.** D. B. Ray and I. M. Singer, *R-torsion and the Laplacian on Riemannian manifolds*, Advances in Math. **7** (1971), 145–210. MR **45** #4447.

**17.** ———, *Analytic torsion for complex manifolds*. M.I.T. (preprint).

**18.** H. Sah and J. Wagoner, *Second homology of Lie groups made discrete*, Comm. Algebra **5** (1977), 611–642.

**19.** I. A. Volodin, *Algebraic K-theory*, Uspehi Mat. Nauk. No. 4 and 5, 1972. MR **52** #14001.

**20.** J. Wagoner, *Diffeomorphisms and the functor* $K_2$, Univ. of California, Berkeley (preprint).

**21.** ———, *Equivalence of Algebraic K-theories*, J. Pure Appl. Algebra (to appear).

**22.** ———, *On* $K_2$ *of the Laurent polynomial ring*, Amer. J. Math. **93** (1971), 123–138. MR **43** #2053.

**23.** ———, *H-cobordisms, pseudo-isotopies, and analytic torsion* (Proc. Internat. Conf. on *K*-theory and $C^*$-algebras, Univ. of Georgia, Athens, Ga., 1975), Lecture Notes in Math., Springer-Verlag, Berlin and New York (to appear).

**24.** C.T.C. Wall, *Classification of hermitian forms*. VI, *Group rings*, Ann. of Math. **103** (1976), 1–80.

UNIVERSITY OF CALIFORNIA, BERKELEY

Proceedings of Symposia in Pure Mathematics
Volume 32, 1978

# ALGEBRAIC $K$-THEORY OF TOPOLOGICAL SPACES. I

FRIEDHELM WALDHAUSEN

This paper is concerned with a functor $A(X)$, from spaces to spaces, which is in some ways analogous to the algebraic $K$-theory functor $K(R)$ which goes from rings to spaces. The two are related by a natural transformation from $A(X)$ to a suitable $K$-theory. The natural transformation itself is induced by a Hurewicz map.

The functor $A(X)$ is of some interest in itself, for example there are about as many definitions of $A(X)$ as there are definitions of $K(R)$. More significantly however it can be used to obtain information about the Whitehead spaces $\mathrm{Wh}^{\mathrm{PL}}(X)$ and $\mathrm{Wh}^{\mathrm{Diff}}(X)$ whose homotopy groups are the PL, resp. Diff, concordance groups, stabilized with respect to dimension.

The plan of the exposition is as follows.

§1 discusses a $K$-theory of simplicial rings. This may be regarded as a model for the study of one aspect of $A(X)$.

§2 gives the quick definition of $A(X)$, via the plus construction.

§3 describes the Whitehead spaces and their relation to $A(X)$.

§4 introduces what by analogy may be called a nonadditive exact-sequence-$K$-theory.

§5 indicates the proof of the main results.

**1. $K$-theory of simplicial rings.** Let $R.$ be a simplicial ring (with unit); then $\pi_0 R.$ is a ring, and $\pi_0\colon R. \to \pi_0 R.$ can be considered as a map of simplicial rings. Let $M_n$ denote $(n \times n)$-matrices, and $\mathrm{GL}_n \to M_n$ the inclusion. Define $\widehat{\mathrm{GL}}_n(R.)$ to be the pullback in the diagram

$$\begin{array}{ccc} \widehat{\mathrm{GL}}_n(R.) & \longrightarrow & M_n(R.) \\ \downarrow & & \downarrow \\ \mathrm{GL}_n(\pi_0 R.) & \longrightarrow & M_n(\pi_0 R.) \end{array}$$

*AMS (MOS) subject classifications* (1970). Primary 55—xx, 57—xx.

and $\widehat{\mathrm{GL}}(R.) = \operatorname{dir\,lim} \widehat{\mathrm{GL}}_n(R.)$. This is a simplicial monoid, and one can form its classifying space $B\widehat{\mathrm{GL}}(R.)$ (to be interpreted, say, as the geometric realization of a certain bisimplicial set). By definition of $\widehat{\mathrm{GL}}(R.)$, the map

$$\pi_1 B\widehat{\mathrm{GL}}(R.) \to \pi_1 B\mathrm{GL}(\pi_0 R.)$$

is an isomorphism, so $\pi_1 B\widehat{\mathrm{GL}}(R.)$ has perfect commutator group, and one can apply Quillen's plus construction to form $B\widehat{\mathrm{GL}}(R.)^+$, together with a natural transformation $B\widehat{\mathrm{GL}}(R.) \to B\widehat{\mathrm{GL}}(R.)^+$ which abelianizes $\pi_1$ and induces an isomorphism on homology. $B\widehat{\mathrm{GL}}(R.)^+$ is a simple space.

REMARK. If $\pi_0 R. = 0$ then $R.$ is contractible (multiply by a path from 1 to 0); hence $B\widehat{\mathrm{GL}}(R.)$ is contractible, and $B\widehat{\mathrm{GL}}(R.)^+$ is not a very interesting space. This shows that our $B\widehat{\mathrm{GL}}(R.)^+$ is very much different, in general, from the $K$-theory of a simplicial ring defined by Anderson **[2]** (one forms $B\mathrm{GL}(R.)^+$ (no ^) by taking the plus degreewise); for example, the Karoubi-Villamayor $K$-theory, which agrees with Quillen's for regular noetherian rings, is the $K$-theory, in Anderson's sense, of a connected simplicial ring **[2]**; as a more immediate example, consider the real numbers. There appears to be only one case where the two definitions agree for general reasons, that of a graded ring considered as a differential graded ring in a trivial way and turned into a simplicial ring by means of the Dold-Kan functor. A question in that context is if there is any relation, in general, between the $K$-theory of a simplicial ring and that of the graded ring of its homotopy groups.

Henceforth we assume that $1 \neq 0$ in $\pi_0 R.$.

DEFINITION. $K(R.) = Z \times B\widehat{\mathrm{GL}}(R.)^+$.

It is a special case of the following proposition that the functor $R. \mapsto K(R.)$ preserves weak homotopy equivalences, that is, if $R. \to R'.$ induces an isomorphism on all homotopy groups then so does the induced map $K(R.) \to K(R'.)$. Thus for example if $R. \xrightarrow{\sim} \pi_0 R.$ then $K(R.)$ gives the Quillen $K$-theory of $(\pi_0 R.)$-modules (in this paper, $K$-theory of a ring will mean that of its free, rather than projective, modules).

Following the usual convention, we will call a map $X \to X'$ *$k$-connected* if, for any basepoint in $X$, the induced map $\pi_j X \to \pi_j X'$ is an isomorphism for $j < k$ and an epimorphism for $j = k$. Bookkeeping with this convention will be easier to follow if one keeps in mind that a $k$-connected map has $(k - 1)$-connected homotopy fibre(s).

PROPOSITION 1.1. *If $R. \to R'.$ is $k$-connected, and $k \geqq 1$, then $K(R.) \to K(R'.)$ is $(k+1)$-connected.*

PROOF. There is a commutative diagram of simplicial sets,

$$\begin{array}{ccc} M_n(R._{(0)}) & \longrightarrow & M_n(R'._{(0)}) \\ \downarrow & & \downarrow \\ \widehat{\mathrm{GL}}_n(R.)_{(e)} & \longrightarrow & \widehat{\mathrm{GL}}_n(R'.)_{(e)} \end{array}$$

where the subscript (0), resp. $(e)$, denotes the zero-, resp. identity-, component, and where the vertical maps are the isomorphisms given by addition of 1. The top map is $k$-connected by hypothesis, and $k \geqq 1$; hence $B\widehat{\mathrm{GL}}(R.) \to B\widehat{\mathrm{GL}}(R'.)$ is $(k+1)$-

connected. Hence $B\widehat{GL}(R.)^+ \to B\widehat{GL}(R'.)^+$, being a map of simple spaces, is also $(k+1)$-connected [6].

The proposition admits the following quantitative refinement.

PROPOSITION 1.2. *Let $R. \to R'.$ be k-connected, where $k \geqq 1$. Then*

$$\pi_{k+1} \mathit{fibre}(K(R.) \to K(R'.)) \xrightarrow{\approx} H_0(\pi_0 R'., \pi_k \mathit{fibre}(R. \to R'.)) \quad (\textit{Hochschild homology})$$
$$\approx \pi_k \mathit{fibre}(R. \to R'.) \,/\, (ar - ra)$$

*where $a \in \pi_k$ fibre$(R. \to R'.)$, and $r \in \pi_0 R'.$*

PROOF. By the Hurewicz theorem, an equivalent assertion is that

$$H_{k+1}\ \mathrm{fibre}(K(R.) \to K(R'.)) \xrightarrow{\approx} H_0(\pi_0 R'., \pi_k \mathrm{fibre}(R. \to R'.)).$$

In order to prove this, we will compare the Serre spectral sequences of the vertical maps in the diagram:

$$\begin{array}{ccc} B\widehat{GL}(R.) & \longrightarrow & B\widehat{GL}(R.)^+ \\ \big\downarrow & & \big\downarrow \\ B\widehat{GL}(R'.) & \longrightarrow & B\widehat{GL}(R'.)^+ \end{array}$$

Let $A$ be an abelian group. We are interested in the cases $A = Z$, and $A = Q$, the rationals, in view of a later application. Then the spectral sequence of the left vertical map is

$$E^2_{p,q} \Rightarrow H_{p+q}(B\widehat{GL}(R.), A)$$

where $E^2_{p,q}$ is given by

$$\mathrm{dir\ lim}_{(n)}\ H_p(B\widehat{GL}_n(R'.), H_q(\mathrm{fibre}(B\widehat{GL}_n(R.) \to B\widehat{GL}_n(R'.)), A)).$$

We have

$$\pi_i\ \mathrm{fibre}(B\widehat{GL}_n(R.) \to B\widehat{GL}_n(R'.)) \xrightarrow{\approx} \pi_{i-1}\ \mathrm{fibre}(\widehat{GL}_n(R.) \to \widehat{GL}_n(R'.))$$
$$\xrightarrow{\approx} \pi_{i-1}\ \mathrm{fibre}(M_n(R.) \to M_n(R'.)) \xrightarrow{\approx} M_n(\pi_{i-1}\mathrm{fibre}(R. \to R'.))$$

for all $i$, this being trivially true for small $i$ because of the connectivity assumption. Furthermore the action of $\pi_1 B\widehat{GL}_n(R'.)$ on the former group corresponds, under the isomorphism, to the conjugation action of $\pi_0 \widehat{GL}_n(R'.) \to^{\approx} GL_n(\pi_0 R'.)$ on the latter. Let $j$ be the smallest number so that $\pi_{j-1}$fibre$(R. \to R'.)$ is nonzero. Then by the Hurewicz theorem and universal coefficient theorem,

$$H_j(\mathrm{fibre}(B\widehat{GL}_n(R.) \to B\widehat{GL}_n(R'.)), A) \approx M_n(\pi_{j-1}\mathrm{fibre}(R. \to R'.)) \otimes A;$$

hence $E^2_{p,j}$ is given by

$$\mathrm{dir\ lim}_{(n)}\ H_p(B\widehat{GL}_n(R'.), M_n(\pi_{j-1}\mathrm{fibre}(R. \to R'.) \otimes A))$$

where the action is the conjugation action.

Let the spectral sequence of the right-hand vertical map be denoted

$$E'^2_{p,q} = H_p(B\widehat{GL}(R'.)^+, B_q) \Rightarrow H_{p+q}(B\widehat{GL}(R.)^+, A)$$

where $B_q = H_q(\mathrm{fibre}(B\widehat{GL}(R.)^+ \to B\widehat{GL}(R'.)^+), A)$. The action in $H_p(B\widehat{GL}(R'.)^+, B_q)$ is trivial. By the fundamental property of the plus construction, the map of spectral sequences is an isomorphism on the abutment, $H_*(B\widehat{GL}(R.), A) \to^{\approx} H_*(B\widehat{GL}(R.)^+, A)$ and on the base, $H_p(B\widehat{GL}(R'.), B_q) \to^{\approx} H_p(B\widehat{GL}(R'.)^+, B_q)$. Since the fibres are connected, we have $A \approx E^2_{0,0} \approx E'^2_{0,0} \approx B_0$; hence $E^2_{p,0} \to^{\approx} E'^2_{p,0}$. Furthermore $E^2_{p,q} \to^{\approx} E'^2_{p,q}$ for $0 < q \leqq k$ since these terms are zero. It follows that we must have

$$E^{\infty}_{0,k+1} \xrightarrow{\approx} E'^{\infty}_{0,k+1}$$

and from this that $E^2_{0.k+1} \to^{\approx} E'^2_{0,k+1}$, i.e., that

$$\operatorname{dir\,lim}_{(n)} H_0(\mathrm{GL}_n(\pi_0 R'_.), M_n(\pi_k \operatorname{fibre}(R. \to R'_.) \otimes A)) \xrightarrow{\approx} H_0(\mathrm{GL}(\pi_0 R'_.), B_{k+1}) \xleftarrow{\approx} B_{k+1}.$$

But clearly the trace map of the coefficients induces an isomorphism of the former term to $H_0(\pi_0 R'_., \pi_k \operatorname{fibre}(R. \to R'_.) \otimes A)$. This completes the proof.

Proposition 1.1 admits a relativization which will be given in the following proposition. To state it properly we need the following notions. Let an $(m, n)$-*connected square* denote a commutative square in which the horizontal maps are $m$-connected, and the vertical maps $n$-connected. Call it *k-homotopy cartesian* if the map of the homotopy fibres of the vertical maps is $(k + 1)$-connected. For example, there is a result in algebraic topology, *homotopy excision*, which says that an $(m, n)$-connected pushout diagram of cofibrations is $k$-homotopy cartesian with $k = (m - 1) + (n - 1)$.

PROPOSITION 1.3. *A $(k - 1)$-homotopy cartesian, $(m - 1, n - 1)$-connected square of simplical rings, with $m, n \geqq 2$ and $k \leqq m + n - 2$, is taken by the functor $K$ to a $k$-homotopy cartesian, $(m, n)$-connected square.*

PROOF. The functor $B\widehat{\mathrm{GL}}$ produces a $k$-homotopy cartesian $(m, n)$-connected square. Let

$$\begin{array}{ccc} A & \longrightarrow & B \\ \downarrow & & \downarrow \\ C & \longrightarrow & D \end{array}$$

be such a square in general. After replacing of $A \to B$ and $C \to D$ by cofibrations, if necessary, their pushout diagram is $(m + n - 2)$-homotopy cartesian, by homotopy excision. From this one sees that $k$-homotopy cartesianness of the square is equivalent, for $k \leqq m + n - 2$, to the map $B \cup_A C \to D$ being $(k + 2)$-connected. In the case at hand we can have $(B \cup_A C)^+ = B^+ \cup_{A^+} C^+$. This shows firstly that the resulting spaces are nilpotent, so **[6]** the connectivity survives under the plus construction. Secondly it shows that the above argument can be traced backward, and the assertion follows.

From Propositions 1.1 and 1.3 one has a spectral sequence relating $K(R.)$ to the $K(\mathrm{Sk}^j R.)$ where $\mathrm{Sk}^j$ denotes the $j$-coskeleton, $E^2_{p,q} \Rightarrow \pi_{p+q} K(R.)$, $p \geqq 0$, $q \geqq 1$, with

$$\begin{aligned} E^2_{p,q} &= \pi_{p+q} \operatorname{fibre}(K(\mathrm{Sk}^{q-1} R.) \to K(\mathrm{Sk}^{q-2} R.)) \quad \text{if } q > 1, \\ E^2_{p,1} &= \pi_{p+1} K(\pi_0 R.) \end{aligned}$$

and

$$E^2_{0,q} \approx H_0(\pi_0 R., \pi_{q-1} R.) \quad \text{if } q > 1$$

by Propsition 1.2

Kiyoshi Igusa has discussed a related spectral sequence in a context of what he calls higher Whitehead groups **[11]**. In this context Igusa points out that by (an analogue of) Proposition 1.3, there is a 'stable range', $p \leqq q-2$, in which $E^2_{p,q}$ can be identified to a certain 'stable' group. We will not enter this discussion here. Rather we will discuss a somewhat different kind of stabilization which will be needed later on.

Let $G(X)$ denote the *loop group* in the sense of Kan **[12]**. $X \mapsto G(X)$ is a functor from pointed connected simplicial sets to free simplicial groups, and $G(X)$ is a loop group for $X$ in the sense that there exists a principal simplicial $G(X)$-bundle over $X$, natural in $X$, with (weakly) contractible total space $X_t \times G(X)$ (a 'twisted cartesian product').

If $R$ is any ring, one may form the simplicial group algebra $R[G(X)]$, hence $K(R[G(X)])$ is defined. For example, Proposition 1.1 says, if $X \to X'$ is $k$-connected then so is $K(R[G(X)]) \to K(R[G(X')])$ provided that $k \geqq 2$.

The functor $K(R[G(X)])$ is related to the usual $K$-theory of a group ring. Indeed, the map $G(X) \to \pi_0 G(X) = \pi_1 X$ induces a natural transformation

$$K(R[G(X)]) \to K(R[\pi_1 X]).$$

This is by no means an equivalence in general. For example,

$$G(X) \to 1 \cdot G(X) \subset R[G(X)]$$

may be considered as a Hurewicz map; and Proposition 1.2 yields

$$\pi_2 \operatorname{fibre}(K(R[G(X)]) \to K(R[\pi_1 X])) \approx (\pi_2 X \otimes R[\pi_1 X])/\pi_1 X.$$

Using a suitable pointed simplicial set as a model, we can decompose the $(n+1)$-sphere, $n \geqq 2$, into its upper and lower hemispheres $D^{n+1}_+$ and $D^{n+1}_-$; this gives for any pointed simplicial set $X$ a pushout diagram

$$\begin{array}{ccc} S^n \wedge (X \cup *) & \longrightarrow & D^{n+1}_+ \wedge (X \cup *) \\ \downarrow & & \downarrow \\ D^{n+1}_- \wedge (X \cup *) & \longrightarrow & S^{n+1} \wedge (X \cup *) \end{array}$$

in which all maps are $n$-connected. By Proposition 1.3, the $(n, n)$-connected square

$$\begin{array}{ccc} K(R[G(S^n \wedge (X \cup *))]) & \longrightarrow & K(R[G(D^{n+1}_+ \wedge (X \cup *))]) \\ \downarrow & & \downarrow \\ K(R[G(D^{n+1}_- \wedge (X \cup *))]) & \longrightarrow & K(R[G(S^{n+1} \wedge (X \cup *))]) \end{array}$$

is therefore $(2n-2)$-homotopy cartesian, consequently the map

$$\begin{aligned} &\Omega^n \operatorname{fibre}(K(R[G(S^n \wedge (X \cup *))]) \to K(R)) \\ &\longrightarrow \Omega^n \operatorname{fibre}(K(R) \to K(R[G(S^{n+1} \wedge (X \cup *))])) \\ &\simeq \Omega^{n+1} \operatorname{fibre}(K(R[G(S^{n+1} \wedge (X \cup *))]) \to K(R)) \end{aligned}$$

is $(n-1)$-connected.

DEFINITION. $K^S(R[G(X)]) = \operatorname{dir\,lim} \Omega^n \operatorname{fibre}(K(R[G(S^n \wedge (X \cup *))]) \to K(R))$.

REMARK. Having to use basepoints and connectivity is somewhat artificial but unfortunately necessary since we are using $G(X)$. For example one may expect that there is a natural transformation

$$K(R[G(X)]) \to K^S(R[G(X)]).$$

This does indeed exist, but it cannot (or at least not obviously) be obtained with the present definition.

Notice that $K^S(R[G(X)])$ is really a functor of two variables, $R$ and $X$. For example if $R = R'[G(Y)]$ there does not appear to be any reason why $K^S(R) = K^S(R[G(*)])$ and $K^S(R'[G(Y)])$ should be the same.

LEMMA 1.4. *The functor* $X \mapsto K^S(R[G(X)])$ *is an (unreduced) homology theory, that is,* $X \mapsto \pi_* K^S(R[G(X)])$ *satisfies the Eilenberg-Steenrod axioms except for the dimension axiom.*

PROOF. The thing to verify is excision. Suppose $X_0$, $X_1$, $X_2$ are simplicial subsets of $X$ so that $X_1 \cup_{X_0} X_2 \to^{\approx} X$. Then the square

$$\begin{array}{ccc} \Omega^n \tilde{K}(R[G(S^n \wedge (X_0 \cup *))]) & \longrightarrow & \Omega^n \tilde{K}(R[G(S^n \wedge (X_1 \cup *))]) \\ \downarrow & & \downarrow \\ \Omega^n \tilde{K}(R[G(S^n \wedge (X_2 \cup *))]) & \longrightarrow & \Omega^n \tilde{K}(R[G(S^n \wedge (X \cup *))]) \end{array}$$

where $\tilde{K}(R[G(Y)]) = \text{fibre}(K(R[G(Y)]) \to K(R))$, is $(n-2)$-homotopy cartesian, by Proposition 1.3. So in the limit the square becomes homotopy cartesian, which is the assertion of excision.

The only use so far of the curious functor $R \mapsto K^S(R)$ is the following result which will be needed later.

PROPOSITION 1.5.

$$\pi_i K^S(Z) \otimes Q \approx \begin{cases} Q, & \textit{if } i = 0, \\ 0, & \textit{if } i > 0. \end{cases}$$

The proof depends on a result of F. T. Farrell and W. C. Hsiang, adapting a technique of Borel [3]. Let $\mathrm{GL}_n(Z)$ act by conjugation on $M_n(Q)$, the rational $(n \times n)$-matrices. The trace $M_n(Q) \to Q$ induces a surjective map

$$\text{dir lim } H_*(\mathrm{GL}_n(Z), M_n(Q)) \to \text{dir lim } H_*(\mathrm{GL}_n(Z), Q).$$

LEMMA (FARRELL AND HSIANG [7]). *This map is an isomorphism.*

PROOF OF PROPOSITION. It suffices to show that, for large $n$,

$$\begin{aligned} B_q &= H_q(\text{fibre}(B\widehat{\mathrm{GL}}(Z[G(S^n)])^+ \to B\,\mathrm{GL}(Z)^+), Q) \\ &\approx \begin{cases} Q & \text{if } q = 0, n, \\ 0 & \text{if } q \leqq 2n-2, \ q \neq 0, n. \end{cases} \end{aligned}$$

This is proved by the method of Proposition 1.2, the comparison of the Serre spectral sequences for rational homology of the vertical maps in the diagram:

$$\begin{array}{ccc} B\widehat{\mathrm{GL}}(Z[G(S^n)]) & \longrightarrow & B\widehat{\mathrm{GL}}(Z[G(S^n)])^+ \\ \downarrow & & \downarrow \\ B\mathrm{GL}(Z) & \longrightarrow & B\mathrm{GL}(Z)^+ \end{array}$$

In the notation set up in the proof of Proposition 1.2, suppose $r$ is the smallest number so that $E^2_{p,r} \to E'^2_{p,r}$ si not an isomorphism, for some $p$. Suppose $r \leqq 2n-2$

and $r \neq n$. The assumption now means that $E'^2_{p,r} \neq 0$. It follows that $E'^2_{0,r} \neq 0$ since this is the coefficient group. Since $E^2_{p,q} \to^{\approx} E'^2_{p,q}$ for $q < r$, and $E^2_{0,r} = 0$, it follows that $E'^2_{0,r}, E'^3_{0,r}, \cdots$, cannot be hit by a nonzero differential. Hence $E'^2_{0,r} \to^{\approx} E'^{\infty}_{0,r}$ must contribute to the abutment, a contradiction.

So suppose the first deviation occurs for $r = n$. The above argument still shows that $E^2_{0,n} \to^{\approx} E'^2_{0,n}$, i.e., that

$$\text{dir lim } H_0(\mathrm{GL}_k(Z),\ M_k(Q)) \xrightarrow{\approx} H_0(\mathrm{GL}(Z),\ B_n) \xleftarrow{\approx} B_n.$$

But

$$\text{dir lim } M_k(Q) \longrightarrow \text{dir lim } H_0(\mathrm{GL}_k(Z),\ M_k(Q)) \xrightarrow{\approx} Q$$

is given by the trace, and it is also the coefficient map, i.e., $H_n(\ , Q)$ applied to the map of fibres. Hence by the lemma of Farrell and Hsiang, $E^2_{p,n} \to^{\approx} E'^2_{p,n}$ for all $p$, and there is no deviation after all. This completes the proof.

**2. The functor** $A(X)$**.** By way of introduction, a definition will be given which is very close to that of the $K$-theory of a simplicial ring. It presupposes the notion of 'ring up to homotopy'.

A *ring up to homotopy* consists of an *underlying space*, $R$, plus a lot of additional structure. This is (i) the additive-group-law-up-to-homotopy, i.e., a homotopy everything $H$-space ($E_\infty$ space) with underlying space $R$ satisfying that $\pi_0 R$ with the induced monoid structure is in fact a group; (ii) the multiplication-law-up-to-homotopy, i.e., a homotopy associative $H$-space with unit and higher coherence conditions ($A_\infty$ space) with underlying space $R$; (iii) homotopy distributivity relating the additive and multiplicative structure, with higher coherence conditions.

The notion of a commutative ring up to homotopy has been successfully codified by May [**14**]. By dropping some of the structure one may expect to obtain from this one workable notion of a ring up to homotopy.

If $R$ is (the underlying space of) a ring up to homotopy, one can form a new ring up to homotopy, the ring of $(n \times n)$-matrices. The underlying space $M_n(R)$ is simply the product of $n^2$ copies of $R$; similarly the additive structure is formed componentwise. However it is the multiplicative structure of $M_n(R)$ in which we are interested.

By restricting to a certain union of components one obtains an $A_\infty$ space with underlying space $\widehat{\mathrm{GL}}_n(R)$, the pullback in the diagram:

$$\begin{array}{ccc} \widehat{\mathrm{GL}}_n(R) & \longrightarrow & M_n(R) \\ \big\downarrow & & \big\downarrow{\scriptstyle \pi_0} \\ \mathrm{GL}_n(\pi_0 R) & \longrightarrow & M_n(\pi_0 R) \end{array}$$

We let $\widehat{\mathrm{GL}}(R) = \text{dir lim } \widehat{\mathrm{GL}}_n(R)$. This is of interest only if $1 \neq 0$ in $\pi_0 R$ which we now assume.

DEFINITION. $K(R) = Z \times B\widehat{\mathrm{GL}}(R)^+$.

There is a 'universal' ring up to homotopy (in the same sense in which the ring of integers is the universal ring with unit). The underlying space of this universal ring up to homotopy is the space

$$\mathrm{Q} = \text{dir lim } \Omega^n S^n.$$

If $G$ is any simplicial group one can form the 'group algebra' $Q[G]$. It has underlying space dir lim $\Omega^n S^n(|G| \cup *)$, the space whose homotopy groups are the stable homotopy groups of $(|G| \cup *)$ (or, what is the same, the framed bordism groups of $|G|$).

If $X$ is a pointed connected simplicial set, let as before $G(X)$ denote the loop group of $X$.

DEFINITION. $A(X) = K(Q[G(X)]) = Z \times B\widehat{GL}(Q[G(X)])^+$.

The notion of 'ring up to homotopy', and the technical problems it entails, enter in this construction of $A(X)$ in the following way: We need that the multiplication on $\widehat{GL}(Q[G(X)])$ is sufficiently associative for the classifying space to exist, and we want a canonical choice for the latter.

On the other hand it turns out that what is $B\widehat{GL}(Q[G(X)])$ for all practical purposes, can also be constructed very directly. This will be described below. It follows that $A(X)$ can also be constructed very directly; and that practically all of the results of the preceding section carry over to $A(X)$, quite independently of a worked out theory of rings up to homotopy, since the proofs do not involve the actual construction of a classifying space of a $\widehat{GL}$.

In order to obtain a more direct definition of $A(X)$, let $G$ be any simplicial group which is a loop group for $X$ in the sense that there exists a principal simplicial $G$-bundle over $X$ with (weakly) contractible total space $X_t \times G$, the latter being a free right simplicial $G$-set. For example, $G = G(X)$, the loop group of Kan.

Define a category $\mathcal{S}(G)$ as follows. The objects are pointed simplicial left $G$-sets $Y$ which are free in the sense that for all $k$, $g \in G_k$, $y \in Y_k$, one has $g(y) = y$ if and only if either $g = 1$ or $y = *$. The morphisms in $\mathcal{S}(G)$ are $G$-maps.

Let $h\mathcal{S}(G)$ be the subcategory of $\mathcal{S}(G)$ of those $G$-maps which are weak homotopy equivalences of the underlying simplicial sets.

Let $\mathcal{S}(G)^n_k$ be the full subcategory of those objects for which

$$|(X_t \times G) \times^G Y| \simeq_{\mathrm{rel}|X|} |X| \vee \bigvee_{j=1,\cdots,k} S^n_j$$

(homotopy equivalence, relative to the subspace $|X|$, to $|X|$ wedge $k$ spheres of dimension $n$), and let

$$h\mathcal{S}(G)^n_k = \mathcal{S}(G)^n_k \cap h\mathcal{S}(G).$$

The latter categories are interrelated by a composition law

$$h\mathcal{S}(G)^n_k \times h\mathcal{S}(G)^n_{k'} \to h\mathcal{S}(G)^n_{k+k'}$$

and a suspension map

$$h\mathcal{S}(G)^n_k \to h\mathcal{S}(G)^{n+1}_k.$$

There is a particular object $(G \cup *)$ in $\mathcal{S}(G)^0_1$. By adding the identity on its $n$-fold suspension, one obtains a stabilization map $h\mathcal{S}^n_k \to h\mathcal{S}(G)^n_{k+1}$.

LEMMA 2.1. *There is a homotopy equivalence*

$$\Omega\ |\text{dir lim}_{(n)}\ h\mathcal{S}(G)^n_k| \simeq \widehat{GL}_k(Q[G(X)])$$

*under which composition of loops in the former space corresponds to matrix multiplication in the latter.*

COROLLARY AND/OR DEFINITION. $A(X) \simeq Z \times |\text{dir lim}_{(n,k)}\ h\mathcal{S}(G(X))^n_k|^+$.

There is a natural transformation

$$A(X) \to K(Z[G(X)])$$

which may be described in two ways. Firstly one may 'linearize' the category $h\mathscr{S}(G)^n_k$, i.e., one applies the functor which sends a free pointed simplicial $G$-set to the free simplicial $Z[G]$-module generated by the nonbasepoint elements; a natural transformation is then given by the Hurewicz map which takes an element to the generator it represents. Secondly, one has the map $B\widehat{\mathrm{GL}}(Q[G(X)])^+ \to B\widehat{\mathrm{GL}}(Z[G(X)])^+$ induced from the ring map $\pi_0\colon Q \to Z$. In view of the linear analogue of Lemma 2.1, the two maps are the same, up to homotopy. As $Q[G(X)] \to Z[G(X)]$ is a rational homotopy equivalence, and 1-connected, the proof of 1.1 (or 1.2) shows:

PROPOSITION 2.2. *The natural transformation $A(X) \to K(Z[G(X)])$ is a rational homotopy equivalence.*

In particular, $A(*) \to K(Z)$ is a rational homotopy equivalence.

Let $\Omega^{\mathrm{fr}}(X)$ (framed bordism) denote the space dir lim $\Omega^n S^n(|X| \cup *)$. The Barratt-Priddy-Quillen-Segal theorem may be stated to say that

$$\Omega^{\mathrm{fr}}(X) \simeq Z \times |\operatorname{dir\,lim} h\mathscr{S}(G(X))^0_n|^+;$$

hence one has a natural transformation $\Omega^{\mathrm{fr}}(X) \to A(X)$, and, clearly, the composite $\Omega^{\mathrm{fr}}(*) \to A(*) \to K(Z)$ is the usual map. In particular, all the elements of $K_*(Z)$ that come from $\pi_*\Omega^{\mathrm{fr}}(*) = \pi^S_*$ also come from $\pi_*A(*)$.

The source of the natural transformation $\Omega^{\mathrm{fr}}(X) \to A(X)$ may be enlarged. For simplicity this will be described only in the case where $X = *$. Notice that in this case, $h\mathscr{S}(G(*))^n_k = h\mathscr{S}(*)^n_k$ is simply the category of pointed simplicial sets, of the weak homotopy type of a certain wedge of spheres, and the maps are pointed maps which are weak homotopy equivalences.

Let these spheres be endowed with orientations. Define $h\bar{\mathscr{S}}(*)^n_1$ to be the subcategory of orientation preserving weak homotopy equivalences. Then

$$\Omega|h\bar{\mathscr{S}}(*)^n_1| \simeq SG_n$$

the space of pointed maps of $S^n$ of degree $+1$, and

$$|\operatorname{dir\,lim} h\bar{\mathscr{S}}(*)^n_1| \simeq BSG.$$

More generally, let $h\bar{\mathscr{S}}(*)^n_k$ consist of those maps in $h\mathscr{S}(*)^n_k$ which are given as a wedge of $k$ maps in $h\bar{\mathscr{S}}(*)^n_1$, followed by some permutation. Then by the Barratt-Priddy-Quillen-Segal theorem,

$$|\operatorname{dir\,lim}_{(n,k)} h\bar{\mathscr{S}}(*)^n_k|^+ \simeq \Omega^{\mathrm{fr}}(BSG)$$

and thus one has a natural transformation $\Omega^{\mathrm{fr}}(BSG) \to A(*)$. If one thinks of $A(*)$ as the $K$-theory of the ring up to homotopy Q, this natural transformation may be pictured as given by the inclusion of the (positive) monomial matrices in Q.

Let $\tilde{\Omega}^{\mathrm{fr}}(BSG)$ be the homotopy fibre of the (naturally split) map $\Omega^{\mathrm{fr}}(BSG) \to \Omega^{\mathrm{fr}}(*)$. Then we have a diagram of fibrations:

$$\begin{array}{ccccc}
\tilde{\Omega}^{\mathrm{fr}}(BSG) & \longrightarrow & \Omega^{\mathrm{fr}}(BSG) & \longrightarrow & \Omega^{\mathrm{fr}}(*) \\
\downarrow & & \downarrow & & \downarrow \\
\mathrm{fibre}(A(*) \to K(Z)) & \longrightarrow & A(*) & \longrightarrow & K(Z)
\end{array}$$

Now $\pi_i\tilde{\Omega}^{\mathrm{fr}}(BSG) = 0$ if $i = 0$ or 1, and $\approx Z_2$ if $i = 2$, and the same is true for $\pi_i$ fibre$(A(*) \to K(Z))$, cf. Corollary 2.7 below. By chasing a representative element through the latter computation, it is not difficult to see that

$$\pi_2\tilde{\Omega}^{\mathrm{fr}}(BSG) \xrightarrow{\approx} \pi_2 \text{ fibre}(A(*) \to K(Z)).$$

Somewhat surprisingly, it appears that $\pi_2$ fibre$(A(*) \to K(Z)) \to \pi_2 A(*)$ is the zero map, or equivalently, that $\pi_3 A(*) \to K_3(Z)$ is not surjective. This comes from an indirect (and involved) argument of Igusa which says that a particular kind of 2-parameter family of cell complexes must exist (cf. the remark after Theorem 3.3 in the next section). An explicit description of this 2-parameter family has not been found so far. It is certain to be complicated, though, since it is closely related to an explicit description of the exotic element of $K_3(Z)$ **[13]**.

Here are the analogues of Propositions 1.1 and 1.3.

PROPOSITION 2.3. *If $X \to X'$ is n-connected, $n \geqq 2$, then so is $A(X) \to A(X')$.*

PROPOSITION 2.4. *The functor A preserves k-homotopy cartesian $(m, n)$-connected squares, provided that $m, n \geqq 2$, $k \leqq m + n - 2$.*

The analogue of Proposition 1.2 is not quite a quantitative statement since it involves computing framed bordism in a fibration. One can do better in special cases. Specifically, one can get spectral sequences from Postnikov towers. The case of the Postnikov tower of $X$ itself seems to be of least interest here, so we will not deal with it.

A ring up to homotopy, with underlying space $R$, has a Postnikov tower. The $j$th term has underlying space $\mathrm{Sk}^j R$, the $j$-coskeleton of $R$. We define $K^j(R) = K(\mathrm{Sk}^j R)$. In the case of $R = Q[G(X)]$, the functor

$$K^j(Q[G(X)]) = Z \times B\widehat{GL}(\mathrm{Sk}^j(Q[G(X)]))^+$$

can again be defined directly, in a way that avoids the general notion of ring up to homotopy. We have, in this case,

$$K^0(Q[G(X)]) = K(Z[\pi_1 X]).$$

The analogue of Proposition 1.2 is

PROPOSITION 2.5. *Let $j \geqq 1$. Then $K^j(R) \to K^{j-1}(R)$ is $(j + 1)$-connected, and*

$$\pi_{j+1}\text{fibre}(K^j(R) \to K^{j-1}(R)) \approx H_0(\pi_0 R, \pi_j R).$$

*In particular if $R = Q[G(X)]$, this is $\approx H_0(\pi_1 X, \pi_j \Omega^{\mathrm{fr}} G(X))$ which, for $j = 1$, is $H_0(\pi_1 X, (Z_2 \oplus \pi_2 X) \otimes Z[\pi_1 X])$.*

The tower of maps $K(R) \cdots \to K^j(R) \to \cdots$, gives rise to a spectral sequence $E^2_{p,q} \Rightarrow \pi_{p+q}K(R)$, $p \geqq 0$, $q \geqq 1$, with

$$E^2_{p,q} = \pi_{p+q}\text{fibre } (K^{q-1}(R) \to K^{q-2}(R)) \quad \text{if } q \geqq 2,$$

where the term $E^2_{0,q}$ is given by the proposition, and

$$E^2_{p,1} = \pi_{p+1}K^0(R) = \pi_{p+1}K(\pi_0 R).$$

Similarly, we may consider the Postnikov tower on the 'coefficient ring' Q, and define

$$A^j(X) = K((\mathrm{Sk}^j Q)[G(X)]).$$

Then $A^0(X) = K(Z[G(X)])$, and the analogue of Proposition 1.2 is

PROPOSITION 2.6. *Let $j \geqq 1$. Then $A^j(X) \to A^{j-1}(X)$ is $(j+1)$-connected, and*

$$\begin{aligned}\pi_{j+1}\mathrm{fibre}(A^j(X) \to A^{j-1}(X)) &\approx H_0(\pi_1 X, \pi_j^S \otimes Z[\pi_1 X]) \\ &\approx \pi_j^S \otimes H_0(\pi_1 X, Z[\pi_1 X]) \quad (\textit{conjugation action}).\end{aligned}$$

Again the tower of maps $A(X) \cdots \to A^j(X) \to \cdots$, gives rise to a spectral sequence $E^2_{p,q} \Rightarrow \pi_{p+q}A(X)$, $p \geqq 0$, $q \geqq 1$, with

$$\begin{aligned}E^2_{p,q} &= \pi_{p+q}\mathrm{fibre}(A^{q-1}(X) \to A^{q-2}(X)) \quad \text{if } q \geqq 2, \\ E^2_{p,1} &= \pi_{p+1}K(Z[G(X)]).\end{aligned}$$

In the special case $X = *$, either of the two towers of maps above specializes to

COROLLARY 2.7. *There is a tower of maps $A(*) \cdots \to A^j(*) \to \cdots$ approximating $A(*)$, with*

(i) $A^0(*) = K(Z)$,
(ii) $A^j(*) \to A^{j-1}(*)$ *is $(j+1)$-connected,*
(iii) $\pi_{j+1}$ *fibre*$(A^j(*) \to A^{j-1}(*)) \approx \pi_j^S$.

Lastly we have to consider the stabilization of the functor $A(X)$. Arguing as in the preceding section we may define it as

$$A^S(X) = \mathrm{dir\,lim}\ \Omega^n\ \mathrm{fibre}(A(S^n \wedge (X \cup *)) \to A(*)).$$

In detail, the $n$th map in the system is the map

$$\Omega^n\ \mathrm{fibre}(A(S^n \wedge (X \cup *)) \to A(*)) \to \Omega^n\ \mathrm{fibre}(A(*) \to A(S^{n+1} \wedge (X \cup *)))$$

which is the map of fibres in the diagram:

$$\begin{array}{ccc} A(S^n \wedge (X \cup *)) & \longrightarrow & A(D^{n+1}_+ \wedge (X \cup *)) \\ \downarrow & & \downarrow \\ A(D^{n+1}_- \wedge (X \cup *)) & \longrightarrow & A(S^{n+1} \wedge (X \cup *)) \end{array}$$

Anticipating from §5 that $A$ is really a functor on simplicial sets, not necessarily pointed nor connected, we see that $A(S^0 \wedge (X \cup *))$ makes sense, and, clearly,

$$A(X) \xrightarrow{\sim} \mathrm{fibre}(A(S^0 \wedge (X \cup *)) \to A(*)).$$

Therefore there is a natural transformation

$$A(X) \to A^S(X).$$

The analogue of Lemma 1.4 is

LEMMA 2.8. *The functor $X \mapsto A^S(X)$ is a homology theory.*

The 'coefficients' of this homology theory is the space $A^S(*)$. In view of Proposition 2.2, the natural transformation

$$A^S(*) \to K^S(Z)$$

is a rational homotopy equivalence. Hence Proposition 1.5 gives

PROPOSITION 2.9.

$$\pi_i A^S(*) \otimes Q \approx \begin{cases} Q & \text{if } i = 0, \\ 0 & \text{if } i > 0. \end{cases}$$

**3. The Whitehead spaces and their relation to** $A(X)$**.** If $X$ is a PL manifold, denote $\mathscr{C}_0^n(X)$ the groupoid in which an object is a PL $h$-cobordism whose lower face is a compact codimension zero submanifold of $X \times I^n$ (where $I^n$ is the $n$-cube); a morphism in $\mathscr{C}_0^n(X)$ is a PL isomorphism which is the identity on $X \times I^n$, the isomorphism need not preserve the upper face. More generally, let $\mathscr{C}_k^n(X)$ be the groupoid of PL $k$-parameter families of such $h$-cobordisms, the parameter domain being the $k$-simplex. Define $\mathscr{C}^n(X)$ to be the simplicial groupoid which in degree $k$ is $\mathscr{C}_k^n(X)$.

Multiplication with the interval gives a map $\mathscr{C}^n(X) \to \mathscr{C}^{n+1}(X)$, and one defines $C^{\mathrm{PL}}(X) = \text{dir lim } |\mathscr{C}^n(X)|$. The functor $X \mapsto C^{\mathrm{PL}}(X)$ extends, canonically up to homotopy, to a functor from spaces to spaces; this kind of argument is well known, it is described in [9] in one case.

In view of the composition law 'gluing at$X$', $C^{\mathrm{PL}}(X)$ is the underlying space of a $\Gamma$-space in the sense of Segal [17]; hence it is canonically an infinite loop space. In particular there is a canonical (connected) deloop $\mathrm{Wh}^{\mathrm{PL}}(X)$, the PL *Whitehead space*.

More or less by definition of this space, $\pi_1 \mathrm{Wh}^{\mathrm{PL}}(X)$ gives a stable classification of $h$-cobordisms, and $\pi_2 \mathrm{Wh}^{\mathrm{PL}}(X)$ classifies stable concordances. In view of Hatcher's stability theorem [8], $\pi_{i+2} \mathrm{Wh}^{\mathrm{PL}}(X)$ is actually isomorphic to the $i$th concordance group of $X$ if $X$ is a compact PL manifold whose dimension is sufficiently large (depending on $i$). Cf. Hatcher's article [9] for a summary of known results.

Note that $\mathrm{Wh}^{\mathrm{PL}}(*) \simeq *$ in view of the (stable) $h$-cobordism theorem and the Alexander trick.

THEOREM 3.1. *There is a map* $A(X) \to \mathrm{Wh}^{\mathrm{PL}}(X)$, *well defined up to homotopy. Its homotopy fibre, denoted* $h(X, A(*))$, *is a homology theory.*

COROLLARY. *The 'coefficients' of this homology theory is* $A(*)$.

PROOF. $\mathrm{Wh}^{\mathrm{PL}}(*) \simeq *$, so by definition of $h(X, A(*))$ there is a homotopy fibration $h(*, A(*)) \to A(*) \to *$.

Theorem 3.1 is entirely a nonmanifold theorem; the proof starts from Hatcher's 'parametrized $h$-cobordism theorem' [8], a nonmanifold reformulation of $\mathrm{Wh}^{\mathrm{PL}}(X)$, and from that point on, manifolds just are not used anywhere in the argument—except maybe an occasional simplex. The proof will be indicated in §5.

REMARK. As will be apparent later, $A(*)$ is the underlying space of a $\Gamma$-space, hence the coefficients of a homology theory by the recipe of [1]. This homology theory coincides with $h(X, A(*))$. For general reasons again, there is a natural transformation

$$h(X, A(*)) \to A(X)$$

which turns out to coincide with the map of Theorem 3.1. By naturality therefore there is a diagram whose rows are homotopy fibrations:

$$\begin{array}{ccccc} h(X, A(*)) & \longrightarrow & A(X) & \longrightarrow & \mathrm{Wh}^{\mathrm{PL}}(X) \\ \downarrow & & \downarrow & & \downarrow \\ h(X, K(Z)) & \longrightarrow & K(Z[G(X)]) & \longrightarrow & \mathrm{Wh}(G(X)) \\ \downarrow & & \downarrow & & \downarrow \\ h(B\pi_1 X, K(Z)) & \longrightarrow & K(Z[\pi_1 X]) & \longrightarrow & \mathrm{Wh}(\pi_1 X) \end{array}$$

Here the upper row is the fibration of Theorem 3.1, and the lower row is a fibration studied in **[18]**. Concerning the middle row, $K(Z[G(X)])$ is as defined in §1, and the map

$$h(B(G(X)),\ K(Z)) \rightarrow K(Z[G(X)])$$

is defined similarly as the map $h(B\pi_1 X,\ K(Z)) \rightarrow K(Z[\pi_1 X])$ in **[18]**. The term $\mathrm{Wh}(G(X))$ in the middle row can be (and is) defined so that the row is a homotopy fibration. This ends the remark.

While there appears to be no direct way to obtain an analogous result for $\mathrm{Wh}^{\mathrm{Diff}}(X)$, the smooth analogue of $\mathrm{Wh}^{\mathrm{PL}}(X)$, it turns out that one may proceed indirectly, using known results about $\mathrm{Wh}^{\mathrm{Diff}}(X)$ and about its relation to $\mathrm{Wh}^{\mathrm{PL}}(X)$, to obtain a result which is just as good. The argument is as follows.

The stabilization procedure to construct $A^S(X)$ from $A(X)$, and the fact that $A^S(X)$ is a homology theory, presuppose only certain formal properties of $A(X)$ and so carry over to other functors sharing these properties. In particular, we may stabilize the whole fibration of Theorem 3.1, and obtain a diagram of homotopy fibrations

$$\begin{array}{ccccc} h(X, A(*)) & \longrightarrow & A(X) & \longrightarrow & \mathrm{Wh}^{\mathrm{PL}}(X) \\ \downarrow\wr & & \downarrow & & \downarrow \\ h^S(X, A(*)) & \longrightarrow & A^S(X) & \longrightarrow & (\mathrm{Wh}^{\mathrm{PL}})^S(X) \end{array}$$

and the left-hand vertical map is a homotopy equivalence since $h(X, A(*))$ is a homology theory already, and therefore unchanged by stabilization. Hence the right-hand square is homotopy cartesian.

It follows from smoothing theory (Burghelea, Lashof and Rothenberg **[5]**) that $F(X)$, the homotopy fibre of $\mathrm{Wh}^{\mathrm{Diff}}(X) \rightarrow \mathrm{Wh}^{\mathrm{PL}}(X)$, is a homology theory. Hence as before, stabilization gives a diagram of fibrations

$$\begin{array}{ccccc} F(X) & \longrightarrow & \mathrm{Wh}^{\mathrm{Diff}}(X) & \longrightarrow & \mathrm{Wh}^{\mathrm{PL}}(X) \\ \downarrow\wr & & \downarrow & & \downarrow \\ F^S(X) & \longrightarrow & (\mathrm{Wh}^{\mathrm{Diff}})^S(X) & \longrightarrow & (\mathrm{Wh}^{\mathrm{PL}})^S(X) \end{array}$$

in which the left-hand vertical map is a homotopy equivalence. Hence the right-hand square is homotopy cartesian.

Putting these two squares together, we obtain the diagram

$$\begin{array}{ccccc} A(X) & \longrightarrow & \mathrm{Wh}^{\mathrm{PL}}(X) & \longleftarrow & \mathrm{Wh}^{\mathrm{Diff}}(X) \\ \downarrow & & \downarrow & & \downarrow \\ A^S(X) & \longrightarrow & (\mathrm{Wh}^{\mathrm{PL}})^S(X) & \longleftarrow & (\mathrm{Wh}^{\mathrm{Diff}})^S(X) \end{array}$$

in which both squares are homotopy cartesian. Hence the homotopy fibres of the vertical maps are all homotopy equivalent, and are mapped to each other by homotopy equivalence. But $\mathrm{Wh}^{\mathrm{Diff}}(S^n) \to \mathrm{Wh}^{\mathrm{Diff}}(*)$ is a $(2n-2)$-connected map [5] (cf. Hatcher [9] for a more direct argument, using only Morlet's lemma of disjunction). Hence

$$(\mathrm{Wh}^{\mathrm{Diff}})^S(*) \simeq *$$

and (since $(\mathrm{Wh}^{\mathrm{Diff}})^S$ is a homology theory) $(\mathrm{Wh}^{\mathrm{Diff}})^S(X) \simeq *$. Hence the homotopy fibre of the right-hand vertical map is $\mathrm{Wh}^{\mathrm{Diff}}(X)$, and it follows therefore that there is a homotopy fibration $\mathrm{Wh}^{\mathrm{Diff}}(X) \to A(X) \to A^S(X)$.

This leads to a numerical result. Namely by Proposition 2.2, and thanks to Borel [3],

$$\pi_i A(*) \otimes \boldsymbol{Q} \approx \pi_i K(Z) \otimes \boldsymbol{Q} \approx \begin{cases} \boldsymbol{Q} & \text{if } i = 0, \\ \boldsymbol{Q} & \text{if } i = 5, 9, 13, \cdots, \\ 0 & \text{otherwise.} \end{cases}$$

And by Propositions 1.5 and 2.9, and thanks to Farrell and Hsiang [7],

$$\pi_i A^S(*) \otimes \boldsymbol{Q} \approx \pi_i K^S(Z) \otimes \boldsymbol{Q} \approx \begin{cases} \boldsymbol{Q} & \text{if } i = 0, \\ 0 & \text{if } i \neq 0. \end{cases}$$

Hence

THEOREM 3.2.

$$\pi_i \mathrm{Wh}^{\mathrm{Diff}}(*) \otimes \boldsymbol{Q} \approx \begin{cases} \boldsymbol{Q} & \textit{if } i = 5, 9, 13, \cdots, \\ 0 & \textit{otherwise}, \end{cases}$$

COROLLARY. *The smooth Alexander trick fails rationally.*

One way to visualize the map $A(X) \to \mathrm{Wh}^{\mathrm{PL}}(X)$ of Theorem 3.1 is to replace $A(X)$ by another functor, not too far removed from it, and then map the latter. This functor is related to the idea of an elementary expansion. We will refer to it as the *combinatorial Whitehead space* of $X$, denoted $\mathrm{Wh}^{\mathrm{Comb}}(X)$. Its definition, which is rather involved, will now be given.

First we need a very rigid notion of cell complex, in fact we want $k$-parameter families of such. Working in the framework of topological spaces, a diagram

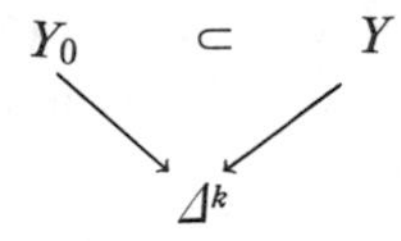

where $\Delta^k$ is the $k$-simplex, will be called a *k-parameter family of cell complexes* from $Y_0$ to $Y$ if it is endowed with the following data:

(i) a finite filtration $Y_0 \subset Y_1 \subset \cdots \subset Y$, over $\Delta^k$,

(ii) for each $j > 0$, an attaching map over $\Delta^k$

$$(S^{n_1} \cup \cdots \cup S^{n_{ij}}) \times \Delta^k \to Y_{j-1}$$

and an isomorphism over $\Delta^k$

$$Y_j \xrightarrow{\approx} Y_{j-1} \cup_{(\cup_i S^{n_i} \times \Delta^k)} \Big( \bigcup_i D^{n_i+1} \times \Delta^k \Big);$$

these data are subject *only* to the equivalence relation of refinement: two cells which are attached simultaneously, may also be attached one after the other (in any order).

Similarly, we define a *k-parameter family of expansions* from $Y_0$ to $Y$, by not attaching disks $D^{n+1}$ along spheres $S^n$, but attaching pairs $(D^{n+1}, D^n_+)$ along pairs $(D^n_-, S^{n-1})$. Again, the structure is supposed to be very rigid, subject only to the equivalence relation of refinement, as above. In particular, we insist here that the particular pairing of cells *is* part of the data, and not subject to change.

DEFINTION. $\mathscr{E}(X)_k$ is the category in which:

(i) an object is a $k$-parameter family of cell complexes from $X \times \Delta^k$ to some $Y$ which is 'acyclic', i.e., the inclusion $X \times \Delta^k \to Y$ is a homotopy equivalence;

(ii) a morphism from $(Y, \cdots)$ to $(Y', \cdots)$ is a $k$-parameter family of expansions from $Y'$ to $Y$ such that the cell structure in $(Y, \cdots)$ coincides with the cell structure induced from the expansion.

We let $\mathscr{E}(X)$ be the simplicial category which in degree $k$ is $\mathscr{E}(X)_k$. Its geometric realization, $E(X) = |\mathscr{E}(X)|$, will be referred to as the *expansion space*.

An interesting question is if the 'two-index-theorem' holds for $E(X)$. That is, if one defines $E^{i,i+1}(X)$ by insisting that all the cells involved have dimension either $i$ or $i + 1$, is it true that dir $\lim_{(i)} E^{i,i+1}(X)$ is homotopy equivalent to $E(X)$? This is far from being obviously true, in fact it might well be wrong.

In view of the composition law 'gluing at $X$', $E(X)$ is the underlying space of a $\Gamma$-space, hence canonically an infinite loop space, and $\mathrm{Wh}^{\mathrm{Comb}}(X)$ is defined as the deloop.

THEOREM 3.3. *There is a map $A(X) \to \mathrm{Wh}^{\mathrm{Comb}}(X)$, well defined up to homotopy. The sequence*

$$\Omega^{\mathrm{fr}}(X) \to A(X) \to \mathrm{Wh}^{\mathrm{Comb}}(X)$$

*is, canonically, the homotopy type of a fibration.*

REMARK. Continuing the discussion of what $\pi_2 A(*)$ is (cf. the material just before Proposition 2.3), we have $\pi_i \mathrm{Wh}^{\mathrm{Comb}}(*) = 0$ if $i = 0, 1$, and

$$\pi_2 A(*) \approx \pi_2 K(Z) \oplus \pi_2 \mathrm{Wh}^{\mathrm{Comb}}(*)$$

and the map

$$Z_2 \approx \pi_2 \tilde{\Omega}^{\mathrm{fr}}(BSG) \xrightarrow{\approx} \pi_2 \text{ fibre } (A(*) \to K(Z)) \longrightarrow \pi_2 \mathrm{Wh}^{\mathrm{Comb}}(*)$$

is surjective. In view of what this map means geometrically, the candidate for a nontrivial element in $\pi_2 \mathrm{Wh}^{\mathrm{Comb}}(*) = \pi_1 E(*)$ is represented by a 'rolling collapse', a circle of cell complexes $S^n \cup_{S^n} D^{n+1}$ where the attaching map is homotopic to the identity, and varies through the nontrivial element of $\pi_1^S$. The aforementioned argument of Igusa is that this element, and hence $\pi_1 E(*)$, must be zero for the following reason. There must exist a 2-parameter family of cell complexes which over the boundary of the parameter domain consists of an odd number of rolling collapses, plus maybe a few circles of expansions ('nonrolling' collapses); for if such a 2-parameter family would not exist, it would follow by a (tricky) geometric argument that $\pi_3^S$ should split off $K_3(Z)$, in contradiction to the result of Lee and Szczarba [**13**]. This ends the remark.

Here is how to map $\mathrm{Wh}^{\mathrm{Comb}}(X)$ to $\mathrm{Wh}^{\mathrm{PL}}(X)$: Take an acyclic cell complex, fatten the cells until a handle decomposed $h$-cobordism (of suitably large dimension) is obtained, and then forget the handle structure. To make this idea work, one redefines the expansion space using handle decomposed $h$-cobordisms instead of cell complexes. Thus if $X$ is a PL manifold, define $\mathscr{E}_0^n(X)$ to be the category in which an object consists of

(i) an object of the category $\mathscr{C}_0^n(X)$ (cf. the beginning of the section),

(ii) a handle decomposition of the $h$-cobordism (including all the necessary data to describe it *completely*), up to an equivalence relation of rearrangement of handles (as in the definition of the expansion space),

and where a morphism is a standard handle cancellation, or composition of such (subject to an equivalence relation of rearrangement). More generally define $\mathscr{E}_k^n(X)$ to be the category whose objects are the PL $k$-parameter families of such handle decomposed $h$-cobordisms, and whose morphisms are the PL $k$-parameter families of standard handle cancellations. $\mathscr{E}^n(X)$ is the simplicial category which in degree $k$ is $\mathscr{E}_k^n(X)$.

Forgetting the handle structure gives a map $\mathscr{E}^n(X) \to \mathscr{C}^n(X)$. This uses that a standard handle cancellation involves a canonical isomorphism of the underlying manifolds.

In order that this define a map $\mathrm{Wh}^{\mathrm{Comb}}(X) \to \mathrm{Wh}^{\mathrm{PL}}(X)$, one needs that $\mathscr{E}^n(X)$ is sufficiently close to $\mathscr{E}(X)$. The latter is seen as follows. One maps $\mathscr{E}^n(X)$ to $\mathscr{E}(X)$ by squeezing each handle to its core. Then a handle-by-handle argument shows that this map is highly connected (depending on the difference of $\dim(X) + n$ and the index of the handle; each test for homotopy equivalence in the limit involves a finite diagram only, so there is a highest handle index).

The map $\mathrm{Wh}^{\mathrm{Comb}}(X) \to \mathrm{Wh}^{\mathrm{PL}}(X)$ so obtained satisfies that

$$\begin{array}{ccc} & A(X) & \\ \swarrow & & \searrow \\ \mathrm{Wh}^{\mathrm{Comb}}(X) & \longrightarrow & \mathrm{Wh}^{\mathrm{PL}}(X) \end{array}$$

commutes up to homotopy.

It appears that a careful wording of the argument actually yields a smooth analogue, a factorization up to homotopy of $\mathrm{Wh}^{\mathrm{Comb}}(X) \to \mathrm{Wh}^{\mathrm{PL}}(X)$ through a map $\mathrm{Wh}^{\mathrm{Comb}}(X) \to \mathrm{Wh}^{\mathrm{Diff}}(X)$. This leads to a startling consequence. Namely in view of the resulting homotopy commutative diagram

$$\begin{array}{ccc} & A(X) & \\ \swarrow & & \searrow \\ \mathrm{Wh}^{\mathrm{Diff}}(X) & \longrightarrow & \mathrm{Wh}^{\mathrm{PL}}(X) \end{array}$$

the homotopy fibre of $A(X) \to \mathrm{Wh}^{\mathrm{Diff}}(X)$ can be identified to the homotopy fibre of

$$\mathrm{fibre}(A(X) \to \mathrm{Wh}^{\mathrm{PL}}(X)) \longrightarrow \mathrm{fibre}(\mathrm{Wh}^{\mathrm{Diff}}(X) \to \mathrm{Wh}^{\mathrm{PL}}(X)),$$

a map of homology theories. Hence $\mathrm{fibre}(A(X) \to \mathrm{Wh}^{\mathrm{Diff}}(X))$ is also a homology theory, and stabilization gives a diagram of fibrations

$$\begin{array}{ccccc} ? & \longrightarrow & A(X) & \longrightarrow & \mathrm{Wh}^{\mathrm{Diff}}(X) \\ \downarrow\wr & & \downarrow & & \downarrow \\ ? & \longrightarrow & A^S(X) & \longrightarrow & (\mathrm{Wh}^{\mathrm{Diff}})^S(X) \end{array}$$

in which the left-hand vertical map is a homotopy equivalence. Because $(\mathrm{Wh}^{\mathrm{Diff}})^S(X) \simeq *$, it follows that $A(X)$ actually splits,

$$A(X) \simeq \mathrm{Wh}^{\mathrm{Diff}}(X) \times A^S(X).$$

It appears that this splitting is hard to reconcile with the computation of $\pi_3 \mathrm{Wh}^{\mathrm{Diff}}(*)$ described in §3 of [9].

Finally a word about the map $A(X) \to \mathrm{Wh}^{\mathrm{Comb}}(X)$. The very description of this map requires the machinery of the following sections, involving a very particular way of constructing simplicial objects. An indication of the nature of this kind of simplicial structure can be seen from the following remarks. For ease of notation we consider only the case $X = *$.

Let $\mathscr{C}$ be the category whose objects are the finite pointed simplicial sets of the homotopy type of, say, a wedge of two spheres of dimension twenty, and whose morphisms are the weak homotopy equivalences which happen to be monomorphisms. Then $\mathscr{C}$ maps to $A(*)$ and hence, by composition, to $\mathrm{Wh}^{\mathrm{Comb}}(*)$. By wishful thinking (a bit too wishful really, but not too far off either) let us insist on the following:

(i) $\mathrm{Wh}^{\mathrm{Comb}}(*)$ can be described as a simplicial object in such a way that

$$\mathrm{nerve}(\mathscr{C}) \to \mathrm{Wh}^{\mathrm{Comb}}(*)$$

can be a simplicial map,

(ii) on 1-simplices, this map is given by

$$(Y_0 \overset{\sim}{\rightarrowtail} Y_1) \mapsto Y_1/Y_0.$$

What does such wishful thinking imply about the simplicial structure of $\mathrm{Wh}^{\mathrm{Comb}}(*)$?

A 2-simplex in $\mathrm{nerve}(\mathscr{C})$ is a sequence $(Y_0 \rightarrowtail^{\sim} Y_1 \rightarrowtail^{\sim} Y_2)$, and its faces are given, respectively, by $(Y_1 \rightarrowtail^{\sim} Y_2)$, $(Y_0 \rightarrowtail^{\sim} Y_2)$, $(Y_0 \rightarrowtail^{\sim} Y_1)$. We conclude that there must be a 2-simplex in $\mathrm{Wh}^{\mathrm{Comb}}(*)$ whose faces are given by $Y_2/Y_1$, $Y_2/Y_0$, $Y_1/Y_0$, a particular arrangement of the terms in the cofibration sequence $(Y_1/Y_0) \rightarrowtail (Y_2/Y_0) \twoheadrightarrow (Y_2/Y_1)$.

What is the general conclusion for $n$-simplices?

**4. An exact sequence $K$-theory in nonadditive categories.** Call a simplicial set finite if the number of nondegenerate simplices is finite. Suppose we want to define $[Y]$, the reduced Euler characteristic of a pointed finite simplicial set $Y$. Then we may take $[Y]$ to be an element of the abelian group eul with generators the finite pointed simplicial sets $Y$, and relations

(i) $[Y] = [Y']$ if $Y \rightarrowtail^{\sim} Y'$,

(ii) $[Y] = [Y'] + [Y'']$ if $Y' \rightarrowtail Y \twoheadrightarrow Y''$ is a cofibration sequence.

This is analogous to the definition of the projective class group $K_0(R)$, the abelian group with generators the finitely generated projective $R$-modules $P$, and relations

(i′) $[P] = [P']$ if $P \approx P'$,

(ii′) $[P] = [P'] + [P'']$ if $P' \rightarrowtail P \twoheadrightarrow P''$ is short exact.

In the latter case, relation (i′) is redundant since it is implied by (ii′). In the former case we might try to find a single type of relation which is equivalent to (i) and (ii) together. However it is clear that the only thing we may gain in doing so is a loss of simplicity.

Hence if we want to interpret $\mathfrak{eul}$ as a low-dimensional homotopy group of some space, this space be better not constructed as a simplicial set (e.g., the nerve of a category); rather we should look for a bisimplicial set (e.g., the nerve of a simplicial category). Also it is hard to imagine that $\mathfrak{eul}$ could be a $\pi_0$ in a reasonably direct way since relation (ii) is a typical $\pi_1$-relation.

On the other hand, $\mathfrak{eul}$ will clearly be the $\pi_1$ of any simplicial category $\mathscr{E}.$ that in low degrees satisfies

(0) $\mathscr{E}_0$ is the trivial category with one object and one morphism,

(1) $\mathscr{E}_1$ is the category of weak equivalences $Y \to^{\sim} Y'$ of finite pointed simplicial sets,

(2) $\mathscr{E}_2$ is a category whose objects are the cofibration sequences $Y' \rightarrowtail Y \twoheadrightarrow Y''$ of finite pointed simplicial sets; and the faces of $(Y' \rightarrowtail Y \twoheadrightarrow Y'')$ are given by the collection $Y'$, $Y$, $Y''$ (in this order, or in reverse order).

This is as far as the structure of $\mathscr{E}.$ can be suggested by the relations (i) and (ii) above. We must now provide our own choice of morphisms for the category $\mathscr{E}_2$. The most natural choice is to let a morphism in $\mathscr{E}_2$ be a weak equivalence of cofibration sequences, that is, a commutative diagram

$$\begin{array}{ccccc} Y' & \rightarrowtail & Y & \twoheadrightarrow & Y'' \\ \downarrow\sim & & \downarrow\sim & & \downarrow\sim \\ \bar{Y}' & \rightarrowtail & \bar{Y} & \twoheadrightarrow & \bar{Y}'' \end{array}$$

where the rows are cofibration sequences, and the vertical arrows weak equivalences.

A moment's reflection shows that the sequence $\mathscr{E}_0, \mathscr{E}_1, \mathscr{E}_2,$ can be continued in a simple way. Namely we may generally define $\mathscr{E}_k$ to be the category in which an object is a sequence of cofibrations

$$* \rightarrowtail Y_1 \rightarrowtail \cdots \rightarrowtail Y_k$$

and where a morphism is a commutative diagram

$$\begin{array}{ccccccc} * & \rightarrowtail & Y_1 & \rightarrowtail \cdots \rightarrowtail & Y_k \\ \downarrow\sim & & \downarrow\sim & & \downarrow\sim \\ * & \rightarrowtail & Y_1' & \rightarrowtail \cdots \rightarrowtail & Y_k' \end{array}$$

in which the vertical arrows are weak equivalences. The new definition of $\mathscr{E}_2$ is equivalent to the old one since by the gluing lemma $Y_2/Y_1 \to Y_2'/Y_1'$ will now also be a weak equivalence.

In order to assemble the $\mathscr{E}_k$ to a simplicial category, we must define face and degeneracy maps. The following rule for face maps from $\mathscr{E}_k$ to $\mathscr{E}_{k-1}$ extends the rule given in (2) above for the case $k = 2$. The $i$th face map, for $i > 0$, just drops $Y_i$ from the sequence

$$* \rightarrowtail Y_1 \rightarrowtail \cdots \rightarrowtail Y_k.$$

But the 0th face map, as it drops $*$, must force $Y_1$ to become a new $*$, that is, the 0th face is given by the sequence

$$* \rightarrowtail Y_2/Y_1 \rightarrowtail \cdots \rightarrowtail Y_k/Y_1.$$

With this rule, the simplicial identities for iterated face maps are satisfied—except possibly for a choice problem with the choice of cokernels.

The degneracy maps from $\mathscr{E}_k$ to $\mathscr{E}_{k+1}$ can be defined in the obvious way, by the insertion of an identity map at the appropriate place of the sequence.

The above choice problem with the face maps is not serious. In fact there is a standard trick to avoid such choice problems, to replace the category in question by an equivalent one to incorporate all the necessary choices. In the case at hand, one may proceed as follows.

Let $\Delta$ denote the category of ordered sets $[0], [1], \cdots, [n] = (0 < 1 < \cdots < n)$, and weakly monotonic maps.

Define $\langle n \rangle$ to be the partially ordered set of pairs $(i, j)$, $0 \leqq i \leqq j \leqq n$, where $(i, j) \leqq (i', j')$ if and only if $i \leqq i'$ and $j \leqq j'$. Considering $\langle n \rangle$ as a category, we may identify it to $\mathfrak{Mor}[n]$, the category whose objects are the morphisms in the ordered set $[n]$ when the latter is considered as a category. The notation emphasizes that $[n] \mapsto \mathfrak{Mor}[n]$ is a covariant functor on the category $\Delta$.

DEFINITION 4.1. $\mathscr{E}'_n$ is the category whose objects are the functors

$$\begin{aligned} Y\colon \mathfrak{Mor}[n] &\to (\text{pointed finite simplicial sets}),\\ (i, j) &\mapsto Y_{(i,j)} \end{aligned}$$

satisfying

(i) for any $i$, $Y_{(i,i)}$ equals the (distinguished) zero object,

(ii) for any triple $i \leqq j \leqq k$, the sequence

$$Y_{(i,j)} \to Y_{(i,k)} \to Y_{(j,k)}$$

is a cofibration sequence,

and whose morphisms are the natural transformations of functors, satisfying that all the maps involved are weak equivalences.

An equivalence from $\mathscr{E}'_n$ to $\mathscr{E}_n$ is given by the forgetful map

$$Y \mapsto (Y_{(0,0)} \rightarrowtail Y_{(0,1)} \rightarrowtail \cdots \rightarrowtail Y_{(0,n)})$$

and it is clear from the definition that the $\mathscr{E}'_n$ assemble to a simplicial category $\mathscr{E}'_{\boldsymbol{\cdot}}$.

We have thus achieved a fairly natural classification of the Euler characteristics of finite pointed simplicial sets by the elements of some homotopy group, namely $\pi_1|\mathscr{E}'_{\boldsymbol{\cdot}}|$. But, after all, one expects Euler characteristics to be classified by a $\pi_0$. So we should really consider the loop space $\Omega|\mathscr{E}'_{\boldsymbol{\cdot}}|$.

THEOREM. $\Omega|\mathscr{E}'_{\boldsymbol{\cdot}}| \simeq A(*)$.

This is a special case of Theorem 5.7 of the next section. To prove it, one has to consider variants of Definition 4.1.

Other variants of Definition 4.1 have to be considered for other purposes. It is therefore desirable to have an abstract version of this definition. The ingredients we need are: a category, and notions of 'cofibration' and 'weak equivalence' in this category. These must satisfy certain conditions, to ensure that Definition 4.1 makes sense. Preferably they should also satisfy other conditions of a general nature which we can expect to hold in cases of interest, and to be useful in proofs.

Thus a *category with cofibrations and weak equivalences* shall mean a category $\mathscr{C}$ together with subcategories $co(\mathscr{C})$ and $w(\mathscr{C})$ satisfying the following three groups of conditions:

(I) $\mathscr{C}$ has a (distinguished) zero object 0;

(II) (1) Isomorphisms in $\mathscr{C}$ are cofibrations (i.e., are morphisms in $co(\mathscr{C})$).

(2) For every object $A$ of $\mathscr{C}$, the morphism $0 \to A$ is a cofibration.

(3) $co(\mathscr{C})$ is closed under cobase change; this means, if in the diagram

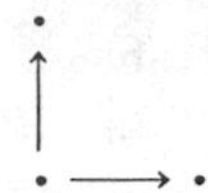

the left vertical arrow is a cofibration, then the pushout of the diagram exists in $\mathscr{C}$, and the right vertical arrow will also be a cofibration.

(III) (1) Isomorphisms in $\mathscr{C}$ are weak equivalences (i.e., are morphisms in $w(\mathscr{C})$).

(2) If in the diagram

$$\begin{array}{ccccc} B & \leftarrowtail & A & \rightarrow & C \\ \downarrow\sim & & \downarrow\sim & & \downarrow\sim \\ B' & \leftarrowtail & A' & \rightarrow & C' \end{array}$$

the left-hand horizontal arrows are cofibrations, and all vertical arrows are weak equivalences, then the map

$$B \cup_A C \to B' \cup_{A'} C'$$

is a weak equivalence.

REMARK. The axioms have been chosen for their simplicity, so that they reproduce easily under chains of constructions. The axioms are sufficiently general to include many uninteresting cases; for example if $\mathscr{C}$ has 0 and colimits, and $co(\mathscr{C}) = \mathscr{C}$; or $w(\mathscr{C}) = \mathscr{C}$. In either of the latter two cases one may expect the homotopy type associated below not to be very interesting either; this is indeed the case.

In practice there will never be any doubt about what the cofibrations are, so the category $co(\mathscr{C})$ will be dropped from the notation in the following definition. By contrast there will, as a rule, be several categories of weak equivalences to choose from, so the category $w(\mathscr{C})$ has to be specified in the notation.

DEFINITION 4.2. $wS_n\mathscr{C}$ is the category whose objects are the functors

$$\begin{aligned} A\colon \mathfrak{Mor}[n] &\to \mathscr{C} \\ (i, j) &\mapsto A_{(i,j)} \end{aligned}$$

satisfying

(i) $A_{(i,i)} = 0$, the distinguished zero object, for every $i$;

(ii) for every triple $i \leqq j \leqq k$, the morphism $A_{(i,j)} \to A_{(i,k)}$ is a cofibration, and

$$\begin{array}{ccc} A_{(i,j)} & \longrightarrow & A_{(i,k)} \\ \downarrow & & \downarrow \\ A_{(j,j)} & \longrightarrow & A_{(j,k)} \end{array}$$

is a pushout;

and whose morphisms are the natural transformations of functors satisfying that all the maps involved are weak equivalences.

$wS.\mathscr{C}$ is the simplicial category which in degree $n$ is $wS_n\mathscr{C}$.

COROLLARY. *$|wS.\mathscr{C}|$ is canonically an infinite loop space.*

PROOF. It follows from the axioms that $(A, A') \to A \vee A' = A \cup_0 A'$ defines a composition law on $|wS.\mathscr{C}|$; therefore $|wS.\mathscr{C}|$ is the underlying space of a $\Gamma$-space, with respect to this composition law, whence the assertion **[17]**.

The present construction generalizes the $Q$-construction of Quillen: Let $\mathscr{A}$ be an exact category in the sense of **[16]**, $co(\mathscr{A})$ the category of admissible monomorphisms, and $w(\mathscr{A})$ the category of isomorphisms. Then there is a natural homotopy equivalence $|wS.\mathscr{A}| \simeq |Q\mathscr{A}|$ (Quillen, unpublished).

Because of the added generality one cannot expect many of the general results on $Q\mathscr{A}$ to carry over directly. In fact, only Theorem 2 of **[16]** has a direct analogue here to which we refer as the *additivity theorem.*

The additivity theorem says, roughly, if $\mathscr{C}$ is a category with cofibrations and weak equivalences, then the category $E(\mathscr{C})$ whose objects are the diagrams $(A \rightarrowtail B)$ in $\mathscr{C}$ can be made into a category with cofibrations and weak equivalences in a natural way, and

$$\begin{gathered} wS.E(\mathscr{C}) \to wS.\mathscr{C} \times wS.\mathscr{C} \\ (A \rightarrowtail B) \mapsto (A, B/A) \end{gathered}$$

is a homotopy equivalence.

That is, cofibration sequences can be replaced by sum diagrams. It thus appears that, philosophically speaking, Definition 4.2 is just another version of the bar construction, applicable in another unusual situation.

**5. The functor $A(X)$, revisited.** Let $X$ be a simplicial set. Denote by $\mathscr{C}(X)$ the category of pairs $(Y, s)$ where $s: X \to Y$ is an injective map, and where a map from $(Y, s)$ to $(Y', s')$ is a map of simplicial sets, $f: Y \to Y'$, such that $fs = s'$.

A map $f: (Y, s) \to (Y', s')$ is called a weak homotopy equivalence, or *$h$-map*, for short, if $|f|: |Y| \to |Y'|$ is a homotopy equivalence relative to the subspace $|X|$;

it is called a *simple map* if $|f|$ has contractible point inverses (and in particular is surjective). Simple maps can be defined without recourse to geometric realization, but this will not be done here. Any simple map is an $h$-map. Let $s\mathscr{C}(X)$ denote the category whose objects are those of $\mathscr{C}(X)$ and whose morphisms are the simple maps.

Denote by $s\mathscr{C}_f(X)$ the full subcategory of $s\mathscr{C}(X)$ of those $(Y, s)$ which are *finite*, i.e., which satisfy that all but finitely many of the nondegenerate simplices of $Y$ are contained in the simplicial subset $s(X)$;

and by $s\mathscr{C}_f^h(X)$ the full subcategory of $s\mathscr{C}_f(X)$ of those $(Y, s)$ which are *acyclic*, i.e., which satisfy that $s: X \to (Y, s)$ is an $h$-map; equivalently, that $|s(X)|$ is a deformation retract of $|Y|$.

Hatcher's main result in **[8]** is that $\Omega\ \mathrm{Wh}^{\mathrm{PL}}(X)$ is homotopy equivalent to $s\mathscr{C}_f^h(X)$. (Actually, Hatcher uses simple PL maps of polyhedra. The translation into simple maps of simplicial sets is nontrivial, but not too surprising either. Also, because of the ambiguity of PL mapping cylinders, the formulation of Hatcher's theorem in terms of simple maps of polyhedra requires additional justification.) It will be indicated below how to prove Theorem 3.1 from the homotopy equivalence between $\Omega \mathrm{Wh}^{\mathrm{PL}}(X)$ and $s\mathscr{C}_f^h(X)$. There is a more direct proof, independent of Hatcher's theorem, but the present proof is simpler at least in the regard that manifolds do not have to be used anymore.

To relate $s\mathscr{C}_f^h(X)$ to $A(X)$, the machinery of the preceding section has to be used. But $\mathscr{C}(X)$ has no cokernels, so we have to modify it.

Let $\mathscr{R}(X)$ be the category in which an object is a triple $(Y, r, s)$ where $r\colon Y \to X$ is a retraction of simplicial sets, and $s$ is a section of $r$; and where a map $f\colon (Y, r, s) \to (Y', r', s')$ is a map $f\colon Y \to Y'$ such that $fs = s'$ and $r = r'f$.

$\mathscr{R}(X)$ has a distinguished zero object $0 = (X, \mathrm{id}, \mathrm{id})$, and it is a category with cofibrations in a natural way: the cofibrations are the maps $(Y', r', s') \to (Y, r, s)$ with $Y' \to Y$ injective. The axioms put down in the preceding section clearly hold. To any cofibration is associated a 'cofibration sequence'

$$(Y', r', s') \rightarrowtail (Y, r, s) \twoheadrightarrow (Y, r, s) \cup_{(Y', r', s')} 0.$$

There are four notions of weak equivalence in $\mathscr{R}(X)$ that we have to be aware of. In either case, the subcategory of weak equivalences will be denoted by prefixing any of the letters $i$, $s$, $h$, $h_X$, respectively, whichever applies.

($i$) $i\mathscr{R}(X)$ is the category of *isomorphisms* in $\mathscr{R}(X)$;

($s$) $s\mathscr{R}(X)$ is the category of *simple maps*, i.e., maps such that $|f|$ has contractible point inverses; or equivalently, where the associated map in $\mathscr{C}(X)$ is simple;

($h$) $h\mathscr{R}(X)$ is the category of *weak homotopy equivalences*, or *h-maps*, for short; by definition, a map $f$ in $\mathscr{R}(X)$ is in $h\mathscr{R}(X)$ if and only if the associated map in $\mathscr{C}(X)$ is in $h\mathscr{C}(X)$;

($h_X$) $h_X\mathscr{R}(X)$ is the category of *hereditary weak homotopy equivalences*, i.e., maps $(Y, r, s) \to (Y', r', s')$ such that for any $X' \subset X$, the induced map $r^{-1}(X') \to r'^{-1}(X')$ is an $h$-map in $\mathscr{R}(X')$; these maps are mainly introduced here for the purpose of making it clear that we will *not* use them.

We have $i\mathscr{R}(X) \subset s\mathscr{R}(X) \subset h_X\mathscr{R}(X) \subset h\mathscr{R}(X)$.

Similarly as before, the subscript '$f$' added to the notation of $\mathscr{R}(X)$, or any of its subcategories, will refer to the full subcategory of finite objects, i.e., those $(Y, r, s)$ satisfying that all but finitely many of the nondegenerate simplices of $Y$ are contained in $s(X)$;

and the superscript '$h$' added to the notation of $\mathscr{R}(X)$ or $\mathscr{R}_f(X)$, or any of their subcategories, will refer to the full subcategory of those $(Y, r, s)$ for which $s\colon X \to (Y, r, s)$ is an $h$-map.

One of the many categories now defined is $s\mathscr{R}_f^h(X)$. We would like to prove that $s\mathscr{R}_f^h(X) \to s\mathscr{C}_f^h(X)$ is a homotopy equivalence. Unfortunately this is wrong in general, simply for the trivial but still annoying reason that if $s\mathscr{R}_f^h(X)$ is not contractible then the two maps

$$s\mathscr{R}_f^h(X) \to s\mathscr{R}_f^h(X \times \Delta^1)$$

are not homotopic.

The counterexample suggests the remedy, namely we must allow things to 'move'. Fortunately there is a simple way to allow for such moving without loss of functoriality, namely, to replace everything in sight by $k$-parameter families, with varying $k$, of the same kind of thing. An organized way of doing so, is to introduce a dummy simplicial direction, in replacing $X$ by the simplicial object of its higher path spaces. To be precise,

DEFINITION. If $K$, $L$ are simplicial sets, let $L^K$ denote the function space, the simplicial set which in degree $n$ is the set of maps

$$K \times \Delta^n \to L.$$

If $F$: (simplicial sets) $\to \mathscr{D}$, $X \mapsto F(X)$, is any functor, define a functor from simplicial sets to the simplicial objects in $\mathscr{D}$, or equivalently, a simplicial object of functors $F.$: (simplicial sets) $\to \mathscr{D}$ by letting $F_k(X) = F(X^{\Delta^k})$.

*Parenthesis.* To understand the meaning of this construction, and why it helps, one should note that a functor $F$: (simplicial sets) $\to$ (simplicial sets) may not respect 'homotopy' in any sense, for example the functor 0-skeleton, $X \mapsto \mathrm{sk}_0(X)$. However there is a natural transformation of functors from simplicial sets to simplicial sets, $F \to \check{F}$ with the following properties:

(i) $\check{F}$ respects simplicial homotopies,

(ii) if $F$ respects simplicial homotopies, then $F(X) \to^{\sim} \check{F}(X)$.

Indeed, one defines

$$\check{F}(X) = \operatorname{diag} F.(X).$$

*Exercise.* What is $\check{\mathrm{sk}}_0$? This ends the parenthesis.

For the purpose of better readability, the notation $F(X^{\Delta^{\cdot}})$ will be used instead of the more precise $F.(X)$.

LEMMA 5.1. $s\mathscr{C}^h_f(X) \to^{\sim} s\mathscr{C}^h_f(X^{\Delta^{\cdot}})$; *if $X$ satisfies the extension condition then also*

$$s\mathscr{R}^h_f(X^{\Delta^{\cdot}}) \xrightarrow{\sim} s\mathscr{C}^h_f(X^{\Delta^{\cdot}}).$$

REMARK. $h_X$-maps are redundant. For if $X$ satisfies the extension condition then

$$s\mathscr{R}^h_f(X^{\Delta^{\cdot}}) \xrightarrow{\sim} h_X\mathscr{R}^h_f(X^{\Delta^{\cdot}}).$$

From now on, $\mathscr{R}_f(X)$ and $\mathscr{R}^h_f(X)$ will be considered as categories with cofibrations and weak equivalences (with several choices for the latter) in the sense of the preceding section. Thus by Definition 4.2 we have, for example, for each $n$ the category $sS_n\mathscr{R}^h_f(X)$ and the simplicial category $sS_n\mathscr{R}^h_f(X^{\Delta^{\cdot}})$, and we have a simplicial category $sS.\mathscr{R}^h_f(X)$ and a bisimplicial category $sS.\mathscr{R}^h_f(X^{\Delta^{\cdot}})$. Henceforth we assume that $X$ satisfies the extension condition.

LEMMA 5.2. *The 'subquotient' map*

$$sS_n\mathscr{R}^h_f(X^{\Delta^{\cdot}}) \to (s\mathscr{R}^h_f(X^{\Delta^{\cdot}}))^n$$
$$(A\colon \mathfrak{Mor}[n] \to \mathscr{R}^h_f(X^{\Delta^k})) \mapsto (A_{(0,1)}, A_{(1,2)}, \cdots, A_{(n-1,n)})$$

*is a weak homotopy equivalence.*

The idea of proof is, if $X \to (Y', r', s')$ is an $h$-map then a cofibration sequence $(Y', r', s') \rightarrowtail (Y, r, s) \twoheadrightarrow (Y'', r'', s'')$ can be 'moved' to a split one because $(Y', r', s')$ is acyclic.

The lemma implies that $|sS.\mathscr{R}^h_f(X^{\Delta^{\cdot}})|$ is homotopy equivalent to the canonical deloop (from the composition law) of $|s\mathscr{R}^h_f(X^{\Delta^{\cdot}})|$. Because of Hatcher's theorem, Lemma 5.1, and the definition of $\mathrm{Wh}^{\mathrm{PL}}(X)$ as a canonical deloop, we have therefore

PROPOSITION 5.3. $|sS.\mathscr{R}_f^h(X^{\Delta^\cdot})| \simeq \mathrm{Wh}^{\mathrm{PL}}(X)$.

Using this transcription, we may compare $\mathrm{Wh}^{\mathrm{PL}}(X)$ to other functors. The following diagram involves the forgetful maps 'simple maps are $h$-maps' (the horizontal arrows) and 'acyclic objects are objects' (the vertical arrows):

$$\begin{array}{ccc} sS.\mathscr{R}_f^h(X^{\Delta^\cdot}) & \longrightarrow & hS.\mathscr{R}_f^h(X^{\Delta^\cdot}) \\ \downarrow & & \downarrow \\ sS.\mathscr{R}_f(X^{\Delta^\cdot}) & \longrightarrow & hS.\mathscr{R}_f(X^{\Delta^\cdot}) \end{array}$$

LEMMA 5.4. (i) *This square is homotopy cartesian*; (ii) $hS.\mathscr{R}_f^h(X^{\Delta^\cdot})$ *is contractible.*

PROOF OF (ii). In each bidegree, the category in this bidegree has a terminal object (in fact, a zero object) and is hence contractible. This implies the assertion, in view of a well-known result on the geometric realization of multisimplicial sets.

Part (i) of the lemma is a special case of a general result which is deduced from the additivity theorem in a similar way as Propositions 7.1—7.3 in **[18]**.

In view of the lemma, the left and bottom arrows in the diagram form a homotopy fibration. The next results below identify the other terms in this fibration. This yields Theorem 3.1.

One may thus say that Theorem 3.1 is obtained by comparison of two notions of weak equivalence in $\mathscr{R}_f(X)$, namely '$h$-map' and 'simple map'. From this point of view, Theorem 3.3 is obtained similarly, by comparison of '$h$-map' and 'isomorphism'.

LEMMA 5.5. *The functor* $X \mapsto sS.\mathscr{R}_f(X^{\Delta^\cdot})$ *is a homology theory.*

To prove the lemma one compares $sS.\mathscr{R}_f(X^{\Delta^\cdot})$ to the functor $X \mapsto sS.\mathscr{R}_f(\mathrm{sk}_0(X^{\Delta^\cdot}))$ which is a homology theory by the recipe of how to associate a homology theory to a $\Gamma$-space **[1]**; cf. **[18]** for a more detailed treatment.

One uses that, for any $X$, the skeleton filtration $\mathrm{sk}_j(X)$ induces a filtration of the identity functor on $\mathscr{R}_f(X)$,

$$\mathrm{sk}_j^*(Y, r, s) = (r^{-1}(sk_j(X)), \cdots).$$

The key fact is that $\mathrm{sk}_j^*$ does induce an endomorphism of $sS.\mathscr{R}_f(X)$, because

(i) simple maps are hereditary (as opposed to $h$-maps),

(ii) the objects are *not* required to be acyclic.

The additivity theorem implies that the identity map on $sS.\mathscr{R}_f(X)$ is homotopic to the map induced from

$$\mathrm{sk}_0^* \vee \mathrm{sk}_1^*/\mathrm{sk}_0^* \vee \cdots \vee \mathrm{sk}_j^*/\mathrm{sk}_{j-1}^* \vee \cdots.$$

This means that for the purposes of $sS.\mathscr{R}_f(X)$, one may restrict to those objects in $\mathscr{R}_f(X)$ which are given as a sum of 'objects with small support'.

LEMMA 5.6. $hS.\mathscr{R}_f(X) \xrightarrow{\sim} hS.\mathscr{R}_f(X^{\Delta^\cdot})$.

THEOREM 5.7. $\Omega\, |hS.\,\mathscr{R}_f(X)\,| \simeq A(X)$.

The main steps in the proof of this theorem will be indicated in the remaining material.

DEFINITION. $\mathscr{R}(X)_k^n$ is the full subcategory of those $(Y, r, s)$ such that $|s|$ is homotopy equivalent, relative to $|X|$, to the inclusion of $|X|$ into ($|X|$ wedge $k$ spheres of dimention $n$).

LEMMA 5.8. $Z \times |\text{dir lim}_{(k,n)}\, h\mathscr{R}(X)^n_k|^+ \simeq A(X)$.

PROOF. The old definition of $A(X)$ was only given for connected pointed $X$. So we assume $X$ is connected and pointed. Let $G$ be any loop group for $X$, for example the $G(X)$ of Kan, and $h\mathscr{S}(G)^n_k$ the category considered in §2, in defining $A(X)$. There are functors

$$h\mathscr{R}(X)^n_k \to h\mathscr{S}(G)^n_k, \quad (Y, r, s) \mapsto r^*(X_t \times G)/(X_t \times G)$$

(and the action changed from right to left), and

$$h\mathscr{S}(G)^n_k \to h\mathscr{R}(X)^n_k, \quad Y \mapsto (X_t \times G) \times^G Y$$

and these functors are adjoint. Hence $h\mathscr{R}(X)^n_k \to^\sim h\mathscr{S}(G)^n_k$; hence

$$Z \times \left|\text{dir lim}_{(k,n)}\, h\mathscr{R}(X)^n_k\right|^+ \xrightarrow{\sim} Z \times \left|\text{dir lim}_{(k,n)}\, h\mathscr{S}(G)^n_k\right|^+$$

and the latter is $A(X)$, by §2.

Let $\mathscr{R}_f(X)^n_k = \mathscr{R}_f(X) \cap \mathscr{R}(X)^n_k$.

LEMMA 5.9. $h\mathscr{R}_f(X)^n_k \to^\sim h\mathscr{R}(X)^n_k$.

The composition law on $\mathscr{R}_f(X)$ induces a composition law on $\mathscr{R}_f(X)^n$, the union of the categories $\mathscr{R}_f(X)^n_k$; hence this is naturally the underlying category of a $\Gamma$-category, and so is the subcategory $h\mathscr{R}_f(X)^n$. Hence a simplicial category $N_\Gamma(h\mathscr{R}_f(X)^n)$, the nerve with respect to the composition law, is defined, cf. **[18]** for details. By Segal **[17]** there is a natural homotopy equivalence

$$\Omega\left|N_\Gamma(h\mathscr{R}_f(X)^n)\right| \xrightarrow{\sim} Z \times \left|\text{dir lim}_{(k)}\, h\mathscr{R}_f(X)^n_k\right|^+;$$

hence one is reduced to showing that

$$\text{dir lim}\left|N_\Gamma(h\mathscr{R}_f(X)^n)\right| \xrightarrow{\sim} \left|hS.\mathscr{R}_f(X)\right|.$$

The category $\mathscr{R}_f(X)^n$ is a category with cofibrations (this requires some care). Hence $hS.\mathscr{R}_f(X)^n$ is defined. There is a map of simplicial categories $N_\Gamma(h\mathscr{R}_f(X)^n) \to hS.\mathscr{R}_f(X)^n$ which sends each sum diagram to a cofibration sequence by forgetting part of the data.

LEMMA 5.10. $\text{dir lim}\; N_\Gamma(h\mathscr{R}_f(X)^n) \to^\sim \text{dir lim}\; hS.\mathscr{R}_f(X)^n$.

Define functors *cone*, $C: \mathscr{R}(X) \to \mathscr{R}(X)$, $C(Y, r, s) =$ (mapping cylinder of $r, \cdots$), and *suspension*, $S = \text{coker}(\text{id} \to C)$. When the cone and suspension are considered as endomorphisms of $hS.\mathscr{R}_f(X)$, one has by the additivity theorem a homotopy $C \to^\sim \text{id} \vee S$. In particular, the suspension represents a homotopy inverse to the identity on $hS.\mathscr{R}_f(X)$, and the map

$$hS.\mathscr{R}_f(X) \to \text{dir lim}_{(S)}\, hS.\mathscr{R}_f(X)$$

is a weak homotopy equivalence.

LEMMA 5.11. $\text{dir lim}_{(n)}\, hS.\mathscr{R}_f(X)^n \to^\sim \text{dir lim}_{(S)}\, hS.\mathscr{R}_f(X)$.

This ends the indication of proof of Theorem 5.7.

## REFERENCES

**1.** D. W. Anderson, *Chain functors and homology theories*, Sympos. Algebraic Topology, Lecture Notes in Math., vol. 249, Springer-Verlag, Berlin and New York, 1971, pp. 1–12. MR **49** #3895.

**2.** ———, *Comparison of K-theories*, Algebraic $K$-Theory, Lecture Notes in Math., vol. 341, Springer-Verlag, Berlin and New York, 1973.

**3.** A. Borel, *Stable real cohomology of arithmetic groups*, Ann. Sci. École Norm. Sup. (4) **7** (1974), 235–272. MR **52** #8338.

**4.** D. Burghelea and R. Lashof, *The homotopy type of the space of diffeomorphisms*. I, II, Trans. Amer. Math. Soc. **196** (1974), 1–50. MR **50** #8574.

**5.** D. Burghelea, R. Lashof, and M. Rothenberg, *Groups of automorphisms of manifolds*, Lecture Notes in Math., vol. 473, Springer-Verlag, Berlin and New York, 1975. MR **52** #1738.

**6.** E. Dror, *A generalization of the Whitehead theorem*, Sympos. Algebraic Topology, Lecture Notes in Math., vol. 249, Springer-Verlag, Berlin and New York, 1971, pp. 13-22.

**7.** F. T. Farrell and W. C. Hsiang, *On the rational homotopy groups of the diffeomorphism groups of disks, spheres, and aspherical manifolds*, these Proceedings, Part I, pp. 325–337.

**8.** A. Hatcher, *Higher simple homotopy theory*, Ann. of Math. (2) **102** (1975), 101–137. MR **52** #4305.

**9.**———, *Concordance spaces, higher simple-homotopy theory, and applications*, these Proceedings, pp. 3–21.

**10.** A. Hatcher and J. Wagoner, *Pseudo-isotopies of compact manifolds*, Astérisque, no. **6** (1973). MR **50** #5821.

**11.** K. Igusa, *Postnikov invariants and pseudo-isotopy* (preprint).

**12.** D. M. Kan, *A combinatorial definition of homotopy groups*, Ann. of Math. (2) **67** (1958), 282–312. MR **22** #1897.

**13.** R. Lee and R. H. Szczarba, *The group $K_3(Z)$ is cyclic of order forty-eight*, Ann. of Math. (2) **104** (1976), 31–60.

**14.** J. P. May (with contributions by F. Quinn. N. Ray, and J. Tornehave), *$E_\infty$ ring spaces and $E_\infty$ ring spectra*, Lecture Notes in Math., vol. 577, Springer-Verlag, Berlin and New York, 1977.

**15.** D. Quillen, *Cohomology of groups*, Actes, Congrès Intern. Math., 1970, tom 2, pp. 47–51.

**16.** ———, *Higher algebraic K-theory*. I, Algebraic $K$-Theory, Lecture Notes in Math., vol. 341, Springer-Verlag, Berlin and New York, 1973, pp. 85–147. MR **49** #2895.

**17.** G. Segal, *Categories and cohomology theories*, Topology **13** (1974), 293–312. MR **50** #5782.

**18.** F. Waldhausen, *Algebraic K-theory of generalized free products*, Ann. of Math. (to appear).

**19.** J. H. C. Whitehead, *Simple homotopy types*, Amer. J. Math. **72** (1950), 1–57. MR **11,** 735.

UNIVERSITÄT BIELEFELD

Proceedings of Symposia in Pure Mathematics
Volume 32, 1978

# METABOLIC AND HYPERBOLIC FORMS

J. P. ALEXANDER, P. E. CONNER AND G. C. HAMRICK

**Introduction.** In our research we have made many detailed computations concerning $W_*(\mathbf{Z}, C)$, $C$ a cyclic group [1], [2], [3]. Several ideals and subgroups of this ring have always intrigued us, particularly the ideal generated by Witt classes represented by forms on projective $\mathbf{Z}[C]$-modules and the image of the Wall groups. This note begins with the question of the relationship between metabolic and hyperbolic forms. This is suggested by the different ways of defining zero elements: in Witt groups metabolics are zero and in the $L$-groups hyperbolics are zero. Theorem 3 recovers in a very special case results of Bak [5], Pardon [8], Wall [9] and others, but we feel an inclusion of the straightforward argument is justified.

Our proofs are not detailed. In fact we only outline the results for the symmetric case. Most of the computations involving group cohomology, while tedious, are elementary.

**Metabolic and hyperbolic forms.** Let $(\pi, V)$ denote a right integral representation of a finite group $\pi$ on a free abelian group $V$. If there is a $\pi$-invariant $\mathbf{Z}$-nonsingular (skew) symmetric inner product $b: V \times V \to \mathbf{Z}$ then we say $(\pi, V)$ is an integral orthogonal (symplectic) representation and denote this by $(\pi, V, b)$.

For some time we have been concerned with relationships between the following concepts.

DEFINITION. (i) $(\pi, V, b)$ is *metabolic* if and only if there is a $\pi$-invariant submodule $N \subset V$ for which $N = N^\perp = \{v \in V | b(v, n) = 0 \text{ for all } n \in N\}$.

(ii) $(\pi, V, b)$ is hyperbolic if and only if there is an integral representation $(\pi, N)$ for which $(\pi, V)$ is equivalent by a $\pi$-equivariant isometry to the natural form on $(\pi, N \oplus N^*)$ given by

$$b_h((n, \varphi), (n', \varphi')) = \varphi(n') + \varphi'(n)$$

*AMS (MOS) subject classifications* (1970). Primary 20C10, 13K05.

where $N^* = \mathrm{Hom}_{\mathbf{Z}}(N, \mathbf{Z})$ with the usual right $\pi$-module action. We denote this hyperbolic form simply by $H(N)$.

Let $W_0(\mathbf{Z}, \pi)$ ($W_2(\mathbf{Z}, \pi)$) denote the Witt group of orthogonal (symplectic) representations of $\pi$ where $(\pi, V, b)$ is zero if it is metabolic. We say that $(\pi, V, b)$ is an *even form* if for each element $g \in \pi$ such that $g^2 = e$ then $b(v, vg)$ is even for all $v \in V$.

THEOREM 1. *If* $\pi = C_p$, *the cyclic group of order* $p$, $p$ *a prime and* $(\pi, V, b)$ *is an even, metabolic form, then* $(\pi, V, b)$ *is hyperbolic.*

To prove this theorem we introduce several new groups. Given $(\pi, N)$ consider all $\mathbf{Z}[\pi]$-embeddings $0 \to (\pi, N) \to^{\rho} (\pi, V, b)$ so that $(\mathrm{im}(\rho))^{\perp} = \mathrm{im}(\rho)$ and $b$ is even. Two embeddings $0 \to (\pi, N) \to^{\rho} (\pi, V, b)$ and $0 \to (\pi, N) \to^{\rho'} (\pi, V', b')$ are *Baer equivalent* if and only if there is a commutative diagram

$$\begin{array}{ccc} & \overset{\rho}{\nearrow} & (\pi, V, b) \\ (\pi, N) & & \downarrow J \\ & \underset{\rho'}{\searrow} & (\pi, V', b') \end{array}$$

where $J$ is a $\mathbf{Z}[\pi]$-module isometry. The sum, $\rho + \rho'$, of two embeddings is described as follows. Consider $L \subset V \perp V'$ where $L = \{(\rho(x), -\rho'(x)) | \, x \in N\}$. ($V \perp V'$ is the orthogonal direct sum of $(\pi, V, b)$ and $(\pi', V', b')$. The associated bilinear form is denoted $b \perp b'$.) Clearly $L$ is summand and $L \subset L^{\perp}$. Let $b''$ denote the natural form induced on $L^{+}/L$ by $b \perp b'$. Then $\rho + \rho'$ is defined by the natural embedding

$$(\pi, N) \xrightarrow{\rho+\rho'} (\pi, L^{+}/L, b'').$$

The hyperbolic form $H(N)$ plays the role of the identity and the inverse of $(\pi, N) \to^{\rho} (\pi, V, b)$ is given by $(\pi, N) \to^{-\rho} (\pi, V, -b)$. The group of embeddings $(\pi, N) \to (\pi, V, b)$ is denoted $\mathrm{Met}_{\pi}(N)$. This is a covariant functor with respect to $\mathbf{Z}[\pi]$-module homomorphisms. A natural transformation $\mathrm{Met}_{\pi}(N) \to \mathrm{Ext}^1_{\mathbf{Z}[\pi]}(N^*, N)$ is given by associating to each embedding $\rho\colon (\pi, N) \to (\pi, V, b)$ the resulting short exact sequence $0 \to N \to V \to N^* \to 0$. If $N$ is projective as a $\mathbf{Z}[\pi]$-module then this map is a *monomorphism*! Furthermore if $\pi$ has odd order then this is a monomorphism for any integral representation $N$.

There are two important constructions involving $\mathrm{Met}_{\pi}(N)$.

CONSTRUCTION 1. Given $(\pi, N_1)$ and $(\pi, N_2)$ there is a homomorphism

$$\mathrm{Ext}^1_{\pi}(N_2^*, N_1) \to \mathrm{Met}_{\pi}(N_1 \oplus N_2)$$

given as follows. If $0 \to N_1 \to^{\alpha} K \to^{\beta} N_2^* \to 0$ is an extension, then its dual is $0 \to N_2 \to^{\beta^*} K^* \to^{\alpha^*} N_1^* \to 0$. Adding these sequences together we get

$$0 \to N_1 \oplus N_2 \xrightarrow{\alpha\oplus\beta^*} K \oplus K^* \to N_2^* \oplus N_1^* \to 0.$$

$K \oplus K^*$ supports the hyperbolic form $H(K)$ and $\mathrm{im}(\alpha \oplus \beta^*)^{\perp} = \mathrm{im}(\alpha \oplus \beta^*)$. This gives us an element of $\mathrm{Met}_{\pi}(N_1 \oplus N_2)$.

CONSTRUCTION 2. If $(\pi, N) \to^{f} (\pi, N')$ is an epimorphism then the induced homomorphism $\mathrm{Met}_{\pi}(N) \to \mathrm{Met}_{\pi}(N')$ can be described as follows. Given $[(\pi, N) \to$

$(\pi, V, b)] \in \mathrm{Met}_\pi(N)$, $\ker f$ is a $\pi$-invariant subspace of $V$ and $\ker f \subset (\ker f)^\perp$. Clearly there is an embedding

$$(\pi, N') \to (\pi, (\ker f)^\perp/\ker f, b').$$

This gives us an element of $\mathrm{Met}_\pi(N')$.

LEMMA 2. *The following is a split exact sequence*

$$0 \to \mathrm{Ext}^1_\pi(N_2^*, N_1) \to \mathrm{Met}_\pi(N_1 \oplus N_2) \to \mathrm{Met}_\pi(N_1) \oplus \mathrm{Met}_\pi(N_2) \to 0. \quad \square$$

This reduces the problem of analyzing $\mathrm{Met}_\pi(N)$ to $N$ an indecomposable $\pi$-representation. Except when $\pi = C_p$, $p$ a prime, our ignorance here is almost complete. Theorem 1 is proved by first using the Reiner-Dietrichsen decomposition for integral $C_p$-representations to show that for an indecomposable representation that $\mathrm{Met}_{C_p}(N) = 0$. This is followed by an application of Lemma 2 and the fact that the elements in the image of Construction 1 are hyperbolic.

By restricting $(\pi, V)$ we can extend this result in the following manner.

THEOREM 3. *If $\pi = C_{p^n}$, $p$ a prime, $C_{p^n}$ the cyclic group of order $p^n$, $V$ is a projective $\boldsymbol{Z}[\pi]$-module and $b$ is a metabolic even form, then $(\pi, V, b)$ is hyperbolic.* $\square$

This theorem is proved by induction on $n$ and uses the following result about group cohomology.

THEOREM 4 [4, PP. 112–113]. *If $\pi$ is a $p$-group, and $A$ is a $\pi$-module without $p$-torsion then the following are equivalent*:

(i) *$A$ is cohomologically trivial,*
(ii) $\hat{H}^q(\pi, A) = \hat{H}^{q+1}(\pi, A) = 0$ *for some* $q \in \boldsymbol{Z}$,
(iii) *$A/pA$ is a free $\boldsymbol{Z}_p[\pi]$-module,*
(iv) *$A$ is a projective $\pi$-module.* $\square$

Theorem 3 is well known for $n = 0$.

Consider $(C_{p^n}, V, B)$. First some notation. $C_p$ will be the subgroup of order $p$ in $C_{p^n}$ and $\tilde{C}_{p^{n-1}} = C_{p^n}/C_p$. $C_p$ is generated by $T^{p^{n-1}}$. Let $\Delta = 1 - T^{p^{n-1}}$ and $\Sigma = 1 + T^{p^{n-1}} + T^{2p^{n-1}} + \cdots + T^{(p-1)p^{n-1}}$. Now $U = \{x \in V \mid \Sigma x = 0\}$ is a projective $\boldsymbol{Z}(\lambda_{p^n})$-module ($\boldsymbol{Z}(\lambda_k)$ are the cyclotomic integers, $\lambda_k = \exp(2\pi i/k)$). Since $V$ is metabolic, so is $U \otimes \boldsymbol{Z}(1/p)$ and hence there is a $C_{p^n}$-invariant summand $W \subset U$ such that $W = W^\perp \cap U$. $W$ is clearly a $\boldsymbol{Z}(\lambda_{p^n})$ projective module.

LEMMA 5. *$W^\perp/W$ is a projective $\boldsymbol{Z}[\tilde{C}_{p^{n-1}}]$-module.*

PROOF. Recall that for a $C_{p^n}$-module $N$ that $H^1(C_p; N) = \ker \Sigma/\mathrm{im}\, \Delta$ and that $H^2(C_p; N) = \ker \Delta/\mathrm{im}\, \Sigma$. Since $V$ is a projective $\boldsymbol{Z}[C_{p^n}]$-module and $W$ is a projective $\boldsymbol{Z}(\lambda_{p^n})$-module, we can use the exact sequence $0 \to W^\perp \to V \to W^* \to 0$ to compute $H^*(C_p; W^\perp)$. Notice that since $\mathrm{im}\, \Delta \subset U$ and $W = W^\perp \cap U$ that $C_p$ acts trivially on $W^\perp/W$. The six-term exact sequence for $0 \to W \to W^\perp \to W^\perp/W \to 0$ reduces to

$$0 \to H^2(C_p; W^\perp) \to H^2(C_p; W^\perp/W) \to H^1(C_p; W) \to 0.$$

Now, $H^2(C_p; W^\perp/W) \approx W^\perp/W \otimes \boldsymbol{Z}_p$ and the first and last terms are free $\boldsymbol{Z}_p[\tilde{C}_{p^{n-1}}]$-modules.

By Theorem 4 this shows $W^\perp/W$ is a projective $\tilde{C}_{p^{n-1}}$-module. □

By induction $W^\perp/W$ is hyperbolic. If $A \subset W^\perp/W$ let $Y = \{x \in W^\perp | x + W \in A\}$ (i. e., $0 \to W \to Y \to A \to 0$ is exact).

LEMMA 6. *There exists* $A \subset W^\perp/W$ *so that*

(i) $W^\perp/W = H(A)$,

(ii) $\delta: H^2(C_{p^n}; A) \to H^1(C_{p^n}; W)$ *is an isomorphism*,

(iii) $Y$ *is a* $\boldsymbol{Z}[C_{p^n}]$-*projective module and* $Y = Y^\perp$,

(iv) $V = H(Y)$.

PROOF. First consider the exact cohomology sequence $0 \to H^2(C_{p^n}; W^\perp) \to^\rho H^2(C_{p^n}; W^\perp/W) \to H^1(C_{p^n}; W) \to 0$. There is a nonsingular finite form on $X = H^2(C_{p^n}; W^\perp/W)$ induced by cup products in cohomology and the bilinear form on $W^\perp/W$. Because $H^2(C_{p^n}; \boldsymbol{Z}_p[\tilde{C}_{p^{n-1}}]) = \boldsymbol{Z}_p$ this form has values in $\boldsymbol{Z}_p$. $W^\perp/W$ is hyperbolic; therefore $X$ is also hyperbolic. There are two ways of describing $X$ as a hyperbolic module. The first comes from choosing $B \subset W^\perp/W$ so that $W^\perp/W = H(B)$. If $\sigma: H^2(C_{p^n}; B) \to H^2(C_{p^n}; W^\perp/W)$ then $X = H(\text{im } \sigma)$. Also $X = H(\text{im } \rho)$. Notice that if $\text{im } \sigma \cap \text{im } \rho = \{0\}$ then the composition $H^2(C_{p^n}; B) \to H^2(C_{p^n}; W^\perp/W) \to H^1(C_{p^n}; W)$ is an isomorphism. Take $A = B$ and the lemma would be proven. In general find $B'$, a projective submodule of $B$, so that $\sigma(H^2(C_{p^n}; B')): m\sigma \cap \text{im } \rho$. Then $B = B' \oplus B''$ and

$$W^\perp/W = B' \oplus B'' \oplus (B')^* \oplus (B'')^* = H(B'' \oplus (B')^*).$$

Now $\text{im}(H^2(C_{p^n}; B'' \oplus (B')^*) \cap \text{im } \rho = \{0\}$. Set $A = B'' \oplus (B')^*$ and $Y = \{w \in W^\perp | w + W \in A\}$. □

**Witt groups and $L$-groups.** If $L_*^h(\pi)$ is the Wall group for surgery to a homotopy equivalence then there is an obvious homomorphism

$$L_0^h(\pi) \to W_0(\boldsymbol{Z}, \pi).$$

One application of Theorem 3 is the following result, well known in much more generality by Bak [**5**], Pardon [**8**], Wall [**9**], and Karoubi [**6**].

COROLLARY 7. *For $p$ a prime the following sequence is exact*:

$$0 \to H^1(C_2, \tilde{K}_0(C_p)) \to L_0^h(C_{p^n}) \to W_0(\boldsymbol{Z}, C_{p^n}).$$

PROOF. The involution of $\tilde{K}_0(C_{p^n})$ is defined by sending a projective module $V$ to its dual $V^*$. An element of $H^1(C_2; \tilde{K}_0(C_{p^n}))$ is represented by a projective module $V$ such that $V \oplus V^*$ is free. The homomorphism $H^1(C_2; \tilde{K}_0(C_{p^n})) \to L_0^h(C_{p^n})$ sends $[V] \mapsto [H(V)]$. Theorem 3 says that if $[U] \in L_0^h(C_{p^n})$ goes to zero in $W_0(\boldsymbol{Z}, C_{p_n})$ then it is hyperbolic, that is $U = H(V)$ where $V$ is projective and $V \oplus V^* = U$ is free. A careful analysis of the units in $\boldsymbol{Z}[C_{p^n}]$ prove the left-hand homomorphism is a monomorphism. □

In fact since $W_*(\boldsymbol{Z}, C_{p^n})$ is torsion free, $p$ odd [**1**], we see that $L_0(C_{p^n})$ is torsion free up to projective kernels. Even though $W_0(\boldsymbol{Z}, C_{2^n})$ is not torsion free, the invariant $H^2(C_{2^n}; V)$ detects the torsion elements. For $V$ projective, this invariant must vanish so in all cases the image of $L_0(C_{p^n}) \to W_0(\boldsymbol{Z}, C_{p^n})$ is torsion free.

It is obvious that the image of the Wall groups in the Witt groups is contained in the ideal generated by elements with representatives $(\pi, V, b)$ where $V$ is a projective $\mathbf{Z}[\pi]$-module. Let $P_*(\mathbf{Z}, \pi)$ denote this ideal.

Given an arbitrary $(C_{p^n}, V, b) \in W_0(\mathbf{Z}, C_{p^n})$ there is a symmetric finite form defined on $H^2(C_{p^k}; V)$, $1 \leqq k \leqq n$, by the composition

$$H^2(C_{p^k}; V) \times H^2(C_{p^k}; V) \to H^4(C_{p^k}; V \otimes V) \to H^4(C_{p^k}; \mathbf{Z}) \to \mathbf{Z}/p^k\mathbf{Z}$$

where the first map is induced by cup products in cohomology, the second by the coefficient pairing $v \otimes v' \mapsto b(v, v')$, and the last by the natural identification of $H^4(C_{p^k}; \mathbf{Z})$ with $\mathbf{Z}/p^k\mathbf{Z}$. Standard arguments involving finite forms [3, 1.7] show that this form can be considered as an element of $W(\mathbf{Z}_p)$. Consider the group homomorphisms $\varphi_k$: $W_0(\mathbf{Z}, C_{p^n}) \to W(\mathbf{Z}_p)$, $p$ odd, $1 \leqq k \leqq n$, given by

$$\varphi_k((C_{p^n}, V, b)) = H^2(C_{p^k}; V) + H^2(C_{p^{k-1}}, V) \in W(\mathbf{Z}_p).$$

PROPOSITION 8. *$\varphi_k$ is a ring homomorphism, $P_0(\mathbf{Z}, \pi) = \bigcap_k \ker \varphi_k$, and*

$$W_0(\mathbf{Z}, C_{p^n})/P_0(\mathbf{Z}, C_{p^n}) \approx \bigoplus_1^n W(\mathbf{Z}_p). \quad \square$$

## BIBLIOGRAPHY


1. J. P. Alexander, P. E. Conner, G. C. Hamrick and J. W. Vick, *Witt classes of integral representations of an abelian p-group*, Bull. Amer. Math. Soc. **80** (1974), 1179–1182. MR **52** #5782.

2. J. P. Alexander and G. C. Hamrick, *The torsion G-signature theorem for groups of odd order* (preprint).

3. J. P. Alexander, G. C. Hamrick and J. W. Vick, *Linking forms and maps of odd prime order*, Trans. Amer. Math. Soc. **221** (1976), 169–185. MR **53** #6600.

4. J. W. S. Cassels and A. Fröhlich, *Algebraic number theory*, Thompson, Washington, D.C., 1967.

5. A. Bak, *The computation of surgery groups of odd torsion groups*, Bull. Amer. Math. Soc. **80** (1974), 1113–1116.

6. M. Karoubi, *Localisations des formes quadratiques* (preprint).

7. J. Milnor and D. Husemöller, *Symmetric bilinear forms*, Ergebnisse der Math. und ihrer Grenzgebiete, Band 73, Springer-Verlag, Berlin and New York, 1973.

8. W. Pardon, *Local surgery theory and applications to the theory of quadratic forms*, Bull. Amer. Math. Soc. **82** (1976), 131–133. MR **53** #1609.

9. C. T. C. Wall, *Surgery on compact manifolds*, Academic Press, New York and London, 1970.


UNIVERSITY OF TEXAS AT AUSTIN

Proceedings of Symposia in Pure Mathematics
Volume 32, 1978

# THE PLUS CONSTRUCTION AND LIFTING MAPS FROM MANIFOLDS

JEAN-CLAUDE HAUSMANN AND PIERRE VOGEL

**Introduction**. The main problem considered in this paper can be roughly described as follows: Define a semi-$s$-cobordism as a cobordism $(W, M, M_-)$ such that the inclusion of $M$ into $W$ is a simple homotopy equivalence (but $\pi_1(M_-)$ can different from $\pi_1(W)$). Let

$$\begin{array}{ccc} & & X \\ & & \downarrow{\scriptstyle\alpha} \\ M & \xrightarrow{f} & Y \end{array}$$

be two maps between connected spaces with $M$ a compact manifold. Does there exist a semilifting of $f$, i.e., a semi-$s$-cobordism $(W, M, M_-)$, an extension $F\colon W \to Y$ of $f$ and a lifting $f_- : M_- \to X$?

The main results are stated in §2, while §1 contains the necessary definitions. Briefly speaking, when $n \geqq 5$, a sufficient condition for the existence of a semilifting is the existence of a lifting of $f$ through $X^+$, where $X^+$ is the space obtained by the Quillen plus construction with respect to the union of all finitely generated perfect subgroups of ker $\pi_1\alpha$. Under certain hypotheses, this condition is also necessary and one can solve the problem of classifying the semiliftings of $f$. These results are proved in §3. In §4, we use them to give an interpretation of the groups $\pi_n(X^+)$ in terms of a certain bordism group of $X$, improving some results of **[H2]**. For instance, one obtains a geometric interpretation of the algebraic $K$-theory groups. Other applications will appear in a subsequent paper.

**1. Definition and statement of the main results**. Let $X$ be a connected topological space, and $N$ a normal perfect subgroup of $\pi_1(X)$. Recall that the Quillen plus construction on $X$ with respect to $N$ **[L]** gives a map $\iota\colon X \to X^+$ characterized by the following two properties:

*AMS* (*MOS*) *subject classifications* (1970). Primary 57D15, 57A70; Secondary 18F25.

(1) $\pi_1(X^+) = \pi_1(X)/N$ and $\pi_1\iota$ is the natural projection,

(2) the induced homomorphism $\iota$: $H(X; \Lambda) \to H(X^+; \Lambda)$ is an isomorphism, where $\Lambda = \boldsymbol{Z}\pi_1(X^+)$ (homology with local coefficients).

Such a map is called a *plus map* (*with respect to* $N$). The space $X^+$ can be obtained by adding an equal quantity of 2- and 3-cells to $X$. The reader will easily check the following lemma:

LEMMA 1.1. *A map* $\alpha: X \to Y$ *between connected spaces is a plus map* (*with respect to* ker $\pi_1\alpha$) *if and only if* $\alpha$ *induces an isomorphism* $H(X; \boldsymbol{Z}\pi_1(Y)) \to H(Y; \boldsymbol{Z}\pi_1(Y))$.

Let $M^n$ be a compact CAT-manifold (CAT = DIFF, PL, or TOP). A *semi-s-cobordism* is a cobordism $(W, M, M_-)$, which is trivial on the boundary of $M$ and such that the inclusion $M \subset W$ is a simple homotopy equivalence. By Poincaré duality (with $\boldsymbol{Z}\pi_1(W)$ as local coefficients) and Lemma 1.1, the inclusion $M_- \subset W$ is a plus map with Whitehead torsion, measured in $\mathrm{Wh}(\pi_1(W))$ equal to zero. If $n \geqq 5$, the classical procedure of elimination of handles of **[Ke]** permits us to describe $W$ as obtained from $M_- \times I$ by attaching handles of index 2 and 3. Finally, observe that if $(W, M, M_-)$ and $(W', M, M_-)$ are two semi-*s*-cobordisms starting from $M$, dim $M \geqq 6$ any CAT-homeomorphism $\beta$: $M_- \to M'$ extends to a CAT-homeomorphism $\bar{\beta}$: $W \to W'$ (it extends to an embedding of the 2- and 3-handles and the remaining parts are *s*-cobordisms).

Let us consider a commutative diagram

$$\begin{array}{ccccc} & & & & X \\ & & & \overset{g}{\nearrow} & \downarrow{\scriptstyle\alpha} \\ \partial M & \subset & M & \xrightarrow{f} & Y \end{array}$$

where $X$ and $Y$ are connected spaces and $M$ is a compact CAT-manifold (our diagrams are strictly commutative, not only up to homotopy). A *semilifting* of the pair $(f, g)$ through $X$ is a semi-*s*-cobordism $(W, M, M_-)$ together with two maps $F$: $W \to Y$ and $f_-$: $M_- \to X$, satisfying $F(x, t) = f(x)$ for $(x, t) \in \partial M \times I \subset W$, $f_-(x) = g(x)$ for $x \in \partial M$ and such that the following diagram is commutative:

$$\begin{array}{ccc} M_- & \xrightarrow{f_-} & X \\ \cap & & \\ W & \overset{F}{\searrow} & \downarrow{\scriptstyle\alpha} \\ \cup & & \\ M & \xrightarrow{f} & Y \end{array}$$

Two semiliftings $((W, M, M_-), F, f_-)$ and $((W', M, M'_-), F', f'_-)$ are called *equivalent* if the map

$$F \cup F'\colon W \cup_{M \cup \partial M \times I} W' \longrightarrow Y$$

is such that the pair $(F \cup F', f_- \cup f'_-)$ admits a semilifting into $X$.

A group $N$ is called *locally perfect* if every finitely generated subgroup of $N$ is contained in a finitely generated perfect subgroup of $N$. A locally perfect group is clearly perfect. The converse is false. For instance, the group presented by $\langle a_k \mid a_k = [a_{k+1}, a_{k+2}]\rangle$, $k \in \boldsymbol{Z}$, is perfect but locally free.

The union $PG$ of all finitely generated perfect subgroups is the largest locally perfect subgroup of $G$. If $N$ is a normal subgroup of $G$, $PN$ is normal in $G$ since it is invariant under any automorphism of $N$.

A statement written for CAT-manifold must be understood, unless otherwise indicated, as a statement valid for DIFF-manifolds as well as for PL- or TOP-manifolds.

**2. Statement of the main theorems.**

THEOREM 2.1. *Let $\iota: X \to X^+$ be a plus map with respect to a locally perfect normal subgroup of $\pi_1(X)$. Let $M$ be a compact* CAT*-manifold of dimension $n \geqq 5$. Let $f: M \to X^+$ and $g: \partial M \to X$ be two maps such that $\iota \circ g = f \mid \partial M$. Then $(f, g)$ admits a semilifting into $X$. Two semiliftings of $(f, g)$ into $X$ are equivalent.*

Theorem 2.1 permits us to prove the more general Theorem 2.2 below. Let us consider a map $\alpha: X \to Y$ and define $N = \ker \alpha$. We have a factorization

$$\begin{array}{ccc} X & \xrightarrow{\iota} & X^+_{PN} \\ {\scriptstyle\alpha}\downarrow & \swarrow{\scriptstyle\alpha^+} & \\ Y & & \end{array}$$

where $X^+_{PN}$ denotes the plus map with respect to the largest locally perfect subgroup $PN$ of $N$. We choose $\iota$ and $\alpha^+$ such that the above diagram commutes strictly and $\alpha^+$ is a Serre fibration.

THEOREM 2.2. *Let $\alpha: X \to Y$ be a map between connected spaces. Let $M$ be a compact* CAT*-manifold of dimension $n \geqq 5$. Let $f: M \to Y$ and $g: \partial M \to X$ be two maps such that $\alpha \circ g = f \mid \partial M$. Suppose that either the largest perfect subgroup of $N$ is equal to $PN$ or $\pi_1(M)$ is finite. Then the set of equivalence classes of semi-liftings of $(f, g)$ into $X$ is bijection with the set of fiber homotopy classes of liftings $F: M \to X^+_{PN}$ such that $F \mid \partial M = \iota \circ g$.*

**3. Proof of the results of** §2. We first prove the following statement:

THEOREM 3.1. *Let $\iota: X \to X^+$ be a plus map with respect to a locally perfect normal subgroup $N$ of $\pi_1(X)$. Suppose that $X^+$ is homotopy equivalent to a finite complex. Let $K$ be a finite complex and $g: K \to X$ be a map. Then, there exist a finite complex $L$ containing $K$ and an extension $\gamma: L \to X$ of $g$ such that $\iota \circ \gamma$ is a plus map with respect to a locally perfect normal subgroup of $\pi_1(L)$.*

PROOF. We will construct a sequence

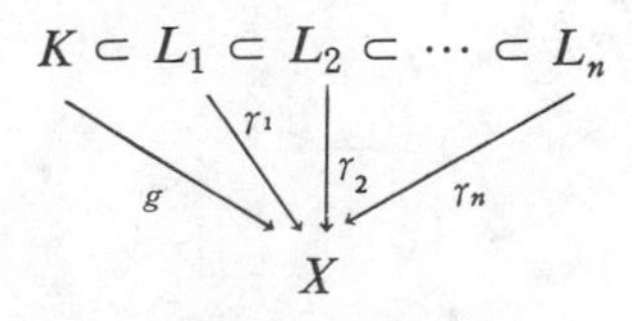


where $L_i$ are finite complexes such that

(i) $H_i(X, L_j; \Lambda) = 0$ for $i \leqq j$, where $\Lambda = \mathbf{Z}\pi_1(X^+)$,

(ii) the kernel of the homomorphism $\pi_1 L_i \to \pi_1(X^+)$ is locally perfect for all $i$,

(iii) $L_n = L$ satisfies the required properties.

Construct first a finite complex $L_0 \supset K$ by adding trivial 1-cells to $K$ in order to get an extension $\gamma_0: L_0 \to X$ of $g$ with $\pi_1(\iota \circ \gamma_0)$ onto. The kernel of $\pi_1(\iota \circ \gamma_0)$ is the normal closure in $\pi_1(L_0)$ of finitely many elements $x_1, \cdots, x_r$. Call $\bar{x}_i \in N$ the image of $x_i$ under $\pi_1\gamma_0$. Since $N$ is locally perfect, there exist $\bar{x}_{r+1}, \cdots, \bar{x}_{r+k} \in N$ such that the subgroup generated by $\bar{x}_1, \cdots, \bar{x}_{r+k}$ in $N$ is perfect. Thus, $\bar{x}_i = C_i$, where $C_i$ is a product of commutators of the $\bar{x}_i$'s. Call $P$ the perfect group defined by the presentation $\{\bar{x}_i \mid \bar{x}_i = C_i\}$, $1 \leqq i \leqq r + k$. One has the following commutative diagram:

$$\begin{array}{ccc} & T = \pi_1(L_0) * P/\{x_i = \bar{x}_i, i \leqq r\} & \\ \nearrow & & \searrow^{t} \\ \pi_1(L_0) & \longrightarrow & \pi_1(X) \end{array}$$

Construct a finite complex $L_1 \supset L_0$ such that $\pi_1(L_1) = T$. The reader will check that ker $\pi_1(\iota \circ \gamma_1)$ is the normal closure of $P$ in $T$ and is thus locally perfect. The space $L_1$ satisfies properties (i) and (ii).

We can now assume inductively that we are able to write $g: K \to X$ as a composition $K \to L_r \to^{\gamma_r} X$ such that $H_i(X, L_r; \Lambda) = 0$ for $i \leqq r \geqq 1$ and $\ker(\pi_1(\iota \circ \gamma_r))$ is locally perfect. Let $z_1, \cdots, z_s$ be a set of generators of the $\Lambda$-module $H_{r+1}(X, L_r; \Lambda) = H_{r+1}(\tilde{X}, \tilde{L}_r; \mathbf{Z})$, where $\tilde{X}$ and $\tilde{L}_r$ are the coverings corresponding to $L_r \to X \to B\pi_1(X^+)$. By [V], each $z_i$ is representable by a map $\zeta_i: (W_i, V_i) \to (\tilde{X}, \tilde{L}_r)$, where $(W_i, V_i)$ is a finite CW-pair with $\tilde{H}_*(W_i, \mathbf{Z}) = 0$, $H_*(V_i; \mathbf{Z}) = H_*(S^r; \mathbf{Z})$ and $W_i$ is of dimension $r - 1$.

Form the finite complex $L_{r+1} = L_r \cup_{\bar{\zeta}_i} (\bigcup_{i=1}^{s} W_i)$ where $\bar{\zeta}_i$ is the composition

$$V_i \xrightarrow{\xi_i | V_i} \tilde{L}_r \longrightarrow L_r.$$

One has a map $\gamma_{r+1}: L_{r+1} \to X$ extending $\gamma_r$ and $H_i(X, L_{r+1}; \Lambda) = 0$ for $i \leqq r + 1$. The kernel of $\pi_1(\iota \circ \gamma_{r+1})$ is the normal closure in $\pi_1(L_{r+1})$ of the images of $*_{i=1}^{s} \pi_1(W_i)$ and of ker $\pi_1(\iota \circ \gamma_r)$. It is thus locally perfect. The space $L_{r+1}$ satisfies conditions (i) and (ii).

Observe that dim $L_r$ = max($r$, dim $K_2$). Therefore, when $n - 1 \geqq$ max(dim $X^+$, dim $K_2$), one has $H_i(X, L_{n-1}; \Lambda) = 0$ for $i \neq n$, and then $H_n(X, L_{n-1}; \Lambda)$ is stably free as a $\Lambda$-module [W, Lemma 2.3]. By adding trivial $(n - 1)$-cells to $L_{n-1}$, we can assume that $H_n(X, L_{n-1}; \Lambda)$ is $\Lambda$-free with basis $z_1, \cdots, z_s$. Using these $z_i$'s to construct $L_n$ as before, one gets $H_*(X, L_n; \Lambda) = 0$. Thus $L_n = L$ and Theorem 3.1 is proved. □

PROOF OF THE THEOREM 2.1. Consider $\iota: X \to X^+$ as a Serre fibration and take the pull-back:

$$\begin{array}{ccc} Y & \longrightarrow & X \\ \downarrow & & \downarrow \\ M & \longrightarrow & X^+ \end{array}$$

Then $Y \to M$ is a plus map.

By Theorem 3.1 (in the case $K = \partial M$) there exist a connected finite complex $L$ and a diagram

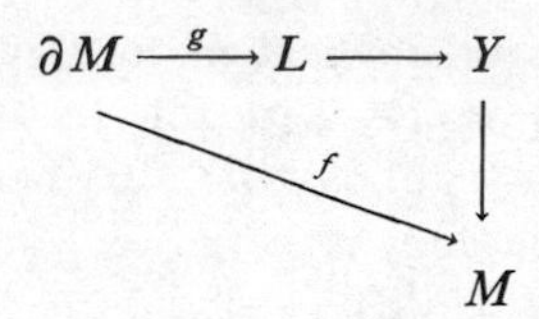

such that the composition $L \to Y \to M$ is a plus map with respect to a locally perfect subgroup of $\pi_1(L)$. Since $\pi_1(L)$ and $\pi_1(M)$ are both finitely presented, $\ker(\pi_1(L) \to \pi_1(M))$ is the normal closure in $\pi_1(L)$ of a finitely generated perfect subgroup of $\pi_1(L)$. Since $L$ is finite an obvious version with boundary of [**H1**, Theorem 5.1] permits us to construct a semi-$s$-cobordism $(W, M, M_-)$ and a diagram:

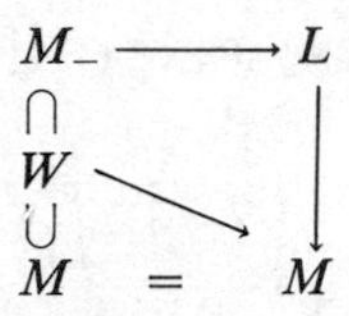

This proves the existence of a semilifting. For the uniqueness, form the manifold $W_0 = W \cup_{M \cup \partial M \times I} W'$ with boundary $\partial W_0 = M_- \cup_{\partial M} M'_-$ and apply the existence of a semilifting we have just proved to the situation:

$$\begin{array}{ccc} \partial W_0 & \xrightarrow{f_- \cup f'_-} & X \\ \cap & & \downarrow \\ W_0 & \xrightarrow{F \cup F'} & X^+ \end{array} \qquad \square$$

PROOF OF THEOREM 2.2. To a semilifting

$$\begin{array}{ccc} M_- & \xrightarrow{f_-} & X \\ \cap & & \\ W & \overset{F}{\searrow} & \downarrow \alpha \\ \cup & & \\ M & \xrightarrow{f} & Y \end{array}$$

of the pair $(f, g)$, we associate a lifting

$$\begin{array}{ccc} & & X^+_{PN} \\ & \overset{\tilde{f}}{\nearrow} & \downarrow \alpha^+ \\ M & \longrightarrow & Y \end{array}$$

as follows: By our hypotheses, the perfect group $\ker(\pi_1(M_-) \to \pi_1(W))$ is mapped by $\pi_1 f_-$ into $PN$. (Either $PN$ is the largest perfect subgroup of $N$ or, if $\pi_1(M)$ is finite, $\ker(\pi_1(M_-) \to \pi_1(W) = \pi_1(M))$ is finitely generated.) Hence, $\iota \circ f_-$ extends in $F'$: $W' \to X^+_{PN}$ where $W'$ is another copy of $W$. Any identification

$$(W \cup_{M_-} W', M) \cong (M \times I, W) \quad (s\text{-cobordism theorem})$$

together with the map $F \cup (\alpha^+ \circ F')$: $W \cup_{M_-} W' \to Y$ gives a map $f_0$: $M \to X^+_{PN}$ and a homotopy from $\alpha^+ \circ f_0$ to $f$. Since $\alpha^+$ is a Serre fibration, this gives a lifting $\tilde{f}$: $M \to X^+_{PN}$ of $f$ whose homotopy class (in the liftings of $f$) is well determined. The bijectivity of this correspondence comes from Theorem 2.1. $\square$

**4. Homology sphere bordism—application to algebraic** $K$**-theory**. Let $X$ be a (pointed) topological space. For $n \geqq 2$, denote by $\Omega_n^{HS}(X)$ (*homology sphere bordism*) the set of equivalence classes of pairs $(\Sigma, f)$, where

– $\Sigma^n$ is an oriented PL-homology sphere, i.e., an oriented $n$-dimensional PL-manifold with $H_*(\Sigma^n; \mathbf{Z}) \cong H(S^n; \mathbf{Z})$,

–$f\colon \Sigma \to X$ is a continuous pointed map.

Two pairs $(\Sigma_1, f_1)$ and $(\Sigma_2, f_2)$ are equivalent if there exists a PL-cobordism $(W, \Sigma_1, \Sigma_2)$ which is an $H_*$-cobordism (i.e., $H(W, \Sigma_i; \mathbf{Z}) = 0$) and a map $F\colon W \to X$ extending the $f_i$'s. ($W$ has a base arc joining the base points of $\Sigma_1$ and $\Sigma_2$ and mapped by $F$ to the base point of $X$.) With the connected sum, $\Omega_n^{HS}(X)$ becomes an abelian group; the zero element is $(S^n$, constant map) and $-(\Sigma, f) = (-\Sigma, f)$. (See **[H2]**.)

REMARK. In **[H2]**, $\Omega_n^{HS}(X)$ is defined with TOP-homology spheres. This gives the same groups if $n \neq 3$ or 4. We use the PL-case which is simpler in places (proof of Proposition 4.4) but all the results of this section are true in the TOP-category.

Throughout this section, we denote by $P$ the largest locally perfect subgroup $P\pi_1(X)$ of $\pi_1(X)$. Observe that, if $P = 0$, then $\Omega_n^{HS}(X) = \pi_n(X)$ when $n \neq 3$ and $\Omega_3^{HS}(X) = \pi_3(X) \oplus \mathscr{H}_3$ where $\mathscr{H}_3$ is the group of PL homology 3-spheres modulo $H^*$-cobordism.

If $f\colon \Sigma \to X$ is a representative of some class of $\Omega_n^{HS}(X)$, one has $\operatorname{Im} \pi_1 f \subset P$. The plus constructions with respect to $\pi_1(\Sigma)$ and $P$ provide a map $f^+\colon \Sigma^+ \cong S^n \to X^+$. (For $n \geqq 5$, $f^+$ is plus-cobordant to the map $i_0 \circ f$, where $\iota_0\colon X \to X^+$ is the plus map with respect to $P$.) This construction gives rise to a well-defined homomorphism $\alpha\colon \Omega_n^{HS}(X) \to \pi_n(X^+)$.

THEOREM 4.1. (a) *For* $n \geqq 5$, $\alpha$ *is an isomorphism.*
(b) *The following sequences are exact*

$$0 \to \Omega_4^{HS}(X) \xrightarrow{\alpha} \pi_4(X^+) \xrightarrow{\delta} \Omega_3^{HS}(F) \xrightarrow{\theta} \Omega_3^{HS}(X) \xrightarrow{\alpha} \pi_3(X^+) \to 0,$$
$$0 \to \Omega_2^{HS}(X) \xrightarrow{\alpha} \pi_2(X^+) \xrightarrow{\beta} H_2(P; \mathbf{Z}) \to 0$$

*where $F$ is the homotopy fiber $F \to^\theta X \to X^+$, $\delta$ is the composed map $\pi_4(X^+) \to \pi_3(F) \to \Omega_3^{HS}(F)$ and $\beta$ is the composed map*

$$\begin{aligned}\pi_2(X^+) &\cong H_2(X^+; \mathbf{Z}\pi_1(X^+)) \\ &\cong H_2(X; \mathbf{Z}\pi_1(X)/P) \to H_2(\pi_1(X); \mathbf{Z}\pi_1(X)/P) = H_2(P; \mathbf{Z}).\end{aligned}$$

Part (a) improves **[H2**, Theorem 5.1], and follows directly from Theorem 2.1. Part (b) will be proved below, after Corollary 4.2.

Recall that the algebraic $K$-theory groups $K_i(A)$ of a ring with unit $A$ were defined by Quillen **[Q]** as $K_i(A) = \pi_i(B\,\mathrm{Gl}(A)^+)$. Thus, Theorem 4.1 gives the following geometric interpretation of $K_i(A)$ (compare with **[H2**, Corollary 5.3]).

COROLLARY 4.2. *For any ring $A$, one has*:
(1) *if* $n \geqq 5$, $K_n(A) \cong \Omega_n^{HS}(B\,\mathrm{Gl}(A))$,
(2) $\Omega_4^{HS}(B\,\mathrm{Gl}(A))$ *is a subgroup of* $K_4(A)$ *and* $K_3(A)$ *is a quotient of* $\Omega_3^{HS}(B\,\mathrm{Gl}(A))$.

PROOF OF COROLLARY 4.2. In view of Theorem 4.1, it suffices to prove that $\ker(\pi_1(B\,\mathrm{Gl}(A)) \to \pi_1(B\,\mathrm{Gl}(A)^+))$ is locally perfect. But this kernel is equal to the subgroup $E(A)$ of $\mathrm{Gl}\,(A)$ generated by the elementary matrices. Now:

$$E(A) = \varinjlim_{m,R} E_m(R)$$

where $R$ ranges over all the subrings of $A$ which are finitely generated as $\boldsymbol{Z}$-algebras. The groups $E_m(R)$ are finitely generated and perfect if $m \geqq 2$ [**B**, Theorem 4.2]. Thus $E(A)$ is locally perfect. □

PROOF OF THEOREM 4.1. In the same way as for $\Omega_n^{HS}(X)$, we define the relative homology sphere bordism $\Omega_n^{HS}(X, A)$, for $n \geqq 2$ and for any pair $(X, A)$ of pointed topological spaces. A representative of a class of $\Omega_n^{HS}(X, A)$ is a pair $(B, f)$, where $B$ is a compact oriented acyclic PL-manifold of dimension $n$ and $f: (B, \partial B) \to (X, A)$ is a continuous pointed map. Two pairs $(B_1, f_1)$ and $(B_2, f_2)$ represent the same class in $\Omega_n^{HS}(X, A)$ if there are a PL $H_*$-cobordism $(T, W)$ between $B_1$ and $B_2$ (i.e., $T$ is acyclic, $\partial T = B_1 \cup W \cup B_2$ and $W$ is an $H_*$-cobordism between $\partial B_1$ and $\partial B_2$) and a map $F: (T, W) \to (X, A)$ extending the $f_i$'s and sending the base arc joining the base points of $\partial B_i$ onto the base point of $A$. With the boundary connected sum, $\Omega_n^{HS}(X, A)$ becomes a group, which is abelian when $n \geqq 3$. One has the following exact sequence:

$$\begin{aligned} \Omega_n^{HS}(A) \longrightarrow \Omega_n^{HS}(X) \longrightarrow \Omega_n^{HS}(X,A) \longrightarrow \Omega_{n-1}^{HS}(A) \\ \to \cdots \to \Omega_2^{HS}(X, A) \longrightarrow \pi_1(A) \longrightarrow \pi_1(X). \end{aligned}$$

We now consider the pair $(X, F)$, where $F$ is the fiber of the map $\iota: X \to X^+$ considered as a Serre fibration. We shall prove the following Lemma and proposition:

LEMMA 4.3. *There is a natural homomorphism* $\mu$: $\Omega_n^{HS}(X, F) \to \pi_n(X^+)$ *such that the following diagram*

$$\begin{array}{ccc} \Omega_n^{HS}(X) & \xrightarrow{\alpha} & \pi_n(X^+) \\ & \searrow \quad \nearrow_{\mu} & \\ & \Omega_n^{HS}(X, F) & \end{array}$$

*is commutative and* $\mu$ *is an isomorphism for all* $n \geqq 2$.

PROPOSITION 4.4. *If $A$ is an acyclic space with $\pi_1(A)$ locally perfect, then* $\Omega_n^{HS}(A) = 0$ *if* $n \neq 3$.

The fiber $F$ is acyclic and $\pi_1(F)$ is the universal central extension of $P$. Since $P$ is locally perfect, $\pi_1(F)$ is also locally perfect [V]. Thus, Theorem 4.1 clearly follows from Lemma 4.3 and Proposition 4.4 except for the right-hand side of the last exact sequence. But $\Omega_2^{HS}(X)$ is a quotient of $\pi_2(X)$ and thus

$$\operatorname{Coker}(\Omega_2^{HS}(X) \to \pi_2(X^+)) = \operatorname{Coker}(\pi_2(X) \to \pi_2(X^+)) = H_2(P; \boldsymbol{Z}). \quad \square$$

PROOF OF LEMMA 4.3. The homomorphism $\mu$ is defined as follows: Let $f$: $(B, \partial B) \to (X, F)$ be a representative of a class of $\Omega_n^{HS}(X, A)$. This defines a map $\bar{f}$: $B/\partial B \to X^+$ and, by plus construction with respect to $\pi_1(B/\partial B)$, one gets a map $\hat{f}$: $S^n \to X^+$. Define $\mu(f)$ by $[\hat{f}] \in \pi_n(X^+)$. The only thing to prove is that $\mu$ is an isomorphism.

By commutativity of the following diagram:

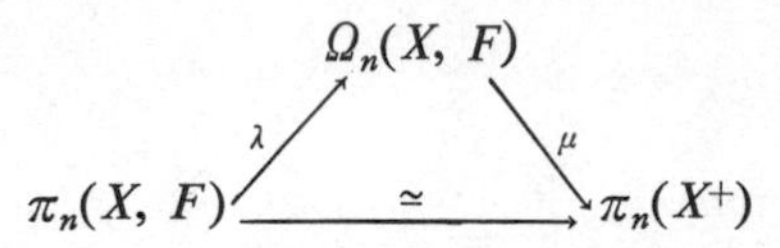

where $\lambda$ is the natural map, one gets that $\mu$ is onto and $\lambda$ is injective. It is thus enough to prove that $\lambda$ is also onto. Let $f: (B/\partial B) \to (X, F)$ be a representative of a class of $\Omega_n^{HS}(X, F)$. The manifold $W = B - \text{int } D^n$, where $D^n$ is an $n$-disk embedded in int $B$, is an $H_*$-cobordism between $\partial B$ and $S^n$. The map $\iota \circ f$ maps $(B, \partial B)$ into $(X^+, \text{pt})$ and is, by obstruction theory, homotopic relative $\partial B$ to a map sending $W$ onto the base point. A lifting of such a homotopy produces an $H_*$-cobordism from $f$ to $f': (D^n, S^{n-1}) \to (X, F)$ and thus $\lambda$ is surjective. □

PROOF OF PROPOSITION 4.4. Let $f: \Sigma^n \to A$ be a representative of a class of $\Omega_n^{HS}(F)$. If $n \geqq 4$, $\Sigma = \partial C$ where $C^{n+1}$ is a contractible manifold. Hence, the assertion follows from Theorem 2.1 applied to the situation:

$$\begin{array}{ccc} \Sigma^n & \xrightarrow{f} & A \\ \cap & & \downarrow \\ C^{n+1} & \longrightarrow & A^+ \simeq \text{pt} \end{array}$$

If $n = 2$, $\Sigma^2 \simeq S^2$ and, by [V] there exists a commutative diagram

$$\begin{array}{ccc} S^2 \simeq \partial W & \subset & W \\ g\downarrow & & \downarrow \\ S^2 & \longrightarrow & A \end{array}$$

where $W$ is an acyclic 3-manifold and this diagram represents the generator of $H_3(A, S^2) \simeq \mathbf{Z}$. Thus $g$ is homotopic to a homeomorphism and $(S^2, f)$ is equivalent to zero. □

REMARK. In order to compute $\Omega_n^{HS}(X)$, it is enough, in view of Theorem 2.1 and Proposition 4.4, to compute $\pi_n(X^+)$ (plus with respect to $P\pi_1(X)$) and $\Omega_3^{HS}(A)$ when $A$ is acyclic with locally perfect fundamental group. Very little is known about the computation of $\Omega_3^{HS}(A)$ but here are some results:

LEMMA 4.5. *Let $g: A \to B$ be a 3-connected map between two acyclic spaces with locally perfect fundamental groups. Then the induced homomorphism $g_*: \Omega_3^{HS}(A) \to \Omega_3^{HS}(B)$ is an isomorphism.*

PROOF. (1) *Surjectivity of $g_*$*: Let $f: \Sigma^3 \to B$ be a map representing a class in $\Omega_3^{HS}(B)$. Let $F$ the homotopy fiber of $g: A \to B$. One has $\pi_1(F) = \pi_2(F) = 0$. Hence, there is no obstruction to lift $f$ through $A$.

(2) *Injectivity of $g_*$*: Let $f: \Sigma^3 \to A$ represent a class in ker $g_*$. There exists an acyclic manifold $C^4$ and an extension $F: C \to B$ of $g \circ f$. The unique obstruction $\theta$ to extend $f$ in a lifting of $F$ through $A$ belongs in

$$H^4(C, \Sigma; \pi_3(F)) = H_0(C; \pi_3(F)) \simeq H_0(\pi_1(C); \pi_3(F))$$

(homology with local coefficients). But $\pi_3(F) = H_3(F)$ and, using the fact that $g$ induces an isomorphism in integral homology and the Serre spectral sequence, one gets $H_0(\pi_1(B); \pi_3(F)) = 0$. Since $\pi_1(B)$ is locally perfect, $H_0(\pi_1(B); \pi_3(F)) = \varinjlim_G H_0(G; \pi_3(F))$ where $G$ ranges over the finitely generated perfect subgroups of $\pi_1(B)$. Thus, there is a finitely generated perfect group $G_0 \subset \pi_1(B)$ and a factorization:

$$\begin{array}{ccc} \pi_1(C) & \xrightarrow{\pi_1 F} & \pi_1(B) \\ & \searrow & \cup \\ & & G_0 \end{array}$$

such that $\theta$ goes to zero in $H_0(G_0; \pi_3(F))$.

Let $K$ be a 2-dimensional acyclic finite complex such that $G_0$ is a quotient of $\pi_1(K)$. (See [**H1**, §2.1] for the construction of such a $K$.) This complex has an embedding in $\boldsymbol{R}^5$ and thus there exists a 4-dimensional homology sphere $T^4$ with a map $T^4 \to K \to B$ sending $\pi_1(T)$ onto $G_0$. The obstruction to get a lifting $\bar{F}$:

$$\begin{array}{ccc} \Sigma & \xrightarrow{f} & A \\ \cap & \overset{\bar{F}}{\nearrow} & \downarrow g \\ C \# T & \longrightarrow & B \end{array}$$

is the image of $\theta$ in $H_0(C\# T; \pi_3(F)) = H_0(G_0; \pi_3(F))$. Hence this obstruction is zero and $\bar{F}$ exists. This proves the injectivity of $g_*$. $\square$

THEOREM 4.6. *Let* $g: X \to Y$ *be a map such that*
(1) $g$ *induces an isomorphism* $\pi_i(X) \to \pi_i(Y)$ *for* $i = 1$ *and* 2.
(2) $g$ *induces an isomorphism*

$$H_*(X; \boldsymbol{Z}\pi_1(Y)/P) \to H_*(Y; \boldsymbol{Z}\pi_1(Y)/P) \qquad (P = P\pi_1(X)).$$

*Then* $g$ *induces an isomorphism from* $\Omega_*^{HS}(X)$ *to* $\Omega_*^{HS}(Y)$.

PROOF. Let $Y \to Y_2$ be the second stage of the Postnikov decomposition of $Y$. Consider the fibration

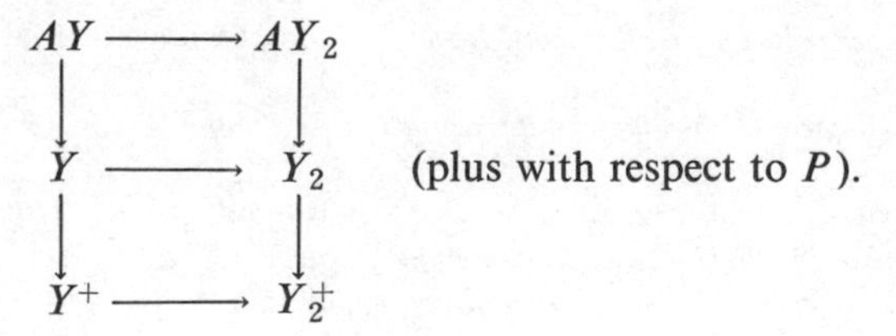

(plus with respect to $P$).

Let $F$ be the fiber of $AY \to AY_2$. Clearly $\pi_1(F) = 0$. We will first prove that $\pi_2(F) = 0$. From the following diagram

$$\begin{array}{ccccccc} & & & & \pi_3(Y^+) & & \\ & & & & \downarrow \delta & & \\ \pi_3(AY_2) & \xrightarrow{\partial} & \pi_2(F) & \xrightarrow{\alpha} & \pi_2(AY) & \longrightarrow & \pi_2(AY_2) \\ & & & & \downarrow & & \downarrow \\ & & & & \pi_2(Y) & \xrightarrow{\simeq} & \pi_2(Y_2) \end{array}$$

one deduces that $\operatorname{Im} \alpha \subset \operatorname{Im} \delta$ and thus $\operatorname{Im} \alpha$ is a trivial $\boldsymbol{Z}\pi$-module ($\pi$ denotes $\pi_1(AY_2)$). On the other hand, $\pi_3(AY_2) \simeq \pi_4(Y_2^+)$ and thus $\operatorname{Im} \partial$ is also a trivial $\boldsymbol{Z}\pi$-module. Hence $H_0(\pi; \operatorname{Im} \alpha) \simeq \operatorname{Im} \alpha$, $H_0(\pi; \operatorname{Im} \partial) \simeq \operatorname{Im} \partial$. Hence $H_1(\pi; \operatorname{Im} \alpha)$

$= 0$ and the five lemma implies that $H_0(\pi;\ \pi_2(F)) \simeq \pi_2(F)$. Since $AY$ and $AY_2$ are both acyclic, the Serre spectral sequence implies that $H_0(\pi;\ \pi_2(F)) = 0$. Therefore $\pi_2(F) = 0$.

Define $\bar{Y}_2$ by the pull-back diagram

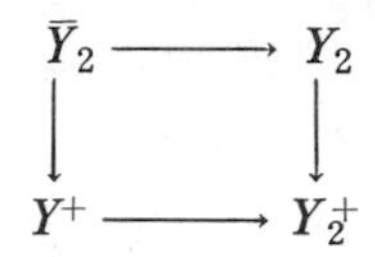

and consider the diagram:

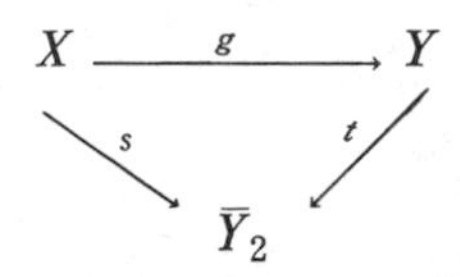

Condition (1) implies that $Y_2$ is also the second stage of the Postnikov decomposition of $X$. Clearly $\bar{Y}_2^+ = Y^+$ and condition (2) makes the map $s^+\colon X^+ \to \bar{Y}_2^+$ a homotopy equivalence. As we saw above, the corresponding maps on the fibers $AX \to AY_2$ and $AY \to AY_2$ are 3-connected and $\pi_1(AX) \simeq \pi_1(AY) \simeq \pi_1(AY_2)$ is locally perfect [V, 5.5]. By Lemma 4.5 and Theorem 4.1, $s_*\colon \Omega_*^{HS}(X) \to \Omega_*^{HS}(\bar{Y}_2)$ and $t_*\colon \Omega_*^{HS}(Y) \to \Omega_*^{HS}(\bar{Y}_2)$ are isomorphisms. Thus $g_*\colon \Omega_*^{HS}(X) \to \Omega_*^{HS}(Y)$ is an isomorphism. □

## REFERENCES


[B] H. Bass, *K-theory and stable algebra*, Inst. Hautes Étude Sci. Publ. Math. **22** (1964), 489–544.

[H1] J.-Cl. Hausmann, *Homological surgery*, Ann. of Math. (2) **104** (1976), 573–584.

[H2] ———, *Homology sphere bordism and Quillen plus construction*, Algebraic $K$-theory (Evanston, 1976), Lecture Notes in Math., vol. 551, Springer-Verlag, Berlin and New York, 1976, pp. 170–181.

[Ke] M. Kervaire, *Le théorème de Barden-Mazur-Stallings*, Comment. Math. Helv. **40** (1965), 31–42. MR **32** #6475.

[L] J.-L. Loday, *K-theorie algebrique et representation de groupes*, Ann. Sci. Ecole Norm. Sup. **9** (1976), 309–377.

[Q] D. Quillen, *Cohomology of groups*, Actes Congrès Internat. Math., Nice **2** (1970), 47–51.

[V] P. Vogel, *Un théorème de Hurewicz homologique*, Comment. Math. Helv. **52** (1977), 393–413.

[W] C. T. C. Wall, *Surgery on compact manifolds*, Academic Press, 1970.



UNIVERSITY OF GENEVA

UNIVERSITY OF NANTES

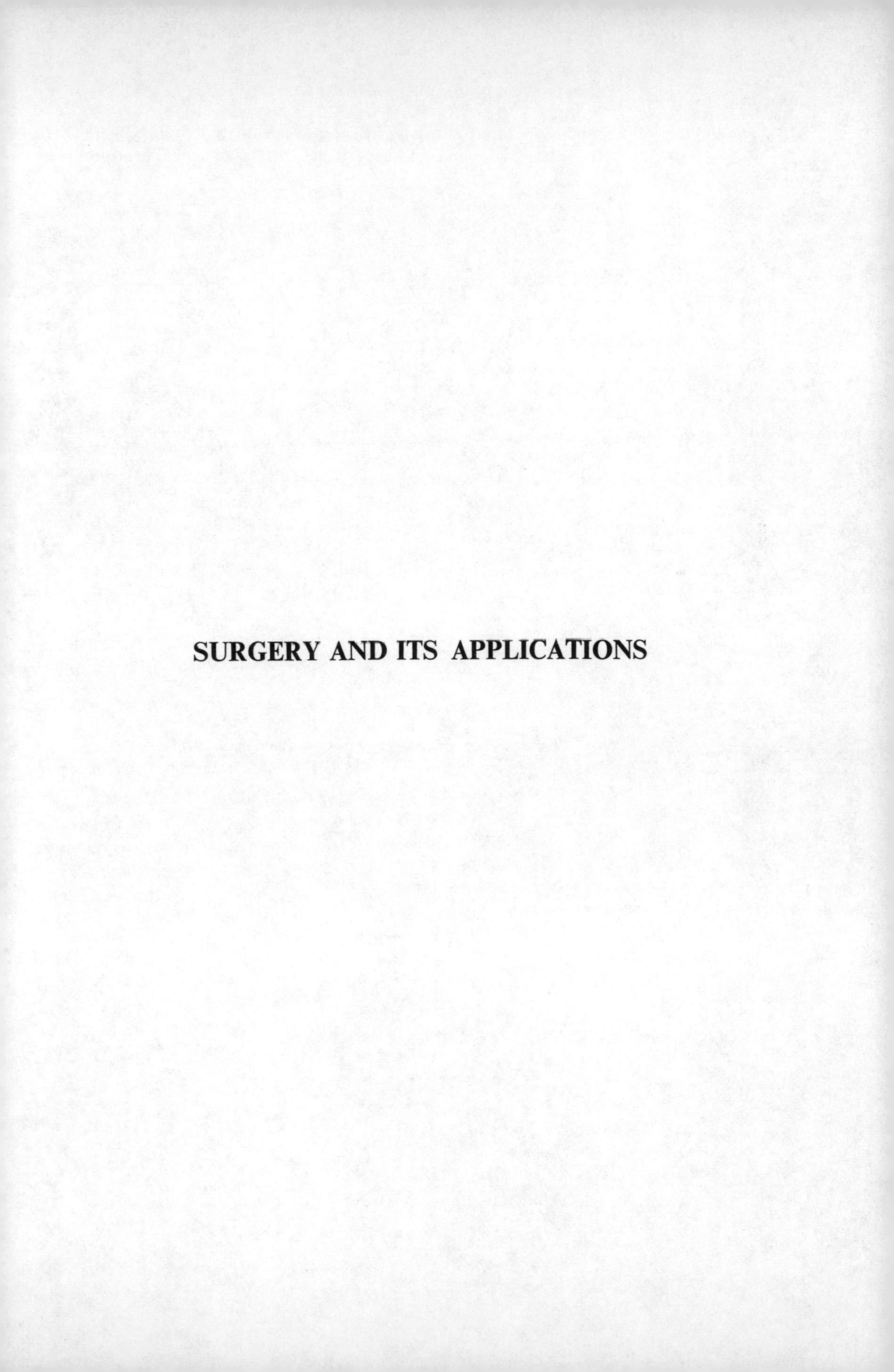

# SURGERY AND ITS APPLICATIONS

Proceedings of Symposia in Pure Mathematics
Volume 32, 1978

# A SURVEY OF THE CLASSIFYING SPACES ASSOCIATED TO SPHERICAL FIBERINGS AND SURGERY

R. JAMES MILGRAM

In this short survey we try to present the main results of the work of G. Brumfiel, Ib Madsen, the author and Dennis Sullivan on the structure of the classifying spaces $B_{\mathrm{PL}}$, $B_{\mathrm{TOP}}$, $G/\mathrm{PL}$, $G/\mathrm{TOP}$, $B_G$ and $B(G/\mathrm{TOP})$ and, where applicable their associated Thom spectra and bordism theories.

It is our feeling (and hope) that the results are in sufficiently precise shape that the reader who understands what they say can apply them to classification problems for manifolds and Poincaré duality spaces without wading through the hundreds of pages of proofs in the literature.

**1. The classifying spaces associated to surgery.** Consider the spaces and maps in the following diagram:

$$
\begin{array}{ccccccc}
\mathrm{PL}/O & & & & & & \\
\downarrow & & & & & & \\
\mathrm{TOP}/O & \longrightarrow & \mathrm{TOP}/\mathrm{PL} & & & & \\
\downarrow & & \downarrow & & & & \\
G/O & \longrightarrow & G/\mathrm{PL} & \longrightarrow & G/\mathrm{TOP} & & \\
\downarrow & & \downarrow & & \downarrow & & \\
B_O & \longrightarrow & B_{\mathrm{PL}} & \longrightarrow & B_{\mathrm{TOP}} & \longrightarrow & B_G \\
 & & & & & & \downarrow \\
 & & & & & & B_{G/O} \longrightarrow B_{G/\mathrm{PL}} \longrightarrow B_{G/\mathrm{TOP}}
\end{array}
\tag{1.1}
$$

where $A/C$ denotes the (homotopy) fiber in the map of classifying spaces $B_C \to B_A$.

---

*AMS* (*MOS*) *subject classifications* (1970). Primary 57C25; Secondary 57C20, 57B10, 57A65, 57A70.

Here, $B_{\mathrm{TOP}}$, $B_G$, $B_{\mathrm{PL}}$, and $B_O$ are stable classifying spaces for various kinds of sphere bundles. Thus, for $X$ a finite complex $[X, B_{\mathrm{TOP}} \times Z]$ is the Grothendieck group of stable isomorphism classes of topological sphere bundles over $X$. Similarly, $[X, B_{\mathrm{PL}} \times Z]$ represents the group of stable sphere bundles over $X$ with "group" the piecewise linear homeomorphisms $S^n \to S^n$. $B_G$ on the other hand represents fiber homotopy sphere bundles or *spherical fiberings* for short. Recall that a spherical fibering is a map $f: E \to X$ whose homotopy fiber is the homotopy type of a sphere. Two spherical fiberings

$$\begin{array}{c} f : E \to X \\ \nearrow \\ f' : E' \end{array}$$

are equivalent if there are maps $g: E' \to E$, $h: E \to E'$ so $f \circ g = f'$, $f = f' \circ h$ and

$$\underset{\text{Fiberwise}}{g \circ h \simeq \mathrm{id}_{E'}}, \qquad \underset{\text{Fiberwise}}{h \circ g \simeq \mathrm{id}_E}.$$

The Whitney sum of two spherical fiberings may be defined by first converting the maps into actual Hurewicz fiberings and then taking fiberwise joins.

Then we can construct the Grothendieck group $K^0_{SF}(X)$ of stable equivalence classes of spherical fiberings over $X$ and $[X, B_G \times Z] = K^0_{SF}(X)$.

Originally the map $B_{0,G} : B_0 \to B_G$ was considered by people such as Atiyah, James and J. F. Adams. The image $(B_{0,G})_*: K^0_0 (X) \hookrightarrow K^0_{SF} (X)$ is called $J(X)$ and was shown to be closely related to the G. W. Whitehead $J$-homomorphism [**2**]. Adams introduced certain operations in $K_0$ theory, the $\psi^k$'s, and conjectured that at certain primes $(p)$ depending on $k$ the composite

$$B_0 \xrightarrow{\psi^k - 1} B_0 \longrightarrow B_{G(p)} \tag{1.2}$$

was homotopically trivial. This was proved by Quillen, Friedlander, and Sullivan [**25**], [**12**], [**28**], though the simplest proof is now due to Becker and Gottlieb [**3**]. The null homotopy in (1.2) gives a lifting

$$\begin{array}{ccc} B_0 & \overset{l}{\dashrightarrow} & G/O_{(p)} \\ & \underset{\psi^k - 1}{\searrow} & \downarrow \\ & & B_0 \end{array} \tag{1.3}$$

and $l$ is seen to inject the homotopy of $B_0$ into a direct summand [**1**].

The resulting connections of $B_G$ and $G/O$ with stable homotopy theory are fundamental and will be covered in more detail later. These spaces and the others in (1.1) also play basic roles in the study and classification of manifolds. Thus, consider the fibering

$$\mathrm{PL}/O \to G/O \to G/\mathrm{PL}. \tag{1.4}$$

The homotopy groups of $\mathrm{PL}/O$ are identified with the groups of diffeomorphism classes of homotopy spheres under connected sum $\Gamma_n = \pi_n (\mathrm{PL}/O)$ and

(1.5)
$$\pi_n(G/\mathrm{PL}) = \begin{cases} Z, & n \equiv 0\,(4), \\ 0, & n \equiv 1,3\,(4), \\ Z/2, & n \equiv 2\,(4), \end{cases}$$

are obtained by interpreting the work of Kervaire and Milnor **[14]**, in dimensions $\geqq 5$ and from Cerf's results in dimension 4 **[10]**. They may be interpreted as surgery obstructions for degree 1 normal maps.

Associated to these spaces Browder-Novikov and Sullivan showed how to analyze the set of simply connected PL-manifolds in a given homotopy type in terms of an exact sequence ($n \geqq 5$) of sets

(1.6)
$$0 \to \mathscr{H}\mathfrak{F}(M^n) \to [M^n, G/\mathrm{PL}] \to \pi_n(G/\mathrm{PL}) \to 0.$$

Here $\mathscr{H}\mathfrak{F}(M^n)$ is the set of all PL normal bordism classes of manifolds homotopy equivalent to $M$. That is, the objects are equivalence classes of homotopy equivalences $f\colon M' \to M$ where $(f, M') \sim (g, M'')$ if there is a manifold $W^{n+1}$, $\partial W = M' \cup M''$ and a homotopy equivalence $H\,|\,W \to I \times M$ so $H|M'$ is $f$ and $H|\,M''$ is $g$.

REMARK 1.7. In particular, two elements in the same class in $\mathscr{H}\mathfrak{F}(M^n)$ are $H$-cobordant so by the PL-$H$-cobordism theorem $n \geqq 5$ are PL-homeomorphic, and we see that there is a surjection of $\mathscr{H}\mathfrak{F}(M)$ onto the set of PL-homeomorphism classes of manifolds in the homotopy type of $M$.

In the differentiable case a similar theory holds but now the sequence analagous to (1.6) is a long exact sequence extending to the left

$$\cdots \to [(I \times M, \partial I \times M); (G/O, X)]$$
$$\to \pi_{n+1}(G/\mathrm{PL}) \to \mathscr{H}\mathscr{D}(M) \to [M^n, G/O] \to \pi_n(G/\mathrm{PL}),$$

where $\mathscr{H}\mathscr{D}(M^n)$ is the set of differentiable $H$-cobordism classes of manifolds homotopy equivalent to $M$. Once more there is a surjection of $\mathscr{H}\mathscr{D}(M^n)$ onto the set of diffeomorphism classes of manifolds homotopy equivalent to $M$.

Work of Sullivan and Kirby-Siebenmann has given us the homotopy types of the spaces $G/\mathrm{PL}$, $G/\mathrm{TOP}$ and $\mathrm{TOP}/\mathrm{PL}$, while Kirby and Siebenmann have also generalized (1.6) to topological manifolds $n \geqq 5$.

THEOREM 1.8 (KIRBY-SIEBENMANN). $\mathrm{TOP}/\mathrm{PL} \simeq K(Z/2, 3)$ *where* $K(Z/2, 3)$ *is the Eilenberg-Mac Lane space with one nontrivial homotopy group* $Z/2$ *in dimension* 3.

THEOREM 1.9 (SULLIVAN). (a) $(G/\mathrm{PL})_{(2)} = \Omega E_3 \times \prod_{i>1} K(Z/2, 4i-2) \times K(Z_{(2)}, 4i)$ *where* $E_3$ *is the 2-stage Postnikov system* $E_3 \to K(Z/2, 2) \to^{\beta Sq^2(\iota)} K(Z_{(2)}, 5)$.

(b) $(G/\mathrm{PL})_{(p)} = B_{O_{(p)}}$ *for p any odd prime.*

COROLLARY 1.10 (KIRBY-SIEBENMANN-SULLIVAN). (a) $(G/\mathrm{TOP})_{(2)} \simeq \prod_{i \geqq 1} K(Z/2, 4i-2) \times K(Z_{(2)}, 4i)$.

(b) $B_{O_{(p)}} \simeq (G/\mathrm{TOP})_{(p)}$ *for all odd p.*

The structure theorems for $G/O$, due primarily to Sullivan, and based on the proof of the Adams conjecture and (1.3) show that

$$G/O_{(p)} \sim B_{O_{(p)}} \times (\mathrm{coker}\ J)\ (p)$$

where $l: B_0 \to G/O(p)$ is injection onto the first factor. Here coker $(J)(p)$ is a $p$-primary space with finite homotopy in each dimension, but the homotopy groups themselves are direct summands of the stable homotopy of spheres.

REMARK 1.11. An important example of how the classification theory works is provided by Sullivan [27] in his study of (1.6) for $M = CP^n$.

He shows that the map $\mathscr{H}\mathfrak{F}(CP^n)$ onto PL-homeomorphism classes of manifolds homotopic to $M$ is an isomorphism $n \geqq 3$. Moreover from 1.9 the set $[CP^n, G/\mathrm{PL}]$ is easily calculated, and the same ideas which gave 1.9 also allowed him to calculate the map $[CP^n, G/\mathrm{PL}] \to \pi_{2n}(G/\mathrm{PL})$. Thus, the classification of homotopy $CP^n$'s is effective.

**2. The lower spaces in (1.1).** In (1.1) the spaces and natural maps

$$B_O \to B_{\mathrm{PL}} \to B_{\mathrm{TOP}} \to B_G \tag{2.1}$$

also play fundamental roles in the classification of manifolds. Currently these are of two types.

First, the Browder-Novikov theorem states that if one has a simply connected Poincaré duality space $X$, then $X$ has the homotopy type of a PL-manifold if and only if its Spivak normal bundle [5, Chapter I] admits a PL-reduction. Now, the Spivak normal bundle, being a (stable) spherical fibering over $X$, is classified by a unique homotopy class of maps $\nu: X \to B_G$ and to say that $\nu$ admits a PL-reduction is the same as being able to compress the map $\nu$ so as to obtain the diagram:

$$\begin{array}{ccc} B_{\mathrm{PL}} & \xrightarrow{B_{\mathrm{PL},G}} & B_G \\ & \nwarrow_{\nu|} \quad \nearrow_{\nu} & \\ & X & \end{array} \tag{2.2}$$

On the other hand, results of Boardman and Vogt show that $B_{\mathrm{PL},G}$ is an infinite loop map [4]. Hence, as is indicated in (1.1) the fibering $G/\mathrm{PL} \to B_{\mathrm{PL}} \to B_G$ admits a classifying map $B_{(G,G/\mathrm{PL})}: B_G \to B_{G/\mathrm{PL}}$ and $\nu$ lifts to $\nu|$ if and only if $B_{(G,G/\mathrm{PL})} \circ \nu \simeq 0$.

THEOREM 2.3 (a) (MADSEN-MILGRAM [18]).

$$B_{(G/\mathrm{PL})\,(2)} \simeq E_3 \times \prod_{i \geqq 2} K(Z/2, 4i - 1) \times K(Z_{(2)}, 4i + 1).$$

(b) (SULLIVAN). $B_{G/\mathrm{PL}(p)} = B^2_{O(p)}$ *for $p$ odd.*

Thus, except for one $K$-invariant in dimension 6 the lifting problem at the prime 2 is purely a cohomology question, and at odd primes is a question in $K$-theory.

REMARK 2.4. The explicit form of the map $B_{(G,G/\mathrm{PL})(2)}$ was also determined in cohomology in [18] and will be explained after we have described $H^*(B_G, Z_{(2)})$. The forms of the map at odd primes will be mentioned below.

REMARK 2.5. For a striking example of how to use 2.3 and the map $B_{(G,G/\mathrm{PL})}$ see the talk of C. T. C. Wall in these PROCEEDINGS.

The situation at odd primes was further explored by Sullivan. He proved that

$$B_{\mathrm{PL}(p)} \simeq B_{O(p)} \times B_{\mathrm{coker}J(p)} \tag{2.6}$$

at each odd prime. Moreover, the solution of the Adams conjecture showed that at odd primes

(2.7) $$B_{G(p)} \simeq B_{\mathrm{im}J(p)} \times \mathrm{B}_{\mathrm{coker}\ J(p)}$$

and the map $B_{(G,G/\mathrm{PL})p}$ restricted to the $B_{\mathrm{coker}\ J}$ part is trivial while restricted to the $B_{\mathrm{im}\ J(p)}$ part is just the natural map in the long exact sequence

(2.8) $$\cdots \to B_{O(p)} \xrightarrow{\psi^q - 1} B_{O(p)} \to B_{\mathrm{im}\ J} \to B_0^2 \xrightarrow{B(\psi^q-1)} B_0^2 \to \cdots.$$

The analysis of the 2-primary obstruction was considerably more involved. To begin we needed the 2-primary cohomology of $B_G$. This was obtained in the following roundabout way.

THEOREM 2.9 (FOLKLORE). *The space of stable homotopy equivalences of the sphere* $\mathscr{H}(S^0)$ *is* $\lim_{n\to\infty} (\Omega^n S^n)_{\pm 1}$ *(where* $(\Omega^n S^n)_i$ *denotes the space of base point preserving degree i maps* $S^n \to S^n$*).*

THEOREM 2.10 (STASHEFF [26]). $B_G = B_{\mathscr{H}(S^0)}$.

Note in particular that $\pi_i(\mathscr{H}(S^0)) = \lim_{n\to\infty} \pi_i(\Omega^n S^n) = \lim_{n\to\infty} \pi_{n+i}(S^n) = \pi_i^s(S^0)$ for $i > 0$, where $\pi_i^s(S^0)$ is the $i$th stable homotopy group of spheres. Thus, the first 6 homotopy groups of $B_G$ are

| 1 | 2 | 3 | 4 | 5 | 6 |
|---|---|---|---|---|---|
| $Z/2$ | $Z/2$ | $Z/2$ | $Z/24$ | 0 | 0 |

and $B_G = RP^\infty \times B_{SG}$ where $SG$ is $\lim_{n\to\infty} (\Omega^n S^n)_{+1}$.

The first two stages of the Postnikov system for $B_{SG}$ are given as follows. The first $K$-invariant must be trivial:

$$K_1 \colon K(Z/2,2) \to K(Z/2,4)$$

since either $K_1^*(\iota_4) = \iota_2^2$ or $K_1^*(\iota_4) = 0$ but the map $B_{SO} \to B_{SG}$ takes the class corresponding to $\iota_2$ back to $w_2$ and $w_2^2 \neq 0$ in $B_{SO}$ so $\iota_2^2 \neq 0$ in $B_{SG}$.

The next $K$-invariant

(2.11) $$K_2 \colon K(Z/2,2) \times K(Z/2,3) \to K(Z/24,5)$$

is $2\beta(\iota_2^2) + 12(Sq^4\iota_3)$.

The term $2\beta(\iota_2^2)$ is forced by the structure of $B_{SO}$ and the fact that the $Z$ in $\pi_4(B_{SO})$ maps surjectively onto the $Z/24$, while the term $12(Sq^2\iota_3)$ is a consequence of the relation $\eta^3 = 12\xi$ in $\pi_3^s(S^0)$ (look at the situation in $\Omega B_{SG} = SG$).

An easy calculation now shows that through dimension 6

$$H^*(B_{SG}, Z/2) = P(w_2, w_3, w_4, w_5, w_6) \otimes E(k_3, Sq^1k_3, Sq^2k_3, Sq^{2,1}k_3)$$

where $j^*(w_4) = \iota_4$ in $H^*(K(Z/24, 4), Z/2)$ in the continuation of (2.11) to the left

$$K(Z/24, 4) \xrightarrow{j} E_3 \xrightarrow{\pi} K(Z/2;2,3) \longrightarrow K(Z/24,5),$$

and $k_3 = \pi^*(\iota_3)$.

Calculating Steenrod operations $Sq^1w_2 = w_3$, $Sq^1w_4 = w_5$ and $Sq^1(Sq^2k_3) = k_3^2 = 0$. A chain level calculation (which is, however, formal) shows that $w_2^2 + 2\{w_4\}$ is a $Z/8$ class with coboundary $8\{Sq^2k_3\}$.

But, in $H^*(B_{SO})$, $w_2^2 + 2\{w_4\}$ is the 2-adic image of the Pontrjagin class $p_4$, which is integral. Thus, we see that the first obstruction to reduction of a PL bundle to a vector bundle is $k_3$, and the second obstruction, the $Z/8$ Bochstein of $(w_2^2 + 2\{w_4\})$ is tied to the first since $\beta_8(w_2^2 + 2\{w_4\}) = Sq^2 k_3$.

Besides these low dimensional calculations we also point out the folklore result, see, e.g., **[6]**.

THEOREM 2.12. *$H^*(B_{SG}, Z/2) = P(w_2, w_3, \cdots, w_n, \cdots) \otimes \mathscr{C}$ as an algebra where the $w_i$ are the universal Stiefel-Whitney classes.*

(The Thom space $M(f)$ of a spherical fibering $f : E \to B$ with fiber $\simeq S^n$ is the mapping cone of $f$. The Thom isomorphism theorem works as usual, and in the universal Thom space over $B_G$ we set $Sq^i U = w_i \cup U$. Thus, since $H^*(B_0) = P(w_1 \cdots w_n \cdots)$ and the $w_i$ are all coming from $H^*(B_G)$, $H^*(B_G)$ contains $P(w_1 \cdots w_n \cdots)$ as well. Finally, $B_0 \to B_G$ is a map of $H$-spaces so $H^*(B_G) \to H^*(B_0)$ is a map of Hopf algebras and the remainder of 2.12 follows from standard facts on Hopf algebras.)

Next, work of Dyer-Lashof **[11]** led to the following general result, which determines $\mathscr{C}$.

THEOREM 2.13 **[22]**. *$H^*(B_G, Z/2) = P(w_1, \cdots, w_n, \cdots) \otimes E(\cdots \sigma(Q_{0,i}) \cdots) \otimes E(\cdots \sigma(Q_I) \cdots)$ where $I$ runs over all sequences of positive integers $(i_1, \cdots, i_n)$, $n \geqq 2$, $\sigma(Q_1)$ has dimension $1 + i_1 + 2i_2 + \cdots + 2^{n-1} i_n$ and $\dim \sigma(Q_{0,i})$ is $2i + 1$ $(i \geqq 1)$.*

In particular, $k_3 = \sigma(Q_{0,1})$, $Sq^1(k_3) = \sigma(Q_{1,1})$, $Sq^2(k_3) = \sigma(Q_{0,2})$.

To obtain the higher two torsion (corresponding to the $Z/8$ in dimension 4 already studied) it was necessary to obtain the structure of the Bochstein spectral sequence for $B_G$. This was possible since the work of Dyer-Lashof made it possible to study the Bochstein spectral sequence for $G$, the phenomenon which led to the $Z/8$ Bochstein generalizes and these two things gave all the differentials.

To describe the answer it is convenient to use homology $H_*(B_G, Z/2)$, which is also a Hopf algebra, using Whitney sum.

THEOREM 2.14 (MILGRAM, UNPUBLISHED, MADSEN **[16]**).

$$H_*(B_G, Z/2) = P(\cdots b_i \cdots) \otimes E(\cdots \sigma(Q_{0,i})_* \cdots) \otimes P(\cdots \sigma(Q_{I'})_* \cdots)$$

*and $i_1 > 1$ in $I'$, otherwise the $I'$ is as in 2.13. Moreover*

$$d_3(\sigma(Q_{0,2i})_*) = b_{2i}^2, \qquad d_1(b_{2i}) = b_{2i-1}.$$

*Finally, on the remaining generators $d_1$ takes generators to generators or their squares and the higher differentials in the Bochstein spectral sequence are formal consequences of these.*

Next, the natural maps $G \to G/\mathrm{PL}$ and $\lambda : B_G \to B_{G/\mathrm{PL}}$ were studied at the prime 2 in **[7]**, **[18]**, **[19]**. The basic results are best expressed in terms of explicit classes in $H^*(B_G, Z_{(2)})$ and $H^*(B_G, Z/2)$.

In particular, if we look at the fundamental classes in dimension $4i - 1$ in 2.3(a) and call them $K_{4i-1}$, then the first result is

THEOREM 2.15. *$\lambda^*(K_{4i-1}) = 0$ unless $i$ is a power of 2, in which case $\lambda^*(K_{4i-1}) = \sigma(Q_{0,2i-1})$.*

Unfortunately, these classes are a bit difficult to describe as a priori intrinsic functions of an explicitly given fiber homotopy sphere bundle. Fortunately, the remaining obstructions turn out to be given in a reasonably explicit way from operations with the Stiefel-Whitney classes, and the classes above are tied to these using the Steenrod operations in the same way that $Sq^2(k_3) = \beta_3(w_2^2 + 2\{w_4\})$.

Specifically, in the subalgebra of $H^*(B_G)$, $P(w_1 \cdots w_n \cdots) \otimes E(\beta_8(w_2^2), \cdots, \beta_8(w_{2n}^2) \cdots)$ we pick out certain special classes which are the Bochsteins of the Newton polynomials in the $w_{2i}^2$. The first nontrivial case is $w_2^4 + 2(w_4^2)$ which is a $Z/16$ class with Bochstein

$$d_{16}(w_2^4 + 2(\omega_4^2)) = w_2^2(Sq^2k_3).$$

In general, the $i$th Newton polynomial $N_{iq}$ has order $2^{3+\nu_2(i)}$ (where $\nu_2(i)$ is the power of 2 occurring in the prime decomposition of $i$) and $N_i$ has dimension $4i$. Then we have

THEOREM 2.16. *Let $\iota_{4j+1}$ be the primitive fundamental class in $H^*(B_{G/\mathrm{TOP}}, Z_{(2)})$; then $\lambda^*(\iota_{4j+1}) = 2^{\alpha(i)-1}\beta(N_i)$ where $\alpha(i)$ is the number of ones in the dyadic expansion of $i$.*

Using 2.15 and 2.16 the 2-adic cohomology structure of $B_{\mathrm{TOP}}$ and $B_{\mathrm{PL}}$ can also be routinely determined. Detailed results along these lines are given in **[19]**.

**3. Some applications.** In Williamson's thesis **[30]** the transversality theorem was proved for PL-bundles and manifolds, while Siebenmann proved transversality for topological bundles except when the fiber has dimension 4. This implies $\pi_*(M(S\,\mathrm{PL})) \cong \Omega_*^{\mathrm{PL}}(\mathrm{pt})$, the cobordism ring of oriented PL manifolds, and except in dimension 4, $\pi_*(M(S\,\mathrm{TOP})) \cong \Omega_*^{\mathrm{TOP}}(\mathrm{pt})$.

Similar results hold for the unoriented cases $\pi_*(M(\mathrm{PL})) \cong \mathcal{N}_*^{\mathrm{PL}}(\mathrm{pt})$ and except in 4, $\pi_*(M(\mathrm{TOP})) = \mathcal{N}_*^{\mathrm{TOP}}(\mathrm{pt})$.

The bordism rings of Poincaré duality spaces $\Omega_*^{\mathrm{PD}}(\mathrm{pt})$ and $\mathcal{N}_*^{\mathrm{PD}}(\mathrm{pt})$ are defined in analogy with the manifold case and Levitt-Morgan **[15]** constructs certain exact sequences which fit together to give the diagram:

$$\begin{array}{ccccccc} \xrightarrow{I} \Omega_i^{\mathrm{PD}}(\mathrm{pt}) & \longrightarrow & \pi_i^s(M(SG)) & \xrightarrow{\partial} & L_{i-1}(0) & \xrightarrow{I} & \Omega_{i-1}^{\mathrm{PD}}(\mathrm{pt}) \\ \downarrow & & \downarrow & & \downarrow J^* & & \downarrow \\ \xrightarrow{I} \mathcal{N}_i^{\mathrm{PD}}(\mathrm{pt}) & \longrightarrow & \pi_i(M(G)) & \xrightarrow{\partial} & L_{i-1}(Z/2,-) & \longrightarrow & \mathcal{N}_{i-1}^{\mathrm{PD}}(\mathrm{pt}) \longrightarrow \end{array} \tag{3.1}$$

Brumfiel-Morgan **[8]**, **[9]** have made extensive calculations of the $\partial$ maps above.

At the prime 2 the result of Browder, Liulevicius, Peterson **[6]** shows that the various Thom spectra above split as wedges of Eilenberg-Mac Lane spectra

$$\begin{aligned} M(SG)_{(2)} &\simeq \bigvee \Sigma K(Z/2^i), \\ M(S\,\mathrm{PL})_{(2)} &\simeq \bigvee \Sigma^{4i}K(Z_{(2)}) \vee \Sigma^s K(Z/2^i), \\ M(S\,\mathrm{TOP})_{(2)} &\simeq \bigvee \Sigma^{4i}K(Z_{(2)}) \vee \Sigma^{s'} K(Z/2^i) \end{aligned}$$

while $M(G)$, $M(\mathrm{PL})$ and $M(\mathrm{TOP})$ are each wedges of suspensions of $K(Z/2)$. In fact we have

$$\pi_*(M(G)) \cong \pi_*(M(0)) \otimes (H_*(B_G)//H_*(B_0))$$

$$(3.2) \qquad \cong \mathcal{N}_*(\mathrm{pt}) \otimes E(\sigma(Q_{0j})) \otimes P(\sigma Q_{I'} \cdots),$$

$$\pi_*(M(\mathrm{PL})) = \mathcal{N}_*(\mathrm{pt}) \otimes (H_*(B_{\mathrm{PL}})//H_*(B_0)).$$

Also, the results of Milnor **[23]** show $\pi_*(M(SG)_p) \simeq \bigvee \Sigma K(Z/p^i)$ for each $p$ while May's results give the exact decomposition.

Returning to (3.1), $\partial\sigma(Q_{0,j}) \neq 0$ **[9]** and since $L_{2i}(Z/2,—) = Z/2$, $L_{2i-1}(Z/2,—) = 0$ it follows that $\mathcal{N}_*^{\mathrm{PD}}(\mathrm{pt})$ injects into $\pi_*(M(G))$. Moreover (3.1) now determines the situation in the oriented case since $L_{4i}(0) = Z$, $L_{4i+2} = Z/2$ with $S_*$ an isomorphism, and $\sigma Q_{(0,2j+1)}$ represents a $Z/2$ class in $\pi_*(M(SG))$.

REMARK 3.3. The $Z$ in $\Omega_{4i}^{\mathrm{PD}}(\mathrm{pt})$ must occur since the index of an oriented P.D. space is a P.D. bordism invariant (the original proof for manifolds used only formal properties of cup product and Poincaré duality in cohomology **[13]**), but (3.1) shows, since $\pi_*(M(SG))$ is torsion, that the index is the only integral invariant of P.D. bordism.

Turning to the PL and TOP bordism the problem breaks into two parts. At the prime 2, (3.2) shows that the essential step is $H_*(B_{\mathrm{PL}})//H_*(B_0)$, $H_*(B_{\mathrm{TOP}})//H_*(B_0)$, which is an involved, but in view of the results of §2, not impossible, calculation. At odd primes Sullivan's splitting theorem and the results of May and Tsuchiya **[21]**, **[29]** make short work of $H_*(B_{\mathrm{PL}}) = H_*(B_{\mathrm{TOP}})$ but there is no simple decomposition of $M(S\,\mathrm{PL})_{(p)}$ so there remain hard problems.

We first discuss the situation at 2.

As was indicated at the end of §2, Madsen-Milgram **[19]** give the structure of $H_*(B_{\mathrm{SPL}}, Z)$ and $H_*(B_{\mathrm{SPL}}, Z/2)//H_*(B_0)$ is given in **[7]**. The general results are not very illuminating and not all the generators have had representative models constructed.

A very important role is played by the geometric construction $\Gamma_n(M) = S^n \times_T M \times M$, which for $\dim(M)$ and $n$ of opposite parity mod (2) is oriented if $M$ is oriented. Basically, by iterating this construction on a comparatively small set of generators one obtains all generators. (This is a geometric interpretation of the Dyer-Lashof operations.)

Perhaps somewhat more interest attaches to the torsion free parts $\Omega_*^{\mathrm{TOP}}/\mathrm{Tor}$ and $\Omega_*^{\mathrm{PL}}/\mathrm{Tor}$. Historically, Milnor's original construction of the exotic spheres and much of the early work in this area centered around the fact (implicit in Milnor's proof) that $\Omega_*/\mathrm{Tor} \to \Omega_*^{\mathrm{PL}}/\mathrm{Tor}$ is not injective onto a direct summand. (More precisely Milnor's work related to spin PL bordism.)

Here we have reasonably complete information.

THEOREM 3.4. (a) $\Omega_*^{\mathrm{PL}}(\mathrm{pt})/\mathrm{Tor} = \Omega_*^{\mathrm{Top}}(\mathrm{pt})/\mathrm{Tor}$ *has generators the differentiable torsion free generators, the Milnor manifolds of index* 8, *and the exotic complex projective spaces* (*discussed at the end of* §1).

(b) $(\Omega_*^{\mathrm{PL}}(\mathrm{pt})/\mathrm{Tor}) \otimes Z_{(p)}$ *is polynomial for each odd* $p$, $P(x_4^{(p)}, \cdots, x_{4p}^{(p)}, \cdots)$.

(c) $(\Omega_*^{\mathrm{PL}}(\mathrm{pt})/\mathrm{Tor}) \otimes Z_{(2)}$ *is a tensor product of a polynomial algebra and a divided power algebra.*

In particular $\Omega_*^{\mathrm{PL}}(\mathrm{pt})/\mathrm{Tor}$ is not globally a polynomial algebra.

Again, related to 3.4 one expects to be able to answer the classical questions of integrability for PL and topological manifolds: which rational polynomials in the Pontrjagin classes are integral for every PL or TOP bundle, or when evaluated for the PL or TOP normal bundle to a manifold $M$ and capped with the orientation class $[M]$.

In principle 3.4 completely solves the second problem above. For the first problem we have given explicit answers in §2. At the prime 2 the general answer seems too complicated to be terribly interesting, so we content ourselves with answering the question of when a *genus* can be integral at (2).

THEOREM 3.5. *The genus $A$ is $Z_{(2)}$ integral on all topological manifolds if and only if its primitive series* $\mathscr{P}(A) = \sum \lambda_i x^i$ *satisfies*

(1) $\lambda_i \in Z_{(2)}$,

(2) $\lambda_{2j} \equiv \lambda_{2j-1}\ (2^j)$,

(3) $\alpha(i) - 4 \geqq \nu_2(i)$ *then* $2^{\alpha(i)-3}$ *divides* $\lambda_i$.

At odd primes the result is somewhat different.

THEOREM 3.6. $H^*(B_{\mathrm{PL}}, Z(\tfrac{1}{2}))/\mathrm{Tor} \cong H^*(B_{\mathrm{TOP}}, Z(\tfrac{1}{2}))/\mathrm{Tor} = P(R_4, R_8, \cdots, R_{4i}, \cdots)$ *where the $R_{4i}$ form the genus with primitive series*

$$P(\mathscr{R}) = \sum (-1)^n (2^{2n-1} - 1)(\mathrm{Num}\ B_{2n}/4n) x^n.$$

(Here $B_{2n}$ is the Bernoulli number and 3.6 was originally conjectured by Brumfiel based on low dimensional calculations.)

Now we review the situation for $M(S\,\mathrm{PL})$ at odd primes. The first results are due to F. Peterson **[24]**, and D. Sullivan.

THEOREM 3.7 (SULLIVAN). *At odd primes there is a $B_0$ orientation* $M(S\mathrm{PL}) \to B_0$.

Next, using the splitting of $B_{\mathrm{PL}}$ at odd primes discussed in §2,

THEOREM 3.8 (SULLIVAN). $M(S\,\mathrm{PL})_p \simeq M(SO)_p \wedge M(\gamma)_p$ *where $\gamma$ is a bundle over* $B(\mathrm{coker}\ J)_{(p)}$.

As a consequence of 3.7, 3.8 one can show that $\pi_*(M(S\,\mathrm{PL}_{(p)}))/\mathrm{Tor}$ is determined by $KO$ characteristic numbers and this leads in particular to 3.5(b).

To obtain information about the torsion structure to date has been considerably harder. F. Peterson made some preliminary calculations in dimensions $< 2p^2(p-1)$, which showed first that there was $p$-torsion, and second that some of these torsion classes are not detected by characteristic numbers in ordinary homology.

Peterson conjectured on the basis of his results that the mapping of the Steenrod algebra $\varphi: \mathscr{A}(p) \to H^*(M(S\,\mathrm{PL}), Z/p)$ defined by $\varphi(\alpha) = \alpha(U)$, maps $\mathscr{A}(p)$ onto the quotient $\mathscr{A}(p)//E(Q_0, Q_1)$. Peterson's conjecture was soon verified by Tsuchiya **[29]**, and we were originally hopeful that his technique would quickly lead to a determination of the $E^2$ term of the Adams spectral sequence converging to $\pi_*(M(S\,\mathrm{PL})) \otimes \hat{Z}_{(p)}$. Unfortunately the techniques were not delicate enough.

More recently, B. Mann and the author **[20]**, and independently H. Ligaard and J. P. May at the University of Chicago, have proved a refined version of Tsuchiya's result.

THEOREM 3.9. *There exists a subalgebra* $\Lambda(A, \underline{W}_3, \cdots, \underline{W}_n, \cdots) \otimes E(W_2, W_3, \cdots)$ *of* $H^*(B_{\mathrm{coker}\ J}; Z/p)$ *satisfying*

$$\begin{aligned} Q_1Q_0(A) &= W_2, \\ Q_1(\lambda(A)) &= Q_0(\underline{W}_3) = W_3, \\ Q_1(\lambda(W_3)) &= Q_0(\underline{W}_4) = W_4, \end{aligned}$$

*etc.* (*where* $\Lambda(\cdots)$ *is a divided power algebra* $\lambda$ *is the divided pth power operator*) *and under the Thom isomorphism* $Q_i(U_\gamma) = W_i \cup U_\gamma$.

Now, this result *does* allow an effective calculation of the entire $E^2$ term of the Adams spectral sequence, and as a corollary we have

THEOREM 3.10. *In the A.S.S. converging to* $\pi_*(M(S\,\mathrm{PL})) \otimes Z_{(p)}$, *there exist differentials of all orders.*

Thus the explicit determination of the odd torsion remains a difficult problem, though in the paper **[20]** we were able to calculate using 3.9, considerably past Peterson's range, and we now feel that all the structure of this A.S.S. is within range of current techniques.

## BIBLIOGRAPHY

**1.** J. F. Adams, *On the groups J(X)*. II, Topology **3** (1965), 137–171. MR **33** #6626.

**2.** M. F. Atiyah, *Thom complexes*, Proc. London Math. Soc. (3) **11** (1961), 291–310. MR **24** #A1727.

**3.** J. C. Becker and D. H. Gottlieb, *The transfer map and fiber bundles*, Topology **14** (1975), 1–12. MR **51** #14042.

**4.** M. Boardman and R. Vogt, *Homotopy invariant algebraic structures on topological spaces*, Lecture Notes in Math., vol. 347, Springer-Verlag, Berlin and New York, 1973.

**5.** W. Browder, *Surgery on simply-connected manifolds*, Ergebnisse der. Math. und ihrer Grenzgebiete, Band 65, Springer-Verlag, New York and Heidelberg, 1972. MR **50** #11272.

**6.** W. Browder, A. Liulevicius and F. Peterson, *Cobordism theories*, Ann. of Math. (2) **84** (1966), 91–101. MR **33** #6638.

**7.** G. W. Brumfiel, Ib Madsen and R. Milgram, PL *characteristic classes and cobordism*, Ann. of Math. (2) **97** (1973), 82–159. MR **46** #9979.

**8.** G. W. Brumfiel and J. W. Morgan, *Quadratic functions, the index modulo* 8, *and a Z/*4-*Hirzebruch formula*, Topology **12** (1973), 105–122. MR **48** #3059.

**9.** ———, *Homotopy theoretic consequences of N. Levitt's obstruction theory to transversality*, Pacific J. Math. **67** (1976), 1–100.

**10.** J. Cerf, *Sur les difféomorphismes de la sphère de dimension trois* ($\Gamma_4 = 0$), Lecture Notes in Math., no. 53, Springer-Verlag, Berlin and New York, 1968. MR **37** #4824.

**11.** E. Dyer and R. Lashof, *Homology of iterated loop spaces*, Amer. J. Math. **84** (1962), 35–88. MR **25** #4523.

**12.** E. Friedlander, Thesis, M.I.T., 1970.

**13.** F. Hirzebruch, *Topological methods in algebraic geometry*, Die Grundlehren der Math. Wissenschaften, Band 131, Springer-Verlag, New York, 1966. MR **34** #2573.

**14.** M. Kervaire and J. Milnor, *Groups of homotopy spheres*. I, Ann. of Math. (2) **77** (1963), 504–537. MR **26** #5584.

**15.** N. Levitt and J. Morgan, *Fiber homotopy transversality*, Bull. Amer. Math. Soc. **78** (1972), 1064–1068.

**16.** Ib Madsen, *On the action of the Dyer-Lashof algebra in* $H_*(G)$, Pacific J. Math. **60** (1975), 235–275.

**17.** Ib Madsen and R. J. Milgram, *On spherical fiber bundles and their* PL *reductions*, New Developments in Topology (Proc. Sympos. Algebraic Topology, Oxford, 1972), Cambridge Univ. Press, London, 1974. MR **49** #8028.

**18.** ———, *The universal smooth surgery class*, Comment. Math. Helv. **50** (1975), no. 3, 281–310. MR **52** #4285.

**19.** ———, Ann. of Math. Studies (to appear).

**20.** B. Mann and R. J. Milgram, *On the action of the Steenrod algebra* $A(p)$ *on* $H^*(MS\,\mathrm{PL}, Z/p)$ *at odd primes* (Mimeo. Stanford Univ.).

**21.** J. P. May, *Homology operations on infinite loop spaces*, Algebraic Topology (Univ. Wisconsin, Madison, Wis., 1970), Proc. Sympos. Pure Math., vol. 22, Amer. Math. Soc., Providence, R.I., 1971, pp. 171–185. MR **47** #7740.

**22.** R. J. Milgram, *The* mod 2 *spherical characteristic classes*, Ann. of Math. (2) **92** (1970), 238–261. MR **41** #7705.

**23.** J. Milnor, *On characteristic classes for spherical fibre spaces*, Comment. Math. Helv. **43** (1968), 51–77. MR **37** #2227.

**24.** F. P. Peterson, *Some results on* PL-*cobordism*, J. Math. Kyoto Univ. **9** (1969), 189–194. MR **40** #4962.

**25.** D. Quillen, *The Adams conjecture*, Topology **10** (1970), 67–80. MR **43** #5525.

**26.** J. Stasheff, *A classification theorem for fibre spaces*, Topology **2** (1963), 239–246. MR **27** #4235.

**27.** D. Sullivan, Thesis, Princeton Univ., 1966.

**28.** ———, *Genetics of homotopy theory and the Adams conjecture*, Ann. of Math. (2) **100** (1974), 1–79.

**29.** A. Tsuchiya, *Characteristic classes for spherical fibre spaces*, Proc. Japan Acad. **44** (1968), 617–622. MR **40** #2115.

**30.** R. E. Williamson, Jr., *Cobordism of combinatorial manifolds*, Ann. of Math. (2) **83** (1966), 1–33. MR **32** #1715.

STANFORD UNIVERSITY

Proceedings of Symposia in Pure Mathematics
Volume 32, 1978

# REMARKS ON NORMAL INVARIANTS FROM THE INFINITE LOOP SPACE VIEWPOINT

IB MADSEN

**0. Introduction.** Consider a Poincaré duality homotopy type $X$. It is well known that the question of realizing $X$ as a manifold (smooth or topological, say) leads to a 2-stage obstruction theory. The primary part is the obstruction in $[X, B(G/O)]$ or $[X, B(G/\mathrm{TOP})]$ to a linear or topological reduction of the Spivak normal fibration $\nu_X$. A reduction of $\nu_X$ implies a 'surgery problem' over $X$ and the secondary obstruction is the surgery obstruction in $L_n(\pi_1 X)$.

A homotopy equivalence $h: M \to X$ (from a manifold $M$) determines a reduction of $\nu_X$ and the set of distinct reductions then becomes naturally identified with $[X, G/O]$ or $[X, G/\mathrm{TOP}]$. Moreover, in this case the surgery obstructions associated to $[f] \in [X, G/O]$ or $[f] \in [X, G/\mathrm{TOP}]$ are given by characteristic numbers which involve certain universal classes $\tilde{L}_{4n} \in H^{4n}(G/\mathrm{TOP}; Z_{(2)})$ and $\tilde{K}_{4n-2} \in H^{4n-2}(G/\mathrm{TOP}; Z/2)$.

In this article I discuss algebraic topological properties of the classifying spaces above from the point of view of infinite loop space theory.

The main new result is the characterization of the Morgan-Sullivan index class $\tilde{L}_{4n}$ in $H^*(G/O; Z_{(2)})$ (Theorem 4.3). This is combined with results of Sullivan and May, Quinn and Ray to give that a $KO$-orientable (stable) spherical fibration $\xi$ admits a topological reduction iff certain characteristic classes (connected with the Arf-invariant one conjecture) $\tau_i(\xi)$ in $H^{2^i-1}$ (base) all vanish.

**1. Transfer spaces.** A half-exact functor $h$ from CW spaces to sets is called a *transfer functor* if to each finite covering $p: \bar{X} \to X$ there is associated a transfer map $p_!: h(\bar{X}) \to h(X)$ such that the assignment $p \to p_!$ is natural under composition and pull-backs of finite coverings (Quillen).

---

*AMS (MOS) subject classifications* (1970). Primary 57C20, 57D65; Secondary 55B20.

*Key words and phrases.* Normal invariants, infinite loop maps, transfer universal surgery classes, Bockstein spectral sequence, spherical fibrations.

The transfer $h(X) \times h(X) \to h(X)$ associated to the trivial 2-sheeted covering of $X$ gives $h(X)$ the structure of an abelian monoid and the transfer maps $p_!$ become homomorphisms.

Let $p\colon \bar{X} \to X$ be an $n$-sheeted covering and $P(\bar{X}) \to X$ the associated principal $\Sigma_n$-covering: $P(\bar{X})$ consists of ordered $n$-tuples $(\bar{x}_1, \cdots, \bar{x}_n)$ of points in $\bar{X}$ over the same point $x$ in $X$, and $\bar{X} = P(\bar{X}) \times_{\Sigma_n} [n]$ where $[n] = \{1, \cdots, n\}$ with the standard $\Sigma_n$-action. There is an obvious map $X \to P(\bar{X}) \times_{\Sigma_n} \bar{X}^n$ and composing with the classifying map $P(\bar{X}) \to E\Sigma_n$ we get the *characteristic map* $X \to^{\bar{p}} E\Sigma_n \times_{\Sigma_n} \bar{X}^n$. If further $q\colon \bar{\bar{X}} \to \bar{X}$ is an $m$-sheeted cover over $\bar{X}$ then the characteristic map $\overline{q \circ p}$ is the composition

$$X \xrightarrow{\bar{p}} E\Sigma_n \times_{\Sigma_n} \bar{X}^n \xrightarrow{1 \times \bar{q}^n} E\Sigma_n \times_{\Sigma_n} (E\Sigma_m \times_{\Sigma_m} \bar{\bar{X}}^m)^n$$

and

$$E\Sigma_n \times_{\Sigma_n} (E\Sigma_m \times_{\Sigma_m} \bar{\bar{X}}^m)^n = E(\Sigma_n \wr \Sigma_m) \times_{\Sigma_n \wr \Sigma_m} \bar{\bar{X}}^{nm},$$

where $\Sigma_n \wr \Sigma_m \subset \Sigma_{nm}$ denotes the wreath product.

Let $H$ be the representing space for $h$, $h(X) = [X, H]$. From the preceding paragraph it follows that $E\Sigma_n \times_{\Sigma_n} H^n$ represents the functor $h_{(n)}(X)$ consisting of pairs $(\bar{X}, \bar{x})$ of an $n$-sheeted covering over $X$ and an element $\bar{x} \in h(\bar{X})$. The transfer gives a natural transformation $h_{(n)} \to h$ which induces certain structure maps

$$d_n\colon E\Sigma_n \times_{\Sigma_n} H^n \to H. \tag{1.1}$$

Moreover, for each $n$, $m$ there is a homotopy commutative diagram

$$\begin{array}{ccc} E\Sigma_n \times_{\Sigma_n} (E\Sigma_m \times_{\Sigma_m} H^m)^n & \xrightarrow{1 \times d_m^n} & E\Sigma_n \times_{\Sigma_n} H^n \\ \downarrow{\scriptstyle \text{incl}} & & \downarrow{\scriptstyle d_n} \\ E\Sigma_{nm} \times_{\Sigma_{nm}} H^{nm} & \xrightarrow{d_{nm}} & H \end{array} \tag{1.2}$$

((1.2) is direct from the required naturality properties of $p \to p_!$.)

A transfer space $H = (H, \{d_n\})$ is a space $H$ with maps $d_n$ as in (1.1) such that the diagrams (1.2) homotopy commute. Clearly, $H$ is a transfer space if and only if $[X, H]$ is a transfer functor; indeed $p_!\colon [\bar{X}, H] \to [X, H]$ is induced from the characteristic map $\bar{p}$ and $d_n$ in the obvious fashion (compare with the original definitions of Kahn and Priddy **[KP]**). An infinite loop space is a transfer space (and in particular each cohomology theory is a transfer functor). One of the basic conclusions of the theory of infinite loop spaces is a partial converse: If for a transfer space $H$ the diagrams (1.2) are *strictly* commutative then $H$ is an infinite loop space (we assume $\pi_0(H)$ is a group) **[BV]**, **[Ma]**, **[S]**.

EXAMPLE 1.3. Denote by $BO^{\oplus}$, $BO^{\otimes}$ the representing spaces for $\widetilde{KO}(X)$ and $(1 + \widetilde{KO}(X), \otimes)$, respectively. Both functors extend to cohomology theories **[S]** and the associated structure maps

$$d^{\oplus}\colon E\Sigma_n \times_{\Sigma_n} (BO^{\oplus})^n \to BO^{\oplus}, \qquad d_n^{\otimes}\colon E\Sigma_n \times_{\Sigma_n} (BO^{\otimes})^n \to BO^{\otimes}$$

classify the virtual vector bundles

$$E\Sigma_n \times_{\Sigma_n} \xi \oplus \cdots \oplus \xi, \qquad E\Sigma_n \times_{\Sigma_n} (1 + \xi) \otimes \cdots \otimes (1 + \xi).$$

For a pair of finite groups $\rho \subset \pi$ we thus have transfers $\mathrm{tr}^{\oplus}$, $\mathrm{tr}^{\otimes}\colon \widetilde{KO}(B\rho) \to$

$\widetilde{KO}(B\pi)$. The additive transfer, $\mathrm{tr}^{\oplus}$, is compatible with the classical induction homomorphism on representation rings under the natural homomorphism $\widetilde{RO}(\pi) \to \widetilde{KO}(B\pi)$. The $\mathrm{tr}^{\otimes}$ is maybe less well known (see **[MST]**).

A mapping $f: H_1 \to H_2$ between transfer spaces is called *transfer commuting* if the induced operation $[—, H_1] \to [—, H_2]$ commutes with transfers. Equivalently the requirement is that $d_1 \circ (1 \times f^n)$ be homotopic to $f \circ d_1$ as maps from $E\Sigma_n \times_{\Sigma_n} H_1^n$ to $H_2$.

Quillen has asked if a (say simply connected) transfer space is an infinite loop space, and similarly one can ask if a transfer commuting map between infinite loop spaces is infinitely deloopable. Note, with respect to the relative question that a map $f: H_1 \to H_2$ between infinite loop spaces is transfer commuting if and only if it commutes with transfers for $p$-fold cyclic coverings for all primes $p$. (This follows from (2.2) because the $p$-Sylow subgroup of $\Sigma p^n$ is the $n$-fold wreath product $\boldsymbol{Z}/p \wr \cdots \wr \boldsymbol{Z}/p$, and because the $p$-Sylow subgroup of $\Sigma_m$ is the product of the $p$-Sylow subgroups in $\Sigma p^{n_1} \times \cdots \times \Sigma p^{n_r}$, $p^{n_1} + \cdots + p^{n_r} = m$.)

Let $X[p]$ denote the localization at $p$ of the space $X$. The next simple result is important for the applications in §§2 and 4.

THEOREM 1.4 **[MST]**. *A transfer commuting map $f: BO^{\oplus}[p] \to BO^{\oplus}[p]$ is an infinite loop map.*

The proof of 1.4 is computational, based on the knowledge of all $H$-endomorphisms of $BO^{\oplus}[p]$. It suffices to show that the completion $\hat{f}$ is deloopable. But $\hat{f} = \sum a_{r,i}(\psi^{p^r+i} - \psi^i)$ $(a_{r,i} = 0$ if $i \geqq p^{r+1} - p^r)$ and evaluating on representation rings one sees that $\hat{f}$ is transfer commuting iff $a_{r,i} = 0$ for $i|p$. In this case $\hat{f}$ is also deloopable since $\psi^k$ is deloopable when $k$ is prime to $p$.

EXAMPLE 1.5. The module of primitive elements in $H^*(BU; \boldsymbol{Z}_{(2)})$ consists of a single copy of $\boldsymbol{Z}_{(2)}$ in each even degree, with generator the Newton polynomial $s_n \in H^{2n}(BU; \boldsymbol{Z}_{(2)})$. We view $s_n$ as a map $s_n: BU[2] \to K(\boldsymbol{Z}_{(2)}, 2n)$. Then $2s_n$ is transfer commuting. We must check that

$$\begin{array}{ccc} E\Sigma_2 \times_{\Sigma_2} BU[2]^2 & \xrightarrow{1\times(2s_n)^2} & E\Sigma_2 \times_{\Sigma_2} K(\boldsymbol{Z}_{(2)}, 2n)^2 \\ \downarrow{\scriptstyle d^{\oplus}} & & \downarrow{\scriptstyle d} \\ BU[2] & \xrightarrow{2s_n} & K(\boldsymbol{Z}_{(2)}, 2n) \end{array}$$

is homotopy commutative. But this follows since $s_n$ is an $H$-map and $(d^{\oplus})^*(2s_n) = i_! i^*(s_n)$ where $i$ is the 2-fold covering $BU[2]^2 \to E\Sigma_2 \times_{\Sigma_2} BU[2]^2$.

On the other hand, from **[A]** we have that $\lambda s_n$ is infinitely deloopable if and only if $2^{\alpha(n)-1} | \lambda$ where $\alpha(n)$ is the number of nonzero terms in the dyadic expansion of $n$.

The preceding example shows that 1.4 (as one might expect) is not a 'general fact' in infinite loop space theory. There are clearly other positive results than 1.4 (e.g., maps between Eilenberg-Mac Lane spaces) but the connection between transfer spaces and infinite loop spaces, and transfer commuting maps and infinite loop maps is obscure at the time of writing.

The category of transfer spaces is not closed under taking (homotopy theoretic) fiber. For example, we show below that the fiber $F$ of $2s_n: BU[2] \to K(\boldsymbol{Z}_{(2)}, 2n)$ is not a transfer space for $\alpha(n) > 2$, but first we recall the definition of mod 2 homology operations.

To each $x \in H_n(X; Z/2)$ there is a sequence of classes $e_i \otimes x \otimes x$ in $H_{i+2n}(E\Sigma_2 \times_{\Sigma_2} X^2; Z/2)$ and if $X$ is a transfer space we set $Q_i(x) = (d_2)_*(e_i \otimes x \otimes x)$. The $a$th (upper) homology operation $Q^a$: $H_n(X; Z/2) \to H_{n+a}(X; Z/2)$ is then defined by $Q^a(x) = Q_{a-n}(x)$. We shall need the following standard facts (see, e.g., [$\mathbf{M}_1$] or [**CLM**]).

(1.6) (i) $Q^a(x) = x \cdot x$ if $a = \deg x$,

(ii) $Sq^a_* Q^b(x) = \sum \binom{b-a}{a-2i} Q^{b-a+i} Sq^i_*(x)$,

(iii) $Q^a Q^b(x) = \sum \binom{i-b-1}{2i-a} Q^{a+b-i} Q^i(x)$,

(iv) $\psi(Q^a(x)) = \sum Q^i(x') \otimes Q^{a-i}(x'')$ where $\psi(x) = \sum x' \otimes x''$,

(v) $Q^a(x \cdot y) = \sum Q^i(x) \cdot Q^{a-i}(y)$.

(In (ii) $Sq^a_*$ denotes the vector space dual of the Steenrod square and in (iv) $\psi$ is the coproduct.)

EXAMPLE 1.7. Let $F$ be the (homotopy) fiber of $2s_7$: $BU[2] \to K(Z_{(2)}, 14)$. Then

$$H_*(F; Z/2) = H_*(BU; Z/2) \otimes H_*(K(Z_{(2)}, 13); Z/2)$$

as Hopf algebras, and $Sq^1_*(s_7) = \iota_{13}$. Homology operations preserve primitive elements (iv) and using (i), (ii) and (v) we have $Q^4(s_1) = s_3$, $Q^8(s_3) = s_7$. Now, $Sq^1_* Q^8 Q^4(s_1) = Q^7 Q^4(s_1)$ and, by (iii), $Q^7 Q^4(s_1) = Q^8 Q^3(s_1)$. But $Q^3(s_1) = 0$ for dimensional reasons. This contradicts the relation $Sq^1_*(s_7) = \iota_{13}$. A similar argument shows that the fiber of $2s_n$ is not a transfer space when $\alpha(n) > 2$.

**2. The surgery spaces.** The fibers $G/O$ and $G/\mathrm{TOP}$ of the natural infinite loop maps from $BO$ and $B\mathrm{TOP}$ to $BG$ classify homotopy trivialized vector bundles and topological bundles, respectively. (If $X$ is a smooth manifold one can also interpret $[X, G/O]$ as normal cobordism classes of surgery problems over $X$, and similarly in the topological case provided $\dim X \neq 4$.)

Let $p$ be any prime and $k$ a natural number prime to $p$ such that $k^{p-1} \not\equiv 1 \pmod p$ if $p$ is odd, and $k \equiv \pm 3 \pmod 8$ if $p = 2$. The affirmed Adams conjecture is equivalent to the existence of a diagram

$$(2.1) \qquad \begin{array}{ccc} & \xrightarrow{A_p} & G/O[p] \\ BSO^{\oplus}[p] & & \downarrow r \\ & \xrightarrow[\psi^k - \psi^1]{} & BSO^{\oplus}[p] \end{array}$$

where $\psi^k$ represents the $k$th Adams operation. (The map $A_p$ is not unique since $[BSO, SG] \neq 0$.)

If $p$ is odd, then $A_p$ can be chosen as an infinite loop map [F]. If $p = 2$, $A_p$ cannot even be chosen as an $H$-map, but for a particular choice of $A_2$ the deviation from 'additivity' has been completely described by Tornehave in [T]: Consider the composition $\Delta$: $BSO \to^{\eta} SO \to SG$ and define $d$: $BO \times BO \to SG$ as

$$d(x, y) = 1 - \frac{1 \pm k}{4}(\Delta(x) - 1) \circ (\Delta(y) - 1)$$

where $-$ indicates loop subtraction in $\Omega^\infty S^\infty$, $\circ$ composition, and 1 is the identity in $SG$. (The sign in $(1 \pm k)/4$ is dictated by $(1 \pm k)/4 \in Z$.) There is a choice of $A_2$ for which

$$(2.2) \qquad A_2(x + y) = A_2(x) A_2(y) d(x, y).$$

Since $G/O = \mathrm{Spin}\, G/\mathrm{Spin}$, a homotopy trivial bundle has two natural Thom classes in $KO$-theory (one comes from the Spin structure and one from the trivialization). Their 'quotient' defines an exponential characteristic class in $KO$(base), and universally an $H$-map

(2.3) $$e\colon G/O \to BSO^{\otimes} \qquad (ph(e) = r^*(\hat{A}\text{-genus})).$$

The composition $e \circ A_p$ is the cannibalistic characteristic class $\rho_{\mathbf{R}}^k\colon BSO^{\oplus}[p] \to BSO^{\otimes}[p]$ [**A**$_1$]. For $p = 2$, $\rho_{\mathbf{R}}^k$ is a homotopy equivalence so if we define cok $J_2$ to be the fiber of $e$ (localized at 2) we have

(2.4) $$G/O[2] \cong BSO[2] \times \operatorname{cok} J_2.$$

At odd primes $p$ one can combine the map $e$ with the natural map $r\colon G/O \to BSO$ to get $e'_p\colon G/O \to BSO[p]$ with $A_p \circ e'$ a homotopy equivalence. Taking cok $J_p$ = Fiber$(e'_p)$ one then again has a splitting $G/O[p] \simeq BSO[p] \times \operatorname{cok} J_p$ (see, e.g., [**MM**$_2$]). (These splittings seem first to have been formulated by D. Sullivan; they can be viewed as reformations of results from [**A**$_1$].)

The homotopy type of $\operatorname{cok} J = \prod \operatorname{cok} J_p$ is at present out of reach. Its homotopy groups are essentially the stable homotopy groups of spheres modulo the (known) image of the $J$-homomorphism. The cohomology groups of cok $J$ have been calculated; they are complicated whereas the $K$-groups are trivial [**HS**]: $\widetilde{KO}^*(\operatorname{cok} J) = 0$.

Adams and Priddy prove in [**AP**] that the homotopy type $BSO[p]$ only supports one infinite loop space structure. Thus from 1.4 a map $f\colon E_1 \to E_2$ between infinite loop spaces with underlying homotopy type $BSO[p]$ is an infinite loop map iff it is transfer commuting. We check in [**MST**] that $\rho_{\mathbf{R}}^k$ and $\psi^k/\psi^1$ are transfer commuting, hence infinite loop maps. Moreover,

THEOREM 2.5 [**MST**]. *The map $e\colon G/O \to BSO^{\otimes}$ is an infinite loop map.*

I shall not give the proof of 2.5 here but just note a few salient points. The space cok $J_p$ can be defined as an infinite loop subspace of $SG[p]$, hence of $G/O[p]$. (This is an outcome of Quillen's work on $K$-theory of finite fields and the characterization of $\Omega^\infty S^\infty$ as the classifying space of the category of finite sets (cf. [**MQRT**]).) Hence there is a fibration of infinite loop spaces $\operatorname{cok} J_p \to G/O[p] \to E$ and, from (2.4), $E \simeq BSO[p]$. Since $\widetilde{KO}^*(\operatorname{cok} J_p) = 0$ both $e$ and the natural map $r\colon G/O[p] \to BSO^{\oplus}[p]$ factor to define maps $\bar{e}, \bar{r}$ from $E$ to $BSO[p]$, and $(\psi^k/\psi^1) \circ \bar{e} \simeq \rho_{\mathbf{R}}^k \circ \bar{r}$. Now, $\psi^k/\psi^1$ and $\rho_{\mathbf{R}}^k \circ \bar{r}$ are infinitely deloopable and 2.5 follows from

LEMMA 2.6 [**MST**]. *Let $E_i$, $i = 1, 2, 3$, be infinite loop spaces, homotopy equjvalent to $BSO[p]$ and suppose given maps $f\colon E_1 \to E_2$, $g\colon E_2 \to E_3$ which are rational equivalences. If two of the three maps $f$, $g$ and $g \circ f$ are infinite loop maps then so is the third.*

We next discuss the classifying space $G$/TOP for topological surgery problems. In contrast to the smooth case the homotopy type has been explicitly determined by Sullivan. (In geometric terms the difference comes from the many smooth manifolds homotopy equivalent to $S^n$.)

Away from 2, $G$/TOP is equivalent to $BSO^{\otimes}$. Moreover, the equivalence $\sigma$: $G/\mathrm{TOP}[\frac{1}{2}] \to BSO^{\otimes}[\frac{1}{2}]$ defined by Sullivan (cf. [**MM**$_2$]) is an infinite loop map [**MST**].

The 2-local structure is more complicated: There are classes $\tilde{K}_{4n-2} \in H^{4n-2}(G/\mathrm{TOP}; \boldsymbol{Z}/2)$, $\tilde{L}_{4n} \in H^{4n}(G/\mathrm{TOP}; \boldsymbol{Z}_{(2)})$ which combine to define a homotopy equivalence

$$(2.7) \qquad \sum \tilde{K}_{4n-2} \times \sum \tilde{L}_{4n}\colon G/\mathrm{TOP}[2] \to \prod_{n=1}^{\infty} K(\boldsymbol{Z}/2,\, 4n-2) \times K(\boldsymbol{Z}_{(2)} 4n).$$

The classes $\tilde{K}$, $\tilde{L}$ are characterized by characteristic class formulas for simply connected surgery problems (of $\boldsymbol{Z}/2^r$-manifolds), but also in the nonsimply connected case do they appear in surgery formulas for normal maps over manifolds **[RS]**, **[MS]**, **[Mi]**. There is only one good definition of $\tilde{K}$ **[RS]** but at least two good choices of $\tilde{L}$, each one with its own merits. We denote by $\tilde{L}$ the class from **[MS]** and by $\tilde{L}'$ the class from **[Mi]**. They differ by an element of order 2:

$$(2.8) \qquad \tilde{L} - \tilde{L}' = \beta_1\Big(\sum_{i \geq 2} Sq^{2^i}\Big)\, Sq^1(\tilde{K})$$

(see **[BM]**).

The classes $\tilde{K}_{4n-2}$ are primitive but $\tilde{L}_{4n}$, $\tilde{L}'_{4n}$ are not, $\psi(\tilde{L}_{4n}) = 1 \otimes \tilde{L}_{4n} + \tilde{L}_{4n} \otimes 1 + 8\sum \tilde{L}_{4i} \otimes L_{4(n-i)}$. To get primitive classes in degree $4n$ we set $\tilde{l}_{4n} = s_n(8\tilde{L}_4, \cdots, 8\tilde{L}_{4n})/8n$, where $s_n$ is the Newton polynomial. Then $\tilde{l}_{4n}$ is an integral polynomial in the $\tilde{L}_{4i}$, congruent to $\tilde{L}_{4n}$ modulo 4·(dec. elements), and it is primitive. Using $\tilde{l}$ (or $\tilde{l}'$) instead of $\tilde{L}$ in (2.7) we have an $H$-space equivalence of $G/\mathrm{TOP}[2]$ with the stated product of Eilenberg-Mac Lane spaces.

The second delooping of $G/\mathrm{TOP}[2]$ again splits as a product of Eilenberg-Mac Lane spaces **[MM$_1$]** but its $H$-structure is twisted: If $\iota_{4n} \in H^{4n}(B^2(G/\mathrm{TOP}); \boldsymbol{Z}/2)$ is the spherical class then $\iota_{4n}^2 = \iota_{8n}$ **[M]**. Hence $B^3(G/\mathrm{TOP}[2])$ has nonzero $k$-invariants.

From the standard description of the $\boldsymbol{Z}/2$-cohomology of Eilenberg-Mac Lane spaces it follows that primitive elements are deloopable to primitive classes. Hence the $\tilde{K}_{4n-2}$ deloop twice to classes in $H^{4n}(B^2(G/\mathrm{TOP}); \boldsymbol{Z}/2)$, and not more than twice by the preceding paragraph. For integral primitive classes the situation is not so simple; the suspension homomorphism

$$\sigma\colon \mathrm{Prim}\, H^*(K(\boldsymbol{Z}_{(2)}, i+1); \boldsymbol{Z}_{(2)}) \to \mathrm{Prim}\, H^*(K(\boldsymbol{Z}_{(2)}, i); \boldsymbol{Z}_{(2)})$$

is not surjective, so it is not even clear that $\tilde{l}_{4n}$ is twice deloopable. In contrast, we proved in **[MM$_1$]** that $\pi^*(\tilde{l}_{4n}) \in H^{4n}(G/O; \boldsymbol{Z}_{(2)})$ is twice deloopable ($\pi\colon G/O \to G/\mathrm{TOP}$) but the proof is calculational and does not show more than stated.

We will see in §4 that $\pi^*(\tilde{l}_{4n})$ is indeed infinitely deloopable, and we make the following

CONJECTURE 2.9. (a) *The primitivized Morgan-Sullivan class $\tilde{l}_{4n} \in H^{4n}(G/\mathrm{TOP}; \boldsymbol{Z}_{(2)})$ is infinitely deloopable.*

(b) *There is an infinite loop space $\boldsymbol{K}$ with underlying homotopy type $\prod K(\boldsymbol{Z}/2, 4n-2)$ such that*

$$G/\mathrm{TOP}[2] \simeq \boldsymbol{K} \times \prod_{n=1}^{\infty} K(\boldsymbol{Z}_{(2)}, 4n) \quad (\textit{as infinite loop spaces}).$$

(c) *The infinite loop space $\boldsymbol{K}$ in* (b) *splits $\boldsymbol{K} \simeq \boldsymbol{K}_A \times \boldsymbol{K}_B$, where $\boldsymbol{K}_A$, $\boldsymbol{K}_B$ have* *underlying homotopy types* $\prod_{i=1}^{\infty} K(Z/2, 2^i - 2)$ *and* $\prod_{n \neq 2^i} K(Z/2, 4n - 2)$, *respectively.*

REMARK 2.10. F. Petersen has conjectured a further splitting of $K_B$, $K_B = \prod_{n \text{ odd}} K_n$, where $K_n = \prod_{i=2}^{\infty} K(Z/2, 2^i n - 2)$, but this does not seem in agreement with the calculations of the homology operations in $G/\mathrm{TOP}$ (cf. [M]). (E.g., for any choice of fundamental class $K_{38}$ in degree 38, $Q_*^{28}(K_{38})$ is a fundamental class in dimension 10, where $Q_*^{28}$ is dual to $Q^{28}$.)

**3. The integral homology of $G/O$ and** cok $J$. For spaces with a complicated torsion structure the most efficient way (if not the only way) to analyze the integral homology structure is via a calculation of the Bockstein spectral sequences.

In the classical description of the higher torsion in the cohomology of Eilenberg-Mac Lane spaces it was convenient to use the Pontrjagin squaring operations to name elements in $H^*(—; Z/p^r)$. These operations which always exist in cohomology can be imitated in homology of transfer spaces.

Let $X$ be a transfer space. There are operations

$$P: H_*(X; Z/2^r) \to H_*(X; Z/2^{r+1})$$

refining the product square, $P(x) = x \cdot x \pmod{2^r}$. These operations have properties analogous to the classical Pontrjagin operations in cohomology (cf. [M$_2$]). Here it suffices to mention that they are natural with respect to transfer commuting maps and that if $X$ is an infinite loop space with classifying space $BX$, and $\sigma_r$: $IH^*(X; Z/2^r) \to H_*(BX; Z/2^r)$ is the homology suspension homomorphism then

$$\sigma_{r+1}P(x) = 2\sigma_r(x) \cdot \beta_r\sigma_r(x). \tag{3.1}$$

(In (3.1) $\beta_r$: $H_*(—; Z/2^r) \to H_*(—; Z_{(2)})$ is the Bockstein associated with the coefficient sequence $Z_{(2)} \overset{2^r}{\rightarrowtail} Z_{(2)} \twoheadrightarrow Z/2^r$ and 2: $H_*(—; Z/2^r) \to H_*(—; Z/2^{r+1})$ is induced from $Z/2^r \subset Z/2^{r+1}$.)

Quite generally one can then get classes in $H_*(X; Z/2^r)$ from $H_*(X; Z/2)$ by iterated use of the Pontrjagin operations. Just how large a part of $H_*(X; Z/2^r)$ one obtains this way can be read off from the Bockstein spectral sequence $\{E_r(X), \partial_r\}$,

$$E_1(X) = H_*(X; Z/2), \qquad E_\infty(X) = H_*(X; Z)/\text{Torsion} \otimes Z/2.$$

If, e.g., $E_r(X)$ is a tensor product of spectral sequences

$$E_{r+1} = P\{\beta^{2^r}\} \otimes E\{\alpha\beta^{2^r-2}\}, \qquad \partial_{r+1}(\beta^{2^r}) = \alpha\beta^{2^r-2} \quad (r \geqq 1) \tag{3.2}$$

then $H_*(X; Z/2^s)/2H_*(X; Z/2^{s-1})$ is generated by classes in $P^{(s-1)}H_*(X; Z/2)$ and their Bocksteins. If $E_r(X)$ is a tensor product of the models (3.2) and models with $E_2 = E_\infty$ then $H_*(X; Z/2^s)/2H_*(X; Z/2^{s-1})$ is generated by classes in $P^{(s-1)}H_*(X; Z/2)$, the $Z/2^s$ reduction of integral torsion classes, their Bocksteins and products of these types. The latter situation occurs for $X = G/O$ and $X = B(G/O)$.

Consider $CP^\infty = BSO(2) \to BSO$ and take $b_{4n} \in H_{4n}(BSO; Z_{(2)})$ to be the image of a generator in $H_{4n}(CP^\infty; Z_{(2)}) = Z_{(2)}$. Let $A_2$: $BSO[2] \to G/O[2]$ be a solution of the Adams conjecture (cf. (2.1)) and set $a_{4n} = (A_2)_*(b_{4n})$.

THEOREM 3.3 [**M**$_2$]. (a) *$H_*(G/O;\ Z/2^s)/2H_*(G/O;\ Z/2^{s-1})$ is multiplicatively generated by classes $a_{4n}$ and $P^{(s-1)}(x)$, $x \in H_*(G/O;Z/2)$, and their Bocksteins.*

(b) *$H_*(B(G/O);\ Z/2^s)/2H_*(B(G/O);\ Z/2^{s-1})$ is multiplicatively generated by classes $\sigma(a_{4n})$ and $P^{(s-1)}(x)$, $x \in H_*(B(G/O);Z/2)$, and their Bocksteins.*

For the subspace cok $J_2 \subset G/O[2]$ the situation is more complicated. Its Bockstein spectral sequence is not quite a tensor product of the models (3.2). One further requires models (one for each $m$ which is not a power of 2).

(3.4) $$E_{r+2} = P\{\kappa_m^{2^r}\} \otimes E\{\alpha_m \cdot \kappa_m^{2^r-1}\}, \qquad \partial_{r+2}(\kappa_m^{2^r}) = \alpha_m \kappa_m^{2^r-1}.$$

Hence $H_*(\text{cok } J;\ Z/2^s)/2H_*(\text{cok } J;\ Z/2^{s-1})$ is not generated by classes $P^{(s-1)}(x)$; one also needs the classes $P^{(s-2)}(\kappa_m)$, $m \neq 2^i$ where $\kappa_m \in H_*(\text{cok } J;\ Z/4)$ has order 4.

For our applications in the next section we require some information on the elements $\kappa_m \in H_{4m}(\text{cok } J;\ Z/4)$ and their (second order) Bocksteins $\beta_2(\kappa_m) \in H_{4m-1}(\text{cok } J;\ Z/2)$. First recall that the space $\Omega^\infty S^\infty$ has two transfer structures: The loop structure and the composition structure. Accordingly there are two sets of products, $*$ and $\circ$, two sets of Pontrjagin squaring operations, $P$ and $\hat{P}$, and two sets of homology operations, $Q^i$ and $\hat{Q}^i$.

Let $u_n \in H_n(SG;\ Z/2)$ be the element in the image of the embedding $RP^\infty \subset SO \subset SG$. Let $i$: cok $J_2 \to G/O[2]$ be the embedding and $j$: $SG \to G/O$ the natural map.

LEMMA 3.5. *For $m$ not a power of 2 there exist elements $\kappa_{4m} \in H_{4m}(\text{cok } J;\ Z/4)$ of order 4 which satisfy*

(i) *$i_*(\kappa_{4m}) \equiv a_{4m}$ modulo $PH_*(G/O;\ Z/2) + 2H_*(G/O;Z/2)$.*

(ii) *$i_*\beta_2(\kappa_{4m}) = j_*((Q^{2m}(u_{2m-1}) + u_{2m} * u_{2m-1}) * [-1]) + \cdots$ where the dots indicate decomposable elements in the composition product.*

(iii) *The classes $\hat{P}^{(r-2)}(\kappa_{4m})$ and $\hat{P}^{(r-1)}(x)$, $x \in H_*(\text{cok } J;\ Z/2)$ and their Bocksteins generate $H_*(\text{cok } J;Z/2^r)/2H_*(\text{cok } J;\ Z/2^{r-1})$ multiplicatively.*

(In 3.5(ii) $[-1]$ denotes the nonzero element of $H_0(\Omega^\infty S^\infty)$ carried by the $-1$ component.) The proof of 3.5 consists of a rather careful study of the Bockstein spectral sequences for the spaces and maps in the fibration cok $J_2 \to G/O[2] \to BSO^\otimes[2]$, using results from [**M**$_2$]. (Similar calculations have been carried out by May.)

REMARK 3.6. It is relatively complicated to give a precise construction of $\kappa_{4m}$; modulo decomposable elements in $H_*(G/O;\ Z/4)$ one has $i_*(\kappa_{4m}) = P(u_{2m}) * [-1]$.

**4. The primitivized Morgan-Sullivan class and topological normal invariants.** The unit $u$: $S^0 \to bO$ in the spectrum for connective orthogonal $K$-theory induces a mapping $\Omega^\infty S^\infty \to^f BO \times Z$ ($f|\Omega^n S^n = \Omega^n(u_n\colon S^n \to bO_n)$) and in particular a mapping of 1-components $f$: $SG \to BO^\otimes$, which is an infinite loop map. The fiber of $Bf$: $BSG \to B(BO^\otimes)$ has been identified by May, Quinn and Ray (see [**MQRT**]) as the space $BKOG$ (originally considered by Sullivan) which classifies pairs $(\xi, \Delta)$ consisting of a stable spherical fibration $\xi$ and a $KO$-orientation $\Delta \in \widetilde{KO}(M(\xi))$.

The infinite loop space $BO^\otimes$ splits in two: $BO^\otimes \simeq RP^\infty \times BSO^\otimes$, and there is a commutative diagram

$$\begin{array}{ccc} SG & \xrightarrow{f} & BO^{\otimes} \\ \downarrow & & \downarrow \\ G/O & \xrightarrow{e} & BSO^{\otimes} \end{array}$$

where $e$ is the infinite loop map from (2.3). In **[MST]** we proved that the triangle

$$\begin{array}{ccc} SG[\frac{1}{2}] & \xrightarrow{i} & G/\mathrm{TOP}[\frac{1}{2}] \\ & {\scriptstyle f}\searrow \quad \swarrow{\scriptstyle 1/\sigma} & \\ & BO^{\otimes}[\frac{1}{2}] & \end{array}$$

and all its deloopings are commutative. Here $i$ is the natural infinite loop map and $\sigma: G/\mathrm{TOP}[\frac{1}{2}] \to BO^{\otimes}[\frac{1}{2}]$ is the homotopy equivalence defined by Sullivan. Hence we have a commutative ladder of homotopy equivalences

$$(4.1) \qquad \begin{array}{ccccccccc} \cdots \to & SG[\frac{1}{2}] & \to & G/\mathrm{TOP}[\frac{1}{2}] & \to & B\mathrm{TOP}[\frac{1}{2}] & \to & BSG[\frac{1}{2}] & \to \cdots \\ & \downarrow{\scriptstyle \mathrm{id}} & & \downarrow{\scriptstyle 1/\sigma} & & \downarrow & & \downarrow{\scriptstyle \mathrm{id}} & \\ \cdots \to & SG[\frac{1}{2}] & \longrightarrow & BO^{\otimes}[\frac{1}{2}] & \to & BKOG[\frac{1}{2}] & \to & BSG[\frac{1}{2}] & \to \cdots \end{array}$$

and we recover Sullivan's result: Away from 2 topological bundle theory is equivalent with $KO$-oriented spherical fibration theory, and we are left with 2-local problems.

In **[BMM]** the mapping $\pi: G/O \to G/\mathrm{TOP}$ was evaluated on homology with $Z/2$ coefficients. For the index classes $\tilde{L}$ and $\tilde{L}'$ the result is (see also **[BM]**),

$$(4.2) \qquad \begin{aligned} \pi^*(L) &= \sum_{i=1}^{\infty} \chi(Sq^{4i})\,(\pi^*\tilde{K}_2^2), \\ \pi^*(\tilde{L}') &= \sum_{i=1}^{\infty} \pi^*(\tilde{K}_2)^{2^i}, \end{aligned}$$

where $\pi^*(\tilde{K}_2)$ is the unique nonzero class in $H^2(G/O; Z/2)$.

The $Z_{(2)}$-integral class $\pi^*(\tilde{L}') \in H^*(G/O; Z_{(2)})$ was completely characterized in **[MM$_1$]**. Somewhat surprisingly it restricts nontrivially to the subspace cok $J$. However, we show below that the Morgan-Sullivan class $\pi^*(\tilde{L})$ does vanish on cok $J$, and in fact

$$\pi^*(\tilde{L}) \in \mathrm{Image}(e^*: H^*(BSO^{\otimes}; Z_{(2)}) \to H^*(G/O; Z_{(2)})).$$

Before presenting the argument, I recall the cohomology of the spectrum $bu$ (for connective $K$-theory) **[A]**.

In each even degree $H^*(bu; Z_{(2)})$ is a single copy of $Z_{(2)}$ and $H^{\mathrm{odd}}(bu; Z_{(2)}) = 0$. The generator of $H^{2n}(bu; Z_{(2)})$ consists of a string of classes $\mathrm{ch}_{i,n} \in H^{2n+2i}(BU[2i, \infty], Z_{(2)})$ where $BU[2i, \infty]$ is the $2i$th space in $bu$, and can be identified with the $2i$th connected cover of $BU$; let $k: BU[2i, \infty] \to BU$ be the projection. Then $\mathrm{ch}_{i,n} \otimes Q = k^*(2^n \cdot \mathrm{ch}_{i+n})$ and $\mathrm{ch}_{i,n} \otimes Z/2 = \chi(Sq^{2n})\,(u_{2i})$, where $\mathrm{ch}_{i+n}$ is the universal Chern character and $u_{2i}$ is the bottom class.

Let $c: BSO^{\oplus} \to BSU^{\oplus}$ be the complexification and define $ph_{1,n-1} = c^*(\mathrm{ch}_{2,2n-2}) \in H^{4n}(BSO^{\oplus}, Z_{(2)})$. The associated map $ph_{1,n-1}: BSO^{\oplus}[2] \to K(Z_{(2)}, 4n)$ is an infinite loop map.

THEOREM 4.3. $\pi^*(\tilde{l}_{4n}) = ((\rho_{R}^3)^{-1} \circ e))^{-1}(ph_{1,n-1})$.

PROOF. The two sides in 4.3 agree on the factor $BSO[2]$ in the splitting $G/O[2] \simeq BSO[2] \times \operatorname{cok} J_2$. This follows from the discussion above (particularly (4.2)) and the rational calculation from **[MM₁]**:

$$(*) \qquad A_2^* \pi^*(\tilde{l}'_{4n}) \otimes Q = 2^{\alpha(n)-1} \cdot s_n.$$

It remains to be seen that $i^*\pi^*(\tilde{l}_{4n}) = 0$ where $i$: $\operatorname{cok} J_2 \to G/O[2]$ is the injection. But $\pi^*(\tilde{l}_{4n})$ is twice deloopable so $i^*\pi^*(\tilde{l}_{4n})$ vanishes on homology classes which are Pontrjagin squares (cf. (3.1)), and by 3.5(iii) it suffices to calculate $i^*\pi^*(\tilde{l}_{4n})$ in $H^*(\operatorname{cok} J; \boldsymbol{Z}/2^r)$ for $r = 1$ and 2.

Both $i^*\pi^*(\tilde{l}_{4n})$ and $i^*\pi^*(\tilde{l}'_{4n})$ vanish in $H^*(\operatorname{cok} J; \boldsymbol{Z}/2)$ since $\operatorname{cok} J_2$ is 5-connected (cf. (4.2)), but in $H^*(G/O; \boldsymbol{Z}/2)$ there is a difference: $\pi^*(\tilde{l}'_{4n}) = 0$ if $\alpha(n) \neq 1$; this is not the case for $\pi^*(\tilde{l}_{4n})$.

Suppose $\alpha(n) > 1$. Then $\pi^*(\tilde{l}'_{4n})$ annihilates the subgroup $2H_*(G/O; \boldsymbol{Z}/2)$ in $H_*(G/O; \boldsymbol{Z}/4)$ and 3.5(iii) shows that $\pi^*(\tilde{l}'_{4n})$ is characterized by its value on $\kappa_{4n}$. From (*) above and 3.5(i),

$$< \pi^*(\tilde{l}'_{4n}), \kappa_{4n} > = 2^{\alpha(n)-1} \quad \text{in } \boldsymbol{Z}/4.$$

Since $\pi^*(\tilde{K}_{4n-2}) = 0$ for $\alpha(n) > 1$ **[BMM]**, $\pi^*(\tilde{l}_{4n}) = \pi^*(\tilde{l}'_{4n})$ if $\alpha(n) > 2$ and $i^*\pi^*(\tilde{l}_{4n}) = 0$. If $4n = 2^i + 2^j$ $(i < j)$ then

$$(**) \qquad \pi^*(\tilde{l}_{4n}) = \pi^*(\tilde{l}'_{4n}) + \beta_1 Sq^{2^i} Sq^1(\pi^* \tilde{K}_{2^j-2})$$

and using 3.5(ii) we have in $\boldsymbol{Z}/4$

$$\begin{aligned}
\langle \pi^*(\tilde{l}_{4n}), i_*(\kappa_{4n})\rangle &= 2 + \langle \beta_1 Sq^{2^i} Sq^1 \pi^*(K_{2^j-2}), i_*(\kappa_{4n})\rangle \\
&= 2 + 2\langle Sq^{2^i} Sq^1 \pi^*(K_{2^j-2}), (Q^{2n}(u_{2n-1}) + u_{2n-1} * u_{2n}) * [-1]\rangle \\
&= 2 + 2\,\langle \pi^*(K_{2^j-2}), u_{2^{j-1}-1} * u_{2^{j-1}-1} * [-1]\rangle
\end{aligned}$$

where the last equation is a consequence of (1.6)(ii), properly dualized. Finally, $\langle \pi^*(K_{2^j-2}), u_{2^{j-1}-1} * u_{2^{j-1}-1} * [-1]\rangle \neq 0$ in $\boldsymbol{Z}/2$ **[BMM,** Theorem 3.3] so altogether $\langle i^*\pi^*(\tilde{l}_{4n}), \kappa_{4n}\rangle = 0$. This completes the proof.

COROLLARY 4.4. *The composite* $G/O[2] \to^{\pi} G/\mathrm{TOP}[2] \to^{l} \Pi K(\boldsymbol{Z}_{(2)}, 4n)$ *is an infinite loop map.*

The primitive elements of $H^{4n+1}(B(G/O))$ injects under suspension in the primitive elements of $H^{4n}(G/O)$. This is true rationally and with $\boldsymbol{Z}/2$ coefficients by direct calculations and then follows with $\boldsymbol{Z}_{(2)}$ coefficients from 3.3(b). Thus we have

COROLLARY 4.5. *The composite*

$$B(G/O)[2] \to B(G/\mathrm{TOP})[2] \to \Pi K(\boldsymbol{Z}_{(2)}, 4n + 1)$$

*factors over* $Be$: $B(G/O[2]) \to B(BSO^{\otimes}[2])$.

The 'Kervaire classes' $\tilde{K}_{4n-2} \in H^{4n-2}(G/\mathrm{TOP}; \boldsymbol{Z}/2)$ desuspends to unique primitive classes $k_{4n-1} \in H^{4n-1}(B(G/\mathrm{TOP}); \boldsymbol{Z}/2)$. We have already used that $\pi^*(\tilde{K}_{4n-2})$ is nonzero in $H^*(G/O; \boldsymbol{Z}/2)$ only if $n$ is a power of 2. In this case on the other hand $\pi^*(\tilde{K}_{4n-2})$ has components in both $BSO[2]$ and $\operatorname{cok} J_2$ under the splitting 2.4. The restriction to $BSO[2]$ was calculated in **[BM]**, but it also follows rather easily from 2.2. The component of $\pi^*(\tilde{K}_{2^i-2})$ in $H^*(\operatorname{cok} J; \boldsymbol{Z}/2)$ is always nonzero, and spherical if and only if there exists an Arf invariant one element in $\pi^s_{2^i-2}(S^0)$.

Since $\sigma$: Prim $H^{\text{odd}}(B(G/O); Z/2) \to$ Prim $H^{\text{ev}}(G/O; Z/2)$ is injective, $(B\pi)^*(k_{4n-1})$ is nonzero only if $\alpha(n) = 1$. The restriction of these classes to $BSG \subset B(G/O)$ define universal characteristic classes $\tau_i \in H^{2^i-1}(BSG; Z/2)$, giving characteristic classes for stable spherical fiber spaces.

COROLLARY 4.6. *A KO-orientable stable spherical fibration $\xi$ admits a topological reduction if and only if the classes $\tau_i(\xi)$ all vanish.*

Finally, results of Ravenel show that if $\xi$ is a spherical fiber space with vanishing Stiefel-Whitney classes then

$$\tau_i(\xi) \cup U = \psi_{i,i}(U) \tag{4.7}$$

where $U \in H^*(M(\xi), Z/2)$ is the Thom class and $\psi_{i,i}$ the secondary operation based on the relation $Sq^{2^{i-1}}Sq^{2^{i-1}} + \sum_{j=0}^{i-2} Sq^{2^i-2^j}Sq^{2^j} = 0$ in the mod 2 Steenrod algebra.

## BIBLIOGRAPHY


A J. F. Adams, *On the Chern characters and the structure of the unitary group*, Proc. Cambridge Philos. Soc. **57** (1961), 189–199. MR **22** #12525.

$A_1$ ———, *On the groups J(X)*. II-III, Topology **3** (1964), 137–171, 193–222. MR **33** #6626; **33** #6627.

AP J. F. Adams and S. Priddy, *Uniqueness of BSO* (preprint).

BV J. M. Boardman and R. Vogt, *Homotopy invariant algebraic structures on topological spaces*, Lecture Notes in Math., vol. 347, Springer-Verlag, Berlin and New York, 1973.

BM G. Brumfiel and I. Madsen, *Evaluation of the transfer and the universal surgery classes*, Invent. Math. **32** (1976), 133–169.

BMM G. Brumfiel, I. Madsen and R. J. Milgram, PL *characteristic classes and cobordism*, Ann. of Math. (2) **97** (1973), 82–159. MR **46** #9979.

CLM F. R. Cohen, T. J. Lada and J. P. May, *The homology of iterated loop spaces*, Lecture Notes in Math., vol. 533, 1976.

F E. Friedlander, *The stable Adams conjecture* (preprint).

HS L. Hodgkin and V. Snaith, *Topics in K-theory*, Lecture Notes in Math., vol. 496, Springer-Verlag, Berlin and New York, 1975.

KP D. S. Kahn and S. B. Priddy, *Applications of the transfer to stable homotopy theory*, Bull. Amer. Math. Soc. **78** (1972), 981–987. MR **46** #8220.

M I. Madsen, *The surgery formula and homology operations*, Proc. Advanced Study Institute on Algebraic Topology (Aarhus, 1970), Vol. II, Math. Inst., Aarhus Univ., Aarhus, 1970. MR **47** #2610.

$M_1$ ———, *On the action of the Dyer-Lashof algebra in $H_*(G)$*, Pacific J. Math. **60** (1975), 235–275. MR **52** #9228.

$M_2$ ———, *Higher torsion in SG and BSG*, Math. Z. **143** (1975), 55–80. MR **51** #11503.

MST I. Madsen, V. Snaith and J. Tornehave, *Infinite loop maps in geometric topology*, Math. Proc. Cambridge Philos. Soc. **81** (1977), 399–430.

$MM_1$ I. Madsen and R. J. Milgram, *The universal smooth surgery class*, Comment. Math. Helv. **50** (1975), 281–310. MR **52** #4285.

$MM_2$ ———, *The oriented bordism of topological manifolds and integrality relations* (to appear).

$Ma_1$ J. P. May, *The geometry of iterated loop spaces*, Lecture Notes in Math, vol. 271, Springer-Verlag, Berlin and New York, 1972.

MQRT ——— (with contributions by F. Quinn, N. Ray and J. Tornehave), *$E_\infty$-ring spaces and $E_\infty$-ring spectra*, Lecture Notes in Math., vol. 577, 1977.

**Mi** R. J. Milgram, *Surgery with coefficients*, Ann. of Math (2) **100** (1974), 194–248. MR **50** #14801.

**MS** J. Morgan and D. Sullivan, *The transversality characteristic class and linking cycles in surgery theory*, Ann. of Math. (2) **99** (1974), 463–544. MR **50** #3240.

**RS** C. Rourke and D. Sullivan, *On the Kervaire obstruction*, Ann. of Math. (2) **94** (1971), 397–413. MR **46** #4546.

**S** G. B. Segal, *Categories and cohomology theories*, Topology **13** (1974), 293–312. MR **50** #5782.

**T** J. Tornehave, *Deviation from additivity of a solution to the Adams conjecture*, Aarhus Univ. (preprint).


Aarhus University

Proceedings of Symposia in Pure Mathematics
Volume 32, 1978

# REMARKS ON SMOOTHING THEORY

GREGORY BRUMFIEL

**1.** If $X$ is a space, we are interested in maps $f$: $M \to X$ where $M$ is a (closed) smooth manifold or a topological manifold and $f$ is a homotopy equivalence, and also where $M$ is a smooth manifold and $f$ is a homeomorphism. Given $X$, both the *existence* and *classification* (modulo suitable equivalence relations) of such maps are fascinating problems which have been quite deeply probed in the last twenty years.

A convenient starting point for the first problem ($f$ a homotopy equivalence) is to assume that $X$ is a Poincaré duality space. The second problem ($f$ a homeomorphism) does not make much sense unless one starts by assuming that $X$ is a topological manifold. Poincaré duality spaces $X$ (P.D. spaces) have a stable normal spherical fibration, classified by a map $X \to BG$, and manifolds $M$ have a stable normal Euclidean space fibre bundle, with structure group $O$ if $M$ is smooth and group TOP in general, classified by a map $M \to BO$ or $M \to B\,\mathrm{TOP}$.

A large part of the existence problems above reduce to the homotopy theoretic lifting problems for the normal bundles with respect to the natural maps of classifying spaces

$$\begin{array}{ccc} & B\,\mathrm{TOP} & \\ \nearrow & & \searrow \\ BO & \longrightarrow & BG \end{array} \tag{1.1}$$

In fact, the problem of existence of a smooth structure on a topological manifold is, for all practical purposes, exactly the lifting problem for the normal bundle with respect to $BO \to B\,\mathrm{TOP}$. In the case of homotopy equivalences, the lifting problem is the first step, followed by analysis of a surgery obstruction.

---

*AMS (MOS) subject classifications* (1970). Primary 57D10, 55F25, 55F35, 55F40; Secondary 57B10, 55D20.

The fibres of the three maps in (1.1) fit into a fibration

(1.2) $$\mathrm{TOP}/O \to G/O \to G/\mathrm{TOP}.$$

The classification problems above (modulo a concordance equivalence relation) reduce to the classification of maps from $X$ to these three spaces, again with a surgery obstruction additional complication in the case of homotopy equivalences.

All the spaces in (1.1) and (1.2) are infinite loop spaces and all the maps are infinite loop maps. While this is not a bad result, it is not good either, at least by itself. To be told that a topological manifold $X$ is smoothable if and only if a certain map $X \to B(\mathrm{TOP}/O)$ is null homotopic leaves the innocent bystander with the feeling that he has been betrayed by both the geometers and the homotopy theorists.

The game, then, is to formulate as many theorems as possible, playing the homotopy theory against the geometry, and vice versa. On the one hand, one wants to describe the spaces and maps in (1.1) and (1.2) in terms of more familiar concepts (cohomology, $K$-theory, stable cohomotopy and associated operations). On the other hand, one wants to find intuitive geometric obstructions and invariants for the existence and classification problems respectively, which a priori give only partial solutions, but which turn out to be complete obstructions or invariants.

Those of us who play this game are beginning to feel rather smug about the homotopy theory of the classification problems (the fibration $\mathrm{TOP}/O \to G/O \to G/\mathrm{TOP}$), and about the lifting problem for $B\mathrm{TOP} \to BG$. The lifting problems for $BO \to B\mathrm{TOP}$ and $BO \to BG$ are pretty well understood "localized at odd primes", due to Sullivan's work, but there are not yet any really satisfactory results about these smoothing problems "localized at 2".

As an example of the state of the art, I will describe the fibration $\mathrm{TOP}/O \to G/O \to G/TOP$ localized at 2, in homotopy theoretic terms, in the following paragraphs. All spaces are assumed localized at 2.

(1.3) $$G/\mathrm{TOP} \cong \prod_{n\geqq 1} K(\boldsymbol{Z}/2, 4n-2) \times \prod_{n\geqq 1} K(\boldsymbol{Z}_{(2)}, 4n).$$

As specific equivalence, we use the Sullivan-Morgan surgery classes $\mathscr{K} = \sum \mathscr{K}_{4n-2}$ and $\mathscr{L} = \sum \mathscr{L}_{4n}$.

(1.4) $$G/O \cong BSO \times \mathrm{Cok}\, J.$$

The splitting is provided by (a) the fibration $\mathrm{Cok}\, J \to G/O \to BSO^{\otimes}$, where $e$: $G/O \to BSO^{\otimes}$, with $ph(e) = \hat{A} \in H^*(G/O, Q)$, is the $KO$ element obtained by dividing the two $KO$-orientations on the Thom spectrum over $G/O$ (one from $M\,\mathrm{Spin}$, one from $S$), and (b) a solution of the Adams conjecture $\alpha$: $BSO \to G/O$, lifting $\psi^3 - \mathrm{Id}$: $BSO \to BSO$ (for definiteness, take the Becker-Gottlieb solution). Then $e \circ \alpha$: $BSO \to BSO^{\otimes}$ is a homotopy equivalence (well known as a $KO$-operation).

The factor $\mathrm{Cok}\, J$ defines a functor which is a canonical direct summand of stable cohomotopy. To obtain $\mathrm{Cok}\, J$ in this form, begin with the fibration of spectra

$$\mathrm{im}\, J \to bO \xrightarrow{\psi^3-\mathrm{Id}} bO\, \langle 4 \rangle.$$

The unit $S \to bO$ lifts uniquely to $S \to \mathrm{im}\, J$. Looping down gives (ring maps)

$QS^\circ \to Z \times \text{Im } J \to Z \times BO$ and $\text{Cok } J \subset SG \subset QS^\circ$ is the fibre of $QS^\circ \to Z \times \text{Im } J$.

The map $G/O \to G/\text{TOP}$ is now described by the $2 \times 2$ cohomology matrix

$$\begin{pmatrix} \mathcal{K}|_{BSO} & \mathcal{K}|_{\text{Cok } J} \\ \mathcal{L}|_{BSO} & \mathcal{L}|_{\text{Cok } J} \end{pmatrix}.$$

The answers are:

(1.5) $$\mathcal{K}|_{BSO} = \sum_{j \geqq 2} \mathcal{K}_{2^j-2}|_{BSO} \quad \text{and} \quad \mathcal{K}_{2^j-2}|_{BSO}$$

is the class in $H^{2^j-2}(BSO, Z/2)$ which is dual to $\sum_{r,s \geqq 0; r+s=2^j-2} x_r x_s \in H_{2^j-2}(BO, Z/2)$ in the usual formula $H_*(BO, Z/2) = Z/2[x_1, x_2, \cdots]$, $x_0 = 1$, $x_i = [RP^i] \subset H_i(BO, Z/2)$. If you prefer formulas in terms of symmetric functions

$$H^*(BO, Z/2) = Z/2[[t_1, t_2, \cdots]]^{\Sigma_\infty},$$

then $\mathcal{K}_{2^j-2}|_{BSO}$ is the restriction to $BSO$ of the class

$$\sum_{u,v; r+s=2^j-2; r,s \geqq 0} \sum t_u^r t_v^s \in H^{2^j-2}(BO, Z/2).$$

If you prefer your formulas in terms of Stiefel-Whitney classes, you will just have to work it out yourself.

(1.6) $$\mathcal{K}|_{\text{Cok } J} = \sum_{j \geqq 2} \mathcal{K}_{2^j-2}|_{\text{Cok } J} \quad \text{and} \quad \mathcal{K}_{2^j-2}|_{\text{Cok } J}$$

can be characterized by giving $\mathcal{K}_{2^j-2}|_{SG}$. This last is complicated because $H_*(SG, Z/2)$ is too much like hard work to describe in a few words. However, one can say that $SG$ has two products, $\cdot$ = composition and $*$ = loop. The homology classes $e_i = (RP^i) \in H_i(SO(i+1)) \subset H_i(SG)$ generate $H_*(SG)$ if both products are used. Then $\mathcal{K}_{2^j-2}|_{SG}$ is uniquely characterized by the properties

(i) $\mathcal{K}_{2^j-2}$ is $\cdot$ -primitive,
(ii) $\langle \mathcal{K}_{2^j-2}, e_{2^j-2} \rangle = 0$ (since $\mathcal{K}|_{SO} = 0$),
(iii) $\langle \mathcal{K}_{2^j-2}, e_r * e_s \rangle = 1$ if $r + s = 2^j - 2$, $r, s > 0$,
(iv) $\langle \mathcal{K}_{2^j-2}, e_{r_1} * \cdots * e_{r_k} \rangle = 0$ if $k > 2$.

This characterization of a cohomology class is a bit incomplete, from the working man's point of view, because one really ought to spell out all the relations these generators $e_i$ satisfy over the two operations $\cdot$, $*$, or better, use a nonredundant basis for $H_*(SG)$. I personally like this characterization (i)—(iv), because of the similarity with (1.5), characterizing $\mathcal{K}|_{BSO}$. This similarity is related to the fact that $SG \to G/O$ can be decomposed as $\text{Im } J \times \text{Cok } J \to BSO \times \text{Cok } J$. The main thing about these calculations is the fact that $\mathcal{K}_{4n-2}|_{G/O} = 0$ if $4n - 2 \neq 2^j - 2$. More explicit formulas than (1.6) can be found in papers of myself, Ib Madsen, and Jim Milgram in various combinations.

$\mathcal{L}|_{BSO}$ is determined by its $Q$- and $Z/2$-reductions. We have

(1.7) $$\rho_Q(\mathcal{L}|_{BSO}) = \tfrac{1}{8}(\psi_H^3 L/L - 1) \in H^*(BSO, Q),$$

where $L = 1 + L_1 + L_2 + \cdots$ is the Hirzebruch genus (for normal bundles, not tangent bundles) and $\psi_H^3$ is the cohomology Adams operation, $\psi^3 z = 3^{2k} z$, $z \in$

$H^{4k}(\ , Q)$. We have $\rho_{Z/2}(\mathscr{L}|_{BSO}) = \chi(S_q^{4*})w_2^2$, where $S_q^{4*} = 1 + S_q^4 + S_q^8 + \cdots$, and $\chi$ is the antiautomorphism of the Steenrod algebra.

(1.8) $$\mathscr{L}|_{\text{Cok } J} = 0.$$

In fact, $\mathscr{L}|_{G/O} \in \text{image}(e^*: H^*(BSO^{\otimes}, Z_{(2)}) \to H^*(G/O, Z_{(2)}))$ and (1.7) identifies precisely which class in $H^*(BSO^{\otimes}, Z_{(2)})$ is involved.

I think (1.8) is a deep theorem. If one makes a calculation, like (1.5) and (1.6), then that is all right, but it is still a calculation. However, if you make a calculation of this order of magnitude of difficulty and get 0, as in (1.8), then you have proved a theorem. You then ought to study the situation so that you can state your theorem properly. All the hard work in this theorem (1.8) is due to Madsen and Milgram.

Putting (1.3), (1.4), (1.5), (1.6), (1.7), (1.8) all together, one has described $G/O \to G/\text{TOP}$. Then the fibre $\text{TOP}/O$ can be described as follows.

Let $\Theta$ be the fibre of $\mathscr{L}|_{BSO}: BSO \to \prod K(Z_{(2)}, 4n)$, which was described in (1.7). Let $K' = \prod K(Z/2, 2^{j+1} - 2)$, $K'' = \prod_{n \neq 2^i} K(Z/2, 4n - 3)$. We have the cohomology classes $\mathscr{K} = \sum \mathscr{K}_{2^{j+1}-2}$, described in (1.6) and (1.5)

$$\text{Cok } J \subset SG \to K', \qquad \Theta \subset BSO \to K'.$$

Form $\mathscr{K} \otimes 1 + 1 \otimes \mathscr{K}: \text{Cok } J \times \Theta \to K'$. Then

(1.9) $$\text{TOP}/O \cong \text{fibre}(\text{Cok } J \times \Theta \to K') \times K''.$$

Unfortunately, (1.9) does not describe the $H$-space structure on $\text{TOP}/O$. The difficulty is that the decomposition $G/O \cong BSO \times \text{Cok } J$ is not additive, because no solution of the Adams conjecture $\alpha: BSO \to G/O$ can be additive due to work of Madsen. In fact, Tornehave and Snaith have independently found formulas for this nonadditivity. These problems are not attacked by obscure homotopy theoretic calculations, by the way. The crucial aspect in understanding the solution of the Adams conjecture, for example, will ultimately reside in better understanding the homogeneous spaces $O(2n)/\mathscr{S}_n \wr O(2)$, and the orbit structure of the left actions of subgroups of $O(2n)$, on $O(2n)/\mathscr{S}_n \wr O(2)$.

For completeness, I will also write down $\text{TOP}/O$, localized at odd primes $p$. This is all due to Sullivan and is in his M.I.T. book. The Adams operation $\psi^3$ is replaced by $\psi^k$, where $k$ generates the group of units in $Z/p^2$. Then $\text{Cok } J_p \subset SG$ is constructed exactly as in the discussion following (1.4). The Adams conjecture again yields $(G/O)_p \cong BSO_p \times \text{Cok } J_p$. Sullivan identifies $(G/\text{TOP})_p \cong BSO_p^{\otimes}$ and then identifies the map $(G/O) \to (G/\text{TOP})_p$ with

$$BSO_p \times \text{Cok } J_p \xrightarrow{\theta^k \times O} BSO_p^{\otimes}.$$

Here $\theta^k: BSO_p \to BSO_p^{\otimes}$ is a $KO$-theory Wu class, associated to the Adams operation $\psi^k$ applied to a $KO \otimes Z[\frac{1}{2}]$ orientation of $MSO$. The operation $\theta^k$ is an $H$-map (from $+$ to $\otimes$) and is characterized by $ph(\theta^k) = \psi_H^k L/L$. These operations were studied long ago by Bott and Adams. (Compare with (1.7).) If $\Theta_p$ is the fibre of $\theta^k: BSO_p \to BSO_p^{\otimes}$, then

(1.10) $$(\text{TOP}/O)_p \cong \text{Cok } J_p \times \Theta_p.$$

In fact, this is an $H$-space decomposition.

Now, if $M$ is a smooth manifold and $M'$ a new smoothing of $M$, classified by some map $M \to \mathrm{TOP}/O$, how is one to find the stable cohomotopy, $KO$-theory, and cohomology invariants provided by (1.9) and (1.10)? This is not at all hopeless, and the geometers and analysts ought now to return to the field of battle, as it were. Roughly, here is what happens.

We interpret a new smoothing of $M$ as a vector bundle $\xi^q$ over $M$, together with a *topological* trivialization $t\colon \xi^q \cong M \times \boldsymbol{R}^q$. This $t$ *defines* a smooth structure on $M \times \boldsymbol{R}^q$, which by the product theorem is $M' \times \boldsymbol{R}^q$ for some unique concordance class of smoothings $M'$ on $M$. The vector bundle $\xi^q$ is canonically a spin bundle, but is also fibre homotopically trivial. This gives *lots* of $KO$-orientations of $T\xi^q$, which can be divided pairwise to give invariants $e\colon M \to BSO^{\otimes}$. We follow by certain equivalences $BSO^{\otimes} \to BSO$ (localized) to get more $KO$-invariants $\gamma\colon M \to BSO$ of our new smoothing. So far, these invariants are just invariants of the composition $M \to \mathrm{TOP}/O \to G/O$. The more delicate *topological* trivialization of $\xi^q$ gives a *reason* why the compositions

$$M \xrightarrow{\gamma} BSO \xrightarrow{\mathscr{L}} \prod K(\boldsymbol{Z}_{(2)}, 4n), \qquad M \xrightarrow{\gamma} BSO \xrightarrow{\theta^k} BSO^{\otimes}$$

are trivial. This reason produces liftings to the fibres $M \to \Theta_2$ and $M \to \Theta_p$ (after localization).

Adams' work on $J(X)$ writes (roughly) $\xi$ as $\psi^k\gamma_p - \gamma_p \in KO(M) \otimes \boldsymbol{Z}_{(p)}$, some $\gamma_p$, $p = 2$ or odd. These are the $\gamma$ occurring above. The more recent Adams conjecture produces a specific fibre homotopy trivialization of $\psi^k\gamma_p - \gamma_p$, suitably localized. This is now compared with the original topological trivialization $t\colon \xi^q \simeq M \times \boldsymbol{R}^q$ to produce a difference element in stable cohomotopy $[M, SG]$. This difference element canonically lies in $[M, \mathrm{Cok}\, J]$.

Finally, the $\boldsymbol{Z}/2$ cohomology data in (1.9) is more subtle and not yet fully understood. I am convinced that the place to look for these invariants, as well as for further information about the $KO$-invariants (with extra structure), is in smooth cobordism of special type from $M$ to $M'$.

**2.** In many ways, it is absolutely crucial to emphasize not only the spaces in (1.1) and (1.2), but the three Thom spectra as well $MO \to M\,\mathrm{TOP} \to MG$. (Sometimes we will want only the oriented Thom spectra $MSO \to MS\,\mathrm{TOP} \to MSG$.) In the first place, the Thom spectra contain all the geometry associated to structures on normal bundles. In the second place, the homotopy theory of the Thom spectra is very closely related to the homotopy theory of the base spaces. In this section I want to state a beautiful geometric "solution" of the lifting problem for $BS\,\mathrm{TOP} \to BSG$, in terms of the Thom spaces.

I have in mind the theorem that a spherical fibre space admits a topological reduction if and only if all maps of manifolds to the Thom space can be made "Poincaré transversal to the zero section". There are at least three approaches to Poincaré transversality available, due to Norman Levitt, Lowell Jones, and Frank Quinn. The theorem I plan to state is but one consequence of the entire theory of transversality (or lack thereof), and surgery on P.D. spaces. All three approaches are quite ingenious and complement each other in many ways. I really do not do any of them justice in the following brief paragraphs.

First, what is a Thom space? If $\pi\colon \xi^q \to B$ is an $S^{q-1}$-spherical fibration, the mapping cylinder of $\pi$ will be called the disc bundle of $\pi$ and denoted $B\xi^q$. The mapping cone of $\pi$ will be called the Thom space, and denoted $T\xi^q$. We thus have

$$\xi^q \subset B\xi^q \subset T\xi^q = B\xi^q \bigcup_{\xi^q} \mathrm{Cone}(\xi^q).$$

Note that $B\xi^q \sim B$ and the inclusion $\xi^q \subset B\xi^q$ is, up to homotopy, just the original map $\pi\colon \xi^q \to B$. Also note that $\xi^q$ is collared (on two sides) in $T\xi^q$, that is, $\xi^q \times [-\frac{1}{2}, +\frac{1}{2}] \subset T\xi^q$ with $\xi^q = \xi^q \times \{0\}, \xi^q \times [-\frac{1}{2}, 0] \subset B\xi^q$, and $\xi^q \times [0, \frac{1}{2}] \subset \mathrm{Cone}(\xi^q)$. If $\pi\colon \xi^q \to B$ is an honest $S^{q-1}$ fibre bundle, then the disc bundle and Thom space above agree with the usual notions.

Secondly, what is a Poincaré duality space $X$? One can certainly give a definition in terms of Poincaré duality. But if $X$ is nonorientable or even nonsimply connected, the precise definition gets a little complicated. So it is better to think of an $n$-dimensional P.D. space as a codimension 0 submanifold of a (closed) manifold $V \subset M^{n+q}$, so that the inclusion $\partial V \to V$ is, up to homotopy type, an $S^{q-1}$-spherical fibration. Then one can prove all the Poincaré duality one wants. If $X$ is a finite complex in $M$ which satisfies Poincaré duality, take a regular neighborhood $V$ of $X$ in $M$. The inclusion $\partial V \to V$ is then a spherical fibration, so the two notions are equivalent.

In particular, we define the normal fibration of $V \subset M$ to be the inclusion $\partial V \to V$. In the case $M = S^{n+q}$ a sphere, we get the Spivak normal spherical fibration of a P.D. space $X$, which we denote $\nu^q_X$.

It is necessary to extend these notions to get P.D. spaces with boundary. For this we use $V \subset M$, where the manifold $M$ has a boundary. The picture below should be sufficient.

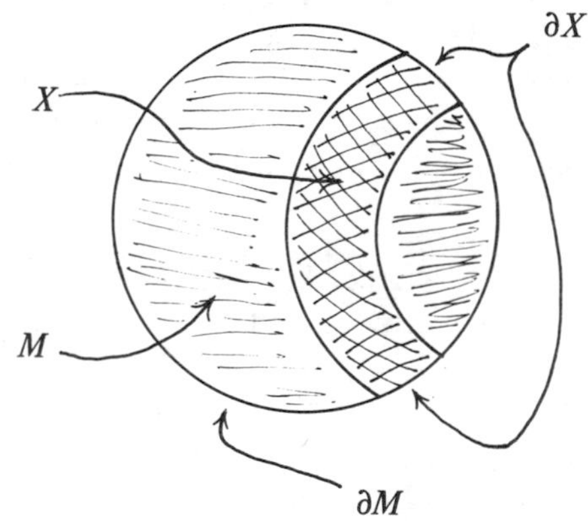


We now define what it means for a map $f\colon N \to T\xi^q$, $N$ a manifold, to be (globally) transversal. We picture $T\xi^q$ as $B\xi^q \cup_{\xi^q} \mathrm{Cone}(\xi^q)$. Since $\xi^q \subset T\xi^q$ is nicely collared, we may assume without real loss of generality that $f^{-1}(B\xi^q)$ is a codimension 0-submanifold of $N$, with nicely collared boundary, $f^{-1}(\xi^q)$. Then $f$ is *transversal* if the diagram

$$\begin{array}{ccc} f^{-1}(\xi^q) & \longrightarrow & \xi^q \\ \downarrow & & \downarrow \\ f^{-1}(B\xi^q) & \longrightarrow & B\xi^q \end{array}$$

is a map of spherical fibrations (which in particular says the left vertical map is a spherical fibration, so that $f^{-1}(B\xi^q) \subset N$ is indeed a P.D. space). The relative version, when $N$ has a boundary, should also be stated, but I will not.

Not all maps can be deformed to transversal maps. One proof is the observation that if they could be so deformed, one could define rational Pontrjagin classes for spherical fibrations. But Norman Levitt discovered that you could set up a (local) obstruction theory to deforming $f\colon N \to T\xi^q_X$ to a transversal map. If $\pi_1(X) = 0$, the obstructions lie in $H^{q+i+1}(N, P_i)$, where $P_i = \pi_i(G/\mathrm{TOP}) = 0, Z/2, 0, Z$ as $i \equiv 1, 2, 3, 4 \pmod 4$ respectively. Certainly this resembles the lifting problem for $B\,\mathrm{TOP} \to BG$. In fact, in an (almost) obvious way, one defines the transversal singular subcomplex $TSG$ of (the singular complex of) $MSG$. John Morgan and Levitt then showed that the lifting problem for $B\xi^q \to BSG$ to $BS\,\mathrm{TOP}$ (and the classification of liftings if one exists) was exactly equivalent to the lifting problem for $T\xi^q \to MSG$ to $TSG$. The fibre of $TSG \to MGS$ is a spectrum which loops down to $G/\mathrm{TOP}$. In particular, they recover not only Sullivan's result that localized at odd primes the obstruction coincides with the obstruction to a $KO \otimes Z[\frac{1}{2}]$ orientation of $T\xi^q$, but they also deduce that the 2 local obstructions are simply cohomology classes $\tilde{\mathscr{K}} \in H^{4*-1}(B\xi^q, Z/2)$ and $\tilde{\mathscr{L}} \in H^{4*+1}(B\xi^q, Z_{(2)})$. These results also come out of Jones' or Quinn's approach.

This result does *not* imply $B(G/\mathrm{TOP})$ localized at 2 is an Eilenberg-Mac Lane space. This latter fact requires more work. But in any application where it is simply sufficient to know the obstructions are cohomology classes, the transversality approach gives enough information.

Actually, quite detailed information is obtained about $\tilde{\mathscr{K}}$ and $\tilde{\mathscr{L}}$. First $\tilde{\mathscr{K}} = V^2\mathscr{E}$, where $\mathscr{E} = \sum_{j\geq 1} e_{2^{j+1}-1}$, and $e_{2^{j+1}-1} \in H^{2^{q+1}-1}(BSG, Z/2)$ is a certain canonical primitive class. So $\mathscr{E}(\xi^q) = 0$ if and only if $\tilde{\mathscr{K}}(\xi^q) = 0$. Secondly, $\tilde{\mathscr{L}} = \beta l$, where $l = L + l_1 + l_2 + \cdots \in H^{4*}(BSG, Z/8)$ is a canonical modulo 8 "Hirzebruch index class", and $\beta\colon H^{4*}(\ \ , Z/8) \to H^{4*+1}(\ \ , Z_{(2)})$ is the Bockstein for $Z_{(2)} \to Z_{(2)} \to Z/8$. We will discuss $l$ further in §4. It is also constructed using the transversality theory.

If one is going to apply the transversality approach to an honest lifting question, perhaps one should not really convert the transversality method into homotopy theory at all. Instead, one should work directly in the machine, geometrically. For example, the question of whether the orbit space $X$ of a finite group $\pi$ acting freely on a space $\tilde{X}$ homotopy equivalent to a sphere $S^{n-1}$ admits a normal invariant might be accessible. It seems to me that this amounts to deciding if the $L(1)$ or $L(Z/2, \text{—})$ part of a surgery obstruction with $\pi_1 = \pi$ vanishes, given that the surgery obstruction lifted to a certain cover vanishes and that the cover is induced from $\tilde{X} \to X$. Or something like that.

**3.** Next I want to discuss a little general nonsense. Let $h^*$, $h_*$ be a generalized cohomology, homology theory. I want to make rather clear that there are two distinct notions of characteristic classes for "manifolds" associated to a bundle theory, the cohomology characteristic classes and the homology characteristic classes.

The cohomology classes are universally given as elements of $h^*(BH)$, where $H$ is $O$, TOP, $G$, or whatever. The homology characteristic classes are universally given as elements of $h^*(MH)$. If $X$ is an "$H$-manifold", then $X$ and $T\nu_X$ are $S$-dual. Thus $h_*(X) \simeq h^*(T\nu_X)$. The classifying map $X \to BH$, covered by $T\nu_X \to MH$, together with $c \in h^*(MH)$, gives a class $c(X) \in h_*(X)$. A warning is in order. The $S$-duality isomorphism $h_*(X) \simeq h^*(T\nu_X)$ is not quite well defined, but can be varied by $H$-bundle automorphisms of $T\nu_X$. For ordinary cohomology, no ambiguity at all occurs. In general, this will not affect what I say below too much, because it is a universal ambiguity commuting with all operations $\mathcal{O}: h_* \to h'_*$.

Once one understands that there are two distinct notions of characteristic classes, one sees why it is that certain obstructions in a lifting problem occur naturally at the Thom space level, while others occur at the classifying space level. The mechanism is this. Given a pair of our classifying spaces $BH \to BH'$, suppose we have a characteristic class $c \in h^*(BH')$ so that $c(X) = 0$ for all $H$-manifolds of some fixed dimension. Then if $X'$ is an $H'$-manifold, the class $c(X') \in h^*(X')$ is an obstruction to an $H$-structure. We call $c$ an *exotic characteristic class* and this is the notion familiar to most. But also, if $s \in h^*(MH')$ is a class which vanishes for $H$-manifolds of some fixed dimension, and $X'$ is an $H'$-manifold, then $s(X') \in h_*(X')$ is an obstruction to an $H$-structure. I will call $s$ a *homological singularity*. Note that just lifting $T\nu_{X'} \to MH'$ to $MH$ will not always imply $s(X') = 0$.

There is a more refined variant of both the notions of exotic characteristic class and homological singularity. We will discuss only the homological version. Suppose $h_*$, $h'_*$ are two homology theories and $\mathcal{O}: h_* \to h'_*$ is a functor. Suppose $c(X) \in h_*(X)$ is a (homology) characteristic class for $H$-manifolds and $c'(X') \in h'_*(X')$ is a characteristic class for $H'$-manifolds where $BH \to BH'$ is one of our lifting problems. Suppose for all $H$-manifolds $X$ it holds that $\mathcal{O}c(X) = c'(X) \in h'_*(X)$. Then if $X'$ is an $H'$-manifold, we obtain an obstruction to an $H$-structure $c'(X') \in h'_*(X')/\text{image}(h_*(X') \xrightarrow{\mathcal{O}} h'_*(X'))$. I call $(h_*, h'_*, \mathcal{O}, c, c')$ a relative homological singularity.

Now, it is one thing to say a relative homological singularity vanishes and quite another to say it vanishes for a reason. For example, suppose we have a map $g: X \to X'$ where $X$ is an $H$-manifold. We say $g: X \to X'$ *resolves* the singularity above if $g_*c'(X) = c'(X')$. This then implies $c'(X') = g_*c'(X) = g_*\mathcal{O}c(X) = \mathcal{O}g_*c(X)$, so the relative singularity certainly vanishes.

Consider the hypotheses:

(A) There is a degree one map $g: X \to X'$, $X$ an $H$-manifold, covered by an $H'$ bundle map $g: \nu_X \to \nu_{X'}$;

(B) There is a lifting of $\nu_{X'}: X' \to BH'$ to $X' \to BH$ (in particular, a bundle lifting $T\nu_{X'} \to MH \to MH'$).

Now, (A) is pretty strong. If $Dg: T\nu_{X'} \to T\nu_X$ is $S$-dual to $g$ and $\hat{g}: T\nu_X \to T\nu_{X'}$ is the bundle map, then $\hat{g} \circ Dg: T\nu_{X'} \to T\nu_X$ is a homotopy equivalence and a short computation gives that $g_*c'(X)$ differs from $c'(X')$ exactly by the automorphism of $h'_*(X')$ induced by $\hat{g} \circ Dg$. But this will not even imply the relative $(c, c')$ singularity vanishes in general. However, if $H$-bundle theory satisfies transversality, then (B) implies (A) and, assuming (B), a computation shows that the relative $(c, c')$ singularity does vanish. That is, $c'(X') \in \mathcal{O}h_*(X')$. However, the map (A) produced by (B) will not necessarily *resolve* the singularity. I leave it to the reader to decide if this general nonsense is complete.

Needless to say, the terminology here is motivated by the intuitive picture of a smoothing of a topological manifold $g: M \to X$ as a "resolution of *all* singularities" phenomena and similarly for a manifold structure on a P.D. space.

EXAMPLES. 3.1. The rational Pontrjagin classes of topological manifolds are integral for smooth manifolds. Also, for smooth $M$, $P_i(m)$ is $w_{2i}^2(M)$ (mod 2). Thus the integrality of Pontrjagin classes can be formulated as a relative singularity associated to the operation $\rho_Q \times \rho_{Z/2}: H_*(\ ,Z) \to H_*(\ ,Q) \times H_*(\ ,Z/2)$.

This example illustrates that if all Thom spaces in sight are orientable for all cohomology theories in sight, then, in fact, there is little difference between the exotic characteristic classes and homological singularities.

3.2. Sullivan's $KO \otimes Z[\frac{1}{2}]$ orientation for $MS$ TOP and lack thereof for $MSG$ can readily be formulated as a relative homological singularity, but not naturally as an exotic characteristic class.

3.3. The transversality machine provides relative homological singularities, since one has (more or less) canonical transversality for $MS$ TOP, which can be interpreted as a lifting $MS$ TOP $\to TSG$ of $MS$ TOP $\to MSG$.

3.4. If $U$: $MS$ TOP $\to BSO$ is Sullivan's $KO \otimes Z[\frac{1}{2}]$ orientation, one obtains $\rho^k \in KO(BS\ \mathrm{TOP}) \otimes Z[\frac{1}{2}]$ as $\psi^k U/U$. Now, $\rho^k: KO \to KO$ is also a $KO$-operation. We then obtain a very important smoothing obstruction, namely for a topological manifold $X$, $\rho^k(X) \in KO(X)/\mathrm{image}(KO(X) \xrightarrow{\rho^k} KO(X))$. This is also a relative singularity, but is a little more complicated because it depends on multiplicative properties and bundle orientations of the cohomology theories involved.

3.5. For simplicity, assume we are working with oriented theories $BSH \to BSH'$. If $X'$ is an $SH'$ manifold, we can ask if there exists a degree one map $X \to X'$, where $X$ is an $SH$-manifold. This can be formulated as a resolution of homological singularities question, souped up even further. Namely, there is the diagram *over* $k(Z, O)$, the spectrum for $Z$-homology

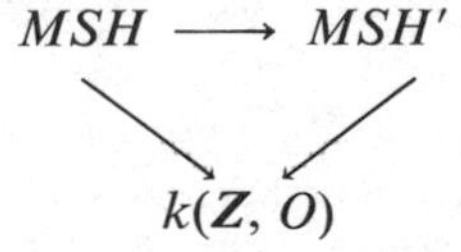


Geometrically, the existence of a homological degree one map seems to be a long way from an $H$-structure on an $H'$-manifold. But it is homotopy theoretically a very strong assumption. For example, if $X'$ is a P.D. space and if there exists any degree one map $X \to X'$, $X$ a topological manifold, then the Thom spectrum $T\nu_X$ is $KO \otimes Z[\frac{1}{2}]$ orientable. In fact the $S$-dual $T\nu_{X'} \to T\nu_X$ plus the $KO \otimes Z[\frac{1}{2}]$ orientation on $T\nu_X$ defines an orientation on $T\nu_{X'}$.

The converse is also true: a $KO \otimes Z[\frac{1}{2}]$ orientation on $T\nu_{X'}$ implies the fundamental homology class of $X'$ is the degree one image of a topological manifold $X \to X'$ (but not necessarily of a smooth manifold). This is really quite amazing, and is a consequence of Sullivan's characterization of PL bundles at odd primes as spherical fibre spaces plus $KO \otimes Z[\frac{1}{2}]$ orientations.

In §4, I will state what additional information is needed about a degree one map $X \to X'$ in order to conclude that $\nu_{X'}$ actually admits a topological bundle structure.

**4.** A topological manifold has a cohomology characteristic class $L = 1 + L_1 + L_2 + \cdots \in H^{4*}(\ , Z_{(2)})$, characterized uniquely by the following universal properties.

4.1. (i) If $\nu$: $M^{4n} \to BS$ TOP is the normal bundle of a $Z/2^k$ manifold (oriented manifold with $2^k$ isomorphic oriented boundary components identified), then $\langle L(M), [M]\rangle = \text{signature}(M) \in \mathbf{Z}/2^k$. (This includes the case $M$ a closed manifold.)

(ii) $\rho_2 L = V^2 \in H^*(\ , \mathbf{Z}/2)$.

(iii) If $\xi$, $\eta$ are topological bundles, $L(\xi + \eta) = L(\xi) \cdot L(\eta)$.

Rationally, $L$ is (the inverse of) the Hirzebruch genus in the Pontrjagin classes. In $H^*(BSO, \mathbf{Z}_{(2)})$, $L$ is not hard to find with these properties, because $H^*(BSO, \mathbf{Z}_{(2)})$ is so well understood. The smooth class is extended to $BS$ TOP by Morgan and Sullivan using transversality and the fact that all homology classes of $MS$ TOP, with $\mathbf{Z}/2^k$ coefficients, are representable by smooth $\mathbf{Z}/2^k$ manifolds mapped to $MS$ TOP.

John Morgan and I have constructed, using Levitt and Morgan's transversality machine, a class $l = 1 + l_1 + l_2 + \cdots \in H^{4*}(BSG, \mathbf{Z}/8)$ with the following properties. (David Frank and Frank Quinn have investigated such classes also.)

4.2. (i) If $\nu: X^{4n} \to BSG$ is the normal fibration of a $\mathbf{Z}/8$ P.D. space, then $\langle l(X), [X]\rangle = \text{signature}(X) \in \mathbf{Z}/8$.

(ii) $\rho_2 l = V^2 \in H^*(\ , \mathbf{Z}/2)$.

(iii) If $\xi$, $\eta$ are spherical fibrations,

$$l(\xi + \eta) = l(\xi) \cdot l(\eta) + (\text{Error term}).$$

The error term is absolutely unavoidable. The class $l$ cannot *possibly* be understood until this error term is pinned down. This requires *proving* some very delicate product formulas for surgery problems on $\mathbf{Z}/2^k$ P.D. spaces, extending the known formulas for manifolds. In addition, some choices must be made as to what cohomology formulas certain transversality obstructions should satisfy. But in the end this can be done and perhaps the nicest choices lead to a *unique* class $l$ for which we have

(iv) (Error term) $= i(\mathscr{K}(\xi) \cdot VS_q^1V(\eta) + VS_q^1V(\xi)\mathscr{K}(\eta))$ where $\mathscr{K} \in H^{4*-1}(BSG, \mathbf{Z}/2)$ is the same class which occurs as a transversality obstruction, mentioned in §2, and $i: H^*(\ , \mathbf{Z}/2) \to H^*(\ , \mathbf{Z}/8)$. Thus the two transversality obstructions $\mathscr{K}$ and $\mathscr{L} = \beta l$ are intimately related.

Another property of $l$ which comes from its construction is

(v) $l|_{BS\,\text{TOP}} = \rho_{\mathbf{Z}/8}(L) \in H^*(BS\,\text{TOP}, \mathbf{Z}/8)$.

So, what we have here is a relative homological singularity for BS TOP $\to BSG$, $(H^*(\ , \mathbf{Z}_{(2)}), H^*(\ , \mathbf{Z}/8), \rho_{\mathbf{Z}/8}, L, l)$.

THEOREM. *The normal bundle of an oriented P.D. space X lifts to BS* TOP *if and only if*

(1) *there is a topological manifold N and a degree one map* $g: N \to X$ *which resolves the* $(L, l)$ *singularity, and*

(2) $l(M \times X) = l(M) \otimes l(X)$ *for all closed oriented smooth manifolds M.*

In fact, in (2) it suffices to test on a single 5-manifold with nonzero de Rham invariant. Alternatively, instead of (2), one could assume that $1 \times g: M \times N \to M \times X$ resolves the $(L, l)$ singularity for $M \times X$.

This theorem contains nothing new certainly. The degree one map gives a $KO \otimes \mathbf{Z}[\frac{1}{2}]$ orientation $U$ of $T\nu_X$, with $ph(U)/U_H = g_*(L(N))$. Since $\rho_{\mathbf{Z}/8}(g_*L(N)) = l(X)$, we have not only $\beta l(X) = \mathscr{L}(X) = 0$, but a specific $\mathbf{Z}_{(2)}$ class reducing to $l(X)$, which is compatible over $Q$ with $ph(U)/U_H$. Finally, (2) guarantees $\mathscr{K}(X) = 0$. In fact, it is unnecessary to pin $l$ down uniquely via 4.2(iii), (iv) for this theorem. That is, 4.2(i), (ii), (v) are adequate since any choice of $l$ with these properties will somehow contain $\tilde{\mathscr{K}}$ in the error term in the diagonal formula. Still, it is good to have a well-defined class $l$ to refer to.

Actually, for the proof of the theorem, it is unnecessary to convert the hypotheses to the $KO \otimes \boldsymbol{Z}[\frac{1}{2}]$ and $H^*(\ \ , \boldsymbol{Z}_{(2)})$, $H^*(\ \ , \boldsymbol{Z}/2)$ statements. One can just work with the picture of $N \subset \nu_X$, using the hypothesis $g_* L(N) \equiv l(X) \pmod 8$, and $l(M \times N) = l(M) \otimes l(N)$ for some suitable 5-manifold. These hypotheses are adequate to prove that all transversality obstructions for manifolds mapped to $T\nu_X$ vanish.

My motivation for emphasizing the transversality approach and the homological singularities language is that perhaps there are very nice geometric obstructions living in the Thom spaces for the lifting problems $BO \to BG$, $BO \to B\mathrm{TOP}$ which can be exploited to derive nontrivial theorems about the smoothing problems. The point is that this works for $B\mathrm{TOP} \to BG$, *without* having to determine $B(G/\mathrm{TOP})$, and, although the smoothing problems are harder, they might not be as hard to unravel in the Thom spaces as it seems to be to deal with $B(G/O)$ and $B(\mathrm{TOP}/O)$. Norman Levitt has, in fact, already attacked the smoothing problem along these lines, but I do not think the positive results are definitive at this time.

STANFORD UNIVERSITY

Proceedings of Symposia in Pure Mathematics
Volume 32, 1978

# FREE ACTIONS OF FINITE GROUPS ON SPHERES

C. T. C. WALL

Just about a century ago it was realised (by Killing, Klein and others) that the sphere and real projective space offered two different global models for geometry which were locally the same. In 1891, Killing formulated (in terminology that now appears vague) the problem of determining all such models. In 1926 Hopf revived the problem, gave a clear statement of its topological setting, and also raised the more general question of studying manifolds covered by spheres. This problem lies deep, and though much was done in the 3-dimensional case serious progress in general is comparatively recent, and can be said to start with Cartan and Eilenberg **[1956]** and Milnor **[1957]**.

I will discuss the problem of classifying the homotopy types of such manifolds; in the first half I outline a reduction of this problem to two others, less complex though perhaps not less deep; then I will discuss the current state of knowledge on these. This includes a virtually complete determination of which groups can act freely on which spheres. This work is part of a collaboration with Charles Thomas and Ib Madsen. To save time, I will omit discussion of the smooth case, except to state one of our main conclusions.

**Examples.** *Orthogonal space-forms.* Since the isometries of the sphere $S^{n-1}$ are all linear, the solution to the original problem is to be sought as follows: Seek linear representations $\phi$ of the given group $\pi$ such that for $1 \neq g \in \pi$, $\phi(g)$ has no fixed points on $S^{n-1}$, or equivalently, 1 is not an eigenvalue of $\phi(g)$. In the present state of representation theory, this is not too hard and a full solution can be found in the book of Wolf **[1967]**, with references to much earlier work.

Let us call representations $\phi$ satisfying the above condition $\mathscr{F}$-representations. Then the sum $\phi_1 \oplus \phi_2$ is an $\mathscr{F}$-representation if and only if $\phi_1$ and $\phi_2$ are (this construction corresponds to the equivariant join of the corresponding spheres), so it is enough to look at irreducible $\phi$. The simplest examples are the (cyclic) sub-

*AMS (MOS) subject classifications* (1970). Primary 57E25; Secondary 57D65, 16A54, 55B25.

groups of $SO_2$: taking sums of these representations yields the lens spaces of de Rham **[1931]**: the 3-dimensional case goes back to Tietze **[1908]**. Another early example was that of binary polyhedral groups, which can be considered as subgroups of the group $S^3$ of unit quaternions, acting on $S^3$ by right translations. This includes the case of generalised quaternion groups $Q_{4k}$ of order $4k$.

Finally if $\pi = \rho\sigma$ is a split extension of a cyclic group $\rho$ by a cyclic group $\sigma$ of coprime order, the faithful irreducible representations are all obtained by taking a faithful 1-dimensional (complex) representation of $\rho$, extending (faithfully) to the centraliser $\zeta$ of $\rho$ in $\sigma$, and inducing up to $\pi$. If $\pi$ admits an $\mathscr{F}$-representation, $\zeta$ meets each Sylow subgroup of $\sigma$; if this holds, all faithful irreducible representations are $\mathscr{F}$-representations (Burnside, **[1905]**). An equivalent condition is that every subgroup of $\pi$ of order $pq$ ($p$, $q$ primes not necessarily distinct) is cyclic. It was also shown by Vincent **[1947]** that any soluble group $\pi$ satisfying this latter condition possesses $\mathscr{F}$-representations.

**Homotopy theory.** We shall follow the basic method of surgery theory (assumed known) and accordingly treat the problem in three stages: homotopy classification, normal invariant and surgery obstruction.

Let us first observe that by the fixed point theorem of Brouwer **[1912]**, an orientation-preserving homeomorphism of $S^{n-1}$ has a fixed point if $n$ is odd. Thus if $\pi$ acts freely, the orientation-preserving subgroup (of index 1 or 2) is trivial. As $\boldsymbol{Z}/2$ can act freely on any sphere, and the orbit space is homotopy equivalent to real projective space, we shall not discuss this case further. So suppose $n$ even. By the same argument, $\pi$ now respects orientation, so the quotient manifold is orientable. We also ignore the trivial case of actions on $S^1$ ($n = 2$), so $n \geqq 4$ throughout. And for our positive results (using surgery) we have to omit the case $n = 4$, and so $n \geqq 6$ is even.

According to Smith **[1944]**, $\boldsymbol{Z}/p \times \boldsymbol{Z}/p$ cannot act freely on any sphere. Thus if $\pi$ can so act, all subgroups of $\pi$ of order $p^2$ ($p$ any prime) are cyclic. Call $\pi$ an $\mathscr{R}$-group if it satisfies this condition. The structure of $\mathscr{R}$-groups is known: the soluble case is due to Zassenhaus **[1935]** and the rest essentially to Suzuki **[1955]**. The Sylow $p$-subgroups of $\pi$ are cyclic, or perhaps (if $p = 2$) generalised quaternionic. If all are cyclic (e.g., if $\pi$ has odd order), $\pi$ is metabelian. In general the quotient of $\pi$ by the maximal normal subgroup $O_{2'}(\pi)$ of $\pi$ of odd order, belongs to one of six types:

I. cyclic,
II. generalised quaternion 2-group,
III. binary tetrahedral,
IV. binary octahedral,
V. $SL_2(p)$ ($p \geqq 5$ prime),
VI. $TL_2(p)$.

We can define $TL_2(p)$ as follows. Choose a nonsquare $\omega \in \boldsymbol{F}_p^\times$. Then $TL_2(p) \cong \{y \in GL_2(p): \det y = 1 \text{ or } \omega\}$ with product given by

$$\begin{aligned} y_1 \circ y_2 &= y_1 y_2 \text{ (product in } GL_2) && \text{if } \det y_1 = 1 \text{ or } \det y_2 = 1, \\ &= \omega^{-1} y_1 y_2 && \text{if } \det y_1 = \det y_2 = \omega. \end{aligned}$$

$\pi$ is soluble for types I—IV, not for V, VI. Types III, IV may be regarded as the case $p = 3$ of V, VI but the detailed descriptions are somewhat different at the prime 3.

A deeper study by Cartan and Eilenberg **[1956]** with later work by Swan **[1960]** gives converse results at the homotopy level. Using Tate cohomology, the cohomology of $\pi$ is *periodic* if $g \in \hat{H}^n(\pi; \boldsymbol{Z})$ is such that for all $a \in \boldsymbol{Z}$ and all $\pi$-modules $A$, cup product with $g$ gives an isomorphism $\hat{H}^a(\pi; A) \to \hat{H}^{n+a}(\pi; A)$. A cohomology class $g$ satisfies this condition if and only if $\hat{H}^n(\pi; \boldsymbol{Z})$ is cyclic, having the same order as $\pi$, and generated by $g$; so we will call such $g$ *generators*. Then $\pi$ has periodic cohomology if and only if $\pi$ is an $\mathscr{R}$-group.

Now define a *$\pi$-polarised space* to consist of a $CW$ complex $X$, dominated by a finite complex, together with an isomorphism $\pi_1(X) \to \pi$ and a homotopy equivalance of the universal cover $\tilde{X}$ on $S^{n-1}$ ($n \geq 3$).

PROPOSITION 1. *Taking the first k-invariant yields a bijection between (polarised) homotopy classes of $\pi$-polarised spaces and generators $g \in \hat{H}^n(\pi; \boldsymbol{Z})$.*

Recall that such a space $X$ determines an obstruction $\theta(X) = \theta(g) \in \tilde{K}_0(\boldsymbol{Z}\pi)$ to its being homotopy equivalent to a finite complex. We will discuss this more fully below.

Note that if $X$ is $\pi$-polarised, there is a natural isomorphism $H^i(\pi; A) \to H^i(X; A)$ for any coefficient module $A$ and integer $i < n - 1$: from now on we will identify these groups. Note also that any polarised space is a Poincaré complex.

**Normal invariants.** A Poincaré complex $X$ has a Spivak normal fibration, classified by a map $X \to BG$. A homotopy class of liftings to $X \to B\,\mathrm{Top}$ is called a *normal invariant*. It determines by transversality a normal cobordism class of normal maps $\phi: M^{n-1} \to X$ ($n \neq 5$) of degree 1.

PROPOSITION 2. *Any polarised space X has a normal invariant.*

PROOF. We need some preliminary remarks. It follows from work of Sullivan that the obstruction corresponding to the top cell of $X$ is zero (for any Poincaré complex), so we may ignore obstructions in $H^{n-1}(X; A)$. Next, since $\mathrm{Top} \subset G$ is an infinite loop map, the obstruction to existence of a normal invariant is a homotopy class of maps $X \to B(G/\mathrm{Top})$, the image of $X \to BG$. It is enough to show this becomes nullhomotopic after localisation at any prime.

At odd primes, $G/\mathrm{Top} \simeq BO$, so we get a class in $KO^{-1}(X; \boldsymbol{Z}_{(\mathrm{odd})})$. This is a summand of $K^{-1}(X; \boldsymbol{Z}_{(\mathrm{odd})})$, which is well known to vanish.

The localisation of $B(G/\mathrm{Top})$ at 2 is an Eilenberg-Mac Lane space, so the obstruction is given by a string of cohomology classes with coefficients $\boldsymbol{Z}_{(2)}$ or $\boldsymbol{Z}/2$. For the covering map $X(\pi_2) \to X$ corresponding to the Sylow 2-subgroup $\pi_2$, the corresponding cohomology groups map injectively, and the obstructions are natural, so it suffices to consider the problem for $\pi_2$.

But if $X(\pi_2)$ is homotopy equivalent to a manifold, it has a normal invariant. And we have orthogonal space-forms in every homotopy type if $\pi_2$ is cyclic, and (at least) in every relevant dimension if $\pi_2$ is generalised quaternionic. This is enough, since we can obtain all generators $g \in \hat{H}^n(\pi_2; \boldsymbol{Z})$ from one such by changing the attaching map of the top cell.

Up to this point, we have been surveying known results. The following, however, is new. We continue the above notations.

PROPOSITION 3. *Any normal invariant of $X(\pi_2)$ coming from a homotopy equivalence with an orthogonal space-form extends to a normal invariant for $X$.*

PROOF. By the above, $X$ does have a normal invariant. Taking this as base point, we can identify normal invariants for $X$ with homotopy classes of maps $X \to G/\mathrm{Top}$. We can use the induced lift also for the covering space $X(\pi_2)$. Then the given normal invariant here determines a map $\alpha: X(\pi_2) \to G/\mathrm{Top}$, which we wish to factor (up to homotopy) through $X$. Since (apart from the top cell) $X(\pi_2)$ is 2-local, it is sufficient to 2-localise throughout.

Now $\alpha$ determines classes $\alpha_{4k} \in H^{4k}(\pi_2; \boldsymbol{Z}_{(2)})$, $\alpha_{4k+2} \in H^{4k+2}(\pi_2; \boldsymbol{Z}/2)$ in dimensions $< n$. Since $H^{4k}(\pi; \boldsymbol{Z}_{(2)})$ maps onto $H^{4k}(\pi_2; \boldsymbol{Z}_{(2)})$ (this is obvious when one recalls that $\pi$ has 2-period 2 or 4), the $\alpha_{4k}$ lead to no obstruction. As to the rest, $H^{4k+2}(\pi_2; \boldsymbol{Z}/2)$ is 0 ($\pi_2$ trivial), $\boldsymbol{Z}/2$ ($\pi_2$ nontrivial cyclic) or $\boldsymbol{Z}/2 \times \boldsymbol{Z}/2$ ($\pi_2$ generalised quaternionic) and—again since the 2-period divides 4—the image of $H^{4k+2}(\pi; \boldsymbol{Z}/2)$ is determined as follows:

*Types* I, II: surjective.

*Types* III, V: zero.

*Types* IV, VI: $\boldsymbol{Z}/2$.

There is thus no obstruction for types I, II.

For the remaining cases, we first reduce to considering the generalised binary tetrahedral groups $T_v^*$ ($v \geqq 1$): defined as the nontrivial (split) extension of a quaternion group of order 8 by a cyclic group of order $\boldsymbol{Z}/3^v$. This has type III, and for any group $\pi$ of type III, $\pi_2$ is contained in a subgroup $T_v^*$. Thus if $\alpha$ extends to $X(T_v^*)$, the obstructions to extending to $X$ must already vanish. If $\pi$ has type IV, there is a subgroup $\pi^0$, of index 2, of type III. Set $\pi_2^0 = \pi^0 \cap \pi_2$. Then $H^{4k+2}(\pi_2; \boldsymbol{Z}/2) \to H^{4k+2}(\pi_2^0; \boldsymbol{Z}/2)$ has image $\boldsymbol{Z}/2$. If the restriction of $\alpha$ to $X(\pi_2^0)$ extends to $X(\pi^0)$, then the restriction of $\alpha_{4k+2}$ to $\pi_2^0$ must vanish. This single mod 2 obstruction must thus coincide with the obstruction to extending $\alpha_{4k+2}$ to $\pi$. So the type IV case reduces to the type III case. Finally for types V, VI we know by [II, Lemma 3.3] that $\alpha$ extends if and only if for each subgroup $Q_8$ of $\pi_2$ lying in a subgroup $T_1^*$ of $\pi$, the restriction of $\alpha$ to $Q_8$ extends to $T_1^*$.

We have thus reduced to the case $\pi = T_v^*$, $\pi_2 = Q_8$. Then the (unique) free orthogonal action of $\pi_2$ on $S^{n-1}$ extends to a free orthogonal action of $\pi$. By the above, the obstruction to extension is a sequence of classes $\alpha_{4k+2} \in H^{4k+2}(\pi_2; \boldsymbol{Z}/2)$. The outer automorphism $c$ of period 3 of $\pi_2$ preserves the orthogonal normal invariant (which is unique) and becomes inner in $\pi$. Thus the classes $\alpha_{4k+2}$ are invariant under $c$. But $c$ permutes the three nonzero elements of $H^{4k+2}(Q_8; \boldsymbol{Z}/2)$: the only invariant class is zero. This completes the proof.

**Surgery.** Instead of going back to first principles, we appeal to [II, Theorem 4.2], which was devised for the present purpose, and now restate it as

PROPOSITION 4. *Let $\phi: M \to Y$ be a normal map of degree 1 from the closed manifold $M$ of odd dimension $m \geqq 5$ to the Poincaré complex $Y$ with finite fundamental group $\pi$. Then we can perform surgery on $\phi$ to obtain a homotopy equivalence if and only if*

(i) *$Y$ is homotopy equivalent to a finite complex,*

(ii) *for every 2-hyperelementary subgroup $\rho$ of $\pi$, the corresponding covering space $Y(\rho)$ of $Y$ is homotopy equivalent to a manifold, and*

(iii) *surgery is possible for the covering normal map* $\phi(\pi_2): M(\pi_2) \to Y(\pi_2)$, $\pi_2$ *the Sylow 2-subgroup of* $\pi$.

I recall that the main ingredients in the proof of this result were the induction theorem due to Andreas Dress **[1975]** and the transfer formula of the author **[1976]**. It is now easy to deduce our first main theorem on homotopy type.

THEOREM 1. *The polarised space $X$ (with $n \geqq 6$) is homotopy equivalent to a manifold if and only if*

(i) *the finiteness obstruction* $\theta(X) \in \tilde{K}_0(Z\pi)$ *is zero, and*

(ii) *for all 2-hyperelementary* $\rho \subset \pi$, $X(\rho)$ *is homotopy equivalent to a manifold.*

PROOF. The conditions are clearly necessary. If they are satisfied, first consider $X(\pi_2)$. This too has zero finiteness obstruction. It follows (by listing cases) that $X(\pi_2)$ has the homotopy type of an orthogonal space-form $Y(\pi_2)$. This homotopy equivalence induces a normal invariant for $X(\pi_2)$, which by Proposition 3 extends to one for $X$. This determines a class of normal maps $\phi: M \to X$ of degree 1. We now apply Proposition 4. Conditions (i) and (ii) are assumed explicitly here, and (iii) holds by the choice of normal invariant. Hence we can do surgery to obtain a homotopy equivalence. This proves the result.

Concerning this result, it seems appropriate to make two remarks. First, we can prove an analogous result in the smooth case, as follows:

*If the polarised space $X$ is homotopy equivalent to a manifold, then it is homotopy equivalent to a smooth manifold whose universal cover is diffeomorphic to* $S^{n-1}$ $(n \geqq 6)$.

Thus we do obtain free smooth actions of $\pi$ on $S^{n-1}$.

Secondly, note that it does not provide an existence theorem for free actions of $\pi$ on spheres. Indeed, the well-known necessary condition due to Milnor has not yet even been mentioned. Accordingly, we now turn to the second part of our problem: namely, the consideration of which homotopy types $X$—or equivalently, generators $g$—satisfy conditions (i), (ii) above. Here, I now have to admit that the only examples of manifolds $X(\rho)$ with $\rho$ 2-hyperelementary that we yet possess come from orthogonal space forms. Since these also yield information about the finiteness obstruction, we now reconsider them.

**Hyperelementary space-forms: homotopy types.**

*Type* I. $\pi$ is an extension of a cyclic group $\rho$ of order $a$ prime to $p$ by a cyclic group $\sigma$ of order $p^r$ ($p$ prime). The extension is split, and is determined by a homomorphism $\alpha: \sigma \to \operatorname{Aut} \rho$. Let $\operatorname{Im} \alpha$ have order $p^s$. Then $\pi$ has an $\mathscr{F}$-representation $\Leftrightarrow$ every subgroup of order $pq$ is cyclic $\Leftrightarrow$ $\alpha$ is *not* injective $\Leftrightarrow s < r$.

The cohomology period is $2p^s$. If $s < r$, the irreducible $\mathscr{F}$-representations all have (complex) degree $p^s$. If one of them has $k$-invariant $g_0$, the others have $k$-invariants $b^{p^s}g_0$, and any $b$ prime to $|\pi|$ can occur. Since adding representations corresponds to multiplying $k$-invariants, it remains only to characterise $g_0$. This we can do via its restrictions to Sylow subgroups (or to $\rho$ and to $\sigma$). These are cyclic, so have cohomology rings which are polynomial on a 2-dimensional generator. It is not hard to see

PROPOSITION 5 (i). *The $k$-invariants which arise from $\mathscr{F}$-representations are those cohomology classes whose restriction to each Sylow subgroup is a nonzero $p^s$th power.*

*Type* II. $\pi$ is an extension of a cyclic group $\rho$ of odd order by a generalised quaternion group $\sigma$ of order $2^k \geqq 8$. Again the extension splits, and we have $\alpha$: $\sigma \to \operatorname{Aut} \rho$. Now $\operatorname{Aut} \rho$ is abelian, so the commutator subgroup of $\sigma$ (in particular the element of order 2) is in $\operatorname{Ker} \alpha$: every subgroup of order $2p$ is automatically cyclic. Moreover, $\sigma^{ab}$ is a four-group, so we can decompose $\rho$ into eigenspaces under $\operatorname{Im} \alpha$. Explicitly, if $\sigma = \langle x, y / x^{2^{k-1}} = 1, y^2 = x^{2^{k-2}}, y^{-1}xy = x^{-1} \rangle$, we can write $\rho = A \times B \times C \times D$ as Cartesian product of (cyclic) groups of coprime orders $a$, $b$, $c$ and $d$ such that

transformation by $x$ centralises $A$ and $B$, but inverts $C$ and $D$;

transformation by $y$ centralises $A$ and $C$, but inverts $B$ and $D$.

In the notation of Milnor [**1957**], this appears as $\boldsymbol{Z}/a \times Q(2^k b; c, d)$. If $k = 3$, $\sigma$ has outer automorphisms permuting $x$, $y$ and $xy$; so $b$, $c$ and $d$ play a symmetrical role. We will suppose $b \geqq c \geqq d$. If $k > 3$, we can only permute $y$ and $xy$, and normalise by $c \geqq d$.

All these groups have cohomological period 4.

Though all have $\mathscr{F}$-representations, the degrees no longer match up. If $c = d = 1$, all irreducible $\mathscr{F}$-representations have real degree 4. Otherwise, all have real degree 8. There is a corresponding dichotomy in determining their homotopy types. As before, we restrict to Sylow subgroups. The part of $H^*(\sigma; \boldsymbol{Z})$ in dimensions divisible by 4 is a polynomial ring. The Chern class (alias $k$-invariant as above) of an irreducible 2-dimensional representation (necessarily an $\mathscr{F}$-representation) yields a generator $\gamma_0 \in H^4$, unique modulo squares.

PROPOSITION 5 (ii). *If $c = d = 1$, the $k$-invariants of $\mathscr{F}$-representations are the classes which restrict to nonzero squares at odd Sylow subgroups, and to nonzero squares times powers of $\gamma_0$ on $\sigma$.*

*If $c > 1$, we have nonzero fourth powers at odd Sylow subgroups, and nonzero fourth powers times powers of $\gamma_0^2$ on $\sigma$. (Note, however, that changing the orientation of $S^{n-1}$ will change the sign of $g$.)*

It is now time to discuss

**Condition (ii) in Theorem 1.** We observed above that the only homotopy types of $X(\pi)$, $\pi$ 2-hyperelementary, currently known to contain manifolds, are those which contain orthogonal space-forms. While I do not expect this to be the complete result, this gives already a fairly large and representative set of examples.

A 2-hyperelementary $\mathscr{R}$-group $\pi$ has an $\mathscr{F}$-representation if every subgroup of order $2p$ is cyclic (and this can only fail for $\pi$ of Type I). But a theorem of Milnor [**1957**] shows that this condition is already necessary for the existence of free topological actions. We may thus impose it.

Next, in 1973 Ronnie Lee reproved Milnor's result by exhibiting an explicit surgery obstruction. He went on to show that if $Q(m; c, d)$ acts freely on $S^{n-1}$, $16 | m$ and $c > 1$, then $8 | n$. Thus 2-hyperelementary groups of type II can be placed in three categories:

K, $c = 1$. Free orthogonal action on any $S^{4r-1}$.

L, $k \geqq 4$, $c > 1$. Free orthogonal action on any $S^{8r-1}$. No free action on any $S^{8r+3}$.

M, $k = 3$, $c > 1$. Free orthogonal action on any $S^{8r-1}$ but on no $S^{8r+3}$. Unknown if free actions exist on any $S^{8r+3}$.

The most obvious outstanding problem in this area is deciding this last case.

**General discussion.** The 'generators' are the units of the Tate cohomology ring $\hat{H}^*(\pi; \boldsymbol{Z})$, and thus form a multiplicative group $\mathscr{G}$, say. Those of degree 0 form a subgroup $\mathscr{G}_0$ isomorphic to the group of units in $(\boldsymbol{Z}/N)^\times$, where $N = |\pi|$. We have an exact sequence $1 \to \mathscr{G}_0 \to \mathscr{G} \to^{\deg} \boldsymbol{Z}$, and the image of "deg" is $d\boldsymbol{Z}$, where $d$ is the period.

The generators of positive degree form a subsemigroup $\mathscr{G}^+$, and by Proposition 1 determine polarised spaces and hence finiteness obstructions. This gives a map $\theta\colon \mathscr{G}^+ \to \tilde{K}_0(\boldsymbol{Z}\pi)$ which, using the algebraic interpretation, is easily seen to be a homomorphism and hence extends to $\mathscr{G}$. The restriction of $\theta$ to $\mathscr{G}_0$ was described by Swan **[1960]**: If $r \in (\boldsymbol{Z}/N)^\times$, then $\theta(r)$ is the class of the projective ideal $\langle r, \sum\{g \in \pi\} \rangle$ in $\boldsymbol{Z}\pi$. Abstractly, $\mathscr{G} \cong \mathscr{G}_0 \oplus \boldsymbol{Z}$: can we choose the splitting so that $\theta(\boldsymbol{Z}) = 0$?

To study this question, first observe that by the induction theorem of Swan **[1960]** the restriction

$$K_0(\boldsymbol{Z}\pi) \to \bigoplus \{K_0(\boldsymbol{Z}\rho)\colon \rho \subset \pi \text{ hyperelementary}\}$$

is injective. This map takes the finiteness obstruction for the (polarised) space $X$ to those for its covering spaces $X(\rho)$. So again it suffices to consider the case where $\pi$ is hyperelementary.

Here again, orthogonal space-forms yield examples of manifolds, and hence of vanishing finiteness obstructions, for $\pi$ of Type II, or of Type I with $s < r$. This gap is filled by our next main result.

THEOREM 2. *Let $\pi$ be $p$-hyperelementary of Type* I *with $s = r$, $g$ a cohomology generator whose restriction to each Sylow subgroup is a $p^s$th power. Then $\theta(g) = 0$.*

I will outline the proof of this result below. The following is an easy consequence of Theorems 1 and 2.

THEOREM 3. *Let $\pi$ be an $\mathscr{R}$-group such that all subgroups of order $2p$ are cyclic. Let $2d_1$ be the cohomology period of $\pi$, and let $d_2 = 2d_1$ if $d_1 \equiv 4 \pmod 8$ and $\pi$ has a hyperelementary subgroup of type* IIL *or* IIM; *$d_2 = d_1$ otherwise.*

*Then if $X$ is a polarised space corresponding to a generator $g$ whose restriction to each cyclic Sylow subgroup is a $d_2$th power, and to each quaternionic Sylow subgroup a $d_2$th power times a power of $\gamma_0$, $X$ is homotopy equivalent to a manifold.*

By Theorem 1, with the induction theorem for $K_0(\boldsymbol{Z}\pi)$, it is enough to consider hyperelementary subgroups. Condition (i) is satisfied by Proposition 5—note that the hypothesis on $g$ is preserved on restriction to subgroups $\rho$, and that the cohomology period of $\rho$ divides that of $\pi$. Condition (ii) is satisfied by Proposition 5 and Theorem 2.

Observe that we can choose $g \in H^{2d_2}(\pi\colon \boldsymbol{Z})$. Thus $\pi$ can act freely—and, in fact, smoothly—on $S^{2d_2-1}$, and on any $S^{2rd_2-1}$ (this holds even if $d_2 = 1$ or 2, by checking Milnor's 1957 list of cases). These are the only spheres on which $\pi$ can act freely, except perhaps in case $d_2 = 2d_1$, and $\pi$ has subgroups of type IIM but none of type IIL. This happens only rarely—for example, if $\pi$ has type I, III or V, then $d_2 = d_1$.

Though the result on dimensions is almost the best possible, that on homotopy types is definitely not.

EXAMPLE 1. $\pi$ $p$-hyperelementary of order $pq$ ($p \neq q$ odd primes). Then $X(g)$ has zero finiteness obstruction—and hence is homotopy equivalent to a manifold—if and only if the restriction of $g$ to the Sylow $q$-subgroup is a $p$th power. This is an easy deduction from the calculation by Galovitch, Reiner and Ullom **[1972]** of $K_0(\boldsymbol{Z}\pi)$.

EXAMPLE 2. $\pi$ the binary tetrahedral group $\mathrm{SL}_2(3)$. In this case, hyperelementary subgroups are cyclic or equal to $\pi_2$ (quaternion of order 8, and normal). Thus $\theta(g) = 0 \Leftrightarrow g|\pi_2$ is a square times a power of $\gamma_0$. For such $g$, the corresponding $X$ is homotopy equivalent to a manifold. The case $g = 7g_0$ does not come from an orthogonal action, or from Theorem 3.

**The projective class group.** PROOF OF THEOREM 2. The localisation square

$$\begin{array}{ccc} A = \boldsymbol{Z}\pi & \to & B = \boldsymbol{Z}\left[\dfrac{1}{p}\right]\pi \\ \downarrow & & \downarrow \\ C = \boldsymbol{Z}_{(p)}\pi & \to & D = \boldsymbol{Q}\pi \end{array}$$

satisfies the conditions (Bass, **[1968]**) for existence of a natural exact sequence

$$K_1B \oplus K_1C \to K_1D \xrightarrow{\partial} K_0A \xrightarrow{i_*} K_0B \oplus K_0C.$$

Suppose we have a finite chain complex $P_*$ of finitely generated projective $A$-modules such that $\chi(P) = \sum(-1)^i[P_i] \in K_0(A)$ is in $\operatorname{Ker} i_* = \operatorname{Im} \partial$. Then $P_* \otimes B$, $P_* \otimes C$ are chain homotopy equivalent to free, based complexes $F_B$, $F_C$ over $B$, $C$ respectively. We now have a preferred homotopy class of chain homotopy equivalences $\eta$: $F_B \otimes D \to F_C \otimes D$ of based free $D$-complexes, defining a torsion element $\tau(\eta) \in K_1(D)$. Then $\chi(P) = \partial\tau(\eta)$. The proof is straightforward, reducing to the definition of the homomorphism $\partial$.

For our problem, the algebra and topology are related as follows (cf. Swan **[1960]**). An exact sequence

$$0 \to \boldsymbol{Z} \to P_{n-1} \to \cdots \to P_0 \to \boldsymbol{Z} \to 0$$

with $P_i$ a finitely generated projective $\boldsymbol{Z}\pi$-module determines a class $g$ in $\operatorname{Ext}^n_{\boldsymbol{Z}\pi}(\boldsymbol{Z}; \boldsymbol{Z}) \cong H^n(\pi; \boldsymbol{Z})$ which is a generator. Conversely, $g$ determines the chain-homotopy type of $P_*$, and $\theta(g) = \chi(P_*)$. Corresponding statements hold on localising $\boldsymbol{Z}$. Though it is difficult to write down $P_*$ explicitly, such complexes can be produced after suitable localisations. We proceed indirectly as follows.

Write $\pi = \rho\sigma$ as usual, with $\alpha\colon \sigma \to \operatorname{Aut} \rho$. Choose an epimorphism $\sigma' \to \sigma$, with $\sigma'$ cyclic of order $p^{s+1}$ (so the kernel $\zeta$ has order $p$), and let the composition $\alpha'$: $\sigma' \to \sigma \to \operatorname{Aut} \rho$ define the split extension $\pi' = \rho\sigma'$. We have an epimorphism $\pi' \to \pi$ with kernel $\zeta$. Since $\alpha'$ has nontrivial kernel, $\pi'$ has $\mathscr{F}$-representations; choose one whose $k$-invariant $g'$ is the image of $g$ at primes other than $p$ (such exist by the conditions on $g$). This yields a *free* $\boldsymbol{Z}\pi'$-chain complex $F'_*$ defining $g'$.

Now $F'_* \otimes_{\boldsymbol{Z}\pi'} \boldsymbol{Z}\pi$ does not give a multiple extension of $\boldsymbol{Z}$ by $\boldsymbol{Z}$, but this fails only at the prime $p$. We can thus take $F_B = F'_* \otimes B$. As to $C$, this contains the idempotent $\varepsilon_\rho = \sum\{g \in \rho\}/|\rho|$, which is central. Hence $C$ splits as $C\varepsilon_\rho \oplus C(1 - \varepsilon_\rho)$, and $C\varepsilon_\rho \cong \boldsymbol{Z}_p\sigma$. Since $\sigma$ is cyclic, with generator $T$, say, we now have the resolution

$$\begin{array}{lcccccccccccc} & \cdots & C\varepsilon_\rho & \xrightarrow{\Sigma T^i} & C\varepsilon_\rho & \xrightarrow{1-T} & C\varepsilon_\rho & \xrightarrow{\Sigma T_i} & C\varepsilon_\rho & \xrightarrow{1-T} & C\varepsilon_\rho & \to \boldsymbol{Z}1/p \to 0 \\ F_C: & & \oplus & & \oplus & & \oplus & & \oplus & & \oplus & \oplus \\ & \cdots & C(1-\varepsilon_\rho) & \xrightarrow{0} & C(1-\varepsilon_\rho) & = & C(1-\varepsilon_\rho) & \xrightarrow{0} & C(1-\varepsilon_\rho) & = & C(1-\varepsilon_\rho) & \longrightarrow 0 \end{array}$$

To compute the torsion, define $F'_C$ for $\pi'$ similarly to $F_C$ for $\pi$, using a generator $T'$ for $\sigma'$ which maps onto $T$. Observe that the corresponding cohomology classes evidently restrict to $p^s$th powers at $\sigma, \sigma'$. We can choose $T'$ so that this restriction agrees with that of $g'$. Then there is a chain homotopy equivalence $F'_* \otimes C' \to F'_C$ whose torsion in $K_1(C')$ clearly maps to 0 in $K_0(A')$. We can thus factor our equivalence $\theta$ as

$$F_B \otimes D = F'_* \otimes D \xrightarrow{\theta_1} F'_C \otimes D \xrightarrow{\theta_2} F_C \otimes D.$$

Now $\partial\tau(\theta_1) = 0$ by an obvious commutative diagram. As to $\theta_2$, observe that $F'_C \otimes C$ differs from $F_C$ only in that each map $\sum T^i$ is replaced by $p\sum T^i$ (since the order of $T'$ is $p$ times that of $T$). Thus $\tau(\theta_2) \in K_1(D)$ is equal to 1 at summands of $D$ corresponding to nontrivial representations of $\pi$, and to $p^{p^s}$ at the trivial component.

Our obstruction is thus $\partial(p^{p^s} \oplus 1)$. Since $\pi$ has Artin exponent $p^s$ (see Bass [**1968**]), this vanishes provided $\partial(p \oplus 1)$ is trivial on cyclic subgroups $\tau$ of $\pi$. But if $\boldsymbol{Z}\tau \subset \boldsymbol{Q}\tau = \boldsymbol{Q} \oplus (\boldsymbol{Q}\tau)'$ has projections $\boldsymbol{Z}$, $(\boldsymbol{Z}\tau)'$, $\partial(p \oplus 1)$ is in $\operatorname{Ker}(K_0(\boldsymbol{Z}\tau) \to K_0((\boldsymbol{Z}\tau)'))$, which is known to vanish for $\tau$ cyclic.

To conclude the proof, it remains to note that any triple $(F_B, F_C, \theta)$ *does* determine a chain-homotopy type of $P_A$ as above—this is a standard argument—so we obtain a complex $P_*$ with $0 = \chi(P_*)$ and hence (after easy modifications) a free complex. The corresponding $g''$ is (by the above) equal to $g$ at Sylow $q$-subgroups with $q \neq p$, and a $p^s$th power on $\sigma$: now by changing $g'$ we see that we can make this any $p^s$th power, and so achieve $g'' = g$. This concludes our outline proof.

The proof of Theorem 3 shows that—apart from troubles with 2-hyperelementary groups of types IIL, IIM—we have succeeded in splitting $\mathscr{G} = \mathscr{G}_0 \oplus \boldsymbol{Z}$ so that $\theta(\boldsymbol{Z}) = 0$. Further progress therefore depends on a study of the ideals $\langle r, \sum\{g \in \pi\} \rangle$ in $\boldsymbol{Z}\pi$ for $\pi$ $p$-hyperelementary. This depends only on $r$ modulo $|\pi|$, and vanishes (as we have seen above) for $p^s$th powers, where $p^s$ is the period. In the opposite direction, our best result so far is

PROPOSITION 6. *If $\theta(g) = 0$, then for each odd prime $p$, the restriction of $g$ to the Sylow $p$-subgroup is a $d_p$th power, where $2d_p$ is the $p$-period of $\pi$.*

But even this (with the analogous result for $p = 2$) is not always sufficient. I hope to give a fuller discussion of this problem at a future date.

## REFERENCES

H. Bass, (1968), *Algebraic K-theory,* Benjamin, New York. MR **40** #2736.

L. E. J. Brouwer, (1912), *Über Abbildung von Mannigfaltigkeiten*, Math. Ann. **71**, 97–115.

W. Burnside, (1905), *On a general property of finite irreducible groups of linear substitutions*, Messenger of Math. **35**, 51–55.

H. Cartan and S. Eilenberg, (1956), *Homological algebra*, Princeton Univ. Press, Princeton, N. J. MR **17**, 1040.

A. W. M. Dress, (1975), *Induction and structure theorems for orthogonal representations of finite groups*, Ann. of Math. (2) **102**, 291–325. MR **52** #8235.

S. Galovich, I. Reiner and S. Ullom, (1972), *Class groups for integral representations of metacyclic groups*, Mathematika **19**, 105–111. MR **48** #4087.

H. Hopf, (1926), *Zum Clifford-Kleinschen Raumproblem*, Math. Ann. **95**, 313–339.

W. Killing, (1891), *Über die Clifford-Klein'schen Raumformen*, Math. Ann. **39**, 257–278.

R. Lee, (1973), *Semicharacteristic classes*, Topology **12**, 183–199. MR **50** #14809.

I. Madsen, C. B. Thomas, and C. T. C. Wall, [II], (1976) *The topological spherical space form problem*. II: *Existence of free actions*, Topology **15**, 375–382.

J. W. Milnor, (1957), *Groups which act on $S^n$ without fixed points*, Amer. J. Math. **79**, 623–630. MR **19**, 761.

G. de Rham, (1931), *Sur l'analysis situs des variétés a n dimensions*, J. Math. Pures Appl. **10**, 115–200.

P. A. Smith, (1944), *Permutable periodic transformations*, Proc. Nat. Acad. Sci. U.S.A. **30**, 105–108. MR **5**, 274.

M. Suzuki, (1955), *On finite groups with cyclic Sylow subgroups for all odd primes*, Amer. J. Math. **77**, 657–691. MR **17**, 580.

R. G. Swan, (1960), *Induced representations and projective modules*, Ann. of Math. (2) **71**, 552–578. MR **25** #2131.

——, (1960), *Periodic resolutions for finite groups*, Ann. of Math. (2) **72**, 267–291. MR **23** #A2205.

H. Tietze, (1908), *Über die topologischen invarianten mehrdimensionaler Mannigfaltigkeiten*, Monatshefte für Math. und Phys. (Wien) **19**, 1–118.

C. B. Thomas and C. T. C. Wall, (1971), *The topological spherical space form problem*. I, Compositio Math. **23**, 101–114.

G. Vincent, (1947), *Les groupes linéaires finis sans points fixes*, Comment. Math. Helv. **20**, 117–171. MR **9**, 131.

C. T. C. Wall, (1976), *Formulae for surgery obstructions*, Topology **15**, 189–210.

J. A. Wolf, (1967), *Spaces of constant curvature*, McGraw-Hill, New York. MR **36**, #829.

H. Zassenhaus, (1935), *Uber endliche Fastkörper*, Abh. Math. Sem. Hamburg **11**, 187–220.

UNIVERSITY OF LIVERPOOL

Proceedings of Symposia in Pure Mathematics
Volume 32, 1978

# FREE ACTIONS BY FINITE GROUPS ON $S^3$

C. B. THOMAS

In 1957 J. Milnor published a list of those finite groups which satisfy the known algebraic conditions for a free action on the standard sphere $S^3$ [7]. The group $\Gamma$ either admits a fixed-point-free representation of real degree 4 (for example, $\Gamma$ is a finite subgroup of the unit quaternions), or $\Gamma$ has cohomological period 4 and admits a fixed-point-free representation of real degree 8. R. Lee [5] subsequently eliminated all groups of the second type from consideration, except for 2-hyperelementary groups of the form $\boldsymbol{Z}/m \times Q(8n; k, l)$, where $k$, $l$, $m$, $n$ are coprime *odd* integers, and the second group has presentation $\{x, y, z | z^{kl} = y^{4n} = 1, y^{2n} = x^2, xyx^{-1} = y^{-1}, yzy^{-1} = z^{-1}, xzx^{-1} = x^r, r \equiv 1\ (k), r \equiv -1\ (l)\}$.

In this paper we examine the implications for this problem of the following conjecture (see [3] for the background).

SMALE CONJECTURE. *The inclusion homomorphism* $i$: SO(4) $\to$ Diff$^+S^3$ *is a homotopy equivalence.*

Diff$^+S^3$ denotes the topological group of orientation preserving $C^\infty$-diffeomorphisms of $S^3$. Subject to this assumption we shall prove the following classification and existence theorems:

THEOREM A. *If the Smale Conjecture is true, and $\Gamma$ is isomorphic to a finite subgroup of* SU(2), *the orbit space of any free action by $\Gamma$ on $S^3$ is homotopy equivalent to a manifold of constant positive curvature.*

Recall that a 3-dimensional manifold of constant positive curvature is the quotient of $S^3$ by a fixed free representation of $\Gamma$ of real degree 4; we shall denote such a space by $S^3/\Gamma$, lin.

THEOREM B. *If the Smale Conjecture is true, there is no free action by the group $Q(8n, k, l)$ on $S^3$.*

---

*AMS (MOS) subject classifications* (1970). Primary 57E25; Secondary 57A10, 55B15.

The proof of both theorems depends on recent work by J. F. Adams on maps between classifying spaces of compact (not necessarily connected) Lie groups. We first note that there is no loss of generality in considering differentiable actions, since the orbit space of a free topological action is first triangulable, and can then be given a smooth structure, which can be lifted to smooth the action of $\Gamma$. With such a smooth action we associate a pair of SU(2)-bundles, which are stably flat. It follows that the first $k$-invariant of the orbit space $S^3/\Gamma$ coincides with the Euler class of a fixed-point-free representation, showing that the free action, if it exists at all, must be homotopically equivalent to a linear action.

There is an alternative approach to homotopy classification, which is independent of the Smale Conjecture. For a particular group $\Gamma$, there is a $(\Gamma, 3)$-polarised complex (see the paper of C. T. C. Wall in these PROCEEDINGS), corresponding to each generator $r$ or $\mathbf{Z}/|\Gamma|$, and for which the finiteness obstruction is the projective ideal $(r, \sum g \in \Gamma)$ generated by $r$ and the sum of the group elements. There exist groups $\Gamma$, for example the binary dihedral group of order 8, for which $(r, \sum)$ is stably free, precisely when the generator $r$ corresponds to a linear action. Indeed we appeal to this result (more for convenience than necessity) in the proof of Theorem A below. However there are exceptions, among them the binary tetrahedral group $T_1^*$ of order 24. In this case there is certainly a finite Poincaré complex, which is homotopically distinct from the unique 3-manifold of constant positive curvature $S^3/T_1^*$, lin; see **[12]** for details of the construction. If the conclusion of Theorem A for this group is independent of the Smale Conjecture, some other geometric hypothesis seems to be necessary.

It is possible to treat those groups $\Gamma$, which can act freely and linearly on $S^3$, and yet which are not isomorphic to subgroups of SU(2) by the methods of **[6]** and the present paper. Neglecting direct products with cyclic groups of coprime order, $\Gamma$ is either an extension of a cyclic group of odd order by a cyclic group of order a power of 2, or a generalised binary tetrahedral group of order $8.3^r$ $(r \geqslant 2)$. In the latter case the finiteness obstruction vanishes only for the homotopy types defined by linear actions; in the former we argue as for the binary dihedral group of Theorem A. Details will appear elsewhere.

The symbol $\mathbf{Z}/r$ denotes the cyclic group of order $r$, $\mathbf{Z}_p^\wedge$ the $p$-adic integers, and $\mathbf{Q}_p^\wedge$ the quotient field $\mathbf{Z}_p^\wedge(1/p)$.

**1. Reduction to unstable bundle theory.** The basic reference for this section is **[12]**.

Let $\Gamma$ act freely and smoothly on $S^3$, and denote the corresponding map of classifying spaces by $Bg: B\Gamma \to B\,\mathrm{Diff}^+S^3$. Assume that there exists a map $f$ making the diagram below commute up to homotopy—such a factorisation of $Bg$ will certainly exist, if the Smale Conjecture is true. For particular groups $\Gamma$, the obstructions to factorisation may well vanish under much weaker hypotheses.

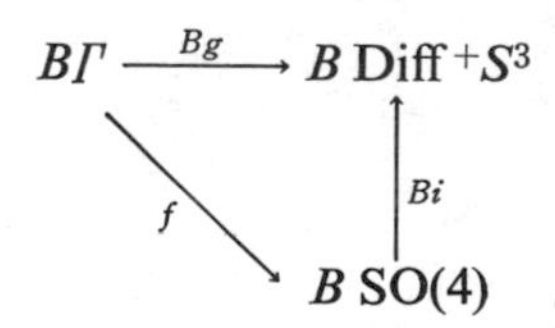


LEMMA 1. *The Euler class $e(f)$ equals the first k-invariant $k_1(S^3/\Gamma)$. In particular $e(f)$ is a generator of $H^4(\Gamma, \mathbf{Z})$.*

PROOF. The $k$-invariant $k_1(S^3/\Gamma)$ is the obstruction to constructing a cross-section to a certain fibration over $B\Gamma$ with fibre $K(Z, 3)$. This is unchanged by restriction to the included fibration with fibre $S^3$, which a short diagram chase shows to be $e(f)$.

LEMMA 2. *The Stiefel-Whitney class* $w_2(f) = 0$, *so that* $f$ *factors through* $f: B\Gamma \to B\,\mathrm{Spin}(4)$.

PROOF. Since $w_2(f)$ is a mod 2 class, $w_2(f) = w_2(f_2)$, where $f_2: B\Gamma_2 \to B\,\mathrm{SO}(4)$ is the lifted map for some 2-Sylow subgroup. Since $\Gamma$ has periodic cohomology, $\Gamma_2$ is either cyclic or binary dihedral, and in either case $S^3/\Gamma_2$ is homotopy equivalent to a linear quotient, see **[6]**, for example. By inspection, $w_2(f_{2,\,\mathrm{lin}}) = 0$, from which it follows by the homotopy invariance of $w_2$ that $w_2(f) = 0$, and that $f$ factors through $B\,\mathrm{Spin}(4)$.

In the proof of the main theorems we shall use the isomorphism $\mathrm{Spin}(4) \cong \mathrm{SU}(2) \times \mathrm{SU}(2)$ to reduce the study of the Spin(4)-bundle over the Eilenberg-Mac Lane space $B\Gamma$ to that of a pair of SU(2)-bundles, classified by maps $f^{(i)} = \pi_i f: B\Gamma \to B\mathrm{SU}(2)$. This is one of the points at which the special 3-dimensional nature of the argument becomes apparent.

**2. Maps between classifying spaces.** The basic reference for this section is **[2]**. For convenience we first give an outline proof of the first result in Adams' theory.

THEOREM 3. *Given a map* $f: BZ/p^t \to BU(n)$, *where* $p$ *is prime, there is a homomorphism* $\phi: Z/p^t \to U(n)$, *such that* $B\phi$ *is stably homotopic to* $f$.

PROOF. Recall from **[1]** that, for an arbitrary compact Lie group, $K(B\Gamma)$ is isomorphic to the $I$-adic completion of the representation ring $R(\Gamma)$, where $I$ is the augmentation ideal. When $\Gamma$ equals $Z/p^t$, it follows that stably $f$ is represented by the sum $a_0 1 + a_1\xi_1 + a_2\xi_2 + \cdots + a_r\xi_r$, where $\xi_1, \cdots, \xi_r$ $(r = p^t - 1)$ are the inequivalent irreducible representations of $Z/p^t$, and $a_i \in Z_p^\wedge$, the ring of $p$-adic integers. Now let $\zeta$ be a primitive $p^t$th root of unity, and map $K(BZ/p^t)$ to the extension field $Q_p^\wedge(\zeta)$ by mapping $\xi_1$ (say) to $\zeta$. If we apply the total Grothendieck-Atiyah characteristic class $\gamma_t$ to the class of $f$, the coefficient of $t^m$ vanishes in dimensions greater than $n$. Passing to $Q_p^\wedge(\zeta)[[t]]$ we have the equation

$$(1 + (\zeta - 1)t)^{a_1}(1 + (\zeta^2 - 1)t)^{a_2} \cdots (1 + (\zeta^2 - 1)t)^{a_r} = 1 + c_1 t + \cdots + c_n t^n.$$

An exercise on power series now shows that, since the right-hand side is polynomial, the coefficients $a_1, a_2, \cdots, a_r$ are ordinary, positive integers. Since $a_0 = n - (a_1 + \cdots + a_r)$, $a_0$ belongs to $Z$, also, and the stable class of $f$ is defined by a homomorphism.

COROLLARY. *If* $p$ *is prime, an* SU(2)*-bundle over* $BZ/p^t$ *is stably equivalent to the flat bundle* $\phi_{p,\,k(p)}$ *associated to the representation matrix*

$$\begin{pmatrix} \zeta^k & 0 \\ 0 & \zeta^{-k} \end{pmatrix} \quad \textit{for some } k = 1, \cdots, r.$$

We next consider the binary dihedral group $D_{4t}^*$ with presentation $\{x, y: y^{2t} = 1, x^2 = y^t, xyx^{-1} = y^{-1}\}^3$.

THEOREM 4. *Given a map $f\colon BD^*_{4t} \to B\,\mathrm{SU}(2)$, where $t$ is odd, there is a homomorphism $\phi\colon D^*_{4t} \to \mathrm{SU}(2)$, such that $B\phi$ is stably homotopic to $f$.*

PROOF. Let $D^*_{4t,p}$ be a Sylow subgroup for each prime $p$ dividing $t$; since $t$ is odd, these are all cyclic. By the corollary to Theorem 3, each lifted map $f_p$ is stably homotopic to $B\phi_{p,k(p)}$. Recall that an irreducible 2-dimensional representation $\phi$ of $D^*_{4t}$ is obtained by transferring a linear representation of the cyclic subgroup generated by $y$ up to the whole group. Inspection of the representing matrices shows that we may choose $\phi$ to restrict to the family $\{\phi_{p,k(p)} | p \text{ dividing } 4t\}$. Since the corresponding result is true for ordinary cohomology, and the Atiyah-Hirzebruch spectral sequence collapses, the restriction map $K(BD^*_{4t}) \to \coprod_p K(BD^*_{4t,p})$ is a monomorphism, and so $f$ is stably homotopic to $B\phi$.

**3. Proof of Theorems A and B.** The proof of Theorem A is by cases:

(i) $\Gamma$ is cyclic. This is classical, see for example [**4**, Chapter IV]; and $S^3/\Gamma$ is homotopy equivalent to a lens space $L^3(r; q, 1)$, since these exhaust the possible homotopy types.

(ii) $\Gamma$ equals $D^*_{4t}$ with $t$ even. This case is discussed in [**6**], and can easily be reduced to $D^*_8$. At the Poincaré complex level, and neglecting orientation, there are two possible homotopy types, one defined by the standard inclusion of $D^*_8$ in SU(2) as $\{\pm 1, \pm i, \pm j, \pm k\}$, and the other by an infinite complex, whose finiteness obstruction generates $\tilde{K}_0(\boldsymbol{Z}D^*_8)$. These first two cases are independent of the Smale Conjecture.

(iii) $\Gamma$ is a binary polyhedral group $\Pi^*_n$ ($n = 24, 48, 120$). This case is solved in [**12**] by an application of ordinary Chern classes to the pair of SU(2)-bundles constructed in §1.

(iv) $\Gamma$ equals $D^*_{4t}$ with $t$ odd. The argument in §1 shows that, if $D^*_{4t}$ acts freely on $S^3$, then the action determines a Spin(4)-bundle, whose Euler class equals $k_1(S^3/D^*_{4t})$. Each of the equivalent pairs of SU(2)-bundles is stably flat by Theorem 4, and it follows that $k_1(S^3/D^*_{4t}) = e(\sigma)$, where $\sigma$ is a fixed-point-free representation of $D^*_{4t}$ in Spin(4). Since the quotient manifolds $S^3/D^*_{4t}$ and $S^3/\sigma$ have the same first $k$-invariant, they are homotopy equivalent, see [**11**, Theorem 1.8] for example, and the theorem is true in this case.

PROOF OF THEOREM B. As in the case of Theorem A this depends on a variant of a proposition in [**2**]. In order to slightly simplify the argument, we shall consider a subgroup of $Q(8n; k, l)$ of the form $Q(8n; k, 1)$ with $n$ and $k$ distinct odd primes. It will clearly be enough to show that the latter group cannot act freely on the standard 3-sphere. We write $Q$ as a semidirect product

$$1 \to Z \to Q(8n; k, 1) \underset{\phi}{\to} D^*_{8n} \to 1,$$

where $Z$ is isomorphic to the cyclic group $\boldsymbol{Z}/k$.

LEMMA 5. *Let $\chi_k$ and $\chi_n$ be characters of $Z$ and $D^*_{8n}$ respectively, such that*
(i) $\chi_k(1) = \chi_n(1) = r$, *and*
(ii) *if $z_1$ and $z_2$ are elements of $Z$, which are conjugate in $Q$, then $\chi_k(z_1) = \chi_k(z_2)$.*
*Then there exists a character $\chi$ of $Q$, which restricts to $\chi_k$ on $Z$ and to $\chi_n$ on $D^*_{8n}$.*

PROOF. Since $Q$ is a semidirect product, each element can be written $(z, d)$, with the law of composition twisted by the action of $D^*_{8n}$ on $Z$. With respect to the induced action on the characters of $Z$, the isotropy subgroup of $\chi_k$ is the whole group $D_{8n}^*$, from which it follows that $\chi_k$ extends to a character $\bar{\chi}_k$ of all of $Q$; compare [10, 9.2].

Thus $\bar{\chi}_k(z, d) = \chi_k(z)$.

Consider the character $\chi = \bar{\chi}_k + \phi^*\chi_n - r$, and check that the restriction condition is satisfied.

Turning now to the proof of the theorem, consider a map of classifying spaces $f: BQ \to B\,\mathrm{SU}(2)$. By Theorems 3 and 4 the lifted maps $f_k$ and $f_n$ are stably defined by homomorphisms with characters $\chi_k$ and $\chi_n$. Condition (i) is obvious ($r = 2$), and $\chi_k$ satisfies condition (ii), since at the *bundle* level it is the lift of a map defined on all of $BQ$, and hence is invariant under $Q$ inner automorphisms. Recall that a character is called positive if it is the character of a homomorphism. Then the character $\chi$ constructed in Lemma 4 is *stably positive* (except perhaps for a 2-Sylow subgroup), since $\chi_k$ is positive (Theorem 3) and $\chi_n$ stably positive (Theorem 5). We thus have the equation $\chi + d = \bar{\chi}_k + \phi^*\chi_n$.

It follows that for odd primes, the Spin(4)-bundle constructed in the first section is stably determined by a pair of homomorphisms into SU(2). The argument in [10, loc. cit.] applied to the group $Q(8n; k, 1)$ shows such a homomorphism cannot be fixed-point-free at both the odd primes $n$ and $k$. Indeed the two-dimensional representations are either constructed from a 1-dimensional representation of $Z$, or are of the form $\phi^*\delta$, for some $\delta: D_{8n}^* \to \mathrm{SU}(2)$. It follows that either the reduction (mod $k$) or the reduction (mod $n$) of the Euler class of the Spin(4)-bundle is trivial, contradicting the condition that $k_1(S^3/Q)$ generate $H^4(Q, Z) \cong Z/8nk$.

**4. Dependence on the Smale Conjecture.** The proofs given in the previous section depend on being able to factorise the classifying map $Bg: B\Gamma \to B\,\mathrm{Diff}^+S^3$ through $B\,\mathrm{SO}(4)$. Even if the Smale Conjecture is false, the homotopy structure of the homogeneous space $F = \mathrm{Diff}^+S^3/\mathrm{SO}(4)$ may be such as to allow factorisation at least for certain groups $\Gamma$. At present all that is known is that $F$ is homotopy equivalent to a countable CW complex [8] and that $\pi_0F$ is trivial [3]. To illustrate the problem, let us consider the groups $D_{4p}^*$, where $p$ is an odd prime.

THEOREM 6. *The classifying map* $Bg: B\,D_{4p}^* \to B\,\mathrm{Diff}^+S^3$ *factorises through* $B\,\mathrm{SO}(4)$ *for any prime p, such that, for all values of i,*

(i) $\pi_iF$ *has no p-torsion, and*

(ii) $\pi_{4i-1}F$ *is p-divisible.*

Recall that the abelian group $A$ is $p$-divisible, if the equation $px = a$ can be solved for all $a$ in $A$.

PROOF. The 2-Sylow subgroup of $D_{4p}^*$ generated by $x$ is cyclic of order 4, and the lifted classifying map $(Bg)_2$: $BZ/4 \to B\,\mathrm{Diff}^+S^3$ factors through $B\,\mathrm{SO}(4)$ by [9]. It follows that, if we write $B\psi$: $BZ/4 \to BD_{4p}^*$ for the finite covering induced by the inclusion, $Bg$ itself will factorise, provided the relative obstructions in $H^{i+1}(\psi^*, \pi_iF)$ vanish. By the long exact sequence in cohomology, the relative groups vanish, provided

(a) $H^{i+1}(D_{4p}^*, \pi_iF) \to H^{i+1}(Z/4, \pi_iF)$ is a monomorphism, and

(b) $H^i(D_{4p}^*, \pi_iF) \to^{\psi^*} H^i(Z/4, \pi_iF)$ is an epimorphism.

Since the cohomological period of $D_{4p}^*$ is four, it is enough to pass to Tate cohomology, and to consider the four cases

$\hat{H}^{-1}$: $\operatorname{Ker}(A \xrightarrow{4p} A) \to \operatorname{Ker}(A \xrightarrow{4} A)$,
$\hat{H}^{0}$: $\operatorname{Coker}(A \xrightarrow{4p} A) \to \operatorname{Coker}(A \xrightarrow{4} A)$,
$\hat{H}^{1}$: $\operatorname{Hom}(D^*_{4p}, A) \to \operatorname{Hom}(\boldsymbol{Z}/4, A)$, and
$\hat{H}^{2}$: $(D^*_{4p})_{ab} \otimes A \to \boldsymbol{Z}/4 \otimes A$.

The abelianised group $(D^*_{4p})_{ab}$ is cyclic of order 4. Hence the last two maps are isomorphisms, as is the first, if $A$ contains no $p$-torsion. The second map is always an epimorphism, and the $p$-divisibility hypothesis on $\pi_{4i-1}F$ is enough to ensure that $H^{4i}(D^*_{4p}, \pi_{4i-1}F)$ restricts monomorphically to $H^{4i}(\boldsymbol{Z}/4, \pi_{4i-1}F)$. Conditions (a) and (b) are thus satisfied, and the result follows.

Loosely speaking, the proof of Theorem A, when the 2-Sylow subgroup $\Gamma_2$ is isomorphic to $\boldsymbol{Z}/4$, does not depend on the 2-torsion in $\pi_*F$. The same kind of argument would apply to Theorem B, provided any free action of the group $D^*_8$ on $S^3$ were conjugate to a linear action. Indeed I feel that any attempt to bypass the Smale Conjecture depends on first solving this geometric problem for the first binary dihedral group.

## REFERENCES


**1.** M. F. Atiyah and G. B. Segal, *Equivariant K-theory and completion*, J. Differential Geometry **3** (1969), 1–18. MR **41** #4575.

**2.** J. F. Adams, *Maps between classifying spaces*, Bull. Amer. Math. Soc. (to appear).

**3.** J. Cerf, *Sur les difféomorphismes de la sphère de dimension trois* ($\Gamma_4 = 0$), Lecture Notes in Math., vol. 53, Springer-Verlag, Berlin and New York, 1968. MR **37** #4824.

**4.** S. Eilenberg and S. Mac Lane, *Homology of spaces with operators*. II, Trans. Amer. Math. Soc. **65** (1949), 49–99. MR **11**, 379.

**5.** R. Lee, *Semicharacteristic classes*, Topology **12** (1973), 183–199. MR **50** #14809.

**6.** R. Lee and C. B. Thomas, *Free actions by finite groups on* $S^3$, Bull. Amer. Math. Soc. **79** (1973), 211–215. MR **47** #4265.

**7.** J. W. Milnor, *Groups which act on* $S^n$ *without fixed points*, Amer. J. Math. **79** (1957), 623–630. MR **19**, 761.

**8.** R. Palais, *Homotopy theory of infinite dimensional manifolds*, Topology **5** (1966), 1–16. MR **32** #6455.

**9.** P. M. Rice, *Free actions of* $Z_4$ *on* $S^3$, Duke Math. J. **36** (1969), 749–751. MR **40** #2064.

**10.** J. P. Serre, *Representations linéaires des groupes finis*, Hermann, Paris, 1967. MR **38** #1190.

**11.** C. B. Thomas, *The oriented homotopy type of compact 3-manifolds*, Proc. London Math. Soc. (3) **19** (1969), 31–44. MR **40** #2088.

**12.** ———, *Poincaré 3-complexes with binary polyhedral fundamental group*, Math. Ann. (to appear).


UNIVERSITY COLLEGE, LONDON

Proceedings of Symposia in Pure Mathematics
Volume 32, 1978

# THE NONSIMPLY CONNECTED CHARACTERISTIC VARIETY THEOREM

LOWELL JONES

**Introduction.** In this paper the Characteristic Variety Theorem (C.V. Theorem) of D. Sullivan [**14**] is proven for the nonsimply connected case. The proof is a fairly routine extension of Sullivan's proof for the simply connected case, the only major new ingredient being the surgery product formula 7.1 in [**9**].

In homotopy terms, the nonsimply connected C.V. Theorem gives a complete analysis of the homotopy type for the surgery classifying spaces $\boldsymbol{L}_k^h(\pi)$ (see Theorem 1.5) defined by F. Quinn [**12**].

The nonsimply connected C.V. Theorem has been most successfully associated with the problem of putting geometric structures on regular neighborhoods. Here are two examples of such applications.

EXAMPLE 0.1 (FROM [7]). $(M, \partial M) \subset (N, \partial N)$ denotes a PL embedding of PL manifold pairs; $(R, R_\partial)$ is a regular neighborhood for $(M, \partial M)$ in $(N, \partial N)$. Given a prime $p$, does $\exists$ a PL group action $\phi\colon \boldsymbol{Z}_p \times (R, R_\partial) \to (R, R_\partial)$ having $(M, \partial M)$ for fixed point set?

This question has a homotopic theoretic formulation. Let $BS\,\mathrm{PL}(S^k)$ be the classifying space for oriented PL block bundles having $S^k$ for fiber and $BS\,\mathrm{PL}(S^k, \boldsymbol{Z}_p)$ be the classifying space for $BS\,\mathrm{PL}(S^k)$-bundles which are equipped with a PL free action of $\boldsymbol{Z}_p$ which preserves the blocks and the fiber orientations. Then $\exists$ a PL action $\boldsymbol{Z}_p \times (R, R_\partial) \to (R, R_\partial)$ iff there exists a lifting of $h$ to $\hat{h}$ making the diagram

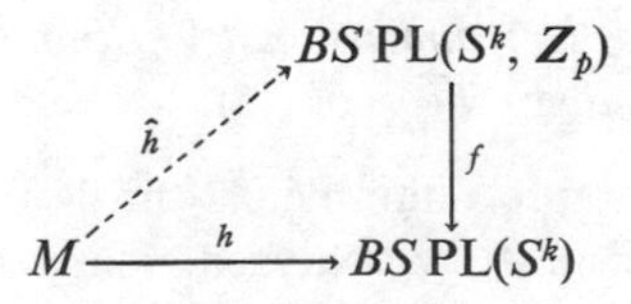


*AMS (MOS) subject classifications* (1970). Primary 57D20.

homotopy commutative. Here $k + 1$ = codimension of $M$ in $N$, $h$ is the classifying map for the regular neighborhood $R$, and $f$ is the forgetful map.

For each connected component $BS\,\mathrm{PL}(S^k, \boldsymbol{Z}_p)_i$ of $BS\,\mathrm{PL}(S^k, \boldsymbol{Z}_p)$ let $G_i$ be the subgroup of $\pi_1(BS\,\mathrm{PL}(S^k, \boldsymbol{Z}_p))$ represented by equivariant bundles $\boldsymbol{Z}_p \times \xi \to \xi$ over $S^1$ such that the twisting map which defines the quotient block bundle $\xi/\boldsymbol{Z}_p$ induces the identity isomorphism on integral homology groups. Let $\rho_i$: $BS\,\mathrm{PL}^\wedge(S^k, \boldsymbol{Z}_p)_i \to BS\,\mathrm{PL}(S^k, \boldsymbol{Z}_p)_i$ be the finite covering map corresponding to the normal subgroup $G_i$. $\rho$: $BS\,\mathrm{PL}^\wedge(S^k, \boldsymbol{Z}_p) \to BS\,\mathrm{PL}(S^k, \boldsymbol{Z}_p)$ is the union of all $\rho_i$.

Using the nonsimply connected C. V. Theorem, plus computations, one can get

PROPOSITION A. *If $k$ is odd and $\geqq 5$, then when localized away from $p$, the mappings $f \circ \rho$: $BS\,\mathrm{PL}^\wedge(S^k, \boldsymbol{Z}_p) \to BS\,\mathrm{PL}(S^k)$ and $p_1$: $BS\,\mathrm{PL}(S^k) \times X \to BS\,\mathrm{PL}(S^k)$ are homotopy equivalent. Here $p_1$ is the first factor projection, and $X$ has the homotopy type of $(\overline{F/\mathrm{TOP}} \otimes \bar{L}^h_{*+k}(\boldsymbol{Z}_p)) \times D$, where $\bar{L}^h_j(\boldsymbol{Z}_p)$ is the kernel of the transfor map $L^h_j(\boldsymbol{Z}_p) \to L^h_j(\{1\})$, $\overline{F/\mathrm{TOP}} \otimes \bar{L}^h_{*+k}(\boldsymbol{Z}_p)$ is defined as in* §1 *below* (*preceding Theorem* 1.5), *and $D$ is a discrete topological space with as many points as there are* PL *$H$-cobordism classes of $k$-dimensional oriented homotopy lens spaces with $\boldsymbol{Z}_p$ for fundamental group.*

EXAMPLE 0.2 (FROM [8]). $(M, \partial M) \subset (N, \partial N)$ and $(R, R_\partial)$ are as before. Assume the codimension of $M$ in $N$ is equal to 2. Is the embedding $(M, \partial M) \subset (N, \partial N)$ PL concordant to a locally flat embedding?

This question has a homotopic theoretic formulation. Let $BSO(2)$ be the classifying space for 2-dimensional oriented vector bundles, and $BS\,\mathrm{PL}^\sim(2)$ be the classifying space for oriented codimension 2 PL manifold thickenings of oriented PL manifolds. $f$: $BSO(2) \to BS\,\mathrm{PL}^\sim(2)$ is the natural inclusion. The embedding $(M, \partial M) \subset (N, \partial N)$ is concordant to a locally flat embedding iff there exists a lifting of $h$ to $\hat{h}$ making the diagram

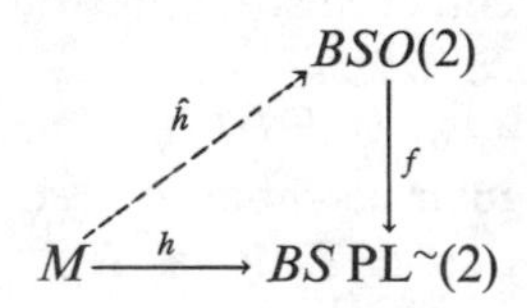


homotopy commutative. Here $h$ is the classifying map for the regular neighborhood $R$ of $M$.

Using the nonsimply connected C.V. Theorem, one can get

PROPOSITION B. *Modulo homotopy difficulties at dimensions* 3 *and* 4, *the map $f$: $BSO(2) \to BS\,\mathrm{PL}^\sim(2)$ is homotopy equivalent to $i_1$: $BSO(2) \to BSO(2) \times Y$. Here $i_1$ identifies $BSO(2)$ with $BSO(2) \times y_0$, $y_0 \in Y$, and $Y$ has the homotopy type of $\overline{F/\mathrm{TOP}} \otimes G_{*-2}$ where $G_j$ is the $j$-dimensional* PL *knot group and $F/\mathrm{TOP} \otimes G_{*-2}$ is defined as in* §1 *below* (*preceding Theorem* 1.5).

To give a complete solution to Example 0.2 the 3- and 4-dimensional difficulties mentioned in Proposition B must be analyzed. This requires the introduction of 4 universal characteristic classes $\theta \in H^2(BS\,\mathrm{PL}^\sim(2), G_3)$, $\beta \in K^0_{F/\mathrm{PL}}(BS\,\mathrm{PL}^\sim(2))$, $\gamma \in K^0_{\overline{F/\mathrm{TOP}}}(BS\,\mathrm{PL}^\sim(2), G_{*+1})$, and $\eta \in H^2(BS\,\mathrm{PL}^\sim(2), Z)$, associated to each $BS\,\mathrm{PL}^\sim(2)$-bundle (see [8]). The classes $\gamma(M, N)$ and $\eta(M, N)$ correspond to Proposition B, and $\theta(M, N)$ and $\beta(M, N)$ pick up the 3- and 4-dimensional complications not handled by Proposition B. A map $h: M \to BS\,\mathrm{PL}^{\sim}(2)$ lifts to $\hat{h}: M \to BSO(2)$ iff $h^*(\theta) = 0, h^*(\beta) = 0, h^*(\gamma) = 0$.

The homotopy type of $BS\,\mathrm{PL}^{\sim}(2)$ is determined by these 4 classes and the universal relationships between them. The class $\eta$ is independent of the classes $\theta, \beta, \gamma$. The relationships among the $\theta, \beta, \gamma$ have been described geometrically in **[8]**, and in the language of Postnikov decompositions in **[2]**.

*Credits.* R. Williamson **[18]** and J. Shaneson **[13]** have given special cases of the surgery product formula, which suffice to prove special cases of the nonsimply connected C.V. Theorem. Browder-Quinn obtained the homotopy type of $\boldsymbol{L}_k^h(\pi)$ localized at 2 (unpublished). Their result was also independently suggested in a remark of **[0**, p. 94], by Burghelea, Lashof and Rothenberg. The first use of the nonsimply C. V. Theorem is due to the author **[6]**, **[7]**. It is the application described in Example 0.1 above. The proof of the nonsimply connected C.V. Theorem contained in §2 below is also taken from **[7]**. Example 0.2 is due to the author **[8]**, with some Postnikov arguments added by Cappell-Shaneson **[2]**. It seems that the results of **[18]** were never correctly proven by Williamson, and that for some time J. Morgan has had a correct proof of the surgery product formulas in **[13]**, **[18]** (IHES preprint 1975).

**1.** $\pi$ denotes any finitely presented group. $\boldsymbol{L}_k^h(\pi)$ are the surgery classifying spaces defined by F. Quinn **[12]**. A good reference for these spaces is pp. 90–94 of **[0]**. Another sketchy reference is the appendix of **[17]**.

$\boldsymbol{L}_k^h(\pi)$ is a huge simplicial complex. A vertex is a framed surgery problem with respect to the fundamental group $\pi$; that is, a framed degree one map $h: (M, \partial M) \to (X, Y)$, where $(M, \partial M)$ is a $k$-dimensional manifold pair and $(X, Y)$ a Poincaré duality pair of formal dimension equal $k$, $h|_{\partial M}: \partial M \to Y$ is a homotopy equivalence, and a homomorphism $\beta: \pi_1(X) \to \pi$ which is the fundamental group set-up. A one-dimensional simplex in $\boldsymbol{L}_k^h(\pi)$ is a surgery cobordism between two vertices, such that the boundary cobordism is a homotopy equivalence and $\beta$: $\pi_1(X) \to \pi$ extends to the range cobordism. The higher dimensional simplices of $\boldsymbol{L}_k^h(\pi)$ carry on this analogy. If $k \geqq 5$, then $\pi_l(\boldsymbol{L}_k^h(\pi)) = L_{k+l}^h(\pi)$, where $L_{k+l}^h(\pi)$ are the surgery groups defined by C. T. C. Wall **[17]**; and $\boldsymbol{L}_k^h(\pi)$, $\boldsymbol{L}_{k+4}^h(\pi)$ are homotopy equivalent. The connected component of $\boldsymbol{L}_k^h(\pi)$, corresponding to the zero in $L_k^h(\pi)$ under the isomorphism $\pi_0(\boldsymbol{L}_k^h(\pi)) \cong L_k^h(\pi)$, is called the *zero component* of $\boldsymbol{L}_k^h(\pi)$ and denoted $\boldsymbol{L}_k^h(\pi)_0$. Every other connected component of $\boldsymbol{L}_k^h(\pi)$ is homotopy equivalent to $\boldsymbol{L}_k^h(\pi)_0$.

$Z|_q$-manifolds $M$, $\delta M$ and $Z|_q$ bordism groups $\Omega_*(X, Z|_q)$ of differentiable oriented $Z|_q$-manifolds are defined in **[11]**. There are also surgery groups $L_k^h(q, \pi)$ "modeled" on the idea of $Z|_q$-manifolds. Each $x \in L_k^h(q, \pi)$ is represented by $(k-1)$-dimensional framed surgery map $f$, together with a surgery null cobordism $F$ of $q$ disjoint copies of $f$. $x = 0 \Leftrightarrow$ there is a completion of surgery for $F$ of the following type: First complete surgery on $f$, then copy this on each of the $q$ copies of $f$ in $\partial F$, finally complete surgery on $F$ away from $\partial F$. There is a short exact sequence

$$(1.1) \qquad 0 \longrightarrow L_k^h(\pi) \otimes Z_q \xrightarrow{i_{q,k}} L_k^h(q, \pi) \xrightarrow{\delta} L_{k-1}^h(\pi)_q \longrightarrow 0.$$

$\delta: L_{k-1}^h(\pi)_q$ is the subgroup of $L_{k-1}^h(\pi)$ of elements with order $q$; if $(f, F)$ represents $x \in L_k^h(q, \pi)$, then $\delta(x)$ is represented by $f$.

$i_{q,k}$: Every framed surgery map of dimension $k$ is a surgery null cobordism of $q$ copies of the empty surgery map. So there is a natural map $h: L_k^h(\pi) \to L_k^h(q, \pi)$. Now note that if $q \cdot y = x$ in $L_k^h(\pi)$, then $h(x) = 0$, so $h$ factors through $L_k^h(\pi) \otimes Z_q$.

Given a map $h: M \to \boldsymbol{L}_k^h(\pi)$ from an $m$-dimensional manifold, or $Z|_q$-manifold, by pulling back the universal surgery problem along $h$, and then amalgamating this pull-back, first over $\delta M$ and then over $M$, we get

$$\sigma(h) \in L_{m+k}^h(\pi) \quad \text{or} \quad L_{m+k}^h(q, \pi). \tag{1.2}$$

Following D. Sullivan **[14]**, **[15]**:

1.3. DEFINITION. A *characteristic variety* for the cell complex $C$ consists of a set of maps $\{f_i: M_i \to C | i \in I\}$ from oriented differentiable manifolds or $Z|_q$-manifolds so that for any mapping $g: C \to \boldsymbol{L}_k^h(\pi)$, with $k \geqq 5$, $g$ is null homotopic iff $\sigma(g \circ f_i) = 0 \; \forall \, i \in I$.

1.4. THEOREM. *If $X$ is a finite cell complex, then there exists a finite characteristic variety, $V$, for $X$. $V$ can be chosen to work simultaneously for all $k \geqq 5$ and all finitely presented groups $\pi$.*

$K_*^{\overline{F/\mathrm{TOP}}}(\ )$ and $K^*_{\overline{F/\mathrm{TOP}}}(\ )$. Let $I: \Omega_*(\mathrm{pt}) \to Z$ be the ring homomorphism gotten by sending each manifold to its index. $Z$ is a $\Omega_*(\mathrm{pt})$-module via $x \cdot n \equiv I(x) \cdot n$. Set

$$K_i^{\overline{F/\mathrm{TOP}}}(X, A) \equiv \sum_{j=-\infty}^{+\infty} \Omega_{i+4j}(X, A) \otimes_{\Omega_*} Z.$$

Then $K_*^{\overline{F/\mathrm{TOP}}}(\ )$ is a well-defined homology functor on the category of finite cell complex pairs and continuous maps. This is proven, localized away from 2, in **[14]**. To check it localized at 2, recall that there is a natural equivalence of homology theories

$$\Omega_n(\ ) \otimes Z_{(2)} \cong \sum_{i+j=n} H_i(\Omega_j \otimes Z_{(2)})$$

(see 15.2, p. 42 in **[4]**). Some elementary computations show that this equivalence can be chosen to preserve the $\Omega_*$-module structure of each functor.

$K^*_{\overline{F/\mathrm{TOP}}}(\ )$ is the functor dual to $K_*^{\overline{F/\mathrm{TOP}}}(\ )$. $\overline{F/\mathrm{TOP}}$ denotes the zeroth element in the loop spectrum for $K^*_{\overline{F/\mathrm{TOP}}}(\ )$. Note that localized at and away from 2 $\overline{F/\mathrm{TOP}}$ is homotopy equivalent to $\prod_{i>0} K(Z, 4i)$, $BO$ respectively.

The notation $\overline{F/\mathrm{TOP}}$ is used because certain results in **[5]**, **[14]** can be reinterpreted as follows: up to homototpy type

$$F/\mathrm{TOP} = \prod_{i>0} K(Z_2; 4i-2) \times \overline{F/\mathrm{TOP}}.$$

$\overline{F/\mathrm{TOP}} \otimes L_{*+k}^h(\pi)$ shall denote the zeroth element in the spectrum for the cohomology theory

$$H^0(\ ) \equiv \sum_{i=0}^{3} K^{k-i}_{\overline{F/\mathrm{TOP}}}(\ , L_i^h(\pi)).$$

$\{C_j \,|\, j \in J\}$ shall denote the collection of all finite subcomplexes of the cell complex $C$. *A premap* from $C$ to a space $X$ is a collection of maps $\{f_j\colon C_j \to X \mid j \in J\}$ satisfying:

$$f_{j_1|C_{j_1} \cap C_{j_2}} \quad \text{and} \quad f_{j_2|C_{j_1} \cap C_{j_2}}$$

are homotopic for all $j_1, j_2 \in J$.

1.5. THEOREM. *For $k \geqq 5$ $\exists$ a premap $pf\colon \boldsymbol{L}_k^h(\pi)_0 \to \overline{F/\mathrm{TOP} \otimes L_{*+k}^h(\pi)}$ from the zero component of $\boldsymbol{L}_k^h(\pi)$, which induces an isomorphism on homotopy groups.*

REMARK 1.6. Theorems 1.4, 1.5 hold much more generally than stated. In fact they hold for any surgery theory I know of. For example:

(1) $L_*^s(\pi)$ = surgery on compact manifolds up to simple homotopy equivalence.

(2) $L_*^h(\pi \to Z_2)$, $L_*^s(\pi \to Z_2)$ = surgery on compact manifolds up to homotopy or simple homotopy equivalence when there are nonorientability complications.

(3) Surgery on noncompact manifolds, developed by Maumary **[10]** and Taylor **[16]**.

(4) Surgery up to homological equivalence, developed by Cappell-Shaneson **[3]**.

(5) Surgery on stratified spaces, developed by Browder-Quinn **[1]**. In this case a vertex of the surgery classifying space would be a surgery problem modeled on a stratified space.

(6) Combinations of (1)—(5).

The proof in all these cases is exactly as for the special case $L_*^h(\pi)$. The hypothesis $k \geqq 5$ in 1.4, 1.5 must be extended so that no surgery problems below dimension five are encountered. For example in (5) above none of the strata lying over a vertex should be allowed below dimension five.

**2.** In this section Theorems 1.4, 1.5 are proven. The proofs assume that each $L_k^h(\pi)$ $(k \geqq 5)$ decomposes into the direct sum of cyclic groups. But these proofs can also be made to work under the assumption that each $L_k^h(\pi)$ is a countable group.

Here is an all purpose "geometric" map induced by a change of coefficients $Z_q \to Z_{q'}$. Let $(V, \partial V)$ be a $Z|_q$-manifold obtained from the manifold pair $(M, \partial M)$ by collapsing to one point each orbit of a group action $\phi\colon \boldsymbol{Z}_q \times \partial M \to \partial M$. For integers $a$, $b$, $d(a, b)$ is their greatest common denominator. $(M', \partial M')$ will denote the union $\bigcup_i (M_i, \partial M_i)$ of $q'/d(q, q')$ disjoint copies of $(M, \partial M)$, and $t_q$, $t_{q'}$ will denote the canonical generators of $\boldsymbol{Z}_q$, $\boldsymbol{Z}_{q'}$. A $\boldsymbol{Z}_{q'}$ action is defined on $\partial M'$ by identifying $\partial_k M_i$ with $\partial_k M_{i+1}$ via the identity map if $i < q'/d(q, q')$ and identifying $\partial_k M_a$ with $(t_q)^a(\partial_k M_1)$ via $(t_q)^a$ for $a = q/d(q, q')$. Collapse the orbits of this action to obtain the $Z|_q$-manifold $V'$. In this way mappings

$$\text{(2.1)} \quad \alpha(q, q')_*\colon L_*^h(q, \pi) \longrightarrow L_*^h(q', \pi), \qquad \beta(q, q')_*\colon \Omega_*(\ , Z|_q) \longrightarrow \Omega_*(\ , Z|_q)$$

are obtained. Note that the second of these homomorphisms is just the usual change of coefficient map for $SO$-bordism theory induced by the homomorphism $\gamma(q, q')\colon Z_q \to Z_{q'}$ which sends $t_q \to t_{q'}^a$, $a \equiv q'/d(q', q)$.

2.2. LEMMA. *For integers $k \geqq 5$, $q \geqq 0$ $\exists$ homomorphisms*

$$r_{q,k} = L_k^h(q, \pi) \longrightarrow L_k^h(\pi) \otimes_Z Z_q$$

*satisfying*:

(a) $r_{q,k}$ *is a left inverse to the* $i_{q,k}$ (*see* 1.1).

(b) *The* $r_{q,k}$ *commute with the periodicity isomorphisms* $L_k^h(q,\pi) \cong L_{k+4}^h(q,\pi)$ *and* $L_k^h(\pi) \otimes_Z Z_q \cong L_{k+4}^h(\pi) \otimes_Z Z_q$.

(c) *All the diagrams*

$$\begin{array}{ccc} \Omega_i\left(\boldsymbol{L}_k^h(\pi), Z|_q\right) & \xrightarrow{r_{k+i,q}\circ\sigma} & L_{k+1}^h(\pi) \otimes_Z Z_q \\ \downarrow{\scriptstyle \beta(q,q')_i} & & \downarrow{\scriptstyle 1\otimes\gamma(q,q')} \\ \Omega_i\left(\boldsymbol{L}_k^h(\pi), Z|_{q'}\right) & \xrightarrow{r_{k+i,q'}\circ\sigma} & L_{k+i}^h(\pi) \otimes_Z Z_{q'} \end{array}$$

*are commutative, where* $\sigma$ *is as in* 1.2.

2.2 will be proven at the end of this section.

PROOF OF THEOREM 1.5. $M(q)$ denotes the space obtained by gluing the disc $D^2$ to the circle $S^1$ along a covering map $\partial D^2 \to S^1$ of degree $q$. For any homology theory $\boldsymbol{H}_*(\ )$, $\boldsymbol{H}_i(X, Z_q) \equiv \boldsymbol{H}_i(X \wedge M(q))$. Clearly $\Omega_*(\ , Z_q) \otimes_{\Omega_*} Z = K_*^{\overline{F/\mathrm{TOP}}}(\ , Z_q)$.

Here is the *universal coefficient information* needed in this proof. There exists a construction $C$ such that for any finite CW complex $X$, $C$ associates a map $C(h, h_q)$: $X \to \overline{F/\mathrm{TOP}_i}$, to any set of homomorphisms $h$: $K_i^{\overline{F/\mathrm{TOP}}}(X) \to Z$, $h_q$: $K_i^{\overline{F/\mathrm{TOP}}}(X, Z_q) \to Z_q$ which commute with the change of coefficient homomorphisms $K_*^{\overline{F/\mathrm{TOP}}}(X) \to K_*^{\overline{F/\mathrm{TOP}}}(X, Z_q)$, $K_*^{\overline{F/\mathrm{TOP}}}(X, Z_q) \to K_*^{\overline{F/\mathrm{TOP}}}(X, Z_{q'})$. Here $\overline{F/\mathrm{TOP}_i}$ is the $i$th element in the loop sepctrum for $K^*_{\overline{F/\mathrm{TOP}}}(\ )$. $C$ satisfies:

2.3.(a) If $g$: $K_i^{\overline{F/\mathrm{TOP}}}(Y) \to Z$, $g_q$: $K_i^{\overline{F/\mathrm{TOP}}}(Y, Z_q) \to Z_q$ is another set of such maps and $f$: $X \to Y$ satisfies $g \circ f_i = h$, $g_q \circ f_i = h_q\ \forall\, i$, then $C(g, g_q) \circ f$ is homotopic to $C(h, h_q)$.

(b) If $x \in K_i^{\overline{F/\mathrm{TOP}}}(X)$ is represented by $\alpha$: $S^{4k+i} \to X$ in $\Omega_{4k+i}(X)$, then $C(h, h_q) \circ \alpha$: $S^{4k+i} \to \overline{F/\mathrm{TOP}_i}$ represents $h(x)$ in $\pi_{4k+i}(\overline{F/\mathrm{TOP}_i}) \cong Z$.

There is also for each integer $n \geqq 2$ a mod $n$ construction $C_n$. $C_n$ associates to each $h_n$ (the other $h_i$, $h$ are not needed) a map $C_n(h_n)$: $X \to \overline{F/\mathrm{TOP}_i} \otimes Z_n$ satisfying 2.3(a) for $q = n$. $C_n$ also satisfies the following mod $n$ version of 2.3(b): If $\bar{x} \in K_i^{\overline{F/\mathrm{TOP}}}(X, Z_n)$ is the image of $x \in K_i^{\overline{F/\mathrm{TOP}}}(X)$ under the change of coefficients $Z \to Z_q$, and $x$ is represented by $\alpha$: $S^{4k+i} \to X$ in $\Omega_{4k+i}(X)$, then $C(h_n) \circ \alpha$: $S^{4k+i} \to \overline{F/\mathrm{TOP}_i} \otimes Z_n$ represents $h_n(\bar{x})$ in $\pi_{4k+i}(\overline{F/\mathrm{TOP}_i} \otimes Z_n) \cong Z_n$.

The existence for $C$, $C_n$ localized at 2 follows from the universal coefficient theorem for ordinary homology. $C$, $C_n$ localized away from 2 are constructed in **[14]** using the universal coefficient theorem for $KO_*(\ ) \otimes Z(\frac{1}{2})$.

Here is how to complete the proof of Proposition B using $C$ and the $C_n$.

Define

$$\mathscr{S}_q^*: K_*^{\overline{F/\mathrm{TOP}}}(X, Z_q) \times [X, \boldsymbol{L}_k^h(\pi)] \longrightarrow L_{*+k}^h(\pi) \otimes Z_q,$$
$$\mathscr{S}^*: K_*^{\overline{F/\mathrm{TOP}}}(X) \times [X, \boldsymbol{L}_k^h(\pi)] \longrightarrow L_{*+k}^h(\pi)$$

as follows. For $x \in K_j^{\overline{F/\mathrm{TOP}}}(X, Z_q)$, $y \in [X, \boldsymbol{L}_k^h(\pi)]$, represent $x$, $y$ by maps $h$: $N \to X \wedge M(q)$, $g$: $X \to \boldsymbol{L}_k^h(\pi)$ where $N$ is a differentiable manifold. Apply the transversality construction discussed on pp. 471–472 of **[11]** to $h$ to get a $Z|_q$ bordism element $f$: $M \to X$. Set $\mathscr{S}_q^j(x, y) \equiv r_{q,4l+j+k} \circ (\sigma(g \circ f))$, where dimension$(M) =$

$4l + j$ (see 2.2, 1.2). If $x \in K_*^{F/TOP}(X)$, then represent $x$ by a map $h\colon N \to X$ from a $(4l + j)$-dimensional manifold and set $\mathscr{S}^j(x, y) \equiv \sigma(g \circ h)$.

Note that $\mathscr{S}_q^*$, $\mathscr{S}^*$ localized away from 2 are independent of the representatives $h_x\colon N \to X \wedge M(q)$ chosen for each $x$ in the above construction: This follows from the surgery product formula 7.1 in **[9]** and from 2.2(b) above.

In more detail here is how the argument goes. We must first show that if $g' = X \to L_k^h(\pi)$, $h'\colon N \to X \wedge M(q)$ are other representatives for $y$ and $x$, and $f'\colon M' \to X$ is constructed from $h'$ as before by transversality, then $\sigma(g' \circ f') = \sigma(g \circ f) +$ (2-primary element) under the periodicity isomorphisms

$$L^h_{4l+j+k}(q, \pi) = L^h_{4l'+j+k}(q, \pi) \qquad \begin{Bmatrix} 4l + j = \dim(M) \\ 4l' + j = \dim(M') \end{Bmatrix}.$$

Then the argument is completed by noting that $\gamma_{q,4l+j+k}$ and $\gamma_{q,4l'+j+k}$ commute with the periodicity isomorphisms (see 2.2(b) above).

To verify that $\sigma(g' \circ f') = \sigma(g \circ f) +$ (2-primary) first note that $h'\colon N' \to X \wedge M(q)$ is cobordant to a disjoint union of maps $\bigcap_{i=1}^r (h_i = X_i \to X \wedge M(q))$, where $X_i = N_i \times A_i$, $q_i\colon N_i \to X \wedge M(q)$, $\rho\colon N_i \times A_i \to N_i$ is the projection, and $h_i = q_i \circ \rho$. The $N_i$, $A_i$ are closed oriented manifolds with $(N_1, q_1)$ equal $(N, h)$, index$(A_1) = +1$, and index$(A_i) = 0$, $i \geqq 2$. Now construct $f_i\colon Y_i \to X$ as follows. Apply the construction on pp. 471–472 in **[11]** to each $q_i\colon N_i \to X \wedge M(q)$ to get $P_i\colon M_i \to X$. Set $Y_i = M_i \times A_i$, and set $f_i$ equal to the composite $M_i \times A_i \xrightarrow{\rho} M_i \xrightarrow{P_i} X$. Note that these same $f_i\colon Y_i \to X$ are also obtained by applying the construction of pp. 471–472 in **[11]** to the $h_i\colon X_i \to X \wedge M(q)$ directly.

Because surgery obstructions remain the same under normal cobordisms, it follows that

(a) $\sigma(g' \circ f') = \sum_{i=1}^r \sigma(g' \circ f_i)$.

I claim that

(b) $\sigma(g' \circ f_i) =$ 2-primary element $i \geqq 2$,

(c) $\sigma(g' \circ f_1) = \sigma(g \circ f) +$ 2-primary element.

Note that by plugging (b), (c) into (a) we get $\sigma(g' \circ f') = \sigma(g \circ f) +$ 2-primary element as desired. So it only remains to verify (b), (c). We shall need a special case of Proposition 7.1 in **[9]** for this. In the notation of that proposition set $\xi^1 =$ point, $\pi^1 = 1$ element group, and substitute for $L_*^h(\xi, \pi)$ the group $L_*^h(q, \pi)$. After these substitutions Proposition 7.1 in **[9]** becomes:

PROPOSITION. *Let $A$ be any closed $t$-dimensional manifold with index$(A) = 0$. Let $L_s^h(q, \pi) \to^{\times A} L_{s+t}^h(q, \pi)$ be the homomorphism obtained by forming the cross product with $A$ of $Z|_q$ normal maps representing elements in $L_s^h(q, \pi)$. Then image$(xA)$ has exponent a divisor of* 8 *if* $s \geqq 5$.

Now to check (b) above, note that $\sigma(g' \circ f_i) = \sigma(g' \circ P_i) \times A_i$. Since index$(A_i) = 0$ for $i \geqq 2$, by the above proposition $\sigma(g' \circ f_i)$ must have order dividing 8. To check (c) above, write $\dim(A_1) = 4x$, and note that index$(A_1 - (CP^2)^x) = 0$. So by the above proposition $\sigma(g' \circ h) \times (A_1 - (CP^2)^x)$ has order dividing $8 \Rightarrow \sigma(g' \circ f) \times A_1 = \sigma(g' \circ f) \times (CP^2)^x +$ (2-primary element). But $\sigma(g' \circ f) \times A_1 = \sigma(g' \circ f_1)$ because $P_1$ can be chosen equal to $f$; and because $g'$ is homotopic to $g$ we have $\sigma(g' \circ f) = \sigma(g \circ f)$, so $\sigma(g' \circ f) \times (CP^2)^x = \sigma(g \circ f)$ under the periodicity isomorphism. Thus $\sigma(g' \circ f_1) = \sigma(g \circ f) +$ (2-primary element) as desired.

This completes the demonstration of why the $\mathscr{S}^i_q$ and $\mathscr{S}^i$ are well defined if localized away from 2.

Localizing at 2, there is a canonical bordism class $h_x: N_x \to X \wedge M_q$ representing each $x \in K_i^{\overline{F/\mathrm{TOP}}}(X, Z_q)$; in fact from $\Omega_*(\ ) \otimes_Z Z_{(2)} \cong H_*(\ , \Omega_*) \otimes Z_{(2)}$ we get

$$K_i^{\overline{F/\mathrm{TOP}}}(-, Z_q) \otimes Z_{(2)} \cong \sum_j H_{1+i+4j}(- \wedge M_q, \Omega_0) \otimes Z_{(2)}.$$

If in the above construction only this bordism class is used (at 2), then $\mathscr{S}^*_q$ will be well defined (at 2). For the same reason $\mathscr{S}^*$ is well defined (at 2).

Now set $X = \boldsymbol{L}^h_k(\pi)$, $I = 1_X$, and apply the constructions $C$, $C_n$ to the homomorphisms

$$K_*^{\overline{F/\mathrm{TOP}}}(X, Z_q) = K_*^{\overline{F/\mathrm{TOP}}}(X, Z_q) \times I \xrightarrow{\mathscr{S}^*_q} L^h_{*+k}(\pi) \otimes Z_q$$

$$K_*^{\overline{F/\mathrm{TOP}}}(X) = K_*^{\overline{F/\mathrm{TOP}}}(X) \times I \xrightarrow{\mathscr{S}^*} L^h_{*+k}(\pi)$$

to get a premap $pf$: $\boldsymbol{L}^h_k(\pi) \to \overline{F/\mathrm{TOP}} \otimes L^h_{*+k}(\pi)$. In more detail, let $L^h_i(\pi) = \bigoplus_j Z_{n(ij)}$ be a direct sum decomposition into cyclic subgroups, commuting with the periodicity isomorphisms $L^h_i(\pi) \cong L^h_{i+4}(\pi)$. $\rho_{ij}: L^h_{ij}(\pi) \to Z_{n(ij)}$ are the projections. For each finite subcomplex $X \subset \boldsymbol{L}^h_k(\pi)$ let $I_X$ denote its inclusion into $\boldsymbol{L}^h_k(\pi)$. If $Z_{n(ij)}$ is finite, apply $C_{n(ij)}$ to

$$h_{n(ij)} \equiv \{K_{i-k}^{\overline{F/\mathrm{TOP}}}(X, Z_{n(ij)}) \times I_X \xrightarrow{\rho_{ij} \circ \mathscr{S}^{i-k}_{n(ij)}} Z_{n(ij)}\}$$

to get $f^X_{ij}: X \to \overline{F/\mathrm{TOP}}_{i-k} \otimes Z_{n(ij)}$. If $Z_{n(ij)} = Z$, apply $C$ to the maps

$$h \equiv \{K_{i-k}^{\overline{F/\mathrm{TOP}}}(X) \times I_X \xrightarrow{\rho_{ij} \circ \mathscr{S}^{i-k}} Z_q\},$$

$$h_q \equiv \{K_{i-k}^{\overline{F/\mathrm{TOP}}}(X, Z_q) \times I_X \xrightarrow{\rho_{ij} \circ \mathscr{S}^{i-k}_q} Z_q\}$$

to get $f^X_{ij}: X \to \overline{F/\mathrm{TOP}}_{i-k}$. Note that all the $h_q$, $h$ commute with the change of coefficients $Z \to Z_q$, $Z_q \to Z_{q'}$, (by 2.2(c)) so $C$ can be applied. By 2.3(a) the $\{f^X_{i,j} | X \subset \boldsymbol{L}^h_k(\pi)\}$ is a premap $pf_{ij}: \boldsymbol{L}^h_k(\pi) \to (\overline{F/\mathrm{TOP}}_{i-k} \otimes Z_{n(ij)})$. By 2.3(b) and $\pi_*(\boldsymbol{L}^h_k(\pi)) \cong L^h_{*+k}(\pi)$, the product of premaps

$$pf \equiv \prod_{i=0}^{3}\Big(\prod_j ; pf_{ij}\Big): \boldsymbol{L}^h_k(\pi) \to \prod_{i=0}^{3}\Big(\prod_j \overline{F/\mathrm{TOP}}_{i-k} \otimes Z_{n(ij)}\Big)$$

induces an isomorphism on homotopy groups.

This completes the proof of Theorem 1.5.

Proof of Theorem 1.4. Any map $g: X \to \boldsymbol{L}^h_k(\pi)$ can be replaced by an inclusion $I: X \to \boldsymbol{L}^h_k(\pi)$ via the mapping cylinder construction. By the proof of Theorem 1.5, $I$ is null homotopic iff all the $\mathscr{S}^*_q$, $\mathscr{S}^*$ vanish on all the $K_*^{\overline{F/\mathrm{TOP}}}(X, Z_q) \times I$, $K_*^{\overline{F/\mathrm{TOP}}}(X) \times I$.

Choose a finite set, $S$, in $\{\bigcup_{q \geq 2} K_*^{\overline{F/\mathrm{TOP}}}(X, Z_q)\} \cup K_*^{\overline{F/\mathrm{TOP}}}(X)$ so that $S$ and all its images under the various change of coefficients $Z \to Z_q$, $Z_q \to Z_{q'}$ and under all the periodicity isomorphisms $K_i^{\overline{F/\mathrm{TOP}}}(X, Z_q) \cong K_{i+4}^{\overline{F/\mathrm{TOP}}}(X, Z_q)$, $K_i^{\overline{F/\mathrm{TOP}}}(X) \cong K_{i+4}^{\overline{F/\mathrm{TOP}}}(X)$ generate $\bigoplus_{q \geq 2} K_*^{\overline{F/\mathrm{TOP}}}(X, Z_q) \oplus K_*^{\overline{F/\mathrm{TOP}}}(X)$ ($X$ is a finite cell complex). For each $s \in S$ choose $v \in \bigcup_{q \geq 2} \Omega_*(X, Z_q) \cup \Omega_*(X)$ that represents $s$. If $v \in \Omega_*(X, Z_q)$ ($q \geq 2$) use the construction on p. 471 in **[11]** to get $v' \in \Omega_*(X, Z|_q)$. Otherwise set

$v' = v$. Let $V$ be the set of all these $v'$. It follows that $V$ is a characteristic variety for $X$.

This completes the proof of 1.4.

PROOF OF 2.2. Decompose each $p$-primary component of $L_i^h(\pi)$ into a direct sum of cyclic groups $\sum_{x \in I_{i,p}} \{t_{i,p,x}\}$ generated by the $t_{i,p,x}$. This can be done so that $I_{i,p} = I_{i+4,p}$ and $t_{i,p,x} \to t_{i+4,p,x}$ defines the periodicity isomorphism $L_i^h(\pi) \cong L_{i+4}^h(\pi)$. $0_{i,x,p}$ shall denote the cardinality of the group $\{t_{i,p,x}\}$. Choose $y_{i,p,x} \in L_{i+1}^h(0_{i,x,p}, \pi)$ so that $\delta y_{i,p,x} = t_{i,p,x}$. These can be chosen so that, for all $i \geqq 5$, $y_{i,p,x} \to y_{i+4,p,x}$ under the periodicity isomorphism $L_{i+1}^h(0_{i,p,x}, \pi) \cong L_{i+5}^h(0_{i,p,x}, \pi)$.

Now, given $y \in L_k^h(q, \pi)$, write $\delta y \equiv \sum_{p,x} a_{p,x} t_{k-1,p,k}$, and set

$$S_{q,k}(y) \equiv y - \sum_{p,x} b_{p,x} \cdot g_{p,x}(y_{k-1,p,x}),$$

where $g_{p,x}$ is the mapping $\alpha(0_{k-1,p,x}, q)_k \colon L_k^h(0_{k-1,p,x}, \pi) \to L_k^h(q, \pi)$, and

$$b_{p,x} = \frac{a_{p,x} \cdot d(0_{k-1,p,x}, q)}{0_{k-1,p,x}}.$$

Note that $\delta(S_{q,k}(y)) = 0$ for all $y$. So there exists $y' \in L_k^h(\pi) \otimes_Z Z_q$ so that $i_{q,k}(y') = S_{q,k}(y)$. Set $r_{q,k}(y) \equiv y'$. $r_{q,k}$ is a well-defined homomorphism iff $S_{q,k}$: $L_k^h(q, \pi) \to L_k^h(q, \pi)$ is. Moreover, if $S_{q,k}$ is well defined, then $\delta y = 0 \Rightarrow S_{q,k}(y) = y \Rightarrow r_{q,k}$ is left inverse to $i_{q,k}$. 2.2(b) is satisfied (assuming $S_{q,k}$ is well defined) because the $\alpha(q, q')_k$ commute with the periodicity isomorphisms and by the choices of the $t_{i,p,x}$, $y_{i,p,x}$. 2.2(c) is satisfied (assuming $S_{q,k}$ is well defined) because $\sigma \circ \beta(q, q')_i = \alpha(q, q') \circ \sigma$ and $S_{q',i} \circ \alpha(q, q')_i = \alpha(q, q')_i \circ S_{q,i}\ \forall\, i, q, q'$.

So 2.2 is proven if $S_{q,k}$ is well defined. Let $y'' = y + y'$ in $L_k^h(q, \pi)$. I claim that for any choice of the $a''_{p,x}$, $a'_{p,x}$, $a_{p,x}$ above, we have $S_{q,k}(y'') = S_{q,k}(y') + S_{q,k}(y)$. From this one deduces that for $y = 0$, $S_{q,k}(y) = 0$ independent of the choice of $a_{p,x}$ (set $y = y' = y''$, $a_{q,x} = a'_{p,x} = a''_{p,x}$ and use additivity); that $S_{q,k}(y)$ is independent of the choice of $a_{p,x}$ (let $a''_{p,x}$ be another choice, set $y'' = y$, $y' = 0$, $a'_{p,x} = a''_{p,x} - a_{p,x}$); and that $S_{q,k}$ is a homomorphism. Now $S_{q,j}(y'') = S_{q,k}(y) + S_{q,k}(y')$ iff $(b''_{p,x} - b_{p,x} - b'_{p,x}) \cdot g_{p,x}(y_{k-1,p,x}) = 0$. Note that $b''_{p,x} - b_{p,x} - b'_{p,x} = 0$ mod $d(0_{k-1,p,x}, q)$ and $g_{p,x}(y_{k-1,p,x})$ has order equal a divisor of $0_{k-1,p,x}$. So $S_{q,k}(y'') = S_{q,k}(y) + S_{q,k}(y')$ if $g_{p,x}(y_{k-1,p,x})$ has order equal a divisor of $q$, which it does by 2.4 below.

This completes the proof of 2.2 (mod 2.4).

LEMMA 2.4. *For $i \geqq 5$, and $q \geqq 2$, any element in $L_i^h(q, \pi)$ has order equal a divisor of $q$.*

PROOF OF 2.4. Let $M(q, n)$ be the space gotten by gluing the $n$-ball $B^n$ to the $(n-1)$-sphere $S^{n-1}$ along a map $\partial B^n \to S^{n-1}$ of degree $q$.

Then the groups $[M(q, n), \boldsymbol{L}_{i-2}^h(\pi)]$ and $L_{i-n}^h(q, \pi)$ are isomorphic whenever $i - n \geqq 5$ : exercise for reader.

Since $\Sigma(M(q, 3)) = M(q, 4)$, $M(q, 4)$ is a co-$H$-space, and $[M(q, 4), M(q, 4)]$ has a group structure. When $q =$ odd, an Eilenberg obstruction agrument and the calculation $\pi_4(S^3) = Z_2$ shows that $1_{M(q,4)}$ has order equal $q$ in $[M(q, 4), M(q, 4)]$. Addition in the set $[M(q, 4), \boldsymbol{L}_k^h(\pi)]$ can be defined in either of two equivalent ways: Use the co-$H$-space structure of $M(q, 4)$ or use the $H$-space structure of $\boldsymbol{L}_k^h(\pi)$.

Since $1_{M(q,4)}$ has order $q$, every element in $[M(q, 4), \boldsymbol{L}_k^h(\pi)]$ must have order dividing $q$. This proves 2.4 if $q$ = odd.

To prove 2.4 if $q = 2^n$ it suffices to show $\exists$ a premap $pf\colon \boldsymbol{L}_k^h(\pi) \to Y$, where $Y$ is a product of Eilenberg-Mac Lane spaces and $pf$ induces an isomorphism (mod odd torsion) on the homotopy groups. To get $pf$ apply the universal coefficient theorem for ordinary homology to the composition of homorphisms

$$H_*(\boldsymbol{L}_k^h(\pi), \Omega_0) \otimes_Z Z_{(2)} \subset \Omega_*(\boldsymbol{L}_k^h(\pi) \otimes_Z Z_{(2)} \xrightarrow{\sigma} L_{*+k}^h(\pi) \otimes Z_{(2)}.$$

Since 2.4 is true for $q$ = odd or $q = 2^n$, it is true for all $q$.

## Bibliography


**0.** D. Burghelea, R. Lashof and M. Rothenberg, *Groups of automorphisms of manifolds*; with an appendix ("The topological category") by E. Pedersen, Lecture Notes in Math., vol. 473, Springer-Verlag, Berlin and New York, 1975. MR **52** #1738.

**1.** W. Browder and F. Quinn, *A surgery theory for G-manifolds and stratified sets*, Manifolds-Tokyo 1973 (Proc. Internat. Conf. on Manifolds and Related Topics in Topology), Univ. Tokyo Press, Tokyo, 1975, pp. 27–36. MR **51** #11543.

**2.** S. E. Cappell and J. L. Shaneson, *Nonlocally flat embeddings, smoothings, and group actions*, Bull. Amer. Math. Soc. **79** (1973), 577–582. MR **47** #9635.

**3.** ———, *The codimension two placement problem and homology equivalent manifolds*, Ann. of Math. (2) **99** (1974), 277–348. MR **49** #3978.

**4.** P. E. Conner and E. E. Floyd, *Differentiable periodic maps*, Ergebnisse der Mathematik und ihrer Grenzgebiete, Band 33, Academic Press, New York; Springer-Verlag, Berlin, 1964. MR **31** #750.

**5.** R. C. Kirby and L. C. Siebenmann, *On the triangulation of manifolds and the Hauptvermutung*, Bull. Amer. Math. Soc. **75** (1969), 742–749. MR **39** #3500.

**6.** L. E. Jones, *Converse to the fixed point theorem of P. A. Smith*, Boulder, Colorado, 1970 (preprint).

**7.** ———, *Combinatorial symmetries of the m-disc*, Berkeley, Calif., 1971 (preprint).

**8.** ———, *Three characteristic classes measuring the obstruction to* PL *local unknotedness*, Bull. Amer. Math. Soc. **78** (1972), 979–980. MR **46** #6368.

**9.** ———, *Patch spaces: A geometric representation for Poincaré spaces*, Ann. of Math. (2) **97** (1973), 306–343. MR **47** #4269.

**10.** S. Maumary, *Proper surgery groups and Wall-Novikov groups*, Algebraic $K$-theory. III: Hermitian $K$-theory and geometric applications (Proc. Conf. Seattle Res. Center, Battelle Memorial Inst., 1972), Lecture Notes in Math., vol. 343, Springer, Berlin, 1973, pp. 526–539. MR **51** #14107.

**11.** J. W. Morgan and D. P. Sullivan, *The transversality characteristic class and linking cycles in surgery theory*, Ann. of Math. (2) **99** (1974), 463–544. MR **50** #3240.

**12.** F. Quinn, Thesis, Princeton Univ., Princton, N.J., 1969.

**13.** J. L. Shaneson, *Product formulas for* $L_n(\pi)$, Bull. Amer. Math. Soc. **76** (1970), 787–791. MR **41** #6232.

**14.** D. Sullivan, *Lecture notes on the characteristic variety theorem*, Princeton Univ., Princeton, N.J., 1967.

**15.** ———, *Geometric periodicity and the invariants of manifolds*, Manifolds—Amsterdam 1970 (Proc. Nuffic Summer School), Lecture Notes in Math., vol. 197, Springer, Berlin, pp. 44–75. MR **44** #2236.

**16.** L. Taylor, Thesis, Univ. of California at Berkeley, 1972.

**17.** C. T. C. Wall, *Surgery an compact manifolds*, Academic Press, New York, 1970.

**18.** R. E. Williamson, *Surgery in* $M \times N$ *with* $\pi_1 M \neq 1$, Bull. Amer. Math. Soc. **75** (1969), 582–585. MR **39** #4859.


State University of New York at Stony Brook

Proceedings of Symposia in Pure Mathematics
Volume 32, 1978

# KERVAIRE'S INVARIANT FOR FRAMED MANIFOLDS*

JOHN JONES AND ELMER REES

**1.** Pontryagin (see **[19]**) set up a correspondence between homotopy classes of maps $S^{n+k} \to S^n$ and bordism classes of framed $k$-dimensional manifolds $M^k$ embedded in $\boldsymbol{R}^{n+k}$. When $k = 2$ and $n \geqq 2$ he constructed a map $q: H_1(M^2) \to \boldsymbol{Z}/2$ (all our homology groups will have $\boldsymbol{Z}/2$ coefficients) which is quadratic with respect to the intersection pairing. Such a quadratic form has an Arf invariant; see **[1]**, **[4]** or **[21]**. Arf($q$) depends only on the framed bordism class of $M$ and defines an isomorphism $\pi_{n+2}S^n \to \boldsymbol{Z}/2$. This procedure can be used in practice to show that certain maps are essential—for example this is done in **[13]**.

In **[10]**, M. Kervaire defined an Arf invariant for $(2l - 2)$-connected, $(4l - 2)$-dimensional closed manifolds which are almost parallelizable and smooth in the complement of a point. The manifolds $S^1 \times S^1$, $S^3 \times S^3$ and $S^7 \times S^7$ may be framed in different ways to have Kervaire invariant one or zero. The Kervaire invariant of any framed $(4l - 2)$-manifold $M$ was defined by first performing surgery to make $M$ $(2l - 2)$-connected. In **[10]**, Kervaire showed that his invariant vanished for closed, smooth 10-dimensional manifolds and constructed a manifold which was smooth in the complement of a point and had Kervaire invariant one. This gave his famous example of a nonsmoothable manifold.

W. Browder **[3]** extended the definition of the Kervaire invariant and also extended Kervaire's result by showing that a framed, smooth $M^{4l-2}$ has Kervaire invariant zero if $l$ is not a power of 2. His definition has since been extended and simplified by E. H. Brown **[5]**. Browder gave necessary and sufficient conditions in terms of the Adams spectral sequence for the existence of a framed $M^{2^{n+1}-2}$ with Kervaire invariant one. These conditions have been verified for $n = 4$ **[17]** and

*AMS* (*MOS*) *subject classifications* (1970). Primary 55E45; Secondary 57D15, 57D90.
*These notes are based on a talk given by Elmer Rees.

$n = 5$ (see [15] for an account of other attempts). An explicit $M^{30}$ was constructed in [7] and directly shown to have a framing of Kervaire invariant one; a sketch of its properties is given in §4 of these notes.

**2.** It is clearly desirable to be able to define the Kervaire invariant of a framed manifold directly, without having to do surgery first. It is also desirable to understand exactly how the invariant depends on the framing. The Browder-Brown approach enables one to define the quadratic form directly on the middle homology of a framed $M^{2k}$ and to study the precise relationship between quadratic forms and framings.

First, consider quadratic forms $q$ defined on a mod 2 vector space $V$ relative to a nonsingular pairing $\langle\ ,\ \rangle$, i.e., $q(x + y) = q(x) + q(y) + \langle x, y\rangle$. Note that the existence of such a form implies $\langle x, x\rangle = 0$ for all $x \in V$. If $q_1$ and $q_2$ are two such forms then their difference $q_1 - q_2$ is linear and so by the nonsingularity of the pairing there is an element $v$ such that $q_1(x) - q_2(x) = \langle x, v\rangle$. This shows that if there is one quadratic form on $V$ then there is a 1-1 correspondence between the set of quadratic forms and $V$ itself. It is easy to show that $\mathrm{Arf}(q_1) + \mathrm{Arf}(q_2) = q_1(v) = q_2(v)$.

Browder and Brown defined the Kervaire invariant of a framed manifold $M^{2k} \subset R^{2k+N}$ as follows. It is the Arf invariant of a quadratic form defined on $H^k(M) \cong H_k(M)$. Let $\nu$ be the normal bundle of $M$, $T(\nu)$ its Thom complex, and let $t$ be the given framing. Then the Pontryagin-Thom construction gives a map $S^{2k+N} \to T(\nu)$ and $t$ gives a homeomorphism $T(\nu) \to \Sigma^n(M_+)$ where $M_+$ denotes $M$ together with a disjoint base point. Given $a \in H^k(M) = [M_+, K_k]$, where $K_k$ denotes an Eilenberg-Mac Lane space $K(Z/2, k)$, we may form the composite $S^{2k+N} \to \Sigma^N M_+ \to^{\Sigma^N a} \Sigma^N K_k$, which is an element of the group $\pi_{2k+N}(\Sigma^N K_k)$. A calculation shows that this group is $Z/2$. So, from our framed manifold $(M^{2k}, t)$, we have constructed a function $q_t\colon H^k(M) \to Z/2$. One may check that $q_t$ is quadratic with respect to the intersection pairing, and that its Arf invariant depends only on the framed bordism class of $(M, t)$. We will denote this invariant by $K(M, t)$; it equals Kervaire's invariant when that is defined. The following theorem, due to E. H. Brown [5], shows the relationship between this quadratic form and those considered by Pontryagin and Kervaire.

THEOREM. *If the Poincaré dual of $a \in H^k(M)$ is represented by an embedded $S^k \subset M^{2k}$ then $q_t(a) = \varepsilon(a) + h(a)$ where*

$$\varepsilon(a) = \begin{cases} 0 & \text{if the normal bundle of } S^k \subset M^{2k} \text{ is trivial,} \\ 1 & \text{otherwise;} \end{cases}$$

$$h(a) = \begin{cases} 0 & \text{if } \varepsilon(a) = 1, \\ \text{the Hopf invariant of the framed embedding} & \\ \quad S^k \subset M^{2k} \subset R^{2k+N} \text{ if } \varepsilon(a) = 0. & \end{cases}$$

The quadratic form clearly depends at most on the framing restricted to the $k$-skeleton of $M$; it is important to understand precisely what it depends on. Browder, in [3], completely analysed this. Browder's work depends on the notion of a Wu-orientation. Note that a framing corresponds to a lift in the diagram

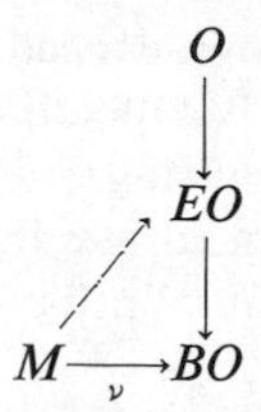


The different choices of framing correspond to maps from $M$ to the fibre, $O$, of this principal bundle. One wants structures, analogous to framings, which give quadratic forms. Since quadratic forms on $H^k(M)$ are in 1-1 correspondence with $H^k(M)$, one would like the structures to be in 1-1 correspondence with $H^k(M)$. Therefore one needs a principal fibration over $BO$ with fibre $K_k$. Such fibrations correspond to $H^{k+1}(BO)$. The element chosen to classify this fibration is the Wu-class $v_{k+1} \in H^{k+1}(BO)$. The total space of this fibration is denoted by $BO\langle v_{k+1}\rangle$ and a lifting to $BO\langle v_{k+1}\rangle$ of the classifying map for the normal bundle of $M$ is called a Wu-orientation of $M$. One good reason for the choice of $v_{k+1}$ is that it is the only $(k + 1)$-dimensional characteristic class that is zero on all $2k$-manifolds. Hence every $2k$-manifold admits a Wu-orientation. A framing gives a Wu-orientation as is easily seen from the commutative diagram

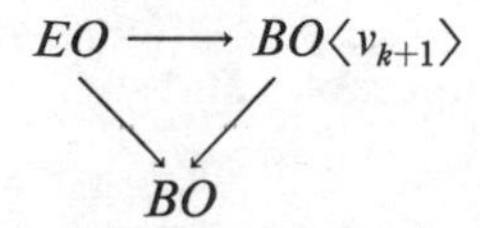


Let $\bar{\gamma}$ be the bundle over $BO\langle v_{k+1}\rangle$ obtained by pulling back the universal bundle over $BO$. A Wu-orientation gives a bundle map $w: \nu \to \bar{\gamma}$. Given $a \in H^k(M) = [M_+, K_k]$ we may form the composite $S^{2k+N} \to T(\nu) \to_{\Delta} M_+ \wedge T(\nu) \to_{a \wedge T(w)} K_k \wedge T(\bar{\gamma})$ where we have taken $\nu$ and $\bar{\gamma}$ to be $N$-dimensional, and $\Delta$ is induced by the diagonal. Hence the Wu-orientation $w$ gives a function $Q_w$: $H^k(M) \to \pi_{2k+N}(K_k \wedge T(\bar{\gamma}))$. If the Wu-orientation arises from a framing then $Q_w$ factors through the map $i: \pi_{2k+N}(K_k \wedge S^N) \to \pi_{2k+N}(K_k \wedge T(\bar{\gamma}))$ induced by the inclusion of a fibre of $\bar{\gamma}$. E. H. Brown [5] shows that the homomorphism $i$ is injective (because $v_{k+1}(\bar{\gamma}) = 0$) and that one may choose an epimorphism $\varepsilon$: $\pi_{2k+N}(K_k \wedge T(\bar{\gamma})) \to \mathbf{Z}/4$ so that the composite $\varepsilon \circ i$ is injective. It is easily shown that the function $q_w = \varepsilon \circ Q_w$ is quadratic in the sense that $q_w(x + y) = q_w(x) + q_w(y) + j(x \cdot y)$ where $j: \mathbf{Z}/2 \to \mathbf{Z}/4$ is the usual inclusion. The choice of $\varepsilon$ ensures that if the Wu-orientation came from a framing then this quadratic form takes values in $\mathbf{Z}/2 \subset \mathbf{Z}/4$ and equals the one previously defined.

The $\mathbf{Z}/4$ here is needed only to study quadratic phenomena on $2k$-manifolds whose $k$th Wu-class is nonzero. For in this case there is a class $x \in H^k(M)$ with $x \cdot x = 1$, and as we have already observed the existence of a $\mathbf{Z}/2$-valued quadratic form associated to the intersection pairing implies that $x \cdot x = 0$ for all $x \in H^k(M)$. Since our main interest is in framed manifolds we will assume our quadratic forms are $\mathbf{Z}/2$-valued.

Suppose now that we have two Wu-orientations, differing by $v \in H^k(M)$, giving quadratic forms $q_1$ and $q_2$. Then it can be shown that the quadratic forms also differ by $v$, that is $q_1(x) + q_2(x) = x \cdot v$. This shows that the relationship between

Wu-orientations and quadratic forms is indeed the simplest possible. So the quadratic form for a framed manifold only depends on the Wu-orientation arising from the framing. We now analyse how a change of framing affects the Wu-orientation, and this then explains the way the quadratic form depends on the framing.

Consider the commutative diagram of principal fibrations

$$\begin{array}{ccc} O & \xrightarrow{x_k} & K_k \\ \downarrow & & \downarrow \\ EO & \longrightarrow & BO\langle v_{k+1}\rangle \\ \downarrow & & \downarrow \\ BO & = & BO \end{array}$$

The fibration $BO\langle v_{k+1}\rangle \to BO$ is induced by $v_{k+1}$ so $x_k$ is the map $\Omega v_{k+1}$ obtained by applying the loop functor to $v_{k+1}\colon BO \to K_{k+1}$. Therefore if two framings of $M$ differ by the map $g\colon M \to O$ then the induced Wu-orientations differ by $g^*x_k$. This leads to the change of framing formula:

THEOREM. *Let $t_1$ and $t_2$ be two framings of $M^{2k}$ differing by $g\colon M \to O$. Then $q_{t_1}(x) + q_{t_2}(x) = x \cdot g^*x_k$ and $K(M, t_1) + K(M, t_2) = q_{t_1}(g^*x_k) = q_{t_2}(g^*x_k)$.*

An easy calculation shows that $v_{k+1}$ is decomposable unless $k + 1$ is a power of 2. The functor $\Omega$ annihilates decomposables and so $x_k = 0$ unless $k + 1$ is a power of 2. Hence one has

COROLLARY. *The quadratic form of a framed manifold $M^{2k}$ is independent of the framing unless $k + 1$ is a power of 2.*

Browder's theorem that the Kervaire invariant of a framed $M^{4l-2}$ vanishes unless $l$ is a power of 2 then follows easily from the following consequence of the Kahn-Priddy theorem due to Nigel Ray [**20**].

THEOREM. *If $\alpha$ is an element of the 2-primary component of $\pi_{n+N}(S^N)$ for $N > n + 1$, then there is a manifold $M^n$ with framings $t_1$ and $t_2$ such that $[M, t_1] = 0$, and $[M, t_2] = \alpha \in \pi_{n+N}(S^N)$.*

The Kervaire invariant of a framed boundary is zero so Browder's theorem follows since we have shown that if $l$ is not a power of 2 then the quadratic form and so certainly the Kervaire invariant of a framed $M^{4l-2}$ is independent of the framing.

Conversely one can use this kind of approach to give sufficient conditions for the existence of framed manifolds $M^{4l-2}$ with Kervaire invariant one when $l$ is a power of 2. Perhaps the simplest is the following result of [**7**].

THEOREM. *There is a framed manifold $M^{2k}$, $k = 2^i - 1$, with Kervaire invariant one if and only if there is an element $\theta \in \pi^s_{2k}(SO)$ detected by $\mathrm{Sq}^{k+1}$, i.e., the class $\mathrm{Sq}^{k+1}x_k$ is nonzero in the cofibre of $\theta$.*

This theorem is also true with $SO$ replaced by $RP^\infty$. It can be used to yield another proof of Browder's result that there is a framed manifold in dimension $2^{i+1} - 2$ with Kervaire invariant one if and only if the element $h_i^2$ in the $E_2$-term of the mod 2 Adams spectral sequence survives to $E_\infty$. The details are worked out in [**7**].

**3.** It is of interest to find criteria under which a manifold will have a framing with Kervaire invariant one. For brevity we will call a manifold "Arf-changeable" if it has framings with different Kervaire invariants. By Ray's result, if there is a framed $M^{4l-2}$ with Kervaire invariant one, then there is an Arf-changeable manifold in the same dimension. Of course the change of framing formula gives a necessary and sufficient condition for a manifold to be Arf-changeable. There is also the following simple criterion for a manifold to be Arf-changeable. Unfortunately we have only been able to use it constructively in simple cases.

THEOREM [7]. *Let $M^{2k}$ be a framed manifold and $g_1, g_2\colon M \to O$ be two maps such that $g_1^*x_k \cdot g_2^*x_k = 1$. Then $M$ is Arf-changeable.*

PROOF. Define $\alpha_i = g_i^*x_k$, $i = 1, 2, 3$, where $g_3(x) = g_1(x)g_2(x)$. Then $\alpha_3 = \alpha_1 + \alpha_2$ and if $q$ is a quadratic form coming from some framing then $q(\alpha_3) = q(\alpha_1) + q(\alpha_2) + 1$. Therefore it follows by the change of framing formula that one of $g_1$, $g_2$ or $g_3$ changes the Kervaire invariant.

COROLLARY. *If $N_1^k$ and $N_2^k$ are framed manifolds with $k = 1, 3$ or $7$, then $N_1 \times N_2$ is Arf-changeable.*

PROOF. There are maps $f_i\colon N_i \to SO$ with $f_i^*x_k \neq 0$. The maps $g_1 = f_1 \circ \pi_1$ and $g_2 = f_2 \circ \pi_2$, where $\pi_i\colon N_1 \times N_2 \to N_i$ is the projection, have the required property.
We will now use a theorem due to Stong [22] to show that highly connected manifolds are not Arf-changeable.

THEOREM. *If $M^{2k}$, $k = 2^r - 1$, is a framed manifold and s-connected where $\phi(s + 1) \geqq r + 1$, then $M$ is not Arf-changeable.*

So for example 8-connected $M^{30}$'s and 9-connected $M^{62}$'s are not Arf-changeable.
PROOF. Stong's theorem says that the map $q_{s+1}\colon BO\langle s + 1\rangle \to BO$ satisfies $q_{s+1}^* w_{2^r} = 0$ if $\phi(s + 1) > r$. Here $BO\langle s + 1\rangle$ is the $(s + 1)$th connected cover of $BO$. The theorem follows immediately.
An interesting class of stably parallelizable manifolds are the hypersurfaces, that is compact codimension one submanifolds of Euclidean space. If they are Arf-changeable then one can prove stronger connectivity results. Suppose $M^{2k} \subset S^{2k+1}$ is a hypersurface. Let $A$, $B$ be the closures of the components of the complement of $M$ in $S^{2k+1}$. The Mayer-Vietoris sequences for cohomology and real $K$-theory show that $i_A^* + i_B^*\colon H^k(A) \oplus H^k(B) \to H^k(M)$ and $i_A^* + i_B^*\colon KO^{-1}(A) \oplus KO^{-1}(B) \to KO^{-1}(M)$ are isomorphisms, where $i_A\colon M \to A$ and $i_B\colon M \to B$ are the inclusions of the boundaries. So given $g\colon M \to O$ write $g = g_1 + g_2$ where $g_1 = i_A \circ g_A$ and $g_2 = i_B \circ g_B$ where $g_A\colon A \to O$, $g_B\colon B \to O$. Therefore $g^*x_k = i_A^*\alpha_A + i_B^*\alpha_B$, where $\alpha_A = g_A^*x_k$ and $\alpha_B = g_B^*x_k$. The natural framing of $M^{2k}$ coming from its embedding in $S^{2k+1}$ extends over both $A$ and $B$. Let $q$ be the quadratic form coming from this framing; then it follows that $q$ vanishes on both $i_A^*H^k(A)$ and $i_B^*H^k(B)$. Therefore

$$q(g^*x_k) = q(i_A^*\alpha_A + i_B^*\alpha_B) = i_B^*\alpha_B \cdot i_A^*\alpha_A = g_1^*x_k \cdot g_2^*x_k.$$

It follows that if a hypersurface is Arf-changeable then one may always use the above theorem from [7] to prove it.

PROPOSITION. *If $M^{30}$ is a 7-connected hypersurface then it is not Arf-changeable.*

PROOF. Let $g_1, g_2: M \to 0$ be maps; then from Stong [22] we know that if $\alpha_i = g_i^* x_{15}$ then $\alpha_i = (\mathrm{Sq}^7 + \mathrm{Sq}^4\mathrm{Sq}^2\mathrm{Sq}^1)\beta_i$ for some $\beta_i \in H^8(M)$ and further $\mathrm{Sq}^2\beta_i = 0$. The result is proved by showing $((\mathrm{Sq}^7 + \mathrm{Sq}^4\mathrm{Sq}^2\mathrm{Sq}^1)\beta_1)\cdot((\mathrm{Sq}^7 + \mathrm{Sq}^4\mathrm{Sq}^2\mathrm{Sq}^1)\beta_2) = 0$ using the parallelizability of $M$ and the Adem relations.

**4.** In [7] the ideas we are describing are used successfully to show that a certain explicitly constructed $M^{30}$ is Arf-changeable. We now describe this manifold. Consider $X_5$ the orientable surface of genus 5; it has a smooth free action of the dihedral group $D_4$ (the symmetries of a square). The quotient space of this action is the non-orientable surface $Y_1$ of Euler characteristic $-1$, the connected sum of the projective plane and the torus. The action is best described by giving a homomorphism $\phi: \pi_1 Y_1 \to D_4$. The group $\pi_1 Y_1$ has generators $A_1$, $A_2$, $B$ and one relation: $A_1 A_2 A_1^{-1} A_2^{-1} = B^2$. The group $D_4$ permutes the four vertices of the square and in this way can be regarded as a subgroup of $S_4$. The homomophism $\phi$ is then given by $\phi(A_1) = (14)(23)$, $\phi(A_2) = (13)(24)$ and $\phi(B) = (13)$. In the following one may assume that $X_5$ is the $D_4$ covering of $Y_1$ associated with $\phi$; the fact that $X_5$ has genus 5 is not necessary.

We now define $M^{30}$ to be $X_5 \times_{D_4} (S^7)^4$ where $D_4$ acts as a permutation group on $(S^7)^4$. It is readily checked that $M^{30}$ is stably parallelizable. Moreover one can construct a map $g: M^{30} \to SO$ as follows:

$$M^{30} \xrightarrow{1\times\omega^4} X_5 \times_{D_4} (SO(8))^4 \xrightarrow{\phi\times 1} ED_4 \times_{D_4} (SO(8))^4 \xrightarrow{D} SO(32).$$

The first map is induced by a map $\omega: S^7 \to SO(8)$ such that $\omega^* x_7$ is nontrivial, $\tilde{\phi}: X_5 \to ED_4$ is the equivariant map that covers $\phi$ and $D$ is a finite version of the Dyer-Lashof map for $SO$ and is described explicitly in [14]. Using the results of [12], it is straightforward to calculate $\alpha = g^* x_{15}$ once one knows enough about $H^*(M)$. It turns out that if $q$ is any quadratic form coming from a framing then $q(\alpha) \neq 0$. The change of framing formula shows that this $M^{30}$ is Arf-changeable. Although the method outlined here only proves that this $M^{30}$ is Arf-changeable, it is in fact possible with a little care to identify a framing on $M$ which has Kervaire invariant one.

In the calculation of $q(\alpha)$ one can use the following lemma which may be of independent interest.

LEMMA [7]. *If $q$ comes from a framing on $M^{4l-2}$ then $q(\mathrm{Sq}^1 a) = a\cdot\mathrm{Sq}^2 a$ for any $a \in H^{4l-2}(M)$.*

PROOF. One considers the map $\mathrm{Sq}^1_*: \pi^s_{4l-2}(K_{2l-2}) \to \pi^s_{4l-2}(K_{2l-1})$ and checks that this is an epimorphism. Moreover an element $\theta$ that maps nontrivially is such that $\theta^*(\iota_{2l-2} \cdot \mathrm{Sq}^2 \iota_{2l-2}) \neq 0$.

The first-named author has recently proved several further results of this kind.

COROLLARY. *If $M^{4l-2}$ is stably parallelizable and $a \in H^{2l-2}(M)$ is such that $\mathrm{Sq}^1 a = 0$ then $a \cdot \mathrm{Sq}^2 a = 0$.*

This corollary for $l$ odd can be deduced from Theorem 1.2 of [16] and for $l = 3$ is given in [18]. It is of interest to note that this corollary is false in general for manifolds all of whose Stiefel-Whitney classes vanish; the manifolds $CP^n$ for $n + 1$ a power of 2 are examples. It is also false for manifolds $M^{4l}$ as the example $SU(3)$ shows. It might be interesting to construct examples in other dimensions.

## REFERENCES

1. C. Arf, *Untersuchungen über quadratische formen in Körpen der charakteristik* 2, J. Reine Angew Math. **183** (1941), 148–167. MR **4,** 237.

2. M. F. Atiyah, *Thom complexes*, Proc. London Math. Soc. (3) **11** (1961), 291–310. MR **24** #A1727.

3. W. Browder, *The Kervaire invariant of framed manifolds and its generalisations*, Ann. of Math. (2) **90** (1969), 157–186.

4. ———, *Surgery on simply connected manifolds*, Springer-Verlag, Berlin and New York, 1972.

5. E. H. Brown, *Generalisations of the Kervaire invariant*, Ann. of Math. (2) **95** (1972), 368–383. MR **45** #2719.

6. F. Hirzebruch and K. H. Mayer, *O(n)-mannigfaltigkieten, exotische sphären und singularitäten*, Lecture Notes in Math., vol. 57, Springer-Verlag, Berlin and New York, 1968.

7. J. Jones, D. Phil. Thesis, Oxford Univ., 1976.

8. J. Jones and E. Rees, *A note on the Kervaire invariant*, Bull. London Math. Soc. **7** (1975), 279–282. MR **52** #6759.

9. D. S. Kahn and S. B. Priddy, *Applications of the transfer to stable homotopy theory*, Bull. Amer. Math. Soc. **78** (1972), 981–987. MR **46** #8220.

10. M. Kervaire, *A manifold which does not admit any differentiable structure*, Comment. Math. Helv. **34** (1966), 256–270.

11. M. Kervaire and J. Milnor, *Groups of homotopy spheres*. I, Ann. of Math. (2) **77** (1963), 504–537.

12. S. O. Kochman, *Homology of the classical groups over the Dyer-Lashof algebra*, Trans. Amer. Math. Soc. **185** (1973), 83–137. MR **48** #9719.

13. K. Y. Lam, *Some interesting examples of non-singular bilinear maps*, Topology **16** (1977), 185–189.

14. I. Madsen, *On the action of the Dyer-Lashof algebra in $H_*G$*, Pacific J. Math. **60** (1975), 235–277. MR **52** #9228.

15. M. E. Mahowald, *Some remarks on the Kervaire invariant problem from the homotopy point of view*, Proc. Sympos. Pure Math., vol. 22, Amer. Math. Soc., Providence, R. I., 1971, pp. 165–169.

16. M. E. Mahowald and F. P. Peterson, *Secondary cohomology operations on the Thom class*, Topology **2** (1963), 367–377. MR **28** #612.

17. M. Mahowald and M. Tangora, *Some differentials in the Adams spectral sequence*, Topology **6** (1967), 349–369.

18. S. Morita, *Smoothability of* PL *manifolds is not topologically invariant*, Manifolds-Tokyo, 1973, Univ. of Tokyo Press, Tokyo, 1975, pp. 51–56. MR **51** #6837.

19. L. Pontrjagin, *Smooth manifolds and their applications in homotopy theory*, Amer. Math. Soc. Transl. (2) **11** (1959), 1–114. MR **22** #5980.

20. N. Ray, *A geometrical observation on the Arf invariant of a framed manifold*, Bull. London Math. Soc. **4** (1972), 163–164.

21. C. P. Rourke and D. P. Sullivan, *On the Kervaire obstruction*, Ann. of Math. (2) **94** (1971), 397–413. MR **46** #4546.

22. R. E. Stong, *Determination of $H^*(BO(k, \cdots, ), Z_2)$ and $H^*(BU(k, \cdots, ), Z_2)$*, Trans. Amer. Math. Soc. **107** (1963), 526–544. MR **27** #1944.

MAGDALEN COLLEGE, OXFORD

ST. CATHERINE'S COLLEGE, OXFORD

Proceedings of Symposia in Pure Mathematics
Volume 32, 1978

# SURGERY ON DIFFEOMORPHISM

YOUN W. LEE

**1. Introduction.** We report a method, "Surgery on diffeomorphism", for the study of $\pi_0(\mathrm{Diff}\ M^n)$ and $I(M^n)$ of a compact, oriented, smooth manifold $M^n$.

"Surgery on diffeomorphism" means the following construction. For an imbedding $\alpha: S^r \times D^{n-r} \to \mathring{M}^n$ and $f \in \mathrm{Diff}\ M^n$, we try to find $g \in \mathrm{Diff}\ M^n$ isotopic to $f$ so that $g$ fixes the points in a neighborhood of $\alpha(S^r \times D^{n-r})$ in $M$. If such a $g$ exists, then define $F \in \mathrm{Diff}\ W$ by $F|_{M\times I} = g \times \mathrm{id}$ and $F|_{D^{r+1}\times D^{n-r}} = \mathrm{id}$, where $W = M \times I \cup_{\alpha_1} D^{r+1} \times D^{n-r}$ for the imbedding $\alpha_1: S^r \times D^{n-r} \to^{\alpha} M \to M \times \{1\}$.

Let $M' = \overline{\partial W - M \times \{0\} - (\partial M) \times I}$, i.e., $M'$ is the result of an ordinary surgery on $M$ using $\alpha$. Then $f' = F|_{M'} \in \mathrm{Diff}\ M'$. We call $f' \in \mathrm{Diff}\ M'$ the result of a surgery on $f \in \mathrm{Diff}\ M$ using $\alpha$.

In this paper, we only consider a manifold with an abelian fundamental group. Let $H_*: \pi_0(\mathrm{Diff}\ M) \to \mathrm{Aut}(H_*(M))$ be the homomorphism defined by $H_*([f]) = H_*(f)$, where $[f]$ denotes the isotopy class containing $f$. We define $\pi_0^+(\mathrm{Diff}\ M) = \mathrm{Ker}(H_*)$.

In §2, we find an obstruction to doing a surgery on $f$, $[f] \in \pi_0^+(\mathrm{Diff}\ M)$. If this obstruction vanishes, define $\phi([f]) = [f'] \in D(M')$, the group of pseudo-isotopy classes of diffeomorphisms of $M'$.

In §3, we find conditions under which $\phi$ is well defined. In this case, $\phi$ is a homomorphism from a normal subgroup of $\pi_0^+(\mathrm{Diff}\ M)$ to $D(M')$.

We state some of the results we can get by this method and short proofs will be given in the corresponding sections.

COROLLARY 2.2. *Let $M^n$ be a simply connected $\pi$-manifold such that $H_{[n/2]}(M) = 0$. Suppose that $\bar{H}_i(M) = 0$, $i \equiv 0, 1, 3, 7$ (Mod 8) and $H_i(M)$ is free otherwise for $i < [n/2]$. Then $I_0(M) = 0$.*

---

*AMS (MOS) subject classifications* (1970). Primary 57D50.

REMARK. $I_0(M)$ is a subgroup of $I(M)$ defined in §2. $I(M)/I_0(M) = 0$ if any automorphism of $H_*(M)$ induced by a PL homeomorphism can be induced by a diffeomorphism of $M$. If $M$ is a product of standard spheres $I_0(M) = I(M)$. (See Schultz [15].) Also compare Corollary 2.2 with the results of Browder [3], Kosinski [8], Schultz [15] and Wall [18].

COROLLARY 2.5. *If* $i + 1 < p$ *and* $M$ *is* $(p - 1)$*-connected, then* $\pi_i(\text{Diff}(S^p \times M^n))$ *contains* $\pi_i(SO_p)$ *as a subgroup.*

PROPOSITION 3.3. *Let $E$ be the total space of an oriented $(p + 1)$-disk bundle over $S^q$ with a characteristic map $b \in \pi_{q-1}(SO_{p+1})$, $p + 2 < q < 2p - 2$. Suppose that $b$ desuspends, i.e., $b \in s_*(\pi_{q-1}(SO_p))$, where $s\colon SO_p \to SO_{p+1}$ is the inclusion and* $p + 1 \equiv 2, 4, 5, 6$ (Mod 8) *if $E$ is not trivial. Then we have an exact sequence,*

$$1 \longrightarrow H \oplus \Gamma^{p+q+1} \longrightarrow \pi_0^+(\text{Diff}\,\partial E) \xrightarrow{\varphi_\alpha} \pi_p(SO) \xrightarrow{\sigma_b} I_0(\partial E) \to 1.$$

*Here $H$ is a quotient group of $\pi_q(SO_{p+1})$, $\varphi_\alpha$ is a homomorphism defined in §2, $I(\partial E)/I_0(\partial E)$ has at most order* 2 *and $\sigma_b$ is closely related to the Milnor's $\sigma$-pairing.*

REMARK. If $E$ is trivial, then Proposition 3.3 is the result of Levine [10], Sato [13] or Terner [17]. Sato [14] studied a similar problem when $p > q$.

**2. Obstruction to surgery.** Let $M^n$ be an $(r - 1)$-connected $(1 \leq r < (2n - 3)/3)$ manifold and $\alpha\colon S^r \times D^{n-r} \to \mathring{M}^n$ be an imbedding. If $[f] \in \pi_0^+(\text{Diff } M)$, by Haefliger [6], the isotopy extension theorem [11] and the uniqueness theorem of tubular neighborhood [11], we can find $g \in [f]$ such that $g|_{\alpha(S^r \times D^{n-r})}$ induces a bundle map $b(g)$ on the trivial bundle $\pi\colon \alpha(S^r \times D^{n-r}) \to \alpha(S^r \times \{0\})$. We regard that $b(g) \in \pi_r(SO_{n-r})$.

Define a subgroup $Q_r^{M(\alpha)}(SO_{n-r})$ of $\pi_r(SO_{n-r})$ by $Q_r^{M(\alpha)}(SO_{n-r}) = \{b(g) \in \pi_r(SO_{n-r})\colon b(g) = g|_{\alpha(S^r \times D^{n-r})}$ and $g$ is isotopic to $\text{id}_M\}$.

Let $\pi_r^{M(\alpha)}(SO_{n-r}) = \pi_r(SO_{n-r})/Q_r^{M(\alpha)}(SO_{n-r})$. Then we can define a homomorphism $\varphi_\alpha\colon \pi_0^+(\text{Diff } M) \to \pi_r^{M(\alpha)}(SO_{n-r})$ by $\varphi_\alpha([f]) = b(g)$.

The following proposition is clear from the construction of $\varphi_\alpha$.

PROPOSITION 2.1. *Let $M^n$ be an $(r - 1)$-connected $(1 \leq r < (2n - 3)/3)$ manifold and $\alpha\colon S^r \times D^{n-r} \to \mathring{M}^n$ be an imbedding. Let $[f] \in \pi_0^+(\text{Diff } M)$, then we can do a surgery on $f$ using $\alpha$ if and only if $\varphi_\alpha([f]) = 0$. If $r < [n/2]$ and* $r \equiv 2, 4, 5, 6$ (Mod 8), *there is no obstruction to doing a surgery.* (*See* [2].)

Now we define $I_0(M^n)$. Let $L = M^n - \mathring{D}^n$. By regarding $I(M^n)$ as a subgroup of $\pi_0(\text{Diff } S^{n-1})$, we have a short exact sequence,

$$1 \longrightarrow \text{Ker}(\partial) \longrightarrow \pi_0(\text{Diff } L) \xrightarrow{\partial} I(M) \longrightarrow 1,$$

where $\partial([f]) = [f|_{\partial L \cong S^{n-1}}]$. Let $A(L) = \text{Im}(H_*)$ and $A_\partial(L) = \text{Im}(H_*|_{\text{Ker}(\partial)})$, where $H_*\colon \pi_0(\text{Diff } L) \to \text{Aut}(H_*(L))$ is the homomorphism mentioned in the introduction.

We have a commuting diagram:

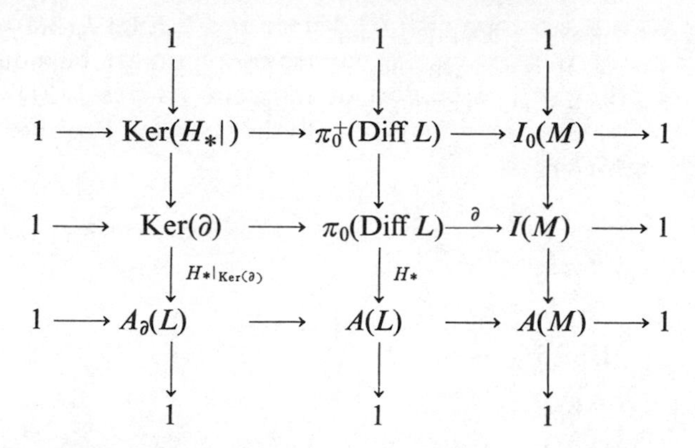

where $I_0(M) = \pi_0^+(\text{Diff } L)/\text{Ker}(H_*|_{\text{Ker}(\partial)})$ and $A(M) = A(L)/A_\partial(L)$.

Now we indicate a proof of Corollary 2.2. Let $[f] \in \pi_0^+(\text{Diff } L)$. Using an induction on the connectivity of $L$ (we need Proposition 2.1 and some facts from Kervaire and Milnor [7] to prove the induction step) and $h$-cobordism theorem [**16**], we get $f' \in \text{Diff } D^n$ as the result of a series of surgeries on $f \in \text{Diff } L$. But $[f'|_{\partial D^n}] = [f|_{\partial L}]$. Hence $I_0(M) = 0$.

We now study the obstruction group.

PROPOSITION 2.3. *If $M^n$ is an $(r-1)$-connected $(r < [n/2])$ $\pi$-manifold, then* $\pi_r^{M(\alpha)}(SO_{n-r}) \cong \pi_r(SO_{n-r})$.

PROOF. Suppose that $b(g) \in Q_r^{M(\alpha)}(SO_{n-r})$ and $g \sim_G \text{id}_M$, i.e., $G$ is an isotopy of $g$ to $\text{id}_M$. Define an imbedding $\beta: S^r \times S^1 \to M \times S^1$ by $\beta((x, [t])) = [G(\alpha(x, 0), t)]$, where we regard $S^1 = I/\{0\} = \{1\}$ and $M \times S^1 = M \times I/(x, 0) = (x, 1)$. Since we may assume that $G$ is a product near the two ends, $\beta$ is an imbedding.

Let $\pi: S^r \times D^{n-r} \to S^r$ be the trivial bundle. Define $\pi': T(b(g)) \to S^r \times S^1$ by $T(b(g)) = S^r \times D^{n-r} \times I/(x, v, 0) = (x, b(g)(x)v, 1)$ and $\pi'([x, v, t]) = (\pi(x, v), [t]) = (x, [t])$.

It is easy to see that the tubular neighborhood $\nu_\beta$ of $\beta$ is bundle isomorphic to $\pi'$. By Lemma 3.5 of [7], $\nu_\beta$ is trivial; thus $\pi' \cong \pi \times \text{id}_{S^1}$. This implies that $b(g)$ is bundle homotopic to the identity. Hence $Q_r^{M(\alpha)}(SO_{n-r}) = 0$.

REMARK. $Q_r^{M(\alpha)}(SO_{n-r}) \neq 0$ in general. For an example, if $M = S^p \times S^{p+i}$, $i = 0, 1$ and $\alpha: S^p \times D^{p+i} \to S^p \times D_+^{p+i} \to S^p \times S^{p+i}$ is the obvious imbedding, then $Q_p^{M(\alpha)}(SO_{p+i}) = \text{Ker}(s_*)$, where $s_*: \pi_p(SO_{p+i}) \to \pi_p(SO_{p+i+1})$ is the homomorphism induced by the inclusion $s: SO_{p+i} \to SO_{p+i+1}$.

For a general manifold, we have the following proposition the proof of which uses Haefliger [**6**]. We omit the proof.

PROPOSITION 2.4. *Let $M^n$ be an $(r-1)$-connected manifold* $(1 \leqq r < (2n-4)/3)$ *such that* (1) $H_{r+1}(M) = 0$ *and* (2) *a set of generators of $H_r(M)$ containing $\alpha$ can be represented by a set of disjoint imbeddings of $S^r \times D^{n-r}$ into $\mathring{M}^n$. Then* $\pi_r^{M(\alpha)}(SO_{n-r}) \cong \pi_r(SO_{n-r})$.

REMARK. The remark after Proposition 2.3 shows that we cannot drop (1) or (2) in Proposition 2.4.

PROOF OF COROLLARY 2.5. Define a map $c : SO_{p+1} \to \text{Diff } S^p$ by $c(v)(x) = v(x)$, $v \in SO_{p+1}$ and $x \in S^p$. It is known that $c_*: \pi_*(SO_{p+1}) \to \pi_*(\text{Diff } S^p)$ is injective. (See [1].) We have the following commuting diagram:

$$\begin{array}{ccccc}
\pi_i(SO_p) & & & & \\
\cong \downarrow (\text{inc})_* & & & & \\
\pi_i(SO_{p+1}) & & & & \\
\downarrow c_* & & & & \\
\pi_i(\text{Diff } S^p) & \xrightarrow{d_1} & \pi_0^+(\text{Diff}(S^i \times S^p)) & \xrightarrow{\varphi_\alpha} & \pi_i(SO_p) \\
\downarrow e_{1*} & & \downarrow e_{2*} & & \cong \downarrow (\text{inc})_* \\
\pi_i(\text{Diff}(S^p \times M)) & \xrightarrow{d_2} & \pi_0^+(\text{Diff}(S^i \times S^p \times M)) & \xrightarrow{\varphi_\beta} & \pi_i(SO_{p+n}).
\end{array}$$

The homomorphisms in the diagram are defined as follows.

$e_1(f)(x, y) = (f(x), y), f \in \text{Diff } S^p$ and $(x, y) \in S^p \times M$,

$e_2(g)(x, y, z) = (g(x, y), z), g \in \text{Diff } (S^i \times S^p)$ and $(x, y, z) \in S^i \times S^p \times M$,

$d_1(u)(x, y) = (x, u(x)(y)), u \in \pi_i(\text{Diff } S^p)$ and $(x, y) \in S^i \times S^p$,

$d_2(v)(x, y, z) = (x, v(x)(y, z)), v \in \pi_i(\text{Diff}(S^p \times M))$ and $(x, y, z) \in S^i \times S^p \times M$.

The range groups of $\varphi_\alpha$ and $\varphi_\beta$ are as in the diagram because of Proposition 2.4, where $\alpha: S^i \times D^p \to S^i \times D^p_+ \to^{\text{inc}} S^i \times S^p$ and $\beta: S^i \times D^{p+n} \to S^i \times D^p \times D^n \to S^i \times D^p_+ \times D^n \to^{\text{inc}} S^i \times S^p \times M^n$ are the obvious imbeddings.

It is clear that $\varphi_\alpha \circ d_1 \circ c_* \circ (\text{inc})_* = \text{id}_{\pi_i(SO_p)}$. Therefore $e_{1_*} \circ c_* \cdot (\text{inc})_*$ is injective.

**3. Homomorphisms induced by surgery.** Let $M^n$ be an $(r - 1)$-connected ($r < (2n - 4)/3$) manifold such that $H_r(M)$ is cyclic and $H_{r+1}(M) = 0$ (we make the last two assumptions to make the discussion simple). Let an imbedding $\alpha: S^r \times D^{n-r} \to \mathring{M}^n$ represent a generator of $H_r(M)$. Define $X = M \times I \cup_{\alpha_0} (D^{r+1} \times D^{n-r})_0 \cup_{\alpha_1} (D^{r+1} \times D^{n-r})_1$, where $\alpha_i: S^r \times D^{n-r} \to^\alpha M \to M \times \{i\}$, $i = 0, 1$, is the obvious imbedding.

Suppose that $f_0 \sim_G f_1 \sim \text{id}_M$ and $f_i$, $i = 0, 1$, fixes the points in a neighborhood of $\text{Im}(\alpha)$. We can define $\bar{G} \in \text{Diff } X$ by $\bar{G}|_{M \times I} = G$ and $\bar{G}|_{X - M \times I} = $ identity. It is easy to see that $X$ is $r$-connected and $[\bar{G}] \in \pi_0^+(\text{Diff } X)$.

By regarding $S^{r+1} = D^{r+1}_+ \cup S^r \times I \cup D^{r+1}_-$, define an imbedding $\bar{\alpha}: S^{r+1} \times \frac{1}{2}D^{n-r} \to \mathring{X}$ as follows.

$$\bar{\alpha}|_{D^{r+1}_+ \times 1/2 D^{n-r}}: D^{r+1}_+ \times \tfrac{1}{2}D^{n-r} \xrightarrow{\text{inc}} (D^{r+1} \times D^{n-r})_1 \subset X,$$

$$\bar{\alpha}|_{S^r \times I \times 1/2 D^{n-r}}: S^r \times I \times \tfrac{1}{2} D^{n-r}$$

$$= S^r \times \tfrac{1}{2} D^{n-r} \times I \xrightarrow{(\alpha|S^r \times 1/2 D^{n-r}) \times \text{id}} M \times I \subset X,$$

$$\bar{\alpha}|_{D^{r+1}_- \times 1/2 D^{n-r}}: D^{r+1}_- \times \tfrac{1}{2} D^{n-r} \xrightarrow{\text{inc}} (D^{r+1} \times D^{n-r})_0 \subset X.$$

We have the homomorphism $\varphi_{\bar{\alpha}}: \pi_0^+(\text{Diff } X) \to \pi_{r+1}^{X(\bar{\alpha})}(SO_{n-r})$. Let $B_{r+1}^{X(\bar{\alpha})}(SO_{n-r}) = \{\varphi_{\bar{\alpha}}([\bar{G}]): f_0 \sim_G f_1 \sim \text{id}_M$ and $f_i$, $i = 0, 1$, fixes the points in a neighborhood of $\text{Im}(\alpha)\}$ and $\bar{B}_{r+1}^{X(\bar{\alpha})}(SO_{n-r}) = \{\varphi_{\bar{\alpha}}([\bar{G}]): \text{id}_M \sim_G \text{id}_M\}$. Define $B_{r+1}^{M(\alpha)}(SO_{n-r}) = B_{r+1}^{X(\bar{\alpha})}(SO_{n-r})/\bar{B}_{r+1}^{X(\bar{\alpha})}(SO_{n-r})$.

PROPOSITION 3.1. *If* $B_{r+1}^{M(\alpha)}(SO_{n-r}) = 0$, *then* $\phi\colon \mathrm{Ker}(\varphi_\alpha) \to D(M')$ *defined by* $\phi([f]) = [f']$ *is a homomorphism. We may replace* $D(M')$ *with* $\pi_0(\mathrm{Diff}\ M')$ *if* $M^n$ *is closed and* $n \geqq 5$.

We show that $\phi$ is well defined. Suppose that $\mathrm{id}_M \sim_G f$ and $f$ fixes the points in a neighborhood of $\mathrm{Im}(\alpha)$. Let $f' \in \mathrm{Diff}\ M'$ be the result of a surgery on $f$ such that $f'|_{M-\mathrm{Im}(\alpha)} = f|_{M-\mathrm{Im}(\alpha)}$ and $f'|_{D^{r+1}\times S^{n-r-1}} =$ identity, where we regard $M' = (M - \mathrm{Im}(\alpha)) \cup D^{r+1} \times S^{n-r-1}$.

Since $B_{r+1}^{M(\alpha)}(SO_{n-r}) = 0$, there exists an isotopy $G_0$ of $\mathrm{id}_M$ to itself such that $\varphi_{\bar{\alpha}}(\bar{G} \circ \bar{G}_0) = 0$. Let $G' \in \mathrm{Diff}\ X'$ be the result of a surgery on $\bar{G} \circ \bar{G}_0 \in \mathrm{Diff}\ X$ using $\bar{\alpha}$. Now $X'$ is diffeomorphic to $M' \times I$. Hence $G'$ is a pseudo-isotopy of $\mathrm{id}_{M'}$ to $f'$.

The second assertion of the proposition is trivial, since $D(M') \simeq \pi_0(\mathrm{Diff}\ M')$ by [5].

REMARK. It is clear that $B_{r+1}^{M(\alpha)}(SO_{n-r}) = 0$ if $r < (n-2)/2$ and $r + 1 \equiv 2, 4, 5, 6$ (Mod 8). If $M = S^p \times S^q$, $p + 1 < q$ and $\alpha\colon S^p \times D^q \to S^p \times D^q_+ \subset S^p \times S^q$ is the obvious imbedding, it is easy to see that $B_{p+1}^{M(\alpha)}(SO_q) = 0$. Under the assumptions of Proposition 3.1, if we define $\phi'\colon \mathrm{Ker}(\varphi_\alpha) \to D(W)$ by $\phi'([f]) = [F]$, then $\phi'$ is a homomorphism.

There are two more homomorphisms induced by surgery in the next proposition.

PROPOSITION 3.2. *Let* $M^n$ *be an* $(r+2)$*-connected* $(r < (2n-5)/3)$ *manifold. For an imbedding* $\alpha\colon S^r \times D^{n-r} \to \mathring{M}$, *there exist homomorphisms* $\psi\colon \pi_0(\mathrm{Diff}\ M) \to D(M')$ *and* $\psi'\colon \pi_0(\mathrm{Diff}\ M) \to D(W)$.

Here we only give the definitions of $\psi$ and $\psi'$. We can assume that there exists an imbedded disk $D^n$ in $M^n$ so that $\alpha(S^r \times D^{n-r}) \subset \mathring{D}^n \subset M^n$. By the disk theorem ([4] and [12]), for any $[f] \in \pi_0(\mathrm{Diff}\ M)$, there exists $g \in [f]$ such that $g$ fixes the points in $D^n$. Let $F = g \times \mathrm{id} \cup \mathrm{id}_{D^{r+1}\times D^{n-r}}$ and $f' = F|_{M'}$.

Define $\psi([f]) = [f'] \in D(M')$ and $\psi'([f]) = [F] \in D(W)$.

Now we give a sketchy proof of Proposition 3.3. We write $\partial E = S^p \times D^q_+ \cup_{f_b} S^p \times D^q_-$, where $f_b \in \mathrm{Diff}(S^p \times S^{q-1})$ is defined by $f_b(x, y) = (b(y)x, y)$ for $(x, y) \in S^p \times S^{q-1}$. Let $\alpha\colon S^p \times D^q \to S^p \times D^q_+ \subset \partial E$ be the obvious imbedding. By Proposition 2.4, $\varphi_\alpha$ is a homomorphism from $\pi_0^+(\mathrm{Diff}\ \partial E)$ to $\pi_p(SO_q) = \pi_p(SO)$.

Let $\sigma\colon \pi_p(SO_{q-1}) \otimes \pi_{q-1}(SO_p) \to \Gamma^{p+q+1}$ be the Milnor's pairing. (See [1] or [9] for the definition.) Then the homomorphism $\sigma_b$ has the property that $\sigma_b(s_*(d)) = \sigma(d, c)$, where $d \in \pi_p(SO_{q-1})$ and $s_*(c) = b$ for some $c \in \pi_{q-1}(SO_p)$. We do not give a precise definition of $\sigma_b$ here.

Proposition 3.1 and the remark after the proposition give a homomorphism $\phi\colon \mathrm{Ker}(\varphi_\alpha) \to D(M')$ but $M' \simeq S^{p+q}$. Thus $D(M') = \Gamma^{p+q+1}$ and it is easy to see that $\phi$ is onto.

Proposition 3.2 gives a homomorphism $\psi\colon \Gamma^{p+q+1} \to \pi_0(\mathrm{Diff}\ \partial E)$ for the imbedding $\alpha'\colon S^{q-1} \times D^{p+1} \to S^{p+q}$ induced by the surgery on $\alpha$. It can be seen that $\mathrm{Im}(\psi) \subset \mathrm{Ker}(\varphi_\alpha)$ and $\phi \circ \psi = \mathrm{id}_{\Gamma^{p+q+1}}$.

Finally, any diffeomorphism in $\mathrm{Ker}(\phi)$ is extendible over $E$ and a further argument shows that $\mathrm{Ker}(\phi)$ is isomorphic to a quotient group of $\pi_q(SO_{p+1})$. If $E$ is trivial, we use the homomorphism $\phi'$ in the remark after Proposition 3.1 and Proposition 2.4 to see that $H = \pi_q(SO_{p+1})$.

## Bibliography

**1.** P. Antonelli, D. Burghelea and P. Kahn, *The non-finite type of some diffeomorphism groups*, Topology **11** (1972), 1–49. MR **45** #1193.

**2.** R. Bott, *The stable homotopy of the classical groups*, Ann. of Math. (2) **60** (1959), 313–337. MR **22** #987.

**3.** W. Browder, *On the action of* $(\theta_n(\partial\pi))$, "*differentiable and combinatorial topology*," A symposium in honor of Marston Morse, Princeton, 1965, 23–36.

**4.** J. Cerf, *Topologie de certains espaces de plongements*, Bull. Soc. Math. France **89** (1961), 227–380. MR **25** #3543.

**5.** ———, *La stratification naturelle des espaces de fonctions différentiables réeles et le théorème de la pseudo-isotopie*, Inst. Haut Études Sci. Publ. Math. No. **39** (1970), 5–173. MR **45** #1176.

**6.** A. Haefliger, *Plongement différentiables de variétès dans varietes*, Comment. Math. Helv. **39** (1961), 47–82. MR **26** #3069.

**7.** M. Kervaire and J. Milnor, *Groups of homotopy spheres*. I, Ann. of Math. (2) **77** (1963), 504–537. MR **26** #5584.

**8.** A. Kosinski, *On the the inertia groups of* $\pi$*-manifold*, Amer. J. Math. **89** (1967), 227–248. MR **35** #4936.

**9.** T. Lawson, *Remarks on the pairings of Bredon, Milnor, Milnor-Munkres-Novikov*, Indiana Univ. J. **22** (1972/73), 833–843. MR **47** #1063.

**10.** J. Levine, *Self equivalences of* $S^n \times S^k$, Trans. Amer. Math. Soc. **143** (1969), 523–543. MR **40** #2098.

**11.** J. Milnor, *Differential structures*, lecture notes, Princeton Univ., 1961.

**12.** R. Palais, *Extending diffeomorphisms*, Proc. Amer. Math. Soc. **11** (1960), 274–277. MR **22** #8515.

**13.** H. Sato, *Diffeomorphlsm groups of* $S^p \times S^q$ *and exotic spheres*, Quart. J. Math. Oxford Ser. (2) **20** (1969), 255–276. MR **40** #6584.

**14.** ———, *Diffeomorphism groups and classification of manifolds*, J. Math. Soc. Japan **21** (1969), 1–36. MR **39** #3525.

**15.** R. Schultz, *On the inertia group of a product of spheres*, Trans. Amer. Math. Soc. **156** (1971), 137–153. MR **43** #1209.

**16.** S. Smale, *Generalized Poincaré conjecture in dimensions greater than four*, Ann. of Math. (2) **74** (1961), 391–406. MR **25** #580.

**17.** E. Terner, *Diffeomorphisms of a product of spheres*, Invent. Math. **8** (1969), 69–82.

**18.** C. T. C. Wall, *The actions of* $\Gamma_{2n}$ *on* $(n-1)$*-connected* $2n$*-manifolds*, Proc. Amer. Math. Soc. **13** (1962), 943–944. MR **26** #783.

University of Utah

Proceedings of Symposia in Pure Mathematics
Volume 32, 1978

# SMOOTH ACTIONS OF SMALL GROUPS ON EXOTIC SPHERES

REINHARD SCHULTZ*

Given a smooth manifold $M^n$ with a great deal of differentiable symmetry and a second smooth manifold $N^n$ homeomorphic to $M$, it is natural to ask how much differentiable symmetry $N$ has. In general, one expects $N$ to be less symmetric than $M$. For example, if $M$ is a sphere and $N$ is not diffeomorphic to $M$, then this is true by classical differential-geometric results. Actually, M. Davis and the Hsiangs have shown $N$ to be far less symmetric than $S^n$ in this case. On the other hand, it is often possible to find smooth actions on $N$ that imitate familiar smooth actions of finite groups, $S^1$, or $S^3$ on $M$ (e.g., [2], [3], [5], [6], [9], [12], [19]). This paper describes some results on the symmetry question in the special case of $M^n = S^n$ (so that $N^n$ is an exotic sphere). In particular, the existence of smooth $Z_p$ actions ($p$ an odd prime) is reduced entirely to homotopy theory (Theorem 4.1 and Proposition 4.3). Unfortunately, this homotopy-theoretic reduction does not give strong global information at this time, but in principle the relevant homotopy problems can be worked out for any specific case.

**1. Motivating examples.** Perhaps the most basic examples of actions on exotic spheres are given using the Brieskorn representations for boundaries of parallelizable manifolds: $\{z_0^{a_0} + \dots z_n^{a_m} = 0\} \cap \{\sum |z_j|^2 = 1\}$. For example, if $a_0 = \dots = a_r = 2$, then the orthogonal group $O_{r+1}$ acts on the manifold. These actions are discussed thoroughly in many places (e.g., [8]); as noted elsewhere in these PROCEEDINGS, there are no similar large actions on exotic spheres not bounding parallelizable manifolds (i.e., *very exotic spheres*). On the other hand, Bredon first showed that $S^1$ or $S^3$ could act smoothly on very exotic spheres [2], and his examples became included in subsequent work of several others [3], [5], [6], [9]. Here is the

*AMS (MOS) subject classifications* (1970). Primary 57E25; Secondary 57E15, 57D55.

*Partially supported by NSF Grants MPS 74–03609 and MCS 76–08794.

basic idea: Suppose we start out with a *semifree linear* $G$-action on $S^n$ (i.e., $G$ has only free orbits and fixed points), with fixed point set $S^k$. Then the equivariant normal bundle of $S^k$ has the form $S^k \times D^{n-k}$, and one constructs a twisted action on some homotopy $\Sigma^m$ by regluing along $S^k \times S^{n-k-1}$ via some equivariant diffeomorphism $k$ homotopic to the identity (note that $G$ acts freely on $S^k \times S^{n-k-1}$). Such diffeomorphisms have been extensively studied using surgery theory; in particular, one can define normal invariants for them valued in $[S^1 \wedge (S^k \times S^{n-k-1}/G), F/O]$ (compare [3], [4], [21]; indeterminacies need not concern us here). This set of normal invariants admits an *additive* surgery obstruction map $\sigma$, and it follows by surgery theory that *all elements in the image of the composite*

$$(1.1)\quad \text{Kernel}\,\sigma \subseteq [S^1 \wedge (S^k \times S^{n-k-1}/G), F/O] \xrightarrow{\pi^*} [S^1 \wedge (S^k \times S^{n-k-1}), F/O] \xrightarrow{(\text{Hopf constr.})^*} [S^{n+k}, F/O]$$

*correspond to exotic spheres with twisted semifree G-actions* in the above sense. As a particular example, we have the following result, half due to Bredon [2] and half due to the author [16]:

THEOREM 1.2. *Suppose* (*for a given n*) *there is a homotopy n-sphere that does not bound a Spin manifold* (*i.e.*, $n \equiv 1$ *or* 2 mod 8, $n \geqq 9$ [20]). *Then some homotopy sphere of this sort supports a smooth semifree* $S^1$ *action.*

Not all exotic spheres admit such twisted semifree actions, however; in particular, the exotic 8-sphere does not if $G = S^1$. On the other hand, a different gluing construction yields a semifree $S^1$ action on the exotic 8-sphere and the elements of $Z_p \cong \Theta_{2p^2-2p-2(p)}$ ($p$ an odd prime) which are the first very exotic spheres of order $p$ [12], [13]. Much of the work described below arose from efforts to understand these examples and fit them into the classification theory of semifree actions on homotopy spheres due to Browder, Petrie, and Rothenberg [5], [9].

Finally, there are *nonexistence* results for actions on exotic spheres that arise. Specifically, the existence of actions with certain representations at fixed points forces some necessary conditions on the ambient exotic spheres' Pontrjagin-Thom invariant; see [10], [11] for some early results of this type.

**2. Differentiable structures and normal invariants.** An understanding of the latter results and a partial clarification of the other results is given by [15], [17], the main results of which we shall describe here. Given a smooth $Z_p$ action on a homotopy sphere $\Sigma^n$ with $k$-dimensional fixed point set, a *knot invariant* $\omega(\Sigma, Z_p)$ was defined in $\pi_k(F_{Z_p}(V)/C_{Z_p}(V))_{(p)}$, where $V$ is the space of normal vectors at a fixed point of $\Sigma$, and the homotopy group classifies pairs $(\xi, \rho)$ where

(i) $\xi$ is a $Z_p$-vector bundle over $S^k$, with free action off the zero section and fiber $V$.

(ii) $\rho$ is an equivariant retraction from the unit sphere bundle of $\xi$ to the unit sphere in $V$.

In particular, this data yields a fiber homotopy trivialization for the associated lens space bundle of $\xi$, and hence a canonical homotopy smoothing of $S^k \times L(V)$ where $L(V)$ is the lens space associated to $V$. Pairs of the form $(\xi, \rho)$, for various $V$, admit an obvious Whitney sum operation; for example, we can add an arbitrary free $Z_p$-module $T$ to get a new object $(\xi \oplus T, \rho \oplus T)$ and a new homotopy smoothing of $S^k \times L(V \oplus T)$. Given *this* homotopy smoothing, we can take its localized normal invariant in

$$(2.1) \qquad [S^k \times L(V \oplus T), F/O]_{(p)}.$$

Actually, we only need a piece of this invariant; using the standard fact that $[S^k L(V \oplus T), F/O]_{(p)}$ is a direct summand of (2.1), we shall take the normal invariant associated to $\omega(\Sigma, Z_p)$ and map it into $[S^k L(V \oplus T), F/O]_{(p)}$.

REMARK. It is more or less implicit in **[5]**, **[9]**, **[14]** that the map $N_*$: $\pi_k(F_{Z_p}(V \oplus T)/C_{Z_p}(V \oplus T))_{(p)} \to [S^k L(V \oplus T), F/O]_{(p)}$ is a homomorphism.

The main result of **[17**, § 3**]** may be restated as follows:

THEOREM 2.2. *Let $c_j \in H^j(S^k L(V \oplus T); \pi_j(F/O)_{(p)})$ be the jth obstruction to making $N_*\omega(\Sigma^n, Z_p)$ nullhomotopic. Then $c_j = 0$ for $j < n$ and $c_n \in H^n(=\pi_n(F/O) \otimes Z_p$ or $\mathrm{Tor}(\pi_n(F/O), Z_p))$ corresponds to minus the Pontrjagin-Thom invariant of $\Sigma$.*

The earlier nonexistence results (and more) are recoverable from Theorem 2.2 and some elementary homotopy-theoretic calculations **[17**, §4**]**.

**3. Normal invariant formulas.** Theorem 2.2 brings us to calculation of the homomorphism

$$(3.1) \qquad \pi_k(F_{Z_p}(V \oplus T)/C_{Z_p}(V \oplus T)) \xrightarrow{\xi} hS(S^k \times L(V \oplus T)) \xrightarrow{\eta} [S^k \times L(V \oplus T), F/O],$$

where $\eta$ is the normal invariant and $\xi$ gives the canonical homotopy smoothing constructed from the vector bundle and fiber retraction. It is convenient to decompose the domain using the exact homotopy sequence of the fibration.

$$(3.2) \qquad F_{Z_p}(V \oplus T) \xrightarrow{i} F_{Z_p}(V \oplus T)/C_{Z_p}(V \otimes T) \longrightarrow BC_{Z_p}(V \oplus T).$$

The first step is to observe that $i * \eta\xi$ may be factored through $[S^k \times L(V \oplus T), F] \cong \{S^k \times L(V \oplus T), S^0\}$; this is true because the homotopy smoothings constructed by $\xi$ are stably tangential, and the factorization is in fact canonical. Denote the canonical factorization of $\eta\xi i_*$ by $\tilde{N}_*$.

Although $\tilde{N}_*$ is defined geometrically, it turns out to have a relatively concise homotopy-theoretic description. Recall from **[1]** that $F_{Z_p}$ is homotopy equivalent to $\Omega^\infty S^\infty(BZ_p^+)$. Let $\bar{\lambda} : F_{Z_p}(V \oplus T) \to \Omega^\infty S^\infty(BZ_p^+)$ be the map constructed there combined with stabilization, and let $\bar{\lambda}_* : \pi_k(F_{Z_p}(V \oplus T)) \to \pi_k^s(BZ_p^+)$ be the induced homomorphism. Also, let $j: L(V \oplus T) \to BZ_p$ be the standard classifying map. Then $\tilde{N}_*$ may be expressed as follows:

THEOREM 3.3. *The restriction of $\tilde{N}_*$ to $\{L(V \oplus T), S^0\}$ is zero, and on complementary direct summand of $\{S^k \times L(V \oplus T), S^0\}$ — identifiable with $\{S^k \wedge (L(V \oplus T)^+), S^0\}$—it is given by the stable cohomotopy slant product $\tilde{N}_*(x) = S^k(j^+)^*[\delta/\bar{\lambda}_*(x)]$, where $\delta \in \{BZ_p^+ \wedge BZ_p^+, S^0\}$ is the composite*

$$BZ_p^+ \wedge BZ_p^+ = B(Z_p \times Z_p)^+ \xrightarrow[\text{diagonal map of } Z_p]{\text{transfer for}} BZ_p^+ \xrightarrow[BZ_p]{\text{collapse}} S^0.$$

There is a corresponding formula for semifree actions with arbitrary groups acting. In particular, this gives an alternate proof for the results of **[12]**, **[13]** mentioned earlier (and greatly simplifies the proofs for the examples announced in **[15**, p. 964]).

Finally, we have to consider $N_*$ on elements that map nontrivially into $\pi_k(BC_{Z_p}(V \oplus T))_{(p)}$. Recall that $C_{Z_p}(V \oplus T)$ is a product $\prod_i U(j_i)$; since $T$ is arbitrary, pick it so large that $\pi_k(BC_{Z_p}(V \oplus T))$ is stable. We may further simplify matters using the following result, essentially due to J. Ewing **[7]**.

PROPOSITION 3.4. *Let $T$ be large as above. Then $\omega(\Sigma, Z_p)$ always maps trivially into $\pi_k(BC_{Z_p}(V \oplus T))_{(p)}$ except if $k = 2$, $p$ is odd, and 2 has odd multiplicative order* mod *p*.

In the exceptional case, however, $\omega(\Sigma, Z_p)$ can map nontrivially (see **[19]**, for example). However, the following result helps us around this point.

THEOREM 3.5. *Let $\xi$ be a real $Z_p$ vector bundle over $S^2$ (trivial action on base), and assume* (i) *$Z_p$ acts freely off the zero section,* (ii) *the unit sphere bundle is equivariantly fiber homotopically trivial. Also suppose 2 has odd multiplicative order* mod *p. Then there are* (a) *an "Adams conjecture splitting" of $F/O_{(p)}$ as $BSO_{(p)} \times \mathrm{Cok}\, J_{(p)}$ and* (b) *an equivariant fiber homotopy trivialization $t$ of $S(\xi)$, such that the normal invariant of $t$ lies in the $KO(S^k \times L(V \oplus T))_{(p)}$ summand of $[S^k \times L(V \oplus T), F/O]_{(p)}$.*

The next statement follows easily from 2.2 and 3.5.

COROLLARY 3.6. *Suppose $Z_p$ ($p$ odd) acts smoothly on a homotopy sphere $\Sigma$ with fixed point set $S^2$. Then there is a possibly different $Z_p$ action on $\Sigma$ with fixed point set $S^2$ and $\omega(\Sigma, Z_p)$ mapping trivially into $\pi_k(BC_{Z_p}(V \oplus T))_{(p)}$.*

In effect, Corollary 3.6 tells us that we can ignore the cokernel of $i_*$ when looking for exotic spheres supporting smooth $Z_p$ actions.

**4. Conditions for existence of $Z_p$ actions.** Theorems 1.2 and 3.3 give rather strict necessary conditions for an exotic sphere $\Sigma^n$ to support a smooth $Z_p$ action with $k$-dimensional fixed point set (see 4.1 for a precise statement). Using the Browder-Petrie-Rothenberg classification of $Z_p$ actions with homotopy spheres as fixed set, together with a few extra arguments, one can prove these necessary conditions are also sufficient if $p \neq 2$.

THEOREM 4.1. *Let $\Sigma^n$ be a homotopy n-sphere (say $n \geqq 7$), let $0 < k < n$ be congruent to $n$* mod 2, *and let $p$ be odd. Then $\Sigma^n$ supports a smooth $Z_p$ action with k-dimensional fixed point set if and only if there is a class $y \in \pi_k^s(BZ_p^+)$ for which the following hold:*

(i) *The class $S^k(j^+)^*[\delta/y] \in \{S^k(L(V \oplus T))^+, S^0\}$ is nullhomotopic on the $(n-1)$-skeleton, and the obstruction to null-homotoping it on the n-skeleton lies in the Pontrjagin-Thom coset of $\Sigma$, reduced* mod *p*.

(ii) *If $\lambda: F_{Z_p} \to Q(BZ_p^+)$ is the homotopy equivalence of* **[1]**, *then the image of $\lambda_*^{-1}(y)$ in $\pi_k(F_{Z_p}/C_{Z_p})_{(p)}$ desuspends to $\pi_k(F_{Z_p}(V)/C_{Z_p}(V))$ for some $(n-k)$-dimensional free $Z_p$ module $V$.*

If $n-k$ is large with respect to $n$, the second condition is automatic. In this "general position" case, there is a more accessible characterization based upon the Atiyah-Hirzebruch spectral sequence

(4.2) $$H_i(BZ_p^{-W}; \pi_j) \Rightarrow \pi_{i+j}^S(BZ_p^{-W}).$$

Here $-W$ refers to the virtual vector bundle determined by $-W \in RO(Z_p)$, and the bottom cell of the spectrum $BG^{-W}$ has dimension $-\dim W$.

PROPOSITION 4.3. *Let $\Sigma^n$, $k$, $p$ be as above, and let $V$ be a fixed $(n-k)$-dimensional free $Z_p$ module. Assume $n > k+2$. Then $\Sigma$ supports a smooth $Z_p$ action with $k$-dimensional fixed point set and normal representation $V$ at fixed points if and only if the class in $E^2_{-(n-k),n}$ $(4.2)_{W=V\oplus T} \cong \pi_n \otimes Z_p$, corresponding to the Pontrjagin-Thom invariant of $\Sigma$, is a boundary in the spectral sequence (here $T$ is some nonzero free $Z_p$-module).*

I believe it is possible to use the above results and some homotopy theory to prove that *every* exotic sphere admits a $Z_p$ action, the homotopy-theoretic contribution resembling the Kahn-Priddy theorem in some way. Unfortunately, at present this is little more than wishful thinking.

**5. Nonsemifree actions and circle actions.** The question considered in the preceding paragraph is a weakening of the following one:

(5.1) Given an exotic sphere $\Sigma^n$ (say $n \geqq 7$), does $\Sigma$ admit a smooth effective $S^1$ action?

One approach to studying this involves looking at the action restricted to a cyclic subgroup, say of prime power order $p^r$. As suggested before (e.g., after Theorem 3.3), everything done until now works perfectly well for semifree $Z_{p^r}$ actions. However, it is possible to find exotic spheres that do *not* admit semifree $Z_{p^2}$ actions (e.g., take the exotic 14-sphere with $p = 2$), and thus any search for circle actions must take nonsemifree actions of $Z_{p^r}$ into account. In **[18]**, **[19]** this is done by introducing the notion of an *ultrasemifree* action: The principal isotropy subgroup is the identity, and the remaining conjugacy classes of isotropy subgroups have a *unique minimal element*, say $H$. This condition is trivial for $Z_{p^r}$ actions because the subgroup lattices are linearly ordered in this case. It follows that $G$ acts freely off the fixed point set of the subgroup $K = \bigcap\{gHg^{-1} \mid g \in G\}$,[1] and this is the key to extending the theory of semifree actions to arbitrary $Z_{p^r}$ actions. All of the above results through Proposition 3.4 have suitable analogs for ultrasemifree actions, and one can combine these with some homotopy theory to get many new $S^1$ actions on exotic spheres. In particular, one can construct smooth $S^1$ actions on arbitrary exotic spheres at least through dimension 17.[2] Furthermore, if $p$ is an odd prime, then every exotic sphere with order a power of $p$ admits a circle action for a reasonably large range of dimensions, through at least dim $\beta_{p-1}$.

Perhaps a few further calculations would be interesting in themselves, but it would be preferable to try for a *global* understanding of (5.1) instead. Algebraically, the above examples reflect the decomposability of elements in $\pi_*$ (represent-

[1]Strictly speaking, the definition of ultrasemifree actions includes an assumption that $K \neq 1$; of course, at any rate this is trivial if $G$ is abelian.

[2]The generators of $\Theta_{10} = Z_6$ require some concepts from Part III of **[17]**, **[18]** not presented here, chiefly because the order is not a prime power.

ing the given exotic spheres) via lower dimensional elements, Hopf maps, and higher order composition; for example, the original actions of [2] correspond to primary composition relations. Clearly some formal framework, for listing and organizing the "admissible" operations that yield circle actions, is an ultimate goal. The results of §4 may be viewed as *partial* answer to the corresponding question of $Z_p$-actions; however, it is to be expected that any analog for $S^1$-actions will be considerably more complicated.

## REFERENCES

**1.** J. C. Becker and R. E. Schultz, *Equivariant function spaces and stable homotopy theory*. I, Comment. Math. Helv. **49** (1974), 1–34. MR **49** #3994.

**2.** G. Bredon, *A $\Pi_*$-module structure for $\Theta_*$ and applications to transformation groups*, Ann. of Math. (2) **86** (1967), 434–438. MR **36** #4570.

**3.** W. Browder, *Surgery and the theory of differentiable transformation groups*, Proc. Conf. on Transformation Groups (New Orleans, La., 1967), pp. 1–46. Springer, New York, 1968. MR **41** #6242.

**4.** ———, *Diffeomorphisms of* 1-*connected manifolds*, Trans. Amer. Math. Soc. **128** (1967), 155–163. MR **35** #3681.

**5.** W. Browder and T. Petrie, *Diffeomorphisms of manifolds and semifree actions on homotopy spheres*, Bull. Amer. Math. Soc. **77** (1971), 160–163. MR **42** #8513.

**6.** G. Brumfiel, *Differentiable $S^1$ actions on homotopy spheres*, mimeographed, University of California, Berkeley, 1968.

**7.** J. Ewing, *Spheres as fixed point sets*, Quart. J. Math. Oxford Ser. (2) **27** (1976), 445–455.

**8.** F. Hirzebruch and K. H. Mayer, *$O(n)$-Mannigfaltigkeiten, exotische Sphären, und Singularitäten*, Lecture Notes in Math., vol. 57, Springer-Verlag, Berlin and New York, 1968. MR **37** #4825.

**9.** M. Rothenberg, *Differentiable group actions on spheres*, Proc. Adv. Study Inst. on Algebraic Topology (Aarhus, 1970), pp. 455–475. Mat. Inst., Aarhus Univ., 1970.

**10.** R. Schultz, *The nonexistence of free $S^1$ actions on certain homotopy spheres*, Proc. Amer. Math Soc. **27** (1971), 595–597. MR **42** #6866.

**11.** ———, *Semifree circle actions and the degree of symmetry of homotopy spheres*, Amer. J. Math. **93** (1971), 829–839. MR **44** #4752.

**12.** ———, *Circle actions on homotopy spheres bounding plumbing manifolds*, Proc. Amer. Math. Soc. **36** (1977), 297–300.

**13.** ———, *Circle actions on homotopy spheres bounding generalized plumbing manifolds*, Math. Ann. **205** (1973), 201–210. MR **52** #1749.

**14.** ———, *Homotopy sphere pairs admitting semifree differentiable actions*, Amer. J. Math. **96** (1974), 308–323.

**15.** ———, *Differentiable $Z_p$ actions on homotopy spheres*, Bull. Amer. Math. Soc. **80** (1974), 961–964. MR **50** #8576.

**16.** ———, *Circle actions on homotopy spheres not bounding spin manifolds*, Trans. Amer. Math. Soc. **213** (1975), 89–98. MR **52** #1750.

**17.** ———, *Differentiable group actions on homotopy spheres*. I. *Differential structure and the knot invariant*, Invent. Math. **31** (1975), 105–128. MR **53** #9264.

**18.** ———, *Differentiable group actions on homotopy spheres*. II. *Ultrasemifree actions and related topics*, mimeographed, Purdue University, 1977.

**19.** ———, *Spherelike G-manifolds with exotic equivariant tangent bundles*, Advances in Math. (to appear).

**20.** R. Stong, *Notes on cobordism theory*, Mathematical Notes No. 7, Princeton Univ. Press, Princeton, N. J., 1968.

**21.** E. C. Turner, *Diffeomorphisms homotopic to the identity*, Trans. Amer. Math. Soc. **186** (1974), 416–498. MR **49** #1532.

PURDUE UNIVERSITY

Proceedings of Symposia in Pure Mathematics
Volume 32, 1978

# AN OBSTRUCTION TO POINCARÉ TRANSVERSALITY

R. J. MILGRAM[1] AND I. HAMBLETON[2]

In [3] an invariant $A(X^{2n}, f)$ in $Z/2$ was defined for a double cover $\pi: \tilde{X} \to X$ of $2n$-dimensional Poincaré duality (PD) spaces classified by a map $f: X \to RP^{2n}$. If the homotopy class of the map $f$ contains a representative which is Poincaré transverse to $RP^{2n-1} \subset RP^{2n}$ [10], we say that $\pi$ is *Poincaré splittable*. The invariant $A(X, f)$ depends only on the bordism class of $(X, f)$ in $\mathcal{N}_{2n}^{\mathrm{PD}}(RP^\infty)$ and vanishes for Poincaré splittable covers. In particular, it vanishes for double covers of PL-manifolds. The authors pointed out that from the map $f: X \to RP^{2n}$, one can construct another obstruction to the existence of a Poincaré splittable double cover bordant to $(X, f)$. Let $\gamma^q \to BG(q)$ be the universal $(q-1)$-spherical fibration and $S^{2n-1} \to RP^{2n-1}$ the double cover (an $S^0$-fibration $\eta$). Then $MG(q) \wedge RP^{2n}$ is the Thom Space of $\gamma \times \eta \to BG(q) \times RP^{2n-1}$ and a Pontrjagin-Thom construction gives a map

$$p(f): S^{q+2n} \to MG(q) \wedge RP^{2n}.$$

If $\pi: \tilde{X} \to X$ is bordant to a Poincaré splittable cover then $p(f)$ is homotopic to a Poincaré transversal map. According to Jones [6], Levitt [7] or Quinn [10], there is one obstruction $\theta p(f)$ (in $Z/2$) to homotoping $p(f)$ to a Poincaré transversal map. In [3] the authors conjectured that $\theta p(f) = A(X, f)$ in all dimension $2n$ ($n \geqq 2$), but when [3] was written there were no known examples for which the invariant $A(X, f)$ was nonzero.

In this note, we construct examples $(X^{2n}, f)$ in all dimensions $2n \geqq 4$, for which $A(X^{2n}, f) = 1$, and outline the proof of the conjecture in dimension 4. This involves using the fact that $A(X, f) \neq 0$ to obtain the exotic characteristic classes of the

---

*AMS* (*MOS*) *subject classifications* (1970). Primary 57B10; Secondary 55G99.

[1]This research was supported in part by MPS 74–07491A01.
[2]Partially supported by the National Research Council of Canada.

Spivak normal bundle to our basic example in dimension 4.

One can establish product formulas for $\theta p(f)$ as is done in **[4]** and for $A(X, f)$ on the basis of our results in §2, and both formulas have the same general shape. Moreover, both $\theta p(f)$ and $A(X, f)$ vanish on $\mathrm{im}(\mathcal{N}_*^{\mathrm{PL}}(RP^\infty)) \subset \mathcal{N}_*^{\mathrm{PD}}(RP^\infty)$; hence, writing $\mathcal{N}_*^{\mathrm{PD}}(RP^\infty)/\mathrm{im}(\mathcal{N}_*^{\mathrm{PL}}(RP^\infty))$ as a module over $\mathcal{N}_*^{\mathrm{PL}}(RP^\infty)$ we must evaluate $A(X, f)$ and $\theta(p(f))$ on generators and show they agree in order to prove their equality. According to Brumfiel and Morgan **[4]**, the Pontrjagin-Thom map $\mathcal{N}_*^{\mathrm{PD}}(RP^\infty) \to \pi_*(RP^\infty \wedge MG)$ is an injection ($* \neq 2$). The problem then is to construct examples to realize enough exotic characteristic numbers. (One could begin by obtaining those which appear in the cohomological formula **[4]** for the transversality obstruction.) On the basis of the results of **[2]** and **[7]**, this program seems feasible but we have not yet attempted it.

The invariant $A(X, f)$ is an Arf invariant based on a quadratic map $q\colon H^n(\tilde{X}; Z/2) \to Z/2$ refining the nonsingular bilinear form $l(a, b) = \langle a \cup T^*b, [\tilde{X}] \rangle$ where $a, b \in H^n(\tilde{X}; Z/2)$ and $T\colon \tilde{X} \to \tilde{X}$ is the free involution. We prove that this map $q$ is the same as the Browder-Livesay map $\psi$ used in **[1]** to define a desuspension obstruction for smooth involutions on homotopy spheres.

Our basic example $X^4$ in dimension 4 is the orbit space of a free simplicial involution on $S^2 \times S^2$. The other examples are obtained from this one by forming the product with suitable smooth manifolds. In each case, the covering space is homotopy equivalent to a smooth manifold. We also indicate some generalizations of the construction using the results of **[9]** on projective homotopy. We sketch some proofs here; full details will appear elsewhere.

**1. A quadratic map for double covers.** In this section, we recall the definition of the quadratic map $q$ and prove that it equals the Browder-Livesay map. All cohomology groups have $Z/2$ coefficients and $[X]$ denotes the fundamental class of a PD space $X$.

Let $\pi\colon \tilde{X} \to X$ be a double cover of $2n$-dimensional PD spaces classified by $f\colon X \to RP^\infty$. We denote the involution on $\tilde{X}$ by $T$ and the map covering $f$ by $\tilde{f}\colon \tilde{X} \to S^\infty$. Form $S^\infty \times_{Z/2} (\tilde{X} \times \tilde{X})$ where $Z/2$ acts on $\tilde{X} \times \tilde{X}$ by interchanging the factors and define $F\colon X \to S^\infty \times_{Z/2} (\tilde{X} \times \tilde{X})$ as the quotient of the equivariant map $\tilde{F}\colon \tilde{X} \to S^\infty \times (\tilde{X} \times \tilde{X})$ given by $\tilde{F}(x) = (\tilde{f}(x), (x, Tx))$. Now if $a_\#$ is a cocycle on $\tilde{X}$ representing $a \in H^n(\tilde{X})$ then $1 \otimes a_\# \otimes a_\#$ is an equivariant cocycle on $S^\infty \times (\tilde{X} \times \tilde{X})$ so represents a class $\alpha \in H^{2n}(S^\infty \times_{Z/2} (\tilde{X} \times \tilde{X}))$.

DEFINITION. $q(a) = \langle F^*(\alpha), [X] \rangle$.

Let $Y = S^\infty \times_{Z/2} \tilde{X}$ and define $\lambda\colon Y \to S^\infty \times_{Z/2} (\tilde{X} \times \tilde{X})$ by $\lambda[u, x] = [u, (x, Tx)]$. If $\rho\colon Y \to X$ is given by $\rho[u, x] = \pi(x)$, then $\rho$ is a homotopy equivalence and $F \circ \rho \simeq \lambda$. We now describe a chain approximation for $\lambda$. Suppose that $T\colon \tilde{X} \to \tilde{X}$ is a simplicial map such that $T\sigma \cap \sigma = \emptyset$ for all simplices $\sigma \in X$ and partially order the simplices so that $T(a \cup_i b) = Ta \cup_i Tb$ where $\cup_i$ denotes the Steenrod cup-sub-$i$-product. We give $S^\infty$ its usual equivariant cellular decomposition with cells $e_i$ and $Te_i$ in each dimension. In the statement below, $\Delta_j\colon C_k(\tilde{X}) \to C_{k+j}(\tilde{X} \times \tilde{X})$ is the $j$th Steenrod map **[11]** and $\tau\colon C_k(\tilde{X} \times \tilde{X}) \to C_k(\tilde{X} \times \tilde{X})$ is defined by $\tau(a \otimes b) = b \otimes a$. We recall the formulas

$$\partial \Delta_j = (1 + \tau)\Delta_{j-1} + \Delta_j \partial \quad \text{and} \quad \Delta_j \cdot T = (T \otimes T)\Delta_j.$$

THEOREM 1. *The map given by*

$$\lambda_\sharp(e_i \otimes c) = \sum_{0 \leq j \leq i} e_j \otimes (1 \otimes T)\tau^j \Delta_{i-j}(c)$$

*and* $\lambda_\sharp(Te_i \otimes c) = (T \otimes \tau)\lambda_\sharp(e_i \otimes Tc)$ *is a chain approximation to* $\lambda$ *where* $c \in C_k(\tilde{X}; Z/2)$.

COROLLARY 2. *For* $a \in H^n(\tilde{X})$,

$$q(a) = \left\langle \sum_{i=0}^{n} e^i \otimes (a_\sharp \cup_i Ta_\sharp), [Y] \right\rangle$$

*where* $a_\sharp$ *is a cocycle representing* $a$, $e^i$ *is dual to* $e_i$ *and* $\rho_*[Y] = [X]$.

With this explicit cochain formula, we can relate $q$ to the Browder-Livesay map $\psi: H^n(\tilde{X}) \to Z/2$. First, we summarize their definition [1].

Let $x$ be a cocycle in $C^n(\tilde{X}; Z/2)$. Then since $x \cup_{n+1} Tx = 0, (1 + T)(x \cup_n Tx) = 0$ so $x \cup_n Tx = (1 + T)\nu^n$. Assuming that $\nu^{n+j}$ are constructed for $0 \leq j \leq i < n$ so that

$$x \underset{n-j}{\cup} Tx + \delta\nu^{n+j-1} = (1 + T)\nu^{n+j}$$

they construct $\nu^{n+i+1}$ satisfying a similar formula. The cochain $\nu^{2n}$ turns out to be determined modulo $\delta C^{2n-1}(\tilde{X}; Z/2) + (1 + T)C^{2n}(\tilde{X}; Z/2)$, and so the class $(1 + T)\nu^{2n}$ represents a cohomology class in $H^{2n}_{Z/2}(C_*(\tilde{X}); Z/2) \cong H^{2n}(X)$. Then if $a = \{x\} \in H^n(X)$ they set $\psi(a) = \langle\{(1 + T)\nu^{2n}\}, [X]\rangle \in Z/2$.

THEOREM 3. *For all* $a \in H^n(\tilde{X})$, $\psi(a) = q(a)$.

PROOF. By construction, $(1 + T)\nu^{2n} = x \cup Tx + \delta\nu^{2n-1}$ where $x$ is a cocycle representing $a$. Set

$$\nu\nu = \sum_{i=0}^{n-1} e^{n-i-1} \otimes \nu^{n+i}$$

and compute

$$\delta\nu = \sum_{i=0}^{n} e^i \otimes \left(x \underset{i}{\cup} Tx\right) + e^0 \otimes (1 + T)\nu^{2n}.$$

Therefore,

$$\delta\nu = \lambda^\sharp(e^0 \otimes x \otimes x) + \rho^\sharp(1 + T)\nu^{2n}$$

so

$$\langle \lambda^\sharp(e^0 \otimes x \otimes x), [Y] \rangle = \langle e^0 \otimes (1 + T)\nu^{2n}, [Y] \rangle$$

and the result follows.

**2. A product formula.** For the construction of the next section, we need to compute $q$ on $\tilde{X} \times N \to^{\pi \times 1} X \times N$ where $N^{2m}$ is a PD space of dimension $2m$. Our main applications are the cases $N = CP^2$ and $N = RP^2$.

THEOREM 4. *Let* $\tilde{X} \times N \to^{\pi \times 1} X \times N$ *be the product covering and* $p + r = n + m$.

*Let $a \in H^p(\tilde{X})$ and $b \in H^r(N)$, then*

$$q(a \otimes b) = \left\langle \sum_{0 \leq j \leq r} F^*(1 \otimes a \otimes a) \cup f^*(u)^j \otimes \mathrm{Sq}_j(b), [X] \otimes [N] \right\rangle$$

*where $u$ generates $H^1(RP^\infty)$.*

We now recall the definition of $A(X, f)$. ($f$: $X \to RP^\infty$ classifies $\pi$: $\tilde{X} \to X$.) According to **[1]** or **[3]**,

$$q(a + b) - q(a) - q(b) = \langle a \cup Tb, [\tilde{X}] \rangle$$

for all $a, b \in H^n(\tilde{X})$. The bilinear form defined by the formula on the right-hand side is nonsingular and even, so there exists a symplectic base for $H^n(\tilde{X})$ with respect to this form. $A(X,f)$ is the Arf invariant associated to any such base. From the definition of $q$ we easily verify that $A(X, f)$ depends only on the class of $(X, f)$ in $\mathcal{N}_*^{\mathrm{PD}}(RP^\infty)$ and vanishes for double covers of PL-manifolds. More generally,

PROPOSITION 5 **[3]**. *If $\pi$: $\tilde{X} \to X$ is a Poincaré splittable double cover of $2n$-dimensional* PD *spaces, then $A(X,f) = 0$ where $f$: $X \to RP^\infty$ classifies $\pi$.*

Using the product formula, we establish

COROLLARY 6. *If $\tilde{X} \times CP^2 \to^{\pi\times 1} X \times CP^2$ is the product covering, $A(X \times CP^2, fp_1) = A(X, f)$ where $p_1$: $X \times CP^2 \to X$ is the projection.*

**3. The examples.** We will now describe the basic example in dimension 4. It is a PD space $X^4$ with fundamental group $Z/2$ and nonzero $A$-invariant.

The complex $X^4$ is among those constructed in **[12**, p. 240**]**. Let $K^3$ be the 3-skeleton of $RP^2 \times S^2$ in a normal cell decomposition and note that $\tilde{K}^3 \simeq S_1^2 \vee S_2^2 \vee S^3$. We obtain $X^4$ by attaching the 4-cell $e^4$ by a different map than that used to get $RP^2 \times S^2$. To describe the map, we need to denote generators of $\pi_2 S_i^2$, $\pi_3 S^3$ and $\pi_3 S_i^2$ by $I_i$, $J$ and $\eta_i$, respectively, for $i = 1, 2$. Then, according to the Hilton-Milnor theorem, $\pi_3 K^3$ is generated by $J$, $\eta_1$, $\eta_2$ and $[I_1, I_2]$. The $Z/2$ action on these is given by

$$T(J) = J - [I_1, I_2], \quad T\eta_i = \eta_i, \quad T[I_1, I_2] = -[I_1, I_2]$$

and the attaching map used to obtain $RP^2 \times S^2$ has class $J$. To construct $X^4$ we use a map in the class $J + \eta_1$ where the notation is chosen so that $S_1^2$ is the sphere covering $RP^2$ in $(RP^2 \times S^2)^{(3)} = \tilde{K}^3$. Since $(1 - T)e^4$ is then attached with class $[I_1, I_2]$, $\tilde{X}^4 \simeq S^2 \times S^2$. Observe that $X^4$ is nonorientable. In fact, there is no orientable example in dimension four.

This PD space has $A(X^4, f) = 1$ where the map $f$: $X \to RP^\infty$ induces the universal covering $\pi$: $\tilde{X} \to X$. To see this we need to describe the generators of $H^2(\tilde{X}^4)$. By construction, $X^4 \simeq (RP^2 \vee S^2) \cup e^3 \cup e^4$. Let $a$ denote the cohomology dual of the class represented by the cover of $RP^2 \subset RP^2 \vee S^2 \subset X^4$, and $b$ denote the dual of the class represented by one cover of $S^2 \subset RP^2 \vee S^2 \subset X^4$. Then $b = \pi^* \bar{b}$ for some $\bar{b} \in H^2(X^4)$. Since $\{a, b\}$ forms a symplectic base, it is enough to show $q(a) = q(b) = 1$.

LEMMA 7. *Let $\pi$: $\tilde{X} \to X$ be a double cover of $2n$-dimensional* PD *spaces and $b \in H^n(X)$. Then*

$$q(\pi^*\bar{b}) = \left\langle \sum_{i=0}^{n} (f^*u)^i \cup \mathrm{Sq}_i(\bar{b}), [X] \right\rangle$$

*where u generates* $H^1(RP^\infty)$.

From this lemma,

$$q(b) = \left\langle \sum_{i=0}^{2} (f^*u)^i \cup \mathrm{Sq}_i(\bar{b}), [X] \right\rangle = \left\langle \bar{a} \cup \bar{b}, [X] \right\rangle = 1$$

where $\bar{a}$ is dual to the class represented by $RP^2 \subset X^4$. To prove $q(a) = 1$ it is necessary to compute $a_\# \cup_i Ta_\#$ where $a_\#$ is the obvious cochain representing $a$. We omit the details.

One can generalize the construction of $X^4$ to higher dimensions in several ways. Here is one direction. Let $K^{n+1}$ be the $(n+1)$-skeleton of $RP^n \times S^n$ in a normal cell decomposition. Since

$$\pi_{n+1} K^{n+1} = \pi_{n+1} RP^n \oplus \pi_{n+1} S^n \oplus \pi_{n+1} S^{n+1},$$

we can construct $K^{n+2}$ by attaching an $(n+2)$-cell to $K^{n+1}$ using a map representing $\eta + \alpha$ where $\eta \in \pi_{n+1} RP^n$ is the nontrivial element and $\alpha \in \pi_{n+1} K^{n+1}$ is the class of the attaching map for the normal $(n+2)$-skeleton of $RP^n \times S^n$.

PROPOSITION 8. *If* $n \equiv 2 \pmod 4$, *then there exists a* PD *space* $X^{2n}$ *with* $\tilde{X} \simeq S^n \times S^n$, $\pi_1 X = Z/2$, $X^{(n+2)} \simeq K^{n+2}$ *in a normal cell decomposition; and* $A(X,f) = 1$.

The point here is that $\eta \in \pi_{n+1} RP^n$ is a projective element if and only if $n \equiv 2\ (4)$ (see [9]). Similarly, by using other projective elements in $\pi_{n+k} RP^n$ for $k < n$, one can construct more examples. For $n = 3$, even though $\eta$ is not projective, we can obtain a PD space $X^6$ with $A(X^6, f) = 1$ by this construction. This is described in [5].

**4. Realization of the transversality obstruction.** Our main result is

THEOREM 9. *In each dimension* $2n \geqq 4$ *there exists a PD space* $X^{2n}$ *and a map* $f\colon X^{2n} \to RP^\infty$ *such that* $A(X^{2n}, f) = 1$, *and* $\tilde{X}$ *has the homotopy type of a smooth manifold.*

PROOF. The method of proof is clear. The example $X^4$ of §2 is crossed with copies of $CP^2$ to obtain examples in dimensions $\equiv 0\ (4)$. From Corollary 6, all these PD spaces have nonzero $A$-invariant. In addition, we note that the above examples $X^{4k}$ provide examples $X^{4k+2}$ $(k \geqq 1)$. Consider $\tilde{X}^{4k} \times RP^2 \to^{\pi\times 1} X^{4k} \times RP^2$. By an argument similar to that of Corollary 6 we see that $A(X^{4k} \times RP^2, fp_1) = 1$ and these give the examples in dimensions $4k + 2$.

**5. The Spivak normal bundle to $X^4$.** Define an injection $\rho\colon \mathcal{N}_*^{\mathrm{PD}}(\mathrm{pt}) \to \mathcal{N}_*^{\mathrm{PD}}(RP^\infty)$ by $\rho\{X^n\} = \{X^n, w_1\}$ where $w_1\colon X^n \to RP^\infty$ classifies the first Stiefel-Whitney class of $X^n$. We need the following

LEMMA 10 [3]. *The Pontrjagin-Thom map* $\mathcal{N}_*^{PD}(RP^\infty) \to \pi_*(RP^\infty \wedge MG)$ *is an injection, so every class in* $\mathcal{N}_*^{PD}(RP^\infty)$ *is detected by characteristic numbers* $(* \neq 2)$.

Consider the class of $\{X^4\}$ in $\mathcal{N}_4^{\mathrm{PD}}(\mathrm{pt})$. We calculate that the Stiefel-Whitney class of $X^4$ is $1 + e^1$. On the other hand, the fact given in §3 that $A(X^4, f) = 1$ together with the fact that $\rho\{X^4\} = (X^4, f)$ shows that $\{X^4\} \neq 0$ in $\mathcal{N}_4^{\mathrm{PD}}(\mathrm{pt})$. This gives

COROLLARY 11. *$K_3(X^4) \neq 0$ and $X^4$ generates the cokernel of* $(\mathcal{N}_4^{\mathrm{Diff}}(\mathrm{pt}) \to \mathcal{N}_4^{PD}(\mathrm{pt}))$. (*Since the only further characteristic classes in dimensions* $\leqq 4$ *are* $K_3$ *and* $\mathrm{Sq}^1 K_3$ [8].)

Let $\kappa$: $X^4 \to^{\tau} S^3 \cup_2 e^4 \to^{\lambda} B_G$ be the composition where $\tau$ is the pinching map and $\lambda$ satisfies $\lambda^*(\kappa_3) \neq 0$, $\lambda^*(w_4) = 0$. If $(\kappa)$ is the induced bundle we have that the Spivak normal bundle of $X^4$ is the Whitney sum $\xi_1 \oplus (\kappa)$ where $\xi_1$ is the nontrivial line bundle.

COROLLARY 12. *In dimension* 4, *$A(X, f)$ coincides with the stable transversality obstruction of* [4].

PROOF. The calculations of [4] show the stable transversality obstruction is given by $e_1 K_3$, and the result follows from Corollary 11.

## REFERENCES

1. W. Browder and R. Livesay, *Free involutions on homotopy spheres*, Tohoku Math. J. **25** (1972), 69–88.

2. G. Brumfiel, I. Madsen and R. J. Milgram, PL *characteristic classes and cobordism*, Ann. of Math. (2) **97** (1973), 83–159.

3. G. Brumfiel and R. J. Milgram, *Normal maps, covering spaces and quadratic functions*, Duke Math. J. (to appear).

4. G. Brumfiel and J. Morgan, *Homotopy theoretic consequences of N. Levitt's obstruction theory to transversality for spherical fibrations*, Pacific J. Math. **67** (1976), 1–100.

5. I. Hambleton, *Free involutions on 6-manifolds*, Michigan Math. J. **22** (1975), 141–149. MR **52** #6762.

6. L. Jones, *Patch spaces*, Ann. of Math. (2) **97** (1973), 306–343. MR **47**#4269.

7. N. Levitt, *Poincaré duality cobordism*, Ann. of Math. (2) **96** (1972), 211–244. MR **47** #2611.

8. R. J. Milgram, *The* mod (2) *spherical characteristic classes*, Ann. of Math. (2) **92** (1970), 238–261. MR **41**#7705.

9. R. J. Milgram, J. Strutt and P. Zvengrowski, *Projective stable stems of spheres* (to appear).

10. F. Quinn, *Surgery on Poincaré and normal spaces*, Bull. Amer. Math. Soc. **78** (1972), 262–267. MR **45** #6014.

11. D. B. A. Epstein and N. Steenrod, *Cohomology operations*, Ann. Math. Studies No. 50.

12. C. T. C. Wall, *Poincaré complexes*. I, Ann. of Math. (2) **86** (1967), 213–245. MR **36**#880.

STANFORD UNIVERSITY

MCMASTER UNIVERSITY

# GROUP ACTIONS

Proceedings of Symposia in Pure Mathematics
Volume 32, 1978

# PSEUDOEQUIVALENCES OF $G$-MANIFOLDS

TED PETRIE*

This paper describes a program for answering *existence* questions in the realm of differentiable transformation groups. Specifically an invariant $I(\cdot)$ is given on the category of smooth $G$-manifolds. We want to know whether there is a $G$-manifold $X$ in the homotopy type of a fixed manifold $Y_0$ so that $I(X)$ takes on a specified value. See Chapter I, §0. This question asks for a method of constructing $G$-manifolds with specified underlying homotopy type and such a method is provided here.

The paper is divided into two chapters. The first introduces the concepts. It defines the role of the three main concepts: quasi-equivalence of $G$ vector bundles—$G$ transversality and $G$ normal cobordism. The second chapter discusses $G$-transversality in greater detail and illustrates via applications some of the questions raised in Chapter I, §0. Each chapter is provided with an introduction outlining its aims.

## Chapter I. The Concepts

**0. Introduction.** Let $G$ be a compact Lie group.

Much of the important work in the subject of differentiable transformation groups deals with the problem of "realization of invariants within a homotopy type". This means an invariant $I(\cdot)$ on the category $C$ of smooth $G$-manifolds is given and a description is sought of the values obtained in the set $\{I(Y) | Y \text{ is homotopy equivalent to some fixed manifold } Y_0, Y \in C\}$. The invariants are generally of two types: (i) a function of the set $\mathcal{S}(Y)$ of conjugacy classes of isotropy groups of $Y$; (ii) a function of the fixed point sets $\{Y^H | H \subseteq G\}$ of subgroups of $G$. For example: (i) $I(Y) = \mathcal{S}(Y)$ or its cardinality $|\mathcal{S}(Y)|$, (ii) $I(Y) = \chi(Y^K)$ or $H^*(Y^K)$, the Euler characteristic, resp. cohomology ring of the fixed point set of $K \subseteq G$.

---

*AMS (MOS) subject classifications* (1970). Primary 57D65, 57E05; Secondary 20C10.

*During the preparation of this paper the author was a Guggenheim fellow. Research partially supported by N. S. F. grant.

This setting encompasses many of the "classical" important problems in differentiable transformation groups. Here are some examples:

(0.1) $I(Y) = \mathscr{S}(Y)$. The question is, which groups $G$ admit $I(Y) = 1 \subseteq G$ if $Y$ is a homotopy sphere? This is the classical question of Cartan-Eilenberg, which groups act freely on spheres? See [**6**].

(0.2) $I(Y) = |\mathscr{S}(Y)|$. The question is, is $I(Y)$ bounded by the dimension of $Y$ for $Y$ a homotopy sphere? This question (in less generality) was raised by Montgomery-Yang for $G = S^1$ in their work on pseudofree actions [**5**]. See [**7**].

(0.3) $I(Y) = Y^G(\chi(Y^G))$ and the question is, can $I(Y) =$ one point (take arbitrary values) for $Y$ a homotopy sphere? This reportedly old question has been popularized by W. C. Hsiang. See [**9**] and §4.

(0.4) $I(Y)$ is the *collection* of representations of $G$ on the tangent space at fixed points of $G$, i.e., $I(Y) = \{TY_p | p \in Y^G\}$. The question is, what are the values of $I(Y)$ if $Y$ is a homotopy complex projective space? See [**11**], [**12**].

In order to answer these questions, we review a constructive method due to the author of producing $G$-manifolds $X$ in the same homotopy type of a given $G$-manifold $Y$.

DEFINITION 0.5. A pseudoequivalence $f: X \to Y$ is a $G$-map $f$ which is also a homotopy equivalence. In the case $X$ and $Y$ are manifolds, we require $X$ and $Y$ to have the same dimension and moreover there exist $\alpha \in KO_G(Y)$ with $TX = f^*(TY - \alpha)$ in $KO_G(X)$.

We describe the three concepts: quasi-equivalence, $G$ transversality and $G$ normal map used to construct pseudoequivalences. In the last section we illustrate how the theory works for treating (0.1)—(0.4) and other questions in differentiable transformation groups.

Within the confines of this space, the author can only state the main results of [**6**], [**9**], and [**10**] which treat the above three concepts. However, accepting the statements of these papers, we illustrate, by way of examples in §4, the elementary manipulations with $G$ vector bundles necessary to apply the theory.

Aside from free actions, semifree actions are the simplest to study. There already exists a large literature on this subject which remains ad hoc and complicated. As the simplest special case of the theory here, we treat some problems in the domain of semifree $\boldsymbol{Z}_p$ actions on homotopy spheres to show how many results in this situation follow from elementary $G$ vector bundle computations. See §4.

In §4 we also present the deepest application of the theory—the construction of smooth actions of some groups on homotopy spheres with just one point fixed by the entire group. This is a problem of some years standing. For this discussion the author restricts himself to a survey of the ideas used to treat this problem. See Theorems 4.10 and 4.20 for precise statements of the results here. This interesting application gives an impression of the full theory presented in §§1–3.

Very briefly we can describe the idea of constructing pseudoequivalences $f: X \to Y$ so the value $I(X)$ solves the problems of (0.1)–(0.4). Start from a $G$-manifold $Y$ with $I(Y)$ an approximation to the sought value of $I(\cdot)$. Construct a proper, fiber preserving $G$-map $\omega$: $N \to M$ between $G$ vector bundles over $Y$ of equal dimension (§1). Find a proper $G$ homotopy of $\omega$ to a map $\theta$ transverse to the zero section $Y \subset M$ (§2). Let $X = \theta^{-1}(Y)$ and $f = \theta|_X$: $X \to Y$. Make $f$ a pseudoequivalence via a $G$ normal cobordism (§3). The resulting $G$-manifold $X$ is a function of $Y$, $M - N \in KO_G(Y)$, and $\omega$ and these data are chosen so that the obstructions to $G$ transversality (§2) and $G$ normal cobordism **[10]** (§3) vanish and so that $I(X)$ is the desired value of $I(\cdot)$. This imposes relations between $Y$ and $\alpha = M - N \in KO_G(Y)$. For some problems it is possible to satisfy these relations and others not. The aim in §4 is to illustrate how one chooses the data to solve the given problem.

This chapter together with the references listed with the questions (0.1)—(0.4) shows that many problems involving $G$ actions on homotopy spheres can be treated in the above manner starting from $Y = S(V)$, the unit sphere of $V$ for a suitable $G$ module $V$. The general theory discussed in §§1–3 for constructing pseudoequivalences $f: X \to Y$ is then reduced to solving "equations or inequalities" in $RO(G)$ or $KO_G(Y)$

It should be emphasized that the application of this theory is by no means restricted to actions on homotopy spheres. The applications here are chosen to illustrate the theory in some familiar situations.

Let us introduce notation. If $Y$ is a $G$-manifold (all $G$-manifolds are smooth $G$-manifolds) and $y \in Y$, the isotropy group at $y$ is $G_y$. If $A \subset Y$ is an invariant submanifold, $\nu(A, Y)$ is its $G$ normal bundle. The $G$ fixed point set of $Y$ is denoted $Y^G$. The component of $Y^G$ containing $y$ is denoted $Y^G_y$. The complex, resp. real, representation ring of $G$ is $R(G)$, resp. $RO(G)$. If $\xi \in KO_G(Y)$, its restriction to $p \in Y$ is denoted by $\xi_p \in RO(G_p)$.

**1. Quasi-equivalence of $G$ vector bundles.** Let $\xi$ and $\eta$ be two real $G$ vector bundles of the same dimension over a $G$-space $Y$. A $G$-map $\omega: \xi \to \eta$ which is proper, fiber preserving and degree one on fibers is called a quasi-equivalence. Let $\alpha = \eta - \xi \in KO_G(Y)$ and define $\alpha \geqq 0$ to mean there exist a $G$ vector bundle $\theta$ over $Y$ and a quasi-equivalence $\omega: \xi \oplus \theta \to \eta \oplus \theta$. In this situation we say $\omega$ is a quasi-equivalence of $\alpha$.

*Problem* 1.1. Given $\alpha \in KO_G(Y)$, give necessary and sufficient conditions for $\alpha \geqq 0$.

When $Y$ is a point and $\alpha \in K_G(Y)$, this is solved by the main theorem of **[6**, Theorem 5.1]. Briefly $\alpha \in R(G)$ satisfies $\alpha \geqq 0$ iff (i) $\dim \alpha^H = 0$ whenever $H \in \mathscr{H}_1$ and (ii) $\dim \alpha^H \geqq 0$ whenever $H \in \mathscr{H}_2$. Here $\dim \alpha^H = \dim_{\boldsymbol{R}} M^H - \dim_{\boldsymbol{R}} N^H$ if $\alpha = M - N \in R(G)$ and $\mathscr{H}_1$ and $\mathscr{H}_2$ are classes of subgroups of $G$, e.g., $\mathscr{H}_1$ contains all $p$ subgroups of $G$. **[6]** also provides a useful sufficient condition for constructing $\alpha \in R(G)$ with $\alpha \geqq 0$.

THEOREM 1.1 **[6**, THEOREM 6.8]. *Let $\chi$ be any complex $G$-module and $p$ and $q$ relatively prime integers prime to the order of the Weyl group of $G$. Set $\alpha = (\psi^p - 1) \cdot(\psi^q - 1)\chi$. Then $\alpha \geqq 0$. (If $G$ is finite, it is its own Weyl group.)*

An extremely useful way of incorporating algebra into our study of quasi-equivalence is through the equivariant cohomology theories $\omega^\alpha_G(Y)$ of **[18]** for $\alpha \in RO(G)$. We are only interested in the case $\dim \alpha = 0$. Let $M$, $N$ and $V$ be real $G$-modules with $\dim N = \dim M$. Let $\alpha = M - N$ and $F_\alpha(V)$ be the space of maps from the unit sphere $S(V \oplus N)$ to the unit sphere $S(V \oplus M)$. Then $F_\alpha(V)$ is a $G$-space. It is acted upon by $G$ via conjugation. The connected component of $F_\alpha(V)$ consisting of maps of degree $i$ is denoted by $F^i_\alpha(V)$. By definition

$$(1.2) \qquad \omega^\alpha_G(Y) = \operatorname*{inj\,lim}_V\, [Y, F_\alpha(V)]^G.$$

Here $[Y, Z]^G$ denotes the $G$ homotopy classes of $G$-maps of $Y$ to $Z$. The zero map of $F_\alpha(V)$ is the map which sends each point of $S(V \oplus N)$ to a fixed point $x \in S(V \oplus M)^G$. The zero map lies in $F_\alpha(V)^G$ and is the base point. In the special case $\alpha = 0$, there is another canonical base point—the identity map $1d$ of $S(V)$. It lies in $F_0^1(V)^G$. This is the more natural base point for quasi-equivalences. The context will dictate which is used in the notation $\omega_G^0(Y, p)$ for $p \in Y^G$.

Quasi-equivalences are included in the setting as the subset ind $\lim_V[Y, F_\alpha^1(V)]^G \subset \omega_G^\alpha(Y)$. For if $\omega \in [Y, F_\alpha^1(V)]^G$, define $\tilde{\omega} : Y \times (N \oplus V) \to Y \times (M \oplus V)$ by $\tilde{\omega}(y, z) = (y, |z|\omega(y)[z/|z|])$. Then $\tilde{\omega}$ is a quasi-equivalence.

Composition of maps defines a multiplication in $\omega_G^0(Y)$ which is a ring. The ring $\omega_G^0(p)$ ($p$ a one point space) is the Burnside ring of the group $G$ [**18**]. It has another more algebraic description as the Grothendieck group of the category of finite $G$ sets [**3**]. In [**3**] we give a complete description of the ring $\omega_G^0(p) = \omega_G^0$. An element $\omega \in \omega_G^0$ is represented as a $G$-map of $S(V)$ to itself for some $G$ module $V$. If $H$ is a subgroup of $G$, define

$$d_H(\omega) = \text{degree } \omega^H, \qquad \omega^H: S(V^H) \to S(V^H). \tag{1.3}$$

Let $\mathscr{S}(G)$ denote the set of conjugacy classes of subgroups of $G$. Define a ring homomorphism $D: \omega_G^0 \to \prod_{\mathscr{S}(G)} \mathbf{Z}$ by setting $D(\omega)_H = d_H(\omega)$ if $H$ is a representative group of a conjugacy class.

THEOREM 1.4 [**3**]. *Let $G$ be finite. Then $D$ is a monomorphism of rings and $D(\omega_G^0)$ is the subring of $\prod_{\mathscr{S}(G)} \mathbf{Z} = \tilde{\mathbf{Z}}$ consisting of $z \in \tilde{\mathbf{Z}}$ satisfying $L_H(z) \equiv O(|W(H)|)$ for each $H \in \mathscr{S}(G)$ with $H \neq G$.*

REMARK 1.5. Each $L_H$ is an element of $\operatorname{Hom}_{\mathbf{Z}}(\tilde{\mathbf{Z}}, \mathbf{Z})$. Its coefficients depend upon the orders of the subgroups of $G$ containing $H$. The order of $W(H) = N(H)/H$ is written $|W(H)|$ and $N(H)$ is the normalizer of $H$ in $G$.

In the remainder of this section we discuss three methods of constructing quasi-equivalences which will be used to construct $G$ actions on spheres in later sections. These are (i) construction of quasi-equivalences over $Y$ from quasi-equivalences over a point $p$, (ii) construction of self quasi-equivalences of a bundle using $\omega_G^0(\cdot)$ and (iii) construction of quasi-equivalences of $\alpha \in KO_G(Y, p)$, $p \in Y^G$, using $G$ homotopy and the equivariant $J$ homomorphism.

EXAMPLE 1.6. *Explicit quasi-equivalences for $Y$ a point.* Let $G = S^1$ and $N$ and $M$ be the complex 2 dimensional $S^1$ modules defined by $N$: $t(z_0, z_1) = (t^p z_0, t^q z_1)$, $M$: $t(z_0, z_1) = (t z_0, t^{pq} z_1)$, where $t \in S^1$, $(z_0, z_1) \in N$ respectively, $M$ and $p$ and $q$ are relatively prime integers. Choose positive integers $a$ and $b$ so that $-ap + bq = 1$ and define $\omega: N \to M$ by

$$\omega(z_0, z_1) = (\bar{z}_0^a \cdot z_1^b,\ z_0^q + z_1^p).$$

One checks that $\omega$ is proper and degree $\omega = 1$ so $\omega$ is a quasi-equivalence of $G$-modules, i.e., $G$ vector bundles over a point. See [**6**].

Note that $\alpha = M - N = (\psi^p - 1)(\psi^q = 1)\chi$ where $\chi$ is the standard complex one dimensional module. Compare Theorem 1.1 above. Of course by restricting to the cyclic subgroups of $S^1$, we achieve explicit quasi-equivalences for these groups too.

As stated above the notion of quasi-equivalence for $Y$ a point is completely understood from [**6**, Theorem 5.1]. Using the functorial properties of $\omega_G^\alpha(\cdot)$ allows use of the information about quasi-equivalences over a point to give information over any space through the trivial observation:

(1.7) $f^*\colon \omega_G^\alpha(Y') \to \omega_G^\alpha(Y)$ preserves quasi-equivalences for any $G$-map $f\colon Y \to Y'$.

Apply this remark to $Y' = p$ and $f$ the unique map of $Y$ to $p$. The observation is exploited in [**8**] and [**13**].

For considering self quasi-equivalences of a $G$ vector bundle $\xi$ over $Y$ we may take $\xi = Y \times V$ for some $G$-module $V$ because of the stabilization property of quasi-equivalences. Then we are dealing with quasi-equivalences in $\omega_G^0(Y)$. For later geometric application we seek a quasi-equivalence $\omega \in \omega_G^0(Y)$ which has specified properties under restriction to the fixed point set. This leads to two problems: (i) Does there exist an $\omega' \in \omega_G^0(Y^G)$ with the specified properties? (ii) Does there exist $\omega \in \omega_G^0(Y)$ with $j^*\omega = \omega'$ where $j\colon Y^G \to Y$ is the inclusion?

To be more specific about the first problem, we give an example with later geometric application. Let $\mathscr{H} \in G$ be a set of subgroups of $G$ closed under conjugation and taking subgroups and let $\mathrm{res}_H\colon \omega_G^0 \to \omega_H^0$ be the restriction homomorphism of rings. Does there exist an element $e \in \omega_G^0$ such that $\mathrm{res}_H(e) = 0$ for all $H \in \mathscr{H}$ and $d_G(e) = 1$? This fits the above question (i) with $Y^G = p$. The answer depends on $G$ and the set $\mathscr{H}$ and can in principle be answered by Theorem 1.4. See [**3**].

The answer to (ii) depends on the properties of (i) but one general technique is to appeal to a localization theorem which says $s^{-1}j^*\colon s^{-1}\omega_G^0(Y) \to s^{-1}\omega_G^0(Y^G)$ is an isomorphism for $j\colon Y^G \to Y$ the inclusion and $s$ a suitable multiplicative subset of $\omega_G^0$. Here is a specific example:

Suppose there is an $e \in \omega_G^0$ with $\mathrm{res}_H(e) = 0$ for $H \in \mathscr{H}$ and $d_G(e) = 1$. Let $s = \{e^\lambda | \lambda = 0, 1, 2, \cdots\} \in \omega_G^0$. Note that each power $\lambda$ of $e$ also satisfies the same conditions as $e$.

LOCALIZATION LEMMA 1.8. *Let $Y$ be a $G$-space such that $H$ is an isotropy group of $Y$ only if $H \in \mathscr{H}$. Then $s^{-1}\omega_G^0(Y) \to s^{-1}\omega_G^0(Y^G)$ is an isomorphism if $s = \{e^\lambda | \lambda = 0, 1, \cdots\}$.*

COROLLARY 1.9. *Suppose $Y$ is a $G$-space with $Y^G = p \cup q$, and if $H$ is an isotropy group of $Y$, then $H \in \mathscr{H}$. Then there is an $\omega \in \omega_G^0(Y)$ with $j^*(\omega) = (1, 1 - e^\lambda)$ for some integer $\lambda$ and $j\colon Y^G \to Y$ the inclusion.*

PROOF. By Lemma 1.8 and the meaning of localization, there is an $x \in \omega_G^0(Y)$ with $j^*(x) = (0, e^\lambda)$ for some $\lambda$. Then $j^*(1 - x) = (1, 1 - e^\lambda)$.

Corollary 1.9 resulted from conversations with tom Dieck.

Let $V$ and $W$ be real $G$-modules. The equivariant $J$ homomorphism $J_G$ is defined via this diagram:

$$\begin{array}{ccc} J_G\colon KO_G(S(W \oplus \boldsymbol{R}), p) & \longrightarrow & \omega_G^0(S(W)) \\ \| & & \| \\ \operatorname*{inj\,lim}_V [S(W), \mathrm{GL}(V)]^G & \longrightarrow & \operatorname*{ind\,lim}_V [S(W), F_0(V)]^G \end{array}$$

induced by the inclusion $\mathrm{GL}(V) \subset F_0^1(V) \subset F_0(V)$. The addition in both groups is defined by composition of maps in $\mathrm{GL}(V)$ and $F_0^1(V)$. Our interest lies in the straightforward

PROPOSITION 1.10. *If $J_G(\alpha) = 0$, there is a quasi-equivalence $\omega$ of $\alpha$.*

PROPOSITION 1.11 **[18]**. *There is a natural isomorphism*

$$\omega_G^0(Y) \otimes Q = \sum_{H \in \mathscr{S}(G)} H^0(Y^H, Q)^{W(H)}.$$

COROLLARY 1.12. *Suppose $W$ is a $G$-module with $W^G \neq 0$. Then*

$$\operatorname{Ker} J_G(S(W \otimes \boldsymbol{R}), p) \oplus Q = KO_G(S(W \oplus \boldsymbol{R}), p) \otimes Q.$$

PROOF. Since $W^G \neq 0$, $S(W \oplus \boldsymbol{R})^H$ is connected for any subgroup $H \subseteq G$; so $\omega_G^0(S(W \oplus \boldsymbol{R}), p) \otimes Q = 0$ for $p \in S(W \oplus \boldsymbol{R})$ by Proposition 1.11. Since $\pi_0(\mathrm{GL}(V)^G) \otimes Q = 0$, the result follows.

COROLLARY 1.13. *Let $W$ be a real $G$-module with $W^G \neq 0$. For every $\alpha \in KO_G(S(W \oplus \boldsymbol{R}), p)$ there is an integer $n$ so that $n\alpha \geqq 0$.*

PROPOSITION 1.14. *Let $W$ be a real $G$-module with* $\dim W^G \equiv 2\ (4)$. *Let $\alpha \in KO_G(S(W \oplus \boldsymbol{R}), p)$ be the difference of two vector bundles $M$ and $N$, $\alpha = M - N$. Then there is an integer $n$ and a quasi-equivalence $\omega\colon nN \to nM$ such that $\omega^G\colon nN^G \to nM^G$ is an isomorphism of vector bundles.*

PROOF. Let $W \oplus \boldsymbol{R} = W'$. $KO(S(W')^G, p) \otimes Q = 0$ because $\dim W^G \equiv 2\ (4)$. By Corollary 1.13, there is an integer $n_1$ and a quasi-equivalence $\omega_1\colon n_1N_1 \to n_1M$ and we may suppose $n_1\alpha^G = 0$ in $KO(S(W')^G, p)$. Using the isomorphism of $n_1N^G$ with $n_1M^G$, we may view $\omega_1^G$ as an element of $\omega^0(S(W')^G)$. Since $\omega^0(S(W')^G, p) \otimes Q = 0$, some multiple $n_2$ of $\omega_1^G$ is the identity, i.e., $n_2\omega_1^G\colon nN^G \to nM^G$ is an isomorphism of vector bundles for $n = n_1n_2$.

**2. $G$ transversality.** A fundamental problem in the theory of differentiable transformation groups is the

2.1. *$G$ Transversality Problem. Let $f\colon N \to M$ be a proper $G$-map between $G$-manifolds $N$ and $M$ and let $Y \subset M$ be an invariant submanifold. When is $f$ properly $G$ homotopic to a map $\theta$ transverse to $Y$? This is written $\theta \pitchfork Y$.*

This problem is solved via an obstruction theory (Chapter II) and **[14]**. In particular see Chapter II, Theorem 4.12. Rather than introduce all the notions involved in this theory, we state some consequences pertinent to constructing $G$-actions on homotopy spheres. The most important situation of application is the case $N$ and $M$ are $G$ vector bundles over $Y$. Then there are rather simple sufficient conditions depending on $\alpha = M - N \in KO_G(Y)$.

First notation: Let $\hat{G}$ denote the set of real irreducible $G$-modules and $1 \in \hat{G}$ the one dimensional trivial $G$-module. Any $\alpha \in RO(G)$ is uniquely expressed as $\alpha = \sum_{\chi \in \hat{G}} \alpha_\chi \chi$ where the $\alpha_\chi$ are integers. Define

$$\langle \chi, \alpha \rangle = \alpha_\chi. \tag{2.2}$$

For each $\chi \in \hat{G}$, let $D_\chi$ be the division algebra $\mathrm{Hom}_{\boldsymbol{R}}(\chi, \chi)^G$ where $G$ acts by conjugation. Let $d_\chi = \dim_{\boldsymbol{R}} D_\chi$ and $\alpha = M - N \in RO(G)$ be the difference of two $G$-modules of equal dimension. Any $\omega \in \omega_G^\alpha(Y)$ is represented as a proper $G$-map $\tilde{\omega}\colon Y \times N \to$

$Y \times M$ preserving fibers over $Y$. We say $\omega$ *is transversal* if $\tilde{\omega}$ is properly $G$ homotopic to a map $\theta$ transverse to $Y \times 0 \subset Y \times M$.

THEOREM 2.3 [8]. *Let $G$ be finite. Let $\alpha \in RO(G)$, $\dim \alpha = 0$ and let $Y$ be a smooth $G$-manifold. Suppose, for each $y \in Y$ and $\chi \in \hat{G}_y - 1$, $d_\chi(\langle \chi, TY_y - \alpha \rangle + 1)$ is greater than or equal to $\dim Y_y^{G_y} - \dim \alpha_y^{G_y}$. Then $\omega \in \omega_G^\alpha(Y)$ is transversal.*

See Chapter II, Theorem 10.1 for an application pertinent to (0.2). Theorem 2.3 is a direct consequence of Chapter II, §§4 and 5. Note the relations between $Y$ and $\alpha$ imposed by $G$ transversality.

THEOREM 2.4 (CHAPTER II). *Let $G$ be a finite group. Each $\omega \in \omega_G^0(Y)$ is transversal.*

The requirement on $G$ in 2.4 is definitely needed. The full obstruction theory of Chapter II, §7 is necessary to analyze the general case.

Theorem 2.3 gives a sufficient condition for the existence of a quasi-equivalence $\tilde{\omega}$ of $\tilde{\alpha} = Y \times M - Y \times N$ to be properly $G$ homotopic to a map transverse to $Y \times 0$. For the general case of arbitrary $\alpha \in KO_G(Y)$, the situation is similar. We state the result somewhat differently and strengthen the hypotheses to achieve a uniqueness result.

THEOREM 2.5. *Let $G$ be finite and let $\alpha = M - N \in KO_G(Y)$ be the difference of two $G$ vector bundles $M$ and $N$. Suppose, for each $y \in Y$ and $\chi \in \hat{G}_y - 1$, $d_\chi(\langle \chi, TY_y - \alpha_y \rangle + 1) \geqq \dim Y_y^{G_y} - \dim \alpha_y^G + 1$. Then any quasi-equivalence $\omega\colon N \to M$ is properly $G$ homotopic to a map $\theta \pitchfork Y$ ($Y \subset M$ is the zero section) and any two such $\theta$ are properly $G$ homotopic via a homotopy transverse to $Y$. See Chapter* II, 4.12–4.14.

COROLLARY 2.6. *Let $G = \boldsymbol{Z}_q$ where $q$ is an odd prime and let $Y = S(W \oplus \boldsymbol{R})$ where $W$ is a real $G$-module with $W^G \neq 0$. Let $\alpha = M - N \in KO_G(Y, p)$ where $p \in Y^G$. If, for each $\chi \in \hat{G} - 1$, $2\langle \chi, W \rangle \geqq \dim W^G - 1$ any quasi-equivalence $\omega\colon N \to M$ is properly $G$ homotopic to a map $\theta$ transverse to $Y$ and $\theta$ is unique up to proper $G$ transverse homotopy.*

PROOF. Since $q$ is odd $d_\chi = 2$. Since $W^G \neq 0$, $Y^G$ is connected; so $\alpha_y = 0$ for any $y \in Y^G$ because $\alpha_p = 0$. The result now follows Theorem 2.5 because $TY_y = W$ and $\dim Y^G = \dim W^G$.

Here are two results on $G$ transversality relevant to finding invariant subspheres of $G$-actions on homotopy spheres.

THEOREM 2.7. *Let $G$ be finite. Let $V \subset W$ be real $G$-modules with $V^G = W^G$ and set $\alpha = W - V \in RO(G)$. Suppose $f\colon \Sigma \to S(W \oplus \boldsymbol{R})$ is a $G$-map where $\Sigma$ is a smooth $G$-manifold. If for all $\sigma \in \Sigma$ and $\chi \in \hat{G}_\sigma - 1$ either* (i) $\langle \chi, \alpha \rangle = 0$ *or* (ii) $d_\chi(\langle \chi, T\Sigma_y - \alpha \rangle + 1) \geqq \dim \Sigma_\sigma^{G_\sigma} + 1$ *then $f$ is $G$ homotopic to a map transverse to $S(V \oplus \boldsymbol{R}) \subset S(W \oplus \boldsymbol{R})$ and the transverse $G$ homotopy class is unique.*

COROLLARY 2.8. *Let $G = \boldsymbol{Z}_q$ where $q$ is an odd prime and let $W$ be a $G$-module with $W^G \neq 0$. Suppose $f\colon \Sigma \to Y = S(W \oplus \boldsymbol{R})$ is a pseudoequivalence and either* (i) $\langle \chi, \alpha \rangle = 0$ *or* (ii) $2\langle \chi, T\Sigma_p - \alpha \rangle \geqq \dim W^G - 1$ *for some $p \in \Sigma^G$ and every $\chi \in \hat{G} - 1$. Then $f$ is $G$ homotopic to a map transverse to $S(V \oplus \boldsymbol{R})$ and the transverse homotopy class is unique.*

PROOF. Since $W^G \neq 0$, $Y^G$ is connected. Since $f$ is a pseudoequivalence, $f^G$ induces an isomorphism in mod $q$ homology by Smith Theory; so dim $\Sigma^G =$ dim $Y^G$ and $\Sigma^G$ is connected. Since $d_x = 2$, the result follows from Theorem 2.7.

REMARKS 2.9. The results of this section were selected with the application of §4 in mind. They are typical in that the inequalities provide relations among the invariants of the $G$-manifold $Y$ and the $G$ vector bundles $\alpha \in KO_G(Y)$ and these relations show up in the properties of the manifolds $X$ arising from $G$ transversality. In any particular case of application it is better to refer to **[8]** because there is invariably more information in a specific problem which serves as additional input to the theory there and one can expect stronger conclusions, e.g., 2.8 is such a case as are the applications of §10 of Chapter II.

**3. $G$ normal maps.** The author warmly thanks his colleagues Madsen, Oliver, Bak and Dress for conversations influencing this section.

The concepts of quasi-equivalence and $G$ transversality are employed to construct $G$ normal maps.

3.1. DEFINITION. A $G$ normal map $f: X \to Y$ consists of two $G$-manifolds $X$ and $Y$ of the same dimension and a $G$-map $f: (X, \partial X) \to (Y, \partial Y)$ of degree 1 together with a stable $G$ vector bundle isomorphism $\hat{f}: TX \to f^*(TY - \alpha)$ where $\alpha \in KO_G(Y)$ and dim $\alpha = 0$.

The specific isomorphism $\hat{f}$ is an important aspect of the structure but is omitted from the notation. An obvious but useful observation is that

$$\dim X_x^H = \dim Y_{f(x)}^H - \dim \alpha_{f(x)}^H \quad \text{if } X_x^H \neq \emptyset \tag{3.2}$$

for the $G$ normal map of 3.1. It is not difficult to prove

PROPOSITION 3.3. *Let $f: X \to Y$ be a $G$ normal map with $TX \cong f^*(TX - \alpha)$, $\alpha \in KO_G(Y)$. Then $\alpha \geqq 0$.*

Thus a $G$ normal map implies a quasi-equivalence of an $\alpha \in KO_G(Y)$. Conversely a quasi-equivalence together with transversality implies a $G$ normal map. Specifically let $\alpha = M - N \in KO_G(Y)$ be the difference of $G$ vector bundles $M$ and $N$ and let $\omega: N \to M$ be a quasi-equivalence. If $\omega$ is properly $G$ homotopic to $\theta \pitchfork Y$ and $X = \theta^{-1}(Y), f = \theta|_X$. Then $f: X \to Y$ is a $G$ normal map with $TX = f^*(TY - \alpha)$. The specific isomorphism $\hat{f}$ is constructed from the transversality hypothesis. In view of this, §§1 and 2 provide conditions on $Y$ and $\alpha \in KO_G(Y)$ for producing $G$ normal maps $f: X \to Y$ with $TX \cong f^*(TY - \alpha)$.

The notion of a $G$ normal map induces a notion of $G$ normal cobordism and leads to the

3.4. *$G$ Normal Cobordism Problem.* Let $f: X \to Y$ be a $G$ normal map. Give necessary and sufficient conditions that $(X, f)$ be $G$ normally cobordant to $(X', f')$ where $f': X' \to Y'$ is a pseudoequivalence.

There is an obstruction theory **[10]**, **[16]** for treating 3.4. Here is a discussion of the rudiments of that theory:

The starting point is Smith Theory which states that if $f: X \to Y$ is a pseudoequivalence, then $f_*^P: H_*(X^P, \mathbf{Z}_p) \to H_*(Y^P, \mathbf{Z}_p)$ is an isomorphism for every $p$-group $P \subset G$ for every prime $p$. This singles out the $p$-subgroups of $G$ for a special role in answering 3.4. First one introduces an abelian group $J_i^{A,B}(R)$,

$i = 0, 1, 2, 3$, attached to a ring $R$ with involution and subgroups $A, B \subset \tilde{K}_0(R)$ of the reduced projective class group of $R$. When $R$ is the integral group ring of $G$ and $A = B = 0$, these are the Wall groups of $G$ written $L_*^h(G)$.

Suppose $f: X \to Y$ is a $G$ normal map. To each $p$ subgroup $P$ of $G$ (including $P = 1$) and each component $\alpha \in \pi_0(Y^P)$ a ring $R_\alpha$ with involution is given depending on the isotropy groups of points of $X^P \cap f^{-1}(\alpha)$. In the case $Y^P$ is connected and $P$ is an isotropy group of $X$, $R_\alpha = \boldsymbol{Z}_{(p)}(\pi_0(W(P)))$ if $\alpha$ is the unique component of $Y^P$. In order to avoid discussing various components of $Y^P$ we henceforth suppose $Y^P$ is connected for each subgroup $P$ of prime power order. To simplify matters even more we assume $G$ finite, though the general case is only slightly more cumbersome. Then to each $p$-subgroup $P$ of $G$ we associate the ring $R_P = \boldsymbol{Z}_{(p)}(W(P))$ for $P \neq 1$ and $R_1 = \boldsymbol{Z}(G)$. We also associate subgroups $A_P$, $B_P \in \tilde{K}_0(R_P)$.

THEOREM 3.5 [**10**], [**16**]. *Suppose $f: X \to Y$ is a $G$ normal map. To each subgroup $P \subset G$ of prime power order, including $P = 1$, there is an obstruction $\sigma_P(f) \in J_*^{A_P, B_P}(R_P)$. If conditions $C_1$, $C_2$ below are satisfied and if all $\sigma_P(f)$ vanish, $(X, f)$ is $G$ normally cobordant to $(X', f')$ with $f': X' \to Y$ a pseudoequivalence. Moreover if $f^P$: $X^P \to Y^P$ induces an isomorphism in* mod *$p$ homology, $\sigma_P(f) = 0$.*

REMARK 3.6. *The obstruction $\sigma_P(f)$ is not a priori defined. It is defined if all $\sigma_{P'}(f) = 0$ for $p$ groups $P'$ strictly containing $P$.*

REMARK 3.7. *Without the assumption $\pi_0(Y^P) = 0$, the obstructions must be indexed by the components of $Y^P$.*

The conditions $C_1$ and $C_2$ are

$C_1$: The Gap Hypothesis on $X$. $\dim X_x^{G_x} > \frac{1}{2} \dim X_x^{G_z}$ whenever $G_z > G_x$ and $\dim X^P > 4$ for every subgroup $P$ of prime power order. This is required for each pair of points $z, x \in X$.

$C_2$: $\pi_1(Y^P) = 0$ for every subgroup $P$ of prime power order.

The condition $C_2$ is probably removable by current technology but the removal of $C_1$ requires the solution of hard unsolved problems in differential topology even for $G = \boldsymbol{Z}_2$.

The appearance of the projective class group in this obstruction theory through the groups $J_*^{A,B}(R)$ results from the conditions of Smith Theory. Specifically, suppose $f: X \to Y$ is a $G$ normal map such that $f^P: X^P \to Y^P$ induces an isomorphism in mod $p$ homology for each $p$ group $P \neq 1$ for all $p$, and

$$\operatorname{Ker}(f_*: H_i(X, C) \to H_i(Y, C)) = K_i(f, C)$$

vanishes for all but one value $n$ of $i$. Here $C$ is any ring.

THEOREM [**16**]. *If $\Lambda$ is the group ring $C(G)$, then* hom $\dim_\Lambda K_n(f, C)$ *is finite and $\chi(f, C) = (-1)^n K_n(f, C)$ represents an element of $\tilde{K}_0(\Lambda)$.*

By modifying the hypothesis and varying the coefficients $C$ depending on the $p$-subgroups of $G$, we obtain invariants $\chi_P(f) \in \tilde{K}_0(R_P)$ and $\chi_1(f) \in \tilde{K}_0(\boldsymbol{Z}(G))$ with $\chi_1(f) = \chi(f, \boldsymbol{Z})$ and $\chi_P(f) = \chi(f^P, \boldsymbol{Z}_{(p)})$.

The obstruction $\sigma_P(f)$ is then fashioned from the $R_P$ valued bilinear form on $\chi_P(f)$ (when $\dim Y^P$ is even) defined by the intersection pairing in $X^P$ and the

$J_*^{A,B}$-groups are made up of all such forms on all $R_P$ projective modules in $B$ modulo the group of hyperbolic forms of projectives in $A$.

Here is a list of some of the properties of the obstruction groups of Theorem 3.5:

(3.8) $J_{2n+1}^{A,B}(\Lambda)$ is finite for $\Lambda = \mathbf{Z}_{(p)}(G)$ or $\Lambda = \mathbf{Z}(G)$ for any finite group $G$.

(3.9) There is a homomorphism Sign: $J_{2n}^{A,B}(\Lambda) \to R(G)$. Its kernel is a finite group of exponent dividing 8 and if $|G|$ is odd, the exponent is at most 2.

(3.10) If $X$ and $Y$ are closed $G$-manifolds, the obstructions $\sigma_P(f)$ are a priori defined modulo torsion in $J_*^{A_P, B_P}(R_P)$. Specifically if we set $\bar{\sigma}_P(f) = \text{Sign}(W(P), X^P) - \text{Sign}(W(P), Y^P)$ where $\text{Sign}(G, X)$ is the Atiyah-Singer $G$-signature of $X$, then $\bar{\sigma}_P(f) \in R(W(P))$ and $\text{Sign}\, \sigma_P(f) = \bar{\sigma}_P(f)$ whenever $\sigma_P(f)$ is defined.

(3.11) In general the groups $(A_P, B_P)$ are $(D(R_P), \tilde{K}_0(R_P))$ where $D(R_P)$ is the kernel of the homomorphism $\tilde{K}_0(R_P) \to \tilde{K}_0(M_P)$ where $M_P$ is a maximal order for $R_P$ in $R_P \otimes Q$.

(3.12) In the case $G$ acts semifreely on $X$ and $Y$, $\sigma_1(f) \in J_*^{0,0}(\mathbf{Z}(G)) = L_*^h(G)$.

In order to use this theory, we need to be able to compute the obstructions and/or give a Zero Theorem, i.e., a criterion for the obstructions to vanish. Using the Atiyah-Singer $G$ Signature Theorem [**1**, p. 582], we can give an explicit formula for the obstructions $\sigma_P(f)$ modulo torsion. Suppose $X$ is a closed $G$-manifold of even dimension. Let $g \in G$ and $L(g, TX^g, \nu(X^g, X)) \in H^*(X^g, \mathbf{C})$ be the class which gives the value $\text{Sign}(g, X)$ of the character $\text{Sign}(G, X)$ at $g$ via evaluation on the orientation class $[X^g]$:

$$L(g, TX^g, \nu(X^g, X))[X^g] = \text{Sign}(g, X).$$

See [**1**, p. 582]. Here $X^g = \{x \in X \mid gx = x\}$. Note that $L(g, a, b)$ is defined for any $a \in KO(X^g)$ and $b \in KO_{G'}(X^g)$ with $b^g = 0$. Here $G'$ is the subgroup generated by $g$.

PROPOSITION 3.13. *Let $f\colon X \to Y$ be a $G$ normal map with $TX = f^*(TY - \alpha)$ for $\alpha \in KO_G(Y)$. Let*

$$L_f(g, \alpha) = L(g, TY^g - \alpha^g, \nu(Y^g, Y) - \alpha_g) \cdot PD^{-1}(f_*^g[X^g]) - L(g, TY^g, \nu(Y^g, Y)).$$

*Here $\alpha_g$ is defined by $j^*(\alpha) = \alpha^g \oplus \alpha_g$ where $j\colon Y^g \to Y$ is the inclusion and $PD\colon H^*(Y^g, \mathbf{C}) \to H_*(Y^g, \mathbf{C})$ is the Poincaré duality isomorphism. Then*

$$L_f(g, \alpha)[Y^g] = \text{Sign}\, \sigma_1(f)(g)$$

*for $g \in G$.*

Clearly this idea can be iterated to give $\text{Sign}\, \sigma_P(f)(\bar{g}) = L_{f^P}(g, \alpha^P)[Y^g]$ for $g \in N(P)$ representing $\bar{g}$ in $W(P)$.

From the formula for $L_f(g, \alpha)$ with $\alpha \in KO_G(Y)$, we see that it only depends on $j^*(\alpha) \in KO_G(Y^g)$ and $f^g$. In particular we can use the formula to define $L_f(g, \beta) \in H^*(Y^g, \mathbf{C})$ for any $\beta \in KO_G(Y^g)$ by replacing $\alpha$ by $\beta$ in the formula.

The function $\mathscr{L}_f(g)\colon KO_G(Y^g) \to \mathbf{C}$ defined by $\mathscr{L}_f(g)(\beta) = \mathscr{L}_f(g, \beta)[Y^g]$ is usually not a homomorphism. But

PROPOSITION 3.14. *Suppose $Y^g$ is a rational homotopy sphere and $f\colon X \to Y$ is a $G$ normal map. Then the function $\mathscr{L}_f(g)$ is a homomorphism.*

PROOF. Since $Y^g$ is a rational homotopy sphere, the rational Pontrjagin and Chern classes give additive functions on bundles over $Y^g$. Since $L(g, a, b)$ is a function of the Pontrjagin classes of $a$ and the Chern classes of the eigenbundles of $b$, the result follows. See p. 582, [**1**].

COROLLARY 3.15. *Suppose $f\colon X \to Y$ is a $G$ normal map with $TX = f^*(TY - \alpha)$. Then for $g \in G$,* Sign $\sigma_1(f)(g) = \mathscr{L}_f(g)(j^*(\alpha))$ *where $j\colon Y^g \to Y$ is the inclusion. If $Y^g$ is a sphere, the function $\mathscr{L}_f(g)$ is additive.*

If we impose the condition $f_*^g(X^g) = [Y^g]$ for all $g \in G$ on the $G$ normal map $f\colon X \to Y$, the class $L_f(g, \beta) \in H^*(Y^g, C)$ does not depend on $f$ but only on $\beta \in KO_G(Y^G, p)$. In fact if

$$(3.16) \qquad L_0(g, \beta) = L(g, TY^g - \beta^g, \nu(Y^g, Y) - \beta_g) - L(g, TY^g, \nu(Y^g, Y)),$$

then $L_f(g, \beta) = L_0(g, \beta)$, and if

$$(3.17) \qquad \mathscr{L}_0(g)(\beta) = L_0(g, \beta)[Y^g]$$

then

COROLLARY 3.18. *Suppose $f\colon X \to Y$ is a $G$ normal map with $TX = f^*(TY - \alpha)$ for $\alpha \in KO_G(Y)$. Suppose $f_*^g[X^g] = [Y^g]$ for all $g \in G$. Then* Sign $\sigma_1(f)(g) = \mathscr{L}_0(g)(j^*(\alpha))$ *where $j\colon Y^g \to Y$ is the inclusion. If $Y^g$ is a sphere, $\mathscr{L}_0(g)$ is an additive function from $KO_G(Y^g)$ to $\boldsymbol{C}$.*

From 3.8–3.15 we have a complete calculation of the obstruction $\sigma_P(f)$ modulo torsion in terms of the explicit data of a $G$ normal map. This is frequently all that is needed for application whenever there is an *additivity* principle for $G$ normal maps to $Y$. If $Y$ is a homotopy sphere with dimension $Y^G > 0$, this is usually the case. For example

PROPOSITION 3.19. *Let $W$ be a $G$-module. Let $X$ be a $G$-space with* dim $X^G > 0$ *and $Y = S(W \oplus \boldsymbol{R})$. If $f\colon X \to Y$ is a $G$ normal map, $2f = f \# f\colon X \# X \to Y \# Y = Y$ is a $G$ normal map and $\sigma_P(2f) = 2\sigma_P(f)$ for all $P$.*

The connected sum is defined with respect to an invariant disk about a point $x \in X^G$ and one about $f(x) \in Y^G$.

COROLLARY 3.20. *Let $W$ be a $G$-module, $Y = S(W \oplus \boldsymbol{R})$ and $f\colon X \to Y$ a $G$ normal map with $TX = f^*(TY - \alpha)$ for $\alpha \in KO_G(Y)$. Suppose* dim $X^G > 0$ *and* Sign$(\sigma_P(f)) = 0$ *for all $P$. Then for some integer $\lambda$, $(2^\lambda X, 2^\lambda f)$ is $G$ normally cobordant to $(X', f')$ with $f'\colon X' \to Y$ a pseudoequivalence and $TX' = f'^*(TY - 2^\lambda\alpha)$.*

PROOF. This follows from 3.15, 3.9 and $T(X \# X) = (f \# f)^*(TY - 2^\lambda\alpha)$.

In situations where the *additivity principle* does not apply we need more detailed knowledge of the groups $J_*^{A,B}(R)$ to obtain a Zero Theorem. It may happen that we know that a $G$ normal map when considered as an $H$ normal map for $H \subset G$ is $H$ normally cobordant to a pseudoequivalence. If this happens for all $H$ in a certain class $\mathscr{H}$ of subgroups of $G$, we may be able to conclude that the original map is $G$ normally cobordant to a pseudoequivalence. This is called an *induction principle* with respect to $\mathscr{H}$.

INDUCTION LEMMA 3.21. *The functors* (i) $J_*^{A(G),B(G)}(\mathbf{Z}_{(p)}(G))$ *and* (ii) $J_*^{A(G),B(G)}(\mathbf{Z}(G))$ *satisfy the induction principle with respect to* (i) *the class of* 2 *hyperelementary subgroups of G and* (ii) *the class of hyperelementary subgroups of G.*

Specifically this means an element in (i) or (ii) is zero iff its restriction to each subgroup $H$ which is 2 hyperelementary, respectively hyperelementary, vanishes.

We now give a geometric criterion for the vanishing of the obstructions $\sigma_P(f)$ where $f: X \to Y$ is a $G$ normal map between closed manifolds. Suppose $X = \partial W$, $Y = \partial Z$ and $F: W \to Z$ is a $G$ normal map with $F|_X = f$—the given $G$ normal map (bundle data included). Let $\hat{\chi}_H(f) = \chi(X^H) - \chi(Y^H)$ be the difference of the Euler characteristics of $X^H$ and $Y^H$.

ZERO THEOREM 3.22. *Let* $P \subset G$ *be a p-group with* $P \neq 1$. *If both* $\sigma_{P'}(f) = 0$, $\sigma_{P'}(F) = 0$ *for* $P' > P$ *and both* $\hat{\chi}_H(f) = 0$, $\hat{\chi}_H(F) = 0$ *for* $H \subset N(P)$ *with* $H/P$ *cyclic then* $\sigma_P(f) \in J_*^{0,\tilde{K}_0}(\mathbf{Z}_{(p)}(W(P))$ *is zero. Here* $\tilde{K}_0 = \tilde{K}_0(\mathbf{Z}_p(W(P)))$.

Let $\mathscr{H}$ the be class of groups for which there is an exact sequence $1 \to P \to H \to Q \to 1$ where $P$ has prime power order and $Q$ is cyclic. Let $\mathrm{ISO}(Z) = \{H \subseteq G | H = G_z \text{ for some } z \in Z\}$.

ZERO THEOREM 3.23. *If* $\sigma_P(f) = 0 = \sigma_P(F)$ *for all groups* $P \subset G$ *of prime power order not* 1 *and if* $\hat{\chi}_H(f) = \hat{\chi}_H(F) = 0$ *for all* $H \in \mathscr{H}$ *and* $\chi_1(F) \in D(G)$, *then* $\sigma_1(f) = 0$ *in* $J_*^{D(G),K_0(\mathbf{Z}(G))}(\mathbf{Z}(G))$.

REMARK 3.24. The preceding two theorems are inductively applied to give a criterion for $(X, F)$ and $(X, f)$ to be $G$ normally cobordant to homotopy equivalences. Fortunately it is possible to give an a priori geometric condition for all $\chi_P(f)$ and $\chi_P(F)$ to vanish. This depends on the work of Oliver (these PROCEEDINGS, pp. 339–346) relating the Burnside ring and the projective class group. Let $\omega_G^0$ be the Grothendieck group of equivalence classes of smooth $G$-manifolds with equivalence defined by $X \equiv Y$ if $\chi(X^H) = \chi(Y^H)$ for all subgroups $H \subseteq G$. Addition is defined by disjoint union. Equivalently $\omega_G^0$ is the Grothendieck group of finite $G$-sets under disjoint union. Viewing a finite $G$-set as a zero dimensional manifold defines a homomorphism of the second group to the first which is an isomorphism. Let $K_0(G, \mathbf{Z})$ be the Grothendieck group under direct sum of finitely generated torsion free $\mathbf{Z}(G)$-modules and $\Delta_G \subset \omega_G^0$ the kernel of the homomorphism to $K_0(G, \mathbf{Z})$ which sends a finite $G$-set $X$ to the permutation representation $\mathbf{Z}(X)$ defined by $X$. If $f: X \to Y$ is a $G$ normal map and $Y - X \in \omega_G^0$ actually lies in $\Delta_G$, then whenever $\chi_P(f)$ is defined in $\tilde{K}_0(R_P)$, it vanishes for $P \neq 1$, $P \in \mathscr{P}$, and lies in $D(G)$ for $P = 1$.

Here is one example of an a priori Zero Theorem whose hypotheses are entirely described by geometric conditions. It is a corollary of the above theorems. Let $\theta_G(Z)$ be the subgroup of $\omega_G^0$ generated by $\{G/H | H \in \mathrm{ISO}(Z)\}$. Let $H_p$ denote a $p$ Sylow subgroup of $H$.

A PRIORI ZERO THEOREM 3.25. *Let* $f: X \to Y$ *be a G normal map and suppose for each subgroup P of prime power order* (a) $\pi_i(Y^P) = \pi_i(\partial Y^P) = 0$, $i = 0, 1$; (b) $H \in \mathrm{ISO}(\partial X)$. *A sufficient respectively necessary condition that G-surgery on X and* $\partial X$ *is possible producing a G normal cobordism of* $(X, f)$ *to a homotopy equivalence is that* $Y - X$ *vanish in* $\omega_G^0/\Delta_G + \theta_G(\partial X)$ *respectively in* $\omega_G^0/\Delta_G + \theta_G(X)$.

REMARK. Both (a) and (b) may be removed by generalizing the Burnside ring and the group $\Delta_G$ and requiring $\pi_0(\partial X) \longrightarrow \pi_0(X)$ to be a surjection of sets.

**4. Applications.** The material of the preceding three sections will now be applied to two situations. The first application is to semifree actions on homotopy spheres. Theorems 4.1 and 4.2 are elementary consequences of the general theory of §§1–3. Theorem 4.2 is the main result of [**17**]. The second application is considerably more difficult—namely the construction of smooth actions on homotopy spheres with only one fixed point. Here we can only give an impression of the ideas and how §3 applies. This is part of the problem posed in (0.3).

First the case of semifree $G = \boldsymbol{Z}_p$ actions on homotopy spheres. Here $p$ is an odd prime. The same considerations apply to arbitrary finite groups acting on homotopy spheres but at the expense of somewhat more complicated hypotheses. The two problems we wish to treat are the existence of a characteristic subsphere and the values of the restriction of the equivariant tangent bundle to the fixed point set $T\Sigma|_{\Sigma^G}$ where $\Sigma$ is a homotopy sphere with semifree $G$-action. Specifically suppose $V \subset W$ are real $G$-modules and $f\colon \Sigma \to S(W \oplus \boldsymbol{R})$ is a pseudoequivalence. If $f$ is transverse to $S(V \oplus \boldsymbol{R})$ with inverse image $\Sigma'$ and $f' = f|_{\Sigma'}$ is a pseudoequivalence, we say $\Sigma'$ is a characteristic subsphere for $V$. The question is under what conditions do characteristic spheres exist. In the second problem we ask for conditions which imply the existence of an infinite number of values of $T\Sigma|_{\Sigma^G} \in KO_G(S(W^G \oplus \boldsymbol{R}))$ where $f\colon \Sigma \to S(W \oplus \boldsymbol{R})$ is a pseudoequivalence and $f^G\colon \Sigma^G \to S(W^G \oplus \boldsymbol{R})$ is a diffeomorphism. This falls into the setting posed in the introduction where $Y = S(W \oplus \boldsymbol{R})$ and $I(\Sigma) = T\Sigma|_{\Sigma^G} \in KO_G(Y^G)$. Both problems make sense for arbitrary finite groups $G$ and we suspect the answers in the general case differ little from those given here for $G = \boldsymbol{Z}_p$. This will be treated in future work and is omitted here for reasons of manuscript deadline.

One useful fact concerning either pseudoequivalences or $G$ normal maps to $S(W \oplus \boldsymbol{R})$ when $\dim S(W \oplus \boldsymbol{R})^G > 0$ is that it is possible to add them. In general some care must be exercised but there is no difficulty in forming $[\Sigma, f] + [\Sigma, f] = 2[\Sigma, f]$ when $f\colon \Sigma \to S(W \oplus \boldsymbol{R})$ is a pseudoequivalence or a $G$ normal map. The addition is formed by taking connected sum in the source and target by removing an invariant disk about a fixed point in each and identifying along the bounding sphere in the standard fashion.

A $G$ normal $h$ cobordism rel fixed set consists of a $G$ normal cobordism $F\colon (W, \partial_0 W, \partial_1 W) \to (Y \times I, Y \times 0, Y \times 1)$ which is a pseudoequivalence and a product on $W^G$, i.e., $W^G \cong Y^G \times I$ and $F^G = F_0^G \times 1d$.

THEOREM 4.1. *Suppose* $V \subset W$, $\dim W - \dim V > 2$, $\dim W^G > 0$, $V^G = W^G$, *and for each* $\chi \in \hat{G} - 1$ *either* $\langle \chi, W - V \rangle = 0$ *or* $2\langle \chi, V \rangle > \dim W^G + 1$. *Suppose* $f\colon \Sigma \to S(W \oplus \boldsymbol{R})$ *is a pseudoequivalence and* $T\Sigma_p = W$ *for* $p \in \Sigma^G$. *Then either* $(\Sigma, f)$ *or* $2(\Sigma, f)$ *is G normally h cobordant to* $(\tilde{\Sigma}, \tilde{f})$ *and* $\tilde{\Sigma}$ *has a characteristic sphere for* $V$. *The former occurs if* $\dim \Sigma$ *is odd and the latter if* $\dim \Sigma$ *is even.*

THEOREM 4.2 [**17**]. *Suppose* $\dim W^G = 2$ *and* $2\langle \chi, W \rangle > \dim W^G + 1$ *for each* $\chi \in \hat{G} - 1$. *There are an infinite number of pseudoequivalences* $f\colon \Sigma \to S(W \oplus \boldsymbol{R})$ *with* $f^G$ *a diffeomorphism and they are distinguished by* $T\Sigma|_{\Sigma^G}$ *in* $KO_G(\Sigma^G) = KO_G(S(W \oplus \boldsymbol{R})^G)$.

We treat Theorem 4.2 first.

Let $Y = S(W \oplus \boldsymbol{R})$ where $W$ satisfies the hypothesis of Theorem 4.2. The strategy is to construct an infinite number of $\alpha \in KO_G(Y, p)$, $p \in Y^G$, with

(i) $\alpha \geqq 0$;

(ii) $\alpha^G = 0$ in $KO(Y^G, p)$;

(iii) $\mathscr{L}_0(g)(j^*(\alpha)) = 0$ for all $g \in G$, $g \neq 1$ for $j: Y^G \to Y$ and $\mathscr{L}_0(1)(\alpha) = 0$ (3.17);

(iv) $j^*(\alpha) \in KO_G(Y^G, p)$ is an infinite set.

Specifically suppose $\alpha = M - N$ and $\omega: M \to N$ is a quasi-equivalence with $\omega^G: N^G \to M^G$ a vector bundle isomorphism (conditions (i) and (ii)). By Corollary 2.6, $\omega$ is properly $G$ homotopic to $\theta$ with $\theta$ transverse to $Y$. Since $\omega^G$ is a vector bundle isomorphism, it is transverse to $Y^G \subset M^G$ and by **[8]** we may suppose $\theta^G = \omega^G$. Then $X = \theta^{-1}(Y)$ and $f = \theta|_X: X \to Y$ is a $G$ normal map with $X^G = Y^G$, $f^G$ is the identity and $TX = f^*(TY - \alpha)$.

Since $f^G$ is a diffeomorphism, the obstruction $\sigma_G(f)$ is zero (3.5). By 3.5 and (3.12), the only obstruction to making $f$ a pseudoequivalence is $\sigma_1(f) \in L^h_{2n}(\boldsymbol{Z}(G))$, $2n = \dim Y$. By 3.18 and (iii), Sign $\sigma_1(f) = 0$. By 3.20, $2^\lambda(X, f)$ is $G$ normally cobordant rel $X^G$ to $(\Sigma, f')$ where $f': \Sigma \to Y$ is a pseudoequivalence with $f'^G$ a diffeomorphism and $T\Sigma = f'*(TY - 2^\lambda\alpha)$. (Actually we may take $\lambda = 1$ if $\dim W \equiv 2$ (4) and 0 if $\dim W \equiv 0$ (4).)

From this discussion, Theorem 4.2 follows once the existence of an infinite number of $\alpha \in KO_G(Y, p)$ satisfying (i)—(iv) has been established. In order to simplify matters, we add the condition $\dim W \equiv 2$ (4) to those of Theorem 4.2. Then the condition $\mathscr{L}_0(1)(\alpha) = 0$ of (iii) is automatically satisfied and (iii) can be expressed as $j^*(\alpha) \in \bigcap_{g \neq 1} \operatorname{Ker} \mathscr{L}_0(g)$ where $\mathscr{L}_0(g)$ is the homomorphism of $KO(Y^G, p)$ to the complex numbers $\boldsymbol{C}$ (3.18).

We establish three facts:

(a) $\bigcap_{g \neq 1} \operatorname{Ker} \mathscr{L}_0(g) \otimes Q \neq 0$,

(b) $j^* \otimes Q: KO_G(Y, p) \otimes Q \to KO_G(Y^G, p) \otimes Q$ is unto and

(c) every $\alpha \in KO_G(Y, p)$ has a multiple $n\alpha$, $n \in \boldsymbol{Z}$, with $n\alpha \geqq 0$, $n\alpha^G = 0$, and in fact if $n\alpha = M - N$, there is a quasi-equivalence $\omega: N \to M$ with $\omega^G: N^G \to M^G$ a vector bundle isomorphism.

This completes the proof because every $\alpha \in KO_G(Y, p)$ with $j^*(\alpha) \in \bigcap_{g \neq 1} \operatorname{Ker} \mathscr{L}_0(g)$ has a multiple $n\alpha$ which gives rise to a pseudoequivalence $f: \Sigma \to Y$ with $f^G$ a diffeomorphism and $f^*(TY - n\alpha) = T\Sigma$.

PROPOSITION 4.3. *Let $Y$ be a $G$-space with $p \in Y^G$ and $KO(Y^G, p) \otimes Q = 0$. Then $j^* \otimes Q: KO_G(Y, p) \otimes Q \to KO_G(Y^G, p) \otimes Q$ is unto.*

PROOF. This follows from the Atiyah-Segal localization theorem for $K_G$ theory **[2]** together with $R(G) \otimes Q \cong Q \times Q(\xi)$ $(\xi^p = 1)$ as a ring and $KO_G(Y, p) \otimes Q \subset K_G(Y, p) \otimes Q$ is the subspace fixed by the homomorphism which sends a $G$ vector bundle to its dual.

COROLLARY 4.4. *Let $W$ satisfy the hypothesis of* 4.2. *Then $j^* \otimes Q$ is unto for $Y = S(W \oplus \boldsymbol{R})$.*

This establishes (b). The condition (c) is a direct consequence of 1.14. It remains to establish (a). We merely quote the fact which is essentially in **[17]**.

PROPOSITION 4.5. *Let* $Y = S(W \oplus \boldsymbol{R})$ *where* $W$ *satisfies* 4.2. *Then* $\bigcap_{g \neq 1} \operatorname{Ker} \mathscr{L}_0(g) \otimes Q \neq 0$.

The key point here is that $Y^G$ is a two sphere. Unfortunately, the statement of 4.5 does not do justice to the excellent paper [**4**] on which the calculation rests.

This completes the proof of Theorem 4.2.

We now turn to the proof of Theorem 4.1. We suppose the data of that theorem. By 2.8 we may suppose $f \pitchfork Y'$ where $Y' = S(V \oplus \boldsymbol{R})$ and set $\Sigma' = f^{-1}Y', f' = f|_{\Sigma'}$. Since $f$ is a $G$ normal map, $f'$ is a $G$ normal map because $f$ is transverse to $Y'$. Moreover $\Sigma^G = \Sigma'^G$ because $Y^G = Y'^G$. Thus $\sigma_G(f') = \sigma_G(f) = 0$. By 3.5 the only obstruction to making $f'$ a pseudoequivalence is $\sigma_1(f')$ and since $\dim(W - V) > 2$, this is the obstruction to the claim of Theorem 4.1. Since $f^G_* = f'^G_*$ is a $\boldsymbol{Z}_p$ homology equivalence, $\Sigma^G$ is a rational homotopy sphere. We have an additivity principle for the obstruction $\sigma_1(f') \in J^{0,0}_v, (\boldsymbol{Z}(G)) = L^h_v(G)$, $v = \dim V$ (3.19). Since the order of $G$ is odd, $L^h_v(G) = 0$ if $v$ is odd, so there is nothing to prove; thus, we suppose $v$ is even. We show Sign $\sigma_1(f') = 0$; so $\sigma_1(f')$ has order at most 2 (3.9). The result then follows from 3.19.

By (3.10)

$$\text{(4.6)} \qquad \operatorname{Sign} \sigma_1(f')(g) = L(g, T\Sigma'^g, \nu(\Sigma'^g, \Sigma'))[\Sigma'^g]$$

because Sign$(g, Y') = 0$ for the sphere $Y'$. We simply list the facts used to conclude (4.6) vanishes for all $g \in G$. Let $i: \Sigma' \to \Sigma$ be the inclusion. Let $\alpha = \Sigma \times W - \Sigma \times V \in KO_G(\Sigma)$. Then

(4.7) (i) $T\Sigma' = i^*(T\Sigma - \alpha)$ by transversality.

(ii) $\Sigma^g = \Sigma'^g$ and $\nu(\Sigma'^g, \Sigma') = \nu(\Sigma^g, \Sigma) - \alpha_g$ for $g \neq 1$.

(iii) The eigenbundles of $\alpha_g$ have trivial Chern classes and the Pontrjagin classes of the bundle $\alpha$ vanish. This is obvious by inspection.

(iv) $L(g, T\Sigma^g, \nu(\Sigma^g, \Sigma))[\Sigma^g] = \operatorname{Sign}(g, \Sigma) = 0$ as $\Sigma$ is a homotopy sphere.

(v) $L(g, T\Sigma^g, \nu(\Sigma^g, \Sigma) - \alpha_g)[\Sigma^g] = \operatorname{Sign}(g, \Sigma) - L(g, T\Sigma^g, \alpha_g)[\Sigma^g]$ for $g \neq 1$ because $\Sigma^g$ is a rational homotopy sphere (3.14).

(vi) $L(1, i^*(T\Sigma - \alpha), 0)[\Sigma'] = 0$ because of (4.7)(iii) and $\Sigma$ is a homotopy sphere. By these remarks (4.6) is (4.7)(v) or (4.7)(vi). These latter expressions vanish because $\Sigma$ is a homotopy sphere and because of (4.7)(iii).

This concludes Theorem 4.1.

The next application is much more difficult and requires the full strength of the preceding theory.

*Problem* 4.8. *Which groups act smoothly on a homotopy sphere* $X$ *with exactly one point* $x$ *fixed by the whole group*?

This problem of some years standing is due to Montgomery-Samelson. Of equal importance is

*Problem* 4.9. *For the groups in* 4.8, *which representations of* $G$ *occur as the isotropy representation on the tangent space of* $X$ *at* $x$?

Suppose $G$ and $X$ are solutions to 4.8. Let $V$ be the $G$-module $TX_x = \nu(x, X)$. The collapsing map of $X$ on the Thom space of $\nu(x, X)$ gives a degree one $G$ map $f: X \to p^{\nu(x,X)} = S(V \oplus \boldsymbol{R})$ which is a homotopy equivalence if $X$ is a sphere; moreover, $V^G = 0$ because $V = TX_x$ and $X^G = x$.

This suggests the following approach to 4.8: Let $V$ be a $G$-module with $V^G = 0$

and $Y = S(V \oplus \mathbf{R})$. Then $Y^G = p \cup q$ consists of two points. Each quasi-equivalence $\omega \in \omega_G^0(Y)$ is transversal by 2.4 (when $G$ is finite). By §3, this gives rise to a $G$ normal map $f: X \to Y$ by representing $\omega$ as a proper $G$-map $\tilde{\omega}: Y \times \Omega \to Y \times \Omega$ for some $G$-module $\Omega$ and then finding a proper $G$-homotopy of $\tilde{\omega}$ to $\theta \pitchfork Y \times 0$ with $X = \theta^{-1}(Y)$ and $f = \theta|_X$. The goal is to choose the data $V$ and $\omega$ so that $X^G =$ one point and the obstructions $\sigma_P(f)$ vanish. Then $f$ may be assumed a pseudoequivalence. The requirement $X^G =$ one point will place restrictions on $G$ and the requirement that all $\sigma_P(f)$ vanish will further restrict $G$ and also restrict $V$.

If $X^G$ is one point $x$ with $f(x) = p \in Y^G$, then a simple equivariant transversality argument using (3.2) shows

(4.10) $$d_G(\omega_q) = 0, \quad \text{where } Y^G = p \cup q.$$

Here $\omega_q \in \omega_G^0(q)$ is the restriction of $\omega$ to $q$. On the other hand

(4.11) $$d_1(\omega_q) = 1, \quad 1 \subset G,$$

because $\omega$ is a quasi-equivalence. The existence of $\omega_q \in \omega_G^0(q)$ satisfying (4.10) and (4.11) already restricts $G$. See **[3]**. For example $G$ cannot be a $p$-group.

We want to apply the Zero Theorems 3.22 and 3.23 to guarantee the vanishing of all the obstructions $\sigma_P(f)$ for all $P$. The conditions of those theorems involve all subgroups of $G$ which are extensions of cyclic groups by $p$-groups. In the case $G$ is abelian these are called hyperelementary. It should not be surprising then that the obstructions $\{\sigma_p(f)\}$ depend on the restrictions $\mathrm{res}_H\, \omega_q \in \omega_H^0(q)$ of $\omega_q$ for all $H \in \mathscr{H}$ where $\mathscr{H}$ is a class of subgroups of $G$ containing the hyperelementary groups. Set

(4.12) $$e = 1 - \omega_q \in \omega_G^0(p),$$

where 1 is the unit of this ring (the identity map of $\Omega$). The above discussion motivates the conditions:

(4.13) $$d_G(e) = 1,$$

(4.14) $$e \in \mathrm{Ker}\Big(\omega_G^0(q) \xrightarrow{\mathrm{res}} \prod_{H \in \mathscr{H}} \omega_H^0(q)\Big),$$

where the homomorphism res is defined by $\mathrm{res}_H$ for $H \in \mathscr{H}$.

We seek an $\omega \in \omega_G^0(Y)$ with

(4.15) $$\omega_p = 1, \quad \omega_q = 1 - e.$$

Then $\omega_q$ satisfies the conditions (4.10) and (4.11) by (4.13) and (4.14). Having such an $\omega$, we can construct a $G$ normal map $f: X \to Y$ with $X^G =$ one point $x$ because we can choose $\theta$ so that its restriction to $p \times \Omega$ is $\omega_p = 1d$ and its restriction to $q \times \Omega^G$ is a self-map of $q \times \Omega^G$ missing $q \times 0$ because degree $\omega_q^G$ is zero.

Let $e_0 \in \omega_G^0(q)$ satisfy (4.13) and (4.14).

LEMMA 4.16. *Let $V$ be a $G$-module with $V^G = 0$ and whose only isotropy groups other than $G$ are in $\mathscr{H}$. Let $Y = S(V \oplus \mathbf{R})$. Then there is an $\omega \in \omega_G^0(Y)$ with $j^*(\omega) = (1, 1 - e_0^\lambda)$ for some positive integer $\lambda$. $j: Y^G \to Y$.*

PROOF. This follows from 1.9.

Let $e = e_0^\lambda$; then the $\omega \in \omega_G^0(Y)$ of 4.16 satisfies (4.15) and we arrive at a $G$ normal map $f: X \to Y$ with $X^G = X$.

Note that 4.16 places restrictions on $V$.

In order to satisfy the hypothesis of the Zero Theorem, we use the fact that $Y$ is the $G$-boundary of the unit disk $Z = D(W \oplus \boldsymbol{R})$ and arrange that, for each $H \in \mathscr{H}$, $f$ has an $H$ extension

$$\begin{array}{ccc} W(H) & \xrightarrow{F(H)} & Z \\ \cup & & \cup \\ X & \xrightarrow{f} & Y \end{array} \tag{4.17}$$

where $W(H)$ is an $H$-manifold with boundary $X$ and $F(H)$: $(W(H), X) \to (Z, Y)$ is an $H$ normal map which restricts to $f$ on $X$. This situation in certain cases allows us to conclude the obstructions for an $H$ normal cobordism of $f$ to (an $H$) pseudoequivalence vanish for all $H \in \mathscr{H}$. The Induction Lemma 3.21 is then used to show that the obstructions to finding a $G$ normal cobordism of $f$ to a $G$ pseudoequivalence vanish also.

Let $i$: $Y \to Z$ be the inclusion and $i_H^*$: $\omega_H^0(Z) \to \omega_H^0(Y)$ the resulting homomorphism. If

$$\omega \in \ker\Big(\omega_G^0(Y) \to \prod_{H \in \mathscr{H}} \omega_0^H(Y)/i_H^* \omega_H^0(Z)\Big) \tag{4.18}$$

then using 2.4, we can achieve (4.17) for each $H \in \mathscr{H}$. Note that (4.18) implies (4.14) because $\omega_p = 1$ and $Z^G$ is connected; thus (4.14) is needed to achieve (4.18). Starting with an $e_0$ satisfying (4.13) and (4.14) and using Localization Lemma 1.8, it is possible to produce $\omega$ satisfying (4.18).

To summarize suppose $G$ is finite. Suppose $\mathscr{H}$ is a class of subgroups of $G$ closed under subgroups and conjugation and which contains all hyperelementary subgroups. Suppose $Y = S(V \oplus \boldsymbol{R})$ where $V$ is a complex $G$-module with $H \subset G$ an isotropy group of $V - 0$ iff $H \in \mathscr{H}$ and $V^G = 0$. Suppose there is an $e \in \omega_G^0(q)$ satisfying (4.13) and (4.14) and an $\omega \in \omega_G^0(Y)$ satisfying (4.15) and (4.18). Then $\omega$ gives rise to a $G$ normal map $f$: $X \to Y$ satisfying the situation of (4.17) and $X^G =$ one point $x$. We now appeal to the Zero Theorems 3.22 and 3.23 and the Induction Lemma 3.21 which applies in the following cases:

THEOREM 4.19. *The following groups act smoothly on homotopy spheres with one fixed point*: (i) $S^3$, $SO_3$, (ii) SL(2, $F$), PSL(2, $F$) *where characteristic $F$ is odd*, (iii) *odd order abelian groups $G$ with the property that no hyperelementary group $H$ in $G$ has prime power index.*

The case SL(2, $\boldsymbol{Z}_5$) was first done by E. Stein in another manner.

The class $\mathscr{H}$ of subgroups is different in each of the cases above, e.g. $\mathscr{H}$ for an odd order abelian group $G$ is $\mathscr{H} = \{H \subseteq G \mid |G|/|H| \text{ is not a prime power}\}$.

For a finite group, let 1 denote the trivial complex one dimensional $G$-module and $C(G)$ the regular representation of $G$. Set $\rho(G) + 1 = C(G)$.

THEOREM 4.20. *For the groups $G$ of* 3.6(ii), *any sufficiently large multiple of $\rho(G)$ occurs as $TX_x$ for some smooth $G$-actions on a homotopy sphere $X$ with $X^G = x$. For the groups* (iii) *any multiple of* $(\rho(G) - \sum_{i=1}^n \rho(P_i))$ *occurs as $TX_x$. Here $G = \prod_{i=1}^n P_i$ is the product of its Sylow subgroups.*

REMARK 4.21. The key fact in treating the groups (iii) is that the relevant groups

$J_n^{A_p,B_p}(Z_{(p)}(W(P)))$ vanish for $n$ odd. This is used in applying the Zero Theorems 3.22 and 3.23 to the obstructions for $F(H)$, $H \in \mathscr{H}$, of (4.17).

The procedure indicated in the preceding applications has been used to treat the problems (0.1)—(0.4). In each of these cases we start with a $G$-module $V$ and $Y = S(V)$ or $(P(V)$ (the complex projective space of $V$). We then choose $\alpha \in R(G)$ and $\omega \in \omega_G^\alpha(Y)$ which is transversal and construct a $G$ normal map $f: X \to Y$ from $\omega$. We arrange $V$ and $\alpha$ so that there is a transversal $\omega$ and all the obstructions $\sigma_p(f)$ vanish. Then $f$ may be assumed to be a pseudoequivalence and $X$ is the sought $G$-manifold realizing the sought value of $I(\cdot)$. The properties of $X$ are functions of $Y$, $\alpha$ and $\omega$ and these data are chosen so that $X$ is the solution of the problem.

It should be remarked that a slight modification is necessary to treat (0.1). Then $\omega \in \omega_G^\alpha(Y)$ mentioned above cannot be a quasi-equivalence. In fact for any $p \in Y$, $d_1(\omega_p) = n$ will in general be a nontrivial divisor of $|G|$. However, we still construct a $G$ map $f: X \to Y$ (of degree $n$) and proceed to use the obstruction theory of §3 to make $X$ a homotopy sphere with free $G$-action except that $f$ will not be a homotopy equivalence. [7] uses these ideas (though not explicitly) for $G$ metacyclic.

## Chapter II. $G$ Transversality and Applications

### 0. Introduction.

0.1. *The G transversality problem.* Given are three smooth $G$-manifolds $N$, $M$ and $Y$ with $Y \subset M$ an invariant submanifold and a proper $G$-map $f: N \to M$. When is there a proper $G$-homotopy of $f$ to a map $\theta$ which is transverse to $Y$?

We give a solution in terms of an obstruction theory—Theorem 4.12. Corollary 4.13 gives a criterion for the vanishing of the obstructions. Two applications are provided: (i) we give a correct proof of the theorem of **[17]** that the Burnside ring of finite $G$-sets is isomorphic to the zeroth equivariant cohomotopy ring of a point $\omega_G^0$. See §6. This is the most elementary application of $G$ transversality and has the virtue of illustrating the ideas of the obstruction theory for 0.1 as well as being crucial for constructing smooth actions of some Lie groups on homotopy spheres with exactly one fixed point **[11]**. (ii) We generalize the work of Montgomery-Yang on pseudofree $S^1$ actions on homotopy spheres [7]. In addition to the inherent interest in pseudofree actions, this setting provides the simplest case of the theory. We are able to treat it without appealing too much to the unpublished results of **[10]** and **[11]**. We give realization theorems for the invariants $I(Y) = |\mathscr{S}(Y)|$ and $I(Y)$—the collection of slice representations of the isotropy groups. See §§9, 10 and 12. A typical result is 12.1. It shows that there are pseudofree actions of $S^1$ on homotopy spheres of dim $4n - 1$ having an arbitrary number of isotropy groups. This generalizes the result of [7] from dimension 7 to arbitrary dimensions.

The equivariant cohomology theory **[17]** $Y \to \omega_G^\alpha(Y)$ for $\alpha \in R(G)$ (the complex representation ring of $G$) gives a fruitful setting for organizing the concepts for dealing with the problem of realizing invariants within a homotopy type, e.g., by varying $G$, $\alpha$ and a representation $V$ of $G$ and letting $Y$ be the unit sphere $S(V)$, we obtain solutions to I(0.1 – 0.3) by choosing a specific $\omega \in \omega_G^\alpha(Y)$ in each case and applying $G$ transversality and $G$ normal cobordism to produce an action of $G$ on a homotopy sphere of the required sort. Application to I 0.1 appeared in **[13]**, I 0.2 appears here (§§9–12) and application to I 0.3 will appear in **[11]**.

Announcements, summaries and discussions of the ideas involved with the obstruction theory for 0.1 appeared in **[14]**, **[15]**, **[9]** and Oberwolfach Lectures as far back as 1973. In particular two ideas were introduced and emphasized: The first is that the problem of equivariant transversality is involved with global phenomena in contrast to the nonequivariant situation where everything is local and trivial. The second is that Shur's Lemma applied to the equivariant vector bundles involved with transversality gives a splitting of the problem into two parts. The first can be solved by traditional means and provides the definition for the transversality obstructions which are concerned with the second part.

More precisely, suppose that $f: N \to M$ is transverse to $Y$ with $X = f^{-1}(Y)$ and $K \subseteq G$. Then $f^K$ is transverse to $Y^K \subset M^K$ and the normal bundle of $X$ in $M$ restricted to $X^K$ has a splitting $\nu(X, N)^K + \nu(X, N)_K$ and $\nu(X, N)^K = \nu(X^K, N^K)$ while $\nu(X, N)_K$ is the restriction of $\nu(N^K, N)$ to $X^K$. The fact that $f$ is transverse to $Y$ at $X^K$ thus is expressed by two equations

(i) $\nu(X^K, N^K) = (f^K)^*\nu(Y^K, M^K)$,

(ii) $\nu(N^K, N)|_{X^K} = (f^K)^*\nu(Y, M)_K$

by Shur's Lemma. The first equation only depends on $f^K: N^K \to M^K$ and is concerned with the action of the normalizer of $K$ mod $K$ on $N^K$ and $M^K$ which (by induction) can be assumed to act freely; so the problem of $f^K$ being transverse to $Y^K$ in $M^K$ is treated by Thom transversality and in particular gives $X^K = (f^K)^{-1}(Y^K)$ as a submanifold of $N$, so the left-hand side of (i) makes sense. It is then equation (ii) which provides the basis for the transversality obstruction theory.

This chapter is organized as follows: §§1—5 deal with the obstruction theory for the $G$ transversality problem. §6 gives the first application—to the Burnside ring of $G$. §7 introduces $\omega_G^\alpha(Y)$ and relates it to quasi-equivalence and $G$ normal cobordism. §8 gives some specific quasi-equivalences for use in the remaining sections. §9 introduces pseudofree $S^1$ manifolds and gives a comparison theorem for invariants of pseudofree manifolds within a homotopy type 9.1. §10 constructs $G$ normal maps which realize the invariants of a pseudofree $S^1$ action. §11 gives a brief discussion of the obstruction theory for converting $G$ normal maps to pseudoequivalences, in particular to the maps of §10. §12 applies §§7—11 to realize the isotropy group and slice representation invariants within the homotopy type of a sphere, i.e., pseudofree $S^1$ actions on homotopy spheres are constructed realizing arbitrary isotropy groups and slice representations subject to 9.1. See 12.1.

**1. The setting.** Let $G$ be a compact Lie group and $H \subset G$ a closed subgroup whose normalizer is designated by $N(H)$. If $Y$ is a $G$-manifold ($G$ acts smoothly on $Y$), $Y^H$ the $H$ fixed point set is an $N(H)$ manifold. If $\xi$ is a $G$ vector bundle over $Y$, then $\xi|_{Y^H}$ the restriction of $\xi$ to $Y^H$, is an $N(H)$ vector bundle over $Y^H$ and has a splitting as $N(H)$ vector bundles

$$\xi|_{Y^H} = \xi^H \oplus \xi_H. \tag{1.1}$$

In fact $\xi_H$ is defined as the $N(H)$ orthogonal complement of $\xi^H$ in $\xi|_{Y^H}$. If $L: \xi \to \eta$ is a $G$ bundle map, then $L|_{\xi|_{Y^H}} = L^H \oplus L_H$ where $L^H$ and $L_H$ are $N(H)$ bundle maps, $L^H: \xi^H \to \eta^H$ and $L_H: \xi_H \to \eta_H$. The splitting of $L|_{\xi|Y^H}$ follows from Schur's Lemma applied to the group $H$.

If $Z \subset Y$ is an invariant submanifold, the $G$ normal bundle of $Z$ in $Y$ is denoted by $\nu(Z, Y)$ and is identified with a $G$ tubular neighborhood of $Z$ in $Y$.

Let $N$ be a $G$-manifold and $\mathscr{L}(N)$ the set of conjugacy classes of isotropy groups of $N$ partially ordered by inclusion. If $H$ is an isotropy group then $\bar{H} \in \mathscr{L}(N)$ denotes its conjugacy class. If the cardinality of $\mathscr{L}(N)$ is $T$, choose a 1-1 function $\alpha$ from $\mathscr{L}(N)$ to the integers 1 through $T$ subject to the relation $\alpha(\bar{K}) > \alpha(\bar{H})$ if $\bar{K} < \bar{H}$. Let

(1.2) $$Z_{k-1} = \bigcup_{\alpha(\bar{H})<k} N^H.$$

Let $N$, $M$ and $Y \subset M$ be $G$-manifolds and $f\colon N \to M$ a proper $G$-map. If $f$ is transverse to $Y$ on $N$, we write $f \pitchfork Y$. We seek to alter $f$ by a proper $G$ homotopy to a map $\theta \pitchfork Y$. This problem is analyzed via an obstruction theory depending on the sets $Z_k$. Suppose

(1.3) $f\colon N \to M$ is transverse to $Y$ on a $G$-neighborhood $W$ of $Z_{k-1}$ and $f^H \pitchfork Y^H$ with $X^H = (f^H)^{-1} Y^H$ whenever $\alpha(\bar{H}) \leqq k$. Let $K \in \mathscr{L}(N)$, $\alpha(\bar{K}) = k$.

For our transversality problem, the hypotheses (1.3) are no more stringent than assuming $f$ is transverse to $Y$ on $Z_{k-1}$. Since transversality is an open condition, the existence of $W$ is guaranteed by the hypothesis $f$ is transverse to $Y$ on $Z_{k-1}$. Since $N(K)/K$ acts freely on $N^K - Z_{k-1}$, we may assume that $f^K \pitchfork Y^K$ by performing a proper $G$ homotopy of $f$ rel $Z_{k-1}$ if necessary. This uses the Thom Transversality Lemma **[6]** for the case of trivial group action and the $G$ homotopy extension lemma **[21**, Lemma 3.2]. For $H$ with $\alpha(\bar{H}) < k$, $f$ is transverse to $Y$ on $N^H$ and this implies $f^H \pitchfork Y^H$.

Applying the discussion of (1.1) to the $N(K)$ vector bundle $\nu(X^K, N)$ gives

(1.4) $$\nu(X^K, N)_K = \nu(N^K, N)|_{X^K} \quad \text{and} \quad \nu(X^K, N)^K = \nu(X^K, N^K).$$

These are elementary and follow from

$$\nu(X^K, N^K) \oplus \nu(N^K, N)|_{X^K} = \nu(X^K, N) = \nu(X^K, X) \oplus \nu(X, N)|_{X^K}.$$

Let $L_f(K)\colon \nu(X^K, N) \longrightarrow \nu(Y^K, M)$ be defined by the composition

(1.5) $$L_f(K)\colon \nu(X^K, N) \subset TN|_{X^K} \xrightarrow{df|X^K} TM|_{Y^K} \xrightarrow{\pi} \nu(Y^K, M).$$

Viewing $\nu(X^K, N)$ and $\nu(Y^K, M)$ as tubular neighborhoods allows us to think of $L_f(K)$ as a map to $M$. Let

(1.6) $$\rho'\colon \nu(Y^K, M) \longrightarrow \nu(Y, M)|_{Y^K}$$

denote the bundle projection. By definition $f$ is transverse to $Y$ at $x \in X^K$ iff $\rho' L_f(K)$ maps the fiber $\nu(X^K, N)_x$ at $x$ surjectively on the fiber $\nu(Y, M)_{f(x)}$.

DEFINITION 1.7. fsbm abbreviates fiberwise surjective bundle map and designates a bundle map which carries each fiber of the source linearly and surjectively on the image fiber.

Let

(1.8) $$X_K = \bigcup_{H>K} X^H, \qquad \bar{H} \in \mathscr{L}(N).$$

Note (1.8) refers to a $G$-space $X$ and (1.5) to a $G$ vector bundle.

Let $V$ be the "germ" of an $N(K)$ open neighborhood of $X_K$ in $X^K$ in the following sense. Let $U$ be a regular neighborhood of $X_K/N(K)$ in $X^K/N(K)$ and $\bar{V}$ the inverse image of $U$ in $X^K$ under the orbit map $O$. We choose $U$ so that $\bar{V} \subset W$ (1.3) and set

(1.9) $$V = \text{interior } \bar{V}, \quad \bar{V} = O^{-1}(U) \quad \text{and} \quad F = X^K - V.$$

Since $f$ is transverse to $Y$ on $\bar{V} \subset W$,

(1.10) $\rho' L_f(K)$ restricted to $\nu(X^K, N)|_{\bar{V}}$ is an $N(K)$ fsbm.

The tangent space to $G$, resp. $N(K)$, at the identity is denoted by $\mathfrak{g}$, resp. $n(k)$. These are $G$-, resp. $N(K)$-modules under the adjoint action and $\mathfrak{g}/n(k)$ is an $N(K)$-module identified with the tangent space of $G/N(K)$ at the identity coset. An invariant metric on $\mathfrak{g}$ allows us to identify $\mathfrak{g}/n(k)$ with the orthogonal complement of $n(k)$ in $\mathfrak{g}$. The exponential mapping [**5**, p. 58] $E: \mathfrak{g} \to G$ is a $G$-map if $G$ acts on itself by conjugation. Define

(1.11) $$\underline{\mathfrak{g}/n(k)} = F \times \mathfrak{g}/n(k).$$

It is an $N(K)$ vector bundle over $F$. In fact it is a subbundle of $\nu(X^K, N)|_F$. Here is an embedding: for $x \in F$ and $v \in \mathfrak{g}/n(k) \subset \mathfrak{g}$, send $(x, v)$ to $(d/dt)E(tv)(x)_{t=0}$, $t \in \boldsymbol{R}$. This is a tangent vector to $N$ at $x \in F$. In fact it is normal to $F$ because $(\mathfrak{g}/n(k))^K = 0$.

(1.12) Define $\nu'$ to be the orthogonal complement of $\underline{\mathfrak{g}/n(k)}$ in $\nu(X^K, N)|_F$.

(1.13) Define $L_f'(K)$: $\nu' \to \nu(Y^K, M)$ by $L_f'(K) = L_f(K) \circ i'$ where $i'$: $\nu' \to \nu(X^K, N)$ is the inclusion.

Observe that a $G$-map $f: N \to M$ carries the curve $E(tv)(x)$ for $x \in F$ into the curve $E(tv)f(x) \subset Y$ as $Y$ is $G$ invariant; so $df_x((d/dt)E(tv)_{t=0}) \in TY_{f(x)}$. This shows $df|_F(\underline{\mathfrak{g}/n(k)}) \subset TY|_{Y^K}$, but $(\mathfrak{g}/n(k))^K = 0$ so in fact $df|_F(\underline{\mathfrak{g}/n(k)}) \subset \nu(Y^K, Y)$. This together with the splitting $\nu(Y^K, M) = \nu(Y^K, Y) \oplus \nu(Y, M)|_{Y^K}$ shows

PROPOSITION 1.14. *$\rho' L_f(K)$ is a fsbm from $\nu_x$ to $\nu(Y, M)_{f(x)}$ for $x \in F$ iff $\rho' L_f'(K)$ is a fsbm from $\nu_x'$ to $\nu(Y, M)_{f(x)}$. In other words, $f$ is transverse to $Y$ at $x \in F$ iff $\rho' L_f'(K)$ is a fsbm from $\nu_x'$ to $\nu(Y, M)_{f(x)}$. Here $\nu_x$ is the fiber of $\nu = \nu(X^K, N)$ over $x$.*

Observe that $G \times_{N(K)} F \cong GF$ is a $G$-submanifold of $N$. The identification is given by sending the point $[g, x]$ of $G \times_{N(K)} F$ determined by $(g, x) \in G \times F$ to $gx$.

PROPOSITION 1.15. $\nu(G \times_{N(K)} F, N) = G \times_{N(K)} \nu'$.

PROOF. $\nu(F, N) = \nu(F, G \times_{N(K)} F) \oplus \nu(G \times_{N(K)} F, N)|_F$. From the fibration $F \to G \times_{N(K)} F \to G/N(K)$, we see that $\nu(F, G \times_{N(K)} F) = \underline{\mathfrak{g}/n(k)}$. Thus $\nu(G \times_{N(K)} F, N)|_F = \nu'$ by definition of $\nu'$. The claim now follows from the observation that two $G$-bundles over $G \times_{N(K)} F$ are equal iff their $N(K)$-restrictions to $F$ are equal.

It follows from 1.15 that the $G$-map $f$ restricted to $\nu(G \times_{N(K)} F, N)$ is completely determined by the $N(K)$-restriction $f'$ of $f$ to $\nu'$; moreover, 1.14 and 1.15 together show that $f$ is transverse to $Y$ on $\nu(G \times_{N(K)} F, N)$ iff $f'$ is transverse to $Y$ on $\nu'$.

Denote the subbundle of $\nu'$ of vectors of norm less than $\delta$ by $\nu_\delta'$ and $S(\nu')$ for the unit sphere bundle of $\nu'$. The strategy is now to modify $f'$ in an $N(K)$ invariant neighborhood of a subset $F_0 \subset F$ with $F - F_0 \subset W$ without changing $f'$ on $\nu'|_{\partial F} \cup S(\nu')$. In fact we seek to modify $f'$ so that it is an $N(K)$ bundle map $L''$ from $\nu_\delta'|_{F_0}$ to $\nu(Y^K, M) \subset M$ for sufficiently small $\delta$. We require $L'' = L_f'(K)$ on $\nu_\delta'|_{\partial F_0}$ where

$\rho' L'_F(K)$ is a fsbm (because $f$ is transverse to $Y$ on $W$). The obstruction to transversality is the obstruction to finding an $N(K)$ extension $L''$ which is a fsbm on $\nu'_\delta|_{F_0}$.

**2. The crunch.** Choose a second open invariant neighborhood $V_1$ of $X_K$ as in (1.9) with $\bar{V} \subset \bar{V}_1 \subset W$ where $\bar{V}_1$ is the closure of $V_1$. Set

(2.0) $$F_0 = X^K - V_1.$$

Let $d$ be an $N(K)$ invariant metric on $\nu'$, so $d(x, y)$ is the distance between $x$ and $y$. Set $\nu'_\delta = \{x \in \nu' | d(x, F_0) \leqq \delta\}$. We may suppose, for any point $v \in \nu'_{2\delta}$, $L'_f(K)(v)$ and $f'(v)$ are joined by a unique geodesic in $M$. This depends on the fact that $L'_f(K)$ and $f'$ agree on $F$. Of course the metric on $M$ defining the geodesics is taken to be $G$ invariant. We abbreviate $L'_f(K)$ by $L'$.

Let $TM$ be the tangent space of $M$ consisting of all pairs $(p, w)$ with $p \in M$, $w \in TM_p$ and $\exp: TM \to M$ the exponential map [5, p. 58] written $\exp(p, w) = \exp_p(w)$.

Since $f'|_F = L'|_F$, $f'(v)$ and $L'(v)$ differ by a small amount if $\|v\|$ is sufficiently small for $v \in \nu'$. This means there is a unique $w(v) \in TM_{L'(v)}$ such that

(2.1) $$\exp_{L'(v)} (w(v)) = f'(v).$$

Choose a smooth invariant function $\alpha$ from $\nu'$ to $[0, 1]$ such that $\alpha = 1$ on $\nu' - \nu'_{2\delta} \cup \nu|\partial F$ and $a = 0$ on $\nu'_\delta | F_0$. Define

(2.2) $$h: \nu' \longrightarrow M \quad \text{as } h(v) = \exp_{L'(v)} (\alpha(v)w(v))$$

for $v \in \nu'_{2\delta}$ and $h = f'$ on $\nu' - \nu'_{2\delta}$ Then

(2.3) $$h = L' \text{ on } \nu'_\delta | F_0 \quad \text{and} \quad h = f' \text{ on } \nu' - \nu'_{2\delta} \cup \nu'|\partial F.$$

Moreover if $L'_z$ denotes the linear map $L|'_{\nu'_z}$, we have

PROPOSITION 2.4. *$\pi dh_z = \pi df'_z = \pi L'_z$ for $z \in F$ (1.13), (1.5).*

PROOF. Let $F$: $TM \to M \times M$ be defined by $F(q, s) = (q, \exp_q s)$. Choose a coordinate chart $U$ in $M$ centered at $L'(z)$. Then $TM|_U \cong U \times U$ and $(u, s) \in U \times U$ denotes a tangent vector to $u = (u, 0)$ of norm $\|s\|$. We may regard $F|_U$ as a map of $U \times U$ to $U \times U$. Then $dF_{(L'(z), 0)} = \left(\begin{smallmatrix} I & I \\ 0 & I \end{smallmatrix}\right)$ [5, p. 58]. Observe that $h(v) = p_2 Fs(v)$ where $s(v) = (L'(v), \alpha(v)w(v))$ and $p_2$: $M \times M \to M$ is projection on the second factor. Since $L'(z) = f'(z)$, $w(z) = 0$ by (2.1). Thus $s(z) = (L'(z), 0)$. Viewing $s(v) = (L'(v), \alpha(v)w(v)) \in U \times U$, we compute

(2.5) $$dh_z = p_2 \circ dF_{(L'(z), 0)} \, ds_z = dL'_z + \alpha dw_z.$$

Repeating (2.5) with $\alpha = 1$ gives $dw_z = df'_z - dL'_z$ so

(2.6) $$dh_z = dL'_z + \alpha(df'_z - dL'_z).$$

Thus 2.4 follows from (2.6) and the fact that $\pi df'_z = \pi L'_z$ for $z \in F$.

Using (2.1) it is easy to give an invariant homotopy between $f'$ and $h$, the homotopy being constant on $\nu'|\partial F$ and on $\nu' - \nu'_{2\delta}$. Replacing $F_0$ by $F$ we may suppose

(2.7) $$f' = L' = L'_f(K) \quad \text{on } \nu'$$

and

(2.8) $\rho' \circ L'$ is a fsbm on $\nu'|_{\partial F}$.

Assume given an $N(K)$ bundle map

(2.9) $L'': \nu' \to \nu(Y^K, M)$ covering $f'|_F$ with $L'' = L'$ on $\nu'|_{\partial F}$ and that $\rho' L''$ is a fsbm.

Replace $f'$ by $L'$ and $L'$ by $L''$. Use $h$ defined by (2.2) for this choice with $\alpha$ now chosen so that $\alpha = 0$ on $\nu'_\delta$ and $\alpha = 1$ on $\nu' - \nu'_{2\delta}$. This gives an $N(K)$-map called $f'': \nu' \to M$ with $f'' = L'$ on $\nu' - \nu'_{2\delta}, f'' = L''$ on $\nu'_\delta$ and $f'' = L'' = L'$ on $\nu'|_{\partial F}$. Thus $f'' = f'$ on $\nu'|_{\partial F} \cup S(\nu')$, $f'' = L''$ on $\nu'_\delta$ and $\rho' L''$ is an $N(K)$ fsbm.

It follows from 1.15 that $f''$ defines a unique $G$-map $\theta$ from $N$ to $M$ which equals the $G$-extension of $f''$ to $\nu(G \times_{N(K)} F, N)$ there and $f$ on its complement; moreover, $\theta$ and $f$ are properly $G$ homotopic and $\theta$ is transverse to $Y$ on $Z_k$. To see this observe that $\theta$ agrees with $f$ outside $\nu(G \times_{N(K)} F, N)$. Inside, $\theta$ is transverse to $Y$ at $gx \in GF = Z_k \cap \nu(G \times_{N(K)} F, N)$ iff $f''$ is transverse to $Y$ at $x \in F$ and this has been achieved by construction. Thus

THEOREM 2.10. *Suppose $f: N \to M$ satisfies* (1.3). *Let $V \subset X^K$ be defined by* (1.9) *and $F = X^K - V$. Define an $N(K)$ vector bundle $\nu'$ over $F$ by* (1.12) *and an $N(K)$ bundle map $L'_f(K)$ of $\nu'$ to $\nu(Y^K, M)$ covering $f|_F$ by* (1.13). *Then $\rho' L'_f(K)$ is an $N(K)$ fsbm* (1.7) *when restricted to $\nu'|_{\partial F}$. There is a $G$-neighborhood $W' \subset W$ of $Z_{k-1}$ and a map $\theta: N \to M$ which is properly $G$ homotopic to $f$* rel *$W' \cup Z_k$ and $\theta$ is transverse to $Y$ on $Z_k$ iff $L'_f(K)$ restricted to $\nu'|_{\partial F}$ extends to an $N(K)$ bundle map $L''$: $\nu' \to \nu(Y^K, M)$ covering $f|_F$ such that $\rho' L'': \nu' \to \nu(Y, M)|_{Y^K}$ is an $N(K)$ fsbm.*

REMARK 2.11. We can choose $\theta$ arbitrarily close to $f$ in the $C^0$ topology and $L'_\theta(K) = L''$. See 2.4. Since $\theta|W' = f|W'$, $L_\theta(H) = L_f(H)$ for $\alpha(\bar{H}) < k$.

Set $L' = L'_f(K)$.

LEMMA 2.12. *$L'$ restricted to $\nu'|_{\partial F}$ extends to an $N(K)$ bundle map $L''$ of $\nu'$ with $\rho' L''$ fiberwise surjective iff $\rho' L': \nu'|_{\partial F} \to \nu(Y, M)|_{Y^K}$ extends to an $N(K)$ fsbm $L'''$: $\nu' \to \nu(Y, M)|_{Y^K}$.*

PROOF. If $L''$ exists, define $L''' = \rho' L''$. Conversely suppose $L'''$ exists. Observe that $\nu(Y^K, M) = \nu(Y^K, Y) \oplus \nu(Y, M)|_{Y^K}$; so $L' = L'_1 \oplus L'_2$ where $L'_2 = \rho' L'$. Define $L'' = L'_1 \oplus L'''$. Then $L'' = L'$ on $\nu'|_{\partial F}$ and $\rho' L'' = L'''$ is an $N(K)$ fsbm.

In view of 2.12, we lose nothing by supposing $L'$ maps $\nu'$ to the subbundle $\nu(Y, M)|_{Y^K}$ and in this context, we seek an extension $L''$ of $L'$ restricted to $\nu'|_{\partial F}$ which is an $N(K)$ fsbm to $\nu(Y, M)|_{Y^K}$.

As before $L'$ splits as $L' = L'^K \oplus L'_K$ where

$$L'^K: \nu'^K \longrightarrow \nu(Y, M)^K = \nu(Y^K, M^K),$$
$$L'_K: \nu'_K \longrightarrow \nu(Y, M)_K. \tag{2.13}$$

LEMMA 2.14. *$L'$ is an $N(K)$ fsbm iff $L'_K$ is an $N(K)$ fsbm.*

PROOF. Since $f^K \pitchfork Y^K$ (1.3), $L'^K = L'_f(K)^K = L'_{f^K}(K)$ is an $N(K)$ fsbm.

Define $\nu'(N^K, N)|_F$ to be the orthogonal complement of the image of $F \times \mathfrak{g}/n(k)$ in $\nu(N^K, N)|_F$. See (1.12). It follows from (1.4) that

$$\nu'_K = \nu'(N^K, N)|_F. \tag{2.15}$$

**3. $G$ transversality and $N(K)$-sections.**

DEFINITION. A $G$ fiber bundle, denoted simply by $E$, consists of $G$-spaces $E$ and $Y$ and a $G$-map $\pi: E \to Y$ such that $|\pi|$ is the projection of a fiber bundle.

The fiber of $E$ is $\pi^{-1}(y) = T$ for $y \in Y$ and the isotropy group of $y$ acts on $\pi^{-1}(y)$. The space of sections $\Gamma(E)$ of $E$ is a $G$-space via $(gs)(y) = gs(g^{-1}(y))$. Set $\Gamma_G(E) = \Gamma(E)^G$ and note that if $K$ is normal in $G$ and acts trivially on $Y$ then

$$\Gamma_G(E) = \Gamma_{G/K}(E^K). \tag{3.1}$$

Suppose that $K$ is normal in $G$ and $A \subset Y^K$ is a $G$ invariant closed subspace such that $G/K$ acts freely on $Y^K - A$. Then $\bar{\pi}^K: E^K|Y^K - A/G \to Y^K - A/G$ is a fibration with fiber $T^K$. This means that if $s_0 \in \Gamma_G(E^K|_A)$, there is an obstruction theory for extending $s_0$ to $s \in \Gamma_G(E^K)$. The obstructions are denoted by

$$O_*(E^K, s_0) \in H^*(Y^K/G, A/G, \pi_{*-1}(T^K)) \tag{3.2}$$

and arise from applying (3.1) and **[18]** to finding a section of $\bar{\pi}^K$. Note that $O_j(E^K, s_0)$ is defined iff $O_i(E^K, s_0) = 0$ for $i < j$ and is the obstruction to extending $s_0$ to $Y_j^K \cup A$ where $Y_j^K$ is the inverse image of the $j$-skeleton of $Y^K/G$.

An example which we need is the $G$ fiber bundle $V_{\xi,\eta} = \mathrm{Hom}^s(\xi, \eta)$ of real surjective homomorphisms of the $G$ vector bundle $\xi$ over $Y$ onto the $G$ vector bundle $\eta$ over $Y$. The action of $G$ is defined by $(gf)(x) = gf(g^{-1}x)$ for $g \in G$, $f \in \mathrm{Hom}^s(\xi,\eta)$ and $x \in \xi$. Then $V_{\xi,\eta}^K$ is a $G/K$ fiber bundle over $Y^K$ if $K$ is normal in $G$. The fiber over $y \in Y^K$ is

$$V(K)_y = \mathrm{Hom}^s_K(\xi_y, \eta_y) \tag{3.3}$$

the space of real surjective $K$-homomorphisms from the fiber $\xi_y$ to $\eta_y$. Note both are real $K$-modules. The function $y \to V(K)_y$ is constant on the components of $Y^K$ and defines a function from the components to topological spaces denoted as $V(K)$.

Replace $G$ by $N(K)$ and $Y$ by $F$ (1.9) and apply the above remarks to $E = V_{\xi,\eta}$ where $\xi = \nu'_K$ and $\eta = f^*|_F \nu(Y, M)_K$ are $N(K)$ vector bundles over $F$. Since $L'_K$ restricted to $\nu'_K|_{\partial F}$ is an $N(K)$ fsbm, it defines a section $s_0 = s_0(f)$,

$$s_0 \in \Gamma_{N(K)}(V_{\xi,\eta}|_{\partial F}). \tag{3.4}$$

To be specific, $L'_K$ defines a unique $N(K)$ bundle map $\hat{L}'_K: \xi \to \eta$ defined by $(f|_F)^* \hat{L}'_K = L'_K$ where $(f|_F)^*: f^*|_F \nu(Y, M)_K \to \nu(Y, M)_K$ is the induced bundle map. Restricting $\hat{L}'_K$ to $\xi|_{\partial F}$ gives the section $s_0$ because an $N(K)$ fsbm of $\xi$ to $\eta$ is the same as an $N(K)$-section of $V_{\xi,\eta}$.

THEOREM 3.5. *Let $f: N \to M$. Assume the hypothesis and notation of* 2.10. *Then $L'_f(K)$ restricted to $\nu'_K|_{\partial F}$ defines a section $s_0 \in \Gamma_{N(K)}(V_{\xi,\eta}|\partial F)$ for $\xi = \nu'_K$, $\eta = (f|_F)^* \nu(Y, M)_K$ and $f$ is properly $G$ homotopic* rel $W' \cup Z_k$ *to a map $\theta$ transverse to $Y$ on $Z_k$ where $W' \subset W$ is a $G$-neighborhood of $Z_k$ iff $s_0$ extends to a section $s \in \Gamma_{N(K)}(V_{\xi,\eta})$.*

PROOF. If $s$ exists, it defines an $N(K)$ fsbm $L'_K: \nu'_K \to \nu(Y, M)_K$ covering $f|_F$. Then $L' = L'^K \oplus L'_K: \nu' \to \nu(Y, M)|_{Y^K}$ is an $N(K)$ fsbm covering $f|_F$ (2.13). Thus $\theta$ exists by 2.10 and $W' = W - \nu(G \times_{N(K)} F, N)$.

Conversely if $\theta$ exists, then $L'_\theta(K): \nu'_K \to \nu(Y, M)_K$ is an $N(K)$ fsbm which equals $L'_f(K)$ on $\nu'|_{\partial F}$ so defines a section $s \in \Gamma_{N(K)}(V_{\xi,\eta})$ extending $s_0$.

Since $N(K)/K$ acts freely on $F$, the obstruction theory of (3.3) applies to the problem of extending $s_0(f)$ and we set

(3.6) $$O_*(f, K) = O_*(V^K_{\xi,\eta}, s_0(f)).$$

By excision the pairs $(F/N(K), \partial F/N(K))$ and $(X^K/N(K), X_K/N(K))$ have the same relative cohomology; so

(3.7) $$O_*(f, K) \in H^*(X^K/N(K), X_K/N(K); \pi_{*-1}(V(K)))$$

(see 3.9).
Putting this together with Theorem 3.5 gives the

THEOREM 3.8. G TRANSVERSALITY THEOREM. *Let $f: N \to M$ be transverse to $Y$ on $Z_{k-1}$ and without loss of generality suppose $f^H \pitchfork Y^H$ for $\alpha(\bar{H}) \leqq k$. Define $X^H = (f^H)^{-1}Y^H$ and if $\alpha(\bar{K}) = k$, $X_K = \bigcup_{H>K} X^H$. Then there is a G invariant neighborhood $W'$ of $Z_{k-1}$ and a proper G-homotopy of f* rel *$W' \cup Z_k$ to a map $\theta \pitchfork Y$ on $Z_k$ iff a sequence of obstructions*

$$O_*(f, K) \in H^*(X^K/N(K), X_K/N(K), \pi_{*-1}(V(K)))$$

*vanish.* (*See* (3.1) *and* 3.5 *for the definition of* $V(K)$.)

REMARK 3.9. The obstruction group should be appropriately interpreted according to the components of $X^K$; namely if $X^K = \bigcup_{j=1}^n X^K_j$ is the union of its connected components, $x_j \in X^K_j$ is any point and $X^j_K = X_K \cap X^K_j$, then

$$O_*(f, K) = \prod_{j=1}^n O_*(f, K)_j \in \prod_{j=1}^n H^*(X^K_j/N(K), X^j_K/N(K), \pi_{*-1}(V(K_{x_j}))).$$

In order not to proliferate notation, we suppress indication of components leaving interpretation of subsequent statements involving this point to the reader.

**4. Analysis of $V(K)$.** Let $G$ be a compact Lie group and $K$ a closed subgroup. The set of irreducible real (complex) $K$-modules is denoted by $\hat{K}$ ($\hat{K}_C$). Then any two $K$-modules $\Gamma$ and $\Omega$ can be expressed as

(4.1) $$\Gamma = \sum_{\psi\in\hat{K}} a_{\psi,\Gamma}\, \psi, \qquad \Omega = \sum_{\psi\in\hat{K}} a_{\psi,\Omega}\, \psi$$

where $a_{\psi,\Gamma}$ and $a_{\psi,\Omega}$ are nonnegative integers. For any $\chi \in \hat{K}$, $D_\chi = \mathrm{Hom}_K(\chi, \chi)$ (the set of real $K$-endomorphisms of $\chi$) is a division algebra over $\boldsymbol{R}$ so is $\boldsymbol{R}$, $C$ or $\boldsymbol{H}$. Let $d_\chi$ be its dimension over $\boldsymbol{R}$.

DEFINITION 4.2. $V^\chi_{\Gamma,\Omega}$ is the subspace of surjective homomorphisms in $\mathrm{Hom}_K(a_{\chi,\Gamma}\chi, a_{\chi,\Omega}\chi)$. Then

(4.3) $$V^\chi_{\Gamma,\Omega} = \mathrm{GL}(a_{\chi,\Gamma}, D_\chi)/\mathrm{GL}(a_{\chi,\Gamma} - a_{\chi,\Omega}, D_\chi)$$

where $\mathrm{GL}(n, D_\chi)$ is the general linear group over $D_\chi$.

We view $\Gamma$ and $\Omega$ as real $K$ vector bundles over a point. Then the $K$-manifold $V_{\Gamma,\Omega}$ of §3 is defined and by Shur's Lemma

(4.4) $$V(K) = V^K_{\Gamma,\Omega} = \prod_{x\in\hat{K}} V^\chi_{\Gamma,\Omega} \quad \text{where } V_{\Gamma,\Omega} = \mathrm{Hom}^s(\Gamma, \Omega).$$

If we define the integers $d_\chi(\Gamma, \Omega)$ and $d(\Gamma, \Omega)$ by

$$(4.5)\qquad \begin{aligned} d_\chi(\Gamma, \Omega) &= (d_\chi(a_{\chi,\Gamma} - a_{\chi,\Omega} + 1) - 1), \qquad \chi \in K, \\ d(\Gamma, \Omega) &= \min_{\chi \in \hat{K}, a_{\chi,\Omega} \neq 0} \{d_\chi(\Gamma, \Omega)\}, \end{aligned}$$

then

LEMMA 4.6. $\pi_i(V(K)) = 0$ *for* $i < d(\Gamma, \Omega)$ *and* $\pi_{d(\Gamma\,\Omega)}(V(K)) \neq 0$.

PROOF. This is immediate from (4.3) and (4.4).

REMARK 4.7. If $d(\Gamma, \Omega) < 0$ we set $\pi_{-1}(V(K)) = \infty$; otherwise its value is 0. This is relevant to the obstruction $O_0(f, K)$ where $\Gamma = \nu'_K|_x\, \Omega = \nu(Y, M)_K|_{f(x)}$ for $x \in X^K$. By convention we set

$$(4.7)\qquad O_0(f, K) = \begin{cases} 0 & \text{if } \pi_{-1}(V(K)) = 0, \\ 1 & \text{if } \pi_{-1}(V(K)) = \infty. \end{cases}$$

This is consistent with the obstruction theory of Theorem 3.10. Also consistent with this theory is

$$(4.8)\qquad O_1(f, K) = 0 \quad \text{if } \pi_0(V(K)) = 0.$$

These are the two cases of ambiguity for the obstructions $O^*(f, K)$ of 3.8.

COROLLARY 4.9. *If the dimension of* $X^K$ *is less than* $d(\Gamma, \Omega) + 1$ *for* $\Gamma$ *and* $\Omega$ *of* (4.7), *then the obstruction groups of* 3.8 *all vanish.*

The statement of 3.8, (4.3), (4.4) and 4.6 express the way in which the transversality obstructions depend on the $G$ homological data of $N$, $M$, $Y$ and $f$.

The product decomposition (4.4) of $V(K)$ gives a product representation for the obstruction groups of 3.8 and in terms of this product representation

$$(4.10)\qquad O_*(f, K) = \prod_{\chi \in \hat{K}} O_*(f)_\chi,$$

$$(4.11)\qquad O_*(f)_\chi \in H^*(X^K/N(K), X_K/N(K); \pi_{*-1}(V^\chi_{\Gamma,\Omega})).$$

If $1 \in \tilde{K}$ denotes the trivial one dimensional $K$-module, then $O_x(f)_1$ is zero because $a_{\Gamma,1} = a_{\Omega,1} = 0$ for $\Gamma = \xi_x, \Omega = \eta_x\, x \in X^K$ (3.5). This means that we can replace the indexings set of (4.10) by $\hat{K} - \{1\}$.

Finally we combine the material of the preceding sections to give the most explicit information on the transversality obstructions.

THEOREM 4.12. $G$ TRANSVERSALITY THEOREM. *Let* $f\colon N \to M$ *be transverse to* $Y$ *on* $Z_{k-1} = \bigcup_{H>K} N^H$ *and without loss of generality suppose* $f^K \pitchfork Y^K$. *Define* $X^H = (f^H)^{-1}(Y^H)$, $H \geqq K$, and $X_K = \bigcup_{H>K} X^H$. *Then there is a* $G$ *invariant neighborhood* $W$ *of* $Z_{k-1}$ *and a proper* $G$*-homotopy of* $f$ rel $W' \cup Z_k$ *to a map* $\theta \pitchfork Y$ *on* $Z_k$ *iff a sequence of obstructions*

$$O_*(f, K) \in H^*(X^K/N(K), X_K/N(K), \pi_{*-1}(V(K)))$$

*vanish. Here* $V(K)$ *is a function of the components of* $X^K$. *The value of* $V(K)$ *at a component* $C$ *of* $X^K$ *is*

$$V(K)_x = \mathrm{Hom}^s_K(\nu'(N^K, N)_x, \nu(Y, M)_{K, f(x)}) \quad \textit{for } x \in C \subset X^K.$$

*Moreover,*

$$V(K)_x = \prod_{\chi \in K} \mathrm{GL}(a_\chi, D_\chi)/\mathrm{GL}(a_\chi - b_x D_\chi)$$

*where* $\nu'(N^K, N)_x = \sum_{x \in \hat{K}} a_\chi \chi$, $\nu(Y, M)_{K, f(x)} = \sum_{\chi \in \hat{K}} b_\chi \chi$. $D_\chi = \mathrm{Hom}_K(\chi, \chi)$ *is a division algebra. Either* $X^K$ *is empty or dimension* $X^K$ *equals dimension* $Y^K$ *– dimension* $M^K$ *+ dimension* $N^K$.

COROLLARY 4.13. *If*

$$\dim Y^K - \dim M^K + \dim N^K < \min_{\chi \in \hat{K}} \{d_\chi(a_\chi - b_\chi + 1) - 1\}$$

*then the obstructions* $O_*(K, f)$ *vanish.*

We state without proof the relative version of 4.12. This in fact follows from what has preceded with only changes in notation required.

THEOREM 4.14. EXTENSION OF TRANSVERSALITY THEOREM. *Let* $N$, $M$ *and* $Y \subset M$ *be G-manifolds such that* $\partial Y \subset \partial M$ *with* $\nu(\partial Y, \partial M) = \nu(Y, M)|_{\partial Y}$. *We allow* $\partial Y = \varnothing$ *and* $\partial M = \varnothing$. *Suppose* $f\colon N \to M$ *is a proper G-map such that* $f\colon \partial N \to \partial M$ *if* $\partial M \neq \varnothing$. *Suppose* $f$ *is transverse to* $Y$ *on a closed G invariant subset* $A \subset X$. *Then for each isotropy group* $K$ *of the action of* $G$ *on* $X$, *there is a sequence of obstructions*

$$O_*(f, K, A) \in H^*(X^K/N(K), X_K \cup A/N(K), \pi_{*-1}(V(K)))$$

*whose vanishing for all* $*$ *and for all* $K$ *imply* $f$ *is properly G homotopic* rel $A$ *to a map* $\theta \pitchfork Y$.

**5. A special case.** In general, 4.12 and its extension 4.14 are the most that can be said about the transversality problem 0.1. However, if $G = T \times F$ where $T$ is a torus and $F$ is a finite group and if $f\colon N \to M$ is a proper $G$-map between $G$ vector bundles over $Y$, these theorems can be considerably improved by expressing the obstructions explicitly in terms of the initial data of $N$, $M$ and $Y$.

Let $N$ and $M$ be $G$ vector bundles over $Y$ and $\alpha = M - N \in KO_G(Y)$. For each $K \subseteq G$ define a function $V'_\alpha(K)$ from components of $Y^K$ to topological spaces by

$$V'_\alpha(K)_y = \mathrm{Hom}^s_K(\nu(Y^K, Y)_y \oplus N_{y,K}, M_{y,K})^1$$

for $y \in Y^K$. Each $K \subseteq G$ contributes a sequence of "obstructions"

$$O_*(\alpha, K) \in H^*(Y^K/N(K), Y_K/N(K), \pi_{*-1}(V'(K))).$$

Specifically $O_j(\alpha, K)$ is defined iff $O_*(\alpha, H) = 0$ for $H > K$ and $O_i(\alpha, K) = 0$ for $i < j$. If $K$ is not an isotropy group of $Y$, the obstructions $O_*(\alpha, K)$ vanish because the obstruction groups vanish; moreover, $O_*(\alpha, gKg^{-1}) = g^*O_*(\alpha, K)$. These two facts imply the obstructions $O_*(\alpha, \cdot)$ are really functions on $\mathscr{L}(Y)$. It is however useful to have $O_*(\alpha, K)$ defined as above for all subgroups of $G$. The relevant properties of these "obstructions" are spelled out in 5.1—5.3.

THEOREM 5.1. *Let* $G = T \times F$ *where* $T$ *is a torus and* $F$ *is finite. Suppose* $f\colon N \to M$ *is a proper G-map between G vector bundles over* $Y$. *The obstructions* $O_*(\alpha, K)$ *have this property: Suppose* $O_*(\alpha, H) = 0$ *for* $H > K$. *Then we may suppose* $O_*(f, H) = 0$ *for* $\bar{H} \in \mathscr{L}(N)$, $H > K$, $f$ *satisfies* (1.3) *and* $O_*(f, K) = (f^K|_{X^K})^* O_*(\alpha, K)$ *where* $X^K = (f^K)^{-1} Y^K$.

[1]Actually $V'_\alpha(K)$ depends on a choice of $N$, $M$ with $M - N = \alpha$.

COROLLARY 5.2. *If all $O_*(\alpha, K)$ vanish, f is properly G homotopic to a map* $\theta \pitchfork Y$

LEMMA 5.3. *The obstructions $O_*(\alpha, K)$ are natural for G bundle maps. Specifically if* $M = h^*M', N = h^*N', h: Y \to Y'$, *then* $O_*(\alpha, K) = h^{K*}O_*(\alpha', K), \alpha' = M' - N'$.

COROLLARY 5.4. *Let* $G = T \times F$, $\Omega$ *any G-module and Y any smooth G-manifold. Then any proper G-map* $f: Y \times \Omega \to Y \times \Omega$ *is properly G homotopic to a map transverse to* $Y \times 0$.

PROOF. Let $N = M = Y \times \Omega$. Then $N = M = h^*\Omega$ where $h$ maps $Y$ to a point and $\Omega$ is regarded as $G$ vector over a point. Thus $O_*(h^*0, K) = h^{K*}O_*(0, K)$ by 5.3. but $O_*(0 ,K) = 0$ because the obstruction groups for $O_*(0, K)$ vanish. Now apply 5.2.

REMARK 5.5. Corollary 5.4 is definitely false for general $G$, e.g., for $G = SO_3$ there are $G$-modules $V$ and $\Omega$ such that for $Y = S(V)$ there is a proper $G$-map $f: Y \times \Omega \to Y \times \Omega$ which is not properly $G$ homotopic to one transverse to 0. See also 7.12.

I do not wish to prove 5.1 and 5.3 but indicate why 5.1 is true. Anyway its proof is much the same as 3.8. If $K \subset G = T \times F$, then $g/n(k) = 0$. This means the vector bundle $\nu'_K$ of (2.15) and 3.8 satisfies

$$\nu'_K = \nu'(N^K, N)|_F = \nu(N^K, N)|_F = f^*|_F(\nu(Y^K, Y) \oplus N_K).$$

Now 3.8 is concerned with finding a certain section $s \in \Gamma_{N(K)}(V_{\xi,\eta})$, $\xi = \nu'_K, \eta = f^*|_F \nu(Y, M)_K$. It suffices to find a certain section $s' \in \Gamma_{N(K)}(V_{\xi',\eta'})$, $\xi' = \nu(Y^K, Y) \oplus N_K$, $\eta' = \nu(Y, M)_K$. Then $s = f^*|_F s'$ is the sought section. The obstructions to finding $s'$ are the $\{O_*(\alpha, K)\}$ above.

**6. Application to the Burnside ring.** The aim of this section is to use the preceding theory of $G$ transversality to give a correct proof of the important theorem **[17]** that the Burnside ring $\Omega(G)$ of a finite group $G$ is isomorphic to a certain equivariant homotopy group $\omega_G^0$. This clears up some of the inaccuracies of **[17]** which mistakenly claims the equivalence of equivariant framed cobordism and equivariant homotopy groups of spheres. The application of $G$ transversality to prove $\Omega(G)$ is isomorphic to $\omega_G^0$ provides an especially illuminating view of the ideas of the obstruction theory. This result and the ideas of this section are important for the construction of smooth actions on homotopy spheres with one fixed point **[11]**. In addition the equivariant cohomology group $\omega_G^g(Y)$ introduced in §7 is a module over $\omega_G^0$.

The category of finite $G$-sets has addition and multiplication defined by disjoint union and cartesian product. The associated Grothendieck group $\Omega(G)$ is therefore a ring, the Burnside ring of $G$.

If $V$ is a $G$-module (real), the space of self-maps of its unit sphere $S(V)$ is denoted by $F(V)$ and inherits an action of $G$ via $(gf)(x) = gf(g^{-1}x)$ for $g \in G, f \in F(V)$ and $x \in S(V)$ and contains GL$(V)$, the real linear automorphisms of $V$ as an invariant subspace.

(6.1) Define $\omega_G^0 = \text{ind lim}_V\, \pi_0(F(V)^G)$.

The direct limit is over all $G$-modules $V$. As long as $V$ contains at least one copy of the trivial $G$-module $\boldsymbol{R}$, $\pi_0(F(V)^G)$ is an abelian group with addition defined as in the homotopy groups of spheres. In particular if $V = V' \oplus \boldsymbol{R}$, the map $\sigma \in F(V)$ defined by $\sigma(v', t) = (v', -t)$ represents $-1$ in $\pi_0(F(V)^G)$. Multiplication in $\omega_G^0$ is defined by composition of maps making it a ring. Our aim is to show

THEOREM 6.2 [17]. *There is a ring isomorphism* $\Phi: \Omega(G) \to \omega_G^0$.

Let $\mathscr{S}(G)$ denote the set of conjugacy classes of subgroups of $G$. It is clear that additively $\Omega(G)$ is the free abelian group with one basis element $G/H$ for each conjugacy class $\bar{H} \in \mathscr{S}(G)$. We define ring homomorphisms $\hat{\chi}$, $\Phi$ and $D$ giving a commutative diagram:

(6.3)
$$\begin{array}{ccc} \Omega(G) & \xrightarrow{\Phi} & \omega_G^0 \\ & {}_{\hat{\chi}}\searrow \quad \swarrow_{D} & \\ & \prod_{H \in \mathscr{S}(G)} \boldsymbol{Z} & \end{array}$$

Theorem 6.2 is completed by showing $\hat{\chi}$ is injective and $\Phi$ is surjective.

(6.4) Define $\hat{\chi} = \prod \chi_H$, $\chi_H(X) =$ Euler characteristic of $X^H$, i.e., $\chi(X^H)$.

(6.5) Define $D = \prod D_H$, $D_H(f) =$ degree $f^H$, $f \in F(V)^G$.

If $X$ is a finite $G$-set, it embeds equivariantly in $S(V \oplus \boldsymbol{R}) = Y$ for some $G$-module $V$. Then its $G$ normal bundle $\nu(X, Y)$ is $X \times V$.

(6.6) Define $\Phi$ as follows: For $X$ a finite $G$-set, $\Phi(X) \in \omega_G^0$ is represented by the composition $Y \xrightarrow{c} X^{\nu(X,Y)} \xrightarrow{h} Y$ where $c$ is the collapsing map onto the Thom space of $\nu(X, Y)$ and $h$ is the map of Thom spaces induced by $(x, v) \to v$. Note $Y$ is the Thom space of $V$ regarded as a vector bundle over a point. $\Phi$ is uniquely extended to the full Burnside ring by additivity. It is independent of the $G$-embedding of $X$ in $Y$.

The fact that $\hat{\chi}$ is injective is an elementary check given the basis $G/H$, $H \in \mathscr{S}(G)$, of $\Omega(G)$. Our interest thus lies in showing $\Phi$ is surjective.

Let $Y = S(V \oplus \boldsymbol{R})$; $p = (0, 1)$. For each $H \subseteq G$ and $y \in Y^H$, the representation of $H$ on the tangent space of $Y$ at $y$ $TY_y$ is $V$. This means that for any self-$G$-map $f$ of $Y$, the differential $df_y$ is a linear $H$-endomorphism of $V$ so splits as $df_y^H \oplus df_{y,H}$ preserving the splitting $V = V^H \oplus V_H$. Let $X$ be a finite $G$-set. Then

$$X = \coprod_{\bar{H} \in \mathscr{S}(G)} G/H \times F(H), \qquad F(H) = X^H - X_H, \tag{6.7}$$

as a $G$-set.

LEMMA 6.8. *Suppose $f$ and $h$ are two self-$G$-maps of $Y$ which are transverse to $p$ with $f^{-1}(p) = h^{-1}(p) = X$ and $df_x = dh_x$ in $\pi_0(\mathrm{GL}(V)^H)$ for each $\bar{H} \in \mathscr{S}(G)$ and $x \in F(H)$. Then $f$ and $h$ are transversally (to $p$) $G$ homotopic.*

PROOF. Using the equality of differentials in $\pi_0$, both $f$ and $G$ are transversally $G$ homotopic rel $X$ to maps $f_0$ and $h_0$ with $f_0 = h_0$ on a tubular neighborhood of $\nu(X, Y)$ of $X$ in $Y$ and both map $Y - \nu(X, Y)$ to $Y - p \cong V$. Since $V$ is $G$ contractible, this means $f_0$ and $h_0$ are $G$ homotopic.

Given any $L \in \mathrm{GL}(V)^H$, let $s_H(L) =$ sign det $(L^H)$. Suppose $f$ is a self-$G$-map of $Y$ transverse to $p$ with $f^{-1}(p) = X$ given by (6.6). For $i = 0, 1$ and $\bar{H} \in \mathscr{S}(G)$, set

$$F_i(H) = \{x \in F(H) | s_H(df_x) = (-1)^i\}, \qquad X_i = \coprod_{\bar{H} \in \mathscr{S}(G)} G/H \times F_i(H).$$

Define $\psi(f) \in \Omega(G)$ by

(6.9) $$\psi(f) = X_0 - X_1 \quad \text{and note } X = X_0 + X_1.$$

LEMMA 6.10. *Any self-map of $Y$ is $G$ homotopic to a map $f \pitchfork p$ such that if $X = f^{-1}(p)$, then $(df_x)_H$ is the identity for each $\bar{H} \in \mathscr{S}(G)$ and $x \in F(H)$.*

PROOF. Use 3.5. Suppose for each $\bar{K} \in \mathscr{S}(G)$ with $K > H$, $f$ is $G$ homotopic to the given map and transverse to $p$ in a $G$-neighborhood of $GY^K$ with $(df_x)_K$ the identity of $V_K$ for each $x \in X^K - X_K$ where $X = f^{-1}(p)$. Moreover suppose $f^H: Y^H \to Y^H$ is transverse to $p$. Set $F = X^H - X_H$. Then $F$ is a finite set so $\partial F = \emptyset$ and by 3.5, any section $s \in \Gamma_{N(H)}(V_{\xi,\eta})$ where $\xi = \eta = F \times V_H$ can be used to extend the transversality hypothesis to the group $H$ with $(df|_F)_H = f^*|_{Fs}$. Take $s$ to be the identity map of $\xi$ to $\eta$. This equation then means that $(df_x)_H$ is the identity.

We now wish to show that $\Phi$ maps $\Omega(G)$ surjectively onto $\omega_G^0$. Given any element of $\omega_G^0$ it is represented by a self-$G$-map $f$ of $Y = S(V \oplus \boldsymbol{R})$ for some $G$-module $V$. By Lemma 6.10 we may suppose $f$ is transverse to $p$ and has the additional properties of that lemma.

LEMMA 6.11. $\Phi\psi(f) = f$ *in* $\omega_G^0$.

PROOF. $\psi(f)$ is the virtual $G$-set $X_0 - X_1$ (6.8) where $X_i$ is equivariantly embedded in $Y$ and these embeddings are used to define $\Phi(X_0 - X_1) = \Phi(X_0) - \Phi(X_1)$. Both $f$ and $\Phi(X_0 - X_1)$ satisfy the hypothesis of Lemma 6.8. This uses the fact that for any $G$-set $Z$ and $z \in Z$, $d\Phi(Z)_z$ is the identity so $s_H d((-1)^i\Phi(Z))_z = (-1)^i$ and $d(\pm \Phi(Z)_{z,H}) =$ identity. These follow from the definition of minus in $\omega_G^0$ and the definition of $\Phi$. The result then follows from 6.8.

COROLLARY 6.12. $\Phi$ *is surjective.*

This completes the proof of 6.2.

Given any element of $\omega_G^0$ it may be represented by some map $f$: $Y \to Y$ transversal to $p$ where $Y = S(V \oplus \boldsymbol{R})$ for some $V$. In general two such representatives will not be transversally $G$ homotopic and this is precisely the reason the equivariant framed cobordism group of framed $G$-manifolds of dimension 0 fails to be $\omega_G^0$.

EXAMPLE 6.13. Take $G = Z_2$ and let $\rho$ be the nontrivial irreducible real $G$-module and $V = \rho \oplus \boldsymbol{R}$. As before $Y = S(V \oplus \boldsymbol{R})$. Let $f$: $Y \to Y$ be the map defined by the matrix

$$\begin{pmatrix} -1 & 0 & 0 \\ 0 & -1 & 0 \\ 0 & 0 & 1 \end{pmatrix}$$

Then $f$ is transverse to $p = (0, 1)$ with $f^{-1}(p) = p$ and $(df_p)_G = -1$.

Since $D_G(f) = -1$, $D_1(f) = 1$ and since $\chi_G(-G/G + G/1) = -1$, and $\chi_1(-G/G + G/1) = 1$, it follows from (6.3) and the fact that $\hat{\chi}$ is injective that $f$ and $\Phi(X)$ are $G$ homotopic where $X = -G/G + G/1$. To define $\Phi(X)$ explicitly we must give a $G$-embedding of $X' = G/G \coprod G/1$ to $Y$. Take $G/G$ to $p$ and $G/1$ to any free orbit.

In order to try to convert the $G$-homotopy $h$: $Y \times I \to Y$ between $f$ and $\Phi(X)$ to a $G$-homotopy transverse to $p$, we may assume $h^G \pitchfork p$ with $(h^G)^{-1}(p) = p \times I \subset$

$Y \times I$. By 4.14 the first obstruction $O_*(h, G, \partial Y \times I)$ to converting the $G$-homotopy $h$ to a transverse $G$-homotopy between $f$ and $\Phi(X)$ lies in

$$H^1(p \times I, \partial p \times I, \pi_0(\mathrm{GL}(V_G)^G)) \cong \pi_0(\mathrm{GL}(V_G)^G) = Z_2.$$

In fact this obstruction is sign $\det(df_p)_G$ / sign $\det(d\Phi(X)_{p,G}) = -1$ as $d\Phi(X)_{p,G}$ is the identity. Here $Z_2$ is written multiplicatively as $\{-1, 1\}$. Thus there is no transverse $G$-homotopy between $f$ and $\Phi(X)$. See also [**21**, p. 137].

**7. Quasi-equivalence, $G$ normal cobordism and $\omega_G^\alpha(Y)$.** Let $\xi$ and $\eta$ be two $G$ vector bundles (complex) of the same dimension over a $G$-space $Y$. A map $\omega: \xi \to \eta$ which is proper, fiber preserving and degree one on fibers is called a quasi-equivalence and abbreviated by q.e. Define $\xi \leqq \eta$ to mean there exist a $G$-bundle $\theta$ and a q.e. $\omega: \xi \oplus \theta \to \eta \oplus \theta$.

*Problem* 7.1 [**4**]. Give necessary and sufficient conditions that $\xi \leqq \eta$.

For $G = 1$, this is solved by the Adams conjecture. Even for $Y$ a point this question is interesting. It is solved in [**4**].

Suppose $\omega: \xi \to \eta$ is a q.e. In addition suppose $\omega$ is properly $G$ homotopic to a map $\theta \pitchfork Y$. Here $Y \subset \eta$ is the zero section. Set $X(\omega) = \theta^{-1}(Y)$ and let $f(\omega)$ be the $G$-map of $X(\omega)$ to $Y$ defined by restricting $\theta$ to $X(\omega)$. Then $G$ transversality implies that $d\theta$ defines a $G$ vector bundle isomorphism $\nu(X(\omega), \xi) \to^{\simeq} f(\omega)^*(\eta)$. Also there are these $G$ vector bundle isomorphisms: $f^*(\omega)(TY \oplus \xi) \cong T\xi|_{X(\omega)} \cong TX(\omega) \oplus \nu(X(\omega), \xi)$. Together they give a stable $G$ vector bundle isomorphism

$$f(\hat{\omega}): TX(\omega) \xrightarrow{\cong} f(\omega)^*(\mu)$$

where $\mu = TY + (\xi - \eta) \in KO_G(Y)$. If $Y$ is a compact oriented manifold, so too is $X(\omega)$ and degree $f(\omega) = 1$. This follows from transversality and that $\omega$ is degree 1 on fibers.

DEFINITION 7.2. A *$G$ normal map $(X, f)$ to $Y$* consists of a $G$-map $f: (X, \partial X) \to (Y, \partial Y)$ between the smooth compact $G$-manifolds $X$ and $Y$ of the same dimension. The map $f$ is to have degree 1 and in addition a stable $G$ vector bundle isomorphism $\hat{f}: TX \to f^*(\xi)$ is given for some $G$ vector bundle $\xi$ over $Y$.

The specific isomorphism $\hat{f}$ is an important part of the data of a $G$ normal map but is omitted from the notation. The notion of a $G$ normal map induces a notion of $G$ normal cobordism.

The above discussion shows that a q.e. with the added assumption of transversality gives rise to a $G$ normal map. This may not be unique up to $G$ normal cobordism (because of the problem of converting a $G$-homotopy to a transverse $G$-homotopy). See Example 6.13. This means the notation $X(\omega)$ is somewhat misleading. We shall restrict attention to those properties of the resulting manifold which only depend on $\omega$.

It is an interesting problem to express the invariants of $X(\omega)$ explicitly in terms of those of $Y$, $\xi$, $\eta$ and $\omega$. One of the key relations for doing this is the equation

$$TX_x \oplus \eta_{f(x)} = TY_{f(x)} \oplus \xi_{f(x)} \quad \text{in } RO(G_x) \tag{7.3}$$

where $X = X(\omega)$, $f = f(\omega)$ and $x \in X$. This equation expresses in an equivariant way the fact that $\theta$ is transverse to $Y$ at $x$. From it we immediately determine two invariants of $X$ in terms of $Y$, $\xi$ and $\eta$.

(7.4) $$\dim X^H = \dim Y^H + \dim_{\mathbf{R}} \xi^H_{f(x)} - \dim_{\mathbf{R}} \eta^H_{f(x)} \quad \text{if } X^H \neq \emptyset.$$

(7.5) $$\nu(X^H, X)_x = \nu(Y^H, Y)_{f(x)} + \xi_{f(x),H} - \eta_{f(x),H} \quad \text{in } RO(H).$$

The discussion in the present generality involves too many meagerly explored areas: $G$ vector bundles, $G$-homotopy and $G$ transversality and relations between them. If we restrict attention to $G$ vector bundles over $Y$ of the form $\xi = Y \times M$, $\eta = Y \times N$ for (complex) $G$-modules $M$ and $N$, we arrive at a more manageable situation primarily because the $G$-homotopy of quasi-equivalences of $\xi$ to $\eta$ is then involved with a group $\omega_G^\alpha(Y)$ where $\alpha = M - N$ and $Y \to \omega_G^\alpha(Y)$ defines an equivariant cohomology theory. This enables us to manipulate $G$-homotopy via algebra.

If $\alpha = M - N \in R(G)$ is the difference of two complex $G$-modules $N$ and $M$ and $V$ is any complex $G$-module, let $F_\alpha(V)$ be the $G$-space of maps of $S(V \oplus N)$ to $S(V \oplus M)$ and define for any compact $G$-space $Y$

(7.7) $$\omega_G^\alpha(Y) = \operatorname*{inj\,lim}_V [Y, F_\alpha(V)]^G.$$

Here $[Y, Z]^G$ denotes the $G$-homotopy classes of $G$-maps of $Y$ to $Z$. For any $\alpha \in RO(G)$, $\omega_G^\alpha(Y)$ is a module over $\omega_G$ and $Y \to \omega_G^\alpha(Y)$ is an equivariant cohomology theory [**17**].

An element $\omega \in \omega_G^\alpha(Y)$ is represented by a proper fiber preserving $G$-map again called $\omega$,

(7.8) $$\omega\colon Y \times N' \to Y \times M'$$

for some $N'$ and $M'$ with $M' - N' = M - N = \alpha$. Specifically $\omega(y, n) = (y, \|n\|\omega(y)(n/\|n\|))$ if $\omega \in [Y, F_\alpha(V)]^G$, $n \in N' = N \oplus V$ and $M' = M \oplus V$.

DEFINITION 7.9. $\omega \in \omega_G^\alpha(Y)$ is *transversal* if its representation (7.8) is properly $G$ homotopic to a map transverse to $Y \times O \subset Y \times M'$.

DEFINITION 7.10. $\omega \in \omega_G^\alpha(Y)$ is a quasi-equivalence if its representation (7.8) is a q.e.

With these definitions our preceding discussion amounts to

LEMMA 7.11. *If $\omega \in \omega_G^\alpha(Y)$ is transversal and a quasi-equivalence, it gives rise to a $G$ normal map $(X(\omega), f(\omega))$ to $Y$.*

In this terminology 5.4 states

THEOREM 7.12. *Let $Y$ be any smooth compact $G$-manifold where $G = T \times F$ with $T$ a torus and $F$ finite. Then any element of $\omega_G^0(Y)$ is transversal. Any quasi-equivalence $\omega \in \omega_G^0(Y)$ gives rise to a $G$ normal map $(X(\omega), f(\omega))$ to $Y$.*

This should be compared to 7.14 for general $G$, e.g., for $G = SO_3$, 7.12 is false. Note explicitly the role of the Lie algebra of $G$.

Example 6.13 shows that the transversal representation of elements of $\omega_G^0(Y)$ need not be unique. For connected $G$ the situation is even more complicated since transversal representation is generally impossible as the next two results show.

Ideally it would be nice to have necessary and/or sufficient conditions for an element $\omega \in \omega_G^\alpha(Y)$ to be transversal for some specific $Y$'s and $\alpha$'s, e.g., for $Y = S(V \oplus \mathbf{R})$ for any $G$-module $V$. Since a complete description of $\omega_G^\alpha(p)$ is available, it is appropriate to begin the study of the problem of transversal representation of

$\omega \in \omega_G^\alpha(Y)$ by expressing conditions depending on its restriction $\omega|_p$ to $\omega_G^\alpha(p)$ for $p \in Y^G$. We remark that $\lambda \in \omega_G^0(p)$ is completely determined by the degrees $D_H(\lambda)$ of $\lambda^H: S(V)^H \to S(V)^H$ for all subgroups $H \subseteq G$. See (6.5) and [3].

Let $H$ be a maximal proper subgroup with $G$ with Lie algebra $h$, $V$ a $G$-module and $j$ the unique $G$-map of $Y = S(V \oplus \boldsymbol{R})$ to $p \in Y^G$.

PROPOSITION 7.13. *If $\mathfrak{g}/h$ is not an $H$-submodule of $V$ and $\omega \in \omega_G^0(Y)$ is any element with $D_H(\omega|_p) \neq D_G(\omega|_p)$ (6.3), then $\omega$ is not transversal.*

PROOF. Suppose $\omega \pitchfork 0$. Let $X = \omega^{-1}(0)$ and $f: X \to Y$ be the composition $X \subset Y \times \Omega \to Y$. The component of $Y^K$ containing $p$ for $K = G, H$ is denoted by $Y_p^K$. Set $X_p^K = (f^K)^{-1} Y_p^K$. Since $\omega$ is a $G$-map transverse to zero, both $\omega^G$ and $\omega^H$ are transverse to zero. This means the degree of the map of $X_p^K$ to $Y_p^K$ is $D_K(\omega|_p)$. Since these degrees differ for $H$ and $K$, $X^H$ strictly contains $X^G$ so there is a point $z$ of $X$ with isotropy group $H$. Since $d\omega_z$ maps $T(Y \times \Omega)_z$ surjectively on $T\Omega_0 = \Omega$ and $T(G/H) = \mathfrak{g}/h$ is mapped to zero, there is an exact sequence $0 \to \mathfrak{g}/h \to V \oplus \Omega \to^{d\omega_z} \Omega \to 0$.

COROLLARY 7.14. *Let $\lambda \in \omega_G^0(p)$ be an element satisfying $D_H(\lambda) \neq D_G(\lambda)$ and suppose $\mathfrak{g}/h$ is not an $H$-submodule of $V$. Then $j^*(\lambda) = \omega \in \omega_G^0(Y)$ is not transversal.*

REMARK 7.15. This section has been concerned with producing $G$ normal maps from quasi-equivalences and $G$ transversality. This process is reversible. Specifically, given a $G$ normal map $(X, f)$ to $Y$, there are real $G$ vector bundles $\xi$ and $\eta$ over $Y$, a q.e. $\omega: \xi \to \eta$ and a proper $G$-homotopy of $\omega$ to a map $\theta$ with $\theta^{-1}(Y) = X$ and $f = \theta|_X$.

REMARK 7.16. Suppose the real $G$ vector bundles $\xi$ and $\eta$ of 7.15 are equivariantly oriented in the sense that, for each $K \subseteq G$, $\xi^K$ and $\eta^K$ are oriented vector bundles over $Y^K$. Let $y \in Y^K$, $Y_y^K$ be the component of $Y^K$ containing $y$ and $X_y^K = (f^K)^{-1} Y_y^K$. Then $X_y^K$ inherits an orientation from $Y_y^K$, $\xi_y^K$ and $\eta_y^K$ by virtue of the fact that $\theta^K$ is transverse to $Y^K$ in $\eta^K$; moreover, whenever $\dim \xi_y^K = \dim \eta_y^K$ we have

$$\text{degree}\, f_y^K = \text{degree}\, \omega_y^K \tag{7.17}$$

if $f_y^K$ is $f$ restricted to $X_y^K$.

If $X_y^K$ is given another orientation from other considerations (7.17) holds up to sign. If $\xi$ and $\eta$ are complex $G$ vector bundles, $\xi^K$ and $\eta^K$ are naturally oriented by their complex structures.

**8. Specific quasi-equivalences.** A useful way of producing elements of $\omega_G^\alpha(Y)$ is to exploit the map $j: Y \to p$ of $Y$ to a point. Clearly $j^*: \omega_G^\alpha(p) \to \omega_G^\alpha(Y)$ preserves quasi-equivalences. For $G$ a cyclic group, we can explicitly construct some quasi-equivalences in $\omega_G^\alpha(p) = \omega_G^\alpha$ without using the full theory of [3] and [4].

Let $G$ be either finite cyclic or $S^1$ and $\hat{G}_C$ the group of complex one dimensional $G$-modules under tensor product. Then $\hat{G}_C \cong G$ if $G$ is finite cyclic and $\hat{G}_C \cong Z$ for $G = S^1$. In both cases $R(G)$ is isomorphic to the group ring of $\hat{G}_C$. Thus if $t \in \hat{G}_C$ is a generator,

$$\begin{aligned} R(G) &= Z[t]/(t^n - 1), \qquad n = |G| \text{ or} \\ R(G) &= Z[t, t^{-1}], \qquad G = S^1. \end{aligned} \tag{8.1}$$

Specifically for $G = S^1$, $t^a$ denotes the complex one dimensional $S^1$ module on which $t \in S^1 \in C$ acts by complex multiplication by $t^a$. By restricting to any cyclic subgroup $H$, $t^a$ becomes an $H$-module.

EXAMPLE 8.2. Let $G = S^1$ and $p$ and $q$ be relatively prime integers. Choose positive integers $a$ and $b$ so that $-ap + bq = 1$. Let $\rho \in \hat{G}_C$, $N = \rho^p + \rho^q$, $M = \rho + \rho^{pq}$. Set $\alpha = M - N \in R(G)$ and let $\omega_\alpha : N \to M$ be the proper $G$-map defined by $\omega_\alpha(z_0, z_1) = (\bar{z}_0^a z_1^b, z_0^q + z_1^p)$ for $(z_0, z_1) \in N$. Then the degree of $\omega_\alpha$, i.e., the degree of the induced map of one point compactification, is 1 **[16]**. (*A complex G-module is oriented by its complex structure.*) Thus $\omega_\alpha \in \omega_G^\alpha$ is a quasi-equivaelnce.

DEFINITION 8.3. For $G = S^1$, $\alpha \in R(G)$ is *special* if either $\alpha = M - N$ is one of the examples of 8.2 or $\alpha = \rho - \rho^{-1}$ for $\rho \in \hat{G}_C$.

If $\alpha = \rho - \rho^{-1}$, there is also a q.e. $\omega_\alpha : \rho^{-1} \to \rho$ defined by $\omega_\alpha(z) = \bar{z}$ for $z \in p^{-1}$.

If $G$ is finite cyclic of order $h$, any complex $n$ dimensional $G$-module $V$ has the form $V = t^{a_1} + \cdots + t^{a_n}$ with $a_i \in Z_h$.

(8.4) Set $\Delta(V) = \prod_{1=i}^{n} a_i \in Z_h$.

Of course $\Delta(V)$ depends on a choice of generator $t \in \hat{G}_C$.

DEFINITION 8.5. $\alpha \in R(G)$ is *elementary* if $\alpha = M - N$ where either $M = \chi$, $N = \chi^{-1}$ or $M = \chi \oplus \chi^{a_1 \cdot a_2}$, $N = \chi^{a_1} \oplus \chi^{a_2}$ for some generator $\chi \in \hat{G}_C$ and units $a_1$ and $a_2$ of $Z_h$.

LEMMA 8.6. *$\alpha$ is elementary iff $-\alpha$ is elementary.*

PROOF. If $\alpha = \chi + \chi^{a_1 \cdot a_2} - (\chi^{a_1} + \chi^{a_2})$, set $\psi = \chi^{a_1}$. Then $-\alpha = \psi + \psi^{a_1^{-1} a_2} - (\psi^{a_1^{-1}} + \psi^{a_2})$. Since $\chi$ is a generator of $\hat{G}_C$ and $a_1$ is a unit of $Z_h$, $\psi$ is a generator of $\hat{G}_C$.

LEMMA 8.7. *Let $G = Z_h$. Suppose $V = t^{a_1} + \cdots + t^{a_{n+1}}$ with each $a_i$ a unit of $Z_h$ and $t \in \hat{G}_C$ a generator. Via a sequence of elementary transformations $V \to V_\alpha = V - \alpha$ with $\alpha$ elementary, $V$ can be transformed into $nt + t^{\pm\Delta(V)} = W$. Similarly $W$ can be transformed into $V$ by these elementary transformations.*

PROOF. This is obvious, e.g., if $\alpha = t^{a_1} + t^{a_2} - (t + t^{a_1 a_2})$, then $\alpha$ is elementary by 8.6 and $V_\alpha = t + t^{a_1 a_2} + t^{a_3} + \cdots + t^{a_{n+1}}$. Replace $a_1$ and $a_2$ by $a_1 a_2$ and $a_3$ and repeat, etc.

COROLLARY 8.8. *Suppose $V$ and $W$ are $n + 1$ dimensional $G$-modules with $\Delta(V) = \pm \Delta(W)$ and $\Delta(V)$ is a unit of $Z_h$. Then each can be transformed to the other via a sequence of elementary transformations.*

LEMMA 8.9. *Let $G = S^1$ and $K_i$, $i = 0, 1, \cdots, l$, be proper closed subgroups with $K_0 \cap K_i = 1$ for $i > 0$. Let $\hat{\alpha} \in R(K_0)$ be elementary. Then there is a special $\alpha \in R(G)$ such that*

(i) $\alpha|_{K_0} = \hat{\alpha}$;

(ii) $\alpha|_{K_i} = 0$ *for $i > 0$ and*

(iii) *there is q.e. $\omega_\alpha \in \omega_G^\alpha$.*

PROOF. Let $\hat{\alpha} = \chi + \chi^{a_1 a_2} - (\chi^{a_1} + \chi^{a_2})$ or $\chi - \chi^{-1}$ be elementary and $h =$ order $K_0$. Choose relatively prime integers $p$ and $q$ with $p \equiv a_1$, $q \equiv a_2 \bmod h$ and choose $\rho \in \hat{G}_C$ so $\rho$ restricts to $\chi$ in $R(K_0)$ and to 1 in $R(K_i)$ for $i > 1$. Then $\alpha = \rho + \rho^{pq} - (\rho^p + \rho^q)$ or $\rho - \rho^{-1}$ is special and satisfies (i) and (ii). Statement (iii) follows from 8.2 and 8.3.

LEMMA 8.10. *Let $G = S^1$ and $K_i$, $i = 0, 1, \cdots, l$, be proper closed subgroups with $K_0 \cap K_i = 1$ for $i > 0$. Suppose $|K_0| = pq$, $(p, q) = 1$. Let $\chi \in (\hat{K}_0)_C$ be a generator. There is a special $\alpha \in R(G)$ such that*

(i) $\alpha|_{K_0} = \chi + 1 - (\chi^p + \chi^q)$;

(ii) $\alpha|_{K_j} = 0, j \geqq 1$, *and*

(iii) *there is a q.e., $\omega_\alpha \in \omega_G^\alpha$ with degree $\omega_\alpha^{Z_p} = q$ and degree $\omega_\alpha^{Z_q} = p$ where $Z_p$ and $Z_q$ are the unique subgroups of $K_0$ of order $p$ and $q$.*

PROOF. Let $\rho \in \hat{G}_C$ be any element which restricts to $\chi$ for $K_0$ and to one for $K_i$, $i > 0$. Then $\alpha = (\rho + \rho^{pq} - (\rho^p + \rho^q))$ and $\omega_\alpha$ of 8.2 provide the required data.

**9. Applications to pseudofree $S^1$-manifolds.** Let $Y$ be a smooth $G$-manifold. Define its singular set ${}^sY$ as $\{y \in Y \mid G_y \neq 1\}$ and ${}^s\bar{Y} = {}^sY/G$. Aside from free actions, the simplest class of $G = S^1$-manifolds are those for which the singular set has dimension 1. This means the orbit space is a manifold except at a finite number of singular points corresponding to "singular orbits". When the underlying manifold is a sphere, this class was studied extensively by Montgomery-Yang in their papers on pseudofree circle actions on homotopy spheres [7], [8]. Their important work was quite exhaustive in dimension 7 and indicates the complexity which can be expected more generally. Specifically their results on *classification* show that the general details of classification will encounter intricate problems in homotopy. We show how their work on *comparison* and *construction* can be generalized as applications of the theory herein. In particular the mysterious results of comparison [7, p. 896], appear as natural consequences (9.1) of the invariants of quasi-equivalence and $G$-transversality. The constructive techniques of $G$-transversality and $G$-surgery provide alternative methods replacing those of [7] which are particular to dimension 7.

For us a pseudofree $S^1$-manifold $Y$ is a closed oriented $G = S^1$-manifold with $G$ acting effectively, $Y^G = \emptyset$ and having singular set of dimension 1. We abbreviate this by $Y \in C$. The orbit of $y \in Y$ determines a point $\bar{y} \in \bar{Y} = Y/G$. Set $G_{\bar{y}} = G_y$ the isotropy group of any point $y$ in the orbit $\bar{y}$. Let $\nu_{\bar{y}}$ be the $G_{\bar{y}}$ module $\nu(Y^{G_{\bar{y}}}, Y)_y$ for any point $y \in \bar{y}$. It is the slice representation at $y \in \bar{y}$. Since $Y$ is oriented, we can assume $\nu_{\bar{y}}$ is a complex $G_{\bar{y}}$-module. Since the complex representation ring of $G_{\bar{y}}$ is $R(G_{\bar{y}}) = Z[t]/(1 - t^m)$ where $m = |G_{\bar{y}}|$ is the order of $G_{\bar{y}}$, $\nu_{\bar{y}} = t^{a_1} + \cdots + t^{a_n}$ for some integers $a_i$ prime to $m$ and $2n - 1 = \dim Y$. There is an indeterminacy in the sign of each $a_i$ because of the choice of complex structure for $\nu_{\bar{y}}$. Set $\Delta(\nu_{\bar{y}}) = \prod_{i=1}^n a_i \in Z_m$ and if $f: X \to Y$ is a $G$-map, $\bar{f}$ denotes the induced map of orbit spaces. Since $y \in C$, $\Delta(\nu_{\bar{y}})$ is prime to $m$. Note that $\Delta(\nu_{\bar{y}})$ is a unit of $Z_m$.

Let $H$ be a cyclic group of order $m$; so $R(H) = Z[t]/(1 - t^m)$ and if $\Sigma = (1 - t^m)/(1 - t) = 1 + t + \cdots + t^{m-1}$, then $(1 - t)$ is a nonzero divisor in $R(H)/(\Sigma)$. The augmentation homomorphism $\varepsilon$ from $R(H)$ to the integers is defined by $\varepsilon f(t) = f(1)$, evaluation at 1. There is a commutative diagram

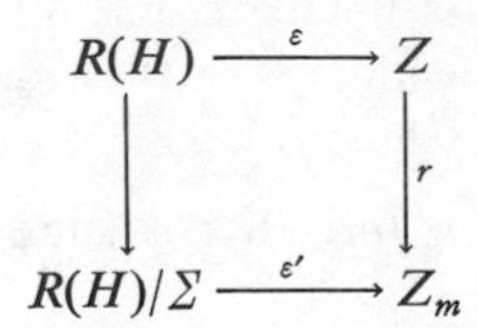

defined by $\varepsilon$ and reduction mod $m$. Any complex $H$-module $A$ defines an element of $R(H)$ and can be written as $A = \sum_{i=1}^{l} t^{a_i}$. Set $\lambda_{-1}(A) = \prod_{i=1}^{l}(1 - t^{a_i})$. If $K \subset H$ we can split the module $A$ as $A = A^K \oplus A_K$. From [1], [2] we know that the equivariant $K$-theories $K_H(A)$ and $K_H(A^K)$ are free $R(H)$-modules with generators $\lambda_A$ and $\lambda_{A^K}$ and the inclusion $A^K \to A$ induces an $R(H)$-homomorphism $s\colon K_H(A) \to K_H(A^K)$ with $s(\lambda_A) = \lambda_{-1}(A_k)\cdot\lambda_{A^K}$.

Here is a comparison theorem which relates the isotropy groups and slice representations of two manifolds $X$ and $Y$ in $C$ which are homotopy equivalent. Compare [7, p. 896.]

THEOREM 9.1 (COMPARISON OF INVARIANTS). *Let $X$, $Y \in C$ and $(X, f, b)$ be a $G$ normal map to $Y$ with $f$ a homotopy equivalence. Then*

(i) *for each $\bar{y} \in \bar{Y}$, $G_{\bar{y}}$ is the direct product of the $G_{\bar{x}}$ with $\bar{x} \in \bar{f}^{-1}(\bar{y})$,*

(ii) $\Delta(\nu_{\bar{f}(\bar{x})}) / \Delta(\nu_{\bar{x}}) \equiv \pm m_{\bar{x}}/m_{\bar{f}(\bar{x})}$ mod $m_{\bar{x}}$ *where* $m_{\bar{x}} = |G_{\bar{x}}|$.

PROOF. (i) The first conclusion is a consequence of the fact that $f^P\colon X^P \to Y^P$ induces a $Z_p$ homology isomorphism for each $p$-group $P$ by Smith Theory. In particular $(f^p)^{-1}Y_\alpha^p$ consists of one component for each component $Y_\alpha^p$ of $Y^p$. Apply this remark to the Sylow $p$-subgroups of $G_{\bar{y}}$ and the component of $Y^P$ containing $y$.

(ii) Suppose without loss of generality (7.15) that $\omega\colon N \to M$ is a quasi-equivalence of real $G$ vector bundles which is properly $G$ homotopic to $\theta \pitchfork Y$ with $\theta^{-1}(Y) = X$ and $f = \theta|_X$. For $y \in Y$ we may suppose $N_y$ and $M_y$ are complex $G_{\bar{y}}$-modules which we *abbreviate by $N$ and $M$*. Then $\omega$ restricted to the fiber over $y$ defines a quasi-equivalence $\omega_y\colon N \to M$ of $G_{\bar{y}}$-modules.

Let $K \subset H$ be subgroups of $G_{\bar{y}}$. Then $\omega_y$ defines a commutative diagram:

$$\begin{array}{ccc} K_H(M) & \xrightarrow{\omega_y^*} & K_H(N) \\ \downarrow s & & \downarrow s \\ K_H(M^K) & \xrightarrow{(\omega_y^K)^*} & K_H(N^K) \end{array}$$

By the remark above, $\omega_y^*(\lambda_M) = a\cdot\lambda_N$ and $(\omega_y^K)^*\lambda_{N^K} = d\lambda_{M^K}$ for some $a$ and $d$ in $R(H)$. Since the degree of $\omega_y$ is 1, $\varepsilon(a) = 1$ and $\varepsilon(d) = d'$ is the degree of $\omega_y^K$. From the diagram, we see that

$$a\cdot\lambda_{-1}(N_K) = d\lambda_{-1}(M_K) \quad \text{in } R(H). \tag{9.2}$$

We take $H$ to be the Sylow $p$-subgroup of $G_{\bar{x}}$ for an $\bar{x} \in \bar{f}^{-1}(\bar{y})$ and $K \subset H$ the cyclic subgroup of order $p$. The order of $H$ is $m$ which is a power of $p$. Since $f^K$ is transverse to $Y^K$ and dim $X^K$ = dim $Y^K$, dim $N^K$ = dim $M^K$ and the degree $d'$ of $\omega^K\colon N^K \to M^K$ is the degree of the map $f_y^K\colon X_y^K \to Y_y^K$ where $Y_y^K$ is the component of $Y^K$ containing $y$ and $X_y^K = (f^K)^{-1}Y_y$ (7.15). Since $Y \in C$, $Y_\alpha^K = G\cdot y = G/G_{\bar{y}}$. By (i), $X_\alpha^K = G/G_{\bar{x}}$ and there is a unique $G$-map of $G/G_{\bar{x}}$ to $G/G_{\bar{y}}$ whose degree is $\pm m_{\bar{y}}/m_{\bar{x}}$ depending on the orientations of these orbits where $m_{\bar{y}}$ and $m_{\bar{x}}$ are the orders of $G_{\bar{y}}$ and $G_{\bar{x}}$ respectively. Thus

$$d' = \pm m_{\bar{y}}/m_{\bar{x}} \quad \text{by 7.16 and (7.17).} \tag{9.3}$$

Since $\Delta(\nu_{\bar{x}})$ and $\Delta(\nu_{\bar{y}})$ are prime to $m_{\bar{x}}$ so prime to $p$, $\nu_{\bar{x}K} = \nu_{\bar{x}}$, $\nu_{\bar{y}K} = \nu_{\bar{y}}$ and (7.5) implies that

(9.4) $$\nu_{\bar{x}} \oplus M_K = \nu_{\bar{y}} \oplus N_K \quad \text{in } R(H).$$

Since $\Delta(M_K)$ and $\Delta(N_K)$ are prime to $p$, (9.4) implies that

(9.5) $$\Delta(\nu_{\bar{x}})/\Delta(\nu_{\bar{y}}) \equiv \Delta(N_K)/\Delta(M_K) \mod m.$$

It follows from (9.2) that

(9.6) $$a/d = \lambda_{-1}(M_K)/\lambda_{-1}(N_K) \quad \text{in } R(H)/(\Sigma).$$

Then $1/d' = \varepsilon(a/d) = \varepsilon(\lambda_{-1}(M_K)/\lambda_{-1}(N_K)) = \Delta(M_K)/\Delta(N_K)$. From (9.3) and (9.5) this gives $m_{\bar{x}}/m_{\bar{y}} \equiv \Delta(\nu_{\bar{x}})/\Delta(\nu_{\bar{y}}) \mod m$. Since this is true for any Sylow $p$-subgroup of $G_{\bar{x}}$, the congruence holds mod $m_{\bar{x}}$.

**10. Constructing $G$ normal maps realizing invariants.** Important invariants of a pseudofree $S^1$-manifold $Y$ are its isotropy groups and their slice representations. We are interested in the values these invariants assume within the specific homotopy type of $Y$. Theorem 9.1 provides relations among these invariants. In this section we show that any values of these invariants are assumed as the isotropy groups and slice representations of a pseudofree $S^1$-manifold $X$ which admits a $G$ normal map $(X, f)$ to $Y$ whenever the given invariants and those of $Y$ satisfy 9.1. In §11 we discuss the problem of making $f$ a homotopy equivalence so that $X$ is in the homotopy type of $Y$.

We say $\alpha \in R(G)$ is positive if $\alpha$ is a $G$-module. In general $\alpha = M - N$ is the difference of two $G$-modules. We set $\dim \alpha = \dim_R M - \dim_R N$.

THEOREM 10.1. *Let $Y \in C$, $\alpha = M - N \in R(G)$. Either all elements of $\omega_G^\alpha(Y)$ are transversal or none are. The necessary and sufficient condition is that for each $\bar{y} \in \bar{Y}$, $\nu_{\bar{y}} - \alpha|_{G_{\bar{y}}}$ be positive in $R(G_{\bar{y}})$.*

PROOF. Represent $\omega \in \omega_G^\alpha(Y)$ by a proper $G$-map $\omega: Y \times N \to Y \times M$ as in (7.8). Since $Y \times N$ and $Y \times M$ are $G$ vector bundles $\xi$ and $\eta$ over $Y$, 5.1 applies to $\omega$. The obstructions $O_*(\xi - \eta, K)$ all vanish because either $Y^K/G$ is empty or zero dimensional and $V'(K)$ in the latter case is nonempty because $\nu_{\bar{y}} - \alpha|_K$ is positive in $R(K)$. See also (4.7).

Suppose $Y \in C$, $\alpha = M - N \in R(G)$ and $\nu_{\bar{y}} - \alpha|_{G_{\bar{y}}}$ is positive for each $\bar{y} \in \bar{Y}$. Then any $\omega \in \omega_G^\alpha(Y)$ is transversal so is represented by a proper $G$-map (7.8) $\omega$: $Y \times N \to Y \times M$ which is properly $G$ homotopic to a map $\theta \pitchfork Y \times 0$.

(10.2) Define $Y_\alpha = \theta^{-1}(Y \times 0)$, $\quad f_\alpha = \theta|_Y$.

Clearly $Y_\alpha$ and $f_\alpha$ are not well-defined functions of $Y$ and $\alpha$. They depend on $\omega$ and a proper $G$-homotopy to $\theta$. Nonetheless some of the $G$-invariants of $Y_\alpha$ are independent of these choices; moreover, our references to $Y$ will be restricted to some specific choice. In particular *we restrict our choice of* $Y_\alpha$ to those constructed from elements of $\omega_G^\alpha(Y)$ of the form $\omega = j^*(\lambda)$, $j$: $Y \to p$, and $\lambda \in \omega_G^\alpha(p)$ is a q.e. If there is some such q.e. $\lambda$ we write $\alpha \geqq 0$. Note for some $\alpha$ there are no such $\lambda$ [**4**, Theorem 5.0].

We are interested in the change in the invariants of $Y$ under the transformation $Y \to Y_\alpha$.

PROPOSITION 10.3. *Suppose* $Y \in C$, $\alpha \in R(G)$, $\alpha \geqq 0$ *and* $\nu_{\bar{y}} - \alpha|_{G_{\bar{y}}}$ *is positive for all* $\bar{y} \in \bar{Y}$; *then* $(Y_\alpha, f_\alpha)$ *is a G normal map to Y and* $Y_\alpha \in C$.

PROOF. $Y_\alpha$ is constructed from a q.e. $\omega = j^*(\alpha)$ so 7.11 applies since $(X(\omega), f(\omega)) = (Y_\alpha, f_\alpha)$.

To see that $Y \in C$, use (7.4) and the fact that if there is a q.e. $\lambda \in \omega_G^\alpha(p)$, then $\dim \alpha^H \geqq 0$ for all $H \subseteq G$ [**4**, Theorem 5.0]. From (7.4) with $\xi = Y \times N$, $\eta = Y \times M$ and $\alpha = M - N$, we have $\dim Y_\alpha^H = \dim Y^H - \dim \alpha^H$. Since $\alpha \in R(G)$, $\dim_{\boldsymbol{R}} \alpha^H$ is even. Thus $Y^H$ is either empty or 1 dimensional for $H \neq 1$ so $Y_\alpha \in C$.

If $X$ and $Y$ are in $C$ and $f$: $X \to Y$ is a $G$-map, then $G_{\bar{x}} \subset G_{\bar{y}}$ if $\bar{f}(\bar{x}) = \bar{y}$. We want to compare $\nu_{\bar{x}}$ and $\nu_{\bar{y}}$ where $X = Y_\alpha$ and $f = f_\alpha$. Note that $\nu_{\bar{x}}$ is positive in $R(G_{\bar{x}})$. This shows the necessity of the following condition.

PROPOSITION 10.4. *Suppose* $Y \in C$, $\alpha \in R(G)$, *and* $\nu_{\bar{y}} - \alpha|_{G\bar{y}} \in R(G_{\bar{y}})$ *is positive for all* $\bar{y} \in \bar{Y}$. *Then for any* $\bar{x} \in \bar{Y}_\alpha$ *and* $\bar{y} = \bar{f}_\alpha(\bar{x})$, *the slice representations satisfy* $\nu_{\bar{x}} = \nu_{\bar{y}} - \alpha|_{G_{\bar{x}}}$ *in* $R(G_{\bar{x}})$.

PROOF. This is immediate from (7.5) where $\xi = Y \times N$ and $\eta = Y \times M$ if $\alpha = M - N$.

Thus the transformation $Y \to Y_\alpha$ effects the transformation $\nu_{\bar{y}} \to \nu_{\bar{y}} - \alpha|_{G_{\bar{x}}}$, $\bar{f}(\bar{x}) = \bar{y}$, of slice representations.

It is now possible to show that Theorem 9.1 gives the complete comparison of relations between slice representations and isotropy groups of $X$, $Y \in C$ when there is a $G$-map $f$: $X \to Y$ which is a homotopy equivalence. The primary steps are:

EXCHANGE LEMMA 10.5. *Let* $Y \in C$ *and* $K_i$, $i = 0, 1, \cdots$, *be its isotropy groups. Suppose* $K_0 \cap K_i = 1$ *for* $i > 0$ *and that* $\nu_{\bar{y}}$ *is a fixed* $K_0$*-module* $V$ *whenever* $G_{\bar{y}} = K_0$. *Given any* $K_0$*-module* $W$ *with* $\dim V = \dim W$ *and* $\Delta(W) = \pm \Delta(V)$, *there is an* $X \in C$ *and a G normal map* $(X, f)$ *to Y with these properties*:

(i) $\bar{f}$; ${}^s\bar{X} \to {}^s\bar{Y}$ *is a homeomorphism*,

(ii) $G_{\bar{x}} = G_{\bar{y}}$ *if* $\bar{x} \in {}^s\bar{X}$, $\bar{y} = \bar{f}(\bar{x})$,

(iii) $\nu_{\bar{x}} = W$ *whenever* $G_{\bar{x}} = K_0$, *and*

(iv) $\nu_{\bar{x}} = \nu_{\bar{y}}$ *whenever* $G_{\bar{x}} \neq K_0$.

SPLITTING LEMMA 10.6. *Let* $Y \in C$ *with* ${}^s\bar{Y} = \bigcup_{i=0}^k \bar{y}_i$. *Suppose the order* $|G_{\bar{y}_0}|$ *of* $G_{\bar{y}_0} = K_0$ *is prime to* $|K_i|$, $i \geqq 1$, *where* $K_i = G_{\bar{y}_i}$. *Given any splitting* $K_0 = K_{01} \times K_{02}$ *and any complex one dimensional* $K_0$*-module* $\chi$ *such that* $\nu_{\bar{y}_1} - \chi$ *is positive, there is an* $X \in C$ *with* ${}^s\bar{X} = \bar{x}_{01} \cup \bar{x}_{02} \bigcup_{i=1}^k \bar{x}_i$ *and a G normal map* $(X, f)$ *to Y such that*

(i) $G_{\bar{x}_{01}} = K_{01}$, $G_{\bar{x}_{02}} = K_{02}$, $\bar{f}^{-1}(\bar{y}_0) = \bar{x}_{01} \cup \bar{x}_{02}$.

(ii) $G_{\bar{x}_i} = G_{\bar{y}_i}$, $\bar{f}^{-1}(\bar{y}_i) = \bar{x}_i$, $i \geqq 1$.

(iii) $\nu_{\bar{x}_{01}} = \nu_{\bar{y}_0} + \chi^q - \chi$, $\nu_{\bar{x}_{02}} = \nu_{\bar{y}_0} + \chi^p - \chi$ *where* $p = |K_{01}|$ *and* $q = |K_{02}|$. *Note* $(p, q) = 1$.

(iv) $\nu_{\bar{x}_i} = \nu_{\bar{y}_i}$ *for* $i \geqq 1$.

The method of proof of both 10.5 and 10.6 is the same; $(X, f)$ is obtained as the result of a series of transformations $Y \to (Y_\alpha, f_\alpha)$ for judicious choices of $\alpha$ resulting from §8. Lemma 10.5 is a corollary of

LEMMA 10.7. *Let* $Y \in C$ *and* $K_i$, $i = 0, 1, \cdots, l$, *be its isotropy groups. Suppose* $K_0 \cap K_i = 1$ *for* $i > 0$, *that* $\hat{\alpha} \in R(K_0)$ *is elementary and* $\nu_{\bar{y}} - \hat{\alpha} \in \boldsymbol{R}(K_0)$ *is positive. Then there is a special* $\alpha \in R(G)$ *with* $\alpha|_{K_0} = \hat{\alpha}$, $\alpha|_{K_i} = 0, i > 0$, *and a* $G$ *normal map* $(Y_\alpha, f_\alpha)$ *to* $Y$ *such that*:

(i) $\bar{f}_\alpha : {}^s\bar{Y}_\alpha \to {}^s\bar{Y}$ *is a homeomorphism.*

(ii) $G_{\bar{x}} = G_{\bar{y}}$ *for* $\bar{y} = \bar{f}_\alpha(\bar{x})$, $\bar{x} \in {}^s\bar{Y}_\alpha$.

(iii) $\nu_{\bar{x}} = \nu_{\bar{y}} - \alpha$ *if* $G_{\bar{x}} = K_0$.

(iv) $\nu_{\bar{x}} = \nu_{\bar{y}}$ *if* $G_{\bar{x}} \neq K_0$.

PROOF. By 8.9, there is a special $\alpha \in R(G)$ with $\alpha \geqq 0$, $\alpha|_{K_0} = \hat{\alpha}$ and $\alpha|_{K_i} = 0$, $i \neq 0$.

By 10.3 there is a $G$ normal map $(Y_\alpha, f_\alpha)$ to $Y$. In order to show (ii), it suffices to show that $\dim X_x^H = \dim Y_y^H$ when $y = f_\alpha(x)$ for each subgroup $H \neq 1$ of $G_{\bar{y}}$. This follows from (7.4) and $\alpha^H = 0$. Now (iii) and (iv) are consequences of 10.4.

The singular set ${}^sY$ is a union of a finite number of orbits $G/K_i$. For an orbit $G/K$, we have an isomorphism [17] $\omega_G^\alpha(G/K) \cong \omega_K^{\alpha|K}(p)$. Under this isomorphism $\omega = j^*(\omega_\alpha)$ restricted to an orbit $G/K_i$ is just $\omega_\alpha|_{K_i}$. For $i > 0$, $\alpha|_{K_i}$ is zero and $\omega_\alpha|_{K_i}$ is the identity of $\omega_{K_i}^0(p)$. This is so because $M|_{K_i} = N|_{K_i}$ are trivial $G$-modules of the same dimension and so the degree of $\omega_\alpha|_{K_i}$ is a complete equivariant homotopy invariant. Thus we may assume $\theta$ restricted to $G/K_i \times N = G/K_i \times M$ is the identity for orbits $G/K_i$, $i > 0$, of $Y$.

The discussion for the orbits $G/K_0$ is even easier because $N^{K_0} = M^{K_0} = 0$ so that $\theta^{K_0}$ is the identity map of $Y^{K_0} \times 0$. Since ${}^sY_\alpha = {}^sY \times {}^sN \cap \theta^{-1}(Y \times 0)$, (i) follows. In fact ${}^sY_\alpha = {}^sY \times 0 \subset Y \times N$.

PROOF OF 10.5. Use Lemma 8.7 to choose elementary $\hat{\alpha}_i \in R(K_0)$, $i = 1, 2, \cdots, r$, so that $W$ is obtained from $V$ by a sequence of transformations $V \to V_{\hat{\alpha}_i}$, $i = 1, 2, \cdots, r$. Use Lemma 8.9 to choose special $\alpha_i \in R(G)$ with $\alpha_i \geqq 0$, $\alpha_i|_{K_0} = \hat{\alpha}_i$ and $\alpha_i|_{K_j} = 0$ for $j \neq 0$.

Use 10.7 to construct $G$ normal maps $(Y_{i+1}, f_{i+1})$ to $Y_i$ inductively by $Y_0 = Y$, $f_0 = \text{identity}$, $Y_{i+1} = (Y_i)_{\alpha_{i+1}}$ and $f_{i+1} = f_{\alpha_{i+1}}$. Then $(X, f) = (Y_r, f_1 \circ f_2 \circ \cdots \circ f_r)$ and $X$ and $f$ have the properties specified by 10.5.

PROOF OF 10.6. Use 8.10 to construct $\alpha \geqq 0$ and $\omega_\alpha \in \omega_G^\alpha$ with the properties stated there. By 10.3 a $G$ normal map $(Y_\alpha, f_\alpha)$ exists. Use (7.4) and 8.10(i) and (ii) to conclude that the isotropy groups of $Y_\alpha$ are $K_{01} = Z_p$ and $K_{02} = Z_q$ and $K_i$, $i \geqq 1$. As in the proof of 10.7, we may suppose ${}^sY_\alpha \cap \bar{f}_\alpha^{-1}(\bar{y}_i) = \bar{x}_i$ for $i \geqq 1$ since $\omega_\alpha$ restricted to an orbit $G/K_i$, $i > 0$, is the identity.

For $L = Z_p$, resp. $Z_q$, $Y_\alpha^L$ is a union of oriented orbits $G/L$ (7.16) and the degree of $f_\alpha^L$ is $q$, resp. $p$, by (7.17) and because these are the degrees of $\omega_\alpha^L$ (8.2 and 8.10). From this we conclude, the degree of the orbit map $\bar{f}_\alpha^L : \bar{Y}_\alpha^L \to \bar{Y}_\alpha^L$ for $L = Z_p$ or $Z_q$ is 1. This is because $Y_\alpha^L$ is a union of oriented copies of $G/L$ and any $G$-map from $G/L$ to $G/K$ has degree $|K| / |L|$ which is either $q$ or $p$.

Since $\bar{Y}_\alpha^L$ consists of a finite set of points (with orientation) and degree $\bar{f}_\alpha^L$ is 1, an elementary argument shows we may suppose $(\bar{f}_\alpha^L)^{-1}(\bar{y}_0)$ consists of one point $\bar{x}_{01}$ for $L = Z_p$ and one point $\bar{x}_{02}$ for $L = Z_p$. Thus $\bar{f}^{-1}({}^s\bar{Y}) \cap {}^s\bar{X}$ is as claimed. Using a homotopy argument similar to [8] we may suppose $\bar{f}^{-1}({}^s\bar{Y}) = {}^s\bar{X}$. Properties (iii) and (iv) are a consequence of 10.4 and 8.10.

**11. Converting $G$ normal maps to pseudoequivalences.** In Chapter I, §3 we discussed the obstruction theory for converting a $G$ normal map to a pseudoequivalence when $G$ was finite. Suppose now $G = T \times F$ where $T$ is a torus and $F$ is finite.

Very briefly here is a discussion of the obstruction theory for converting a $G$ normal map $(X, f)$ to $Y$ into a pseudoequivalence. Each subgroup $P$ of prime power order contributes an obstruction $\sigma_P(f)$ contained in the Wall group $L_*(\Lambda_P(\pi_0(N(P)/P)))$ depending on the indicated group ring where $\Lambda_P$ is a subring of the rationals and $N(P)$ is the normalizer of $P$ in $G$. The obstruction $\sigma_P(f)$ arises from $f^P: X^P \to Y^P$ and is the obstruction to making $f^P$ induce an isomorphism in homology with mod $p$ coefficients if $|P|$ is a power of $p$. This is a necessary condition (from Smith theory) if $f$ is a pseudoequivalence. The obstruction $\sigma_P(f)$ is defined iff $\sigma_{P'}(f) = 0$ whenever $P' > P$ (where $|P'|$ is a power of $p$) and $f^{P'}$ induces an isomorphism in mod $p$ homology.

The trivial group 1 has prime power order and $\sigma_1(f)$ is defined if $f^P$ induces a mod $p$ homology equivalence for each subgroup $P > 1$ of order a power of $p$ for every prime $p$ and $\sigma_1(f) \in J^{D(\pi_0(G)), \tilde{K}_0}(\boldsymbol{Z}(\pi_0(G)))$ where $\tilde{K}_0 = \tilde{K}_0(\boldsymbol{Z}(\pi_0(G)))$.

The case where $G = S^1$ is the simplest illustration of the theory for in that case $\pi_0(N(P)/P)$ is always 1 and the obstructions $\sigma_P(f)$ are elements of $L_*(Z_{(p)})$, the Wall group of the ring of integers localized at $p$. The obstruction $\sigma_1(f)$ lies in $L_*(Z)$ and this group is $Z$, $Z_2$ or 0 for $*$, respectively 0, 2, 1 mod 4, and $\sigma_1(f)$ is a signature or Arf invariant of a quadratic form when $*$ is even. Here $*$ equals $\dim Y - \dim G$.

Applying this discussion to pseudofree $S^1$ manifolds gives a partial converse to 9.1.

THEOREM 11.1. *Let $X, Y \in C$ and $(X, f)$ be a $G$ normal map to $Y$ such that, for each $\bar{y} \in \bar{Y}$, $G_{\bar{y}}$ is the direct product of the $G_{\bar{x}}$ with $\bar{x} \in \bar{f}^{-1}(\bar{y})$. Then all $\sigma_P(f)$ vanish whenever $P$ is a nontrivial subgroup of prime power order. Thus $\sigma_1(f) \in L_*(Z)$ is defined. If it vanishes $(X, f)$ is $G$ normally cobordant* (rel *a $G$-neighborhood of ${}^sX$) to a pseudo-equivalence.*

PROOF. This follows from the above discussion and the fact that the hypothesis implies $f^P: X^P \to Y^P$ induces an isomorphism in mod $p$ homology for each non-trivial $p$-group. This is because $Y^P$ is a union of orbits $G/H$, $H \supset P$, $X^P \cap f^{-1}(G/H)$ consists of one orbit $G/K$ and any $G$-map of $G/K$ to $G/H$ has degree $|H|/|K|$ which is prime to $p$.

**12. Pseudofree actions on homotopy sheres.** The canonical models for pseudofree $G = S^1$-manifolds are the unit spheres $S(W)$ where $W = t^{a_1} + \cdots + t^{a_n}$, $t \in \hat{G}_C$, a generator and $a_i \in Z$ with $(a_i, a_j) = 1$, $i \neq j$. The isotropy groups are $Z_{a_1}$; so we see that the number of isotropy groups of $S(W)$ is bounded by the dimension of $S(W)$ which is $2n - 1$. This observation prompted Montgomery-Yang to ask whether in general for a pseudofree $S^1$ homotopy sphere $Y$, the number of isotropy groups is bounded by the dimension of $Y$. They showed it could be arbitrarily large for $2n - 1 = 7$ using techniques particular to dimension 7 [7]. They also established for homotopy seven spheres a comparison theorem similar to 9.1. In this section we indicate how the results of the preceding sections apply to give generalizations of their results to arbitrary dimensions.

THEOREM 12.1. *Let $Y = S(V)$ be the unit sphere of the $G$-module $V = t^q + nt$ where both $n$ and $q$ are odd. For any product splitting of the unique isotropy group $G_{\bar{y}} = Z_q$, $\bar{y} = {}^s\bar{Y}$,*

(i) $Z_q = Z_{q_1} \times \cdots \times Z_{q_k}$, $\Pi q_i = q$, *and any set of* $Z_{q_i}$ *modules* $V_i$, $i = 1, 2, \cdots k$, *satisfying*

(ii) $\Delta(\nu_{\bar{y}})/\Delta(V_i) \equiv \pm q_i/q \bmod q_i$,

*there is an* $X \in C$ *and a* $G$ *normal map* $(X, f)$ *to* $Y$ *with* $f$ *a homotopy equivalence and for* $i = 1, 2, \cdots k$,

(iii) ${}^s\bar{X} = \bigcup_{i=1}^k \bar{x}_i$, $G_{\bar{x}_i} = Z_{q_i}$, $\bar{f}^{-1}(\bar{y}_i) = \bar{x}_i$,

(iv) $\nu_{\bar{x}_i} = V_i$.

Compare with [7, p. 896] and 9.1.

PROOF. First apply the Splitting Lemma 10.6 to split the isotropy group $Z_q$ of $Y$ into the $k$ isotropy groups $Z_{q_i}$ as required in (iii) for some $G$ normal map $(Z, h)$ to $Y$. This is done by applying 10.6 $k - 1$ times constructing $G$ normal maps $(Z_{i+1}, g_{i+1})$ to $Z_i$ for $i = 1, 2, \cdots, k-1$ with $Z_1 = Y$. The manifold $Z_{i+1}$ is to have $i + 1$ singular points and $i + 1$ isotropy groups $Z_{q_j}$, $j = 1, 2, \cdots, i$, and $Z_{s_{i+1}}$ where $s_{i+1} = (\prod_{j=1}^i g_j)^{-1} q$. Properties 10.6(iii) and (iv) guarantee that for any singular point $\bar{z} \in \bar{Z}_{i+1}$ and $\bar{y} = \bar{g}_{i+1}(\bar{z})$, that $\Delta(\nu_{\bar{z}}) = \Delta(\nu_{\bar{y}})|G_{\bar{y}}|/|G_{\bar{z}}|$. Let $Z = Z_k$ and $h$ be the composition of the $g_i$'s. Then $\bar{Z}$ has $k$ singular points $\{\bar{z}_i | i = 1, 2, \cdots, k\}$ and isotropy groups $Z_{q_i}$, $i = 1, 2, \cdots, k$, and $\Delta(\nu_{\bar{z}_i}) = |G_{\bar{y}}|/|G_{\bar{z}_i}| \Delta(\nu_{\bar{y}}) \equiv q/q_i \bmod q_i$ where $\bar{h}(\bar{z}_i) = \bar{y}$.

Since $\Delta(\nu_{\bar{y}})/\Delta(V_i) \equiv \pm q_i/q \bmod q_i$, $\Delta(V_i) \equiv \pm \Delta(\nu_{\bar{z}_i}) \bmod q_i$. This is the condition which is needed to apply the Exchange Lemma 10.5 to exchange the slice representations $\nu_{\bar{z}_i}$ with the $V_i$ for $i = 1, 2, \cdots, k$. This is done inductively as above with $k$ steps and produces $(X', f')$.

By Theorem 11.1 the only obstruction to converting $(X', f')$ to $(X, f)$ with $f$ a pseudoequivalence is $\sigma_1(f) \in L_*(Z) = Z_2$ because $* = \dim Y - 1$ is congruent to 2 mod 4 as $n$ is odd. Because $q$ is odd, all isotropy groups have odd order. If $q$ were one, $G$ would act freely on $X$ and $Y$ and $\sigma_1(f)$ would be the Arf invariant of the normal maps $\bar{f}: \bar{X} \to \bar{Y}$ which could be computed from the Sullivan formula [19]. In fact for any odd $q$, the computation of $\sigma_1(f)$ can be reduced to the case $q = 1$ and the data arranged so that $\sigma_1(f) = 0$.

Since $(X', f')$ satisfies (iii) and (iv) and since $(X, f)$ is $G$ normally cobordant to $(X', f')$ rel ${}^sX'$, $(X, f)$ satisfies (iii) and (iv) and in addition $f$ is a pseudoequivalence.

## CHAPTER I REFERENCES


**1.** M. F. Atiyah and I. M. Singer, *The index of elliptic operators*. III, Ann. of Math. (2) **87** (1968), 546–604. MR **38** #5245.

**2.** M. F. Atiyah and G. B. Segal, *The index of elliptic operators*. II, Ann. of Math. (2) **87** (1968), 531–545. MR **38** #5244.

**3.** A. Dress, T. Petrie and T. tom Dieck, *Geometric modules over the Burnside ring* (to appear).

**4.** J. Ewing, *Characters, Dirichlet series, discriminants and periodic maps*, Topology (to appear).

**5.** D. Montgomery and C. T. Yang, *Differentiable pseudo-free circle actions*, Proc. Nat. Acad. Sci. U.S.A. **68** (1971), 894–896. MR **43** #4066.

**6.** A. Myerhoff and T. Petrie, *Quasi equivalence of G modules*, Topology **15** (1976), 69–75. MR **52** #11963.

**7.** T. Petrie, *The existence of free metacyclic actions on homotopy spheres*, Bull. Amer. Math. Soc. **76** (1970), 1103–1106. MR **41** #9247.

**8.** ———, *G transversality and applications*, Notes, Rutgers University.

**9** ———, *Smooth G actions on spheres with one fixed point* (to appear).

**10.** ———, *G normal maps* (to appear).

**11.** T. Petrie, *A setting for smooth $S^1$ actions with applications to real algebraic actions on $P(C^{4n})$*, Topology **13** (1974), 363–374. MR **51** #4294.

**12.** ———, *Exotic $S^1$ actions on $CP^3$ and related topics*, Invent. Math **17** (1972), 317–327. MR **47** #7759.

**13.** ———, *Equivariant quasi-equivalence, transversality and normal cobordism*, Proc. Internat. Congr. Math. (Vancouver, 1974), Vol. 1, Canad. Math. Congr., Montreal, 1975, pp. 537–541.

**14.** ———, *G transversality*, Bull. Amer. Math. Soc. **81** (1975), 721–722. MR **51** #6858.

**15.** ———, *Obstructions to transversality for compact Lie groups*, Bull. Amer. Math. Soc. **80** (1974), 1133–1136. MR **51** #4300.

**16.** ———, *G maps and the projective class group*, Comment. Math. Helv. Jan. (1977).

**17.** R. Schultz, *Spherelike G manifolds with exotic equivariant tangent bundles*, Advances in Math. (to appear).

**18.** G. Segal, *Equivariant stable homotopy theory*, Proc. Internat Congr. Math. (Nice, 1970), Gauthier-Villars, Paris, 1971, pp. 59–63.

## Chapter II References

**1.** M. Atiyah, *K theory*, Benjamin New York, 1967. MR **36** #7130.

**2.** M. Atiyah and G. Segal, *The index of elliptic operators*. II, Ann. of Math. (2) **87** (1968), 531–545. MR **38** #5244.

**3.** T. tom Dieck and T. Petrie, *Geometric modules over the Burnside ring* (to appear).

**4.** A. Meyerhoff and T. Petrie, *Quasi-equivalence of G modules*, Topology **15** (1976), 69–75.

**5.** J. Milnor, *Morse theory*, Ann. of Math. Studies, no. 51, Princeton Univ. Press, Princeton, N.J., 1963. MR **29** #634.

**6.** ———, *Differential topology*, Lectures on Modern Math., Vol. 2, Wiley, New York, 1964, pp. 165–183. MR **31** #2731.

**7.** D. Montgomery and C. T. Yang, *Differentiable pseudo-free circle actions on homotopy spheres*, Proc. Nat. Aci. Sci. U.S.A. **68** (1974), 894–896.

**8.** ———, *Homotopy equivalence and a differentiable pseudo-free circle actions on homotopy spheres*, Michigan Math. J. **20** (1973), 145–159. MR **48** #3069.

**9.** T. Petrie, *Equivariant quasi-equivalence, transversality and normal cobordism*, Proc. Internat. Congr. Math. (Vancouver, 1974), Canad. Math. Congr., Montreal, 1975, pp. 537–541.

**10.** ———, *G surgery* I–IV (to appear).

**11.** ———, *Smooth actions on spheres with one fixed point* (to appear).

**12.** ———, *A setting for smooth $S^1$ actions with applications to real algebraic actions on $P(C^{4n})$*, Topology **13** (1974), 363–374. MR **51** #4294.

**13.** ———, *The existence of free metacyclic actions on homotopy spheres*, Bull. Amer. Math. Soc. **76** (1970), 1103–1106. MR **41** #9247.

**14.** ———, *G transversality*, Mimeographed notes, Rutgers Univ., New Brunswick, N.J., 1975.

**15.** ———, *G transversality*, Bull. Amer. Math. Soc. **81** (1975), 721–722. MR **51** #6858.

**16.** ———, *Exotic $S^1$ actions on $CP^3$ and related topics*, Invent. Math. **17** (1972), 317–327. MR **47** #7759.

**17.** G. Segal, *Equivariant stable homotopy*, Proc. Internat. Congr. Math. (Nice, 1970), Gauthier-Villars, Paris, 1971, pp. 59–63.

**18.** N. Steenrod, *The topology of fiber bundles*, Princeton Univ. Press, Princeton, N.J., 1951. MR **12**, 522.

**19.** D. Sullivan, *Geometric topology seminar notes*, Mimeographed notes, Princeton Univ., Princeton, N.J., 1967.

**20.** C. T. C. Wall, *Surgery on compact manifolds*, Academic Press ,New York, 1970.

**21.** A. Wasserman, *Equivariant differential topology*, Topology **8** (1969), 127–150. MR **40** #3563.

Rutgers University

Proceedings of Symposia in Pure Mathematics
Volume 32, 1978

# $G$-SMOOTHING THEORY

R. LASHOF AND M. ROTHENBERG

**Introduction.** If $G$ is a finite group then smoothing theory **[L3]** for topological manifolds can be carried over to $G$-smoothing theory for $G$-manifolds with only small changes.

1. *Replace vector bundles by G-vector bundles.* A $G$-vector bundle is a vector bundle $p: E \to X$ such that $E$ and $X$ are $G$-spaces and $p$ and the 0-section $s$ are $G$-maps, and $G$ acts on $E$ through vector bundle maps. The tangent bundle of a smooth $G$-manifold is a $G$-vector bundle.

2. *Replace $R^n$ bundles by G-$R^n$ bundles.* If $M$ is a $G$-manifold then the tangent microbundle is a $G$-microbundle. However to prove a $G$-Kister theorem that $G$-microbundles contain $G$-$R^n$ bundles unique up to equivalence one needs to know that the $G$-microbundle is locally equivalent to a $G$-vector bundle. This will be true if $M$ is locally $G$-smoothable in the sense of Bredon **[B1]**. It is easy to show that a $G$-manifold $M$ is locally smoothable if its tangent microbundle is locally linear. A standard construction gives classifying spaces for $G$-vector bundles and $G$-locally linear $R^n$ bundles.

3. *G-isotopy extension theorem.* Let $M$ be a locally smoothable $G$-manifold and $K \subset M$ a $G$-invariant compact subspace. Consider the semisimplicial complex $E(K, M)$ of $G$-embeddings $f: \Delta^i \times U \to \Delta^i \times M$, $f$ commuting with projection on the $i$-simplex $\Delta^i$, when $U$ is a $G$-neighborhood of $K$ and we identify $f$ and $f'$: $\Delta^i \times U' \to \Delta^i \times M$ if they agree on some smaller neighborhood of $K$. Then $r$: $\mathrm{Homeo}(M) \to E(K, M)$ is a fibration, where $H(M)$ is the *ss*-complex of $G$-homeomorphisms of $M$. This may be proved using the work of Siebenmann **[S1]**.

4. *G-immersion theorem.* Let $M$ be a locally smoothable $G$-manifold. For $H \subset G$ let $M_{(H)} = (x \in M | G_x$ is conjugate to $H)$. Let $M^H =$ fixed point set under $H$. $M^H$ is a locally flat submanifold. Let $M^{(H)} = GM^H$.

LEMMA. *Let $(H_j)_{j \in J}$ be the orbit types of $M$. For each $j \in J$, let $[M_i^j | \, i \in I(j)]$ be the*

*AMS (MOS) subject classifications* (1970). Primary 57E10, 57E15, 57D10; Secondary 57A30, 57A35, 57A55, 55F35.

*G-components of $M^{(H_j)}$. The minimal elements of this partially ordered set under inclusion are the topologically closed G-components of the $M_{(H_j)}$, and hence are G-submanifolds.*

DEFINITION (BIERSTONE). A smooth $G$-manifold is said to have a good handle-bundle decomposition if there is a normal direction in each handle bundle on which $G$ acts trivially.

THEOREM (BIERSTONE). *A smooth G-manifold has a good handle-bundle decomposition if and only if the minimal elements of the* ($M_i^j$) *are nonclosed* (*as manifolds*).

DEFINITION. If a locally smooth $G$-manifold $M$ has nonclosed minimal elements in ($M_i^j$), we say that $M$ satisfies the *Bierstone condition*. Bierstone **[B2]** proves a $G$-Gromov Theorem for smooth $G$-manifolds satisfying the Bierstone condition. In particular, he proves

THEOREM. $I(M^n, N^n) \to^d R(TM, TN)$ is *a homotopy equivalence when $M^n$ satisfies the Bierstone condition, $I(M, N)$ is the space of G-immersions and $R(TM, TN)$ is the space of G-vector bundle maps.*

A topological version of this is true for locally smoothable $G$-manifolds where $R(TM, TN)$ is $G$-$R^n$ bundle maps and the spaces are treated semisimplicially. This uses the $G$-isotopy extension theorem above.

5. *G-smoothing up to sliced concordance.* As in the case $G = (e)$, the immersion theory has the immediate consequence that if $M$ is a locally smoothable $G$-manifold satisfying the Bierstone condition, then the sliced concordance classes of $G$-smoothings of $M$ are in 1-1 correspondence with the isotopy classes of reductions of its tangent bundle to a $G$-vector bundle. Two smoothings $M_\alpha$, $M_\beta$ of $M$ are said to be *sliced concordant* if there exists a smoothing $(M \times I)_\gamma$ of $M \times I$ such that the projection $\Pi_2$ on $I$ is a submersion (and hence $\Pi_2^{-1}(t)$ is a smooth submanifold for each $t \in I$) and $(M \times 0)_\gamma = M_\alpha$, $(M \times 1)_\gamma = M_\beta$.

Gerald Anderson **[A1]** also outlines a proof of the result along the same lines but he does not mention the $G$-isotopy extension theorem and may have a different argument in mind.

6. *G-engulfing theorem.* In order to pass from a smoothing theory up to sliced concordance to a general smoothing theory up to isotopy, one uses an engulfing lemma (this is where the restriction $n \neq 4$ comes into ordinary smoothing theory).

THEOREM. *Let $M^n$ be a compact smooth G-manifold and $h: M \times R \to V^{n+1}$ be a G-homeomorphism onto a smooth G-manifold such that $h|M \times 0$ and $h|\partial M \times R$ are smooth. Then if for each $H \subset G$,* dim $V^H \neq 4$, *there is a G-diffeomorphism $f: M \times R \to V$ such that $f|M \times 0 = h|M \times 0$ and $f|\partial M \times R = h|\partial M \times R$.*

7. *G-smoothing theorem.*

THEOREM. *Let $M^n$ be a locally smoothable G-manifold such that for any $H \subset G$,* dim $M^H \neq 4$. *Then the G-isotopy classes of G-smoothings of M are in* 1-1 *correspondence with the isotopy classes of reductions of the tangent bundle to a G-vector bundle.* (*If $\partial M \neq \emptyset$, we need* $\dim(\partial M)^H \neq 4$.)

In terms of classifying spaces we have

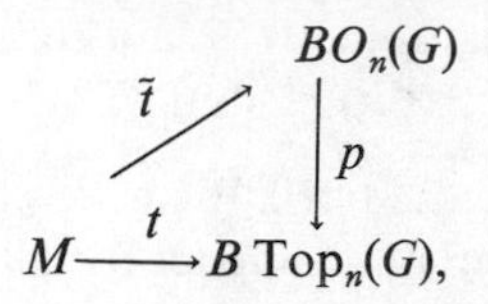

where $t$ and $\tilde{t}$ are $G$-maps. Thus our problem reduces to $G$-obstruction theory. This comes down to studying the obstructions to lifting the map on fixed point sets for $H \subset G$.

$[B\,\mathrm{Top}_n(G)]^H$ is the disjoint union of $B(\mathrm{Top}_n^\rho)$, where $\rho: H \to O(n)$ is a representation and $\mathrm{Top}_n^\rho \subset \mathrm{Top}_n$ is the subgroup of homeomorphisms commuting with $\rho(h)$, $h \in H$. The sum is over one $\rho$ from each $\mathrm{Top}_n$ equivalence class of representations. The fibre of $p_n^H: [BO_n(G)]^H \to [B\,\mathrm{Top}_n(G)]^H$ above $x \in B\,\mathrm{Top}_n^\rho$ is the disjoint union of $\mathrm{Top}_n^{\rho^1}/O_n^{\rho^1}$ when $\rho^1$ is $\mathrm{Top}_n$ equivalent to $\rho$ and we pick one $\rho^1$ from each $O_n$ equivalence class. For groups $H$ of odd prime power order, Schultz has shown that $\mathrm{Top}_n$ equivalence implies $O_n$ equivalence and hence $(p_n^H)^{-1}(x) = \mathrm{Top}_n^\rho/O_n^\rho$. In any case, our obstructions lie in the homotopy groups of $\mathrm{Top}_n^\rho/O_n^\rho$, $\rho: H \to O(n)$.

8. $\pi_i \mathrm{Top}_n^\rho/O_n^\rho$, *$\rho$ semifree.* We consider a stable representation $\alpha: H \to O_n \subset O_{n+1} \subset \cdots \subset O_{n+k}$, where the action of $H$ on $S^{n-1}$ is free.

THEOREM. *There are fibrations (up to homotopy equivalence):*

(1) $P^\alpha(S^{n+k}) \to \mathrm{Top}_{n+k}^\alpha/O_{n+k}^\alpha \to \mathrm{Top}_{n+k+1}^\alpha/O_{n+k+1}^\alpha$,

(2) $P^\alpha(S^{n+k} \bmod S^k) \to \mathrm{Top}_{n+k}^\alpha/\mathrm{Top}_k \to \mathrm{Top}_{n+k+1}^\alpha/\mathrm{Top}_{k+1}$,

(3) $P^\alpha(S^{n+k} \bmod S^k) \to P^\alpha(S^{n+k}) \to P(S^k)$,

*where $P^\alpha(S^{n+k}) \subset P(S^{n+k})$ is the subgroup of pseudoisotopies (or concordances) which commute with $\alpha$.*

THEOREM (D. ANDERSON AND W. C. HSIANG).

$$P^\alpha(S^{n+k} \bmod S^k) \simeq P_b(L \times R^{k+1}),$$

*where $P_b(L \times R^{k+1})$ is the group of bounded pseudoisotopies of $L \times R^{k+1}$, $L = S^{n-1}/\alpha$.*

THEOREM (D. ANDERSON AND W. C. HSIANG [**A2**]). *For $n + k \geqq 6$, $\Pi = \Pi_1(L)$.*

$$\Pi_i P_b(L \times R^{k+1}) = \begin{cases} K_{-k+1+i}Z(\Pi), & 0 \leqq i < k-1, \\ \tilde{K}_0[Z(\Pi)], & i = k-1, \\ Wh_1(\Pi), & i = k, \\ \Pi_{i-k-1}P(L \times D^{k+1}), & i \geqq k+1. \end{cases}$$

REMARKS. For $\Pi$ Abelian, $K_{-S}[Z(\Pi)] = 0$ for $S > 1$.

For $\Pi$ Abelian + prime power order, $K_{-1}[Z(\Pi)] = 0$.

For $\Pi$ cyclic of order $p$, $\tilde{K}_0[Z(\Pi)] =$ class group of $Q(e^{2\pi i/p})$.

For $\Pi$ finite $\tilde{K}_0[Z(\Pi)]$ is finite.

For $\Pi$ free Abelian $\tilde{K}_0[Z(\Pi)] = 0$.

The above results show that to compute $\Pi_i \mathrm{Top}_{n+k}^\alpha/O_{n+k}^\alpha$, at least up to extensions, one can either compute $\Pi_i \mathrm{Top}^\alpha/O^\alpha$ and work backwards or begin at $\Pi_i \mathrm{Top}_n^\alpha/O_n^\alpha$ and work up. We have had some success in both approaches, but before stating the results we first note an immediate consequence for stability of $G$-smoothings.

9. *Stability theorems.*

DEFINITION. Let $M$ be a $G$-manifold. $M$ is said to have a spine of codim $\geqq r$ if the cohomological dimension of $M^H/G \leqq \dim M^H - r$.

EXAMPLE 1. $M \times R^r$, where $G$ acts trivially on $R^r$, has a spine of codim $r$.

EXAMPLE 2. If $M^H$ and $\partial M^H$ are noncompact for all $H \subset G$, then $M$ has a spine of codim $\geqq 1$.

DEFINITION. Let $\mathscr{S}_G(M)$ be the isotopy classes of $G$-smoothings of $M$. Let $\bar{\mathscr{S}}_G(M)$ be the isotopy classes of stable $G$-smoothings, i.e., of $M \times R^s$, $s$ arbitrarily large. Let $M_0 = M - \partial M$.

Consider the following statements:

$A_r$: $M$ spine codim $r$ then $\mathscr{S}_G(M) \to \bar{\mathscr{S}}_G(M)$ is epi,

$B_r$: $M$ spine codim $r$ then $\mathscr{S}_G(M) \to \bar{\mathscr{S}}_G(M)$ is bijective,

$C_r$: $M$ spine codim $r$ then $\mathscr{S}_G(M) \to \bar{\mathscr{S}}_G(M_0)$ is epi.

THEOREM. *Let $M$ be a locally smoothable $G$-manifold, $G$ finite and acting semifreely. Then if* $\dim M - \dim M^G \neq 2$ *and* $\dim M^G \geqq 5$:

(1) *If $G$ is finite Abelian, then* $A_2$, $B_3$, $C_2$ *are true.*

(2) *If further $G$ is of prime power order, then* $A_1$, $B_2$, $C_1$ *are true.*

(3) *If further $G$ is of prime order and the class group of* $Z[e^{2\pi i/o(G)}] = 0$ *then* $A_0$, $B_1$, $C_0$ *are true.*

(4) *If* $G = Z_2, Z_3$ *then* $A_0$, $B_0$, $C_0$ *are true.*

10. *Computation of* $\Pi_i \operatorname{Top}^\alpha_{n+k}$. First note that

$$\Pi_i \operatorname{Top}^\alpha_{n+k} \simeq \Pi_i \operatorname{Top}_k \oplus \Pi_i \operatorname{Top}^\alpha_{n+k}/\operatorname{Top}_k.$$

Write $T^\alpha_{n,k} = \operatorname{Top}^\alpha_{n+k}/\operatorname{Top}_k$. Then

(a) One has an exact sequence, $n + k \geqq 6$, $k \geqq 0$:

$$\begin{aligned} 0 &\to \Pi_{k+1}\tilde{A}(L) \to \Pi_{k+1}T^\alpha_{n+1,\,k+1} \to Wh_1(\Pi) \to \Pi_k T^\alpha_{n,\,k} \to \Pi_k T^\alpha_{n+1,\,k+1} \\ &\to \tilde{K}_0 Z(\Pi) \to \Pi_{k-1}T^\alpha_{n,\,k} \to \Pi_{k-1}T^\alpha_{n+1,\,k+1} \to K_{-1}[Z(\Pi)] \to \cdots. \end{aligned}$$

Here $\tilde{A}(L)$ = block automorphisms of $L = S^{n-1}/\alpha$. Also

$$\Pi_1(L) = \begin{cases} H & \text{if } n \geqq 3, \\ Z & \text{if } n = 2, \\ e & \text{if } n = 1. \end{cases}$$

(b) For $n + i \geqq 5$, $i \geqq 1$, there is an exact sequence **[A3]** ($L_* = L \cup$ point)

$$\begin{aligned} &\to [\Sigma^{n+1}(L_*), G/\operatorname{Top}] \to \mathscr{L}^s_{n+1}(\Pi) \to \Pi_i[\mathscr{H}(L)/A(L)] \\ &\to [\Sigma^i(L_*), G/\operatorname{Top}] \to \dots \to [\Sigma(L_*), G/\operatorname{Top}] \to \mathscr{L}^s_n(\Pi), \end{aligned}$$

where $\mathscr{H}(L)$ is the space of homotopy equivalences of $L$ and $\mathscr{L}^s_n(\Pi)$ is the Wall group. Thus (a) and (b) determine $\Pi_i \operatorname{Top}^\alpha_{n+k}/\operatorname{Top}_k$, $0 \leqq i \leqq k$, up to extensions, $n + k \geqq 6$. (For $i = 0$, one needs a special argument.) For $i \geqq k + 1$, we have

$$\Pi_i \operatorname{Top}^\alpha_{n+k}/\operatorname{Top}_k \simeq \Pi_{i-k-1} A(L \times D^{k+1}),$$

where $A(L \times D^{k+1})$ = homeomorphism of $L \times D^{k+1}$ fixed on $L \times S^k$. Results of Hatcher and Wagoner give results on $\Pi_0 A(L \times D^{k+1})$. The higher homotopy groups are still unknown. However, for $G$-smoothing, $G$ acting semifreely, this is sufficient.

11. *Computation of* $\Pi_i$ $\mathrm{Top}^\alpha/O^\alpha$, $o(H)$ *prime power.*

THEOREM. *If* $n + k \geqq 8$ *and* $k \geqq 5$, *there exists an exact sequence*

$$\to H_{n+k}(\tilde{K}_0) \to C_\alpha^{n+k} \to [S_\alpha^{n+k}, \mathrm{Top}/O]_H \to H_{n+k-1}(K_0) \to \cdots.$$

Here

(a) $[S_\alpha^{n+k}, \mathrm{Top}/O]$ = equivariant homotopy classes of base-pointed maps = stable $H$-smoothings of $S_\alpha^{n+k}$ which give standard $\alpha$ action on $D_\alpha^{n+k} \subset S_\alpha^{n+k}$.

(b) $W \in C_\alpha^{n+k}$ if $W$ is a smooth $H$-manifold, $W$ homeomorphic to $S^{n+k}$, $W^H$ homeomorphic to $S^k$. If $x \in W^H$, the action of $H$ on $W_x$ is given by $\alpha$. We can take the $H$-connected sum along the fixed point set. Identify $W = 0$ if $W = \partial V$ where $\partial V$ is homeomorphic to $D^{n+1}$ and $V^H$ is homeomorphic to $D^{m+1}$. $C_\alpha^{n+k}$ is determined up to extensions by homotopy groups of spheres and Wall groups, etc. (Rothenberg [**R1**]). $C_\alpha^{n+k}$ is finitely generated, rank is known if $H$ is cyclic.

(c) A differential $d_i\colon \tilde{K}_0 \to \tilde{K}_0$ can be defined so that $d_i d_{i+1} = 0$. If $o(H)$ is odd, it is conjectured that $d_i(x) = x + (-1)^i x$. If this is the case then

$$H_{\mathrm{even}}(\tilde{K}_0) = \text{elements of order 2},$$
$$H_{\mathrm{odd}}(\tilde{K}_0) = \tilde{K}_0/2\tilde{K}_0.$$

THEOREM. *There is an exact sequence*

$$\to [(D^{n+k+1} \times L, S^{n+k} \times L), (\mathrm{Top}/O, *)] \to [S_\alpha^{n+k}, \mathrm{Top}/O]_H \to [S^k, \mathrm{Top}^\alpha/O^\alpha]$$
$$\to [(D^{n+k} \times L, S^{n+k-1} \times L), (\mathrm{Top}/O, *)] \to \cdots.$$

*Thus* $\Pi_k(\mathrm{Top}^\alpha/O^\alpha)$ *is finitely generated with rank* = *rank* $C_\alpha^{n+k}$.

**1. *G*-bundles.** Let $p\colon E \to X$ be a locally trivial bundle with fibre $F$ and group $A$. $p$ is called a *G-bundle*, or more precisely a *G-A bundle* if $E$ and $X$ are $G$-spaces, $p$ is a $G$-map, and $G$ acts on $E$ through $A$-bundle maps. Two $G$-$A$ bundles over $X$ are called *G-A equivalent* if they are $A$-equivalent via a $G$-equivariant map.

EXAMPLE 1. A $G$-vector bundle of dimension $n$ is simply a $G$-$L_n$ bundle, $L_n$ the group of linear isomorphisms of $R^n$.

If $p\colon E \to X$ is a $G$-$A$ bundle, $G$ acts on the associated principal $A$-bundle $P$ through bundle maps. That is, $G$ acts on the left and $A$ acts on the right of $P$ and these actions commute. Conversely, if $p\colon P \to X$ is a principal $G$-$A$ bundle and $A$ acts on the left of $F$, then $E = P \times_A F$ is a $G$-$A$ bundle with fibre $F$. Two $G$-$A$ bundles with fibre $F$ are $G$-$A$ equivalent if and only if their associated principal $G$-$A$ bundles are $G$-$A$ equivalent.

In order to prove a covering homotopy property or to produce a classifying space for ordinary bundles the local triviality property is essential. For $G$-$A$ bundles we will need a $G$-local triviality condition for the same purpose. Before defining this condition we recall the local structure of a completely regular $G$-space $X$ (see Bredon [**B1**]): For any $x \in X$ there is a $G_x$-invariant subspace $V_x$ containing $x$, called a *slice* through $x$, such that $\psi_x\colon G \times_{G_x} V_x \to X$, $\psi_x[g, v] = gv$ is a homeomorphism onto a $G$-invariant neighborhood of the orbit $Gx$. The $G$-invariant neighborhood, $GV_x$, is called a *tube* about $Gx$. For any $G$-space $X$ we define a *G-chart* to be a pair $(V, H)$ where $H \subset G$, $V$ is an $H$-invariant subspace of $X$ and the map $\psi\colon G \times_H V \to X$, $\psi[g, v] = gv$, is a homeomorphism onto an open set. (Note that $V$ need not be a slice, i.e., $H$ may not be $G_x$ for any $x \in V$.) A *G-atlas* is a family $\{(V_\alpha, H_\alpha)\}$ of $G$-

charts such that $\{GV_\alpha\}$ covers $X$. If $X$ is paracompact, then any cover by $G$-invariant open sets has a refinement which is a $G$-atlas.

We note that the preimage under a $G$-map of a $G$-chart is a $G$-chart. That is, if $(V, H)$ is a $G$-chart in $X$ and $f\colon Y \to X$ is a $G$-map, then $f^{-1}(V, H) = (f^{-1}(V), H)$. This means $f^{-1}(V)$ is $H$-invariant and $G \times_H f^{-1}(V) \to^{\psi} Gf^{-1}(V) = f^{-1}(GV)$ is a homeomorphism.

We now describe the appropriate generalization of product bundle: Let $H \subset G$ and $\rho\colon H \to A$ be a representation. If $A$ acts on the left of $F$ and $H$ on the left of $V$, then $H$ acts on $V \times F$ by $h(v, y) = (hv, \rho(h)y)$. We denote by $1_\rho^H(V)$ the $G$-$A$ bundle over $G \times_H V$ with fibre $F$, given by $p\colon G \times_H (V \times F) \to G \times_H F$, $p[g, (v, y)] = [g, v]$. (Note that $p$ is trivial as an $A$-bundle.)

DEFINITION. A $G$-$A$ bundle $p\colon E \to X$ with fibre $F$ is called *$G$-$A$ locally trivial* (or simply *$G$-locally trivial* if $A$ is fixed) if there is a $G$-atlas $\{(V_\alpha, H_\alpha)\}$ on $X$ such that $E|GV_\alpha$ is $G$-$A$ equivalent to $1^{H_\alpha}(V_\alpha)$ for some representation $\rho_\alpha\colon H_\alpha \to A$ (under the identification $\psi_\alpha\colon G \times_{H_\alpha} V_\alpha \to GV_\alpha$).

If $X$ is completely regular this is equivalent to Bierstone's definition **[B2]**: For each $x \in X$, there is a $G_x$-invariant neighborhood $U_x$ such that $p^{-1}(U_x)$ is $G_x$-$A$ equivalent to $U_x \times F$ with $G_x$ action $h(u, y) = (hu, \rho_x(h)y)$, where $u \in U_x$, $h \in G_x$, $y \in F$ and $\rho_x\colon G_x \to A$ is a representation.

If $p\colon E \to X$ is a $G$-$A$ bundle and $f\colon Y \to X$ is a $G$-map the induced bundle $f^*(p)\colon f^*E \to Y$ is a $G$-$A$ bundle. Further if $p$ is $G$-$A$ locally trivial, then $f^*(p)$ is $G$-$A$ locally trivial.

A $G$-$A$ bundle is $G$-$A$ locally trivial if and only if the associated principal bundle is $G$-$A$ locally trivial.

The following is essentially due to Bierstone and Wasserman.

THEOREM 1. *Any $G$-$L_n$ bundle over a completely regular $X$ is $G$-locally trivial.*

PROOF. Let $p\colon E \to X$ be the associated $G$-vector bundle. Let $x \in X$ and let $\bar\varphi\colon \bar U_x \times R^n \to p^{-1}(\bar U_x)$ be a local trivialization.

We can assume $\bar U_x$ is $G_x$-invariant, since any neighborhood contains a $G_x$-invariant one as an $L_n$-bundle. Define $\rho_x\colon G_x \to L_n$ by $\rho_x(h)y = \bar\varphi_x^{-1}h\bar\varphi_x y$, $h \in G_x$, $y \in R^n$. Now let $\varphi\colon \bar U \times R^n \to p^{-1}(\bar U)$ be the map obtained by averaging over $G_x$; i.e.

$$\varphi_u y = \frac{1}{|G_x|} \sum h^{-1}\bar\varphi_{hu}(\rho_x(h)y), \qquad u \in \bar U_x.$$

Then $\varphi_u$ is linear, $\varphi_x = \bar\varphi_x$ and $g\varphi(u, y) = \varphi(gu, \rho_x(g)y)$, $g \in G_x$. Since $\varphi_x = \bar\varphi_x$ is an isomorphism, $\varphi_u$ is an isomorphism for $u$ in some smaller $G_x$-invariant neighborhood $U_x \subset \bar U_x$. Hence $\varphi|U_x \times R^n\colon U_x \times R^n \to p^{-1}(U_x)$ is a $G_x$-$L_n$ trivialization, and $p$ is $G$-$L_n$ locally trivial. Q.E.D.

Just as for ordinary bundles, one may ask if a $G$-$A$ bundle reduces to a $G$-$B$ bundle, $B \subset A$. If $A/B$ has local cross-section in $A$, this is true if and only if the associated $G$-$A$ bundle with fibre $A/B$ has a $G$-cross-section. A $G$-vector bundle over a paracompact base can always be given a $G$-invariant Riemannian metric, and so reduces to a $G$-$O_n$ bundle, $O_n$ the orthogonal group. This reduction is unique up to equivalence. In particular, if follows from Theorem 1 that $G$-$O_n$ bundles over paracompact spaces are $G$-$O_n$ locally trivial.

DEFINITION 2. A $G$-$A$ bundle $p\colon E \to X$ is called *numerable* if there is a trivializing

$G$-partition of unity, i.e., there exists a $G$-partition of unity subordinate to a $G$-atlas $\{(V_\alpha, H_\alpha)\}$ such that $E|GV_\alpha$ is equivalent to $1^{H_\alpha}(V_\alpha)$, some $\rho_\alpha\colon H_\alpha \to A$.

Bierstone [**B3**] proves that a $G$-locally trivial bundle over a paracompact base satisfies the $G$-covering homotopy property. We generalize this slightly to:

THEOREM 2. *A numerable G-A bundle satisfies the G-covering homotopy property.*

To show this one follows the proof for ordinary numerable bundles as given in Husemoller [**H1**], simply substituting $G$-partitions of unity for ordinary partitions of unity. This comes down to showing that a $G$-bundle $E$ over $X \times I$, when $G$ acts trivially on the $I$-factor, is $G$-equivalent to $E_0 \times I$, where $E_0 = E/X \times (0)$. To see that Husemoller's argument works one needs to observe the following lemma and corollary.

LEMMA 3. *Let $X$ be a G-space and suppose $X \times I = (W, H)$, i.e., $X \times I$ may be identified with $G \times_H W$ via $[g, w] \to gw$. Then $W = W_0 \times I$, where $W_0 = W \cap (X \times (0))$.*

PROOF. The map $W_0 \times I \subset G \times_H W \xrightarrow{\pi} G/H$ has image $eH$ since $G/H$ is discrete and $I$ is connected. Hence $W_0 \times I \subset W$. To see that $W \subset W_0 \times I$ first note that since $X \times (0)$ is $G$-invariant, $X \times (0) = G \times_H W_0$ and $X \times I = G \times_H (W_0 \times I)$. Hence if $w \in W$, $w = g(w_0, t)$, some $g \in G$, $w_0 \in W_0$, $t \in I$. Since $(w_0, t) \in W$, we must have $g \in H$; and $g(w_0, t) = (gw_0, t)$, $gw_0 \in W_0$. Hence $w \in W_0 \times I$ and $W = W_0 \times I$.

COROLLARY 4. *With $X \times I = (W, H)$ as in the above lemma, $1^H_\rho(W) = 1^H_\rho(W_0) \times I$ as a G-A bundle.*

Of course, Theorem 2 has the:

COROLLARY 5. *If $p\colon E \to X$ is a numerable G-A bundle and $f_i\colon Y_i \to X$, $i = 1, 2$, are G-homotopic G-maps, then $f_1^*(p)$ and $f_2^*(p)$ are G-A equivalent.*

DEFINITION 3. A universal $G$-$A$ bundle $p\colon P \to X$ is a $G$-$A$ numerable bundle such that $G$-$A$ equivalence classes of $G$-$A$ numerable bundles over any $G$-space $Y$ correspond to $G$-homotopy classes, $[Y, X]_G$, of $G$-maps of $Y$ into $X$, the correspondence being given by induced bundles.

Now for ordinary bundles a universal $A$-bundle is characterized as a numerable contractible principal $A$-bundle [**D1**]. By a slight refinement of this argument one may prove:

THEOREM 6. *A numerable principal G-A bundle $p\colon P \to X$ is universal if and only if it satisfies: For each $H \subset G$ and representation $\rho\colon H \to A$ consider $P$ as an H-space under the action $p \to hp\rho(h)^{-1}$, $h \in H$, $p \in P$. Then*

(1) $P^H \neq \emptyset$.

(2) *For any $p_H \in P^H$, $P$ is H-contractible to $p_H$.*

PROOF. We first show the conditions are necessary:

(1) Consider the $G$-$A$ bundle $\pi\colon G \times_H A \to G/H$, $H$ acting through $\rho\colon H \to A$. Let $e_1 \in G$, $e_2 \in A$ be the unit elements. Then $[e_1, e_2] \in G \times_H A$ and $h[e_1, e_2] = [h, e_2] = [e_1, \rho(h)e_2] = [e_1, e_2]\rho(h)$. Since $p$ is universal, there is a $G$-$A$ bundle map $\varphi\colon G \times_H A \to P$. Then

$$h\varphi[e_1, e_2] = \varphi(h[e_1, e_2]) = \varphi([e_1, e_2]\rho(h)) = (\varphi[e_1, e_2])\rho(h).$$

Hence $\varphi[e_1, e_2] \in P^H$, and (1) is satisfied.

(2) Consider the $G$-$A$ bundle $\pi: G \times_H (P \times A) \to G \times_H P$, $H$ acting on $P$ as above. Define the $G$-$A$ bundle maps $\lambda_i: G \times_H (P \times A) \to P$, $i = 0, 1$, by $\lambda_0[g, p, a] = gpa$ and $\lambda_1[g, p, a] = gp_H a$, $p_H \in P^H$. By the universality of $P$ there is a $G$-$A$ homotopy $\lambda_t: G \times_H (P \times A) \to P$, $0 \leqq t \leqq 1$, between $\lambda_0$ and $\lambda_1$. In particular, $p = \lambda_0[e_1, p, e_2]$ is deformed to $\lambda_1[e_1, p_H, e_2] = p_H$. But

$$\lambda_t[e_1, hp\rho(h)^{-1}, e_2] = \lambda_t[h, p, \rho(h)^{-1}] = h\lambda_t[e_1, p, e_2]\rho(h)^{-1}.$$

Hence $\lambda_t$ defines an $H$ contraction of $P$ to $p_H$.

The proof of sufficiency is just the same as in Dold **[D1]**, except that we replace the section extension property by the equivariant section property. In particular, if $p': P' \to X'$ is a $G$-$A$ bundle, $G$-$A$ bundle maps of $P'$ into $P$ correspond to $G$-sections of the associated bundle $P' \times_A P$ over $X'$ with fibre $P$. If $p'$ is numerable and $(V, H)$ is a chart, $P' \times_A P|GV \simeq G \times_H (V \times P)$. But $G$-sections of a bundle of the form $\pi: G \times_H (V \times P) \to G \times_H V$ are in 1-1 correspondence with $H$-sections of $\pi: V \times P \to V$. The sufficiency follows from the $H$-contractibility of $P$. Q.E.D.

If we restrict our attention to a class of numerable $G$-$A$ bundles for which the local trivializations are given by representations $\rho$ in some designated subset $S$ of all representations, then a numerable $G$-$A$ bundle with local trivializations in $S$ will be universal for the class of bundles if and only if (1) and (2) are satisfied with respect to representations in $S$. In particular, we are interested in the case where $S$ consists of representations into some subgroup $B \subset A$. These bundles will be denoted as $G$-$(A, B)$ *bundles*. We have in mind the case where $A = \mathrm{Top}_n$, the group of homeomorphisms of $R^n$ fixing $0 \in R^n$ and $B = O_n$. The associated bundles with fibre $R^n$ will be called locally linear $G$-$R^n$ bundles. In fact, a $G$-$R^n$ bundle $p: E \to X$, with $X$ paracompact, will be locally linear if and only if for each $x \in X$ there exists a $G$-invariant neighborhood $U$ of $x$ such that $E|U$ is $G$-$R^n$ equivalent to a $G$-vector bundle.

THEOREM 7. *Universal $G$-$(A, B)$ bundles exist (see Lu* **[L1]***).*

PROOF. We first note some properties of infinite joins.

(1) If $\pi_i: P_i \to X_i$, $i = 1, 2, 3, \ldots$, is any sequence of principal $G$-$A$ bundles then the infinite join $P = *_i P_i$ is a $G$-$A$ bundle.

In fact a point in $P$ is of the form $\sum \lambda_i p_i$, where $\lambda_i$ are the join coordinates, $\lambda_i \geqq 0$ with only a finite number nonzero, and $\sum \lambda_i = 1$. The actions of $G$ and $A$ on $P$ are given by $g(\sum \lambda_i p_i) = \sum \lambda_i(gp_i)$ and $(\sum \lambda_i p_i)a = \sum \lambda_i (p_i a)$. Let $\pi: P \to X$ be the quotient map under the $A$-action, i.e., $X = P/A$.

Then as shown in Husemoller **[H1]**, $\pi$ is a locally trivial, in fact numerable, $A$-bundle. We will show

(2) If each $\pi_i$ is a numerable $G$-$(A, B)$ bundle, so is $\pi$.

Consider a chart $(V, H) \subset X_j$. Let $\tilde{W} = \{\sum \lambda_i p_i | \lambda_j > 0 \text{ and } p_j \in \pi_j^{-1}(V)\}$. Then $\tilde{W}$ is $A$-invariant and $G\tilde{W} = \{\sum \lambda_i p_i | \lambda_j > 0 \text{ and } p_j \in \pi_j^{-1}(GV)\}$ is $A$-invariant. Also $\pi(G\tilde{W}) = G\pi(\tilde{W})$. Let $W = \pi(\tilde{W})$. Now $r_j: G\tilde{W} \to p_j$, $r_j(\sum \lambda_i p_i) = p_j$, is a $G$-$A$ map and induces a $G$-map $\bar{r}_j: GW \to GV \subset X_j$, with $\bar{r}_j^{-1}(V) = W$. Hence $(W, H)$ is a chart in $X$.

If $P_j|GV = 1_\rho^H(V)$, we claim $P|GW = 1_\rho^H(W)$. This is equivalent to showing that if $\varphi_j: V \times A \to \pi_j^{-1}(V)$ is an $H$-$A$ equivalence with $H$-action $h(v, a) = (hv, \rho(h)a)$, then there is an $H$-$A$ equivalence $\varphi: \tilde{W} \times A \to \tilde{W} = \pi^{-1}(W)$ with $H$-action $h(w, a) = (hw, \rho(h)a)$. Following Husemoller, define $\tilde{s}: \tilde{W} \to \tilde{W}$, $\tilde{s}(\sum \lambda_i p_i) = \sum \lambda_i p_i a^{-1}$, where $\varphi_j^{-1}(p_j) = (v, a)$. Then $s$ is $A$-invariant and induces a cross-section $s: W \to \tilde{W}$. Define $\varphi: W \times A \to \tilde{W}$ by $\varphi(w, a) = s(w)a$. Then $\varphi$ is an $A$-equivalence. To see that $\varphi$ is an $H$-$A$ equivalence note that

$$\begin{aligned} hs(\pi(\textstyle\sum \lambda_i p_i)) &= h\tilde{s}(\textstyle\sum \lambda_i p_i) = \textstyle\sum \lambda_i hp_i a^{-1} \\ &= \textstyle\sum \lambda_i hp_i \rho(h^{-1})\rho(h)a^{-1} = (\tilde{s}(\textstyle\sum \lambda_i hp_i \rho(h)^{-1}))\rho(h) \\ &= (sh\pi(\textstyle\sum \lambda_i p_i))\rho(h), \end{aligned}$$

i.e., $hs(w) = (s(hw))\rho(h)$ and $h\varphi(w, a) = s(hw)\rho(h)a = \varphi(hw, \rho(h)a)$. Hence $P|GW = 1_\rho^H(W)$.

Taking an atlas of charts in each $X_j$, the reunion of all the corresponding charts in $X$ will be an atlas in $X$. Now the $\lambda_i: P \to [0, 1]$, being $A$-invariant, induce $\bar{\lambda}_i: X \to [0, 1]$. Let $\mu_i(x) = \max(0, \lambda_i(x) - \sum_{j<i}\bar{\lambda}_j(x))$. Then $\{\mu_i^{-1}(0, 1]\}$ is locally finite and $\nu_i = \mu_i/\sum_j \mu_j$ is a locally finite partition of unity subordinate to $\{\bar{\lambda}_i^{-1}(0, 1]\}$. Let $\{\alpha_i^k\}$ be a partition of unity on $X_i$ subordinate to the trivializing atlas. Let $\beta_i^k = \alpha_i^k \circ \bar{r}_i: \bar{\lambda}_i^{-1}(0, 1] \to [0, 1]$, $\bar{r}_i$ as above. Then $\{\beta_i^k \nu_i\}$ is a partition of unity subordinate to the trivializing atlas on $X$. Hence $\pi$ is a numerable $G$-$(A, B)$ bundle.

(3) For any representation $\rho: H \to B$, $P^H = *_i P_i^H$, assuming $P_i^H \neq \varnothing$, all $i$.

Now choose a representative $H$ from each conjugacy class of subgroups of $G$ and a representative $\rho: H \to B$ from each $A$-equivalence class of representations of $H$ in $B$. Let $\{\rho_\beta\}$ be this set. Let $E_\beta = G \times_{H_\beta} A$, $\rho_\beta: H_\beta \to B$, and let $E = \coprod_\beta E_\beta$. Let $p_\beta: G \times_{H_\beta} A \to G/H_\beta$ be the projection and $p: E \to \coprod_\beta G/H_\beta$ be $\coprod p_\beta$. Then $E$ is a numerable $G$-$(A, B)$ bundle. Further for any subgroup $H$ of $G$ and representative $\rho: H \to B$, $E^H \neq \varnothing$. In fact, $[e_1, e_2] \in G \times_{H_\beta} A$ satisfies $h[e_1, e_2] = [e_1, e_2]\rho_\beta(h)$ and $[g, e_2]$ satisfies $ghg^{-1}[g, e_2] = [g, e_2]\rho_\beta(h)$, $h \in H$. Finally $[e_1, a]$ satisfies $h[e_1, a] = [e_1, a]a^{-1}\rho_\beta(h)a$. So all subgroups and representatives have fixed points.

For each $i$, let $P_i = E$ and $\pi_i = p$. Then $P = *_i P_i$ satisfies property (1) of Theorem 6. Now let $p_H \in P^H$. Then $p_H$ belongs to some finite join $*_{i \leq k} P_i^H \subset P$. Let $P^k = \{\sum \lambda_i p_i \in P | \lambda_i = 0 \text{ for } i \leq k\}$. Then it is easy to construct (see **[H1]**) a $G$-$A$ deformation of $P$ into $P^k$. For any representation $\rho: H \to B$ this deformation will be in particular an $H$-deformation. Now $p_H * P^k \subset P$ and since $p_H$ is a fixed point, the contraction of this cone to the vertex $p_H$ will be an $H$-deformation. Hence $P$ satisfies (2) of Theorem 6, and $P$ is universal. Q.E.D.

REMARK. If $X$ is a $G$-space and $(V, H)$ is a chart (i.e., $G \times_H V \to GV$ is a homeomorphism) then $(gV, gHg^{-1})$ is a chart with the same image (i.e., $G \times_{gHg^{-1}} V \to GgV = GV$ is a homeomorphism). Further, the trivial $G$-$A$ bundle $1_\rho^H(V)$ over $GV$ is equivalent to $1_{\rho \circ c(g^{-1})}^{gHg^{-1}}(gV)$, where $c(g)$ means conjugation by $g$.

The next theorem gives information on the universal base space for $G$-$(A, B)$ bundles. In order to state it we need some notation: Let $H \subset G$. For any representation $\rho: H \to B$, let $A^\rho = \{a \in A | \rho(h)a\rho(h)^{-1} = a\}$. Then $A^\rho$ is a closed subgroup of $A$. Let $R$ be a collection of representations of $H$ in $B$ containing exactly one representation from each $A$-equivalence class.

THEOREM 8. *Let* $\pi: P \to X$ *be a universal* $G$-$(A, B)$ *bundle. Then* $X^H = \coprod_{\rho \in R} BA^\rho$ *(disjoint union), where* $BA^\rho$ *is a universal base space for the topological group* $A^\rho$.

PROOF. Let $P^\rho = \{p \in P \mid hp = p\rho(h), h \in H\}$.

(1) $P^\rho$ is nonempty and contractible (by Theorem 6).

(2) For any $p \in P^\rho$, $\pi^{-1}(\pi(p)) \cap P^\rho = pA^\rho$, i.e., $h(pa) = (pa)\rho(h) \Leftrightarrow p\rho(h)\, a = pa\rho\,(h) \Leftrightarrow \rho(h)a = a\rho(h)$.

(3) $\pi(P^\rho) \subset X^H$.

(4) If $x \in X^H$, $x \in \pi(P^\rho)$, some $\rho \in R$.

In fact, by the remark above, there is a trivializing chart $(V, H')$ with $x \in V$ and $P|GV = 1_{\rho'}^{H'}(V)$ $\rho': H' \to B$. Now $H \subset G_x \subset H'$. Since $\pi^{-1}(V) \simeq V \times A$ with $H'$ action, $h'(v, a) = (h'v, \rho'(h')a)$; $(x, e_2) \in V \times A$ and $h(x, e_2) = (x, e_2)\rho'(h)$. Hence $x \in \pi(P^\rho)$ where $\rho = \rho' \mid H$. If $\rho \notin R$, then $c(a) \circ \rho \in R$, some $a \in A$. But if $p \in P^\rho$, $pa^{-1} \in P^{c(a)\circ\rho}$. Hence $x \in \pi(P^\rho)$, some $\rho \in R$.

(5) $\pi \mid P^\rho$ is a locally trivial principal $A^\rho$-bundle and $\pi(P^\rho)$ is open in $X^H$.

If $x \in \pi\,(P^\rho)$, then we may choose a trivializing chart $(V, H')$ such that $x \in V$ and $P|GV = 1_{\rho'}^{H'}$, $\rho': H' \to B$ and $\rho'|H' = \rho$. Now if $U = V \cap X^H$ we claim $\pi^{-1}(U) \cap P^\rho = U \times A^\rho \subset V \times A$. In fact, $\pi^{-1}(U) = U \times A$ with $h'(u, a) = (h'u, \rho(h')a)$. If $(u, a) \in P^\rho$, $h(u, a) = (u, a)\,\rho(h) = (hu, \rho(h)a) = (u, \rho(h)a)$ by (3). Hence $\rho(h)a = a\rho(h)$ and $a \in A^\rho$. Since $V$ is open in $X$, $U$ is open in $X^H$. But $(u, e_2) \in P^\rho$ and $U \subset \pi(P^\rho)$. Hence $P^\rho$ is locally trivial and $\pi(P^\rho)$ is open in $X^H$.

(6) $\pi(P^\rho) \cap \pi(P^{\rho'}) = \emptyset$, $\rho, \rho' \in R$, $\rho \neq \rho'$.

If $x \in \pi(P^{\rho'}) \cap \pi(P^{\rho'})$ and $p \in \pi^{-1}(x) \cap P^\rho$, then $pa \in P^{\rho'}$, some $a \in A$. Hence $\rho = c(a) \circ \rho'$. Contradiction.

By (1), (4), (5) and (6), $\pi(P^\rho) = BA^\rho$ and $X^H = \coprod_{\rho \in R} BA^\rho$.

*Notation*. We will denote the classifying space for $G$-$O(n)$ bundles by $BO_n(G)$ and for $G$-$(\mathrm{Top}_n, O_n)$ bundles by $B\,\mathrm{Top}_n(G)$. This last will cause no confusion because only locally linear $R^n$-bundles will be used. Note that since $G$-vector bundles are locally linear $R^n$-bundles, there is a $G$-map of $BO_n(G)$ into $B\,\mathrm{Top}_n(G)$ defined up to $G$-homotopy. More explicitly:

LEMMA 9. *If* $\pi: E\,\mathrm{Top}_n(G) \to B\,\mathrm{Top}_n(G)$ *is a universal* $G$-$(\mathrm{Top}_n, O_n)$ *bundle then the quotient map* $q: E\,\mathrm{Top}_n(G) \to E\,\mathrm{Top}_n(G)/O_n$ *is a universal* $G$-$O_n$ *bundle and we can take* $BO_n(G) = E\,\mathrm{Top}_n(G)/O_n$. *With this choice* $BO_n(G)$ *is a numerable* $G$-$(\mathrm{Top}_n, O_n)$ *bundle over* $B\,\mathrm{Top}_n(G)$ *with fibre* $\mathrm{Top}_n/O_n$.

PROOF. $q$ is a principal $G$-$O_n$ bundle. To see that it is $G$-locally trivial let $(V, H)$ be a trivializing chart for $\pi$. Then $\pi^{-1}(GV) \simeq G \times_H (V \times \mathrm{Top}_n)$, $\rho : H \to O_n$, and $\pi^{-1}(GV)/O_n \simeq G \times_H (V \times \mathrm{Top}_n/O_n)$. Now $\mathrm{Top}_n \to \mathrm{Top}_n/O_n$ has local cross-sections since $O_n$ is compact Lie. Since $\mathrm{Top}_n$ and hence $\mathrm{Top}_n/O_n$ is metrizable, $\mathrm{Top}_n$ is an $H$-$O_n$ locally trivial bundle over $\mathrm{Top}_n/O_n$ and hence has a trivializing $H$ partition of unity. It is easy to see that this implies that $q$ is a numerable $G$-$O_n$ bundle. Since $E\,\mathrm{Top}_n(G)$ satisfies (1) and (2) of Theorem 6 as a $G$-$(\mathrm{Top}_n, O_n)$ bundle it satisfies (1) and (2) as a $G$-$O_n$ bundle and $q$ is universal.

THEOREM 10. *Let* $p_n: BO_n(G) \to B\,\mathrm{Top}_n(G)$ *be the* $G$-*bundle of Lemma* 9. *For any* $H \subset G$, *let* $p_n^H: [BO_n(G)]^H \to [B\,\mathrm{Top}_n(G)]^H$ *be the restriction to fixed point sets. Then* $p_n^H$ *is a numerable bundle such that the fibre over the component* $B\,\mathrm{Top}_n^\rho$, $\rho: H \to$

*$O_n$, of $[B\,\mathrm{Top}_n(G)]^H$ is $\coprod_{\rho' \in S} \mathrm{Top}_n^{\rho'}/O_n^{\rho'}$, where $S$ consists of one representation $\rho'\colon H \to O_n$ from each $O_n$-equivalence class which is $\mathrm{Top}_n$ equivalent to $\rho$.*

PROOF. If $z \in E\,\mathrm{Top}_n(G)$ is such that $q(z) \in [BO_n(G)]^H = [E\,\mathrm{Top}_n(G)/O_n]^H$, then $hz = z\rho'(h)$, $\rho' : H \to O_n$ a representation. But if $\pi(z) \in B\,\mathrm{Top}_n^{\rho}$, $\rho'$ is $\mathrm{Top}_n$ equivalent to $\rho$ (see step (6) of the proof of Theorem 8). Thus

$$p_n^{-1}(B\,\mathrm{Top}_n^{\rho}) \cap [BO_n(G)]^H = \coprod_{\rho' \in S} BO_n^{\rho'},$$

where $BO_n^{\rho'} = (E\,\mathrm{Top}_n(G))^{\rho'}/O_n^{\rho'}$, $(E\,\mathrm{Top}_n(G))^{\rho'} = \{z \in E\,\mathrm{Top}_n(G) \mid hz = z\rho'(h)\}$. Since for each $\rho' \in S$, $B\,\mathrm{Top}_n^{\rho} = B\,\mathrm{Top}_n^{\rho'} = (E\,\mathrm{Top}_n\,(G))^{\rho'}/\mathrm{Top}_n^{\rho'}$ and the quotient map is a numerable bundle (see proof of Theorem 8), $p_n^H | B\,\mathrm{Top}_n^{\rho}$ is a numerable bundle with fibre $\coprod_{\rho' \in S} \mathrm{Top}_n^{\rho'}/O_n^{\rho'}$.

**2. *G*-microbundles.** Before defining $G$-microbundles we recall the definition of microbundles without $G$-action.

DEFINITION 1. An $R^n$-microbundle $\mu$ over $X$ is a diagram of maps and spaces $X \xrightarrow{s} E \xrightarrow{p} X$ such that $ps = \mathrm{id}_X$ and there exist an open covering $\{U_\alpha\}$ of $X$ and open embeddings $\varphi_\alpha\colon U_\alpha \times R^n \to p^{-1}(U_\alpha)$ such that

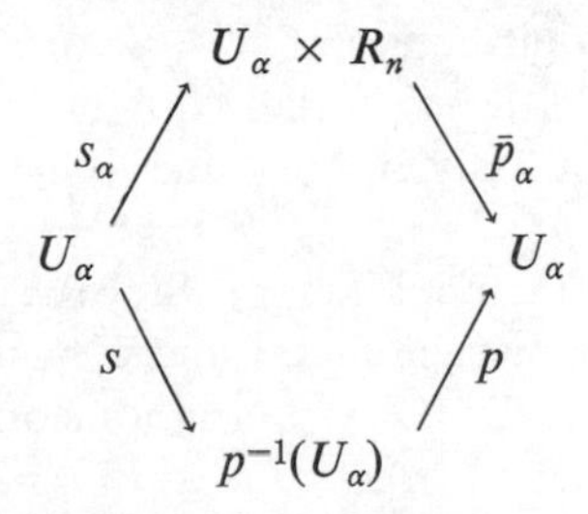


commutes, where $s_\alpha(x) = (x, 0)$ and $p_\alpha(x, y) = x$. $\mu$ is called *numerable* if there exists a partition of unity subordinate to $\{U_\alpha\}$.

If $E^0$ is a neighborhood of $sX$ in $E$, $X \xrightarrow{s} E^0 \xrightarrow{p} X$ is again a microbundle. However, $E^0$ may not be a numerable microbundle even if $E$ is, as the following example from **[H2]** shows:

EXAMPLE 1. Let $X$ be a denumerable set and let $x_0 \in X$ be a fixed element. Define a topology on $X$ by requiring $U \subset X$ to be open if $U$ is empty or contains $x_0$. Then $X$ is connected and so any continuous function $f\colon X \to R$ is constant. Consider the trivial microbundle $\varepsilon\colon X \xrightarrow{s} X \times R \xrightarrow{p} X$, $s(x) = (x, 0)$, $p(x, r) = x$. Writing $X = \{x_i, i = 0, 1, 2, \cdots\}$, let $E^0 \subset X \times R$ be the set $E^0 = \bigcup_i x_i \times (-1/i, 1/i)$. Then $E^0 = \bigcup_i U_i \times (-1/i, 1/i)$, where $U_i \subset X$ is the open set $U_i = \{x_j, 0 \leqq j \leqq i\}$. Hence $E^0$ is an open neighborhood of $s(X) = X \times 0$. We claim $E^0$ is not a numerable microbundle. Suppose there existed a partition of unity $\{\lambda_\alpha\}$ and open embeddings $\varphi_\alpha\colon W_\alpha \times R \to p^{-1}(W_\alpha) \cap E^0$, $W_\alpha = \lambda_\alpha^{-1}(0, 1]$, such that $p\varphi_\alpha = p_\alpha$ and $\varphi_\alpha s_\alpha = s$ as above. Define $f_\alpha\colon W_\alpha \to R$ by $f_\alpha(x) = pr_2\varphi_\alpha(x, 1)$. Then $f_\alpha$ is a continuous positive function and so is $f = \sum \lambda_\alpha f_\alpha\colon X \to R$. Then on the one hand, $f$ is a constant function, and on the other hand $\{(x, r) \in X \times R \mid r \leqq f(x)\}$ must be in $E^0$. Contradiction.

We will say that a neighborhood $E^0$ of $sX$ in the numerable microbundle $\mu$: $X \xrightarrow{s} E \xrightarrow{p} X$ is a microbundle neighborhood if $X \to E^0 \xrightarrow{p} X$ is a numerable microbundle. Two numerable microbundles $\mu_i: X \xrightarrow{s_i} E_i \xrightarrow{p_i} X$, $i = 1, 2$, are called equivalent if there exist microbundle neighborhoods $E_i^0$ of $s_i X$ and a homeomorphism $\varphi$ of $E_1^0$ onto $E_2^0$ such that $\varphi s_1 = s_2$ and $p_2 \varphi = p_1$.

The fact that every microbundle neighborhood $E^0$ of $sX$ is not a microbundle neighborhood leads to some technical complications. For example, it is not immediately obvious that the above definition is actually an equivalence relation. If $X$ is paracompact, every neighborhood of $sX$ is a microbundle neighborhood and these problems disappear.

An $R^n$-bundle with 0-section is a microbundle. Since the category of numerable $R^n$-bundles and the category of numerable $R^n$-bundles with 0-section are equivalent we will always assume our $R^n$-bundles have 0-section. Equivalent $R^n$-bundles define equivalent $R^n$-microbundles.

In **[H2]** it is claimed that every numerable $R^n$-microbundle contains a numerable $R^n$-bundle. Unfortunately this is false if $X$ is not Hausdorff, as the following example shows. (The difficulty in Per Holm's argument is that in 1.9 of **[H2]** and in §2 he is always working inside a bundle, whereas in applying these results to the main theorem he has to work inside a microbundle.)

EXAMPLE 2. Let $X$ and $E^0 \subset X \times R$ be as in Example 1. Let $\tilde{E}$ be the space obtained by taking two disjoint copies of $X \times R$ and identifying corresponding points of $E^0$. To be explicit, let $\bar{R}$ be another copy of $R$, then $\tilde{E} = X \times R \coprod X \times \bar{R}/\sim$, $(x_i, r_i) \sim (x_j, r_2)$ if and only if $i = j$ and $r_1 = r_2 \in (-1/i, 1/i)$. Define $\tilde{s}$: $X \to \tilde{E}$ by $\tilde{s}(x) = (x, 0)$ and $\tilde{p}$: $\tilde{E} \to X$ by $\tilde{p}(x, r) = x$, $\tilde{p}(x, \bar{r}) = x$. Let $\tilde{\pi}: \tilde{E} \to X \times R$, $\tilde{\pi}(x, r) = (x, r)$, $\tilde{\pi}(x, \bar{r}) = (x, r)$. Since $X \times R \coprod X \times \bar{R} \to_{/\sim} \tilde{E} \to_{\tilde{\pi}} X \times R$ is a homeomorphism on each factor and $\tilde{E}$ has the quotient topology, the inclusion $i$: $X \times R \to \tilde{E}$ and $\bar{i}$: $X \times \bar{R} \to \tilde{E}$ are homeomorphisms onto open subsets. Now let $E \subset \tilde{E} \times I$ be $\{((x, r), t) \in \tilde{E} \times I \mid t \neq 1\} \cup \{((x, \bar{r}), t) \in \tilde{E} \times I \mid t \neq 0\}$. Define $s$: $X \times I \to E$ by $s = \tilde{s} \times \mathrm{id}$, and $p$: $E \to X \times I$ by $p = \tilde{p} \times \mathrm{id} \mid E$. Then over $X \times [0, \frac{3}{4})$ we have the trivialization $\varphi_1$: $X \times [0, \frac{3}{4}) \times R \to E$, $\varphi_1(x, t, r) = ((x, r), t)$, and over $X \times (\frac{1}{4}, 1]$ the trivialization $\varphi_2 : X \times (\frac{1}{4}, 1] \times R \to E$, $\varphi_2(x, t, r) = ((x, \bar{r}), t)$. Thus $E$ is a microbundle. By taking $\lambda_1, \lambda_2$: $X \times I \to [0, 1]$, $\lambda_1(x, t) = \max(0, \frac{3}{4} - t)$, $\lambda_2(x, t) = \max(0, t - \frac{1}{4})$ and normalizing, we get a trivializing partition of unity. Hence $E$ is numerable.

Now if $E$ contains a numerable line bundle, this bundle must clearly be orientable and hence trivial. Let $\varphi$: $X \times I \times R \to E$ be the embedding (preserving projection and 0-section). Now as in Example 1, $f_0(x) = pr_3 \varphi_1^{-1} \varphi(x, 0, 1)$ and $f_1(x) = pr_3 \varphi_2^{-1} \varphi(x, 1, 1)$ must be constant functions. Let $\pi$: $E \to X \times I \times R$, $\pi((x, r), t) = (x, t, r)$, $\pi((x, \bar{r}), t) = (x, t, r)$. Then $f(x, t) = pr_3\, \pi \varphi(x, t, 1)$ is continuous and positive and constant on each slice $X \times t$, $t \in I$. Further $f_0(x) = f(x, 0)$, $f_1(x) = f(x, 1)$. For any $i$, $\varphi(x_i, 0, 1) = ((x_i, f_0(x_i)), 0)$, $\varphi(x_i, 1, 1) = ((x_i, \overline{f_1(x_i)}), 1)$ and $\varphi(x_i, t, r) = ((x_i, f(x_i, t)), t)$ or $((x_i, \overline{f(x_i, t)}), t)$. Since $\varphi$ is continuous we must have $f(x_i, t) \in (-1/i, 1/i)$ for some $t \in I$. Since $f(x, t)$ is independent of $x$, say $f(x, t) = g(t)$, we have $|g(t)| < 1/i$ for some $t = t_i$, each $i$. Since $g$ is continuous, there exists a $t_0 \in I$ such that $g(t_0) = 0$. Hence $f(x, t_0) \equiv 0$, contradicting the fact that $\varphi$ is a homeomorphism.

Of course the base space $X$ of Example 2 is not Hausdorff. We will show (in particular) that every numerable $R^n$-microbundle with Hausdorff base space contains a numerable $R^n$-bundle as a microbundle neighborhood of $s(X)$, unique up to equivalence. The argument will be only a slight modification of Per Holm's argument.

DEFINITION 2. A *G-$R^n$ microbundle* is a microbundle $\mu: X \xrightarrow{s} E \xrightarrow{p} S$ such that $X$ and $E$ are $G$-spaces and $s$ and $p$ are $G$-maps. $\mu$ is called *G-locally linear* if there exists a $G$-atlas $(V_\alpha, H_\alpha)$ on $X$ such that we have a commutative diagram

$$\begin{array}{ccccc} G \times V_\alpha & \xrightarrow{s_\alpha} & G \times_{H_\alpha} (V_\alpha \times R^n) & \xrightarrow{p_\alpha} & G \times_{H_\alpha} V_\alpha \\ \downarrow{\scriptstyle \psi_\alpha} & & \downarrow{\scriptstyle \varphi_\alpha} & & \downarrow{\scriptstyle \psi_\alpha} \\ GV_\alpha & \xrightarrow{s} & p^{-1}(GV_\alpha) & \longrightarrow & GV_\alpha \end{array}$$

where $p_\alpha$ is the $G$-vector bundle $1^{H_\alpha}_{\rho_\alpha}(V_\alpha)$, some representation $\rho_\alpha: H_\alpha \to O_n$ and $\varphi_\alpha$ is a $G$-homeomorphism onto an open subset.

$\mu$ is called *numerable* if there exists a $G$-partition of unity subordinate to $\{(H_\alpha, V_\alpha)\}$.

If $E^0$ is a $G$-invariant neighborhood of $sX$ in $E$, $X \xrightarrow{s} E^0 \xrightarrow{p} X$ is again a $G$-microbundle. $E^0$ will be called a *G-microbundle neighborhood* of $sX$ if it is numerable.

Two numerable locally linear $G$-microbundles $\mu_i: X \xrightarrow{s_i} E \xrightarrow{p_i} X$, $i = 1, 2$, are called *equivalent* if there exist $G$-microbundle neighborhoods $E_i^0$ of $s_i X$ and a $G$-homeomorphism $\varphi$ of $E_1^0$ onto $E_2^0$ such that $\varphi s_1 = s_2$ and $p_2 \varphi = p_1$.

Every numerable locally linear $G$-$R^n$ bundle is a numerable locally linear $G$-microbundle and $G$-equivalent bundles define $G$-equivalent microbundles.

In order to avoid excessive notation, a $G$-$R^n$ microbundle will mean, for the remainder of §2, a numerable locally linear $G$-$R^n$ microbundle with Hausdorff base space.

THEOREM 1. *G-equivalence of G-$R^n$ microbundles is an equivalence relation.*

For the proof it will be useful to make the

DEFINITION 3. Let $\mu_i: X_i \xrightarrow{s_i} E_i \xrightarrow{p_i} X_i$, $i = 1, 2$, be two $G$-$R^n$ microbundles. A *G-microbundle embedding* $\varphi: E_1 \to E_2$ over $\psi: X_1 \to X_2$ is a $G$-embedding $\varphi: E_1 \to E_2$ such that $s_2 \psi = \varphi s_1$ and $\psi p_1 = p_2 \varphi$. Two $G$-microbundle embeddings $\varphi'$: $E_1' \to E_2$, $\varphi''$: $E_1'' \to E_2$ of $G$-microbundle neighborhoods of $s_1 X$ in $E_1$ into $E_2$ over $\psi: X_1 \to X_2$ are in the same *germ* if they agree on a neighborhood of $s_1 X$.

The key lemma is:

LEMMA 2. *Let $\mu: X \xrightarrow{s} E \xrightarrow{p} X$ be a G-$R^n$ microbundle and $\varphi_i: G \times_{H_i} (V_i \times R^n) \to E$, $i = 1, 2$, be G-microbundle embeddings over $\psi_i$: $G \times_{H_i} V_i \to GV_i = X$, $\psi_i(g, v_i) = gv_i$, $v_i \in V_i \subset X$. If* Im $\varphi_1$ *is open, there is a G-microbundle embedding $\varphi: G \times_{H_1} (V_1 \times R^n) \to G \times_{H_1} (V_1 \times R^n)$ over the identity such that*:

(1) Im $\varphi$ *is open,*

(2) Im$(\varphi_1 \varphi) \subset$ Im $\varphi_2$,

(3) *$\varphi$ is in the germ of the identity and $\varphi$ is G-isotopic to the identity through open G-microbundle embeddings in the identity germ. (As usual, $H_i$ acts in $R^n$ through $\rho_i: H_i \to O_n$.)*

PROOF. Let $x \in X$, and let $D^n$ be the unit disc in $R^n$. For any $(g_1, v_1) \in G \times V_1$ with $g_1 v_1 = x$, $\varphi_1 [g_1, v_1 \times D^n] \subset p^{-1}(x)$ depends only on $x$ and we denote it by $\varphi_{1x}(D^n)$. Similarly for $\varphi_2$. Let $p_1: G \times_{H_1} (V_1 \times R^n) \to G \times_{H_1} V_1$ be the projection.

For any $(g_1, v_1)$ with $g_1v_1 = x$, the map $\pi_{(g_1,v_1)} : p_1^{-1}[g_1, v_1] = [g_1, v_1 \times R^n] \to R^n$, $\pi_{(g_1,v_1)}[g_1, (v_1, y)] = y$ is well defined up to an orthogonal transformation. Hence $\min(1, \text{radius of } \pi_{(g_1,v_1)}(\varphi_1^{-1}\varphi_2(D^n)))$ depends only on $x$ and will be denoted $r(x)$.

Note that since $\varphi_1$ is open, $\varphi_1[g_1, v_1 \times R^n]$ is open in $p^{-1}(x)$ and $\varphi_1[g_1, v_1 \times R^n] \cap \varphi_{2x}(\mathring{D}^n)$ is homeomorphic to an open set in $R^n$. By invariance of domain, $\pi_{(g_1,v_1)}\varphi_1^{-1}\varphi_{2x}(\mathring{D}^n)$ is open in $R^n$, and it contains $0 \in R^n$. The radius is the distance from 0 to the closed complement; so $r(x) > 0$.

We claim $r(x)$ is continuous. First it is sufficient to check this on the components of $X$. A component $X_0$ of $X$ is of the form $g_1V_1' = g_2V_2'$, $V_1'$, $V_2'$ components of $V_1$, $V_2$ respectively, $g_1, g_2$ fixed. Thus $p_i^{-1}\psi_i^{-1}(X_0) = [g_i, V_1' \times R^n] \simeq X_0 \times R^n$, and we can consider the embeddings $\varphi_i|\psi_i^{-1}(X_0)$ to be microbundle embeddings $\varphi_i: X_0 \times R^n \to p^{-1}(X_0)$. Let $x \in X_0$; then $\varphi_{1x}(D(s)) \subset \varphi_{2x}(D^n)$, where $D(s)$ is the disc of radius $s < r(x)$ centered at $0 \in R^n$. Let $B(s) = \varphi_{2x}^{-1}\varphi_{1x}(D(s)) \subset R^n$. Then $B(s)$ is a compact neighborhood of $0 \in R^n$ homeomorphic to $D(s)$. Since $\varphi_1$ is open, $\varphi_{2x'}(B(s)) \subset \varphi_{1x'}(R^n)$ for $x'$ near $x$. Hence by the continuity of $\varphi_1^{-1} \circ \varphi_2|B(s)$, $r(x') \geqq r(x) - \varepsilon$ for $x'$ near $x$.

Since $\varphi_1$ is open, $\varphi_{1x}(R^n) \cap \varphi_{2x}(R^n)$ is open in $\varphi_{2x}(R^n)$ and hence open in $\varphi_{1x}(R^n)$. Thus $\varphi_{1x}^{-1}\varphi_{2x}(R^n)$ is an open neighborhood of $D(r(x))$. Suppose there exist $x'$ arbitrarily near $x$ such that $r(x') > r(x) + \varepsilon$. Then for any $g \in D(r(x) + \varepsilon) - D(r(x))$, $\varphi_{1x'}(y) \in \varphi_{2x'}(D^n)$. Choose such a $y$ so that $y \in \varphi_{1x}^{-1}\varphi_{2x}(R^n)$. By choosing $x'$ in arbitrarily small neighborhoods of $x$ we get a set $\{\varphi_{1x'}(y)\} \in \varphi_2(X_0 \times R^n)$. Since $\{\varphi_{2x'}^{-1}\varphi_{1x'}(y)\} \subset D^n$ has a convergent subnet and $\{x'\} \subset X_0$ converges to $x$, we can assume $\varphi_{1x'}(y)$ converges to a point in $\varphi_2(x \times D^n)$. On the other hand, $\{\varphi_{1x'}(y)\}$ converges to $\varphi_{1x}(y) \in \varphi_2(x \times R^n)$. But $\varphi_{1x}(y) \notin \varphi_2(x \times D^n)$. Since $X_0 \times R^n$ and hence $\varphi_2(X_0 \times R^n)$ is Hausdorff, this is a contradiction. Hence $r(x') \leqq r(x) - \varepsilon$ for $x'$ near $x$, and $r$ is continuous.

Thus $r(x)$ is well defined, continuous, positive and $G$-invariant.

Define $\gamma_x: [0, \infty) \to [0, \infty)$ to be the homeomorphism which sends $[r(x)/2, 1]$ linearly onto $[r(x)/2, r(x)]$, $[1, 2]$ linearly onto $[r(x), 2]$ and is the identity on $[0, r(x)/2]$ and $[2, \infty)$. Then $\gamma_x$ and $\gamma_x^{-1}$ depend continuously on $x$. Let $\bar{\gamma}_x$ be the homeomorphism of $R^n$ given by $\bar{\gamma}_x(t, z) = \gamma_x(t)z$, $t \in [0, \infty)$, $z \in S^{n-1}$. Also let $\mu: [0, \infty) \to [0, 1)$ be a homeomorphism which is the identity on $[0, \frac{1}{2}]$, and $\bar{\mu}: R^n \to D^n$ the homeomorphism $\bar{\mu}(tz) = \mu(t)z$. Define $\hat{\gamma}: G \times_{H_1}(V_1 \times R^n) \to G \times_{H_1}(V_1 \times R^n)$ by $\hat{\gamma}[g, (v_1, y)] = [g, (v_1, \bar{\gamma}_x(y))]$, $x = gv_1$, and $\hat{\mu}: G \times_{H_1}(V_1 \times R^n) \to G \times_{H_1}(V_1 \times \mathring{D}^n)$ by $\hat{\mu}[g, (v_1, y)] = [g, (v_1, \bar{\mu}(y))]$. Then $\hat{\gamma}$ and $\hat{\mu}$ are homeomorphisms. Let $\varphi = \hat{\gamma} \circ \hat{\mu}$. Since $\mu$ is open, $\varphi$ is open. By construction $\operatorname{Im} \varphi_1\varphi \subset \operatorname{Im} \varphi_2$.

Since $\gamma_x\mu$ is the identity in $[0, r(x)/2]$, $\varphi$ is the identity on the neighborhood $\{[g, (v_1, y)] | y \in D^n(r(gv_1)/2)\}$ of the 0-section. Since $\mu$ is isotopic to the identity rel $[0, \frac{1}{2}]$ and $\lambda_x$ is continuously $G$-isotopic to the identity rel $[0, r(x)/2]$, $\varphi$ is $G$-isotopic to the identity through open $G$-microbundle embeddings in the identity germ.

REMARK. We only used the fact that $X$ was Hausdorff in proving $r(x)$ continuous. If $\mu$ were actually a numerable $G$-$R^n$ bundle, no hypothesis on $X$ is needed. In fact, as above, the argument comes down to looking at two $G$-embeddings $\varphi_i: X_0 \times R^n \to p^{-1}(X_0)$. By cutting down $X_0$, we can assume $p^{-1}(X_0) = X_0 \times R^n$. Thus we have two families of embeddings $\varphi_{ix}: R^n \to R^n$. It is then obvious from the continuity of $\varphi_i$, $i = 1, 2$, that $r(x)$ is continuous.

COROLLARY 3. *If* $\varphi: G \times_H (V \times R^n) \to G \times_H (V \times R^n)$ *is a G-microbundle embedding*, $\operatorname{Im} \varphi$ *is a neighborhood of the* 0-*section.*

COROLLARY 4. *If* $E^0$ *is a G-microbundle neighborhood of the* 0-*section in* $G \times_H (V \times R^n)$, *there is a G-microbundle embedding* $\varphi: G \times_H (V \times R^n) \to G \times_H (V \times R^n)$ *in the identity germ such that* $\operatorname{Im} \varphi \subset E^0$.

COROLLARY 5. *If* $E^0$ *is a G-microbundle neighborhood of sX in the G-microbundle* $\mu: X \xrightarrow{s} E \xrightarrow{p} X$, *then* $E^0$ *has local trivializations which are open in E and hence define local trivializations of E.*

COROLLARY 6. *Let* $\mu_i: X \xrightarrow{s_i} E_i \xrightarrow{p_i} X$, $i = 1, 2$, *be G-microbundles, and* $E_1^0$ *a G-microbundle neighborhood of* $s_1X$ *in* $E_1$. *If* $\varphi: E_1^0 \to E_2$ *is a G-microbundle embedding, then* $\varphi(E_1^0)$ *is a G-microbundle neighborhood of* $s_2X$ *in* $E_2$ *and hence* $\varphi$ *defines a G-microbundle equivalence.*

Corollary 3 follows trivially. Corollaries 4, 5, 6 follow by first reducing the problem to a local trivialization.

PROOF OF THEOREM 1. $G$-equivalence is clearly reflexive and symmetric; we show it is transitive. Let $E_i$, $i = 1, 2, 3$, be $G$-microbundles over $X$, and $\varphi': E_1' \to E_2'$ and $\varphi'': E_2'' \to E_3''$ be $G$-equivalences. We may choose a $G$-partition of unity $\{\lambda_\alpha\}$ which is trivializing for both $E_1'$ and $E_2''$. Let $W_\alpha = \lambda_\alpha^{-1}(0, 1]$, and $\varphi_\alpha': G \times_{H_\alpha} (V_\alpha' \times R^n) \to E_1'/W_\alpha$ and $\varphi_\alpha'': G \times_{H_\alpha''} (V_\alpha'' \times R^n) \to E_3''/W_\alpha$ be the trivializations. Then $\varphi_1 = \varphi' \circ \varphi_\alpha'$ and $\varphi_2 = (\varphi'')^{-1} \circ \varphi_\alpha''$ are two trivializations of $E_2/W_\alpha$. By Corollary 5, we can assume $\varphi_1$ is open. By Lemma 2, there exists $\varphi_2': G \times H_\alpha'(V_\alpha' \times R^n) \hookleftarrow$ such that $\operatorname{Im} \varphi_1\varphi_\alpha \subset \operatorname{Im} \varphi_2$ or $\operatorname{Im} \varphi' \circ \varphi_\alpha \subset E_2''$. Now $E_1^0 = \bigcup_\alpha \operatorname{Im} \varphi_\alpha \subset E_1'$ is a microbundle neighborhood of $s_1X$ and $\varphi'(E_1^0) \subset E_2''$. Hence $\varphi'' \circ \varphi': E_1^0 \to E_3$ is a $G$-microbundle embedding and defines a $G$-equivalence by Corollary 6. Q.E.D.

REMARK. The proof shows that the intersection of two $G$-microbundle neighborhoods of the 0-section in a $G$-microbundle contains a $G$-microbundle neighborhood.

The key result needed to prove that an $R^n$-microbundle contains an $R^n$-bundle is Kister's theorem that there is a continuous map $\lambda: \operatorname{Emb}(R^n, R^n)_0 \to \operatorname{Top}(n)$ from the space of embeddings of $(R^n, 0)$ in $(R^n, 0)$ into the space of homeomorphisms of $(R^n, 0)$, with the $C$-$O$ topology such that $\lambda(f)|D^n = f|D^n$. The Alexander trick then implies that $\operatorname{Top}(n)$ is a deformation retract of $\operatorname{Emb}(R^n, R^n)_0$ rel $D^n$. We need a $G$-version of this.

Let $\rho_i: G \to O(n)$, $i = 1, 2$, be representations. Define an action of $G$ on $\operatorname{Emb}(R^n, R^n)_0$ by $g(f) = \rho_2(g)^{-1} f \rho_1(g)$. Then $\operatorname{Emb}(R^n, R^n)_0^G = \{f \in \operatorname{Emb}(R^n, R^n)_0 | f\rho_1(g) = \rho_2(g) f\}$.

THEOREM 7 (G-KISTER THEOREM). *There is a continuous map* $\lambda: \operatorname{Emb}(R^n, R^n)_0^G \to \operatorname{Top}(n)^G$ *such that* $\lambda(f)|D^n = f|D^n$.

The proof given in **[K1]** for Kister's theorem goes over without change to give the $G$-Kister theorem since all the radial contractions and expansions used in the proof commute with the orthogonal actions of $G$.

Let $\operatorname{Homeo}(D^n, \partial)$ be the space of homeomorphisms of $D^n$ fixed on the boundary with the $C$-$O$ topology. The Alexander trick proves that this space is contractible to the identity. Exactly the same argument proves:

THEOREM 8 ($G$-ALEXANDER TRICK). *Homeo*$(D^n, \partial)^G$ *is contractible to the identity.*

Theorems 7 and 8 then give

COROLLARY 9. $\mathrm{Top}(n)^G$ *is a deformation retract of* $\mathrm{Emb}(R^n, R^n)_0^G$ rel $D^n$.

THEOREM 10. *Let $\mu$ be a numerable locally linear $G$-$R^n$ microbundle over a Hausdorff space $X$. Let $A$ be a $G$-invariant subspace of $X$ and $W$ a $G$-halo of $A$. Then*

(a) *If $\eta$ is a numerable locally linear $G$-$R^n$ bundle over $W$ contained in $\mu|W$ there is a numerable locally linear $G$-$R^n$ bundle $\mathscr{E}$ in $\mu$ such that $\mathscr{E}|A = \eta|A$.*

(b) *If $\mathscr{E}_1$, $\mathscr{E}_2$ are numerable locally linear $G$-$R^n$ bundles in $\mu$ and $\mathscr{E}_1|W = \mathscr{E}_2|W$, there is a $G$-bundle equivalence $\varphi\colon \mathscr{E}_1 \simeq \mathscr{E}_2$ such that $\varphi|A$ is the identity. Further the inclusion of $\mathscr{E}_1$ in $\mu$ is $G$-isotopic through $G$-microbundle embeddings* rel $A$ *to $\varphi$.*

The theorem follows from Lemma 2 (and corollaries) and the $G$-Kister Theorem just as in **[H2]**, with the obvious modifications necessary to take the $G$-action into account. We will not repeat the details here.

THEOREM 11. *If $\mathscr{E}$ is a numerable locally linear $G$-$R^n$ bundle then there is a $G$-homotopy $H_t\colon \mathscr{E} \to \mathscr{E}$, $H_0 =$ identity, $H_1 =$ projection of $\mathscr{E}$ into the* 0*-section, and $H_t$ a $G$-microbundle embedding for $t < 1$.*

This follows from Theorem 10 (b) just as in **[H2]**. Because we are always inside a $G$-$R^n$ bundle, it is not necessary to assume $X$ is Hausdorff. (See remark following Lemma 2.)

DEFINITION 4. Let $\mu_i\colon X_i \xrightarrow{s_i} E_i \xrightarrow{p_i} X_i$, $i = 1,2$, be two $G$-$R^n$ microbundles. A $G$-microbundle map $\varphi\colon \mu_1 \to \mu_2$ is a $G$-map $\varphi : E_1^0 \to E_2$, $E_1^0$ a $G$-microbundle neighborhood of $s_1(X_1)$, commuting with 0-section and projection, such that with $\psi\colon X_1 \to X_2$ the $G$-map defined by $\varphi$, the induced map $\varphi_*\colon E_1^0 \to \psi^* E_2$ is a $G$-microbundle embedding (and hence an equivalence). Two such are identified if they agree on some $G$-microbundle neighborhood of $s_1(X_1)$.

More generally, we will say that a $G$-map $\varphi\colon E_1^0 \to E_2$, commuting with 0-section and projection, *defines a $G$-microbundle map* if there is a $G$-microbundle neighborhood $E_1' \subset E_1^0$ such that $\varphi|E_1'$ is a $G$-microbundle map.

Note that a $G$-microbundle map is an embedding in each fibre. A partial converse is:

LEMMA 12. *If either* (1) *$\mu_2$ above is a numerable locally linear $G$-$R^n$ bundle, or* (2) *$X_1$ is paracompact, then a $G$-map $\varphi\colon E_1^0 \to E_2$, commuting with* 0*-section and projection, which is an embedding in each fibre, defines a $G$-microbundle map.*

PROOF. (1) The problem reduces to the case where $E_1^0 = X \times R^n$, $\psi^* E_2 = X \times R^n$ and $\varphi_*\colon X \times R^n \to X \times R^n$ is an embedding in each fibre. By Theorem 7, there is a $G$-$R^n$ bundle equivalence $\lambda(\varphi_*)\colon X \times R^n \to X \times R^n$ such that $\lambda(\varphi_*)|X \times D^n = \varphi_*|X \times D^n$, i.e., $\varphi_*|X \times \mathring{D}^n$ is open.

(2) If $E_2^0$ is a $G$-$R^n$ bundle in $E_2$, then $\varphi^{-1}(E_2^0)$ is a $G$-microbundle neighborhood since $X_1$ is paracompact, and (2) follows from (1).

**3. Locally smooth $G$-manifold.**

DEFINITION 1. A $G$-manifold is a $G$-space $M$ such that $M$ is a paracompact topological manifold.

A *smooth* $G$-manifold is a $G$-manifold such that $M$ is smooth and the action $\mu: G \times M \to M$ is smooth. (Since $G$ is finite this just means $\mu_g: M \to M$ is smooth, $g \in G$.)

A *locally smooth* $G$-manifold is a $G$-manifold $M$ such that each $x \in M$ has a $G$-invariant neighborhood $U_x$ which admits a smoothing such that $\mu : G \times U_x \to U_x$ is smooth.

If $M$ is a locally smooth $G$-manifold and $V \subset M$ is a topological submanifold, $V$ is called a *locally smooth submanifold* if for each $x \in V$ there is a $G$-invariant neighborhood $U_x$ of $x$ in $M$ admitting a smoothing on which $G$ acts smoothly and such that $U_x \cap V$ is a smooth submanifold of $U_x$. Note in particular, $V$ is then a locally flat submanifold of the topological manifold $M$.

The following can be found in **[B1]**:

LEMMA 1. *Let $M$ be a (locally) smooth $G$-manifold, then $M^G$ is a (locally) smooth submanifold.*

LEMMA 2 (SLICE THEOREM). *Let $M$ be a (locally) smooth $G$-manifold and $x \in M$. Then the orbit $Gx$ has a $G$-invariant neighborhood isomorphic to a vector bundle $G \times_{G_x} V_x$. That is, $V_x$ is a Euclidean space embedded in $M$ so that $x$ corresponds to the origin, $G_x$ acts orthogonally on $V_x$, and $\psi_x: G \times_{G_x} V_x \to M$, $\psi_x [g, v] = gv$, is a (homeomorphism) diffeomorphism onto an open neighborhood of $Gx$.*

If $V_x$ is as above, $V_x$ is called a *smooth slice* and $GV_x$ is called a *smooth normal tube* about $Gx$.

A $G$-manifold $M^n$ is locally smoothable if and only if there is a $G$-atlas $\{(V_\alpha, H_\alpha)\}$ such that $V_\alpha = R^n$ and $H_\alpha$ acts orthogonally.

THEOREM 3 ($G$-TUBULAR NEIGHBORHOOD THEOREM). *Let $M$ be a (locally) smooth $G$-manifold and $x \in M$. Any two smooth normal tubes of $Gx$ are isotopic. That is, if $\psi_i: G \times_{G_x} V_x^i \to M$, $i = 1, 2$, are two smooth tubes about $Gx$, then $\psi_1$ is $G$-(isotopic) diffeotopic* rel $Gx$ *to a ($G$-$R^n$ bundle) $G$-vector bundle map $\psi_1'$ onto $\psi_2 (G \times_{G_x} V_x^2)$. (That is, $\psi_2^{-1}\psi_1'$ is a bundle map.)*

In the smooth case this is proved in **[B1]**, the locally smooth case is a consequence of

PROPOSITION 4. *Let $X$ be any completely regular $G$-space and $x \in X$. Any two tubes about $Gx$ have isomorphic germs. That is, if $\psi_i: G \times_{G_x} V_x^i \to X$, $i = 1, 2$, are two tubes about $Gx$, we can replace $V_x^1$ by a sufficiently small $G_x$-invariant open neighborhood of $x$ in $V_x^1$ so that $\psi_1$ becomes an open $G$-embedding*

$$\psi_1: G \times_{G_x} V_x^1 \to \psi_2 (G \times_{G_x} V_x^2)$$

*and $\psi_2^{-1}\psi_1$ commutes with projection onto $G/G_x$.*

PROOF. Since $G$ is finite, $V_x^i$ is a neighborhood of $x$. Replace $V_x^1$ by Int $V_x^1 \cap$ Int $V_x^2$. Then $\psi_1$ becomes an open $G$-embedding into $\psi_2$ $(G \times_{G_x} V_x^2)$. Further $\psi_2^{-1}\psi_1$ commutes with projection since it is a $G$-map.

PROOF OF THEOREM 3. Since $V_x^1 = R^n$, we can choose a small open disc $D^n$ contained in the $G_x$ invariant neighborhood of $x$ satisfying the conclusions of Proposition 4. Since $R^n$ may be radially deformed into $D^n$ rel $0 \in R^n$, the identity map of $G \times_{G_x} V_x'$ may be deformed via $G$-microbundle embeddings to a $G$-homeomorphism $\psi_1^0$ into $G \times_{G_x} D^n$ rel $Gx$. Hence by the $G$-Kister Theorem of §2, $\psi_2^{-1}\psi_1^0$ is $G$-isotopic through $G$-embeddings rel $Gx$ to a $G$-$R^n$ bundle map.

REMARK. If $M$ is a (locally) smooth $G$-manifold with boundary, and $x \in \partial M$, then a smooth slice $V_x = H^n = R^{n-1} \times [0, \infty]$ with $Gx$ acting orthogonally on $R^{n-1}$ and trivially on $[0, \infty)$. If we define an $H^n$ bundle to be a $\mathrm{Top}_{n-1}$ bundle with fibre $H^n$, $\mathrm{Top}_{n-1}$ acting trivally on $[0, \infty)$, then the analogue of Theorem 3 holds, where if $x \in \partial M$ we substitute $G$-$H^n$ bundle map for $G$-$R^n$ bundle map and $G$-$O_{n-1}$ bundle map for $G$-vector bundle map. The proof requires straightening out a local $G$-collar. This is done by applying Brown's argument in the orbit space. This technique also leads to the global $G$-collaring theorem below.

DEFINITION 2. Let $M$ be a (topological) smooth $G$-manifold with boundary. A *$G$-collar* of $\partial M$ in $M$ is a (topological) smooth $G$-embedding $\varphi: \partial M \times [0, 1) \to M$ where $\varphi(x, 0) = x$, $x \in \partial M$, and $G$ acts trivially on $[0, 1)$.

THEOREM 5 (BREDIN [**B1**]). *If $M$ is a (locally) smooth $G$-manifold with boundary, $\partial M$ has a $G$-collar in $M$, unique up to $G$-(isotopy) diffeotopy.*

REMARK. It follows that any (locally) smooth $G$-manifold $M$ with boundary can be extended to a (locally) smooth $G$-manifold $M$ without boundary by adding an exterior $G$-collar.

If $M$ is a $G$-manifold its tangent $G$-microbundle is $\tau M$: $M \xrightarrow{\Delta} M \times M \xrightarrow{pr_1} M$, where $G$ acts on $M \times M$ by the diagonal action.

THEOREM 6. *A $G$-manifold $M^n$ (without boundary) is a locally smooth $G$-manifold if and only if its tangent $G$-microbundle is a numerable locally linear $G$-$R^n$ microbundle.*

PROOF. If the tangent microbundle of a $G$-manifold is locally linear we have for each chart $(V, H)$ of some atlas the commutative diagram:

$$\mathrm{D}_1 \qquad \begin{array}{ccccc} G \times_H V & \xrightarrow{s} & G \times_H (V \times R^n) & \xrightarrow{p} & G \times_H V \\ \downarrow \psi & & \downarrow \varphi & & \downarrow \psi \\ GV & \xrightarrow{\Delta} & GV \times M & \xrightarrow{pr_1} & GV \end{array}$$

where $\varphi$ is an open $G$-embedding.

Since $M$ is completely regular, for each $x \in V$ there is an $H$-slice $V_x$ in $V$. $V_x$ is also a $G$-slice and $G \times_H (H \times_{H_x} V_x) = G \times_{G_x} V_x$ is a tube about $Gx$, where $G_x = H_x$ (see [**B1**]). Further, $p^{-1}(G \times_{G_x} V_x) \simeq G \times_{G_x} (V_x \times R^n)$. Thus $\mathrm{D}_1$ restricts to

$$\mathrm{D}_2 \qquad \begin{array}{ccccc} G \times_{G_x} V_x & \xrightarrow{s} & G \times_{G_x} (V_x \times R^n) & \xrightarrow{p} & G \times_{G_x} V_x \\ \downarrow \psi & & \downarrow \varphi & & \downarrow \psi \\ GV_x & \xrightarrow{\Delta} & GV_x \times M & \xrightarrow{pr_1} & GV_x \end{array}$$

Now $pr_2 \circ \varphi: G \times_{G_x} (x \times R^n) \to M$ is a $G$-embedding into an open neighborhood of

$x \in M$, and $G_x = H_x$ acts on $R^n$ by the restriction of the orthogonal action of $H$. Thus $M$ is locally smoothable.

Conversely, if $M$ is locally smoothable, let $\psi: G \times_H V \to M$ be a chart, $V = R^n$, $H$ acting orthogonally. Then $\tau M|GV$ contains $\tau(GV)$: $GV \xrightarrow{\Delta} GV \times GV \xrightarrow{pr_1} GV$. Define $\varphi: G \times_H (V \times R^n) \to GV \times GV$ by $\varphi[g, (v, y)] = (gv, gv + gy)$. With this definition of $\varphi$, $D_1$ above commutes and $\tau M$ is locally linear. Since $M$ is paracompact, $\tau M$ is numerable.

REMARK. If $M$ is a $G$-manifold with $G$-collared boundary, applying Theorem 6 to $\tilde{M}$ and $\partial M$ we get: A $G$-manifold $M$ with $G$-collared boundary is locally smooth if and only if both $\tau\tilde{M}$ and $\tau(\partial M)$ are numerable locally linear $G$-$R^n$ microbundles.

DEFINITION 3. If $X$ is a $G$-space and $q: X \to X/G = X^*$ is the quotient map, we define a *G-component* of $X$ to be $q^{-1}(X_0^*)$ where $X_0^*$ is a component of $X^*$.

$G$-components are closed sets. If $M$ is a locally smooth $G$-manifold then $M^*$ is locally connected and hence the $G$-components of $M$ are both open and closed.

DEFINITION 4. Let $X$ be a $G$-space and $x \in X$. The conjugacy class $(G_x)$ of $G_x$ is called the *type* of the orbit $Gx \simeq G/G_x$ The set. of all conjugacy classes of subgroups of $G$ is partially ordered as follows: $(H_1) \leqq (H_2)$ if some conjugate of $H_2$ is a subgroup of $H_1$. The union $X_{(H)}$ of all orbits of type $H$ is a $G$-invariant subspace of $X$. The saturation $X^{(H)} = GX^H$ of $X^H$ is a closed $G$-invariant subspace.

REMARKS. (1) $X^{(H)} = \{x \in X | (G_x) \leqq (H)\}$ and $X_{(H)} \subset X^{(H)}$.

(2) If $(K) \leqq (H)$, $X^{(K)} \subseteq X^{(H)}$.

(3) If $x \in X$ and $V_x$ is a slice, then for any $x' \in GV_x = \psi_x(G \times_{G_x} V_x)$, $(G_x) \leqq (G_{x'})$. Thus any $x'$ sufficiently near $x$ satisfies $(G_x) \leqq (G_{x'})$. (See **[B1]**.)

LEMMA 7. *Let $X$ be a completely regular G-space. For any $H \subset G$, $X_{(H)}$ is open in $X^{(H)}$.*

PROOF. If $x \in X_{(H)}$ and $V_x$ is a slice through $x$, let $x' \in (G \times_{G_x} V_x) \cap X^{(H)}$. Then $(G_{x'}) = (G_x)$ by the above remarks. Hence $X_{(H)}$ is open in $X^{(H)}$.

The following are proved in **[B1]**.

THEOREM 8. *Let $M$ be a (locally) smooth G-manifold. Then*

*(a) $M_{(H)}$ is a (locally) smooth G-submanifold of $M$.*

*(b) $M^H \cap M_{(H)}$ is a (locally) smooth submanifold of $M$.*

*Now let $\varphi: G \times_{N(H)} M^H \to M^{(H)}$, $\varphi[g, x] = gx$, $N(H)$ the normalizer of $H$.*

*(c) $\varphi$ sends $G \times_{N(H)} (M^H \cap M_{(H)})$ (homeomorphically) diffeomorphically onto $M_{(H)}$.*

*(d) $G \times_{N(H)} (M^H \cap M_{(H)}) \simeq G/H \times_{N(H)/H} (M^H \cap M_{(H)})$.*

*(e) $N(H)/H$ acts freely on $M^H \cap M_{(H)}$ and $M_{(H)}$ is a bundle over $M_{(H)}/G \simeq (M^H \cap M_{(H)})/(N(H)/H)$ with fibre $G/H$.*

THEOREM 9. *Let $M$ be a (locally) smooth G-manifold with $M^* = M/G$ connected. There exists a maximal orbit type $(H)$ for $M$. (That is, $(H) \geqq (G_x)$ for $x \in M$ and equality holds for some $x \in M$.) $M_{(H)}$ is open and dense in $M$ and $M_{(H)}/G$ is connected.*

LEMMA 10. *Let $M$ be a (locally) smooth G-manifold. Then*

*(a) $M^{(H)}/G$ is locally arcwise connected.*

*(b) The components of $M^{(H)}/G$ are open and arcwise connected.*

(c) *If $M_i^{(H)}$ is a G-component of $M^{(H)}$, then $M_i^{(H)} \cap M_{(H)}$ is both open and closed in $M_{(H)}$ and hence is a union of G-components of $M_{(H)}$.*

PROOF. (a) Let $x \in M^H$ and let $K = G_x$. Let $G \times_K V$ be a normal tube of $G_x$; i.e. $V$ is an orthogonal $K$-space through $x$. Then $K \supset H$. Note that $gM^H = M^{gHg^{-1}}$. Now $M^{(H)} \cap (G \times_K V) = G \times_K (V \cap M^{(H)})$. Hence if $V \cap M^{(H)}$ is arcwise connected, $M^{(H)}/G$ will be locally arcwise connected. But $V \cap M^{(H)} = \bigcup_g (V \cap gM^H)$; and $V \cap gM^H = V \cap M^{gHg^{-1}}$ is a linear subspace of $V$ if $gHg^{-1} \subset K$ and empty otherwise. Thus $V \cap M^{(H)}$ is a union of subspaces and hence arcwise connected.

(b) This is an immediate consequence of (a).

(c) Since $M^{(H)}$ is closed in $M$, $M_i^{(H)}$ is closed in $M$. Since $M_i^{(H)}$ is open in $M^{(H)}$ and $M^{(H)} \supset M_{(H)}$, $M_i^{(H)} \cap M_{(H)}$ is open and closed in $M_{(H)}$. Since $M_{(H)} \to M_{(H)}/G$ is both an open and closed map, (c) follows.

LEMMA 11. *Let $M^i_{(H)}$ be a G-component of $M_{(H)}$; then*

(a) *$M^i_{(H)}/G$ is a connected manifold.*

(b) *Let $\overline{M^i_{(H)}}$ be the closure of $M^i_{(H)}$ in M. Then $\overline{M^i_{(H)}} - M^i_{(H)}$ = union of some G-components of $M_{(K)}$'s, $(K) < (H)$.*

(c) *If $M^i_{(H)}$ is closed in M, then $M^i_{(H)}$ is a G-component of $M^{(H)}$.*

(d) *M is the disjoint union of the $M^i_{(H)}$ as (H) runs over the orbit types of M and i over the G-components of $M_{(H)}$.*

(e) *Only a finite number of the $M^i_{(H)}$ meet any compact set.*

PROOF. (a) This follows from Theorem 8.

(b) Let $x \in M^i_{(H)} - \overline{M^i_{(H)}}$; then $G_x = K$, $(K) < (H)$. Let $V$ be a linear slice through $x$ and let $y \in M^i_{(H)} \cap V$. Then $(G_y) = (H)$ and $G_y \subset K$. Let $z \in V^K$, then for $k \in K$, $k(z + y) = (z + y)$ if and only if $ky = y$. So $G_{z+y} = g_y$. Now $z + y \in M^i_{(H)} \cap V$ since $tz \in V^K, 0 \leqq t \leqq 1$. Since also $sy \in M^i_{(H)} \cap V, 0 < s \leqq 1, z + sy \in M^i_{(H)} \cap V$ and $z = z + 0 \cdot y \in \overline{M^i_{(H)}}$. Thus $V^K \subset \overline{M^i_{(H)}}$. Now $(G \times_K V)_{(K)} = G \times_K V^K$ and hence $(G \times_K V)_{(K)} \subset \overline{M^i_{(H)}}$.

It follows that $\overline{M^i_{(H)}} \cap M_{(K)}$ is open in $M_{(K)}$, $(K) < (H)$. Since it is also closed, it is the union of $G$-components.

(c) Under the hypothesis, $M^i_{(H)}$ is open and closed in $M^{(H)}$. Since it is $G$-connected, it is a $G$-component of $M^{(H)}$.

(d) This is immediate.

(e) We proceed by induction on dim $M$. The result is trivial for dim $M = 0$. Assume for dimensions $< n$ and suppose dim $M = n$. If $C$ is a compact set it is covered by a finite number of normal tubes. So it is sufficient to prove for $M = G \times_K V$ that $M_{(H)}$ has only a finite number of $G$-components. If $S$ is the unit sphere in $V$, let $M_0 = G \times_K S$. Now the number of components of $M_{(H)}$ equals the number of components of $(M_0)_{(H)}$, except possibly if $(K) = (H)$. But in this last case $M_{(H)}$ has just one $G$-component. Since $M_0$ is compact and dim $M_0 < n$, the result follows from the induction hypothesis.

Let $\{(H_j), j \in J\}$ be the orbit types of $M$, and for each $j \in J$, let $\{M^j_i, i \in I(j)\}$ be the $G$-components of $M^{(H_j)}$. Partially order the $\{M^j_i\}$ by inclusion. If $(K) \leqq (H)$, $M^{(K)} \subseteq M^{(H)}$ and the $G$-components of $M^{(K)}$ are contained in the $G$-components of $M^{(H)}$. Thus a minimal element of $\{M^j_i\}$ can be written in the form $M_i^{(H)}$, where $M_j^{(K)} \subseteq M_i^{(H)}$, $(K) \leqq (H)$, implies $(K) = (H)$ and $j = i$. Then $M_i^{(H)} \subset M_{(H)}$ and

$M_i^{(H)}$ is a $G$-component of $M_{(H)}$ which is a closed subset of $M$. Conversely, if $M^i_{(H)}$ is a $G$-component of $M_{(H)}$ which is a closed subset of $M$, then by Lemma 11(c) and (d), $M^i_{(H)}$ is a $G$-component of $M^{(H)}$ which is minimal in $\{M^i_i\}$.

THEOREM 12 (BIERSTONE). *Let $M$ be a (locally) smooth $G$-manifold. The minimal elements of the partially ordered (under inclusion) set $\{M^i_i\}$ are the topologically closed $G$-components of the $M_{(H_j)}$. In particular, they are (locally) smooth $G$-submanifolds.*

DEFINITION 5. Let $H \subset G$ and $\rho: H \to O(k)$ be a representation. Let $E(\rho)$ be the $G$-vector bundle $G \times_H R^k$ and $D(\rho) = G \times_H D^k$, $S(\rho) = G \times_H S^{k-1}$ the associated disc and sphere bundles. The Whitney sum $D(\rho_1) \oplus D(\rho_2)$, $\rho_1: H \to O(k)$, $\rho_2: H \to O(l)$ is called a *handle bundle* or *G-handle* of *index* $k$ and total dimension $k + l$ over $G/K$. ($D(\rho_1) \oplus D(\rho_2) = G \times_H (D^k \times D^l)$ where $H$ acts by the diagonal action, $h(y, z) = (\rho_1(h)y, \rho_2(h)z)$, $y \in D^k$, $z \in D^l$.)

Now let $M^n$ be a $G$-manifold and $N^n$ be a $G$-invariant submanifold in Int $M$. We will say that $M$ is obtained from $N$ by adding a $G$-handle of index $k$ if

$$M = \hat{N} \bigcup_f G \times_H (D^k \times D^{n-k}),$$

where $\hat{N} = N$ union a closed $G$-collar and $f$ is a $G$-embedding of $G \times_H (S^{k-1} \times D^{n-k})$ in $\partial\hat{N}$. $M^n$ is said to have a $G$-handle decomposition if $M = \bigcup M_i$, $M_i$ a $G$-invariant submanifold $M_i \subset \text{Int } M_{i+1}$, and $M_{i+1}$ is obtained from $M_i$ by adding a $G$-handle.

If $M$ is a smooth $G$-manifold, a $G$-handle decomposition will be called smooth if each $M_i$ is a smooth submanifold and the $G$-embedding $f_i$ of the boundary of the $G$-handle in $\partial\hat{M}_i$ is smooth.

THEOREM 13 (WASSERMAN **[W1]**). *Every smooth G-manifold has a smooth G-handle decomposition.*

Wasserman's proof uses $G$-invariant Morse functions. As pointed out by Bierstone **[B2]**, another approach is by induction up the orbit types. Let $M^i_{(H)}$ be a minimal element of $\{M^i_k\}$. $M^i_{(H)}/G$ is a smooth manifold and hence has a handle decomposition. This defines a $G$-handle decomposition of $M^i_{(H)}$; i.e., the preimage of a handle $D^p \times D^q \subset M^i_{(H)}/G$ is a $G$-handle $G/H \times D^p \times D^q \subset M^i_{(H)}$. Using a $G$-invariant normal tube of $M^i_{(H)}$ to thicken this up we get $G$-handles in $M$ covering a normal tube of $M^i_{(H)}$. Let $M_1 = M -$ minimal elements. Then $M_1$ is a smooth $G$-manifold and a minimal element of $M_1$ is of the form $M^i_{(K)}$. Now $M^i_{(K)}$ minus the open normal tubes surrounding the initial set of minimal elements is a $G$-manifold with boundary. (Note that the boundary of the normal tubes meets $M^i_{(K)}$ transversally.) Its $G$-quotient is a smooth $G$-manifold with boundary and hence has a handle decomposition built up from the boundary giving a $G$-handle decomposition of the $G$-manifold with boundary in $M^i_{(K)}$, which is again thickened up using a normal tube of $M^i_{(K)}$, etc. (At the higher levels we get smooth $G$-manifolds with corners, which have to be rounded.)

DEFINITION 6. A $G$-handle $G \times_H (D^k \times D^l)$ is called *good* if $H$ acting on $R^l$ leaves some one-dimensional subspace fixed; i.e., $\rho_2: H \to O(l)$ factors through some $O(l - 1) \subset O(l)$.

From the description above we see that good $G$-handles correspond to handles of index less than the dimension of the corresponding $M^j_{(K)}/G$. Thus to have a good $G$-handle decomposition it is necessary that the minimal elements of the $\{M^j_i\}$ be nonclosed manifolds. This turns out to be sufficient; i.e., that the manifolds with boundary that we successively construct in the $M^j_{(K)}/G$ also have handle decompositions in the boundary without handles of top index. Explicitly:

THEOREM 14 (BIERSTONE [**B2**]). *Let $M$ be a smooth $G$-manifold (with or without boundary). The following conditions are equivalent:*

(a) *All minimal elements of the partially ordered set $\{M^j_i\}$ are nonclosed (as manifolds).*

(b) *$M$ has a good smooth $G$-handle decomposition.*

(c) *All $M^i_{(H)}$ are nonclosed.*

DEFINITION 7. A (locally) smooth $G$-manifold is said to satisfy the Bierstone condition if (a) above holds.

REMARKS. 1. It follows from Lemma 11(c) that if $M$ is a compact smooth $G$-manifold then $M$ has a $G$-handle decomposition with only finitely many $G$-handles. The same holds for a good $G$-handle decomposition if $M$ also satisfies the Bierstone condition.

2. If $M$ satisfies the Bierstone condition, then so does any open $G$-submanifold.

3. If $M$ is any (locally) smooth $G$-manifold, then $M \times R$ (trivial $G$-action on $R$) satisfies the Bierstone condition. Also if we let $M_0 = M -$ one $G$-orbit from each minimal element of the $\{M^j_i\}$, then $M_0$ satisfies the Bierstone condition.

LEMMA 15. *If $M$ is a (locally) smooth $G$-manifold (without boundary) satisfying the Bierstone condition then for any $x_* \in M^i_{(H)}/G$ there is a proper map $f$: $[0, \infty] \to M/G$ such that $f(0) = x_*$ and $\mathrm{Im}(f) \subset M^i_{(H)}/G$.*

PROOF. If $M^i_{(H)}$ is a minimal element of $\{M^j_k\}$ then it is a closed subset of $M$ and cannot be contained in any compact set since it is a nonclosed manifold. Hence $M^i_{(H)}/G$ is not contained in any compact set of $M/G$. Since $M^i_{(H)}/G$ is a connected manifold it must be connected to $\infty$; i.e., for any $x_* \in M^i_{(H)}/G$ there is such an $f$.

Now for any $M^i_{(H)}$ which is not minimal, $\overline{M^i_{(H)}}$ contains an $M^j_{(K)}$ by Lemma 11(b). We can assume by induction that $M^j_{(K)}/G$ is connected to $\infty$. It follows that $M^i_{(H)}/G$ must also be connected to $\infty$.

**4. $G$-engulfing and $G$-isotopy extension theorems.**

DEFINITION 1. Let $M$ be a locally smooth $G$-manifold. A $G$-string ($G$-half-string) is a string (or half-string) $L$ is some $M_{(H)}/G$, i.e., a locally flat embedded line (or half-line).

Equivalently, a $G$-string (half-string) is a locally smooth $G$-invariant submanifold of $M$-orbit type preserving $G$-homeomorphic to $G/H \times L$, where $L$ is a line (or half-line), and $G$ acts trivially on $L$. That is, the submanifold must lie in $M_{(H)}$ and the quotient map $q: M_{(H)} \to M_{(H)}/G$ provides a one-to-one correspondence.

THEOREM 1. (a) *A $G$-string (or half-string) in $M^n$ has a trivial normal $G$-vector bundle, i.e., a neighborhood $G$-homeomorphic to $(G \times_H R^{n-1}) \times (-\infty, \infty)$, where $(G \times_H 0) \times (-\infty, \infty) = G/H \times L$ (or $(G \times_H 0) \times [0, \infty) = G/H \times L$) and $H$ acts orthogonally on $R^{n-1}$.*

(b) *A $G$-string (or half-string) which is closed in $M$ has a trivial normal $G$-disc* *bundle which is closed in $M$, i.e., a closed neighborhood $G$-homeomorphic to $(G \times_H D^{n-1}) \times (-\infty, \infty)$, where $(G \times_H 0) \times (-\infty, \infty) = G/H \times L$ (or $G$-homeomorphic to $(G \times_H D^{n-1}) \times [-1, \infty)$ where $(G \times_H 0) \times [0, \infty) = G/H \times L$).*

PROOF. (a) Since any half-string may always be extended to a string by the definition of locally flat, we may assume $L$ is a string in part (a).

Now a normal vector (disc) bundle neighborhood of a $G$-string in $M$ corresponds under the quotient map to an open (closed) cone bundle neighborhood of $L$ in $M/G$; the base of the cone being $S^{n-2}/H$. By Lacher's theorem **[L2]**, $L$ has a trivial normal bundle $R^{k-1} \times (-\infty, \infty)$ in $M_{(H)}/G$ where $k = \dim (M_{(H)}/G)_i$, $(M_{(H)}/G)_i$ the component containing $L$. This lifts to the $G$-normal bundle $G/H \times R^{k-1} \times (-\infty, \infty)$ of $G/H \times L$ in $M_{(H)}$. Since $M_{(H)}$ is a locally smooth $G$-submanifold of $M$, each $x \in M_{(H)}$ has a $G$-invariant neighborhood $U_x$ such that $U_x$ has a normal $G$-vector bundle in $M$. Thus each $t \in L$ has a $G$-invariant neighborhood in $M$ of the form

$$\left(G \underset{H}{\times} R^{n-1}\right) \times (t - \varepsilon, t + \varepsilon) = \left(G \underset{H}{\times} R^{n-k}\right) \times R^{k-1} \times (t - \varepsilon, t + \varepsilon).$$

Thus in $M/G$ each, $t \in L$ has a product cone bundle neighborhood, $\mathring{C} \times (t-\varepsilon, t + \varepsilon)$, where $\mathring{C}$ is the open cone in $S^{n-2}/H$. Note also that this product structure preserves the strata, i.e., if $(c, t) \in M^i_{(K)}/G$, then $(c, t') \in M^i_{(K)}/G$, $t' \in (t - \varepsilon, t + \varepsilon)$, $c \in \mathring{C}$.

Following ideas in **[L2]**, we will engulf $L \subset M/G$ in such a product cone bundle by a strata preserving engulfing. This product cone bundle will then lift to a trivial $G$-normal vector bundle of the $G$-string in $M$, by Theorem II.7.1. of **[B1]**.

Choose a sequence of points $t_i$, $i \in I$, in $L$ such that $t_i \to \pm\infty$ as $i \to \pm\infty$ and such that each $[t_i, t_{i+1}]$ has a product open cone bundle neighborhood as above. For simplicity, reparametrize $L$ so that $t_i = i$.

The argument goes as follows:

*Step* 1. There is an ambient strata preserving isotopy $h^i_t$ of $M/G$ sending $L$ into $L$ and fixed outside any given neighborhood of $[-i-1, i] \cup [i, i+1]$ and such that $h^i_1[-i-1, i+1] = [-i, i]$.

*Step* 2. For each $i$ there is a product closed cone bundle neighborhood $N_i$ of $[-i, i]$ such that $N_i \subset \operatorname{Int} N_{i+1}$ and $N_i$ is a subclosed cone bundle of $N_{i+1}|[-i-\varepsilon, i+\varepsilon]$. (That is, if $N_{i+1}|[-i-\varepsilon, i+\varepsilon] = C \times [-i-\varepsilon, i+\varepsilon]$, then $N_i = aC \times [-i-\varepsilon, i+\varepsilon]$, where $aC$ is the cone of radius $a$, $0 < a < 1$.)

*Step* 3. $\bigcup_i N_i$ is a product open cone bundle neighborhood of $L$ in $M/G$.

PROOF OF STEP 1. Let $V_i$ be any product closed cone bundle neighborhood of $[i, i+1]$, over say $[i-\varepsilon, i+1+\varepsilon]$, and $V_{-i}$ one for $[-i-1, -i]$ over $[-i-1-\varepsilon, -i+\varepsilon]$. (Obviously any open cone bundle neighborhood contains a closed cone bundle neighborhood.) Using the product structure, one easily constructs an ambient strata preserving isotopy $h^i_t$ of $V_i$, fixed on $bV_i$, sending $[i-\varepsilon, i+1+\varepsilon]$ into itself such that $h^i_1[i, i+1] = [i-\varepsilon/2, i]$. Similarly for $V_{-i}$. Extend to $M/G$ by the identity. Since every neighborhood of $[i, i+1]$ contains such a $V_i$, and similarly for $[-i-1, -i]$, Step 1 is proved.

PROOF OF STEP 2. Assume by induction that we have constructed a product closed cone bundle neighborhood $N_i$ of $[-i, i]$ for $1 \leqq i \leqq j$ such that

(a) For $i < j$, $N_i \subset \operatorname{Int} N_{i+1}$.

(b) For $i < j$, $N_i$ is a subclosed cone bundle of $N_{i+1}|[-i-\varepsilon, i+\varepsilon]$, some $\varepsilon$.

(c) For $i \leqq j$, if $N_i$ is defined over $[i-\varepsilon, i+\varepsilon]$, there is a product open cone bundle $N_i^+$ defined over some neighborhood of $[i-\varepsilon, i+\varepsilon]$ such that $N_i$ is a closed subcone bundle of $N_i^+|[i-\varepsilon, i+\varepsilon]$.

Note that the existence of $N_1$ satisfying (c) is obvious. We prove $j+1$ with $h_t^j$ the ambient isotopy of $M/G$ of Step 1, $h_1^j N_j$ is a product closed cone bundle neighborhood of $[-j-1, j+1]$. Since $h_t^j$ is fixed on $N_j|[-j+\varepsilon, j-\varepsilon]$ for a suitable choice of $h_t^j$, if we take a subclosed cone bundle $N_j^-$ of $N_j|[-j+\varepsilon, j-\varepsilon]$, the pairs $(N_j, N_j^-)$ and $(h_1^j N_j, N_j^-)$ satisfy the properties analogous to (a) and (b) above. Now $N_j^+ \supset N_j \supset N_j^-$ and we can take an ambient isotopy $g_t^j$ of $M/G$, fixed outside a neighborhood of $N_j$ in $N_j^+$, strata preserving and with $g_t^j(L) = L$ and with $g_1^j N_j = N_j$. Take $N_{j+1} = g_1^j h_1^j N_j$ and $N_{j+1}^+ = g_1^j h_1^j N_j^+$. Then (a), (b) and (c) are satisfied for $j+1$.

Proof of Step 3. By construction the bundle projections of the $N_i$ fit together to make $\bigcup_i N_i$ a space over $L$. Further, over each interval $[j, j+1]$, $\bigcup_i N_i|[j, j+1]$ is obviously a product open cone bundle. It follows that $\bigcup_i N_i$ is a product open cone bundle neighborhood of $L$.

This completes the proof of part (a).

(b) Note first that given a $G$-vector bundle $E$ over $X$ and any $G$-invariant function $f\colon X \to R_+$ there is a $G$-$R^n$ bundle equivalence $\varphi\colon E \to E$ such that if $D_x$ is the unit disc bundle over $x$, $\varphi_x|D_x$ is multiplication by $f(x)$. Applying this to the $G$-vector bundle neighborhood of part (a), we choose $f\colon L \to R_+$ going sufficiently rapidly to zero as we go out to $\pm\infty$ so that $\varphi|$Disc bundle will be a proper embedding in $M$. The image will be the desired closed neighborhood. (For $L$ a half-string we first restrict the vector bundle to $[-1, \infty)$.)

Corollary 2. *Let $M$ be a locally smooth $G$-manifold (without boundary) satisfying the Bierstone condition and let $x_i^*$, $i = 1, 2, 3, \cdots$, be a sequence of points in $M/G$ with only a finite number in any compact subset. Then there is a $G$-isotopy $h_t\colon M \to M$ with $h_0$ the identity and $h_1(M) \subset M - \bigcup_i q^{-1}(x_i^*)$.*

Proof. By Lemma 3.15 (see remark following), there is for each $i$ such that $x_i^* \in M_{(H)}/G$ a proper map $f_i\colon [0, \infty) \to M/G$ such that $f_i(0) = x^*$ and $\operatorname{Im}(f_i) \subset M_{(H)}/G$. Since the $\{x_i^*\}$ are discrete, we can assume the $f_i$ are locally flat embeddings. By Theorem 1, they have closed product neighborhoods which we can choose sufficiently small to be disjoint. The isotopy is then obvious.

Our next goal is to prove a $G$-isotopy extension theorem. This is a direct consequence of Siebenmann's isotopy extension theorem for stratified sets.

Definition 2 (Siebenmann). A *stratified set* $X$ is a metrizable space with a filtration $X \supset \cdots \supset X^{(k)} \supset X^{(k-1)} \supset \cdots \supset X^{(-1)} = \emptyset$ by closed subsets such that for each $k \geqq 0$, the components of $X^{(k)} - X^{(k-1)}$ are open in $X^{(k)} - X^{(k-1)}$. It is called a Top stratified set if $X^{(k)} - X^{(k-1)}$ is a topological $k$-manifold without boundary, called the *$k$-stratum*.

A stratified set $X$ is *locally cone-like* if for each $x \in X$, say $x \in X^{(k)} - X^{(k-1)}$, there is an open neighborhood $U$ of $x$ in $X^{(k)} - X^{(k-1)}$, a compact stratified set of finite dimension $L$ and an isomorphism $U \times \mathring{C}(L)$ onto an open neighborhood of $x$ in $X$. ($\mathring{C}(L)$ is the open cone in $L$. $U$ is regarded as stratified with $U = U^{(k)} - U^{(k-1)}$.)

A *CS set* is a locally cone-like Top stratified set.

EXAMPLE. If $M^n$ is a locally smooth $G$-manifold, $M/G$ can be given a CS structure as follows. First assume $\partial M = \emptyset$. Let $(M/G)^{(k)}$ = union of the $\overline{M^i_{(H)}/G}$ with dim $M^i_{(H)}/G = k$ (see §3). This makes $M/G$ into a Top stratified set. It is locally cone-like because if $x^* \in M^i_{(H)}/G$, $q^{-1}(x)$ has a $G$-invariant neighborhood $U = G/H \times R^k$ in $M^i_{(H)}$ with normal $G$-vector bundle $(G \times_H R^{n-k}) \times R^k$ by local smoothability. Hence $U^* = R^k$ is an open neighborhood of $x^*$ with product neighborhood $U \times C(S^{n-k-1}/H)$ in $M/G$.

If $\partial M \neq \emptyset$, take $(M/G)^{(k)}$ to be the union of $(\text{Int } M/G)^{(k)}$ and $(\partial M/G)^{(k)}$.

THEOREM 3 (SIEBENMANN **[S1]**). *Let $X$ be a Hausdorff, locally compact, locally connected CS set and $K \subset V \subset M$ a $G$-invariant compact set and $G$-invariant open neighborhood. Let $f_t: V \to M$, $t \in I^n$, be a continuous family of open $G$-embeddings, with $f_{t_0}$ = inclusion for some $t_0 \in I^n$. If $f_t$ respects strata (where $V^{(k)} = V \cap X^{(k)}$), then there exists a continuous family $F_t: X \to X$ of homeomorphisms fixed outside a compact set such that $F_t$ respects strata and $F_t|K = f_t|K$.*

*Addendum.* If in addition $f_t$ leaves fixed $V \cap Y_i$, where $\{Y_i\}$ is a family of closed substratified sets with $Y_i^{(k)} - Y_i^{(k-1)}$ open in $X^{(k)} - X^{(k-1)}$, then $F_t$ leaves the $Y_i$ fixed.

REMARKS. (1) $\partial M/G$ is a closed substratified set of $M/G$ with open strata as above. (2) By substituting $F_t \circ F_{t_0}^{-1}$ for $F_t$, we see that we may assume $F_{t_0}$ = identity.

THEOREM 4 ($G$-ISOTOPY EXTENSION THEOREM). *Let $M^m$ be a locally smooth $G$-manifold and $K \subset V \subset M$ a $G$-invariant compact set and $G$-invariant open neighborhood. Let $f_t: V \to M$, $t \in I^n$, be a continuous family of open $G$-embeddings, with $f_{t_0}$ = inclusion for some $t_0 \in I^n$. Then there exists a continuous family $F_t: M \to M$ of $G$-homeomorphisms such that $F_t|K = f_t|K$ and $F_t$ is fixed outside a compact set.*

*Relative version.* If in addition $f_t$ leaves $V \cap \partial M$ fixed, then $F_t$ leaves $\partial M$ fixed.

PROOF. Note that since $f_t$ is an open $G$-embedding it preserves orbit type and hence $f_t^*: V^* \to M^*$ preserves strata. Let $K$ be a compact $G$-invariant neighborhood of $K$ in $V$. Let $F_t^*: M^* \to M^*$ be the strata preserving extension of $f_t^*|K^*$. Then $F_t^*$ lifts to a continuous family $F_t$ of $G$-homeomorphisms by Theorem II.7.1 (see also 7.3) of **[B1]**, with $F_{t_0}$ = identity. The proof of II.7.1 of **[B1]** is the usual neighborhood-by-neighborhood argument using a $G$-partition of unity. Hence one may construct the lift so that $F_t|K = f_t|K$. If $f_t$ leaves $V \cap \partial M$ fixed, then $f_t^*$ leaves $V^* \cap \partial M^*$ fixed and we may insist that the extension $F_t^*$ leave $\partial M^*$ fixed. By a similar refinement of the lifting argument we get $F_t$ extending $f_t$ with $F_t|\partial M$ = identity. (Actually, for $G$ finite there is a unique lift such that $F_{t_0}$ = identity since each strata is a covering space, and hence automatically $F_t|K = f_t|K$.)

Consider the s.s. complex $E(K, M)$ of $G$-embeddings $f: \Delta^i \times U \to \Delta^i \times M$, commuting with projection in $\Delta^i$, where $U$ is a $G$-invariant open neighborhood of $K$ and we identify $f$ and $f': \Delta^i \times U' \to \Delta^i \times M$ if they agree in a smaller neighborhood of $K$. Also let Homeo($M$) be the s.s. complex of $G$-homeomorphisms of $M$. Then Theorem 4 implies

THEOREM 4′. *The restriction map $r$*: Homeo($M$) $\to E(K, M)$ *is a Kan fibration.*

Let $V^{n+1}$ be a smooth $G$-manifold, $M^n$ a smooth compact invariant submanifold and $h: M \times R \to V$ a $G$-homeomorphism extending the inclusion on $M \times 0$.

We would like to conclude that there is a $G$-diffeomorphism $f\colon M \times R \to V$ extending the inclusion. Let us recall the situation for $G$-trivial: Let $M^n$ be a smooth compact submanifold of the smooth manifold $V^{n+1}$. Then

THEOREM 5. (a) *Let $h\colon M \times R \to V$ be a homeomorphism such that $h|\partial M \times R$ is smooth and $h|M \times 0$ is the inclusion. If $n + 1 \neq 4$, there is a diffeomorphism $f\colon M \times R \to V$ with $f = h$ on $M \times 0 \cup \partial M \times R$.*

(b) *Let $h\colon M \times R \to V$ be a proper homotopy equivalence such that $h|\partial M \times R$ is a diffeomorphism onto $\partial V$ and $h|M \times 0$ is the inclusion. If $n + 1 \neq 4$, $f$ as in* (1) *exists provided that when $n + 1 = 3$, $M \neq RP^2$ and $V^3$ contains no fake* 3*-discs.*

PROOF. When $n + 1 > 4$ both (a) and (b) follow from engulfing. When $n + 1 \leqq 3$ and $h$ is a homeomorphism (a) follows from the uniqueness of smoothing in these dimensions. When $n + 1 \leqq 3$ and $h$ is a proper homotopy equivalence this follows from **[H3]** and **[B4]**. That is, Husch and Price show $V$ has a collar neighborhood of each end and Brown shows that the region in between is a product.

We use the results for $G$-trivial and induction up the strata of $M/G$ to give:

THEOREM 6. *Let $V^{n+1}$ be a smooth $G$-manifold, $M^n$ a smooth compact invariant submanifold and $h\colon M \times R \to V$ a $G$-homeomorphism* (*trivial action on $R$*) *such that $h|\partial M \times R$ is smooth and $h|M \times 0$ is the inclusion. If no component of $V^H$ has dimension* 4 *for any $H \subset G$, there exists a $G$-diffeomorphism $f\colon M \times R \to V$ such that $f = h$ on $M \times 0 \cup \partial M \times R$.*

PROOF. Let $\{M_i^{\tau}\}$ be the partially ordered set of $G$-components of the $M^{(H)}$, $H \subset G$ (see above and §3). Note that for each $M_i^j$ there is a corresponding $V_i^j$, $G$-homeomorphic to $M_i^j \times R$. If $M_i^{(H)}$, say, is a minimal element, then $M_i^{(H)}/G = M^i_{(H)}/G$ is a smooth manifold and $V_i^{(H)}/G$ is a smooth manifold with $h^*\colon M_i^{(H)}/G \times R \to V_i^{(H)}/G$, $h^*$ the map induced by $h$, a homeomorphism such that $h^* |\partial (M_i^{(H)}/G) \times R$ is smooth and $h^* \mid M_i^{(H)}/G \times 0$ the inclusion. Thus by Theorem 5, if $\dim V_i^{(H)}/G \neq 4$, there is a diffeomorphism $f^*\colon M_i^{(H)}/G \times R \to V_i^{(H)}/G$ with $f^* = h^*$ in $M_i^{(H)}/G \times 0 \cup \partial(M_i^{(H)}/G) \times R$. $f^*$ lifts to a $G$-diffeomorphism $f\colon M_i^{(H)} \times R \to V_i^{(H)}$ with $f = h$ on $M_i^{(H)} \times 0 \cup \partial M_i^{(H)} \times R$. Further if $N$ is the union of disjoint normal tubes of the minimal elements in $M$, $f$ extends to a smooth proper $G$-embedding of $N \times R$ in $V$.

For the inductive step, let now $M_i^{(H)}$ be such that if $N$ is the union of normal tubes of the lower strata in $M_i^{(H)} = \overline{M^i_{(H)}}$, there exists a smooth proper $G$-embedding $f\colon N \times R \to V$ such that $f = h$ on $N \times 0 \cup (N \cap \partial M) \times R$. In general, $N$ is a smooth $G$-manifold with corners which we round off invariantly. $M^i_{(H)}$ is a smooth $G$-manifold meeting $N$ transversally and $M^i_{(H)}$ is $K = \overline{M^i_{(H)} - N}$ union a smooth open $G$-collar attached along $\partial_1 K = \partial K \cap \partial N$. Likewise $V^i_{(H)}$ meets $f(N \times R)$ transversally and $V^i_{(H)}$ is $L = \overline{V^i_{(H)} - f(N \times R)}$ union a smooth open $G$-collar in $\partial_1 L = \partial L \cap \partial f(N \times R)$ extending the collar on $\partial_1 K$. On the other hand, $h\colon M^i_{(H)} \times R \to V^i_{(H)}$ is a $G$-homeomorphism, and $h^*$ defines the proper homotopy equivalence:

$$h_1^*\colon K/G \times R \to (M^i_{(H)}/G) \times R \xrightarrow{h^*} V^i_{(H)}/G = L/G \cup \partial_1 L/G \times [0, 1) \xrightarrow{r} L/G,$$

$r$ the retraction defined by the collar. Further $h_1^*$ is properly homotopic to $h_2^*$ where $h_2^*$ agrees with $f^*$ on $\partial_1 K/G \times R$ and with $h^*$ on $\partial_2 K/G \times R$, $\partial_2 K = \partial M \cap \partial K$, and $h_2^* \mid K/G \times 0$ is inclusion. Thus $K/G$ is a smooth compact submanifold of $L/G$ and $h_2^*$ is a proper homotopy equivalence such that $h_2^* \mid \partial K/G \times R$ is a diffeomorphism onto $\partial V$ with $h_2^* \mid K \times 0$ the inclusion. By Theorem 5(b), if $\dim M^i_{(H)}/G \neq 4$, $h_2^* \mid \partial K/G \times R$ extends to a diffeomorphism $f_2^*: K/G \times R \to L/G$, $f_2^* \mid K/G \times 0$ the inclusion. (Note that if $n + 1 = 3$, $K/G \neq RP^2$ since $\partial K/G \neq \emptyset$. Further, since $L/G \subset V^i_{(H)}/G \simeq M^i_{(H)}/G \times R$, $L/G$ contains no fake 3-discs.) Thus there is a $G$-diffemorphism $f_2: K \times R \to L$ such that $f_2$ agrees with $f$ on $\partial_1 K \times R$ and with $h$ on $\partial_2 K \times R$, and $f_2 \mid K \times 0$ is inclusion. Adjoining to $N$ a normal disc bundle $N'$ of $K$, we can extend $f$ to a smooth proper $G$-embedding of $(N \cup N') \times R \to V$, the inclusion on $(N \cup N') \times 0$ and agreeing with $h$ on $((N \cup N') \cap \partial M) \times R$. This completes the induction step and proves Theorem 6.

**5. $G$-immersion theory.** Let $M^n$ and $Q^n$ be smooth $G$-manifolds of the same dimension, $\partial Q = \emptyset$. Recall that $M$ has a good smooth $G$-handle decomposition if and only if $M$ satisfies what we have called the Bierstone condition, i.e., the minimal elements of the $\{M^i_i\}$ are nonclosed manifolds (§3).

THEOREM 1 (BIERSTONE **[B2]**). *If $M$ has a good smooth $G$-handle decomposition the differential $D: \mathscr{I}(M, Q) \to \mathscr{R}(TM, TQ)$ is a weak homotopy equivalence.*

Here $\mathscr{I}(M, Q)$ is the space of $G$-immersions with the $C^\infty$-topology and $\mathscr{R}(TM, TN)$ is the space of $G$-vector bundle maps with the $C^0$-topology. We will prove an analogous result for $G$-immersions of locally smooth $G$-manifolds satisfying the Bierstone condition. We will work semisimplicially; so to compare smooth immersions with topological we replace the above spaces with their singular complexes $I^d(M, Q)$ and $R^d(TM, TQ)$, respectively. (The superscript $d$ is used to indicate the Diff category.) Then Theorem 1 becomes:

THEOREM 1′. *If $M^n$ and $Q^n$ are smooth $G$-manifolds with $\partial Q = \emptyset$ and $M$ satisfying the Bierstone condition, the s.s. map $D: I^d(M, Q) \to R^d(TM, TQ)$, defined by the differential, is a homotopy equivalence.*

DEFINITION 1. Let $M^n$ and $Q^n$ be topological $G$-manifolds of the same dimension with $\partial Q = \emptyset$. First assume $\partial M = \emptyset$. A *$G$-immersion* of $M$ in $Q$ is a $G$-map $f: M \to Q$ which is a local homeomorphism. Define the s.s. complex $I^t(M, Q)$ by taking a simplex to be a $G$-immersion $f: \Delta \times M \to \Delta \times Q$, commuting with projection on the simplex $\Delta$. That is, giving $\Delta$ the trivial $G$-action, $f$ is a $G$-map and a local homeomorphism.

If $\partial M \neq \emptyset$, there are four essentially equivalent ways of defining $I^t(M, Q)$:

$I^t(M, Q)_1$. Set $I^t(M, Q)_1 = I^t(\tilde{M}, Q)$, $\tilde{M} = M$ union an open $G$-collar.

$I^t(M, Q)_2$. $I^t(M, Q)_2$ is $I^t(\tilde{M}, Q)/\sim$, where $f_1, f_2: \Delta \times \tilde{M} \to \Delta \times M$ are declared equivalent if they agree on $\Delta \times U$, $U$ some neighborhood of $M$ in $\tilde{M}$.

$I^t(M, Q)_3$. A simplex of $I^t(M, Q)_3$ is a $G$-immersion $f: \Delta \times U \to \Delta \times Q$ commuting with projection, where $U$ is a $G$-invariant open neighborhood of $M$ in $\tilde{M}$; and two such $f_i: \Delta \times Q \to \Delta \times Q$, $i = 1, 2$, being equivalent if they agree on $\Delta \times U$, $U \subset U_1 \cap U_2$ a smaller neighborhood of $M$ in $\tilde{M}$.

$I^t(M, Q)_4$. A simplex of $I^t(M, Q)_4$ is a $G$-map $f: \Delta \times M \to \Delta \times Q$ commuting with projection on $\Delta$, such that for each $(s, x) \in \Delta \times M$, there is an open neighborhood $\tilde{U}_x$ of $x$ in $\tilde{M}$, and $V$ of $s$ in $\Delta$, and a $G$-embedding $\tilde{f}_x$: $V \times \tilde{U}_x \to V \times Q$ commuting with projection on $V$ satisfying $\tilde{f}_x \mid V \times (\tilde{U}_x \cap M) = f \mid V \times (\tilde{U}_x \cap M)$.

We have obvious s.s. maps: $I^t(M, Q)_1 \to I^t(M, Q)_2 \to I^t(M, Q)_3 \to I^t(M, Q)_4$. It is not difficult to see that each map is surjective and a homotopy equivalence. (This is essentially the existence and uniqueness of $G$-collars.) Of course, analogues $I^d(M, Q)_i$, $i = 1, 2, 3, 4$, of all these complexes exist in the Diff category and we have the forgetful maps $F$: $I^d(M, Q)_i \to I^t(M, Q)_i$. Also note that $I^d(M, Q) = I^d(M, Q)_4$.

DEFINITION 2. Let $\mu_i$: $X_i \xrightarrow{s_i} E_i \xrightarrow{p_i} X_i$, $i = 1, 2$, be $G$-$R^n$ microbundles. Define $R^t(\mu_1, \mu_2)$ to be the s.s. complex whose simplices are $G$-microbundle maps $\varphi$: $\Delta \times E_1^0 \to \Delta \times E_2$, commuting with projection on $\Delta$; two such $\varphi_i$: $\Delta \times E_1^i \to \Delta \times E_2$, $i = 1, 2$, being identified if they agree on some smaller $G$-microbundle neighborhood $E_1^0 \subset E_1^1 \cap E_1^2$ of $s_1(X_1)$.

DEFINITION 3. Let $M^n$ and $Q^n$ be $G$-manifolds without boundary and $\tau M$ and $\tau Q$ their tangent $G$-microbundles. If $f$: $M \to Q$ is a $G$-immersion, we define the *differential Df* of $f$ to be the $G$-map $Df = f \times f$: $M \times M \to Q \times Q$. Since $Df$ restricted to a neighborhood of the 0-section is an embedding on each fibre, it defines a unique $G$-microbundle map (see 2.12). Define $D$: $I^t(M, Q) \to R^t(\tau M, \tau Q)$ as follows: If $f$: $\Delta \times M \to \Delta \times Q$ is a simplex in $I^t(M, Q)$, then $Df$: $\Delta \times \tau M \to \Delta \times \tau Q$ is the simplex defined by $Df_s = f_s \times f_s$, $s \in \Delta$.

If $\partial M \neq \emptyset$, let $\tau M = \tau\tilde{M} \mid M$. Note that if $U$ is a $G$-invariant open neighborhood of $M$ in $M$, then $\tau U \mid M$ is a $G$-microbundle neighborhood of the 0-section in $\tau\tilde{M} \mid M$. Also note that the restriction $r$: $R(\tau\tilde{M}, TQ) \to R(\tau M, \tau Q)$ is a homotopy equivalence since $\tilde{M}$ is a $G$-deformation retract of $M$. Define $D_1 : I^t(M, Q)_1 \to R(\tau M, \tau Q)$ by $D_1 = r\,D$, $D$: $I^t(\tilde{M}, Q) \to R(\tau\tilde{M}, \tau Q)$. Then $D_1$ defines $D_2$: $I^t(M, Q)_2 \to R(\tau M, \tau Q)$ by passage to the quotient, and $D_3$: $I^t(M, Q)_3 \to R(\tau M, \tau Q)$ similarly. Note however, that in contrast to the smooth case, no natural $D_4$ is defined. Also we have the commutative diagram:

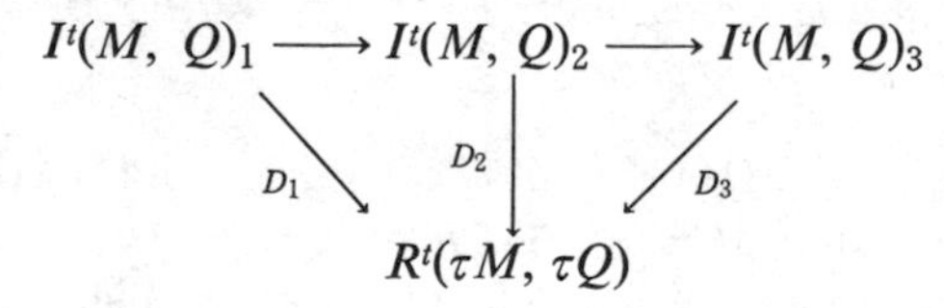


In the Diff category we get a commutative diagram with all four $D_i$'s.

If $M^n$ is smooth and we take a $G$-invariant Riemannian metric on $M^n$ (or $\tilde{M}^n$ if $\partial M \neq \emptyset$) we can identify $TM$ with a $G$-microbundle neighborhood of the 0-section in $\tau M$ by the exponential map. Then if $\partial M = \emptyset$,

$$\begin{array}{ccc} I^d(M, Q) & \xrightarrow{D} & R^d(TM, TQ) \\ \downarrow F & & \downarrow F \\ I^t(M, Q) & \xrightarrow{D} & R^t(\tau M, \tau Q) \end{array}$$

commutes up to homotopy. Similarly, if $\partial M \neq \emptyset$ we get a homotopy commutative diagram for each $D_i$, $i = 1, 2, 3$.

If $\partial M \neq \emptyset$, we will set $I^t(M, Q) = I^t(M, Q)_2$ and $D = D_2$.

THEOREM 2. *If $M$ and $Q$ are $G$-manifolds with $\partial Q = \emptyset$ and $M$ a locally smooth $G$-manifold satisfying the Bierstone condition, then* $D: I^t(M, Q) \to R^t(\tau M, \tau Q)$ *is a homotopy equivalence.*

The proof is formally almost identical to the $G$-trivial case **[L3]**, and we will only indicate the changes needed to make the proof work here.

First assume $M$ is a smoothable compact $G$-manifold. Then $M$ has a good $G$-handle decomposition. The proof of Theorem 2 in this case comes down to the following three points (see **[B2]** and **[L3]**), which hold for $M$ locally smoothable:

(1) If $M$ is a 0-handle, then $D: I^t(M, Q) \to R^t(\tau M, \tau Q)$ is a homotopy equivalence.

(2) If $M$ is obtained from $M_0$ by adding a closed $G$-collar and then a good $G$-handle, the restriction map $r: I^t(M, Q) \to I^t(M_0, Q)$ is a Kan fibration.

(3) Let $M$ and $M_0$ be as in (2); then $r: R^t(\tau M, \tau Q) \to R^t(\tau M_0, \tau Q)$ is a Kan fibration.

The proof of (1) is formally identical to that in **[L3]**; we merely substitute $G$-deformations for the deformations in **[L3]**. The proof of (2) is also formally identical to that in **[L3]**, using the $G$-isotopy extension theorem in place of the isotopy extension theorem. Moreover, in the last step of the proof in **[L3]** where one puts the handle in good position by pushing the core of the handle away from itself along a normal direction (this is where one requires no handles of top dimension in **[L3]**), it is necessary here to have a $G$-trivial normal direction in order to get a $G$-deformation. This is where we need a good $G$-handle. (3) follows from the $G$-covering homotopy theorem for $G$-locally linear $R^n$ bundles (§1), and the fact that any neighborhood of the 0-section in $\tau M$ contains such a bundle (§2).

The inductive argument now takes the form: Assume the result for all $M^n$ having good $G$-handle decompositions with handles of index $\leqq k$, $k < n - 1$. The result holds for $k = 0$ by (1). Now if $M$ is obtained from $M_0$ by adding a good handle of index $k + 1$ and the result is true for $M_0$, then consider the map of fibrations (2) and (3):

$$\text{(A)}\qquad \begin{array}{ccc} I^t(M, Q) & \xrightarrow{D} & R^t(\tau M, \tau Q) \\ {\scriptstyle r}\downarrow & & \downarrow{\scriptstyle r} \\ I^t(M_0, Q) & \xrightarrow{D} & R^t(\tau M_0, \tau Q) \end{array}$$

The result for $M$ will hold if we can show the map on fibres is a homotopy equivalence.

But if the $k + 1$ handle is $N = G \times_H (D^{k+1} \times D^{n-k+1})$, and we let $\partial_1 N = G \times_H (S^k \times D^{n-k+1})$ and $\partial_1 N \times I \subset N$ be a closed $G$-collar, then as in **[L3]**, the fibrations in (A) have the same fibres and map of fibres (up to homotopy equivalence) as the fibrations

$$\text{(B)}\qquad \begin{array}{ccc} I^t(N, Q) & \longrightarrow & R^t(\tau N, \tau Q) \\ {\scriptstyle r}\downarrow & & \downarrow{\scriptstyle r} \\ I^t(\partial_1 N \times I, Q) & \longrightarrow & R^t(\tau(\partial_1 N \times I), \tau Q) \end{array}$$

Now $N = G \times_H (D^{k+1} \times D^{n-k+1}) = G \times_H D^n$ is a 0-handle and the top map is a homotopy equivalence by (1). On the other hand, any $G$-handle decomposition of $G \times_H S^k$ gives a good handle decomposition of $\partial_1 N \times I = G \times_H (S^k \times D^{n-k+1}) \times I$ with handles of index $\leqq k$; so the bottom map is a homotopy equivalence by the induction assumption. Thus the map on fibres is a homotopy equivalence, and the result holds for $M$.

Note that the above argument shows that if a locally smooth $G$-manifold $M$ is built up from a submanifold $M_0$ by adding good handles, then if the result holds for $M_0$ it holds for $M$.

Finally, the general case follows as in **[L3]** from (B), by covering $M$ by a countable locally finite family of smoothable $G$-neighborhoods. As in **[L3]**, we have to throw out a countable number of points $\{x_i^*\}$ from $M/G$ (i.e., orbits from $M$) so as to insure we are only adding good handles at each step. But if $M$ satisfies the Bierstone condition we can push $M$ off these points by 4.2, and the result for $M$ follows. Q.E.D.

One may also prove various relative immersion theorems by the same methods. For example, if $M$ is a locally smooth $G$-manifold with boundary we will say that $M^n$ satisfies the *relative Bierstone condition* if the minimal elements of the $\{M_i^j\}$ are noncompact (see §3). Given a $G$-immersion $f\colon \partial M \times (-1, 1) \to Q^n, \partial Q = \emptyset$, let $I_f^t(M, Q)$ be the subcomplex of $I^t(M, Q)$ of simplices of immersions which agree with $f$ on some neighborhood of $\partial M$, where we identify $\partial M \times (-1, 1)$ with a $G$-bicollar of $\partial M$ in $\tilde{M}$. Also let $R_{Df}^t(\tau M, \tau Q)$ be the subcomplex of $R^t(\tau M, \tau Q)$ of simplices of $G$-microbundle maps which agree on $\tau M \mid \partial M$ with $Df$. Then we have

THEOREM 3. *Let $M$ be a locally smooth $G$-manifold with boundary, and $f\colon \partial M \times (-1, 1) \to Q$ a $G$-immersion, $\partial Q = \emptyset$. If $M$ satisfies the relative Bierstone condition, then $D\colon I_f^t(M, Q) \to R_{Df}^t(\tau M, \tau Q)$ is a homotopy equivalence.*

**6. *G*-smoothing theory.**

DEFINITION 1. A topological $G$-manifold $M$ admits a *$G$-smoothing* if there exist a smooth $G$-manifold $V$ and a $G$-homeomorphism $\alpha$ of $V$ onto $M$. Two such $\alpha_i\colon V_i \to M$, $i = 1, 2$, are identified if $\alpha_2^{-1}\alpha_1$ is a diffeomorphism. Frequently one identifies the underlying topological $G$-manifold of $V$ with $M$ via $\alpha$, and then one calls $\alpha$ a *smooth $G$-structure on $M$* and denotes it by $M_\alpha$.

Two $G$-smoothing $\alpha_i\colon V_i \to M$, $i = 1, 2$, are called *isotopic* if $\alpha_1$ is $G$-isotopic through onto $G$-homeomorphisms to a $G$-homeomorphism $\bar{\alpha}_1$ such that $\alpha_2^{-1}\bar{\alpha}_1$ is a diffeomorphism. Viewed as smooth $G$-structures $M_{\alpha_i}$, $i = 1, 2$, this is equivalent to saying the identity map of $M$ is $G$-isotopic to a $G$-diffeomorphism of $M_{\alpha_1}$ onto $M_{\alpha_2}$.

The lemma below follows from the $G$-trivial case **[L3]**:

LEMMA 1. *Let $f\colon M^n \to Q^n$ be a topological $G$-immersion, $\partial M = \partial Q = \emptyset$. If $\alpha$ is a smooth $G$-structure on $Q$, there is a unique smooth $G$-structure $f^*\alpha$ on $M$ such that $f\colon M_{f^*\alpha} \to Q_\alpha$ is a smooth $G$-immersion. ($f^*\alpha$ is called the induced $G$-smoothing.)*

DEFINITION 2. A *$G$-vector bundle reduction* of a $G$-microbundle $\mu$ over $X$ is a $G$-vector bundle $\mathscr{E}$ over $X$ and a $G$-microbundle embedding $\varphi\colon E(\mathscr{E}) \to E(\mu)$ over $1_X$. Two such $\varphi_i\colon E(\mathscr{E}_i) \to E(\mu)$, $i = 1, 2$, are called *isotopic* if $\varphi_1$ is $G$-isotopic through $G$-microbundle embeddings over $1_X$ to a $G$-microbundle embedding $\bar{\varphi}_1$ such that there exists a $G$-vector bundle equivalence $\theta: E(\mathscr{E}_1) \to E(\mathscr{E}_2)$ with $\varphi_2 \circ \theta = \bar{\varphi}_1$. Caution: 'Isotopic' is called 'equivalent' in **[L3]**.

We can make similar definitions of isotopy classes of $G$-$R^n$ bundle reductions of a $G$-microbundle and isotopy classes of $G$-vector bundle reductions of $G$-$R^n$ bundles. The reductions and isotopies in the first case will be $G$-microbundle embeddings and isotopies over $1_X$ and in the second case $G$-$R^n$ bundle maps and isotopies over $1_X$. By §3, a locally linear $G$-$R^n$ microbundle has a unique isotopy class of $G$-$R^n$ bundle reductions. Now, an isotopy class of $G$-vector bundle reduction of a $G$-$R^n$ bundle contained in a $G$-$R^n$ microbundle defines a unique isotopy class of $G$-vector bundle reduction of the $G$-microbundle. Conversely, given a locally linear $G$-$R^n$ bundle in a $G$-microbundle, a $G$-vector bundle reduction of the $G$-microbundle is isotopic to one that factors through a reduction of the $G$-$R^n$ bundle, and an isotopy class of $G$-vector bundle reductions of the $G$-microbundle defines a unique isotopy class of $G$-vector bundle reductions of the $G$-$R^n$ bundle.

Since we frequently identify a $G$-microbundle with some $G$-microbundle neighborhood of the 0-section, we often identify a locally linear $G$-microbundle with a particular locally linear $G$-$R^n$ bundle. Also given a $G$-vector bundle reduction of the $G$-$R^n$ bundle we can identify the underlying $G$-$R^n$ bundles and speak of a *$G$-vector bundle structure* on the $G$-$R^n$ bundle, and, by abuse of notation, of a $G$-vector bundle structure on the $G$-microbundle.

(These identifications may seem somewhat confusing, but without them the notation is hopelessly cumbersome.)

If $V$ is a smooth $G$-manifold with $\partial V = \emptyset$, any $G$-invariant metric on $V$ defines a *smooth* $G$-microbundle embedding of $TV$ in $\tau V = V \times V$, via the exponential map. On the other hand, it follows from the $G$-tubular neighborhood argument **[B1]**, that any two smooth $G$-vector bundle reductions of $\tau V$ are isotopic through smooth $G$-vector bundle reductions. From this it follows that if $M$ is a $G$-manifold (with or without boundary):

PROPOSITION 2. *An isotopy class $(\alpha_0)$ of $G$-smoothings of $M$ induces a unique isotopy class $(\varphi_0)$ of $G$-vector bundle reductions of $\tau M$, such that if $\alpha: V \to M$ is in $(\alpha_0)$ and $\psi: T\tilde{V} \to \tilde{V} \times \tilde{V}$ is any smooth $G$-microbundle embedding then $\alpha \times \alpha \circ \psi \circ (\alpha^{-1})_*: (\alpha^{-1})^* TV \to \tau M$ is in $(\varphi_0)$. (Here $\tilde{V} = V$ if $\partial M = \partial V = \emptyset$.)*

Let $\rho: H \to O(n)$ be a representation and let $R^n_\rho$ be Euclidean $n$-space with this action. Let $K^n \subset R^n_\rho$ be a compact $H$-invariant topological submanifold with an $H$-collar neighborhood in $R^n_\rho$. The following is the key lemma for piecing together local $G$-smoothings to get a global $G$-smoothing of a $G$-manifold:

LEMMA 3. *Suppose $\alpha$ is an $H$-smoothing of $K$ such that the isotopy class $(\varphi)$ of $H$-vector bundle reductions of $\tau K$ induced by $(\alpha)$ extends to $(\bar{\varphi})$ in $\tau(R^n_\rho)$. Then $\alpha$ extends to an $H$-smoothing $\bar{\alpha}$ of $R^n_\rho - P$, $P$ a finite set of $H$-orbits in $R^n_\rho - K$, such that $(\bar{\alpha})$ induces $(\bar{\varphi}) \mid R^n_\rho - P$.*

PROOF. Identify $\tau R^n_\rho$ with the underlying $H$-$R^n$ bundle of the $H$-vector bundle $TR^n_\rho$ and $\tau K$ with $\tau R^n_\rho | K$. Then the hypothesis says we have an $H$-vector bundle structure $(\tau R^n_\rho)_{\bar{\varphi}} | K$, which (as a reduction of $\tau K = (\tau R^n_\rho) | K$) is in the isotopy class induced by $\alpha$. We identify $TK_\alpha$ with $(\tau R^n_\rho)_{\bar{\varphi}} | K$ by this equivalence.

Since $R_\rho^n$ is $H$-contractible to $0 \in R_\rho^n$, the $G$-covering homotopy property (§1) for $G$-vector bundles implies $(\tau R_\rho^n)_{\bar\varphi}$ is $H$-vector bundle equivalent to $pr_1: R_\rho^n \times R_{\rho'}^n \to R_\rho^n$ with the diagonal $H$-action. (Note $\rho'$ is Top($n$) equivalent to $\rho$, but we do not know that it is linearly equivalent in general.) Let $\lambda$: $R_\rho^n \times R_{\rho'}^n \to TR_{\rho'}^n$ be the $H$-vector bundle map $\lambda(x, y) = (0, y)$, and let $\bar\lambda$: $(\tau R_{\rho'}^n)_{\bar\varphi} \to TR_{\rho'}^n$ be the composition of $\lambda$ with the equivalence $(\tau R_{\rho'}^n)_{\bar\varphi} \to R_\rho^n \times R_{\rho'}^n$.

Now $\bar\lambda | TK_\alpha$: $TK_\alpha \to TR_{\rho'}^n$ is an $H$-vector bundle map and by Bierstone's immersion theorem (5.1), there is a smooth $H$-immersion $f$: $K_\alpha \to R_{\rho'}^n$ such that the smooth differential $Df$ is $G$-homotopic to $\bar\lambda | TK_\alpha$. (Note that $\tilde K_\alpha$ and hence $K_\alpha$ satisfies the Bierstone condition since $R_\rho^n$ does.) This implies that if $\tilde f$: $\tilde K_\alpha \to R_{\rho'}^n$ is any smooth $H$-immersion extending $f$, then $\tilde f \times \tilde f | \tau K$: $\tau K \to \tau R_{\rho'}^n$ is also isotopic to $\bar\lambda | \tau K$.

If $\overline{R_\rho^n - K}$ satisfied the relative Bierstone condition we could extend $\tilde f$ |neighborhood of $K$ to a topological $H$-immersion $\bar f$: $R_\rho^n \to R_{\rho'}^n$ such that $\bar f \times \bar f$ is $H$-homotopic to $\bar\lambda$: $\tau R_\rho^n \to \tau R_{\rho'}^n$ (where we take $\tau R_{\rho'}^n$ to be the underlying $H$-$R^n$ bundle $\tau R_{\rho'}^n$). Further the smoothing $\bar\alpha$ on $R_\rho^n$ induced by $\bar f$ would satisfy $\bar\alpha | K = K_\alpha$, and $D\bar f$: $T(R_\rho^n)_{\bar\alpha} \to TR_{\rho'}^n$ is $H$-homotopic to $\bar\lambda$ (where $T(R_\rho^n)_{\bar\alpha}$ is identified with an $H$-vector bundle structure on $\tau R_\rho^n$), and thus $(\bar\alpha)$ would induce $(\bar\varphi)$.

But it is only necessary to remove one orbit from each compact minimal element of the $\overline{(R_\rho^n - K)_i^j}$ to make it satisfy the relative Bierstone condition. To see that this set is finite, note that $(R_\rho^n)^{H'}$, $H' \subset H$, is a linear subspace, say $R^l$. By Alexander duality $R^l - K = R^l - K^{H'}$ has only a finite number of components. It follows that there are only a finite number of minimal elements in $\{(R_\rho^n - K)_i^j\}$. Since $K$ has an $H$-collar neighborhood, each $(R_o^n - K)_i^j$ is contained in exactly one $\overline{(R_\rho^n - K)_i^j}$. Thus $R_\rho^n - P$, $P$ a finite set of orbits in $R_\rho^n - K$, satisfies the relative Bierstone condition, and $\alpha$ extends to $\bar\alpha$ satisfying the conclusion of the lemma.

Corollary 4. *Let $K^n$ be a compact $G$-invariant topological submanifold of $E(\rho) = G \times_H R^n$, $\rho: H \to O(n)$, with a $G$-collar neighborhood. Suppose $\alpha$ is a $G$-smoothing of $K$ such that the isotopy class $(\varphi)$ of $G$-vector bundle reductions of $\tau K$ induced by $(\alpha)$ extends to $(\bar\varphi)$ on $\tau E(\rho)$. Then $\alpha$ extends to a $G$-smoothing $\bar\alpha$ of $E(\rho) - P$, $P$ a finite set of $G$-orbits in $E(\rho) - K$, such that $(\bar\alpha)$ induces $(\bar\varphi) | (E(\rho) - P)$.*

Proof. If $K^0 = K \cap [e, R^n]$, $G \times_H K^0 = K$. Hence Corollary 4 follows from Lemma 3 applied to $K^0$.

By an argument completely analogous to the $G$-trivial case **[L3]** (using the $G$-engulfing result 4.2 in place of the corresponding $G$-trivial result) we get:

Theorem 5. *Let $M$ be a locally smooth $G$-manifold without boundary, satisfying the Bierstone condition. If $\varphi$ is a $G$-vector bundle reduction of $\tau M$, there exists a $G$-smoothing $\alpha$ of $M$ such that $(\alpha)$ induces $(\varphi)$.*

Definition 3. If $M$ is an arbitrary topological $G$-manifold without boundary, we say that $M$ satisfies the Bierstone condition if $G$-component of $M^{(H)}$ is compact, any $H \subset G$. If $M$ is locally smooth this definition is equivalent to the original definition.

From the fact that if $M$ has a $G$-vector bundle reduction of $\tau M$, then $M$ is locally smooth (3.6) we get:

Theorem 5′. *Let $M$ be a topological $G$-manifold without boundary, satisfying the*

*Bierstone condition. If $\varphi$ is a G-vector bundle reduction of $\tau M$, there exists a G-smoothing $\alpha$ of M such that* $(\alpha)$ *induces* $(\varphi)$.

DEFINITION 4. Define the s.s. complex $VR(\tau M)$ of $G$-vector bundle reductions of $\tau M$ by taking a simplex to be a $G$-vector bundle over $\Delta \times M$ and a $G$-microbundle embedding $\varphi: E(\mathscr{E}) \to \Delta \times \tau M$. Two such $\varphi_i: E(\mathscr{E}_i) \to \Delta \times \tau M$ are identified if there exists a $G$-vector bundle equivalence $\theta: E(\mathscr{E}_1) \to E(\mathscr{E}_2)$ with $\varphi_2\theta = \varphi_1$. This is obviously a Kan complex.

DEFINITION 5. Define the s.s. complex $\mathscr{S}(M)$ of $G$-smoothings of $M$ by taking a simplex to be a smooth $G$-manifold $V$ (with corners) and a $G$-homeomorphism $\alpha: V \to \Delta \times M$ such that $p = pr_1 \circ \alpha: V \to \Delta$ is a submersion. Two such $\alpha_i: V_i \to \Delta \times M$ are identified if there exists a $G$-diffeomorphism $\lambda: V_1 \to V_2$ such that $\alpha_2 \circ \lambda = \alpha_1$. Unfortunately, $\mathscr{S}(M)$ as it stands may not be a Kan complex. What we need is that $\alpha$ be a product near the faces. For example, if $\Delta$ is the 1-simplex $I$, we need $p^{-1}[0, \varepsilon] = [0, \varepsilon] \times V_0$, $p^{-1}[1 - \varepsilon, 1] = [1 - \varepsilon, 1] \times V_1$, where $V_i = p^{-1}(i)$, $i = 0, 1$ and $\alpha | p^{-1}[0, \varepsilon] = 1 \times \alpha_0$ and $\alpha | p^{-1}[1 - \varepsilon, 1] = 1 \times \alpha_1$. For a precise definition of the condition on higher simplices see the Appendix 1 of **[B5]**. We assume this condition is imposed so that $\mathscr{S}(M)$ is a Kan complex.

Note that for any $s \in \Delta$, $\alpha_s: V_s \to M_s$ is a $G$-smoothing of $M$, where $V_s = p^{-1}(s)$ and $M_s = s \times M = pr_1^{-1}(s)$.

A 1-simplex of $\mathscr{S}(M)$ is called a *sliced concordance* **[L3]**, and a $k$-simplex is called a $k$-parameter *sliced family* of smoothings **[K2]**.

Since the slices $V_s$, $s \in \Delta$, are leaves of a foliation of $V$, the bundle of tangent vectors along the slices, $T_s(V)$, is locally trivial and a $G$-vector bundle over $V$.

By induction up the skeletons of $\mathscr{S}(M)$ we can define, for any $\alpha: V \to \Delta \times M$, a family $\psi_s$ of smooth $G$-microbundle embeddings, $\psi_s: TV_s \to V_s \times V_s$, so that $\alpha_s \times \alpha_s \circ \psi_s: TV_s \to M_s \times M_s = \tau M_s$ defines a $G$-microbundle embedding $\varphi(\alpha)$: $T_s V \to \Delta \times \tau M$, and $\varphi: \alpha \to \varphi(\alpha)$ defines a s.s. map $\varphi: \mathscr{S}(M) \to VR(\tau M)$.

REMARK. $\varphi$ of course depends on the choices made, but any two are homotopic. (If we defined smooth $G$-microbundles and smooth $G$-microbundle reductions of a topological $G$-microbundle as in **[L3]** or **[K2]**, then one would get a natural s.s. map from $\mathscr{S}(M)$ to the s.s. complex of such reductions. One would then show that the latter was homotopy equivalent to $VR(\tau M)$.)

Under suitable assumptions on $M$, $\varphi$ will be a homotopy equivalence. But we first note the more obvious:

PROPOSITION 6. *Let M be a G-manifold without boundary satisfying the Bierstone condition. Then isotopy classes of G-vector bundle reductions are in one-to-one correspondence with sliced concordance classes of G-smoothings.*

PROOF. Note that we are to prove that $\varphi$ is bijective on components. Theorem 5 states that $\varphi$ is surjective on components. To show it is injective, let $M_\alpha$ and $M_\beta$ be two $G$-smoothings and consider $TM_\alpha$ and $TM_\beta$ to be $G$-vector bundle reductions of an underlying $G$-$R^n$ bundle $\tau_0 M \subset \tau M$. To say these are isotopic reductions means the identity map of $\tau_0 M$ is homotopic through $G$-$R^n$ bundle maps of $\tau_0 M$ onto itself to one which is a $G$-vector bundle equivalence of $TM_\alpha$ onto $TM_\beta$. Let $f: M_\alpha \to M_\beta$ be a corresponding smooth immersion. Then as a topological $G$-

immersion $f$ is regularly homotopy to the identity. That is, we have a $G$-immersion $F: I \times M \to I \times M$ such that $F_0 = f$ and $F_1$ = identity. Let $(I \times M)_\gamma$ be the smoothing induced by $F$ from $I \times M_\beta$. Since $f$ is smooth $(0 \times M)_\gamma = M_\alpha$ and $(1 \times M)_\gamma = M_\beta$. Thus $(I \times M)_\gamma$ is the sliced concordance.

To show $M_\alpha$ and $M_\beta$ are isotopic smoothings we need to show that $(I \times M)_\gamma$ is $G$-diffeomorphic over $I$ to $I \times M_\alpha$.

Note that if $(I \times M)_\gamma$ was a sliced concordance between closed manifolds one need merely pick a $G$-invariant metric and follow the normal lines to the slices straight across to show $(I \times M)_\gamma$ is $G$-diffeomorphic to a product. When $M$ is not closed, the normal lines may wander out to infinity, so it is necessary to use engulfing to keep that from wandering too far. By a proof formally the same as in the $G$-trivial case (see **[B6]** and **[K2]**), using the $G$-engulfing result of 4.6, one gets:

THEOREM 7. *Let $M$ be a locally smooth $G$-manifold (with or without boundary) such that for any $H \subset G$ no component of $M^H$ has dimension four. Then any $k$-parameter sliced family of $G$-smoothings of $M$ is $G$-diffeomorphic over $\Delta^k$ to a product.*

Using Theorem 7, the argument of Proposition 6 easily generalizes (see **[K2]**) to prove:

THEOREM 8. *Let $M$ be a locally smooth $G$-manifold without boundary satisfying the Bierstone condition and such that for any $H \subset G$ no component of $M^H$ has dimension four. Then $\varphi: \mathscr{S}(M) \to VR(\tau M)$ is a homotopy equivalence.*

We now come to our main result:

THEOREM 9. *Let $M^n$ be a locally smooth $G$-manifold without boundary, such that for any $H \subset G$ no component of $M^H$ has dimension four. Then the isotopy classes of $G$-smoothings of $M$ are in one-to-one correspondence with the isotopy classes of $G$-vector bundle reductions of $\tau M$.*

PROOF. (a) *Existence.* Remove one orbit from each compact minimal element of the $\{M_i^j\}$ (§3), so that $M_0 = M - P$, $P$ a countable set of orbits with only a finite number in any compact set, satisfies the Bierstone condition. If $\tau M$ has a $G$-vector bundle reduction, $M_0$ has a $G$-smoothing $(M_0)_\alpha$ by Theorem 5.

Now take a collection of disjoint normal tubes of the orbits in $P$. Such a normal tube is of the form $G \times_H R^n$, $H$ acting orthogonally on $R^n$. $G \times_H R^n$ satisfies the Bierstone condition and so the reduction of $\tau M$ induces a $G$-smoothing $(G \times_H R^n)_\beta$. Thus $(G \times_H (R^n - 0))_\alpha$ and $(G \times_H (R^n - 0))_\beta$ are two $G$-smoothings corresponding to the same $G$-vector bundle reduction of $\tau(G \times_H (R^n - 0))$. By Theorem 7, these smoothings are isotopic, i.e., the identity map of $G \times_H (R^n - 0)$ is $G$-isotopic to a diffeomorphism of $(G \times_H (R^n - 0))_\beta$ into $(G \times_H (R^n - 0))_\alpha$. Attach $(G \times_H R^n)_\beta$ to $(M_0)_\alpha$ by this diffeomorphism. Doing this for each orbit in $P$, we get a $G$-smoothing of $M$ corresponding to the given reduction of $\tau M$.

(b) *Uniqueness.* Let $M_\alpha$ and $M_\beta$ be two $G$-smoothings such that $\varphi(\alpha)$, $\varphi(\beta)$ are isotopic reductions of $\tau M$. We again look at $M_0 = M - P$; then $(M_0)_\alpha$ and $(M_0)_\beta$ are sliced concordant by Proposition 6 or Theorem 8. Similarly, if $G \times_H R^n$ is a normal tube of an orbit in $P$, $(G \times_H R^n)_\alpha$ and $(G \times_H R^n)_\beta$ are sliced concordant. Further, the two sliced concordances of $(G \times_H (R^n - 0))_\alpha$ to $(G \times_H (R^n - 0))_\beta$ correspond to simplices of $VR(\tau(G \times_H (R^n - 0)))$ which are homotopic rel endpoints. By Theorem 8 these two sliced concordances are themselves sliced concordant rel endpoints. By Theorem 7, all these sliced concordances are actually isotopies. Thus we may extend the isotopy of $(M_0)_\alpha$ to $(M_0)_\beta$ to one from $M_\alpha$ to $M_\beta$.

Thus an isotopy class of $G$-vector bundle reductions induces an isotopy class of $G$-smoothings. The converse was already proved in Proposition 2.

The argument of Theorem 9 clearly extends to $k$-isotopy classes, or equivalently in view of Theorem 7, we get:

THEOREM 8′. *Let $M^n$ be a locally smooth G-manifold without boundary such that for any $H \subset G$ no component of $M^H$ has dimension four. Then $\varphi: \mathcal{S}(M) \to VR(\tau M)$ is a homotopy equivalence.*

The above theorems have relative versions. We only give the relative version of Theorem 9: If $M$ is a locally smooth $G$-manifold with boundary then (§3), $\partial M$ has a $G$-collar. A $G$-smoothing $(\partial M)_\alpha$ of $\partial M$ defines a product smoothing $\partial M_\alpha \times (-1, 1)$ of the bicollar of $\partial M$ in $M$. Since any two $G$-collars are $G$-isotopic, this gives a well-defined isotopy class of $G$-vector bundle reductions of $\tau M|\partial M = \tau \tilde{M}|\partial M$. The arguments above are easily modified to prove:

THEOREM 9 REL. *Let M be a locally smooth G-manifold with boundary, such that for any $H \subset G$ no component of $M^H$ has dimension four. Then the isotopy classes of G-smoothings $\alpha$ on $\partial M$ are in one-to-one correspondence with the isotopy classes of G-vector bundle reductions of $\tau M$ such that the restriction to $\tau M|\partial M$ is a fixed reduction in the isotopy class defined by $\alpha$.*

REMARKS ON THE PL-CATEGORY. (1) If $M$ is a smooth $G$-manifold then a slight extension of Whitehead's $C^1$ triangulation theorem gives a $C^1$ triangulation of $M$ such that $G$ acts by PL homeomorphism.

(2) If $\mathcal{E}$ is a $G$-vector bundle over a PL-space $X$, then $E(\mathcal{E})$ may be triangulated so that it becomes a $G$-PL $R^n$ bundle over $X$.

(3) The analogues of Theorem 9 and Theorem 9 rel are true with locally smooth $G$-PL manifolds replacing locally smooth $G$-manifolds.

(4) The analogues of Theorems 9 and 9 rel are true with locally smooth $G$-PL manifold structure replacing $G$-smoothing and locally linear $G$-PL $R^n$-bundle reduction replacing $G$-vector bundle reduction.

(5) The $G$-PL Kister Theorem is true for locally linear $G$-PL $R^n$-microbundles.

(6) We do not have a theory going from $G$-manifolds to $G$-PL manifolds which are not locally smooth.

**7. $G$-obstruction theory.** Theorem 6.9 states that if $\dim M^H \neq 4$, then the isotopy classes of $G$-smoothings of $M$ are in one-to-one correspondence with the isotopy classes of $G$-vector bundle reductions of $\tau M$. If $M$ is locally smooth this can be stated in terms of classifying spaces, i.e., $\tau M$ is locally linear (§3) and hence contains a locally linear $G$-$R^n$ bundle (§2) which is represented by a $G$-homotopy class of maps $t: M \to B\,\mathrm{Top}_n(G)$ (§1), and isotopy classes of reductions correspond to $G$-homotopy classes of lifts $\bar{t}: M \to BO_n(G)$ of $t$.

In order to enumerate $G$-homotopy classes of lifts we use an equivariant obstruction theory for extending an equivariant lift over successive skeleta of an equivariant $CW$-complex. The definition of an *equivariant CW-complex* or *G-CW-complex* is obtained from the definition of an ordinary $CW$-complex as follows: Instead of adjoining $n$-cells $D^n$ by maps from $S^{n-1}$, we adjoin $G$-spaces of the form $G/H \times D^n$ (equivariant $n$-cells) by equivariant maps from $G/H \times S^{n-1}$. The standard elementary properties of $CW$-complexes remain valid in the equivariant case (with the obvious definitions):

(1) $G$-homotopy extension property,

(2) $G$-cellular approximation theorem,

(3) $G$-cofibration property for $L \subset K$ a $G$-$CW$ complex pair.

A $G$-map $f: X \to Y$ between arbitrary $G$-spaces is an *n-equivalence* of $G$-spaces, $n \geqq 1$, if for each $H \subset G$ the map $f^H: X^H \to Y^H$ is an $n$-equivalence. If $f$ is an $n$-equivalence for all $n$, $f$ is called a *weak G-homotopy equivalence*. The following are proved in [**M1**]:

(4) If $f: X \to Y$ is an $n$-equivalence of $G$-spaces, and $K$ is a $G$-$CW$ complex, then $f_*: [K, X]_G \to [K, Y]_G$ is injective if $\dim K < n$ and surjective if $\dim K \leqq n$. Here $[K, X]_G = G$-homotopy classes of $G$-maps.

(5) A $G$-map $f: K \to L$ between $G$-$CW$ complexes is a weak $G$-homotopy equivalence if and only if it is a $G$-homotopy equivalence.

EXAMPLE 1. A smooth $G$-manifold $M^n$ is an $n$-dimensional equivariant countable $CW$-complex [**M1**].

EXAMPLE 2. A topological $G$-manifold $M^n$ is a $G$-retract of a smooth $G$-manifold $N^{n+k}$ and hence a $G$-retract of a finite dimensional equivariant $CW$-complex.

In fact, by [**B1**], $M^n$ can be $G$-embedded in a representation space $R_\rho^{n+k}$ as a closed subspace. By [**S3**], $M$ is a $G$-$ANR$ and hence there is a $G$-invariant open neighborhood $N^{n+k}$ of $M$ in $R_\rho^{n+k}$ which retracts onto $M$.

Next we show that a locally smooth $G$-manifold is a $G$-deformation retract of a smooth $G$-manifold. First we need the existence of inverse bundles:

DEFINITION 1. Let $\mu$ be a locally linear $G$-$R^n$ bundle over a $G$-space $X$. A locally linear $G$-$R^n$ bundle $\nu$ over $X$ is *inverse* to $\mu$ if $\mu \oplus \nu \simeq 1_\rho^G(X)$, some $\rho: G \to O(n+m)$. (Recall that $1_\rho^G(X) = G \times_G (X \times R_\rho^{n+m}) = X \times R_\rho^{n+m}$ with diagonal $G$-action.)

PROPOSITION 1. *A locally linear G-$R^n$ bundle over a finite dimensional G-CW complex has an inverse.*

The proof will follow the lines of Milnor's proof for the $G$-trivial case [**M2**]:

First we consider a $G$-space $B$ constructed by taking a $p$-cell $G/H \times D^p$ and identifying $G/H \times S^{p-1}$ to a fixed point $a$. Taking the standard $D^{p-1}$ in $D^p$, the reflection $r$ of $D^p$ across $D^{p-1}$ induces a $G$-map $\bar{r}: B \to B$ by $\bar{r}\,[g, x] = [g, \bar{r}x]$, $\bar{g} \in G/H$, $x \in D^p$. Note that $\bar{r}(a) = a$. If $\mathscr{E}$ is a locally linear $G$-$R^n$ bundle over $B$ then the fibre $\mathscr{E}_a$ over $a$ may be identified with $R_\rho^n$, some $\rho: G \to O(n)$.

LEMMA 2. $\mathscr{E} \oplus \bar{r}^*\mathscr{E} \simeq 1_{\rho\oplus\rho}^G(B)$.

PROOF. Let $B \vee B$ be the one point union with respect to $a$. Since $\bar{r}(a) = a$, we can identify $(\bar{r}^*\mathscr{E})_a$ with $\mathscr{E}_a$ and we can form $\mathscr{E} \vee \bar{r}^*\mathscr{E}$ over $B \vee B$ by this identification. Also we have $\mathscr{E} \vee 1_\rho$ and $1_\rho \vee \mathscr{E}$ over $B \vee B$ by the identification of $\mathscr{E}_a$ with $(1_\rho(B))_a$.

Let $\bar{c}: B \to B \vee B$ be the $G$-map induced from the standard map $c: S^p \to$

$S^p \vee S^p$ of degree one in each factor and collapsing $S^{p-1}$ to $a$. Let $f\colon B \vee B \to B$ be the identity map in the first factor and $\bar{r}$ on the second. Then $fc$ is $G$-homotopic to the constant map $B \to a$. Thus we have:

(1) $\bar{c}^*(\mathscr{E} \vee 1_\rho) \simeq \bar{c}^*(1_\rho \vee \mathscr{E})$,

(2) $f^*\mathscr{E} \simeq \mathscr{E} \vee \bar{r}^*\mathscr{E}$ and $c^*f^*\mathscr{E} \simeq 1_\rho(B)$, i.e., $c^*(\mathscr{E} \vee \bar{r}^*\mathscr{E}) \simeq 1_\rho(B)$.

(3) $(1_\rho \oplus 1_\rho)(B) \simeq c^*((\mathscr{E} \vee r^*\mathscr{E}) \oplus (1_\rho \vee 1_\rho)) \simeq c^*(\mathscr{E} \vee 1_\rho) \oplus c^*(1_\rho \vee \bar{r}^*\mathscr{E}) \simeq \mathscr{E} \oplus \bar{r}^*\mathscr{E}$.

Proof of Proposition 1. Let $\mu$ be a locally linear $G$-$R^n$ bundle over a finite dimensional $G$-$CW$ complex $K$. The proof is by induction on $\dim K$. If $\dim K = 0$, $K$ is the disjoint union of 0-cells, $G/H_i \times D_j^0$, $H_i \subset G$. First consider a single 0-cell $G/H \times D^0$; then $\mu|(G/H \times D^0) \simeq 1_\rho^H(D^0) \simeq G \times_H R_\rho^n$, some $\rho\colon H \to O(n)$. If $\rho$ is a subrepresentation of $\bar{\rho}\colon G \to O(n+k)$, i.e., $\bar{\rho}|H = \rho \oplus \varphi$, $\varphi\colon H \to O(k)$, $1_{\bar{\rho}}^G(G/H) \simeq \mu \oplus \nu$, where $\nu \simeq 1_\varphi^H(D^0) \simeq G \times_H R_\varphi^k$, and $\mu \oplus \nu \simeq G \times_H (R_\rho^n \times R_\varphi^k)$. In fact, the $G$-equivalence $\psi\colon \mu \oplus \nu \to 1_{\bar{\rho}}^G(G/H) \simeq G/H \times R_{\bar{\rho}}^{n+k}$ with diagonal $G$-action, is given by $\psi[g, (x, y)] = (\bar{g}, g(x+y))$.

If $K = \coprod G/H_i \times D_j^0$, then $\mu|(G/H_i \times D_j^0) = 1_{\rho_{i,j}}^{H_i}(D_j^0)$, $\rho_{i,j}\colon H_i \to O(n)$. Since there are only a finite number of $H \subset G$ and a finite number of $n$-dimensional representations of $H$ in $O(n)$, there is a representation $\bar{\rho}\colon G \to O(n+k)$, some $k$, such that each $\rho_{i,j}$, is a subrepresentation of $\bar{\rho}$ (see **[B1]**). Thus $1_{\bar{\rho}}^G(K) = \mu \oplus \nu$, where $\nu|(G/H_i \times D_j^0) \simeq 1_{\varphi_{i,j}}^{H_i}(D_j^0)$.

Now assume the result holds for dimensions less than $p$, and $\dim K = p$. Then $\mu|K^{p-1}$ has an inverse $\eta$. If $\sigma\colon G/H \times D^p \to K$ is a $p$-cell, $\sigma^*\mu \simeq 1_\rho^H(D^p) \simeq (G \times_H R_\rho^n) \times D^p$ and if $\bar{\rho}\colon G \to O(n+k)$ with $\bar{\rho}|H = \rho \oplus \varphi$, then $\eta \oplus 1_{\bar{\rho}}(K^{p-1})$ satisfies

$$(\partial\sigma)^*(\eta \oplus 1_{\bar{\rho}}) \simeq (\partial\sigma)^*\eta \oplus (\partial\sigma)^*\mu \oplus 1_\varphi^H(S^{p-1}) \simeq 1_\lambda^G(G \times_H S^{p-1}) \oplus 1_{\bar{\rho}}^H(S^{p-1}),$$

where $\mu \oplus \eta = 1_\lambda^G(K^{p-1})$. That is, $\eta \oplus 1_{\bar{\rho}}$ extends over the $p$-cell $\sigma$. Again we can choose a single $\bar{\rho}$ which works for all $p$-cells, and hence $\eta \oplus 1_{\bar{\rho}}$ extends to $\eta'$ over $K$.

Consider $K \cup \operatorname{cone}(K^{p-1})$. Since $\mu \oplus \eta'|K^{p-1}$ is trivial, it extends to some bundle $\mathscr{L}$ over $K \cup \operatorname{cone}(K^{p-1})$. But $K \cup \operatorname{cone}(K^{p-1})$ has the $G$-homotopy type of a wedge of $G$-spaces of the form of $B$ in Lemma 2. Hence $\mathscr{L}$ has an inverse $\mathscr{L}^*$, and $\mu \oplus \eta' \oplus \mathscr{L}^*|K$ is trivial.

Corollary 3. *Let $M^n$ be a locally smooth G-manifold; then $M^n$ is a G-deformation retract of a smooth G-manifold $N^{n+k}$. In particular, $M^n$ has the G-homotopy type of a finite dimensional G-CW complex.*

Proof. Since $M^n$ is a $G$-retract of a $G$-neighborhood $V$ in $R_\rho^{n+k}$, $r^*\tau M$ over $V$ has an inverse and hence $\tau M \simeq \iota^*r^*\tau M$ has an inverse $\nu$. Now just as in Milnor **[M2]**, $\tau E(\nu) \simeq p^*\tau M \oplus p^*\nu$, $p\colon E(\nu) \to M$. Hence $\tau E(\nu)$ is trivial, and in particular reduces to a $G$-vector bundle. Thus by Theorem 6.5, $E(\nu)$ is $G$-smoothable, and by Theorem 2.11, $M$ is a $G$-deformation retract of $N = E(\nu)$.

Remark. The existence of an inverse bundle $\nu$ implies that given a $G$-embedding of $M$ in a locally smooth $G$-manifold $N$, then $M$ has a normal locally linear $G$-$R^n$ bundle $(\tau N|M) \oplus \nu$ in $N \times R_\rho^q$, where $\tau M \oplus \nu = 1_\rho^q$. See **[M2]**.

Following Bredon **[B7]**, Illman **[I1]**, and Bierstone **[B2]**, we describe equivariant singular and cellular cohomology theory: A *contravariant coefficient system* $\lambda$ for

$G$ is a contravariant functor from the category of $G$-spaces of the form $G/H$ and $G$-maps to the category of modules over a ring $R$ with unit. Then Illman defines an equivariant singular cohomology with coefficients in $\lambda$ on the category of $G$-spaces and $G$-maps, satisfying the following analogues of the Eilenberg-Steenrod axioms: The first five axioms are as in **[E1]**, simply substituting $G$-maps and $G$-homotopy. The excision axiom is satisfied in the following strong sense:

EXCISION AXIOM. *An inclusion of the form* $i: (X - U, A - U) \to (X, A)$, *where* $\bar{U} \subset \operatorname{Int} A$ *(U and A are G-subsets of the G-space X) induces isomorphisms* $i^*: H^n(X, A; \lambda) \to H^n(X - U, A - U; \lambda)$.

The dimension axiom takes the form

DIMENSION AXIOM. *For any* $H \subset G$, $H^n(G/H; \lambda) = 0$ *for* $n \neq 0$ *and there exists an isomorphism* $\gamma_H: H^0(G/H, \lambda) \to \lambda(G/H)$ *which commutes with homomorphisms induced by G-maps* $G/H \to G/K$.

Illman proves that any two equivariant cohomology theories satisfying the above axioms (with the excision axiom weakened by the assumption that $U$ is open in $X$) and defined on the category of finite dimensional $G$-$CW$ complexes are isomorphic in that category.

Bredon defines a cellular equivariant cohomology theory with coefficients in $\lambda$ satisfying the axioms and defined on the category of all $G$-$CW$ complexes. We follow Bierstone's description: Define the $n$th cellular cochain module $\bar{C}^n(K, L; \lambda)$ as follows. Fix a characteristic map $\sigma: (G/H \times D^n, G/H \times S^{n-1}) \to (K^n, K^{n-1})$ for each equivariant $n$-cell of $(K, L)$. Denote by $s(\sigma)$ the subgroup $H \subset G$, and let $P = \bigoplus_{H \subset G} \lambda(G/H)$. Let $F_n$ be the free abelian group generated by distinguished characteristic maps $\sigma$ for the $n$-cells of $(K, L)$, and define $\bar{C}^n(K, L; \lambda)$ as the $R$-module of (abelian group) homomorphisms $f: F_n \to P$ such that $f(\sigma) \in \lambda(G/s(\sigma))$. $\delta: \bar{C}^n(K, L; \lambda) \to \bar{C}^{n+1}(K, L; \lambda)$ is given by $(\delta f)(\tau) = \sum [\tau; \sigma] f(\sigma)$, where the sum is over the $n$-cells $\sigma$, and $[\tau, \sigma]$ is a homomorphism $\lambda(G/s(\sigma)) \to \lambda(G/s(\tau))$ generalizing the usual incidence number. Let $\bar{H}^n(K, L; \lambda)$ be the cellular cohomology modules.

Now since we are dealing with obstructions to a $G$-cross-section in a $G$-bundle it is necessary to consider equivariant cohomology with local coefficients. To understand this first recall the properties of a $G$-fibration:

DEFINITION 2. A $G$-map $f: X \to Y$ of $G$-spaces is a *G-fibration* if $f$ satisfies the $G$-homotopy lifting property for $G$-$CW$ complexes.

Bredon **[B7]** shows $f$ is a $G$-fibration if and only if $f^H: X^H \to Y^H$ is a Serre fibration for each $H \subset G$.

Thus the obstruction to extending a $G$-cross-section over the $n$-skeleton of a $G$-$CW$ complex, assuming one over the $n-1$ skeleton, comes down to extending the lift over each $G$-$n$ cell $G/H \times D^n$ given it over $G/H \times S^{n-1}$. This in turn comes down to extending the lift over $S^{n-1}$ in $f^H: X^H \to Y^H$ to $D^n$.

Recall that in the $G$-trivial case, a local coefficient bundle over $X$ is a bundle $\mathscr{B}$ with fibre an $R$-module and with structure group $\operatorname{Aut}(F)$ with discrete topology. Such a bundle is determined by a characteristic homomorphism $\chi: \pi_1(X, x_0) \to \operatorname{Aut}(F)$ up to equivalence.

DEFINITION 3. Let $X$ be a $G$-space. A *local coefficient system* $\mathscr{B} = \mathscr{B}_X$ over $X$ is a family of local coefficient bundles $\mathscr{B}_H^i$ over each component $X_i^H$ of $X^H$, $H \subset G$, such that if we set $\mathscr{B}_X(G/H) = \bigcup_i \mathscr{B}_H^i$ (the fibres over different components may be different) $\mathscr{B}_X$ is a contravariant functor from the category of $G$-spaces $\{G/H\}$ and $G$-maps to the category of local coefficient bundles and bundle homomorphisms. Explicitly, if $\rho$: $G/H_1 \to G/H_2$ is a $G$-map and $\rho(eH_1) = gH_2$, then $\rho$ defines $\rho^*$: $X^{H_2} \to X^{H_1}$ by $\rho^*(x) = gx$ and $\mathscr{B}_X(\rho)$: $\mathscr{B}_X(G/H_2) \to \mathscr{B}_X(G/H_1)$ is a map over $\rho^*$ which is a homomorphism on each fibre.

Further, we assume $\mathscr{B}_X$ is locally trivial: For each $x \in X$, there is a slice $V_x$ and trivializations $\psi_H$: $\mathscr{B}_X(G/H)|V_x^H \to V_x \times \mathscr{B}_X(G/H)_x$, $H \subset G_x$, such that for $\rho$: $G/H_1 \to G/H_2$ and $H_1, H_2 \subset G_x$, we have a commutative diagram

$$\begin{array}{ccc} \mathscr{B}(G/H_2)|V_x^{H_2} & \xrightarrow{\psi_{H_2}} & V_x^{H_2} \times \mathscr{B}(G/H_2)_x \\ \downarrow{\scriptstyle \mathscr{B}(\rho)|V_x^{H_2}} & & \downarrow{\scriptstyle \rho^* \times \mathscr{B}(\rho)_x} \\ \mathscr{B}(G/H_1)|V_x^{H_1} & \xrightarrow{\psi_{H_1}} & V_x^{H_1} \times \mathscr{B}(G/H_1)_x \end{array}$$

when $\rho(eH_1) = gH_2$, $g \in G_x$.

REMARKS. (1) If $G = (e)$, then $\mathscr{B}_x$ is just an ordinary local coefficient bundle.

(2) If each $\mathscr{B}_H^i$ is trivial, then $\mathscr{B}_X$ is just a contravariant coefficient system in the sense of Illman.

(3) If $f$: $Y \to X$ is a $G$-map and $\mathscr{B}_X$ is a local coefficient system over $X$, $f^*\mathscr{B}_X$ is a local coefficient system over $Y$.

To define equivariant singular cohomology with coefficients in $\mathscr{B}_X$, recall **[I1]** that a singular $G$-$n$-simplex is a $G$-map $T$: $G/H \times \Delta_n \to X$. Note that $T(\bar{e} \times \Delta_n) \in X^H$ and $(T|\bar{e} \times \Delta_n)^*\mathscr{B}(G/H) \simeq \Delta_n \times \mathscr{B}(G/H)_{T(0)}$, $0 \in \Delta_n$ the 0-vertex. Write $t(T) = G/H$. Let $\hat{C}^n(X; \mathscr{B})$ be the group of functions $c(T)$ on the singular $G$-$n$-simplices such that $c(T) \in \mathscr{B}(tT)_{T(0)}$. Define $C^n(X; \mathscr{B})$ as the subgroup of $\hat{C}^n(X; \mathscr{B})$ satisfying: If $T$: $G/H \times \Delta_n \to X$ and $T'$: $G/H' \times \Delta_n \to X$ and $h$: $G/H \times \Delta_n \to G/H' \times \Delta^n$ is a $G$-map which covers $\mathrm{id}_{\Delta_n}$ with $T = T'h$, then $c(T) = h^*c(T')$. Here $h^*$: $\mathscr{B}(G/H') \to \mathscr{B}(G/H)$ is the map defined by $h|G/H$: $G/H \to G/H'$, where $G/H$ is identified with $G/H \times 0$.

Define $\delta$: $\hat{C}^n(X; \mathscr{B}) \to \hat{C}^{n+1}(X; \mathscr{B})$ by $(\delta c)(T) = \sum_{i=0}^n (-1)^i p_i^* c(T^{(i)})$, where $T^{(i)}$ is the $i$th face of $T$ and $p_i^*$: $\mathscr{B}(t(T^{(i)}))_{T^{(i)}(0)} \to \mathscr{B}(tT)_{T(0)}$ is defined by the above trivialization of $(T|\Delta_{n+1})^*\mathscr{B}(tT)$. Then $\delta$ restricts to $\delta$: $C^n(X; \mathscr{B}) \to C^{n+1}(X; \mathscr{B})$ and we define $H^n(X; \mathscr{B}) = H^n(C(X; \mathscr{B}))$.

If $A \subset X$ is $G$-invariant, define $C^n(X, A; \mathscr{B})$ as the subgroup of $C^n(X; \mathscr{B})$ of functions $c$ which vanish on simplices in $A$. Define $H^n(X, A; \mathscr{B}) = H^n(C(X, A; \mathscr{B}))$. Also define $H^n(A; \mathscr{B}) = H^n(A; \mathscr{B}|A)$.

Of course when $G = (e)$, this is simply cohomology with local coefficients, and when the $\mathscr{B}_H^i$ are trivial this is just Illman's equivariant cohomology.

Consider the category of pairs $(X; \mathscr{B}_X)$, $\mathscr{B}_X$ a local coefficient system in $X$, and bundle maps $(f, \varphi)$; $f$: $X \to Y$ a $G$-map, $\varphi$: $\mathscr{B}_X \to \mathscr{B}_Y$ a bundle map over $f$, i.e., for each component $X_i^H$ of $X^H$, $\varphi|\mathscr{B}_X(G/H)^i$: $\mathscr{B}_X(G/H)^i \to \mathscr{B}_Y(G/H)^j$ is a bundle map over $f|X_i^H$: $X_i^H \to Y_j^H$, and $(\varphi|X^{H_2})\mathscr{B}_X(p) = \mathscr{B}_Y(\rho)(\varphi|X^{H_1})$ where $\rho$: $G/H_1 \to G/H_2$. Then one may prove, just as in Illman, that $H^*(X; \mathscr{B}_X)$ is a cohomology theory on this category: For example:

HOMOTOPY AXIOM. *If* $(f, \varphi), (g, \psi)\colon (X, A; \mathscr{B}_X) \to (Y, B; \mathscr{B}_Y)$ *are G-homotopic, then* $(f, \varphi)^* = (g, \psi)^*\colon H^*(Y, B; \mathscr{B}_Y) \to H^*(X, A; \mathscr{B}_X)$.

EXCISION AXIOM. *If* $\bar{U} \subset \operatorname{Int} A$, *U G-invariant, then* $i^*\colon H^*(X, A; \mathscr{B}_X) \to H^*(X-U, A-U; \mathscr{B}_X|X-U)$ *is an isomorphism.*

DIMENSION AXIOM. *For any* $H \subset G$ *and* $\mathscr{B}$ *over* $G/H$, $H^n(G/H; \mathscr{B}) = 0$ *for* $n \neq 0$ *and there exists an isomorphism* $\gamma_H^{\mathscr{B}}\colon H^0(G/H; \mathscr{B}) \to \mathscr{B}(G/H)_{\bar{e}}$ *such that if* $(\rho, \varphi)\colon (G/H_1, \mathscr{B}_1) \to (G/H_2, \mathscr{B}_2)$, *then* $\varphi \circ \gamma_{H1}^{\mathscr{B}_1} \circ (\rho, \varphi)^* = \mathscr{B}_2(\rho)\gamma_{H2}^2$.

Any two such theories satisfying these axioms (with $U$ open in the excision axiom) on the category $\{(X; \mathscr{B}_X)\}$, with $X$ a finite dimensional $G$-$CW$ complex, are naturally equivalent.

Further, we can define a cellular theory in the category of local coefficient systems over $G$-$CW$ complexes as follows: Let $\bar{C}^n(X, \mathscr{B}_X)$ be the group of functions on the $G$-$n$ cells such that if $\sigma\colon G/H \times D^n \to X$, $c(\sigma) \in \mathscr{B}(G/H)_{\sigma(0)}$, $0 \in D^n$. Since $(\sigma|D^n)^*\mathscr{B}(G/H) \simeq D^n \times \mathscr{B}(G/H)_{\sigma(0)}$, we again have a well-defined coboundary. The resultant theory, $\bar{H}^n(X, A; \mathscr{B}_X)$ will satisfy the axioms.

REMARK. The cellular theory may be defined from the singular theory as in Dold [D2]. One gets that $\bar{H}^n(X; \mathscr{B}) = H^n(X^{(n+k)}; \mathscr{B})$ for $k > 0$. (However, one does not get that $\bar{H}^n(X; \mathscr{B}) = H^n(X; \mathscr{B})$ unless $X$ is finite dimensional.)

Now we wish to prove:

THEOREM 4. *Let* $M^n$ *be a locally smooth G-manifold and* $\mathscr{B}$ *a local coefficient system over* $M$; *then* $H^i(M^n; \mathscr{B}) = 0$ *for* $i > n$, *and if no component of M is compact,* $H^n(M^n; \mathscr{B}) = 0$.

Before giving the proof let us recall some facts from the $G$-trivial case: We begin with the sheaf duality theorem [S4]. Let $A$ be a subspace of the paracompact space $X$ and $S$ a sheaf over $X$. Then

(1) If $X - A = M^n$ is a topological $n$-manifold and $A$ is closed in $X$

$$\check{H}^i(X, A; S) \simeq H_{n-i}^{\varphi M}(M; T \otimes S|M).$$

Here $\check{H}(X, A; S)$ is Čech cohomology with coefficients in $S$. $H_{n-i}^{\varphi M}(M, T \otimes S/M)$ is singular homology based on locally finite chains with supports in $\varnothing_M$ and coefficients in $T \otimes S\,|\,M$, where $\varphi_M = \{B \subset M|B$ is closed in $X\}$ and $T$ is the orientation sheaf.

(2) If $X$ is homologically locally connected (e.g., locally contractible) and $A$ is taut in $X$ (e.g., $A$ open or $A$ a closed $ANR$) then $\check{H}^i(X, A; S) \simeq H^i(X, A; S)$ for $S$ a local coefficient bundle, where the right side is singular cohomology.

In particular, if $X = M^n$ and $A = \varphi$, $S$ a local coefficient bundle we have:

(3) $H^i(M; S) \simeq H_{n-i}(M; T \otimes S)$, where the left is singular cohomology and the right is singular homology based on locally finite chains. Thus $H^i(M; S) = 0$ for $i > n$. If no component of $M$ is compact, then every 0-simplex is the boundary of an infinite 1-chain and $H_0(M; T \otimes S) = 0$, and thus $H^n(M; S) = 0$.

(4) More generally, for any family $\varphi$ of closed supports and any sheaf $S$ on a topological $n$-manifold $M$

$$\check{H}_\varphi^i(M; S) \simeq H_{n-i}^\varphi(M; T \otimes S).$$

(5) In fact, (1) follows from (4) and the isomorphisms

$$\check{H}^i(X, A; S) \simeq \check{H}^i(X; S_M) \simeq \check{H}^i_{\varphi M}(M; S|M),$$

$A$ closed in $X$.

(6) Let $\varphi$ be a family of closed supports in $X$ and $S$ a sheaf over $X$. Then $\check{H}^i(X; S) = \text{Inj Lim } H^i(X, S_U)$ over open $U$ such that $\bar{U} \in \varphi$. ($S_U$ is $S|U$ extended by zero to $X$.)

This implies

(7) Let $A$ be closed in $X$, $S$ a sheaf over $X$, $\{U\}$ open neighborhoods of $A$. Then

$$\check{H}^i(X, A; S) = \lim_U \check{H}^i(X, \bar{U}; S) = \lim_U \check{H}^i(X, U; S).$$

(8) Let $A$ be closed in $X$, $M = X-A$, $S$ a sheaf over $M$. Then

$$\check{H}^i_{\varphi M}(M;S) = \lim_U \check{H}^i(M, \bar{U} \cap M; S) = \lim_U \check{H}^i(M, U \cap M; S),$$

over open neighborhoods of $A$.

(9) In particular, if $S$ is a coefficient bundle over $M^n = X - A$, $A$ closed in $X$, then from (8), (4), and (2)

$$H^{\varphi M}_{n-i}(M, T \otimes S) = \lim_U H^i(M, U \cap M; S)$$

over open neighborhoods of $A$. This is zero for $i > n$, and for $i = n$ if $\bar{M}$ has no compact components.

Proof of Theorem 4. Let $M^n$ be a locally smooth $G$-manifold and let $b\bar{M}_i^{(H)} = M_i^{(H)} - M^i_{(H)}$. Recall (§2) that $bM_i^{(H)}$ is a union of $M_j^{(K)}$ where $(K) < (H)$. Inductively, we will prove that $H_q(M_i^{(H)}; \mathscr{B}|M_i^{(H)} = 0$ for $q > \dim M_i^H$.

First note that it follows from Corollary 3 that $bM_i^{(H)}$ is a $G$-$ANR$, since for smooth $G$-manifolds it is actually an equivariant subcomplex.

For a minimal element, $bM_i^{(H)} = \emptyset$ and $M_i^{(H)}/G = M^i_{(H)}/G = M_i^H/(N(H)/H)$ is a manifold of dim $M_i^H$. Also note that any $T: G/K \times \Delta_n \to M_i^{(H)}$ may be factored through $T': G/H \times \Delta_n \to M_i^{(H)}$. Thus we need look at only simplices of the form $G/H \times \Delta_n$; but these correspond to singular simplices of $M_i^{(H)}/G$. Further, since $N(H)/H$ acts freely on $M_i^H$, $\mathscr{B}(G/H)^i$ is induced from a unique local coefficient bundle $S$ on $M_i^{(H)}/G$. Thus we see that

$$C^*(M_i^{(H)}; \mathscr{B}|M_i^{(H)}) = C^*(M_i^{(H)}/G; S)$$

and hence $H^q(M_i^{(H)}; \mathscr{B}|M^{(H)}) \simeq H^q(M_i^{(H)}/G; S)$. Thus the result holds for minimal elements by the sheaf duality theorem above.

It follows by induction and the Mayer-Vietoris sequence that if the result is proved for $M_j^{(K)}$, $(K) < (H)$, then $H^q(bM_i^{(H)}; \mathscr{B}|bM_i^{(H)}) = 0$ for $q \geqq \dim M_i^H$. Now consider $H^q(M_i^{(H)}, bM_i^{(H)}; \mathscr{B}|M_i^{(H)})$. Since $bM_i^{(H)}$ is a $G$-$ANR$, it follows that this last group is equal to $\text{Lim } H^q(M_i^{(H)}, U, \mathscr{B}|M_i^{(H)})$ over $G$-invariant open neighborhoods of $bM_i^{(H)}$. By excision,

$$H^q(M_i^{(H)}, U; \mathscr{B}|M_i^{(H)}) = H^q(M^i_{(H)}, U \cap M^i_{(H)}; \mathscr{B}|M^i_{(H)}).$$

But by the same argument as for the minimal elements, this last may be identified with $H^q(M^i_{(H)}/G, U \cap M^i_{(H)}/G; S)$ for a unique local coefficient bundle over $M^i_{(H)}/G$. By (9) above, $\text{Lim } H^q(M^i_{(H)}/G, U \cap M^i_{(H)}/G; S)$ over $(U \cap M^i_{(H)})/G =$

$U/G \cap M^i_{(H)}/G$ must be zero for $q > \dim M^H_i$. Hence $H^q(M_i^{(H)}, bM_i^{(H)}; \mathscr{B})$ is zero for $q > \dim M_i^H$, and hence $H^q(M_i^{(H)}; \mathscr{B}|M_i^{(H)}) = 0$ for $q > \dim M_i^{(H)}$. Finally, if $(H)$ is the principal orbit type, and $\overline{M^i_{(H)}}/G = M_i^{(H)}/G = M/G$ will be noncompact if $M$ has no compact components, and $H^n(M_i^{(H)}; \mathscr{B}|M_i^{(H)}) = H^n(M; \mathscr{B}) = 0$, completing the proof.

REMARK. The above argument actually shows that there is a sheaf $S$ defined on $M/G$ so that $H^q(M/G; S) \simeq H^q(M; \mathscr{B})$.

*Question.* Does Theorem 4 imply $M$ has the homotopy type of an $n$-dimensional $G$-$CW$ complex ($n - 1$ dimensional if no component is compact)?

Let $p: E \to K$ be a $G$-fibration over a $G$-$CW$ complex $K$. Let $F_x = p^{-1}(x)$. For each $x \in K$ we *assume* $F_x^H$ is connected and $q$-simple. For each $q > 0$ define a local coefficient system $\mathscr{B} = \mathscr{B}(\pi_q) = \tilde{\pi}_q$ on $K$ by taking $\mathscr{B}^i_H$ to be the bundle whose fibre over $x \in K_i^H$ is $\pi_q(F_x^H)$. It is easy to check that $\mathscr{B}$ satisfies the definition of local coefficient system. In particular, the local triviality of $\mathscr{B}$ follows from the fact that any $x \in K$ is a strong $G_x$ deformation retract of a $G_x$ neighborhood, and the equivariant lifting property.

By the usual arguments we get (see **[B7]**)

THEOREM 5. *Let $L \subset K$ be a $G$-subcomplex.*

(a) *Let $f$ be a $G$-cross-section of $p$ over $L \cup K^{q-1}$. Then $f$ determines $c_f \in \bar{C}^q(K, L; \tilde{\pi}_{q-1})$ such that*

(1) $\delta c_f = 0$. *Hence $c_f$ determines $[c_f] \in \bar{H}^q(K, L; \tilde{\pi}_{q-1})$.*

(2) *$f$ extends to a cross-section over $L \cup K^q$ if and only if $c_f = 0$.*

(3) $[c_f] = 0$ *if and only if there exists a $G$-cross-section $f'$ over $L \cup K^q$ with $f'|L \cup K^{q-2} = f|L \cup K^{q-2}$.*

(b) *Given two $G$-cross-sections $f_0, f_1$ which coincide on $L$ and a $G$-homotopy $h$* rel *$L$ over $L \subset K^{q-1}$ of $f_0$ to $f_1$, then there is a class $[d(f_0, h, f_1)] \in \bar{H}^q(K, L; \tilde{\pi}_q)$ such that $h|L \cup K^{q-2}$ is extendable* rel *$L$ over $L \cup K^q$ if and only if $[d(f_0, h, f_1)] = 0$.*

**8.** $\mathrm{Top}_n(G)$. From the $G$-obstruction theory of §7 and Theorem 1.10, we see that $G$-smoothings require the study of $\mathrm{Top}_n^\alpha$ and $\mathrm{Top}_n^\alpha / O_n^\alpha$, $\alpha: H \to O(n)$ a representation of $H \subset G$. $\mathrm{Top}_n^\alpha(O_n^\alpha)$ is the subgroup of $\mathrm{Top}_n(O_n)$ commuting with $\alpha(h)$, $h \in H$. We first study $\mathrm{Top}_n^\alpha$ under stabilization.

Let $\alpha: H \to O_n \subset O_{n+1} \subset \cdots \subset O_{n+k} \subset \cdots$. We assume $H$ acts on $S^{n-1}$ via $\alpha$ without fixed points. The action of $H$ via $\alpha$ on $R^{n+s} = R^n \times R^s$, some $s$, induces actions on invariant subspace such as $D^{n+k} = D^n \times D^k$, $S^{n+k} = S^{n-1} * S^k$, $S^{n+k} \times D^l = (S^{n-1} * S^k) \times D^l$, $R^{n+k} \times D^l = R^n \times R^k \times D^l$, etc. By $A^\alpha(D^{n+k} \bmod \partial)$, $A^\alpha(D^{n+k} \bmod D^k \cup \partial)$, etc., we mean the group of homeomorphisms commuting with the action of $H$ and fixed on the indicated subspace. Actually, we take the singular complex of these groups and treat them semisimplicially. We also need the complex $A_\gamma^\alpha(R^{n+k} \times D^l \bmod R^k \times D^l \cup \partial)$ of automorphism germs, i.e., an $r$-simplex is an equivalence class of $H$-embeddings $\Delta_r \times R_\varepsilon^{n+k} \times D^l \to \Delta_r \times R^{n+k} \times D^l$ which commute with projection onto $\Delta_r$ and are the inclusion on $\Delta_r \times (R_\varepsilon^k \times D^l \cup \partial)$, where $R_\varepsilon^{n+k}$ is an open $\varepsilon$-ball about $0 \in R^{n+k}$, and two such embeddings defined on $R_{\varepsilon_1}^{n+k} \times D^l$ and $R_{\varepsilon_2}^{n+k} \times D^l$ are identified if they agree on $R_{\varepsilon_3}^{n+k} \times D^l$, $\varepsilon_3 < (\varepsilon_1, \varepsilon_2)$.

Using the $G$-isotopy extension theorem of §4, the arguments below are analogous to those for $G$-trivial **[B5]**. First note that $A^\alpha(D^{n+k} \bmod D^k \cup \partial)$ is contractible by the Alexander trick. Also $A^\alpha(R^{n+k} \times D^l \bmod D_\varepsilon^{n+k} \times D^l \cup \partial)$, $D_\varepsilon^{n+k}$ the closed $\varepsilon$-ball, is contractible similarly. Thus the restriction map

$$A^\alpha(R^{n+k} \times D^l \bmod R^k \times D^l \cup \partial) \to A_\gamma^\alpha(R^{n+k} \times D^l \bmod R^k \times D^l \cup \partial)$$

is a Kan fibration with contractible fibre and hence a homotopy equivalence.

Let $P^\alpha(S^{n+k} \bmod S^k)$ be the group of pseudoisotopies (or concordances), i.e.,

$$P^\alpha(S^{n+k} \bmod S^k) = A^\alpha(S^{n+k} \times D^l \times I \bmod S^{n+k} \times D^l \times (0) \cup S^{n+k} \times \partial D^l \times I).$$

PROPOSITION 1. *For* $k \geqq -1$,

$$P^\alpha(S^{n+k} \bmod S^k) \xrightarrow{r} A^\alpha(S^{n+k} \bmod S^k) \xrightarrow{\Sigma} A^\alpha(S^{n+k+1} \bmod S^{k+1})$$

*is a fibration (up to homotopy), where $r$ is restriction to $S^{n+k} \times (1)$ and $\Sigma$ is suspension.*

PROOF. This follows from the fibration:

$$P^\alpha(S^{n+k} \bmod S^k) \xrightarrow{i} A^\alpha(D^{n+k+1} \bmod D^{k+1}) \xrightarrow{r_1} A_\gamma^\alpha(R^{n+k+1} \bmod R^{k+1}),$$

and the commutative diagram:

$$\begin{array}{ccc} A^\alpha(D^{n+k+1} \bmod D^{k+1}) & \xrightarrow{r_1} & A_\gamma^\alpha(R^{n+k+1} \bmod R^{k+1}) \\ \simeq \uparrow c & & \simeq \uparrow r_2 \\ A^\alpha(S^{n+k} \bmod S^k) & \xrightarrow{\Sigma} & A^\alpha(S^{n+k+1} \bmod S^{k+1}) \end{array}$$

where $c$ is coning, $r_1$ and $r_2$ are restrictions to germs about 0. That $c$ and $r_2$ are homotopy equivalences follows again from the Alexander trick.

*Note.* $r_1$ and hence $\Sigma$ may not be surjective.

Let $\mathrm{Top}_{n+k}^\alpha/\mathrm{Top}_k$ be the fibre of $B\,\mathrm{Top}_k \to B\,\mathrm{Top}_{n+k}^\alpha$.

PROPOSITION 2. $A^\alpha(S^{n+k} \bmod S^k) \simeq \mathrm{Top}_{n+k}^\alpha/\mathrm{Top}_k$, $k \geqq 0$.

PROOF. This follows from the fibration:

$$A^\alpha(S^{n+k} \bmod S^k) \to A^\alpha(S^{n+k} \bmod S^0) \xrightarrow{r} A(S^k \bmod S^0)$$

and the commutative diagram

$$\begin{array}{ccc} A^\alpha(S^{n+k} \bmod S^0) & \xrightarrow{r} & A(S^k \bmod S^0) \\ \simeq \downarrow & & \downarrow \simeq \\ \mathrm{Top}_{n+k}^\alpha & \underset{i}{\overset{r}{\rightleftarrows}} & \mathrm{Top}_k \end{array}$$

Note that because $S^k$ is the fixed point set of the action it is invariant, similarly $R^k = 0 \times R^k \subset R^{n+k}$.

From Propositions 1 and 2 we have

PROPOSITION 3. *For* $k \geqq 0$,

$$P^\alpha(S^{n+k} \bmod S^k) \xrightarrow{r} \mathrm{Top}_{n+k}^\alpha/\mathrm{Top}_k \xrightarrow{i} \mathrm{Top}_{n+k+1}^\alpha/\mathrm{Top}_{k+1}$$

*is a fibration.*

PROPOSITION 4. *For $k \geqq 0$, there is a fibration*

$$P^\alpha(S^{n+k}) \to \mathrm{Top}^\alpha_{n+k}/O^\alpha_{n+k} \xrightarrow{i} \mathrm{Top}^\alpha_{n+k+1}/O^\alpha_{n+k+1}.$$

In fact, by the same arguments as above we have:

PROPOSITION 1'. *For $k \geqq -1$ we have a fibration*

$$P^\alpha(S^{n+k}) \xrightarrow{r} A^\alpha(S^{n+k}) \xrightarrow{\Sigma} A^\alpha(S^{n+k+1} \bmod S^0).$$

PROPOSITION 2'. *For $k \geqq 0$, $A^\alpha(S^{n+k} \bmod S^0) \simeq \mathrm{Top}^\alpha_{n+k}$.*

PROPOSITION 3'. *For $k \geqq -1$ we have a fibration*

$$P^\alpha(S^{n+k}) \to A^k(S^{n+k}) \to \mathrm{Top}^\alpha_{n+k+1}.$$

PROOF OF PROPOSITION 4. Now note that $O^\alpha_{n+k+1}/O^\alpha_{n+k} = O_{k+1}/O_k = S^k$ and that we have a commutative diagram:

$$\begin{array}{ccccc} A^\alpha(S^{n+k} \bmod S^0) & \longrightarrow & A^\alpha(S^{n+k}) & \longrightarrow & S^k \\ \uparrow & & \uparrow & & \uparrow \\ O^\alpha_{n+k} & \longrightarrow & O^\alpha_{n+k+1} & \longrightarrow & S^k \end{array}$$

Hence

$$A^\alpha(S^{n+k})/O^\alpha_{n+k+1} \simeq A^\alpha(S^{n+k} \bmod S^0)/O^\alpha_{n+k} \simeq \mathrm{Top}^\alpha_{n+k}/O^\alpha_{n+k}.$$

Since also

$$\begin{array}{ccc} O^\alpha_{n+k+1} & \longrightarrow & \mathrm{Top}^\alpha_{n+k+1} \\ & \searrow \quad \nearrow & \\ & A^\alpha(S^{n+k}) & \end{array}$$

commutes, we get that $\mathrm{Top}^\alpha_{n+k}/O^\alpha_{n+k} \simeq A^\alpha(S^{n+k})/O^\alpha_{n+k+1} \to \mathrm{Top}^\alpha_{n+k+1}/O^\alpha_{n+k+1}$ also has homotopy fibre $P^\alpha(S^{n+k})$.

Now let $\tilde{A}^\alpha(S^{n+k})$, $\widetilde{\mathrm{Top}}^\alpha_{n+k}$, etc., be the complex of block automorphisms commuting with the action [B5]. As pointed out by Morlet in the $G$-trivial case, $\tilde{P}^\alpha(S^{n+k})$, etc., is always contractible [**B5**]. Now all the arguments above go through for block automorphisms and lead to:

PROPOSITION $\tilde{1}$. *For $k \geqq -1$, $\tilde{A}^\alpha(S^{n+k} \bmod S^k) \simeq^\Sigma \tilde{A}^\alpha(S^{n+k+1} \bmod S^{k+1})$.*

PROPOSITION $\tilde{2}$. *$\tilde{A}^\alpha(S^{n+k} \bmod S^k) \simeq \widetilde{\mathrm{Top}}^\alpha_{n+k}/\widetilde{\mathrm{Top}}_k$, $k \geqq 0$.*

PROPOSITION $\tilde{3}$. *$\widetilde{\mathrm{Top}}^\alpha_{n+k}/\widetilde{\mathrm{Top}}_k \simeq \widetilde{\mathrm{Top}}^\alpha_{n+k+1}/\widetilde{\mathrm{Top}}_{k+1}$, $k \geqq 0$.*

PROPOSITION $\tilde{4}$. *$\widetilde{\mathrm{Top}}^\alpha_{n+k}/O^\alpha_{n+k} \simeq \widetilde{\mathrm{Top}}^\alpha_{n+k+1}/O^\alpha_{n+k+1}$, $k \geqq 0$.*

PROPOSITION $\tilde{1}'$. *For $k \geqq -1$, $\tilde{A}^\alpha(S^{n+k}) \simeq \tilde{A}^\alpha(S^{n+k+1} \bmod S^0)$.*

PROPOSITION $\tilde{2}'$. *$\tilde{A}^\alpha(S^{n+k} \bmod S^0) \simeq \widetilde{\mathrm{Top}}^\alpha_{n+k}$, $k \geqq 0$.*

PROPOSITION $\tilde{3}'$. *For $k \geqq -1$, $\tilde{A}^\alpha(S^{n+k}) \simeq \widetilde{\mathrm{Top}}^\alpha_{n+k+1}$.*

PROPOSITION 5. *$\Omega^{k+1}A^\alpha(S^{n+k} \bmod S^k) \simeq A^\alpha(S^{n-1} \times D^{k+1} \bmod \partial)$ and $\Omega^{k+1}P^\alpha(S^{n+k} \bmod S^k) \simeq P^\alpha(S^{n-1} \times D^{k+1} \bmod \partial)$. Further, the equivalences commute with the restriction map.*

PROOF. We have the fibrations

(a) $$A^\alpha(D^{n+k} \bmod D^k \cup \partial) \to A^\alpha(S^{n+k-l} \times D^l \bmod S^{k-l} \times D^l \cup \partial) \to A^\alpha_\gamma(R^{n+k-l} \times D^l \bmod R^{k-l} \times D^l \cup \partial).$$

(b) $$A^\alpha(S^{n+k-l-1} \times D^{l+1} \bmod S^{k-l-1} \times D^{l+1} \cup \partial) \to A^\alpha(D^{n+k} \bmod D^k \cup \partial) \to A^\alpha_\gamma(R^{n+k-l} \times D^l \bmod R^{k-l} \times D^l \cup \partial).$$

Since $A^\alpha(D^{n+k} \bmod D^k \cup \partial)$ is contractible we get for $l \leqq k$,

(c) $$\Omega A^\alpha(S^{n+k-l} \times D^l \bmod S^{k-l} \times D^l \cup \partial) \simeq A^\alpha(S^{n+k-l-1} \times D^{l+1} \bmod S^{k-l-1} \times D^{l+1} \cup \partial).$$

Hence the first statement follows by induction. Since the identical fibrations hold for $P^\alpha$ and the restriction $r: P^\alpha \to A^\alpha$ induces a map of fibrations, the second and third statements follow.

For $\tilde{A}^\alpha$ we have from Proposition $\tilde{1}$, that $\tilde{A}^\alpha(S^{n+k} \bmod S^k) \simeq \tilde{A}^\alpha(S^{n-1})$. On the other hand, we have from the fibration:

$$\tilde{A}^\alpha(S^{n-1} \times D^{l+1} \bmod \partial) \to \tilde{P}^\alpha(S^{n-1} \times D^l \bmod \partial) \to \tilde{A}^\alpha(S^{n-1} \times D^l \bmod \partial).$$

PROPOSITION $\tilde{5}$. $\tilde{A}^\alpha(S^{n-1} \times D^{k+1} \bmod \partial) \simeq \Omega^{k+1}\tilde{A}^\alpha(S^{n-1})$.

Also we note that by definition we have an exact sequence

$$\pi_0 P^\alpha(S^{n-1} \times D^{k+1}) \to \pi_0 A^\alpha(S^{n-1} \times D^{k+1} \bmod \partial) \to \pi_0 \tilde{A}^\alpha(S^{n-1} \times D^{k+1} \bmod \partial) \to 0.$$

Hence by taking $\Omega^{k+1}$ of the fibration of Proposition 4, we get

PROPOSITION 6. *There is an exact sequence* ($k \geqq 0$)

$$0 \to \pi_{k+1} \tilde{A}^\alpha(S^{n-1}) \to \pi_{k+1} \mathrm{Top}^\alpha_{n+k+1}/\mathrm{Top}_{k+1} \to \pi_k P^\alpha(S^{n+k} \bmod S^k) \to \pi_k \mathrm{Top}^\alpha_{n+k}/\mathrm{Top}_k \to \pi_k \mathrm{Top}^\alpha_{n+k+1}/\mathrm{Top}_k \to \pi_{k-1} P^\alpha(S^{n+k} \bmod S^k) \to \cdots \to \pi_0 \mathrm{Top}^\alpha_{n+k}/\mathrm{Top}_k \to \pi_0 \mathrm{Top}^\alpha_{n+k+1}/\mathrm{Top}_{k+1}.$$

Further we have a commutative diagram

$$\begin{array}{ccccc} & & \pi_i \mathrm{Top}^\alpha_{n+i}/\mathrm{Top}_i & \to & \pi_i \mathrm{Top}^\alpha_{n+i+r}/\mathrm{Top}_{i+r} \\ & \nearrow & \downarrow & & \downarrow \\ \pi_i \tilde{A}^\alpha(S^{n-1}) & & & & \\ & \underset{\simeq}{\searrow} & & & \\ & & \pi_i \widetilde{\mathrm{Top}}^\alpha_{n+i}/\widetilde{\mathrm{Top}}_i & \underset{\simeq}{\to} & \pi_i \widetilde{\mathrm{Top}}^\alpha_{n+i+r}/\widetilde{\mathrm{Top}}_{i+r} \end{array}$$

Hence we get

PROPOSITION 7. $\pi_i \tilde{A}^\alpha(S^{n-1}) \to \pi_i \mathrm{Top}^\alpha_{n+i+r}/\mathrm{Top}_{i+r}$ *is split injective*, $r \geqq 0$.

From Proposition 6, using the fact that

$$\pi_i(\mathrm{Top}^\alpha_{n+k+1}/\mathrm{Top}_{k+1}, O^\alpha_{n+k+1}/O_{k+1}) \simeq \pi_i(\mathrm{Top}^\alpha_{n+k+1}/O^\alpha_{n+k+1}, \mathrm{Top}_{k+1}/O_{k+1})$$

and that $O^\alpha_n \simeq O^\alpha_{n+k+1}/O_{k+1}$ all $k$, we get

PROPOSITION 8. *There is an exact sequence* ($k \geqq 0$)

$$0 \to \pi_{k+1}(\tilde{A}^\alpha(S^{n-1}), O_n^\alpha) \to \pi_{k+1}(\mathrm{Top}^\alpha_{n+k+1}/O^\alpha_{n+k+1}, \mathrm{Top}_{k+1}/O_{k+1})$$
$$\to \pi_k P^\alpha(S^{n+k} \bmod S^k) \to \pi_k(\mathrm{Top}^\alpha_{n+k}/O^\alpha_{n+k}, \mathrm{Top}_k/O_k)$$
$$\to \pi_k(\mathrm{Top}^\alpha_{n+k+1}/O^\alpha_{n+k+1}, \mathrm{Top}_{k+1}/O_{k+1}) \to \pi_{k-1} P^\alpha(S^{n+k} \bmod S^k)$$
$$\to \cdots \to \pi_0(\mathrm{Top}^\alpha_{n+k}/O^\alpha_{n+k}, \mathrm{Top}_k/O_k) \to \pi_0(\mathrm{Top}^\alpha_{n+k+1}/O^\alpha_{n+k+1}, \mathrm{Top}_{k+1}/O_{k+1}).$$

Also, $\pi_i(\tilde{A}^\alpha(S^{n-1}), O_n^\alpha) \to \pi_i(\mathrm{Top}^{\ \alpha}_{n+i+r}/O^\alpha_{n+i+r}, \mathrm{Top}_{i+r}/O_{i+r})$ is split injective, $r \geqq 0$.

PROPOSITION 9. *There is a fibration* $P^\alpha(S^{n+k} \bmod S^k) \to P^\alpha(S^{n+k}) \to P(S^k)$. *Further, for* $k \geqq 5$, $\pi_i P(S^k) = 0$ *for* $i \leqq k+1$. *Hence for* $k \geqq 5$ *we have an exact sequence*

$$\pi_k P^\alpha(S^{n+k} \bmod S^k) \to \pi_k(\mathrm{Top}^\alpha_{n+k}/O^\alpha_{n+k}) \to \pi_k(\mathrm{Top}^\alpha_{n+k+1}/O^\alpha_{n+k+1})$$
$$\to \pi_{k-1} P^\alpha(S^{n+k} \bmod S^k) \to \pi_{k-1}(\mathrm{Top}^\alpha_{n+k}/O^\alpha_{n+k}) \to \pi_{k-1}(\mathrm{Top}^\alpha_{n+k+1}/O^\alpha_{n+k+1})$$
$$\to \cdots \to \pi_0(\mathrm{Top}^\alpha_{n+k}/O^\alpha_{n+k}) \to \pi_0(\mathrm{Top}^\alpha_{n+k+1}/O^\alpha_{n+k+1}).$$

EXAMPLES. Let $Z_2$ act on $R^1$ by reflection and let $\alpha\colon \bigoplus Z_2 \to O(n)$ be the product action on $R^n$. It is not difficult to see that $\mathrm{Top}^\alpha_{n+k} \simeq \mathrm{Top}_k$ and $O^\alpha_{n+k} = O_k$.

We now assume that the action of $H$ on $S^{n-1}$ via $\alpha$ is *free*. We let $L^{n-1} = S^{n-1}/H$. Then the following are obvious.

PROPOSITION 10. (a) $A^\alpha(S^{n-1} \times D^{k+1} \bmod \partial) = A(L \times D^{k+1} \bmod \partial)$, $k \geqq 0$.
(b) $P^\alpha(S^{n-1} \times D^{k+1} \bmod \partial) = P(L \times D^{k+1} \bmod \partial)$, $k \geqq -1$.
(c) $A(L) = A^\alpha(S^{n-1})/C$, *$C$ the center of $H$ acting by covering transformations.*

Following Anderson and Hsiang [**A2**], define $P(D^l \times (S^{k-l-1} * \mathring{c}L), \partial)$ as the complex of pseudoisotopies of $D^l \times (S^{k-l-1} * \mathring{c}L)$ that are the identity over $\partial D^l \times (S^{k-l-1} * \mathring{c}L) \cup D^l \times (S^{k-l-1} * c)$, where $\mathring{c}L$ is the open cone on $L$ and $c$ is the vertex. Note that

$$S^{k-l-1} * \mathring{c}L = (R^{k-l} \times \mathring{c}L) \cup (S^{k-l-1} \times c).$$

Let $P(D^l \times R^{k-l} \times \mathring{c}L, \partial)$ be the complex of pseudoisotopies of $D^l \times R^{k-l} \times \mathring{c}L$ which are the identity over $(\partial D^l \times R^{k-l} \times \mathring{c}L) \cup (D^l \times R^{k-l})$. Then we have obviously

(a) The restriction $r\colon P(D^l \times (S^{k-l-1} * \mathring{c}L), \partial) \to P(D^l \times R^{k-l} \times \mathring{c}L, \partial)$ is an isomorphism.

(b) $P^\alpha(R^{n+k-l} \times D^l \bmod R^{k-l} \times D^l \cup \partial) \simeq P(D^l \times R^{k-l} \times \mathring{c}L, \partial)$.

As pointed out in our introductory remarks

(c) $P^\alpha(R^{n+k-l} \times D^l \bmod R^{k-l} \times D^l \cup \partial) \simeq P^\alpha_\gamma(R^{n+k-l} \times D^l \bmod R^{k-l} \times D^l \cup \partial)$.

By (a) of Proposition 5

(d) $P^\alpha(S^{n+k-l} \times D^l \bmod S^{k-l} \times D^l \cup \partial) \simeq P^\alpha_\gamma(R^{n+k-l} \times D^l \bmod R^{k-l} \times D^l \cup \partial)$.

By (c) of Proposition 5

(e) $\Omega^l P^\alpha(S^{n+k} \bmod S^k) \simeq P^\alpha(S^{n+k-l} \times D^l \bmod S^{k-l} \times D^l \cup \partial)$.

Thus we have:

PROPOSITION 11. $\Omega^l P^\alpha(S^{n+k} \bmod S^k) \simeq P(D^l \times (S^{k-l-1} * \mathring{c}L), \partial)$.

REMARKS. For any closed manifold $L$ one may prove:

$$P_b(L \times D^l \times R^{k-l+1}, \partial) \simeq P((L \times S^{k-l}) \times D^l, \partial) \simeq P(D^l \times (S^{k-l} * \mathring{c}L), \partial).$$

Anderson and Hsiang give a rather difficult and lengthy proof. Actually the first equivalence is true for any compact *space* $L$ and follows from a direct Kister-Mazur type argument. The second is essentially the Alexander trick for cones and some fibrations analogous to those in Proposition 5 (using Siebenmann's isotopy extension theorem for stratified spaces). We will not need these here, but one should note that in our case:

$$P^\alpha(S^{n+k} \bmod S^k) \simeq P_b(L \times R^{k+1}) \simeq P(L * S^k \bmod S^k).$$

THEOREM 12 (ANDERSON AND HSIANG). *If* $\dim L + k \geqq 5$,

$$\pi_0 P(D^l \times (S^{k-l-1} * \mathring{c}L), \partial) \simeq \begin{cases} K_{-k+1+l}(Z\pi_1 L), & l < k-1, \\ \tilde{K}_0(Z\pi_1 L), & l = k-1, \\ Wh_1(\pi_1 L), & l = k. \end{cases}$$

COROLLARY 13. *If* $n - 1 + k \geqq 5$

$$\pi_l P^\alpha(S^{n+k} \bmod S^k) \simeq \begin{cases} K_{-k+1+l}(Z\pi_1 L), & l < k-1, \\ \tilde{K}_0(Z\pi_1 L), & l = k-1, \\ Wh_1(\pi_1 L), & l = k. \end{cases}$$

Using Corollary 13 and Proposition 10 in Proposition 6, we get:

THEOREM 14. *There is an exact sequence* ($k \geqq 0$, $\dim L + k = n - 1 + k \geqq 5$):

$$\begin{aligned} 0 &\to \pi_{k+1} \tilde{A}(L) \to \pi_{k+1} \mathrm{Top}^\alpha_{n+k+1}/\mathrm{Top}_{k+1} \to Wh_1(\pi_1 L) \to \pi_k \mathrm{Top}^\alpha_{n+k}/\mathrm{Top}_k \\ &\to \pi_k \mathrm{Top}_{n+k+1}/\mathrm{Top}_{k+1} \to \tilde{K}_0(Z\pi_1(L)) \to \pi_{k-1} \mathrm{Top}^\alpha_{n+k}/\mathrm{Top}_k \\ &\to \pi_{k-1} \mathrm{Top}^\alpha_{n+k+1}/\mathrm{Top}_{k+1} \to K_{-1}(Z\pi_1 L) \to \pi_{k-2} \mathrm{Top}^\alpha_{n+k}/\mathrm{Top}_k \\ &\to \cdots \to K_{-k+1}(Z\pi_1 L) \to \pi_0 \mathrm{Top}^\alpha_{n+k}/\mathrm{Top}_k \to \pi_0 \mathrm{Top}^\alpha_{n+k+1}/\mathrm{Top}_{k+1}. \end{aligned}$$

One may, of course, make similar substitutions in Propositions 8 and 9.

REMARKS. Using results of Bass and others, one has:

For $\pi$ abelian, $K_{-s}(Z\pi) = 0$ for $s > 1$.

For $\pi$ abelian and prime power order, $K_{-1}(Z\pi) = 0$.

For $\pi$ cyclic of order $p$, $\tilde{K}_0(Z\pi) =$ class-group of $Q(e^{2\pi i/p})$.

For $\pi$ finite, $\tilde{K}_0(Z\pi)$ is finite.

For $\pi$ free abelian, $\tilde{K}_0(Z\pi) = 0$.

Let $\mathscr{H}(L) =$ space of homotopy equivalences of $L^{n-1}$; then from **[A3]**, we have the surgery exact sequence, with $\mathscr{L}^s_i(\pi)$ the Wall surgery obstruction group for simple homotopy equivalences, $\pi = \pi_1(L)$; $\dim L + i \geqq 6$, $L^* = L \cup$ (point):

$$\begin{aligned} &\to [\Sigma^{i+1} L^*, G/\mathrm{Top}] \to \mathscr{L}^s_{n+i}(\pi) \to \pi_i(\mathscr{H}(L)/\tilde{A}(L)) \\ &\to [\Sigma^i L^*, G/\mathrm{Top}] \to \cdots \to \pi_1(\mathscr{H}(L)/\tilde{A}(L)) \qquad (15) \\ &\to [\Sigma L^*, G/\mathrm{Top}] \to \mathscr{L}^s_n(\pi). \end{aligned}$$

Thus for example, if $n + k \geqq 6$ one may in theory compute $\pi_i \mathrm{Top}^\alpha_{n+k}/\mathrm{Top}_k$, $0 \leqq i \leqq k$, up to extensions, from (14) and (15).

For $i = 0$, we begin with Proposition 3′ to give a fibration $P^\alpha(S^{n-1}) \to A^\alpha(S^{n-1}) \to \mathrm{Top}^\alpha_n$. Note that $\mathrm{Top}^\alpha_n$ is isomorphic to $A^\alpha_0(S^{n-1} \times R)$, where the subscript means end preserving. Hence we have a map of fibrations

$$\begin{array}{ccccc} & & C & = & C \\ & & \downarrow & & \downarrow \\ P^\alpha(S^{n-1}) & \longrightarrow & A^\alpha(S^{n-1}) & \longrightarrow & A_0^\alpha(S^{n-1} \times R) \\ \simeq\downarrow & & \downarrow & & \downarrow \\ P(L) & \longrightarrow & A(L) & \longrightarrow & A_0(L \times R) \end{array}$$

Note that we have an exact sequence, dim $L \geqq 5$,

$$\pi_0 A(L) \to \pi_0(A_0(L \times R)) \xrightarrow{\lambda} Wh(\pi_1 L),$$

where $\lambda$ gives the obstruction to the $h$-cobordism between $g(L \times 0)$ and $L \times t$, $t$ large, being a product. For Lens spaces [**A2**], this is always trivial.

Since we also have a map of fibrations

$$\begin{array}{ccccc} A^\alpha(S^{n-1} \times I \bmod \partial) & \longrightarrow & P^\alpha(S^{n-1}) & \longrightarrow & A^\alpha(S^{n-1}) \\ \simeq\downarrow & & \simeq\downarrow & & \downarrow \\ A(L \times I \bmod \partial) & \longrightarrow & P(L) & \longrightarrow & A(L) \end{array}$$

$\pi_1 A_0^\alpha(S^{n-1} \times R) \simeq \pi_1 A_0(L \times R)$ since this is the same as $\pi_0 A^\alpha(S^{n-1} \times I \bmod \partial) \to \pi_0 A^\alpha(L \times I \bmod \partial)$. Thus we have exact sequences

$$0 \to \pi_0 \tilde{A}(L) \to \pi_0 A_0(L \times R) \to Wh_1(\pi_1 L)$$

and

$$0 \to C \to \pi_0 \mathrm{Top}_n^\alpha \to \pi_0 A_0(L \times R) \to 0.$$

Therefore $\pi_0 \mathrm{Top}_n^\alpha$ is determined up to extension from $\pi_0 \tilde{A}(L)$. From (15) we can determine $\pi_1(\mathscr{H}(L)/\tilde{A}(L))$ up to extension. On the other hand, we have $\pi_1(\mathscr{H}(L)/\tilde{A}(L)) \to \pi_0 \tilde{A}(L) \to \pi_0 \mathscr{H}(L)$ exact. Finally, we have

PROPOSITION 16. *$\pi_0 \mathrm{Top}_{n+k}^\alpha/\mathrm{Top}_k \to \pi_0 \mathrm{Top}_{n+k+1}^\alpha/\mathrm{Top}_{k+1} \xrightarrow{\tau} K_{-k}(\pi_1 L)$ is exact. In particular, if $\pi_1 L$ is abelian, the first map is surjective, $k > +1$.*

PROOF. Write $L * S^{k+1} = (L * S^k) * S^0 \simeq (S^0 * L) * S^k$. Consider the fibration

$$F \to A(L * S^{k+1} \bmod S^{k+1}) \xrightarrow{r} A_\gamma(\mathring{c}L * S^k \bmod S^k).$$

The fibre $F$ consists of homeomorphisms of $S^0 * L * S^k$ which are the identity on $S^0 * S^k$ and near $c * L * S^k$. $F$ is contractible by the Alexander trick. Hence $[h] \in \pi_0 A(L * S^{k+1} \bmod S^{k+1})$ will be in the image of $\pi_0 A(L * S^k \bmod S^k)$ if $rh$ is the restriction of $k \in A(cL * S^k \bmod S^k)$.

Now following Anderson and Hsiang, an embedding $f\colon \mathring{c}L * S^k \to cL * S^k$ fixed on $c * S^k$ (i.e., $f \in A_\gamma(\mathring{c}L * S^k \bmod S^k)$) determines an embedding $\bar{f}\colon T^{k+1} \times c_\varepsilon L \to T^{k+1} \times cL$ agreeing with $f$ on $D_\varepsilon^{k+1} \times c_\varepsilon L$ and fixed on $T^{k+1} \times c$, $T^{k+1}$ the torus. Then closure$(T^{k+1} \times cL - T^{k+1} \times c_\varepsilon L)$ is an $h$-cobordism with base $T^{k+1} \times L$. Let $\tau(f) \in Wh_1(\pi_1(L) \times Z^{k+1})$ be the Whitehead torsion. As shown by Anderson and Hsiang following Siebenmann, $\tau$ actually lies in $K_{-k}(\pi_1 L)$.

Now if $\tau$ is trivial we get a homeomorphism $g$ of $T^{k+1} \times cL$ agreeing with $f$ on $D_\varepsilon^{k+1} \times c_\varepsilon L$ and fixed on $T^{k+1} \times c$. Then $g$ is covered by a bounded homeomorphism $\tilde{g}$ on $R^{k+1} \times cL$ agreeing with $f$ on a neighborhood of $D_\varepsilon^{k+1} \times c_\varepsilon L$ and fixed on $R^{k+1} \times c$. This in turn extends to a homeomorphism of $D^{k+1} \times cL = cL * S^k$ fixed on $S^k$ and agreeing with $f$ on a neighborhood of $0 \times c = c_\varepsilon L * S^k$.

*Addendum.* The same argument shows that if $\alpha_1, \alpha_2: H \to O_n$ are free on $S^{n-1}$, then the obstruction to showing that $\alpha_1, \alpha_2$ are $\text{Top}_{n+k}$ equivalent when they are $\text{Top}_{n+k+1}$ equivalent lies in $K_{-k}(Z(H))$.

REMARK. This means the exact sequence of (14) may be extended to $\to \pi_0 \text{ Top}^{\alpha}_{n+k}/\text{Top}_k \to \pi_0 \text{ Top}^{\alpha}_{n+k+1}/\text{Top}_{k+1} \to K_{-k}(Z\pi_1 L)$.

DEFINITION. Let $M$ be a locally smooth $G$-manifold. $M$ is said to have a spine of codimension $\geqq r$ if the cohomological dimension of $M_i^{(H)}/G \leqq \dim M_i^H - r$, $M^H \neq \emptyset$. Similarly for $\partial M^H$. All $H \subset G$.

EXAMPLE 1. $M \times R^r$ where $G$ acts trivially on $R^r$ has a spine of codim $\geqq r$.

EXAMPLE 2. If $M^H$ and $\partial M^H$ are noncompact for all $H \subset G$ then $M$ has a spine of codim $\geqq 1$.

EXAMPLE 3. If there exist a $G$-$CW$ complex $(K, \partial K)$ and a $G$-homotopy equivalence $\psi: (K, \partial K) \to (M, \partial M)$ and $\dim M^H - \dim K^H \geqq r$ and similarly for $\partial M^H$, then $M$ has a spine of codim $\geqq r$.

Let $\mathscr{S}_G(M)$ be the isotopy classes of $G$-smoothings of $M$.

Let $\bar{\mathscr{S}}_G(M)$ be the isotopy classes of stable $G$-smoothings, i.e., of $M \times R^s$, $s$ arbitrarily large.

Let $M_0 = M - \partial M$. Consider the following statements:

$A_r$: $M$ spine codim $r$ then $\mathscr{S}_G(M) \to \bar{\mathscr{S}}_G(M)$ is epi,

$B_r$: $M$ spine codim $r$ then $\mathscr{S}_G(M) \to \bar{\mathscr{S}}_G(M)$ is bijective,

$C_r$: $M$ spine codim $r$ then $\mathscr{S}_G(M) \to \bar{\mathscr{S}}_G(M_0)$ is epi.

THEOREM 17. *Let $M$ be a locally smoothable $G$-manifold, $G$ finite and acting semi-freely. Then if* $\dim M - \dim M^G \neq 2$, *and* $\dim M^G \geqq 5$:

(1) *If $G$ is finite abelian,* $A_2$, $B_3$, $C_2$ *are true.*

(2) *If further $G$ is of prime power order,* $A_1$, $B_2$, $C_3$ *are true.*

(3) *If further $G$ is of prime order $p$ and the class group of* $Q(e^{2\pi i/p}) = 0$, *then* $A_0$, $B_1$, $C_0$ *are true.*

(4) *If* $G = Z_2, Z_3$ *then* $A_0$, $B_0$, $C_0$ *are true.*

PROOF. This follows from Proposition 9, Corollary 13 and the remarks following Theorem 14, and by the obstruction theory of §7.

Finally, we note a result of R. Schultz **[S2]**:

THEOREM 18 (SCHULTZ). *If $G$ is of odd prime power order, then two representations* $\rho_1: G \to O_n$, $\rho_2: G \to O_n$ *are* $\text{Top}_n$ *equivalent if and only if they are* $O_n$ *equivalent.*

**9.** In §8 we developed some exact sequences which give us a hold on computing the obstruction groups for equivariant smoothing. In this section we develop some further techniques which allow us to tackle the stable obstruction groups. In view of the stabilization Theorem 8.17, these stable groups yield interesting information concerning smoothings of actions.

Recall the notation: We let $B \text{ Top}_n(G)$ be the basespace of universal $G$-$(\text{Top}_n, O_n)$ bundles, and $BO_n(G)$ the basespace of universal $G$-$O_n$ bundles. We consider the natural map $p_n: BO_n(G) \to B \text{ Top}_n(G)$, which we can take as a $G$-fibration. For $K$ a locally smoothable $n$-dimensional $G$-manifold, we have the tangent map

$t\colon K \to B\,\mathrm{Top}_n(G)$. We are looking for lifts of $t$, $\bar{t}\colon K \to BO_n(G)$. By §7 the obstructions to having such lifts lie in the homotopy groups of the fibres of $p_n^H, p_n^H\colon BO_n(G)^H \to B\,\mathrm{Top}_n(G)^H$, where $H$ is a subgroup of $G$. These fibres depend on the point $x \in B\,\mathrm{Top}_n(G)^H$ one is looking at. This dependency is as follows:

We consider *stable* representations $\alpha\colon H \to O_k \subset O_{k+1} \subset \cdots \subset 0$. Such a stable representation has a smallest dimensional representative of dimension $i(\alpha)$. For $L_m = \mathrm{Top}_m$ or $O_m$, $m \geqq i(\alpha)$, $H$-acts on $L_m$ via $\alpha$ by $hl(x) = hlh^{-1}(x)$, $h \in H$, $x \in R^m$, $1 \in L_m$. The fixed point set of $L_m$ under this action will be denoted by $L_m^\alpha$. It is the centralizer of $\alpha(H)$ in $L_m$.

For a given $\alpha$, let $T_n(\alpha)$ be a complete set of representatives of all representations topologically equivalent to the $n$-dimensional representative of $\alpha$. That is, $T_n(\alpha)$ is a family of $n$-dimensional representations of $H$, no two distinct ones being linearly equivalent, all being topologically equivalent, and if given any representation $\alpha'$ topologically equivalent to $\alpha$, there is one (and only one) representation $\alpha''$ in $T_n(\alpha)$ linearly equivalent to $\alpha'$.

Let $EO_n(G) \xrightarrow{\lambda_n} BO_n(G)$ be the universal $n$-dimensional $G$-vector bundle. For $y \in (BO_n(G))^H$, $\lambda_n^{-1}(y)$ is an $H$-vector space and this defines, uniquely up to linear equivalence, a representation $\alpha_y\colon H \to O_n$. Then by 1.14,

$$(p_n^H)^{-1}(x) \cong \sum_{\alpha' \in T(\alpha)} \mathrm{Top}_n^{\alpha'}/O_n^{\alpha'},$$

where $\alpha = \alpha_y$ for any $y \in (p_n^H)^{-1}(x)$, and $\sum$ represents disjoint union.

We can now consider the following commutative diagram:

$$\begin{array}{ccccc} BO_n(G) & \xrightarrow{\psi_n} & \overline{BO}_n(G) & \xrightarrow{\iota} & BO_{n+1}(G) \\ & \searrow^{p_n} & \downarrow{\scriptstyle \bar{p}_n} & & \downarrow{\scriptstyle p_{n+1}} \\ & & B\,\mathrm{Top}_n(G) & \xrightarrow{\iota} & B\,\mathrm{Top}_{n+1}(G) \end{array}$$

where $p_n$ is the induced bundle over $B\,\mathrm{Top}_n(G)$ induced from the bundle $p_{n+1}$ and the natural stabilization map $\bar{\iota}\colon B\,\mathrm{Top}_n(G) \to B\,\mathrm{Top}_{n+1}(G)$. We can take $\psi_n$ also to be a $G$-fibration. We consider the fibre of $\psi_n^H\colon (BO_n(G))^H \to (\overline{BO}_n(G))^H$. For $z \in (\overline{BO}_n(G))^H$, let $F_{n,z}^H = (\psi_n^H)^{-1}(z)$. Then by definition

$$\pi_r(F_{n,z}^H) = \pi_{r+1}(\mathrm{Top}_{n+1}^\alpha; O_{n+1}^\alpha; \mathrm{Top}_n^\alpha),$$

where $\alpha = \alpha_{\iota(z)}$, $\iota(z) \in BO_{n+1}(G)$, and we have the exact sequence

$$\pi_r(F_{n,z}^H) \to \pi_r(\mathrm{Top}_n^\alpha, O_n^\alpha) \to \pi_r(\mathrm{Top}_{n+1}^\alpha, O_{n+1}^\alpha) \to \pi_{r-1}(F_{n,z}^H) \cdots .$$

We also have the exact sequence

$$\pi_r(F_{n,z}^H) \to \pi_r(O_{n+1}^\alpha, O_n^\alpha) \to \pi_r(\mathrm{Top}_{n+1}^\alpha, \mathrm{Top}_n^\alpha) \to \pi_{r-1}(F_{n,z}^H).$$

Let $m = m(z) = n - i(\alpha)$. Then $O_{n+1}^\alpha/O_n^\alpha = S^m$ and thus we have an exact sequence

$$\pi_r(F_{n,z}^H) \to \pi_r(S^m) \to \pi_r(\mathrm{Top}_{n+1}^\alpha, \mathrm{Top}_n^\alpha) \to \pi_{r-1}(F_{n,z}^H).$$

For any space $X$, recall $P(X)$ is the space of topological pseudoisotopies. If $X$ is an $H$-space, for some group $H$, so is $P(X)$, and if the actions of $H$ on $X$ is denoted by $\alpha$, then we write $P^\alpha(X)$ for $(P(X)^H)$.

PROPOSITION 1. *$\pi_r(F^H_{n,z}) \cong \pi_r(P^\alpha(S^n))$ for $r > 0$ and $\pi_0(P^\alpha(S^n)) \to \pi_0(F^H_{n,z})$ is onto.*

PROOF. This follows from 8.4 since both are fibers of the same maps.

Observe that we have the fibration $P^\alpha(S^n, S^m) \to P^\alpha(S^n) \to P(S^m)$, and by **[B5]** if $m \geqq 5$, $\pi_r(P(S^m)) = 0$ for $r \leqq m + 1$. Thus for $m \geqq 5$, we have $\pi_r(P^\alpha(S^n, S^m)) = \pi_r(P^\alpha(S^n))$, $r \leqq m$, and $\pi_{m+1}(P^\alpha(S^n, S^m)) \to \pi_{r+1}(P^\alpha(S^n))$ is epi. For these values of $r$ the homotopy groups are computed by 8.13 when the representation $\alpha$ defines a relatively free action of $H$.

We now consider a $G$-manifold $M$ and the following diagram:

$$\begin{array}{ccc} & & BO_\infty(G) = BO(G) \\ & & \downarrow \qquad\quad \downarrow \\ M & \xrightarrow{\bar{t}} & B\,\mathrm{Top}_\infty(G) = B\,\mathrm{Top}(G) \end{array}$$

We are looking for lifts of $\bar{t}$. As we noted the obstructions lie in $\pi_r(\mathrm{Top}^\alpha, O^\alpha)$, where $\alpha$ is a stable representation of a subgroup $H$ of $G$. For $H = (e)$ these homotopy groups are closely related to the stable homotopy groups of spheres and are known for small values of $r$. For $H \neq (e)$ the problem is more difficult. $\pi_0(\mathrm{Top}^\alpha, O^\alpha)$ is not known and probably not $(e)$ (see however 8.15 and 8.16), even for groups $Z/pZ$. We will derive results for the higher homotopy groups using certain indirect techniques which we sketch now.

Suppose $M$ is stably $G$ parallelizable. That is, we can take $\bar{t}(M) = x \in B\,\mathrm{Top}(G)^G$. Then $\pi^{-1}(x) = \mathrm{Top}/O$ on which $G$ acts via some stable representation $\alpha: G \to O_n \subset 0$ and $\bar{\mathcal{S}}_G(M)$, the stable $G$ smoothings of $M$, can be identified with the $G$ homotopy classes of mappings of $M$ into $\mathrm{Top}/O$. The obvious example of such a manifold is $S^n_\alpha$, a sphere on which $G$ acts linearly via a stable representation $\alpha$: $G \to O_{n+1} \subset O$.

For the remainder of this section we assume the representation is *relatively free*, that is, $G$ acts relatively freely on $S^n$ via $\alpha$. Consider the tangent diagram:

$$\begin{array}{ccc} & & BO_n(G) \\ & & \downarrow \\ S^n_\alpha & \xrightarrow{t} & B\,\mathrm{Top}_n(G) \end{array}$$

$t$ has a canonical lifting $t_0$ since $S^n_\alpha$ is a smooth $G$-manifold. By the exact $G$ homotopy sequence for $G$ fibrations, the homotopy classes of liftings of $t$ agreeing with $t_0$ at the basepoint $p \in S^n_\alpha$ correspond exactly to $[[S^n_\alpha, \mathrm{Top}_n/O_n]]$, where we use $[[\;\;]]$ to denote basepointed $G$ equivariant homotopy classes. (We are now assuming $n > i(\alpha)$, so there is a basepoint, and $G$ is action on $\mathrm{Top}_n/O_n$ via $\alpha$.) On the other hand, by our smoothing theory such homotopy classes of liftings also correspond to the slice concordance classes of smoothings of $S^n_\alpha, f\colon S^n_\alpha \to V$, where $f \mid D^n_\alpha$, $D^n_\alpha$ a disc neighborhood of $p$, is a diffeomorphism. Thus $\mathcal{S}_G(S^n_\alpha, p) = [[S^n_\alpha, \mathrm{Top}_n/O_n]]$.

We now introduce some geometrically defined groups studied in **[R1]** and **[R2]** which will give us useful approximations to $\mathcal{S}_G(S^n_\alpha, p)$. We recall some definitions from those papers. A smooth $G$-manifold is called a $(G, \alpha)$-manifold if for each $x \in M^G$ the linear action of $G$ on $M_x$ is stably equivalent to $\alpha$.

For $M$ a $(G, \alpha)$-manifold, the group of the bundle $t(M) \mid M^G \to M^G$ has a canonical reduction from $GL(n)$ to $O^\alpha_n$. An $\alpha$ *orientation* for $M$ is a further reduction of

$t(M)|M^G$ to the component of the identity of $O_n^\alpha$. If $M^G$ has simply connected components such a reduction always exists. If further $M$ and each component of $M^G$ is orientable and $G$ is of odd order, then such an orientation exists. Two $\alpha$ oriented $(G, \alpha)$-manifolds are called $(G, \alpha)$ *isomorphic* if there exists a $G$-diffeomorphism between them preserving the orientations.

As in **[R1]** we define $\Sigma_\alpha^n$ as the set of $(G, \alpha)$ isomorphism classes of $n$-dimensional $\alpha$ oriented $(G, \alpha)$-manifolds $M$ such that $G$ acts relatively freely on $M$ and such that $M$ is homeomorphic to $S^n$ and $M^G$ is homeomorphic to $S^m$. (Here and hereafter $m = n - i(\alpha)$.)

For the remainder of this section, unless otherwise stated, we assume $m \geqq 1$, and for the order of $G$ even, $m \neq 1$. Since we wish to apply $G$ engulfing, we assume $i(\alpha) \neq 2$. Let $f\colon S_\alpha^n \dashrightarrow V$ represent an element of $\mathscr{S}_G(S_\alpha^n, p)$. Then $V$ is a $(G, \alpha)$-manifold and is $\alpha$ orientable. In fact, assuming the orientation of $V$ restricts to the canonical one on $f(D_\alpha^n)$ determines a unique $\alpha$ orientation for $V$ with $f|D_\alpha^n \to V$ an orientation preserving embedding. Thus $V$ determines an element of $\Sigma_\alpha^n$ and thus we have $\psi\colon \mathscr{S}_G(S_\alpha^n, p) \to \Sigma_\alpha^n$ given by $\psi(f) = V$.

Proposition 2. *For $m \geqq 5$, $\psi$ is surjective.*

Proof. Let $V$ represent an element of $\Sigma_\alpha^n$. Let $V_0 = V - (p_1)$, where $p_1$ is a point of $V^G$. Then $R_\alpha^n$ is $G$-diffeomorphic to a neighborhood of $p_0$ in $V_0$, where $p_0$ is in $V_0^G$. Applying $G$ engulfing **[C3]** yields a $(G, \alpha)$-isomorphism $f_1\colon R_\alpha^n \cong V_0$ and compactifying yields an equivariant homeomorphism $f\colon S_\alpha^n \cong V$. Then $f^{-1}$ determines an element of $\mathscr{S}_G(S_\alpha^n, p)$ with $\psi(f^{-1}) = V$. Q.E.D.

We now recall that by **[R1]** for $m \geqq 1$ we can define an oriented $(G, \alpha)$ connected sum which puts a group structure on $\Sigma_\alpha^n$ and $\mathscr{S}_G(S_\alpha^n, p)$ and it is easily checked that with respect to these group structures $\psi$ is a homomorphism. The group structure on $\mathscr{S}_G(S_\alpha^n, p)$ agrees with the usual homotopically defined group structure on $[[S_\alpha^n, \mathrm{Top}_n/O_n]] \cong \mathscr{S}_G(S_\alpha^n, p)$.

As in **[R1]** we let $C_\alpha^n = \Sigma_\alpha^n / B_\alpha^n$, where $V$ is in $B_\alpha^n$ if $V = \partial W$ and $W$ is a (relatively free) $\alpha$ oriented $(G, \alpha)$-manifold whose orientations restrict to that of $V$, and $W$ is homeomorphic to $D^{n+1}$, and $W^G$ is homeomorphic to $D^{m+1}$. $C_\alpha^n$ is a quotient group of the group $\Sigma_\alpha^n$.

Proposition 3. *For $m \geqq 5$, there is a monomorphism $\lambda\colon C_\alpha^n \to [[S_\alpha^n, \mathrm{Top}_{n+1}/O_{n+1}]]$ such that the following diagram commutes.*

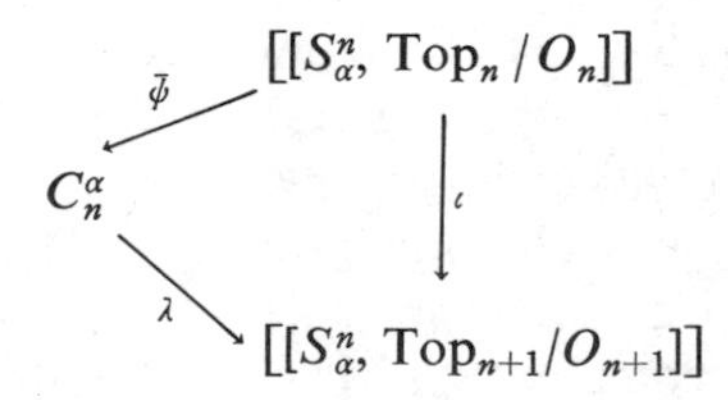


*where $\bar\psi$ is induced from $\psi$.*

Proof. Since $\psi$ is epi, we need only show that for any $f\colon S_\alpha^n \to V$ representing an element of $\mathscr{S}_G(S_\alpha^n, p) = [[S_\alpha^n, \mathrm{Top}_n/O_n]]$, $\bar\psi(f) = 0$ if and only if $\iota(f) = 0$. Let $V = \partial W$, where $W$ and $W^G$ are topological discs. Let $x \in W^G - \partial W^G$. Applying $G$

engulfing we get a $(G, \alpha)$-isomorphism $g: V \times [0, \infty) \cong W - (x)$. Then $g(f \times \mathrm{Id})$: $D_\alpha^{n+1} - (0) \to W - (x)$ is a $G$-homeomorphism and compactifying yields $h: D_\alpha^{n+1} \to W$ extending $f$. $\iota(f)$ represents a smoothing of $S_\alpha^n \times R$ which extends to a smoothing of $R_\alpha^{n+1}$ via $h$. Hence $\iota(f)$ is the zero element of $[[S_\alpha^n, \mathrm{Top}_{n+1}/O_{n+1}]]$. Conversely, if $\iota(f) = 0$ then $f$ extends to $f: D_\alpha^{n+1} \to W$, for some $W$ as above. Q.E.D.

The above constructions permit us to derive results on the equivariant smoothing obstruction groups. The above constructions are useful when $m \geqq 5$, and we assume this for the remainder of this section. The groups $C_\alpha^n$ are not computable for $n - m = 2$, and the case $n - m = 1$ is only interesting for $G = Z/2Z$ and has very special characteristics. *For simplicity we will assume for the remainder of this section that G is cyclic of prime power order.* Analogous results can be derived for more complicated groups but they are more complicated to state.

Recall that $F_n$ is the fiber of the $G$ fibration $\mathrm{Top}_n/O_n \to \mathrm{Top}_{n+1}/O_{n+1}$. From **[B5]** we know that $\pi_r(F_n) = 0$ for $r \leqq n$, and by 9.1, 8.13, and the remarks following 9.1, we have $\pi_r(F_n^G) = 0$ for $r < m - 1$, $\pi_{m-1}(F_n^G) = \widetilde{KO}(Z(G))$, and $\pi_m(F_n^G) = Wh(G)$.

Let $W = S^m \times R_\alpha^{n-m}$ be a $G$ tubular neighborhood of $S^m = (S_\alpha^n)^G$ in $S_\alpha^n$. Then the $G$ Puppe exact sequence yields an exact sequence

$$[S^m, F_n^G] \xleftarrow{u_1} [[S_\alpha^n, F_n]] \xleftarrow{u_2} [[S_\alpha^n; W, F_n; p]].$$

PROPOSITION 4. *$u_1$ is an isomorphism.*

PROOF. (a) $u_1$ is epi. Let $\psi: S^m \to F_n^G$ and $\psi_0: W \to F_n$ be any equivariant extension. The coefficients for the obstruction cocycles to extending $\psi_0 | \partial W$ to $S_\alpha^n - \mathrm{Int}\, W$ lie in $\pi_k(F_n)$ since $G$ acts freely on $S_\alpha^n - \mathrm{Int}\, W$, and these are all 0 for $k < n + 1$. Thus $\psi_0$ extends.

(b) $u_2$ is 0. We will show $[[S_\alpha^n; W, F_n; p]]$ is 0. An element of this group is represented by $\lambda: S_\alpha^n - \mathrm{Int}\, W \to F_n$ such that $\lambda | \partial W = p$. Let $V$ be the cone over $S_\alpha^n - \mathrm{Int}\, W$. Then $\lambda$ is defined on the base of $V$ and extend it by sending the cone over $\partial W$ to $p$. Then since $G$ acts freely on $V -$ (cone over $\partial W$) and $\pi_r(F_n) = 0$ for $r < n + 1$, by the argument of (a), $\lambda$ extends to $V$. Thus $\lambda$ represents 0 in $[[S_\alpha^n; W, F; p]]$ and since $\lambda$ was arbitrary this group is 0.

*Addendum.* The argument of Proposition 4 shows also that $[[S_\alpha^{n-1}; F^n]] \cong [S_\alpha^{m-1}, F_n^G]$ and $[[S_\alpha^{n-k}, F_n]] = 0$ for $k > 1$, that these results are independent of any restrictions on $m$.

By the definition of $F$ we have the following exact sequence: $[[S_\alpha^n, F_n]] \to [[S_\alpha^n, \mathrm{Top}_n/O_n]] \to [[S_\alpha^n, \mathrm{Top}_{n+1}/O_{n+1}]] \to [[S_\alpha^{n-1}, F_n]] \to$ . Using this and Propositions 3 and 4 yields the exact sequence

$$\begin{aligned} 0 \to C_\alpha^n \to [[S_\alpha^n, \mathrm{Top}_{n+1}/O_{n+1}]] &\xrightarrow{k_n} \widetilde{KO}(Z(G)) \xrightarrow{j_{n-1}} [[S_\alpha^{n-1}, \mathrm{Top}_n/O_n]] \\ &\to [[S_\alpha^{n-1}, \mathrm{Top}_{n+1}/O_{n+1}]] \to 0 \end{aligned}$$

and further we have $[[S_\alpha^{n-1}, \mathrm{Top}_{n+1}/O_{n+1}]] \cong [[S_\alpha^{n-1}, \mathrm{Top}/O]]$. Let $d_n: \tilde{K}_0(Z(G)) \to \tilde{K}_0(Z(G))$ be given by $d_n = k_n j_n$. Then $d_{n-1} d_n = 0$ where we set $d_n = 0$ for $n < 5$. We then have a chain complex $(\tilde{K}_0, d)$ given by

$$\cdots \tilde{K}_0(Z(G)) \xrightarrow{d_n} \widetilde{KO}(Z(G)) \xrightarrow{d_{n-1}} \tilde{K}_0(Z(G)) \cdots .$$

Then the previous exact sequence and some diagram chasing yield an exact sequence

$$(1)\qquad \begin{aligned} &H_n(\tilde{K}_0) \to C^n_\alpha \to [[S^n_\alpha, \mathrm{Top}/O]] \to H_{n-1}(\tilde{K}_0) \\ &\to C^{n-1}_\alpha \to [[S^{n-1}_\alpha, \mathrm{Top}/O]] \to \cdots. \end{aligned}$$

This sequence is exact for $m = m(n) = n - i(\alpha) \geqq 6$.

The groups $C^n_\alpha$ are known modulo homotopy groups of spheres and extension problems [**R1**]. For $G = Z/pZ$, $p$ a prime, $\tilde{K}_0(Z(G))$ is the class group of the ring of integers of the cyclotomic field $Q(e^{2\pi i/p})$, a finite abelian group which is, in theory, known. We conjecture that for $p$ an odd prime, $d_n(x) = x + (-1)^n(x)$. If this is true, then $H_n(\tilde{K}_0)$ is the 2-torsion of $\tilde{K}_0$ for $n$ even and $K_0/2K_2$ for $n$ odd. In that case, we would have a good hold on $[[S^n_\alpha, \mathrm{Top}/O]]$.

We now use the above results to determine $\pi_r(\mathrm{Top}^\alpha/O^\alpha)$ for $r$ large enough. Observe that, for $m > 0$,

$$[[S^m, \mathrm{Top}/O]] = [S^m, (\mathrm{Top}/O)^\alpha] = [S^m, \mathrm{Top}^\alpha/O^\alpha],$$

where, as usual $S^m = (S^n_\alpha)^G$. The Puppe sequence applied to the following $G$ cofibration $S^m \to S^n_\alpha \to S^{n-m-1}_\alpha \times D^{m+1}/S^{n-m-1}_\alpha \times S^m$ yields the exact sequence

$$(2)\qquad K_{n-1} \leftarrow [S^m, \mathrm{Top}^\alpha/O^\alpha] \leftarrow [[S^n_\alpha, \mathrm{Top}/O]] \leftarrow K_n \leftarrow$$

defined for $m > 0$, where $K_n = [[S^{n-m-1}_\alpha \times D^{m+1}; S^{n-m+1} \times S^m, \mathrm{Top}/O; *]]$. Let $f\colon (S^{n-m-1}_\alpha \times D^{m+1}, S^{n-m+1}_\alpha \times S^m) \to (\mathrm{Top}/O, *)$ represent an element of $K_n$. Since $G$ acts freely on $S^{n-m-1}_\alpha \times D^{m+1}$, there exists a $G$-map $\lambda$ unique up to $G$-homotopy with $\lambda\colon S^{n-m-1}_\alpha \to E_G$, where $E_G$ is a contractible free $G$-space. Let $\tilde{\lambda}$ be the composition of $\lambda$ with the projection $S^{n-m-1}_\alpha \times D^{m+1} \to S^{n-m-1}_\alpha$. Let $\hat{f}$ be

$$f \times \lambda\colon (S^{n-m-1}_\alpha \times D^{m+1}, S^{n-m-1}_\alpha \times S^m) \to (E_G \times \mathrm{Top}/O, E_G \times *).$$

Let $L^\alpha = S^{n-m-1}_\alpha/G$ be the lens space of $\alpha$. We then have

$$\hat{f}/G\colon L^\alpha \times D^{m+1}, L^\alpha \times S^m \to (E_G \times_G \mathrm{Top}/O, B_G),$$

where $B_G = E_G/G$. It follows easily that $K_n$ consists of homotopy classes of maps $h\colon L^\alpha \times D^{m+1} \to E_G \times_G \mathrm{Top}/O$ which satisfy the following conditions.

(1) If $\pi$ is the bundle map $E_G \times_G \mathrm{Top}/O \to B_G$, then $\pi h\colon L^\alpha \times D^{m+1} \to B_G$ is $\tilde{\lambda}/G$.

(2) $h(L^\alpha \times S^m) \subset B_G = E_G \times_G * \subset E_G \times_G \mathrm{Top}/O$ and $h|L^\alpha \times S^m = \tilde{\lambda}/G|L^\alpha \times S^m$.

The homotopies are required to preserve both (1) and (2).

Another way to look at this is to consider the fibration induced over $L^\alpha$ by $\lambda/G$. We then have the following map of fibrations

$$\begin{array}{ccc} \mathrm{Top}/O & & \mathrm{Top}/O \\ \downarrow & & \downarrow \\ E & \longrightarrow & E_G \times_G \mathrm{Top}/O \\ \downarrow{\scriptstyle \pi_2} & & \downarrow{\scriptstyle \pi} \\ L^\alpha & \xrightarrow{\lambda/G} & B_G \end{array}$$

The basepoint of Top/$O$ determines a canonical cross-section for $\pi$ and hence for $\pi_2$, call it $j$. Then if $\Gamma(E)$ is the space of cross-sections of $\pi_2$ with basepoint $j$, we have $K_n = \pi_{n+1}(\Gamma(E), j)$.

PROPOSITION 5. $\pi_{n+1}(\Gamma(E), j) = \pi_{n+1}((\mathrm{Top}/O)^{L^\alpha})$, $n \geqq 0$.

PROOF. $G$ acts on $\mathrm{Top}_r/O_r$ via $\alpha$ for $r \geqq n - m = i(\alpha)$ and Whitney sum induces a $G$-equivariant map $\mathrm{Top}_r/O_r \times \mathrm{Top}_s/O_s \to \mathrm{Top}/O$, with $G$ acting trivially on the second factor $\mathrm{Top}_s/O_s$. Let $*$ be a basepoint of $\mathrm{Top}_r/O_r$. Then

$$\rho: * \times \mathrm{Top}_s/O_s \to \mathrm{Top}_r/O_r \times \mathrm{Top}_s/O_s \to \mathrm{Top}/O$$

is $G$-equivariant and $\pi_k(\rho) = 0$ for $k \leqq s$. We then have the bundle map $j_s$:

$$\begin{array}{ccc} L^\alpha \times \mathrm{Top}_s/O_s = S_\alpha^{n-m-1} \times_G \mathrm{Top}_s/O_s & \xrightarrow{j_s} & S_\alpha^{n-m-1} \times_G \mathrm{Top}/O = E \\ \searrow & & \swarrow \\ & L^\alpha & \end{array}$$

with $\pi_k(j_s) = 0$ for $k \leqq s$. Proposition 5 follows.

Proposition 5 and the exact sequence (2) yield the exact sequence

$$\text{(3)} \qquad \pi_n((\mathrm{Top}/O)^{L^\alpha}) \leftarrow [S^m, \mathrm{Top}^\alpha/O^\alpha] \leftarrow [[S_\alpha^n, \mathrm{Top}/O]] \leftarrow \pi_{n+1}((\mathrm{Top}/O)^{L^\alpha}).$$

Note that since $\pi_i(\mathrm{Top}/O)$ is finite for all $i$, $\pi_i(\mathrm{Top}/O^{L^\alpha})$ is also finite for all $i$ and thus $[S^m, \mathrm{Top}^\alpha/O^\alpha] \otimes Q \simeq [[S_\alpha^n, \mathrm{Top}/O]] \otimes Q$ and thus from the exact sequence (1) and the known properties of $K_0$ and $C_\alpha^n$ we conclude that $[S^m, \mathrm{Top}^\alpha/O^\alpha]$, $m \geqq 5$, is finitely generated. We can say more. For $G = Z/p^rZ$, $p \neq 2$, $p$ a prime, we can apply results of Ewing **[E2]** and the exact sequences of **[R1]** to conclude the following

PROPOSITION 6. (1) *For $n$ even, $n \neq 2i(\alpha) - 2$, $C_\alpha^n \otimes Q = 0$.*
(2) *For $n$ and hence $m$ odd and $n \neq 2i(\alpha) - 1$, there is an exact sequence*

$$0 \to \pi_m(O_{i(\alpha)}^\alpha) \otimes Q \to \tilde{L}_{n+1}^h(G) \otimes Q \to C_\alpha^n \otimes Q \to 0,$$

*where $\tilde{L}^h(G)$ is the reduced Wall surgery group, and $O_{i(\alpha)}^\alpha$ is the centralizer of the representation $\alpha: G \to O_{i(\alpha)}$ which is the product of unitary groups $U_{k_1} \times U_{k_2} \times \cdots \times U_{k_l}$, where $k_1 + k_2 + \cdots + k_l = i(\alpha)/2$.*

## BIBLIOGRAPHY


**A1**. G. Anderson, *Classifications of structures on abstract manifolds* (preprint).

**A2**. D. Anderson and W. C. Hsiang, *The functor $K_{-i}$ and pseudo-isotopies of polyhedra* (preprint).

**A3**. P. L. Antonelli, D. Burghelea and P. J. Kahn, *The concordance-homotopy groups of geometric automorphism groups*, Lecture Notes in Math., vol. 215, Springer-Verlag, Berlin and New York, 1971. MR **50** #11293.

**B1**. Glen Bredon, *Introduction to compact transformation groups*, Academic Press, New York, 1972.

**B2**. E. Bierstone, *Equivariant Gromov theory*, Topology **13** (1974), 327–345. MR **50** #5818.

**B3**. ———, *The equivariant covering homotopy property for differential G-fibre bundles*, J. Differential Geometry **8** (1973), 615–622. MR **49** #6260.

**B4**. E. M. Brown, *Unknotting in $M^2 \times I$*, Trans. Amer. Math. Soc. **123** (1966), 480–505. MR **33** #6640.

**B5.** D. Burghelea, R. Lashof and M. Rothenberg, *Groups of automorphisms of manifolds*, Lecture Notes in Math., vol. 473, Springer-Verlag, Berlin and New York, 1975. MR **52** #1738.

**B6.** D. Burghelea and R. Lashof, *The homotopy type of the space of diffeomorphisms*. I, Trans. Amer. Math. Soc. **196** (1974), 1–36. MR **50** #8574.

**B7.** Glen Bredon, *Equivariant cohomology theories*, Lecture Notes in Math., no. 34, Springer-Verlag, Berlin and New York, 1967. MR **35** #4914.

**C1.** E. H. Connell, D. Montgomery, C. T. Yang, *Compact groups in* $E^n$, Ann. of Math. (2) **80** (1964), 94–103; Correction, Ann. of Math. (2) **81** (1965), 194. MR **29** #189.

**D1.** A. Dold, *Partitions of unity in the theory of fibrations*, Ann. of Math. (2) **78** (1963), 223–256. MR **27** #5264.

**D2.** ———, Lectures on Algebraic Topology, Springer, N. Y., 1972.

**E1.** S. Eilenberg and N. E. Steenrod, *Foundations of algebraic topology*, Princeton Univ. Press, Princeton, N. J., 1952. MR **14** #398.

**E2.** J. Ewing, *Characters, Dirichlet series, discriminants and periodic maps* (preprint).

**H1.** D. H. Husemoller, *Fibre bundles*, McGraw-Hill, New York, 1966. MR **37** #4821.

**H2.** Per Holm, *The microbundle representation theorem*, Acta Math. **117** (1967), 191–213. MR 34 #8427.

**H3.** L. S. Husch and T. M. Price, *Finding a boundary for a 3-manifold*, Ann. of Math. (2) **91** (1970), 223–235. MR **41** #9269.

**I1.** S. Illman, *Equivariant singular homology and cohomology*. I, Mem. Amer. Math. Soc. **1** (1975), issue 2, no. 156. MR **51** #11482.

**K1.** N. H. Kuiper and R. Lashof, *Microbundles and bundles*. I, Invent. Math. **1** (1966), 1–17. MR **35** #7339.

**K2.** R. C. Kirby and L. C. Siebenmann, *Essays on topological manifolds, smoothings and triangulations* (to appear).

**L1.** S. J. Luh, *Classifying spaces of* $G$-$\Gamma$ structures, Thesis, University of Chicago, 1975.

**L2.** R. C. Lacher, *Locally flat strings and half-strings*, Proc. Amer. Math. Soc. **18** (1967), 299–304. MR **35** #3670.

**L3.** R. Lashof, *The immersion approach to triangulation and smoothing*, Proc. Sympos. Pure Math., vol. 22, (Univ. of Wisconsin, Madison, Wis., 1970), Amer. Math. Soc., Providence, R. I., 1971, pp. 131–164. MR **47** #5879.

**M1.** T. Matumoto, *On G-CW complexes and a theorem of J. H. C. Whitehead*, J. Fac. Sci. Univ. Tokyo Sect. I A Math. **18** (1971–72), 363–374. MR **49** #9842.

**M2.** J. Milnor, *Microbundles*. I, Topology **3** (1964), suppl. 1, 53–80. MR **28** #4553b.

**R1.** M. Rothenberg, *Differentiable group actions on spheres*, Proc. Advanced Study Inst. on Algebraic Topology (Aarhus Univ., 1970), Vol. 2, Mat. Inst., Aarhus Univ., Aarhus, 1970. MR **46** #919.

**R2.** M. Rothenberg and J. D. Sondow, *Nonlinear smooth representations of compact Lie groups* (preprint).

**S1.** L. C. Siebenmann, *Deformations of homeomorphisms on stratified sets*. I, II, Comment. Math. Helv. **47** (1972), 123–136; ibid. **47** (1972), 137–163. MR **47** #7752.

**S2.** R. Schultz, *On the topological classification of linear representations* (preprint).

**S3.** ———, *Homotopy decompositions of equivariant function spaces*. II, Math. Z. **132** (1973), 69–80.

**S4.** R. G. Swan, *The theory of sheaves*, University of Chicago Press, 1964.

UNIVERSITY OF CHICAGO

Proceedings of Symposia in Pure Mathematics
Volume 32, 1978

# TORSION INVARIANTS AND FINITE TRANSFORMATION GROUPS

MEL ROTHENBERG*

The introduction of torsion invariants to study actions of finite groups on simplicial complexes goes back 40 years to the work of Reidemeister, Franz, and de Rham. In 1950, Whitehead generalized these notions in defining the torsion of a homotopy equivalence on nonsimply connected complexes, and showed that this torsion has striking geometric content. This work of Whitehead, after laying fallow for many years, was rediscovered in the early 1960's and became the starting point of an incredibly rich development in algebra and geometric topology. At the same time, these developments tended to bypass the field of transformation groups, their source and inspiration. This happened in spite of the beautiful survey article of Milnor [M] (a complete bibliography prior to 1966 can be found there), which clearly demonstrated the power of these torsion invariants as a tool for the study of transformation groups.

In this paper we shall extend the definition of torsion invariants and apply these invariants to the study of the actions of finite groups on spheres and disks. Since classical torsion is defined only for free actions (an enormous restriction for transformation group theory), our first task is to extend the domain of definition of these invariants to broader and more natural categories of actions. Our main category of interest is smooth actions, but we get there via pl actions, and thus our approach is very close to that of the founders of the theory. In principle, the method of extending the domain of our actions beyond free actions is simple. If $X$ is a space on which $G$ acts and $H \subset G$ is a subgroup, then $X^{(H)} = \{x \in X \mid G_x = H\}$ is the space of a free $N(H)/H$ action, where $G_x = \{g \in G \mid gx = x\}$ and $N(H)$ is the normalizer of $H$ in $G$. Hence $X = \bigcup_{H \subset G} X^{(H)}$ is the union of "strata", the spaces of free group actions. This decomposition provides a method of extending invariants de-

*AMS* (*MOS*) *subject classifications* (1970). Primary 57E25; Secondary 57E15, 54H15, 18F25.
*Supported in part by NSF Grant MCS-75-08280.

fined for free group actions to invariants defined for much more general types of group actions.

In §1 we carry out this extension in a fairly general categorical context. In doing so we start from scratch and "rediscover" the classical theory as a specialization of a quite categorical and abstract construction. The major technical result is the Product Theorem 1.27 which is crucial to the geometric applications in the succeeding sections.

In §2 we develop some constructions on $G$-$CW$ complexes. These complexes are the main device through which we relate the abstract algebra to the geometry and topology of transformation groups. We show that we can realize many of our algebraic invariants topologically and give a topological interpretation to their vanishing.

§3 is the heart of this paper. Here we come to smooth $G$-manifolds, and formulate and prove an equivariant $s$-cobordism theorem. Under certain hypotheses we can relate this theorem to the invariants defined in Definitions 3.9 and 3.10. With this theorem we can classify some fairly general families of smooth actions of $G$ on disks (Theorem 3.13). The methods developed here enable us to construct systematically infinite families of differentiably distinct actions of $G$ on spheres and disks for rather general finite groups $G$ (Theorems 3.20, 3.23). The problem and hence the method is considerably simpler and more elementary than the problem of constructing free actions, for which a much more developed theory and method exist. The results of this section are a relatively complete generalization (in the category of finite groups) of some earlier results of J. Sondow and myself on relatively free actions **[RS]**.

§4 gives two pretty applications of the torsion invariants to the study of orthogonal actions on spheres. The first is a reformulation and proof of a theorem of de Rham which says that, for a compact Lie group $G$, two orthogonal actions of $G$ on $S$ are smoothly equivalent if and only if they are linearly equivalent. Our proof is rather close to the original proof of de Rham **[D]** whose paper foreshadows the construction of the generalized torsion invariants. A second application is a theorem giving necessary and sufficient conditions, in terms of the invariants, for two orthogonal actions of $G$ on $S^n$ to be $G$ homotopically equivalent.

In §5 we extend our results to the category of piecewise linear $G$-manifolds. We show that the theory goes through there, and is in certain ways even simpler, provided we restrict ourselves to pl manifolds on which the action of $G$ is locally linear.

In §6 we reformulate our theory of actions on spheres and disks as a stable theory. Here, at the cost of some sharpness, results can be stated very simply and generally. In §7 we speculate on possible generalizations and further routes to explore. The main theme is the application of Smith theory to get invariants under very weak hypotheses. The task of computing the groups in which these more general invariants live remains a major one, and the computation is obviously a precondition for the invariants to be useful.

This paper is an outgrowth of investigations begun with Jon Sondow a number of years ago on relatively free actions, and is conceived as the first of two; the second will extend the results here to compact Lie groups. In spite of some serious technical hurdles most of these results do so extend.

We wish to point out that the influence of Milnor's article [M] on the research reported here is enormous.

**1. Some elementary categorical algebra.** In this section we will define $K_1$ and Wh as invariants of certain types of categories and functors. Our point of view is that these invariants measure lack of additivity, in a sense to be made clear. Our definitions yield the classical objects for rings and group rings. Beyond this, whether, or in what respect, they compare with the various other generalizations studied remains to be settled. Our definitions are designed for the study of transformation groups which is our main interest. Possibly they may also be interesting or useful for other purposes and with this in mind we present them in this self-contained section. The substance of this section lies primarily in the definitions. Given the definitions, most of the proofs and arguments are fairly simple and transparent. We leave the proofs of some of the assertions which depend on familiarity with the classical definitions to the readers. Those familiar with the classical definitions should have no difficulty in supplying the proofs.

We consider only categories $A$ with specified initial-terminal objects, always denoted by 0. All functors are covariant and required to preserve 0. The unique map $0 \to 0$ will also be denoted by 0. The unique map $X \to 0 \to Y$ will be denoted by $\mathbf{0}$. Our categories are to be small in the sense that the class of isomorphism classes of objects form a set. For $X_1, X_2 \in A$, $X_1 \vee X_2$ will always denote a coproduct. That is, we assume we are given $\iota_i: X_i \to X_1 \vee X_2$, $i = 1, 2$, such that $(\iota_1 \times \iota_2)^*: A(X_1 \vee X_2, Z) \to A(X_1, Z) \times A(X_2, Z)$ is a bijection for every $Z \in A$. We then have $p_i \in A(X_1 \vee X_2, X_i)$ defined by the conditions that $(\iota_1 \times \iota_2)^*(p_1) = (\mathrm{Id}, \mathbf{0})$ and $(\iota_1 \times \iota_2)^*(p_2) = (\mathbf{0}, \mathrm{id})$. If $X_1 \vee' X_2$ is another coproduct there exists a unique isomorphism $f: X_1 \vee X_2 \to X_1 \vee' X_2$ commuting with the various $\iota_i, p_i, \iota_i', p_i'$.

1.1. DEFINITION. An *ES structure* on a category $A$ consists of a family of diagrams $ES(A)$, $ES(A) = \{X \to^i Y \to^j Z\}$, $X, Y, Z \in A$, $i, j$ maps of $A$, where $0 \to 0 \to 0$ is a member of the family. A category with a given $ES$ structure is called an *ES category*. A functor $F: A \to B$ such that $F(ES(A)) \subset ES(B)$ is called an *ES functor*. To an Abelian category we always associate the Abelian $ES$ structure $ES(A) = \{X \to^i Y \to^j Z\}$ such that $0 \to X \to^i Y \to^j Z \to 0$ is exact. To any category we have the *wedge ES* structure given by $\{X_1 \to^{\iota_1} X_1 \vee X_2 \to^{p_2} X_2\}$. They agree for an Abelian category if and only if all exact sequences split.

An $ES$ category is the most general type of category for which it makes sense to define a $K_0$ group.

1.2. DEFINITION. Let $A$ be an $ES$ category. Let $K_0(A)$ be the Abelian group generated by isomorphism classes of objects of $A$, subject to the relations $\{X\} = \{Y\} + \{Z\}$ for each $Y \to^i X \to^j Z$ in $ES(A)$. Note that $K_0(\ )$ is a functor on the category of $ES$ categories and $ES$ functors.

We wish to define $K_1$ vis a vis such categories. It turns out that the thing to look at is the category of finite complexes of $A$. A finite complex $(X, d)$ of $A$ is just a sequence $\{X_i, d_i\}$, where $X_i \in A$, $d_i: X_i \to X_{i-1}$, such that $X_i = 0$ for all but a finite number of $i$, and $d_{i-1}d_i = \mathbf{0}$. The finite complexes of $A$ with the obvious maps form a category, $C(A)$. Given a functor $F: A \to B$ there is an obvious functor $F^*$: $C(A) \to C(B)$. (We do not use $C(F)$ for this since we have another use in mind for

$C(F)$.) For any functor $F$: $A \to B$, $F(A)$ will denote the full subcategory of $B$ generated by the objects of $A$ in the image of $F$.

1.3. DEFINITION. If $F$: $A \to B$ is a functor we denote by $C(F)$ the category whose objects are the objects of $C(F(A))$ and whose maps are those maps in $C(F(A))$ which are in the image of $F^*$, $F^*: C(A) \to C(F(A))$. If $A$ has an $ES$ structure, then $C(F)$ inherits an induced one—to wit, $(X, d) \to^{\psi} (X', d') \to^{\psi'} (X'', d'')$ is in $ESC(F)$ if for each $i$ there exists $Y \to^{f} Y' \to^{f'} Y''$ in $ES(A)$ with $F(f) = \psi_i$ and $F(f') = \psi'_i$. Since $C(A) = C(\mathrm{Id})$, $C(A)$ has an $ES$ structure.

For a category $A$ and any map $f \in A(X, Y)$ and $i \in Z$ we have the object $i(f)$ of $C(A)$ defined by $i(f) = \{X_j, d_j\}$, $X_j = 0$ for $j \neq i, i-1$, $X_i = X$, $X_{i-1} = Y$, and $d_i = f$. This defines a functor from the category of maps of $A$, $M(A)$ to $C(A)$, $i$: $M(A) \to C(A)$, which embeds $M(A)$ as a full subcategory. Let $I(A)$ be the full subcategory of $M(A)$ of isomorphisms of $A$. We can embed $A$ as a full subcategory of $I(A)$ by sending $X$ to $\mathrm{Id}_X$. Hence we have $A \to I(A) \to M(A) \to^{i} C(A)$ where each of the arrows represents an embedding as a full subcategory.

1.4. DEFINITION. If $A$ is an $ES$ category, a *closed* subcategory $D$ of $A$ is a full subcategory such that if $\{X \to^{i} Y \to^{j} Z\} \in ES(A)$ and if two of these objects are in $D$, so is the third. Given any full subcategory $D$ of $A$ there is clearly a smallest closed subcategory of $A$ containing $D$, denoted by $\bar{D}$, the *closure* of $D$. In particular, let $F: A \to B$ be a functor, $f$ an isomorphism in $F(A)$. Then $i(f) \in C(F)$. Let $C_0(F)$ be the closure of the subcategory of $C(F)$ generated by all $i(f)$, $f \in I(F(A))$, $i \in Z$. If $A$ is an $ES$ category, then so is $C(F)$ and $C_0(F)$ inherits this structure.

1.5. DEFINITION. Let $F: A \to B$ be a functor, $A$ an $ES$ category. Then the induced map $\hat{F}$: $K_0(C_0(A)) \to K_0(C_0(F))$ is determined. We let $K_1(F) = \text{cokernel } \hat{F}$. Alternatively, we observe $I(A) \to^{i} C(A) \to C(F)$ factors through $I(A) \to^{i} C_0(F)$. We could define $K_1(F) = K_0(C_0(F))/\bigcup_{i \in Z} i(f)$.

1.6. EXAMPLE. Let $R$ be a ring, $A$ the category whose objects are $R^n$, $n \geqq 0$, $R^0 = (0)$, and whose maps $\psi: R^{n_1} \to R^{n_2}$ are given by $\psi(r_1, r_2, \cdots, r_{n_1}) = (r_k, r_{k+1}, \cdots, r_{n_1}, 0, 0, \cdots, 0)$, where $1 \leqq k$, and $n_1 - k \leqq n_2$. If $k = 1$, $\psi$ is called an injection; if $n_1 - k = n_2$, $\psi$ is called a projection. Define an $ES$ structure on $A$ by diagrams $\{R^{n_1} \to^{i} R^{n_1+k} \to^{p} R^k\}$, where $i$ is an injection and $p$ a projection. Let $B$ be the category of left $R$ modules and $F$ the obvious functor. Then $K_1(F) = K_1(R)$, where $K_1(R)$ denotes the usual functor.

1.7. EXAMPLE. Let $G$ be a group. A $G$-set $X$ is called $G$-finite if there exist only a finite number of $G$-orbits in $X$. A basepointed $G$-set is a $G$-set with a basepoint $*$ fixed under $G$. Let $A(G)$ be the category of basepointed $G$-finite $G$-sets such that $G$ acts freely off the basepoint. The maps are basepoint preserving $G$-maps. We consider the wedge $ES$ structure on $A(G)$. For any ring consider the functor from $A(G)$ to left $R(G)$-modules given by $F(X) = R \otimes S_0(X)$, where $S_0(X)$ are the singular 0-chains of the discrete pair $(X, *)$. Then $\bar{K}_1(F) = K_1(F)/1(-\mathrm{Id}_X) = \mathrm{Wh}_R(G)$, the classical Whitehead group of $G$ over $R$, where $-\mathrm{Id}$: $R(G) \to R(G)$. If $G$ is a topological group then each object of $A(G)$ has a natural topology such that $G$ acts continuously. Consider the functor from $A(G)$ to left $R(G)$-modules given by $F(X) = H_*(X, *; R)$, the singular homology with coefficients in $R$. It is reasonable to define the topological Whitehead group $\mathrm{Wh}^t_R(G) = K_1(F)/1(-\mathrm{Id})$. If $G$ is discrete this is the classical functor.

The examples above serve to justify the notation. What we wish to study is an obvious generalization of the previous example. Namely, replace in the above example the category $A(G)$ with the category $B(G)$, the Burnside category, whose objects are all basepointed $G$-finite $G$-sets and whose maps are basepointed $G$-maps. The functor $F$ is sometwhat more subtle. We develop it through the following universal construction.

1.8. *The universal additive extension.* Let $A$ be a category, $R$ a ring. An $R$-extension of $A$ is a functor $\iota: A \to B$ such that

(1) $\iota$: Objects($A$) $\cong$ Objects($B$) is a bijection so that we can identify the objects of $A$ and $B$ via $\iota$.

(2) $\iota$ preserves coproducts.

(3) For $X, Y \in B$, $B(X, Y)$ is an $R$-module and composition is $R$ bilinear.

A category satisfying (3) will be called an *$R$-category*.

To construct a kind of universal $R$-extension of $A$ we proceed as follows: First we define the *indecomposables* of $A$, IND($A$). For $X$ in $A$, $X \in$ IND($A$) if, for any $Y, Z \in A$ and $f$: $X \to Y \vee Z$. $f$ factors through $X \to^{f_1} Y \to^{\iota_1} Y \vee Z$ or through $X \to^{f_2} Z \to^{\iota_2} Y \vee Z$.

1.9. DEFINITION. Define the category $R(A)$ as follows: The objects of $R(A)$ = objects of $A$. $R(A)(X, Y)$ is the free $R$-module generated by the elements of $A(X, Y)$ modulo the one relation $\mathbf{0} = 0$. Composition is given by extending composition in $A$ in the obvious way. The obvious functor $j$: $A \to R(A)$ does not in general preserve coproducts. Let $K(X, Y) = \{w \in R(A)(X, Y) \mid \text{all } u, u \in R(A)(Z, X), \text{where } Z \in \text{IND}(A), wu = \mathbf{0}\}$. Let $R \otimes A$ be the category with the same objects as $A$ and $R(A)$ and where $R \otimes A(X, Y) = R(A)(X, Y)/K(X, Y)$. It is easily verified that $R \otimes A$ satisfies (1), (2) and (3) of 1.8 with respect to the natural functor $\iota: A \to R \otimes A$. Observe that for $G$ a discrete group and $A = A(G)$, $R \otimes A \cong F(A)$, where $F$ is the functor of 1.7. Observe also that if $A$ is an additive category, then IND($A$) consists only of the 0's of $A$ and $R \otimes A(X, Y) = (\mathbf{0})$.

1.10. PROPOSITION. *For any category A and a class W of objects of A, we say W generates A if, for any $f \in A(X, Y)$, $f \neq \mathbf{0}$, there exists $w \in W$ and $j \in A(W, X)$ with $fj \neq \mathbf{0}$. Let B be any R-category and let $T$: $A \to B$ be any functor such that $T$(IND($A$)) generates $T(A)$. Then T extends uniquely to a $\hat{T}: R \otimes A \to B$ with $\hat{T}: R \otimes A(X, Y) \to B(T(X), T(Y))$ an R morphism and $T = \hat{T}\iota$.*

1.11. EXAMPLE. Let $U_n = \{*, 1, 2, \cdots, n\}$, for a nonnegative integer $n$. Let $U$ be the category with the sets $U_n$ as objects and whose maps are basepoint preserving functions. We put the wedge *ES* structure on $U$. For any ring $R$, let $\bar{R}$: $U \to R \otimes U$ be the natural functor. Then $K_1(\bar{R}) = K_1(R)$.

We say that a category has coproducts if every pair of objects of $A$ has a coproduct. Observe that if $A$ has coproducts and $X \in A$ there is a functor $F$: $U \to A$ determined uniquely up to isomorphism of functors by the condition that $F(U_1) = X$ and $F$ preserves coproducts. The result of the previous example plus this universal property of $U$ justify the following definition.

1.12. DEFINITION. Let $A$ be any category with wedge *ES* structure. For any ring $R$ we define $K_1(A; R) = K_1(\iota)$, where $\iota$: $A \to R \otimes A$ is the natural functor. Note that this is functorial with respect to ring homomorphisms as well as coproduct preserving functors. With this notation $K_1(U; R) = K_1(R)$, where $U$ is the category of the previous example. We let $\bar{K}_1(A; R) = K_1(A; R)/1(-\text{Id}_X)$, where $X \in$ Obj($R \otimes A$).

Observe that if $A$ is additive, IND($A$) consists only of 0 elements, and thus $K_1(A; R) = (0)$. Hence $K_1(A; R)$ is a measure of the lack of additivity of $A$. Our definition of $K$ is not directly comparable to that of Bass [**B1**] since he begins with a category with a functorial product operation, while we do not assume this. The following observations are intended to show that the two definitions are close in spirit.

Recall that for an isomorphism $f: X \to Y$ in $R \otimes A$ and $i \in Z$, we have the complexes $i(f)$, and by definition these complexes generate $K_1(A; R)$. Suppose $X \vee Y$ exists in $A$. Consider the complex $Z$ given by $Z_j = 0, j \neq i, i+1, i-1$. $Z_{i+1} = X$, $Z_i = X \vee Y$, $Z_{i-1} = Y$, $d_{i+1} = \mathrm{Id}_X - f$, $d_i = \mathrm{Id}_Y + f$. (We have suppressed, and will in the future suppress, denoting the various injection and projection maps associated to the coproduct.) Then we have two elements of $ES(C_0(\iota))$:

(1) $i(f) \to Z \to i+1\,(-f)$,

(2) $i(\mathrm{Id}_Y) \to Z \to i+1\,(\mathrm{Id}_X)$.

The second shows that $Z = 0$ in $K_1(A; R)$; the first shows that then $-i(f) = (i+1)(-f)$ in $K_1(A; R)$. Hence if $A$ has coproducts, it follows that $K_1(A; R)$ is generated by $1(f)$ as $f$ runs over the isomorphisms of $R \otimes A$. Suppose $f_j: X_j \to Y_i$, $j = 0, 1$, are isomorphisms in $R \otimes A$. Then clearly $f_0 \vee f_1: X_0 \vee X_1 \to Y_0 \vee Y_1$ is an isomorphism in $R \otimes A$ and $1(f_0 \vee f_1) = 1(f_0) + 1(f_1)$ in $K_1(A; R)$. The following relations are central.

1.13. PROPOSITION. *If $A$ has coproducts, then for any two composable isomorphisms $f_1, f_2$ in $R \otimes A$, $1(f_1 f_2) = 1(f_1) + 1(f_2)$ in $K_1(A; R)$.*

PROOF. Let $f_i: X_i \to X_{i-1}$, $i = 1, 2$, be isomorphisms in $R \otimes A$. Let $Z$ be the following complex in $R \otimes A$:

$$\begin{aligned} &Z_j = 0, \qquad j \neq 0, 1, 2; \\ &Z_2 = X_2 \vee X_1; \qquad Z_1 = X_1 \vee X_2 \vee X_0 \vee X_1; \\ &Z_0 = X_0 \vee X_1; \\ &d_2 = f_2 + f_2^{-1} + -f_1 f_2 - \mathrm{Id}_{X_1};\ d_1 = f_1 + f_2 + \mathrm{Id}_{X_0} + \mathrm{Id}_{X_1}. \end{aligned}$$

Then we have an element $1(f_1 \vee f_2) \to Z \to 2(-f_1 f_2 \vee \mathrm{Id}_{X_1})$ of $ES(C_0(\iota))$. Thus in $K_1(A; R)$, $Z = 1(f_1 \vee f_2) - 1(f_1 f_2 \vee \mathrm{Id}_{X_1})$. On the other hand, we have $2(f_2 \vee f_2^{-1}) \to Z \to 1(\mathrm{Id}_{X_0} \vee \mathrm{Id}_{X_1})$. Hence $Z = 2(f_2 \vee f_2^{-1})$ in $K_1(A; R)$. Thus $2(f_2 \vee f_2^{-1}) = 1(f_1 \vee f_2) - 1(f_1 f_2 \vee \mathrm{Id}_{X_1})$ in $K_1(A; R)$. This is true for every $f_1$, hence for $f_1 = \mathrm{Id}_{X_1}$. But then the right side is 0; hence $2(f_2 \vee f_2^{-1}) = 0$. Hence $1(f_1) + 1(f_2) = 1(f_1 \vee f_2) = 1(f_1 f_2 \vee \mathrm{Id}_{X_1}) = 1(f_1 f_2)$. Q.E.D.

We can sum up previous results as follows. Let $K_1^*(A; R)$ be the free Abelian group generated by isomorphisms $f$ in $R \otimes A$ modulo the following relations:

(1) $f = 0$ if $f = i\,(f_0)$, $f_0$ an isomorphism in $A$;

(2) $f_1 \vee f_2 - f_1 - f_2 = 0$, where $\vee$ denotes coproduct coming from coproduct of objects in $A$;

(3) if $f_1, f_2$ are composable, then $f_1 f_2 - f_1 - f_2 = 0$.

If $A$ has coproducts then we have an epimorphism $K_1^*(A; R) \to K_1(A; R)$. The group $K_1^*$ is very similar to Bass's construction. In fact, a slightly more refined analysis shows:

1.14. THEOREM. *If $A$ has coproducts, $K_1^*(A; R) \cong K_1(A; R)$.*

We now come to our central object of study.

1.15. DEFINITION. Let $G$ be a group and $B(G)$ the category of all basepointed $G$-finite $G$-sets whose maps are basepointed $G$-maps. We define $\mathrm{Wh}(G; R) = \bar{K}_1(B(G); R)$. Note that this is different from $\mathrm{Wh}_R(G)$ which we considered earlier.

We wish to compute $\mathrm{Wh}(G; R)$ in terms of "known" objects, $\mathrm{Wh}_R(H)$, where $H$ runs over quotients of subgroups of $G$. Toward this end, we introduce the following definition:

1.16. DEFINITION. Let $A$ be a category, $A_0$, $A_1$ full subcategories. We say $A = A_0 \vee A_1$ if the following three conditions are satisfied:

(1) If $X \in A$, then $X = X_0 \vee X_1$, $X_i \in A_i$; and further the decomposition is unique in the following sense: If $f: X_0 \vee X_1 \to Y_0 \vee Y_1$ is an isomorphism in $A$, with $X_i$, $Y_i$ in $A_i$, then $f = f_0 \vee f_1$ with $f_i: X_i \to Y_i$ an isomorphism in $A_i$. Note that this implies that $\mathrm{Object}(A_0) \cap \mathrm{Object}(A_1) = 0$ objects of $A$.

(2) $A(X_0, X_1) = (\mathbf{0})$ for $X_0 \in A_0$, $X_1 \in A_1$.

(3) Let $\iota: A \to R \otimes A$; then $\iota(\mathrm{IND}(A_0))$ generates $\iota(A_0)$.

1.17. THEOREM. *If $A$ has coproducts and $A = A_0 \vee A_1$, then the inclusion functors $A_i \to A$ induce an isomorphism $K_1(A_0; R) + K(A_1; R) = K_1(A; R)$.*

PROOF. Conditions (1) and (2) and the fact that $A$ has coproducts imply that the coproduct of two elements of $A_0$ is in $A_0$. Thus $A_0$ has coproducts. Hence $R \otimes A_0$ and $R \otimes A$ are additive categories, and coproducts are also products. For $X_i \in A_i$, $A(X_0, X_1) = (\mathbf{0})$. Thus $R \otimes A(X_0, X_1) = (\mathbf{0})$, and thus for $Y_i \in A_i$, $R \otimes A(X_0, Y_0) \to R \otimes A(X_0, Y_0 \vee Y_1)$ is a bijection. Let $(X, d)$ be a complex in $R \otimes A$. Then $X_j = Y_j \vee Z_j$, $Y_j \in A_0$, $Z_j \in A_1$. Then we have $\bar{d}_j: Y_j \to Y_{j-1}$, $\bar{d}_j$ in $R \otimes A_0$, such that the following diagram commutes:

$$\begin{array}{ccc} Y_j & \xrightarrow{\iota_j} & Y_j \vee Z_j \\ \big\downarrow{\scriptstyle \bar{d}_j} & & \big\downarrow{\scriptstyle d_j} \\ Y_{j-1} & \xrightarrow{\iota_{j-1}} & Y_{j-1} \vee Z_{j-1} \end{array}$$

Since $\bar{d}_j = p_{j-1} d_j \iota_j$, where $p_{j-1}: Y_{j-1} \vee Z_{j-1} \to Y_{j-1}$, $\bar{d}_j$ is determined uniquely once we have chosen the representation of $X_j$ as $Y_j \vee Z_j$. By the uniqueness condition (1) above, the complex $(Y, \bar{d})$ is determined uniquely up to isomorphism in $C(\iota_0)$, $\iota_0: A_0 \to R \otimes A_0$.

Now define the complex $(Z, \bar{\bar{d}})$ as follows:

$$\bar{\bar{d}}_j: Z_j \to Z_{j-1}, \qquad \bar{\bar{d}}_j = p'_{j-1} d_j \iota'_j,$$

where $\iota'_j: Z_j \to Y_j \vee Z_j$ and $p'_{j-1}: Y_{j-1} \vee Z_{j-1} \to Z_{j-1}$ are the associated injections and projections. Using the fact that $\mathrm{Id}_{X_j} = \iota'_j p'_j + \iota_j p_j$, we have

$$\begin{aligned} \bar{\bar{d}}_{j-1} \bar{\bar{d}}_j &= p'_{j-2} d_{j-1} \iota'_{j-1} p'_{j-1} d_j \iota'_j = p'_{j-2} d_{j-1} (\mathrm{Id}_{X_{j-1}} - \iota_{j-1} p_{j-1}) d_j \iota'_j \\ &= -\, p'_{j-2} d_{j-1} \iota_{j-1} p_{j-1} d_j \iota'_j = -\, p'_{j-2} \iota_{j-2} \bar{d}_{j-1} p_{j-1} d_j \iota'_j = 0 \end{aligned}$$

since $p'_{j-2} \iota_{j-2} = 0$. Hence $(Z, \bar{\bar{d}}) \in C(\iota_1)$, $\iota_1: A_1 \to R \otimes A_1$, determined uniquely up to isomorphism in $C(\iota_1)$. Further, $p'_j: X_j \to Z$ induces $p': (X, d) \to (Z, \bar{\bar{d}})$, a map in $C(\iota)$. Thus we have an element $\mu(X, d)$ of $ESC(\iota)$, $(Y, \bar{d}) \to^{j} (X, d) \to^{p'} (Z, \bar{\bar{d}})$, determined by $(X, d)$, where the two end complexes and the maps $j$, $p'$ are determined uniquely up to isomorphism in $C(\iota)$.

Since the assignments $(X, d) \to (Y, \bar{d})$ and $(X, d) \to (Z, \bar{\bar{d}})$ preserve elements of $ESC(\iota)$, it follows that if $(X, d) \in C_0(\iota)$, then $(Y, \bar{d}) \in C_0(\iota_0)$ and $(Z, \bar{\bar{d}}) \in C_0(\iota_1)$. Hence the function $(X, d) \to ((Y, \bar{d}), (Z, \bar{\bar{d}}))$ induces $\lambda: K_1(A; R) \to K_1(A_0; R) + K_1(A_1; R)$. On the other hand, the inclusion functors induce $\rho: K_1(A_0; R) + K_1(A_1; R) \to K_1(A; R)$. The composition is clearly the identity. On the other hand, the element $(Y, \bar{d}) \to^{j} (X, d) \to^{p'} (Z, \bar{\bar{d}})$ of $ESC(\iota)$ shows that $\rho$ is an epimorphism. Thus $\lambda$, $\rho$ are isomorphisms and inverses to one another. Q.E.D.

1.18. THEOREM. $\mathrm{Wh}(G; R) = \sum \mathrm{Wh}_R(N(H)/H)$, *where* $N(H)$ *is the normalizer in* $G$ *of the subgroup* $H$ *of* $G$ *and the sum runs over all conjugacy classes of subgroups of* $G$.

PROOF. Let $\mathrm{Conj}(G)$ be the set of all conjugacy classes of subgroups of $G$. Let $W$ be the family of finite subsets of $\mathrm{Conj}(G)$. $W$ is a directed set ordered by inclusion. For each $\alpha \in W$, let $B(\alpha)$ be the category of $G$-finite $G$-basepointed sets $X$ such that for all $y \in X - (*)$, $[G_y] \in \alpha$. Each $B(\alpha)$ is a full subcategory of $B(G)$ closed under taking coproducts. For $\alpha \leqq \alpha'$, $B(\alpha) \subset B(\alpha') \subset B(G)$, and it follows directly from the definitions that $\mathrm{Wh}(G; R) = \bar{K}_1(B(G); R) = \lim_\alpha \bar{K}_1(B(\alpha); R)$.

Let $\alpha_0 \in W$ and $H \subset G$ such that for $H_0 \subset H$, $[H_0] \notin \alpha_0$. Let $\alpha_1 = ([H])$ and $\alpha = \alpha_0 \cup \alpha_1$. Then $A = B(\alpha)$, $A_0 = B(\alpha_0)$, $A_1 = B(\alpha_1)$ satisfy the hypotheses of 1.17. Hence $K_1(B(\alpha); R) = K_1(B(\alpha_0); R) + K_1(B(\alpha_1); R)$. Consider the functor $\lambda: A(N(H)/H) \to B(\alpha_1)$ given by $\lambda(X) = (G \times_{N(H)} (X - *)) \cup *$. $\lambda$ induces an isomorphism of categories;

$$\mathrm{Wh}_R(N(H)/H) = \bar{K}_1(A(N(H)/H); R) = \bar{K}_1(B(\alpha_1); R).$$

Hence

$$\bar{K}_1(B(\alpha); R) = \bar{K}_1(B(\alpha_0); R) + \mathrm{Wh}_R(N(H)/H).$$

An easy induction then shows that, for all $\alpha \in W$, $\bar{K}_1(B(\alpha); R) = \sum_{[H] \in \alpha} \mathrm{Wh}_R(N(H)/H)$, and the theorem follows. Q.E.D.

Unfortunately, the notion of $K_1$ defined above is not quite general enough for our purposes. We must introduce a more general construction. As we did earlier, consider an $ES$ category $A$ and a functor $F: A \to B$. Let $g: B \to D$ be a functor, where $D$ is an Abelian category. $(X, d) \in C(F)$ is called $g$-acyclic if its image $g(X, d)$ in $C(D)$ is acyclic as a complex in the Abelian category $D$.

1.19. DEFINITION. The set of all $g$-acyclic complexes generates a full subcategory $C(F; g) \subset C(F)$ which inherits an $ES$ structure from $C(F)$. We define $K_1(F; g) = K_0(C(F; g))/\bigcup_i i(f)$, where $f \in I(A)$. Let $f$ be a map in $B$ and let $\mathrm{Ker}(g) = \{f: X \to X \mid g(f) = \pm \mathrm{Id}_{g(X)}\}$. We define $\bar{K}_1(F; g) = K_1(F; g)/\bigcup_i i(f)$ such that $f \in \mathrm{Ker}(g)$.

Observe that if $gF: A \to D$ preserves $ES$ structure, $C(F; g)$ is closed in $C(F)$. In particular, $C_0(F) \subset C(F; g)$ for $gF$ an $ES$ functor. Now let $A$ be a category, $R$ a ring, $D$ an Abelian $R$ category. A functor $T: A \to D$ will be called *nice* if $T$ preserves coproducts, i.e., $T(X \vee Y) = T(X) + T(Y)$, and $T(\mathrm{IND}(A))$ generates $T(A)$. Then we have a unique extension $\hat{T}: R \otimes A \to D$ of $T$ and as usual, with $\iota: A \to R \otimes A$, we define $K_1(A; T) = K_1(\iota; \hat{T})$, and $\bar{K}_1(A; T) = \bar{K}_1(\iota; \hat{T})$. These constructions are functorial in $A$ and $T$.

If further we have a transformation $f: T \to T'$ of functors such that $f: T(X) \cong T'(X)$ for each $X$ in $A$, then $C(\iota; T) = C(\iota; T')$ and thus $K_1(A; T) = K_1(A; T')$.

Since $C_0(\iota) \subset C(\iota; T)$ for any nice functor $T$, we have a homomorphism $K_1(A; R) \to K_1(A; T)$. Is there a $T$ which makes this map an isomorphism?

1.20. DEFINITION. Let $X \in A$. We say $X$ is *$T$-projective* if given any diagram

$$\begin{array}{ccc} & & Z \\ & & \downarrow j \\ X & \xrightarrow{i} & Y \end{array}$$

where $i$ and $j$ are maps in $R \otimes A$ and $T(j)$ is surjective, there exists a map $k: X \to Z$ in $R \otimes A$ with $i = jk$. We say that $T$ is a *projective* functor if every $X$ in $A$ is $T$-projective.

1.21. THEOREM. *Let $A$ be a category with coproducts. If $T$ is a nice projective functor, then $C_0(\iota) = C(\iota; T)$ and thus $K_1(A; T) = K_1(A; R)$.*

PROOF. Let $\alpha: X \to Y$ be a map in $R \otimes A$ with $\hat{T}(\alpha)$ an isomorphism. Since $Y$ is $T$-projective, there exists an $f: Y \to X$ with $\alpha f = \mathrm{Id}_Y$. Since $X$ is $T$-projective, there exists a $g: X \to Y$ with $fg = \mathrm{Id}_X$. Thus $\alpha = \alpha f g = g$. Hence $f = \alpha^{-1}$ and $\alpha$ is an isomorphism.

Let $(C, d) \in C(\iota; T)$, $(C, d) = \to 0 \to 0 \to C_{n+i} \to C_{n+i-1} \to \cdots \to C_i \to 0 \to 0$. We wish to show $(C, d) \in C_0(\iota)$. We prove it by induction on $n$. By the previous paragraph we have proved it for $n = 1$.

Since $C_i$ is $T$-projective, there exists $\lambda: C_i \to C_{i+1}$, $d_{i+1}\lambda = -\mathrm{Id}$. Define a complex $(C_\lambda, d_\lambda)$ as follows:

$$\begin{aligned} (C_\lambda)_j &= C_j, \quad j \neq i, i+1, i+2, \; d_j^\lambda = d_j, \\ (C_\lambda)_{i+2} &= C_{i+2} \vee C_i, \\ (C_\lambda)_{i+1} &= C_{i+1}, \\ (C_\lambda)_i &= 0, \quad d_{i+2}^\lambda = d_{i+2} + \lambda. \end{aligned}$$

Then $(C_\lambda, d_\lambda) \in C(\iota; T)$, and hence by induction $(C_\lambda, d_\lambda) \in C_0(\iota)$. Now consider the complex $(\bar{\bar{C}}, \bar{\bar{d}})$, where

$$\begin{aligned} \bar{\bar{C}}_j &= C_j, \quad j \neq i+2, i+1, \\ \bar{\bar{C}}_{i+1} &= C_i \vee C_{i+1}, \\ \bar{\bar{C}}_{i+2} &= C_i \vee C_{i+2}, \quad \bar{\bar{d}}_{i+2} = \mathrm{Id} + \lambda + d_{i+2}, \bar{\bar{d}}_{i+1} = \mathrm{Id} + d_{i+1}. \end{aligned}$$

Then we have the elements of $ESC(\iota)$:

$$\begin{gathered} (C, d) \to (\bar{\bar{C}}, \bar{\bar{d}}) \to i + 2(C_i), \\ i + 1(C_i) \to (\bar{\bar{C}}, \bar{\bar{d}}) \to (C_\lambda, d_\lambda). \end{gathered}$$

Hence $(C_\lambda, d_\lambda) \in C_0(\iota)$ implies $(\bar{\bar{C}}, \bar{\bar{d}}) \in C_0(\iota)$ implies $(C, d) \in C_0(\iota)$. Q.E.D.

1.22. EXAMPLE. Let $G$ be a group. For each $\alpha \in \mathrm{Conj}(G)$, pick out a representative $H_\alpha$ and consider the functor $\lambda^{H_\alpha}: B(G) \to D^{H_\alpha}$, where $D^{H_\alpha}$ is the category of left $R(N(H_\alpha)/H_\alpha)$-modules, and $\lambda^{H_\alpha}(X) = R \otimes S_0(X^{H_\alpha})$, where $S_0(X^{H_\alpha})$ are the singular 0 chains of the discrete space $X^{H_\alpha}$ modulo $*$, and $\lambda^{H_\alpha}$ on maps being the induced maps. Then the functor $\lambda = \times_\alpha \lambda^{H_\alpha}: B(G) \to \times_\alpha D^{H_\alpha}$ is a nice projective functor. Observe that if we select new representatives $H'_\alpha$ of $\alpha$, there is a natural identification of $X^{H_\alpha}$ and $X^{H'_\alpha}$. Then there is a natural transformation of the corresponding functors which is an isomorphism on objects. As a result, $C(\iota; \lambda) = C(\iota; \lambda')$. If we consider $W \subset \mathrm{Conj}(G)$, and restrict $\alpha$ to lie in $W$, we get a corresponding functor $\lambda^W = \times_{\alpha \in W} \lambda^{H_\alpha}: B(G) \to \times_{\alpha \in W} D^{H_\alpha}$. This is a nice, projective functor when restricted to $B(W)$, the full subcategory of $B(G)$ whose objects are those $X \in B(G)$ with $[G_y] \in W$ for all $y \in X - (*)$.

We are now ready to introduce one of the most important invariants of transformation group theory, the Reidemeister torsion. We follow the approach of Milnor [M] which is based on the following observation. Let $G$ be a finite group. Let $R_1(G)$ be an $R(G)$-module, where now $R$ is assumed to be a field of characteristic 0 or prime to the order of $G$, and suppose we are given an $R(G)$ epimorphism $\varepsilon: R(G) \to R_1(G)$. Since $R(G)$ is semisimple, $\varepsilon$ splits, and we have a decomposition $R(G) = R_1(G) + R_0(G)$ of $R(G)$. Here $R_0(G)$ is the kernel of $\varepsilon$, and both summands are two-sided ideals of $R(G)$ which annihilate one another. Then for any $R(G)$-module $M$, we have a decomposition $M = R_0(G)M + R_1(G)M$. Thus given any $R(G)$ complex $(M, d)$, we have a decomposition $M_i = R_0(G)M_i + R_i(G)M_i$ and $d_i$ splits into $d_i^0 + d_i^1$. Thus $H_i(M) = H_i^0(M) + H_i^1(M) = R_0(G)H_i(M) + R_1(G)H_i(M)$. We have then that the action of $R(G)$ on $H_*(M)$ factors through an action of $R_1(G)$ iff $R_0(G)$ annihilates $H_*(M)$ iff $(R_0(G)M, d)$ is acyclic.

If $H$ and $H'$ are conjugate subgroups of $G$ with $H' = gHg^{-1}$ then $N(H') = gN(H)g^{-1}$ and thus we have an isomorphism $\hat{g}: R(N(H)/H) \cong R(N(H')/H')$. If $R_0(N(H)/H)$ is a two-sided ideal of $R(N(H)/H)$, then $\hat{g}(R_0(N(H)/H)) = R_0(N(H')/H')$ is a two-sided ideal of $R(N(H')/H')$ which is independent of the choice of $g$.

1.23. DEFINITION. Let $G$ be a group and $R$ a field of characteristic prime to the order of $G$, or of characteristic 0. Let $W \subset \mathrm{Conj}(G)$. A *Reidemeister torsion functor*, $\tau$, is given by assigning to each $[H]$ in $W$, a two-sided ideal $R_0^\tau(N(H)/H)$ of $R(N(H)/H)$. By the remark above this is determined on each representative of $[H]$ when it is given on one. This induces functors $\tau_H: B(G) \to D_H$ given on objects by $\tau_H(X) = R_0^\tau(N(H)/H)S_0(X^H)$, where $S_0(\ )$ is the functor of 1.22. We then set $\tau = \times_{[H] \in W} \tau_H: B(G) \to \times_{[H] \in W} D_H$ where we have picked out representatives of $H$ for each $[H]$ in $W$. As in 1.22 if we select different representatives for $[H]$ we get a natural transformation between the corresponding functors which is an isomorphism on the objects. Thus $C(\iota; \tau) = C(\iota; \tau')$, where $\tau, \tau'$ are associated with same ideals $R_0(N(H)/H)$.

From our previous discussion we can characterize $C(\iota, \tau_H)$ as follows when $G$ is finite. For any element $(X, d)$ of $C(\iota)$ we have the subcomplexes $(X^H, d^H = d \mid X^H)$. We then have the complex

$$S_H(X) = \cdots \to R \otimes S_0(X_i^H) \xrightarrow{(d_i^H)^*} R \otimes S_0(X_{i-1}^H) \to \cdots$$

which is a complex of $R(N(H)/H)$ modules. Then $(X, d) \in C(\iota; \tau_H)$ iff

$$\cdots \to R_0^\tau(N(H)/H)(R \otimes S_0(X_i^H)) \to R_0^\tau(N(H)/H)(R \otimes S_0(X_{i-1}^H)) \to \cdots$$

is acyclic which is true iff $R_0^\tau(N(H)/H)$ annihilates $H_*(S_H(X))$.

The groups $K_1(B(G); \tau)$ yield very interesting invariants of $G$-actions. We would like to have a decomposition theorem of type 1.18 for this group, but we do not know how to prove it for a general group $G$. This is because the argument of 1.17 which assigns to $(X, d)$ the pair $((Y, d), (Z, d))$ argues that for $(X, d)$ "universally" acyclic, the same is true for $Y$ and $Z$. The corresponding result when $(X, d)$ is $T$-acyclic does not necessarily follow, although it does follow that if $(X, d)$ is $T$-acyclic and if one of the two of $Y$ or $Z$ is $T$-acyclic, the other is. We can show the following

1.24. THEOREM. *Let $W$ be the family of all subgroups of $G$; let $G$ be finite Abelian, $\tau$ a Reidemeister torsion functor such that, for the natural map $p\colon R(G) \to R(G/H)$, $p(R_0^\xi(G)) = R_0^\xi(G/H)$. Then*

$$K_1(B(G); \tau) = \sum_{H \in W} K_1(A(G/H); \tau_H)$$

*and*

$$\bar{K}_1(B(G); \tau) = \sum_{H \in W} \bar{K}_1(A(G/H); \tau_H),$$

*where $A(G/H)$ is the full subcategory of $B(G)$ of those $X$ such that $G_x = H$ for all $x \in X - *$.*

PROOF. We prove the first equation, the proof of the other is similar. Let $V$ be the family of subsets of $W$. Given $\alpha \in V$, $X \in B(G)$, let $X^\alpha = \{x \in X \mid G_x \in \alpha\} \cup (*)$ and $B(\alpha)$ the full subcategory of $B(G)$ whose objects are those $X$ such that $X = X^\alpha$. Observe that $X^\alpha \in B(\alpha)$. Let $\iota_\alpha\colon B(\alpha) \to R \otimes B(\alpha)$, $\iota\colon B(G) \to R \otimes B(G)$ be the canonical functors. Given $(X, d) \in C(\iota)$, we have the complex $(X^\alpha, d^\alpha) \in C(\iota_\alpha)$ and this induces a functor $C(\iota_{\alpha'}) \to C(\iota_{\alpha \cap \alpha'}) \subset C(\iota_\alpha)$.

For any $H \subset G$, let $\bar{H} = \{H' \subset G \mid H \subset H'\}$. $\bar{H} \in V$. Let $V^0 \subset V$, $V^0 = \{\alpha \in V \mid H \in \alpha$ and $H \subset H'$ implies $H' \in \alpha\}$. Note that $\bar{H} \in V^0$ for every $H \subset G$. Further if $\alpha \in V^0$ and $H \in \alpha$ is a smallest element then $\alpha - (H) \in V^0$. Now Theorem 1.24 will follow exactly as 1.18 if in $\alpha \in V^0$ we can show $K_1(B(\alpha); \tau) = K_1(B(\alpha - (H)); \tau) + K_1(B((H)); \tau)$ and this will follow from the argument of 1.17 if we can show that, for $(X, d) \in C(\iota_\alpha)$ $\tau$-acyclic, $(X^{\alpha-(H)}, d^{\alpha-(H)}) \in C(\iota_{\alpha-(H)})$ is $\tau$-acyclic. Now by assumption, $(X, d)$ is $\tau$-acyclic if and only if $R_0^\xi(G)$ annihilates $H_*(S_H(X))$ for all $H \subset G$. Since $(X^H)^{H'} = X^{H \cup H'}$, it follows that $(X, d)$ $\tau$-acyclic implies $(X^H, d^H)$ $\tau$-acyclic for every $H$. Now let $H_1, H_2, \ldots, H_m$ be minimal elements of $\alpha - (H)$. Then $\alpha - (H) = \bigcup_j \bar{H}_j$ and $(X, d) \in C(\iota_\alpha; \tau)$ implies $(X, d)$ $\tau$-acyclic implies $(X^{H_j}, d^{H_j})$ $\tau$-acyclic. Let $\alpha_k = \bigcup_{j=1}^k \bar{H}_j$, $\bar{\alpha}_k = \bigcup_{j=1}^k \overline{H_j \cap H_{k+1}}$. For any $H' \subset G$, we have an exact sequence $0 \to S_{H'}(X^{\bar{\alpha}_k}) \to S_{H'}(X^{\alpha_k}) + S_{H'}(X^{H_{k+1}}) \to S_{H'}(X^{\alpha_{k+1}}) \to 0$. By induction on $k$ it follows that $(X^H)$ $\tau$-acyclic for all $H$ implies $(X^{\alpha_m})$ $\tau$-acyclic implies $(X^{\alpha-(H)}, d^{\alpha-(H)})$ $\tau$-acyclic. Q.E.D.

We wish to show that $K_1(A(G/H); \tau)$ is a familiar object and thus locate where the Reidemeister torsion lives. Let $K = G/H$. The torsion is determined by $R_0^\xi(K)$, a two-sided ideal of $R(K)$. Since the group $K$ acts on $R_0^\xi(K)$, we have a homomorphism $j\colon K \to \bar{K}_1(R_0^\xi(K))$, and we denote the cokernel of $j$ by $\tau(K; R)$. Since $R_0^\xi(K)$ is a ring with identity (because $R(K)$ is semisimple), multiplication by that identity yields an epimorphism of rings $s\colon R(K) \to R_0^\xi(K)$, and thus $j$ factors $K \to \bar{K}_1(R(K)) \to \bar{K}_1(R_0^\xi(K))$. Hence we have a map $\mathrm{Wh}_R(K) \to \tau(K; R)$. For a field of characteristic 0 this is known to be surjective.

1.25. THEOREM. *If* $R_0^\tau(K) \neq 0$, $\bar{K}_1(A(K); \tau_H) = \tau(K; R)$. *If* $R_0^\tau(K) = 0$, *then* $\bar{K}_1(A(K); \tau_H) \cong K_0(A(K)) \cong Z$, *the isomorphism given by the Euler map* $\sum (X, d) = \sum (-1)^i X_i$.

PROOF. The second statement is trivial; we prove the first. To construct the isomorphism we first observe that $R \otimes A(K)$ is isomorphic to a full subcategory of the category of free left $R(K)$-modules with coproducts corresponding to direct sums. With this identification the functor $\hat{\tau}: R \otimes A(K) \to D_H$ can be identified with the functor $M \to R_0^\tau(K) M$. For $M \in R \otimes A(K)$, let $s_M: M \to M$ be given by $s(x) = sx$ where $s$ is the identity in $R_0^\tau(K)$. Then $s_M^2 = s_M$ and $\hat{\tau}(s_M) = \mathrm{Id}$.

Further, given a diagram in $R \otimes A(K)$ with $\hat{\tau}(i)$ an epimorphism

$$\begin{array}{ccc} & & N_1 \\ & & \downarrow i \\ M & \xrightarrow{j} & N_2 \end{array}$$

there exists a $\lambda$, $\lambda: M \to N_1$, with $i\lambda = js_M$.

An element of $K_1(R_0^\tau(K))$ is given by a finite free, based, acyclic $R_0^\tau(K)$ complex $(M, d)$. The operation $(M, d) \to (R(K) \otimes_{R_0^\tau(K)} M, \mathrm{Id} \otimes d)$ yields a universally acyclic complex of $R \otimes A(K)$ and hence an element of $C(\iota; \tau)$. On the other hand, given $(N, d) \in C(\iota; \tau)$, $(R_0^\tau(K)N, d)$ is a finite, free, acyclic $R_0^\tau(K)$ complex. It is not quite based, but a class of bases is determined up to multiplication by elements of $K$. Thus the two operations are inverses, carry over the required relations, and yields

$$\tau(K; R) \xrightarrow{\psi} \bar{K}_1(A(K); \tau) \xrightarrow{\lambda} \tau(K; R)$$

with $\lambda\psi = \text{Identity}$.

To show $\psi$ is an isomorphism it will suffice to show it is surjective. Suppose at first we have an element of $\bar{K}_1(A(K); \tau)$ represented by a two term complex:

$$0 \to 0 \to \cdots \to M \xrightarrow{d} N \to 0 \to 0. \tag{1}$$

Since $\tau(d)$ is an isomorphism, and $M$, $N$ free $R(K)$-modules, it follows by dimensionality that $M = N = \bigoplus_n R(K)$. Let $R_1(K)$ be a complementary subalgebra to $R_0^\tau(K)$ in $R(K)$. Thus $R(K) = R_0^\tau(K) + R_1(K)$ and $R_0^\tau(K) \cdot R_1(K) = 0$. Then we have a decomposition

$$d = d_0 + d_1: R_0^\tau(K)M + R_1(K)M \to R_0^\tau(K)M + R_1(K)M,$$

(where we have identified $M$, $N$ via an isomorphism in $A(K)$).

Consider $d': M \to M$ given by $d' = d_0 + \mathrm{Id}_{R_1(K)M}$. Then

$$\cdots 0 \to 0 \to M \xrightarrow{d'} M \to 0 \to 0 \cdots, \tag{2}$$

$$\cdots 0 \to 0 \to M \xrightarrow{d(d'^{-1})} M \to 0 \to 0 \cdots \tag{3}$$

are elements of $C(\iota; \tau)$. By the argument of 1.13 the difference of the first two complexes equals the third in $K_1(A(K); \tau)$, but the third is 0 since $\tau(d(d'^{-1})) = \mathrm{Id}$. The second complex is just $\mathrm{Id} \otimes d_0$ and hence in the image of $\psi$. Now consider a complex in $C(\iota; \tau)$ of length $i$, $(C, d)$

$$0 \to 0 \to 0 \to C_{n+i} \to C_{n+i-1} \cdots C_n \to 0 \to 0,$$

and assume all complexes of length less than $i$ are represented in $\bar{K}_1(A(K); \tau)$ by elements in the image of $\psi$. We now repeat the construction of 1.21.

Choose $\lambda: C_i \to C_{i+1}$ with $d_{i+1} = -s_{C_i}$. Define $(C_\lambda, d_\lambda)$ as follows: $(C_\lambda)_j = C_j$, $j \neq i, i+2$, $(C_\lambda)_i = 0$, $(C_\lambda)_{i+2} = C_i \vee C_{i+2}$, $d_j^\lambda = d_j$ for $j \neq i, i+1, i+2$, $d_{i+2}^\lambda = d_{i+2} + \lambda$, otherwise $d_j^\lambda = 0$.

Now define $(\bar{\bar{C}}, \bar{\bar{d}})$, with $\bar{\bar{C}}_j = C_j, \bar{\bar{d}}_j = d_j$ for $j \neq i+2, i+1$, $\bar{\bar{C}}_{i+2} = C_j \vee C_{i+2}$, $\bar{\bar{C}}_{i+1} = C_i \vee C_{i+1}$, $\bar{\bar{d}}_{i+2} = s_{C_i} + \lambda + d_{i+2}$, $\bar{\bar{d}}_{i+1} = \mathrm{Id}_{C_i} + d_{i+1}$. Since $d_{i+1}\lambda = -s_{C_i}$ this is a complex.

We now have elements in $ESC(\iota)$

$$(C, d) \to (\bar{\bar{C}}, \bar{\bar{d}}) \to i + 2(s_{C_i}), \qquad i + 1(\mathrm{Id}_{C_i}) \to (\bar{\bar{C}}, \bar{\bar{d}}) \to (C_\lambda, d_\lambda).$$

Since $(C, d)$ is $\tau$-acyclic it follows that $(\bar{\bar{C}}, \bar{\bar{d}})$ and $(C_\lambda, d_\lambda)$ are $\tau$-acyclic. Since $i + 2(s_{C_i})$ and $i + 1(\mathrm{Id}_{C_i})$ are 0 in $\bar{K}_1(A(K); \tau)$, $(C, d) = (C_\lambda, d_\lambda)$ in $\bar{K}_1(A(K); \tau)$ and by induction $\psi$ is epi. Q.E.D.

1.26(a). Definition. Classical Reidemeister torsion comes from taking $R_0^\xi(N(H)/H)$ as the kernel of the augmentation $R(N(H)/H) \to R$. This is the ideal generated by $((x-1))$ for $x \in N(H)/H$. We will call the classical Reidemeister torsion the *oriented R-torsion*. For actions of groups on manifolds the following generalization is sometimes convenient. Let $\alpha: G \to O_n$ be a representation of $G$, and let $S_\alpha$ be the unit sphere in $R^n$ with the induced action. For each $H \subset G$, $N(H)/H$ acts on $S_\alpha^H = S^k$, for some smaller dimensional sphere. This yields a homomorphism

$$\rho: N(H)/H \to \mathrm{Aut}(H_k(S_\alpha^H)) = \underset{\substack{0\\ \text{or}}}{\mathrm{Aut}(Z)} = \underset{\substack{0\\ \text{or}}}{Z/2Z}.$$

Let $R_0^\xi(N(H)/H) = \mathrm{kernel}(R(N(H)/H)) \to^{\rho^*} R(Z/2Z)$ or $R(0) = R$. This obviously depends only on $[H]$ and thus defines a Reidemeister torsion functor which we denote by $\tau^\alpha$. Such torsion will be called *unoriented R-torsion*.

Suppose $G_0 \subset G$ a subgroup. Then the map $X \to G \times_{G_0} (X - *) \cup (*)$ defines a coproduct preserving functor $e: B(G_0) \to B(G)$ and extends canonically to $e$: $R \otimes B(G_0) \to R \otimes B(G)$ and thus induces $e^*: K_1(B(G_0); R) \to K_1(B(G); R)$. Further, if $g: R \otimes B(G) \to D$ is any functor to an Abelian category we have an induced map $K_1(B(G_0); ge) \to K_1(B(G); g)$. In particular, for a nice functor $T$, we have $e^*: K_1(B(G_0); Te) \to K_1(B(G); T)$.

If $G_0$ is of finite index in $G$ the restriction of actions yield a coproduct preserving functor $r: B(G) \to B(G_0)$. Thus for a nice functor $j: B(G_0) \to D$, we have $r^*$: $K_1(B(G); jr) \to K_1(B(G_0); j)$. Now letting $T$ of the previous paragraph be equal to $jr$, we have

$$K_1(B(G_0); jre) \xrightarrow{e^*} K_1(B(G); jr) \xrightarrow{r^*} K_1(B(G_0); j).$$

If $G_0$ is in the center of $G$, then $jre = mj$, where $m = o(G/G_0)$. Thus a complex in $R \otimes B(G_0)$ is $j$-acyclic if and only if it is $jre$-acyclic. More generally, for $G_0$ normal in $G$, $K_1(B(G_0); jre) = K_1(B(G_0); j)$ and we have

$$K_1(B(G_0); j) \xrightarrow{e^*} K_1(B(G); jr) \xrightarrow{r^*} K_1(B(G_0); j),$$

where $r^*e^*$ = multiplication by $o(G/G_0)$ if $G_0$ is in the center of $G$. In general, $K_1$ behaves as a Frobenius module **[B1]** which allows us to use induction techniques, and should facilitate computation a great deal. Analogous results hold for $\bar{K}_1$.

1.26(b). DEFINITION (THE HIGHER $K_i$ GROUPS). Suppose $A$ is a category with coproducts such that, for any $Z \in A$,

$$A(Z, X \vee Y) \xrightarrow{p_1^* \times p_2^*} A(Z, X) \times A(Z, Y)$$

is 1-1, and suppose $T$ is a nice functor on $A$. Then we have a functor from $C_0(A) = C_0(\mathrm{Id}_A)$ to $C(\iota; T)$, where $\iota: A \to R \otimes A$. We can also perform the Quillen construction **[Q]** on both categories, which leads to classifying spaces and a continuous $j$: $BQC_0(A) \to BQC(\iota; T)$. Denote the homotopy fiber of this map by $F(A; T)$. It is then reasonable, in view of **[Q]**, to define $K_i(A; T) = \pi_i(F(A; T), 0)$, where 0 is the obvious basepoint of $F(A; T)$. As in Quillen **[Q]** one can prove that $K_1$ is the group defined earlier. It would be interesting to know whether $K_2$, for the appropriate $A$, is the Milnor-Steinberg group.

There is a fairly easy generalization of the Kwun-Szczarba product theorem **[KS]** to our situation which we now wish to give. For $X_1, X_2 \in A$, a product of $X_1$ and $X_2$ is an object $X_1 \times X_2$ of $A$ along with maps $\pi_i: X_1 \times X_2 \to X_i$, $i = 1, 2$, such that

$$\pi_1^* \times \pi_2^*: A(Z, X_1 \times X_2) \to A(Z, X_1) \times A(Z, X_2)$$

is a bijection for all $Z \in A$. By $X_1 \times X_2$ we always mean a product of $X_1$ and $X_2$. Products are unique up to canonical isomorphism. We say products exist in $A$ if any two objects have a product. We have $X_1 \vee X_2 \to^{p_1 \times p_2} X_1 \times X_2$. We write $Y = X_1 \wedge X_2$ and say $Y$ is a smash product of $X_1$ and $X_2$ if $X_1 \times X_2 = (X_1 \vee X_2) \vee Y$ with the coproduct inclusion of $X_1 \vee X_2 \to X_1 \times X_2$ given by $p_1 \times p_2$. These products are unique up to isomorphism, and we say $A$ has smash products if any two objects of $A$ have such a product.

As usual, let $\iota: A \to R \otimes A$ be the canonical functor and suppose that $(X, d)$ and $(X', d')$ are in $C(\iota)$. We say that an object $(X \otimes X', d \otimes d')$ of $C(\iota)$ is a *tensor product* of $(X, d)$ and $(X', d')$ if $(X \otimes X')_j = \bigvee_{k+m=j} (X_k \wedge X'_m)$ and

$$(d \otimes d')_j = \sum_{k+m=j} (d_k \wedge \mathrm{Id}_{X'_m} + (-1)^k \mathrm{Id}_{X_k} \wedge d'_m).$$

Any two tensor products of $(X, d)$ and $(X', d')$ are isomorphic in $C(\iota)$. The symbol $(X \otimes X', d \otimes d')$ will always be used for a tensor product. If $A$ has both smash products and coproducts, tensor products always exist in $C(\iota)$. Observe that in $A$ there is a natural map $\lambda$: $(X_1 \wedge X_2) \vee (X_1 \wedge X_3) \to X_1 \wedge (X_2 \vee X_3)$ which is an isomorphism in $R \otimes A$. We say $A$ is a *nice category if*

(1) $A$ has coproducts,

(2) $A$ has smash products,

(3) $\lambda$ is an isomorphism in $A$ for all $X_1, X_2, X_3$ in $A$.

For the remainder of the discussion of products, we assume $A$ is a nice category.

If $A$ has the wedge $ES$ structure and $(X \to Y \to Z)$ in $ES(A)$, then, for any $W$ in $A$, $(W \wedge X \to W \wedge Y \to W \wedge Z)$ is in $ES(A)$. It follows immediately that for $(X, d) \to (Y, d') \to (Z, d'')$ in $ESC(\iota)$ and $(W, d''')$ in $C(\iota)$,

$$(W \otimes X, d''' \otimes d) \to (W \otimes Y, d''' \otimes d') \to (W \otimes Z, d''' \otimes d'')$$

is in $ESC(\iota)$. We now consider $A$ a full subcategory of $C(\iota)$ by identifying $X$ in $A$ with $0 \to 0 \to X \to 0 \to 0$ in $C(\iota)$. Also, for any $(X, d)$ in $C(\iota)$ we have the suspension $(S^k(X), S^k d)$, where $S^k(X)_i = X_{i-k}$, $S^k d_i = (-1)^k d_{i-k}$. Note that $C_0(\iota)$ and $C(\iota; T)$ for a nice functor $T$ are closed under the operation $S^k$.

For $(X, d) \in C_0(\iota)$, we say $(X, d)$ is of *depth* 1 if $(X, d) = i(\rho)$, for some $\rho$ an isomorphism of $R \otimes A$. We say $(X, d)$ is of *depth n*, if it is not of depth $n-1$, but is one term of an element of $ESC_0(\iota)$, whose other terms are of depth $\leqq n - 1$. Every object of $C_0(\iota)$ is of some unique depth. An object $(X, d)$ of $C(\iota)$ is called of *length n* if exactly $n$ of the terms $(X_i)$ are nonzero. By inducting simultaneously on length and depth we get: If $(Y, d') \in C_0(\iota)$ of depth $\leqq n$, and $(X, d) \in C(\iota)$ of length $\leqq m$, then $(X \otimes Y, d \otimes d') \in C_0(\iota)$ of depth $\leqq n + m$. It follows that we have a pairing

$$K_0(C(\iota)) \otimes K_1(A; R) \to K_1(A; R). \tag{1}$$

Suppose $T: A \to D$ is a nice functor. $T$ is called *multiplicative* if for every $(Y, d') \in C(\iota; T)$ and $X \in A$, $(X \otimes Y, \mathrm{Id}_X \otimes d') \in C(\iota; T)$. For such a $T$ it follows by an easy induction on length that $(X \otimes Y, d \otimes d') \in C(\iota^\circ, T)$ if $(X, d) \in C(\iota)$ and $(Y, d') \in C(\iota T)$. Hence for a multiplicative functor we get a pairing:

$$K_0(C(\iota)) \otimes K_1(A; T) \to K_1(A; T). \tag{2}$$

Since $A$ is a subcategory of $C(\iota)$ we get induced pairings.

$$K_0(A) \otimes K_1(A; T) \to K_0(C(\iota)) \otimes K_1(A; T) \to K_1(A; T), \tag{3}$$

$$K_0(A) \otimes K_1(A; R) \to K_0(C(\iota)) \otimes K_1(A; R) \to K_1(A; R). \tag{4}$$

Finally, given $(X, d)$ in $C(\iota)$ we have the *Euler character* $\varepsilon(X, d) = \sum_i (-1)^i X_i$, an element of $K_0(A)$. This induces the *Euler map* $\varepsilon: K_0(C(\iota)) \to K_0(A)$.

1.27. PRODUCT THEOREM. *For $T$ a multiplicative functor and $(X, d)$ in $C(\iota)$, $v$ in $K_1(A; T)$, $w$ in $K_1(A; R)$ we have*
*$(X, d) \cdot v = \varepsilon(X, d) \cdot v$ and $(X, d) \cdot w = \varepsilon(X, d) \cdot w$,*
*where $\cdot$ denotes the pairings of* (2), (3), *and* (4) *above.*

PROOF. The proof is an easy induction on the length of $(X, d)$ (see **[KS]**).

Observe that (2) induces a pairing $K_1(A; R) \otimes K_1(A; R) \to K_1(A; R)$. However, an easy induction on depth shows that for $(X, d) \in C_0(\iota)$, $\varepsilon(X) = 0$ and hence by 1.26 this induced pairing is 0.

One disadvantage of 1.27 is that many interesting functors are not multiplicative. In particular, Reidemeister torsion is not multiplicative. What is true for many such functors is that if $(X, d)$ and $(Y, d')$ are in $C(\iota; \tau)$, then $(X \otimes Y, d \otimes d')$ is in $C(\iota; \tau)$. This induces a pairing $K_1(B(G); \tau) \otimes K_1(B(G); \tau) \to K_1(B(G); \tau)$. In general, this pairing is very mysterious. We will give some results in special cases, and even here we find the product theory involves some rather subtle formulas.

Recall that for $K$ a group, $R(K)$ has a natural diagonal $\Delta: R(K) \to R(K \times K) = R(K) \otimes R(K)$ induced by the diagonal homomorphism $K \to K \times K$. $M \subset R(K)$ is called a *co-ideal* if $\Delta(M) \subset R(K) \otimes M + M \otimes R(K)$. For the remainder of this section, we consider only torsion functors $\tau$ such that $R^\tau(N(H)/H)$ is both an ideal and a co-ideal of $R(N(H)/H)$, for each $[H) \in \mathrm{Conj}(G)$. For such torsion functors we have that if $(X, d), (X', d') \in C(\iota; \tau)$ then $(X \otimes X', d \otimes d') \in C(\iota; \tau)$ and hence we have a pairing $K_1(B(G); \tau) \otimes K_1(B(G); \tau) \to K_1(B(G); \tau)$.

We will now derive a product formula for $\tau$ under the additional hypothesis of 1.24. For the remainder of this discussion of products, we assume

(1) $G$ is Abelian,

(2) $W$ is the family of all subgroups of $G$,

(3) under the natural map $p: R(G) \to R(G/H)$, $R_0^\tau(G/H) = p(R_0^\tau(G))$.

Under these hypotheses, the condition of the previous paragraph is equivalent to the assumption that $R_0^\tau(G)$ is a co-ideal, and in fact, we can forget about the ideals $R_0^\tau(G/H)$ and consider only the ideal $R_0^\tau(G)$, which is independent of $H$.

Let $R_0^\tau(G \times G) = R_0^\tau(G) \otimes R(G) \cup R(G) \otimes R_0^\tau(G) \subset R(G \times G) = R(G) \otimes R(G)$, an ideal. By hypothesis $\Delta(R_0^\tau(G)) \subset R_0^\tau(G \times G)$. Let $R_1^\tau(G \times G)$ be a complementary ideal.

For $H_1, H_2 \subset G$ the natural embedding $G/H_1 \to G/H_1 \times G/H_2$ yields a ring homomorphism $R(G/H_1) \to R(G/H_1) \otimes R(G/H_2)$. On the other hand, the diagonal induces a monomorphism $\Delta: G/H_1 \cap H_2 \to G/H_1 \times G/H_2$. Thus $R(G/H_1 \times G/H_2)$ is a module over $R(G/H_1 \cap H_2)$ and in fact $R(G/H_1 \times G/H_2) \cong c(H_1, H_2)R(G/H_1 \cap H_2)$, as a module over $R(G/H_1 \cap H_2)$, where $c(H_1, H_2) = o(G)o(H_1 \cap H_2)/o(H_1)o(H_2)$. This construction yields a homomorphism $\hat{\lambda}_{H_1, H_2}: K_1(R(G/H_1)) \to K_1(R(G/H_1 \cap H_2))$, which we call a transfer.

We now consider $H_1$, with $R_0^\tau(G/H_1) \neq 0$. We let $R_0^\tau(G/H_1 \times G/H_2) = R_0^\tau(G/H_1) \otimes R(G/H_2) \cup R(G/H_1) \otimes R_0^\tau(G/H_2)$ an ideal of $R(G/H_1) \otimes R(G/H_2) = R(G/H_1 \times G/H_2)$. The map $a \to a \otimes 1$ induces an algebra monomorphism $j$: $R_0^\tau(G/H_1) \to R_0^\tau(G/H_1 \times G/H_2)$. Hence we have $j^*: K_1(R_0^\tau(G/H_1)) \to K_1(R_0^\tau(G/H_1 \times G/H_2))$. We wish to define a transfer homomorphism $k^*: K_1(R_0^\tau(G/H_1 \times G/H_2)) \to K_1(R_0^\tau(G/H_1 \cap H_2))$.

The map $G/H_1 \cap H_2 \to G/H_1 \times G/H_2$ makes $R_0^\tau(G/H_1 \times G/H_2)$ an $R_0^\tau(G/H_1 \cap H_2)$-module (but not a unitary module). By assumption $R_0^\tau(G/H_1 \times G/H_2) = p^*(R_0^\tau(G \times G))$ where $p: G \times G \to G/H_1 \times G/H_2$. It follows that $R(G/H_1 \times G/H_2) = R_0^\tau(G/H_1 \times G/H_2) + R_1^\tau(G/H_1 \times G/H_2)$, where $R_0^\tau$, $R_1^\tau$ are mutually annihilating ideals and $R_1^\tau(G/H_1 \times G/H_2) = p^*R_1^\tau(G \times G)$. We have the equality:

$$\text{(a)} \quad R_0^\tau(G/H_1 \cap H_2)(R(G/H_1 \times G/H_2)) = R_0^\tau(G/H_1 \cap H_2)(R_0^\tau(G/H_1 \times G/H_2)).$$

This follows because $R_0^\tau(G/H_1 \cap H_2)(R_1^\tau(G/H_1 \times G/H_2)) = 0$, and this follows because $R_0^\tau(G)(R_1^\tau(G \times G)) = 0$ and this follows because $R_1^\tau(G \times G) = R_1^\tau(G) \otimes R_1^\tau(G)$ and $R_0^\tau(G)$ is a co-ideal. We know that, as an $R(G/H_1 \cap H_2)$-module, $R(G/H_1 \times G/H_2) = c(H_1, H_2)R(G/H_1 \cap H_2)$. Thus $R_0^\tau(G/H_1 \cap H_2)R(G/H_1 \times G/H_2) = c(H_1, H_2)R_0^\tau(G/H_1 \cap H_2)$. Thus from (a) we get $R_0^\tau(G/H_1 \cap H_2)R_0^\tau(G/H_1 \times G/H_2) = c(H_1, H_2)R_0^\tau(G/H_1 \cap H_2)$. This induces a homomorphism $k_*: K_1(R_0^\tau(G/H_1 \times G/H_2)) \to K_1(R_0^\tau(G/H_1 \cap H_2))$. Composing this with the map $j^*: K_1(R_0^\tau(G/H_1)) \to K_1(R_0^\tau(G/H_1 \times G/H_2))$ yields $k^*j^*: K_1(R_0^\tau(G/H_1) \to K_1(R_0^\tau(G/H_1 \cap H_2))$. This in turn induces a homomorphism $\lambda_{H_1, H_2}: \tau(G/H_1; R) \to \tau(G/H_1 \cap H_2; R)$ (see 1.25). If $R_0^\tau(G/H_1) = 0$ we set $\lambda_{H_1, H_2} = 0$. Recall that under our hypothesis $\tilde{K}_1(B(G); \tau) = \sum_{H \subset G} \tau(G/H; R)$. Thus $a \in \tilde{K}_1(B(G); \tau)$, $a = \sum_{H \subset G} a_H$, $a_H \in \tau(G/H; R)$. We can now state and prove our product theorem:

1.28. Torsion Product Theorem. *Let $G$ be finite Abelian, $W$ = set of sub-*

*groups of $G$, $\tau$ a Reidemeister torsion functor with $R_0^\xi(G)$ a co-ideal and such that $R_0^\xi(G/H) = pR_0^\xi(G)$ for all $H \in W$. If $X$ and $Y$ are $\tau$-acyclic complexes of $R \otimes B(G)$, then so is $X \otimes Y$, and $\tau(X \otimes Y) \in \tilde{K}_1(B(G); \tau)$ is given as follows:*

(a) *If $R_0^\xi(G/H) \neq 0$ then*

$$\tau(X \otimes Y)_H = \sum_{H_i, H_j \subset G; H_i \cap H_j = H} \varepsilon_{H_j}(Y)\, \lambda_{H_i, H_j}(\tau(X)_{H_i}) + \varepsilon_{H_i}(X)\, \lambda_{H_j, H_i}(\tau(Y)_{H_j}),$$

*where, for any complex $(Z, d)$ of $R \otimes B(G)$ and $H' \subset G$, $\varepsilon_{H'}(Z) = \sum_i (-1)^i \varepsilon_{H'}(Z_i)$, $\varepsilon_{H'}(Z_i) =$ the number of orbits of type $H'$ in $Z_i - (*)$, $*$ is the basepoint. Note that $\varepsilon(Z) = \sum_{H'} \varepsilon_{H'}(Z) G/H'$ in $K_0(B(G))$.*

(b) *If $R_0^\xi(G/H) = 0$, then*

$$\tau(X \otimes Y)_H = \sum_{H_i \cap H_j = H} c(H_i, H_j)\, \varepsilon_{H_i}(X)\, \varepsilon_{H_j}(Y).$$

PROOF. Let $H_1$ be a minimal isotopy subgroup of $\bigcup_i X_i$ and $H_2$ a minimal isotopy subgroup of $\bigcup_j Y_j$. We then have the elements of $ESC(\iota; \tau)$:

$$X^0 \to X \to \bar{X}, \quad \text{and} \quad Y^0 \to Y \to \bar{Y},$$

where

(i) if $x \in X_i^0$ then $G_x \neq H_i$; if $x \in \bar{X}_i - (*)$, then $G_x = H_1$,

(ii) if $y \in Y^0$, then $G_y \neq H_2$; if $y \in \bar{Y} - (*)$, then $G_y = H_2$.

Starting with $X^0 = Y^0 =$ the trivial complex we can assume by induction that the theorem is true for $X^0 \otimes \bar{Y}$, $\bar{X} \otimes Y^0$, and $X^0 \otimes Y^0$. Let $V = X^0 \otimes Y \cup X \otimes Y^0 \subset X \otimes Y$. Then $X^0 \otimes Y^0 \to V \to (X^0 \otimes \bar{Y}) \cup (\bar{X} \otimes Y^0)$ and $V \to X \otimes Y \to \bar{X} \otimes \bar{Y}$ are in $ESC(\iota; \tau)$. For $H \neq H_1 \cap H_2$, $\tau(X \otimes Y)_H = \tau(V)_H = \tau(X^0 \otimes Y^0)_H + \tau(X^0 \otimes \bar{Y})_H + \tau(\bar{X} \otimes Y^0)_H$ and the formula follows by induction. For $H = H_1 \cap H_2$ we have

$$\begin{aligned}\tau(X \otimes Y)_H &= \tau(V)_H + \tau(\bar{X} \otimes \bar{Y})_H \\ &= \tau(X^0 \otimes Y^0)_H + \tau(X^0 \otimes \bar{Y})_H + \tau(\bar{X} \otimes Y^0)_H + \tau(\bar{X} \otimes \bar{Y})_H.\end{aligned}$$

By induction then, the theorem reduces to showing:

(1) If $R_0^\xi(G/H) \neq 0$, then

$$\tau(\bar{X} \otimes \bar{Y}) = \varepsilon_{H_2}(\bar{Y})\, \lambda_{H_1, H_2}(\tau(X))_{H_1} + \varepsilon_{H_1}(\bar{X})\, \lambda_{H_2, H_1}(\tau(Y))_{H_2}.$$

(2) If $R_0^\xi(G/H) = 0$ and thus $R_0^\xi(G/H_1) = R_0^\xi(G/H_2) = 0$, then

$$\varepsilon_H(\bar{X} \otimes \bar{Y}) = c(H_1, H_2)\, \varepsilon_{H_1}(\bar{X})\, \varepsilon_{H_2}(\bar{Y}).$$

(2) is true by inspection, so it remains to prove only (1). $\tau(\bar{X} \otimes \bar{Y})$ is determined as an element of a quotient of $K_1(R_0^\xi(G/H))$ by means of an acyclic $R_0^\xi(G/H)$ free chain complex $R_0^\xi(G/H)(S_0(\bar{X}) \otimes S_0(\bar{Y}))$, where $S_0(\bar{X})$ and $S_0(\bar{Y})$ are the free $R(G/H_1)$, $R(G/H_2)$ complexes of 1.22. Let $R_0^\xi(G/H_1 \otimes G/H_2) = R_0^\xi(G/H_1) \otimes R(G/H_2) \cup R(G/H_1) \otimes R_0^\xi(G/H_2)$, the ideal of $R(G/H_1) \otimes R(G/H_2)$ defined earlier. Then the diagonal $G \to G \times G$ induces $R_0^\xi(G/H) \to R_0^\xi(G/H_1 \times G/H_2)$ and gives the action of $R_0^\xi(G/H)$ on $(S_0(\bar{X}) \otimes S_0(\bar{Y}))$. Now $R_0^\xi(G/H)(S_0(\bar{X}) \otimes S_0(\bar{Y}))$ is the image under $k^*$: $K_1(R_0^\xi(G/H_1 \times G/H_2)) \to K_1(R_0^\xi(G/H))$ of $R_0^\xi(G/H_1 \times G/H_2)$ $(S_0(\bar{X}) \otimes S_0(\bar{Y}))$. We have

$$R_0^\tau(G/H_1 \otimes G/H_2) = R_0^\tau(G/H_1) \otimes R_0^\tau(G/H_2) + R_0^\tau(G/H_1) \otimes R_1^\tau(G/H_2)$$
$$+ R_1^\tau(G/H_1) \otimes R_0^\tau(G/H_2)$$
$$= A_1 + A_2 + A_3,$$

three mutually annihilating ideals of $R_0^\tau(G/H_1 \times G/H_2)$. If follows that

$$K_1(R_0^\tau(G/H_1 \times G/H_2)) = K_1(A_1) + K_1(A_2) + K_1(A_3);$$

the element $R_0^\tau(G/H_1 \times G/H_2)(S_0(\bar{X}) \otimes S_0(\bar{Y}))$ has components $A_1(S_0(\bar{X}) \otimes S_0(\bar{Y})) + A_2(S_0(\bar{X}) \otimes S_0(\bar{Y})) + A_3(S_0(\bar{X}) \otimes S_0(\bar{Y}))$. We can now apply the product theorem of **[KS]** and deduce that

$$A_1(S_0(\bar{X}) \otimes S_0(\bar{Y})) = 0 \quad \text{in } K_1(A_1),$$
$$A_2(S_0(\bar{X}) \otimes S_0(\bar{Y})) = \varepsilon_{H_2}(Y)j^*(S_0(\bar{X})),$$

and

$$A_3(S_0(\bar{X}) \otimes S_0(\bar{Y})) = \varepsilon_{H_1}(\bar{X})j^*(S_0(\bar{Y})).$$

The theorem now follows by the definitions. Q.E.D.

The formula of 1.28 looks very complicated and is so, yet often one can simplify by the following observation. If $X$ is $\tau$-acyclic and if $R_0^\tau(G/H) \neq 0$, then we have the acyclic complex $R_0^\tau(G/H)S_0(X(^{(H)}))$, and this forces $\varepsilon_H(X) = 0$ by the usual elementary argument. Thus we have

1.28 (a). COROLLARY. *For $\tau$-oriented torsion, and hence $R_0^\tau(G/H) \neq 0$ for $H \neq G$, we have*

$$\tau(X \otimes Y)_H = \varepsilon_G(Y)\tau(X)_H + \varepsilon_G(X)\tau(Y)_H,$$
$$\tau(X \otimes Y)_G = \varepsilon_G(X)\varepsilon_G(Y).$$

More generally, for $X \in B(G)$, let $l(X) = \sum_{H \subset G} o(G/H)\varepsilon_H(X) =$ number of elements of $X - (*)$. For $(X, d)$ a finite complex of $R \otimes B(G)$, we let $l(X) = \sum_i (-1)^i\, l(X_i)$. Then $l(X) = \sum_{H \subset G} o(G/H)\, \varepsilon_H(X)$. An easy induction argument shows, for $H \subset G$, $\varepsilon_{H'}(X) = 0$ for all $H' \supset H$, if and only if $l(X^{H'}) = 0$ for all $H' \supset H$. Note that under our hypothesis, if $R_0^\tau(G/H) = 0$ then $R_0^\tau(G/H') = 0$ for all $H' \supset H$. Putting this all together yields

1.28(b). COROLLARY. *Suppose that for $X, Y \in C(\iota; \tau)$, $R_0^\tau(G/H) = 0$ implies $l(X^H) = l(Y^H) = 0$. Then $\tau(X \otimes Y) = 0$.*

The way we pass from homotopy equivalences to acyclic complexes in algebraic topology is through the mapping cone construction. This construction has an obvious extension to our situation. Let $(C, d)$ and $(C', d')$ be elements of $C(R \otimes A)$ and $\alpha: C \to C'$ a map in $C(R \otimes A)$. A *mapping cone* $(C_\alpha, d_\alpha)$ for $\alpha$ is an object in $C(R \otimes A)$ such that $(C_\alpha)_i = C_{i-1} \vee C_i'$ and $(d_\alpha)_i = (-d_{i-1} + \alpha_{i-1}) \vee d_i'$. If $A$ has coproducts, mapping cones always exist and any two are isomorphic as objects in $C(\iota)$. Further, $(C', d') \to (C_\alpha, d_\alpha) \to S(C, d)$ is in $ESC(\iota)$. For the remainder of this section we will assume $A$ has coproducts.

We say $\alpha$ is an *equivalence* if $(C_\alpha, d_\alpha) \in C_0(\iota)$. Suppose $T: A \to D$ is a nice functor. We say $\alpha$ is a *T-equivalence* if $T(\alpha)$: $H(T(C, d)) \to H(T(C', d'))$ is an isomorphism. Then the exact sequence $0 \to T(C', d') \to T(C_\alpha, d_\alpha) \to T(S(C, d))$

$\to 0$ and the fact that $T(C_\alpha, d_\alpha) = (C_{T(\alpha)}, d_{T(\alpha)})$ implies $\alpha$ is a $T$-equivalence if and only if $H(T(C_\alpha, d_\alpha)) = 0$ if and only if $(C_\alpha, d_\alpha)$ is $T$-acyclic.

Suppose $\alpha\colon (C, d) \to (C', d')$ is such that $\alpha_i\colon C_i \to C'_i$ is an isomorphism in $R \otimes A$ for each $i$. Then $1(\alpha_i) \in C_0(\iota)$.

1.29. PROPOSITION. *$(C_\alpha, d_\alpha) \in C_0(\iota)$ and $(C_\alpha, d_\alpha) = \sum(-1)^i\, 1(\alpha_i)$ in $K_1(A; R)$.*

PROOF. The proof follows by induction on the length of $C$ and the fact that if we are given a commutative diagram

$$\begin{array}{ccccc} (A', d'_{A'}) & \longrightarrow & (B', d'_{B'}) & \longrightarrow & (C', d'_{C'}) \\ \uparrow{\scriptstyle \alpha_1} & & \uparrow{\scriptstyle \alpha_2} & & \uparrow{\scriptstyle \alpha_3} \\ (A, d_A) & \longrightarrow & (B, d_B) & \longrightarrow & (C, d_C) \end{array}$$

where the rows are in $ESC(\iota)$, then $(C_{\alpha_1}, d_{\alpha_1}) \to (C_{\alpha_2}, d_{\alpha_2}) \to (C_{\alpha_3}, d_{\alpha_3})$ is in $ESC(\iota)$.

We say that $\alpha$ and $\alpha'$ are *chain homotopic* if for each $i$ there exist maps $\delta_i\colon C_i \to C'_{i+1}$, in $R \otimes A$, such that $\alpha_i - \delta_{i-1}\, d_i = \alpha'_i - d'_{i+1}\, \delta_i$. If $\alpha$ and $\alpha'$ are chain homotopic then $\mathrm{Id} + \delta_{i-1}\colon (C_\alpha)_i \to (C_{\alpha'})_i$ induces an isomorphism in $C(R \otimes A)$ with inverse given by $\mathrm{Id} - \delta_{i-1}$. We need the following proposition.

1.30. PROPOSITION. *Let $h\colon X \to Y$ be a map in $R \otimes A$ and consider $\mathrm{Id} + h\colon X \vee Y \to X \vee Y$. It has an inverse $\mathrm{Id} - h$. Then $1(\mathrm{Id} + h) = 0$ in $K_1(A; R)$.*

PROOF. Consider $\psi = (\mathrm{Id} + h) \vee \mathrm{Id}\colon X \vee Y \vee Y \to X \vee Y \vee Y$. It suffices to show $1(\psi) = 0$ in $K_1(A; R)$. Let $t\colon Y \vee Y \to Y \vee Y$ be the twist, i.e., $t = i_2 p_1 + i_1 p_2$. Let $\rho = (i_1 + i_2) \vee i_2$, $\rho\colon Y \vee Y \to Y \vee Y$. Then $\psi = (\mathrm{Id} \vee t)^{-1} \rho^{-1} \psi^{-1} \cdot (\mathrm{Id} \vee \rho)\, \psi (\mathrm{Id} \vee t)$, and by the composition formula and the commutativity of $K_1(A; R)$, $1(\psi) = 1(\mathrm{Id}) = 0$. Q.E.D.

1.30. COROLLARY. *Suppose $\alpha$ and $\alpha'$ are chain homotopic equivalences ($T$-equivalences). Then $(C_\alpha, d_\alpha)$ and $(C_{\alpha'}, d_{\alpha'})$ represent the same element of $K_1(A; R)$ ($K_1(A; T)$).*

PROOF. Let $f = \mathrm{Id} + \delta\colon C_\alpha \to C_{\alpha'}$. Then $(C_{\alpha'}, d_{\alpha'}) \to (C_f, d_f) \to S(C_\alpha, d_\alpha)$ is in $ESC(\iota)$. Thus $C_{\alpha'} + S(C_\alpha) = C_f$ in $K_1(A; R)$ $(K_1(A; T))$. By 1.29 and 1.30, $C_f$ is 0 in $K_1(A; R)$ (and thus in $K_1(A; T)$). Thus $C_{\alpha'} = -S(C_\alpha)$ in $K_1(A; R)$ $(K_1(A; T))$. Since $\alpha$ is certainly chain homotopic to itself, we have $C_\alpha = -S(C_\alpha) = C_{\alpha'}$. Q.E.D.

Now consider $\alpha\colon C \to C'$ and $\alpha'\colon C' \to C''$ maps in $C(R \otimes A)$.
We have:

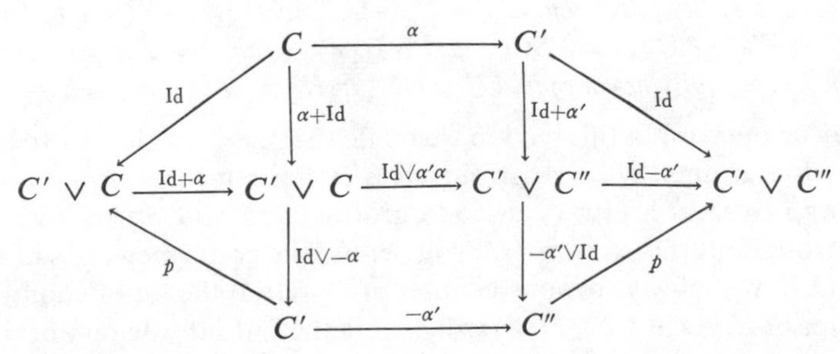

It follows that if $C_\alpha$, $C_{\alpha'}$ are in $C_0(\iota)$ $(C(\iota; T))$ so is $C_f$, where $f$ is the composition

$$C' \vee C \xrightarrow{\mathrm{Id}+\alpha} C' \vee C \xrightarrow{\mathrm{Id}\vee\alpha'} C' \vee C'' \xrightarrow{\mathrm{Id}-\alpha'} C' \vee C''$$

and $C_\alpha + C_{\alpha'} = C_f$ in $K_1(A; R)$ $(K_1(A; T))$.

On the other hand, we have an isomorphism $g\colon C_f \to C_{\mathrm{Id}\vee\alpha'\alpha}$, where $1(g_i)$ represents 0 in $K_1(A; R)$. Thus by 1.29, $C_f = C_{\mathrm{Id}\vee\alpha'\alpha} = C_{\alpha'\alpha}$ in $K_1(A; R)$ $(K_1(A; T))$. Hence we have shown

1.32. PROPOSITION. *If $\alpha\colon C \to C'$ and $\alpha'\colon C' \to C''$ are equivalences (T-equivalences), then so is $\alpha'\alpha$ and $C_{\alpha'\alpha} = C_\alpha + C_{\alpha'}$ in $K_1(A; R)$ $(K_1(A; T))$.*

Suppose $\alpha\colon C \to C'$ is a chain homotopy equivalence. That is, there exists $\alpha'\colon C' \to C$ such that $\alpha\alpha'$ and $\alpha'\alpha$ are chain homotopy equivalent to the respective identities. Then for any nice functor $T$, $C_\alpha \in C(\iota; T)$. Thus if there exists a nice projective functor on $A$, $C_\alpha \in C_0(\iota)$.

*Question.* Is it true that for any chain homotopy equivalence $\alpha$, $C_\alpha \in C_0(\iota)$?

1.33. PROPOSITION. *Suppose $C \to^\alpha C' \to^\gamma C''$ is in $ESC(\iota)$, and $\alpha$ is a T-equivalence (equivalence). Then $C''$ is in $C(\iota; T)$ $(C_0(\iota))$ and $C'' = C_\alpha$ in $K_1(A; T)$ $(K_1(A; R))$.*

PROOF. We have a commutative diagram;

$$\begin{array}{ccccc} C & \xrightarrow{\alpha} & C' & \xrightarrow{\gamma} & C'' \\ \uparrow{\scriptstyle \mathrm{Id}} & & \uparrow{\scriptstyle \alpha} & & \uparrow{\scriptstyle 0} \\ C & \xrightarrow{\mathrm{Id}} & C & \longrightarrow & 0 \end{array}$$

Hence $C_{\mathrm{Id}} = C_{\mathrm{Id}} + C_0 = C_0 = C''$ in $K_1(A; T)$ $(K_1(A; R))$.

We sum up previous propositions in the following theorem:

1.34. THEOREM. *Let $T$ be a nice functor on $A$ and $\alpha$ a map in $C(R \otimes A)$ with $\alpha$ a T-equivalence (equivalence); then*

(1) $C_\alpha \in C(\iota; T)$ $(C_0(\iota))$.

(2) *If $\alpha'$ is chain-homotopic to $\alpha$, $C_{\alpha'} = C_\alpha$ in $K_1(A; T)$ $(K_1(A; R))$.*

(3) *If $\alpha$ and $\alpha'$ are composable T-equivalences (equivalences), then $\alpha'\alpha$ is a T-equivalence (equivalence) and $C_{\alpha'\alpha} = C_{\alpha'} + C_\alpha$ in $K_1(A; T)$ $(K_1(A; R))$.*

(4) *If we have a commutative diagram*

$$\begin{array}{ccccc} C & \longrightarrow & C' & \longrightarrow & C'' \\ \uparrow{\scriptstyle \alpha} & & \uparrow{\scriptstyle \alpha'} & & \uparrow{\scriptstyle \alpha''} \\ B & \longrightarrow & B' & \longrightarrow & B'' \end{array}$$

*where the rows are in $ESC(\iota)$, and any two of the vertical maps are T-equivalences (equivalences) so is the third, and $C_{\alpha'} = C_\alpha + C_{\alpha''}$ in $K_1(A; T)$ $(K_1(A; R))$.*

(5) *If $C \to^\alpha C' \to^p C''$ is in $ESC(\iota)$ and $\alpha$ is a T-equivalence (equivalence) then $C''$ is T-acyclic (universally acyclic) and $C_\alpha = C''$ in $K_1(A; T)$ $(K_1(A; R))$.*

**2. *G-CW* complexes.** In this section we relate the previous algebra to questions of transformation groups. We restrict ourselves to finite groups $G$, although most of the results go over to arbitrary discrete groups, and still others to compact Lie groups. Throughout this section, $[H]$ will denote the conjugacy class of a subgroup $H$ of $G$, and $W$ will always denote a subset of $\mathrm{Conj}(G)$, the set of conjugacy classes of subgroups of $G$. As in 1.22, $B(W)$ will denote the full subcategory of the Burnside category $B(G)$ of $G$ determined by $W$, i.e., all $G$-finite sets whose isotropy subgroups lie in $W$. Categories of the form $B(W)$ with the wedge $ES$ structure will be denoted by $\mathbf{B}$. $T$ will denote a nice functor on $\mathbf{B}$ (see 1.19). The special case of a Reidemeister torsion functor will be denoted by $\tau$. Recall (1.23) that these are given by selecting two-sided ideals $R_0^{\xi}(N(H)/H)$ of $R(N(H)/H)$ for $[H] \in W$, where $R$ is a field of characteristic 0 or prime to the order of $G$. We can also permit an arbitrary commutative ring $R$ with unit when $R_0^{\xi}(N(H)/H) = R(N(H)/H)$. Associated to this last functor, we get, when $W = \mathrm{Conj}(G)$, the $R$ Whitehead torsion $\mathrm{Wh}(G; R)$, and for $R = Z$ the generalized Whitehead torsion $\mathrm{Wh}(G; Z)$.

The link between the previous algebra and geometry is the notion of $G$-$CW$ complex. We consider the category $CW(G)$ of finite $G$-$CW$ complexes and $G$-maps as defined in **[B2]** or **[I1]**. For $W \subset \mathrm{Conj}(G)$ we consider the full subcategory $CW(W)$ given by all finite $G$-complexes $X$ such that $[G_x] \in W$ for all $x \in X$. These categories satisfy the natural analogues of the elementary homotopy properties of the category of finite $CW$ complexes. In particular, in these categories, $G$-subcomplex inclusions are $G$-cofibrations, every $G$-map is homotopic to a $G$-cellular map, unique up to cellular $G$-homotopy, etc.

If $X$ is an element of $CW(W)$, $Y$ a $G$-subcomplex, then the $n$-cells of $X - Y \cup (*)$, $*$ a basepoint, are a $G$-finite set, and in fact, an element of $B(W)$. We denote this element by $(X, Y)_n^\wedge$. The incidence relation determines a boundary map on cellular $n$-chains, $\partial_n\colon C_n(X, Y) \to C_{n-1}(X, Y)$. $\partial_n$ determines a map, denoted by $\hat{\partial}_n$, from $(X, Y)_n^\wedge \to (X, Y)_{n-1}^\wedge$ in $Z \otimes B(W)$. Thus we have determined an element $(X, Y)^\wedge$ of $C(Z \otimes B(W))$.

If $f\colon (X, Y) \to (X', Y')$ is a *cellular* $G$-map, then $f$ determines $\hat{f}\colon (X, Y) \to^\wedge (X', Y')^\wedge$ where $\hat{f}$ is a map in $C(Z \otimes B(W))$. Further, if $R$ is a commutative ring with identity, then $Z \to R$ induces $Z \otimes B(W) \to R \otimes B(W)$. Hence the complex $R \otimes (X, Y)^\wedge$ and the map $R \otimes \hat{f}\colon R \otimes (X, Y)^\wedge \to R \otimes (X', Y')^\wedge$ are well defined in $C(R \otimes B(W))$.

For any nice functor $T$ we can consider those $f$ such that $R \otimes \hat{f}$ is a $T$-equivalence. For example, if $\tau$ is a Reidemeister torsion functor, $R \otimes \hat{f}$ is a $\tau$-equivalence if and only if

$$f_*\colon R_0^{\xi}(N(H)/H)H_*(X^H, Y^H; R) \cong R_0^{\xi}(N(H)/H)H_*(X'^H, Y'^H; R) \quad \text{for each } [H] \in W.$$

If $R \otimes \hat{f}$ is a $T$-equivalence, we say $f$ is a $T$-equivalence. When $f$ is a $T$-equivalence, we let $T(f)$ be the class of $R \otimes \hat{f}$ in $\tilde{K}_1(B; T)$. By 1.33, $T(f)$ depends only on the $G$-homotopy class of $f$. Thus the notion of $T$-equivalence can be extended to, not necessarily cellular, $G$-maps $f$ and the torsion $T(f)$ of such maps is well defined.

The following two questions present themselves.

(1) Given $a \in \tilde{K}_1(B(W); T)$ and $X \in CW(W)$, when does there exist an $X'$ in $CW(W)$ and a $T$-equivalence $f\colon X \to X'$ such that $T(f) = a$? (We write $X$ for the pair $(X, \emptyset)$.)

(2) Given a $T$-equivalence $f\colon X \to X'$, under what geometric condition is $T(f) = 0$?

These seem to be difficult questions. We have some partial answers, particularly in the case when $T$ is the generalized Whitehead torsion. Since there are natural homomorphisms $\mathrm{Wh}(G; Z) \to \mathrm{Wh}(G; R) \to K_1(B(G); T)$ for any nice functor $T$, this special case provides some information in the general case.

First consider question (2). Here we follow the work of Illman **[I1]**. Let $X$ and $X'$ be elements of $CW(W)$, each containing $Y$ as a $G$-subcomplex. By an obvious modification of Illman's definition, we can define the notion of a formal deformation in $W$ of $X$ to $X'$ rel $Y$. Let $f_i$: $Y \to X$ be the inclusions. Then

2.1. THEOREM. *If $f_1$ is a $T$-equivalence and $X$ a formal deformation of $X'$ in $W$* rel $Y$, *then $f_2$ is a $T$-equivalence and $T(f_1) = T(f_2)$.*

PROOF. This follows from the definition of deformation and repeated application of (3) and (5) of Theorem 1.34.

2.2. COROLLARY. *If $f$: $X \to X'$ is a formal deformation in $W$, then $T(f) = 0$. In particular, if $f$: $X \to I \times X$ is one of the natural inclusions then $T(f) = 0$.*

One of the main results of Illman is a partial converse of 2.1.

2.3. THEOREM. *Let $X$, $X'$ be $G$-$CW$ complexes, $G$ abelian. Let $t_0$ be generalized Whitehead torsion. Let $f$: $X \to X'$ be a $t_0$-equivalence and assume all the components of $X^H$ and $X'^H$ are simply connected for every $H \subset G$. Then $f$ is $G$-homotopic to a formal deformation of $X$ to $X'$ if and only if $t_0(f) = 0$.*

PROOF. The theorem follows from Theorem 1.4 of Illman and our result 1.18.

We wish to prove an invariance under subdivision theorem. The following will suffice for our purposes although stronger results are undoubtedly possible.

2.4. DEFINITION. A cellular $G$-map $f$: $X \to X'$ is called a *subdivision* if $f$ is a homeomorphism and to each cell $\lambda$: $D^n \to X$ of $X$ we have $f\lambda(D^n) = \bigcup_i \lambda_i(D^n)$, where $\lambda_i$: $D^n \to X'$ are cells of $X'$.

2.5. THEOREM (INVARIANCE UNDER SUBDIVISION). *Let $f$: $X \to X'$ be a subdivision, with $X$ and $X'$ in $CW(W)$. Then for any nice functor $T$, $f$ is a $T$-equivalence and $T(f) = 0$.*

PROOF. Since $f$ is a homeomorphism, $f$: $X^H \cong X'^H$ for each $H \subset G$. Hence $f$ is a $t_0$-equivalence for the generalized Whitehead torsion functor $t_0$. Since $t_0$ is a projective functor (1.21, 1.22), $f$ is a $T$-equivalence for any nice functor $T$ and $T(f)$ is the image of $t_0(f)$ under the composite homromophism

$$\tilde{K}_1(B(W); Z) \to \tilde{K}_1(B(W); R) \to \tilde{K}_1(B(W); T).$$

Thus if suffices to show $t_0(f) = 0$.

Observe that if $Y$ is a subcomplex of $X$, then $Y' = f(Y)$ is a subcomplex of $X'$ and $f|\ Y$: $Y \to Y'$ is also a subdivision. Hence in $C_0(\iota)$ we have the commutative diagram where the rows are in $ESC_0(\iota)$:

$$\begin{array}{ccccc} \hat{Y} & \longrightarrow & \hat{X} & \longrightarrow & (X, Y)^\wedge \\ \downarrow{\scriptstyle f|} & & \downarrow{\scriptstyle f} & & \downarrow{\scriptstyle \tilde{f}} \\ \hat{Y}' & \longrightarrow & \hat{X}' & \longrightarrow & (X', Y')^\wedge. \end{array}$$

By induction on the number of $G$-orbits of $X$, we can reduce the problem to showing $t_0(\tilde{f}) = 0$ where $(X, Y)^\wedge$ is a single $G$-orbit. Thus as a complex, $(X, Y)^\wedge_n = G/H \cup (*)$ for some $H \subset G$ and some chosen $n$ and $(X, Y)^\wedge_j = *$ for $j \neq n$. Since $f$ is a homeomorphism, we have $G_x$ conjugate to $H$ for $x \in X' - Y'$. Let $\lambda$: $D_n \to X$ be the cell corresponding to $[e]$ in $(X, Y)^\wedge_n = G/H \cup (*)$. Let $(X' - Y')^0_j$ be all the $j$-cells of $X' - Y'$ which are contained in $f(\lambda(D_n))$. Since $f$ is a homeomorphism, it follows easily that $(X', Y')^{\wedge}_j = (G/H \times (X' - Y')^0_j) \cup (*)$ and $\tilde{f} = \mathrm{Id} \times \tilde{f}_0$. Thus $t_0(\tilde{f})$ is in the image of $j$: $\tilde{K}_1(B(e); Z) \to \tilde{K}_1(B(W); Z)$, where $j$: $B((e)) \to B(W)$ is given by $j(X) = G/H \times X$. But $\tilde{K}_1(B((e)); Z) = 0$. Thus $t_0(f) = 0$ and hence by induction, $t_0(f) = 0$.

2.6. COROLLARY. *Let $f$: $X \to X'$ be a T-equivalence. $f$ remains a T-equivalence if either $X$ or $X'$ or both are replaced by subdivisions. Furthermore, $T(f)$ is unaltered by such a replacement.*

REMARK. As we shall see later, for $CW$ complexes with more than one orbit type it is not true that $T(f) = 0$ for $f$ a $G$-homeomorphism. Thus our invariants depend on the cellular structure and not just the topological type of the spaces involved.

We now consider (1), the realization problem. We will show how to realize elements in $\tilde{K}_1(B(W); Z) = \tilde{K}_1(B(W); t_0)$, where $t_0$ is generalized Whitehead torsion.

2.7. DEFINITION. We say $X \in CW(W)$ is *rich* if, for each $H \subset G$, with $[H] \in W$, there exists an $x \in X^H$ such that the orbit of $x$ under $N(H)$ (the normalizer of $H$ in $G$), is contained in a connected component of $X^H$. For example, if $X^G \neq \emptyset$, $X$ is rich.

2.8. REALIZATION THEOREM. *Let $a \in \tilde{K}_1(B(W); Z)$. Let $X$ be rich and $k \geqq 2$. Then there exists $Y \in CW(W)$, with $X \subset Y$ and $Y$ is formed from $X$ by attaching $k$-cells by trivial maps along the boundaries and then attaching $(k + 1)$-cells to the result, such that the inclusion $\iota$: $X \hookrightarrow Y$ is a $t_0$-equivalence and $t_0(\iota) = a$.*

PROOF. The argument of 1.18 shows that

$$\tilde{K}_1(B(W); Z) = \sum_{[H] \in W} \mathrm{Wh}(N(H)/H).$$

Using the composition formula 1.34(3), we can assume $a \in \mathrm{Wh}(N(H)/H)$ for some $H \subset G$. Since $X$ is rich we can find a component $X^H_0$ of $X^H$ invariant under $N(H)$. Let $x \in X^H_0$.

Let $a$ be given by the $m \times m$ nonsingular matrix $(a_{ij})$, with $a_{ij} \in Z(N(H)/H)$. Then $a$ as a transformation takes $mZ(N(H)/H)$ into itself. Let $W = N(H)/H \times \bigvee_m S^k$ with induced left $N(H)$-action. Then the orbit of the basepoint $e \times p$ is identified with $N(H)/H$ and we have $H_*(W, N(H)/H) = H_k(W, N(H)/H) = mZ(N(H)/H)$. We have the $N(H)$-map $\rho$: $N(H)/H \to X^H_0$, given by $[g] \to gx$ and can form the $N(H)$-complex $W \cup_\rho X^H_0$. Observe that

$$H_*\Big(W \underset{\rho}{\cup} X^H_0, X^H_0\Big) = H_k(W, N(H)/H) = mZ(N(H)/H).$$

We have the following commutative diagram with exact rows and columns:

$$\begin{array}{ccccccc}
 & & 0 & & & & \\
 & & \uparrow & & & & \\
 & \overset{\gamma}{\nearrow} & H_k\Big(W \underset{\rho}{\cup} X^H_0, X^H_0\Big) \simeq mZ(N(H)/H) & & & & \\
 & & \uparrow & & & & \\
\pi_k\Big(W \underset{\rho}{\cup} X^H_0\Big) & \xrightarrow{\tau} & \pi_k\Big(W \underset{\rho}{\cup} X^H_0, X^H_0\Big) & \longrightarrow & \pi_{k-1}(X^H_0) & \xrightarrow{\lambda} & \pi_{k-1}\Big(W \underset{\rho}{\cup} X^H_0\Big)
\end{array}$$

Since $W \cup_\rho X_0^H$ has $X_0^H$ as a retract, the map $\lambda$ is clearly monic. Hence $\tau$ is epi. Thus $\gamma$ is epi.

Now let $v_1, v_2, \ldots, v_m$ be the natural basis of $mZ(N(H)/H)$ over $Z(N(H)/H)$. Then $a(v_j) = \sum_i a_{ij}v_i$. Let $b_j: S^k \to W \cup_\rho X_0^H$ with $\gamma(b_j) = a(v_j)$. Let $V_0 = N(H)/H \times S_1^k + S_2^k + \cdots + S_m^k$ which is contained in $V = N(H)/H \times D_1^{k+1} + D_2^{k+2} + \ldots + D_m^{k+1}$. We have the $N(H)$-map $\psi: V_0 \to W \cup_\rho X_0^H$ given by $\psi(g, y) = gb_j(y)$ when $y \in S_j^k$. We can then form $Y_0 = V \cup_\psi (W \cup_\rho X_0^H)$. It follows immediately from the construction that, for all $H' \subset N(H)$, the inclusion $\iota: X_0^H \hookrightarrow Y_0$ induces a homotopy equivalence $(X_0^H)^{H'} \simeq Y^{H'}$ and $\mathrm{Wh}(\iota) = a$.

We now enlarge $Y_0$ to a $G$-complex by setting $Y_1 = G \times_{N(H)} Y_0$ with the $G$-action coming from the action of $G$ on the first factor. The inclusion $X_0^H \hookrightarrow Y_0$ induces an inclusion $G \times_{N(H)} X_0^H \hookrightarrow Y_1$. We also have the natural $G$-map $f$: $G \times_{N(H)} X_0^H \to X$ given by $f(g, x) = gx$. Let $Y = Y_1 \cup_f X$. Then the inclusion $X \hookrightarrow Y$ satisfies the conclusion of the theorem.

**3. Smooth *G*-manifolds.** In this section we develop our results about the ultimate objects of our interest, smooth $G$-manifolds. The basic result is that in certain cases the invariants we have introduced form a complete set of obstructions for a $G$-manifold $N$ to be $M \times I$, for some $G$-manifold $M$. This provides us with a method for classifying nice actions of finite groups on disks, and some information about such actions on spheres. In particular it permits us to construct infinite families of nonlinear actions of almost any finite group on spheres and other manifolds.

In this section $G$ will be a compact Lie group and all $G$-manifolds will be smooth manifolds with $C^\infty$-actions. It has long been expected that an $s$-cobordism theorem holds in this category and indications of how such a theorem might be formulated and proven can be found in Browder-Quinn **[BQ]**. Since the formulation there is somewhat incomplete and imprecise we will formulate it more adequately. Once this has been done, it is relatively straightforward to link it with our invariants when $G$ is finite and thus derive the result mentioned above. One would expect an analogous theory in the PL category. However, unless one makes some restrictions on the actions, the theory definitely fails, and Whitehead torsion invariants do not measure the failure of an $h$-cobordism to be a product cobordism. In fact, there are examples of such nontrivial $h$-cobordisms in the PL category for the group $Z/2Z$. In §4, we take up the question of overcoming these problems by restricting ourselves to locally linear actions. Finally, we will see by examples in this section that our invariants are not topological invariants, and thus leave the problem of distinguishing distinct topological actions untouched.

For a $G$-space $X$ and a *closed* subgroup $H$ of $G$, we write $X^{(H)} = \{x \in X \mid G_x = H\}$, and $X^{[H]} = \{x \in X \mid [G_x] = [H]\}$. If $X$ is a $G$-manifold (always assumed 2nd countable), then $X^{(H)}$ is a smooth $N(H)$-submanifold of $X^H$, which is open and dense in $X^H$, and $X^{[H]}$ is a smooth $G$-submanifold of $X$. We are allowing here smooth manifolds to have different components of different dimensions. Further, let $X_H = X^{(H)}/N(H) = X^{[H]}/G$. Then the maps $\rho_1: X^{(H)} \to X_H$ and $\rho_2: X^{[H]} \to X_H$ are projections of fibre bundles, where $\rho_1$ is a principal bundle with group $N(H)/H$ and $\rho_2$ is the associated bundle with fibre $G/H$. $X_H$ is a smooth manifold, and $X_H = \sum_i X_H^i$, where each $X_H^i$ is a component of $X_H$ of which there are a countable number. $X_H$ depends only on the conjugacy class of $H$, i.e., if $H$ and $H'$ are conjugate, $X_H = X_{H'}$.

Let $\iota: X \to \partial Y \subset Y$ be a $G$-embedding of compact $G$-manifolds with $\dim X = \dim Y - 1$. $\iota$ is called a $G$-cobordism of $X$ to $\overline{\partial Y - X}$.

3.1. DEFINITION. $\iota$ is called a *G h-cobordism* if for each $[H] \in \operatorname{Conj}(G)$ (the family of conjugacy classes of *closed* subgroups of $G$) the induced maps $\iota^{[H]}: X^{[H]} \to Y^{[H]}$ and $\iota'^{[H]}: \overline{(\partial Y - X)}^{[H]} \to Y^{[H]}$ are homotopy equivalences.

Using an inductive procedure, we will define a family of elements $\mathrm{Wh}^i_H(\iota)$ in $\mathrm{Wh}(\pi_1(X^i_H))$ for $G$ $h$-cobordisms $\iota$, which depend only on the conjugacy class of $H$ in $G$.

3.2. DEFINITION OF THE ELEMENTS $\mathrm{Wh}^i_H(\iota)$. Suppose we have defined these elements for all $G$ $h$-cobordisms whose domains have at most $n - 1$ orbit types. Suppose $\iota: X \to Y$ is a $G$ $h$-cobordism such that $X$ has exactly $n$ orbit types. Let $H$ be a maximal isotopy subgroup that actually occurs. Then $Y^{[H]}$ is a *compact* $G$-submanifold of $Y$ with $Y^{[H]} \cap X = X^{[H]}$. We can then choose a closed $G$-tubular neighborhood $E$ of $Y^{[H]}$ in $Y$, where $E$ is a $G$-disk bundle over $Y^{[H]}$ such that $E \cap X = E \mid X^{[H]}$ is a tubular neighborhood of $X^{[H]}$ in $X$ and $E \cap \overline{\partial Y - X} = E \mid \overline{(\partial Y - X)}^{[H]}$ is a tubular neighborhood of $\overline{(\partial Y - X)}^{[H]}$ in $\overline{\partial Y - X}$. Let $S(E)$ be the associated sphere bundle to $E$, $S(E) \subset E$ and $\mathring{E} = E - S(E)$. Set $\mathring{Y} = Y - \mathring{E}$, $\mathring{X} = X - \mathring{E}$. Then $\mathring{X} = X \cap \mathring{Y}$ and the inclusion map $\iota_0: \mathring{X} \to \mathring{Y}$ is also a $G$ $h$-cobordism and $\mathring{X}$ has one less orbit type than $X$. We then define for $[H'] \neq [H]$,

$$\mathrm{Wh}^i_{H'}(\iota) = \mathrm{Wh}^i_{H'}(\iota_0),$$

which is permissible since $X_{H'}$ is homotopically equivalent to $\mathring{X}_{H'}$ for $[H'] \neq [H]$. Further $\lambda_H: X_H \to Y_H$ is a homotopy equivalence of compact smooth manifolds. Hence the Whitehead torsion of $\lambda_H$ is defined and sits in $\sum_i \mathrm{Wh}(\pi_1(X^i_H))$, and these components are by definition $\mathrm{Wh}^i_H(\iota)$.

An easy induction argument shows

3.3. THEOREM. *The invariants* $\mathrm{Wh}^i_H(\iota)$ *are well defined and independent of the choice made in* 3.2 *of tubular neighborhoods and of the maximal isotopy subgroups. In particular, if* $\phi: (Y, X) \to (Y', X')$ *a G-diffeomorphism, where* $\iota: X \to Y$ *and hence* $\iota': X' \to Y'$ *are G h-cobordisms, then* $\phi\,(\mathrm{Wh}^i_H(\iota)) = \mathrm{Wh}^i_H(\iota')$ *for each H and i.*

REMARK. The invariants $\mathrm{Wh}^i_H(\iota)$ should really be denoted $\mathrm{Wh}^i_{[H]}(\iota)$ to indicate they depend only on the conjugacy class of $H$. However, in the interest of avoiding cumbersome notation we do not do this.

We are now ready to formulate the $s$-cobordism theorem.

3.4. EQUIVARIANT $s$-COBORDISM THEOREM. *Let* $\iota: X \to Y$ *be a G h-cobordism such that* $Y'_H = X'_H \times I$ *for* $\dim X'_H \leqq 4$. *Then the pair* $(Y, X)$ *is G-diffeomorphic to* $(X \times I, X \times (0))$ *if and only if* $\mathrm{Wh}^i_H(\iota) = 0$ *for all H, i.*

PROOF. This follows from the usual $s$-cobordism theorem and induction on the number of distinct orbit types.

We also have a realization theorem.

3.5. REALIZATION THEOREM. *Let $X$ be a compact $G$-manifold. Suppose, for each $[H] \in \mathrm{Conj}(G)$ and each component $X_H^i$ of $X_H$, we have selected an element $w_H^i$ of $\mathrm{Wh}(\pi_1(X_H^i))$ such that $w_H^i = 0$ when $\dim X_H^i \leqq 4$. Then there is a $G$ h-cobordism $\iota: X \to Y$ such that $Y_H^i = I \times X_H^i$, where $\dim X_H^i \leqq 4$ and $\mathrm{Wh}_H^i(\iota) = w_H^i$ for all $H$, $i$.*

PROOF. This follows from induction on the number of orbit types and the argument of Theorem 11.1 of [M].

From 3.4 and 3.5 and the argument of the proof of 11.3 in [M] we get

3.6. THEOREM. *Let $\iota_j: X \to Y_j$ be $G$ h-cobordisms, $j = 1, 2$. Suppose that for $\dim X_H \leqq 4$, $Y_{j,H}^i = X_H^i \times I$ and further suppose $\mathrm{Wh}_H^i(\iota_1) = \mathrm{Wh}_H^i(\iota_2)$ for all $H$, $i$. Then there exists a $G$-diffeomorphism $\lambda: Y_1 \to Y_2$ with $\iota_2 = \lambda\iota_1$*

Thus the invariants $\mathrm{Wh}_H^i(\iota)$ are quite powerful. However they are generally difficult to compute. We wish to link them to our more accessible invariants defined earlier. The key to this is the existence of a unique nice equivariant triangulation of the smooth manifold. Precisely:

3.7. DEFINITION. Let $G$ be a finite group. $\lambda : K \to X$ is called a *$G$ $C^\infty$-triangulation* of the $C^\infty$ $G$-manifold $X$ if $K$ is a $G$-simplicial complex and $\lambda$ is an equivariant homeomorphism which is a $C^\infty$-triangulation in the usual sense. (See [W].)

3.8. THEOREM. *Every $C^\infty$ $G$-manifold $X$ admits a $GC^\infty$-triangulation $\lambda: K \to X$. If we are given another $\lambda': K' \to X$, then $\lambda'^{-1}\lambda$ is $\varepsilon$-$G$-isotopic to a $G$ PL-equivalence* [I2].

Thus the theory developed in §2 for $G$-$CW$ complexes can be applied directly to the category of smooth $G$-manifolds when $G$ is a finite group. In particular, given $\iota: X \to Y$ a $G$ $h$-cobordism, we have a $t_0$-equivalence, where $t_0$ is the generalized Whitehead torsion functor. Thus $t_0(\iota) \in \mathrm{Wh}(G; Z) = \sum_{[H] \in \mathrm{Conj}(G)} \mathrm{Wh}(N(H)/H)$ is defined, and we can write $t_0(\iota) = \sum_{[H] \in \mathrm{Conj}(G)} t_0^H(\iota)$, where $t_0^H(\iota) \in \mathrm{Wh}(N(H)/H)$. We have the principal bundle $N(H)/H \to X^{(H)} \to X_H$. Thus for each component of $X_H$, $X_H^i$ we have the induced principal bundle $N(H)/H \to X_i^{(H)} \to X_H^i$. This gives us a well-defined homomorphism $\psi_H^i: \pi_1(X_H^i) \to N(H)/H$ defined up to inner automorphism of $N(H)/H$. This defines a unique homomorphism $\bar{\psi}_H^i: \mathrm{Wh}(\pi_1(X_H^i)) \to \mathrm{Wh}(N(H)/H)$. Chasing through the various definitions and inducting on the number of orbit types yields

3.9. THEOREM. $t_0^H(\iota) = \sum_i \bar{\psi}_H^i(\mathrm{Wh}_H^i(\iota))$.

3.10. COROLLARY. *If $X^{(H)}$ is connected and simply connected for all $H \in \mathrm{Conj}(G)$ then we have the identity $t_0^H(\iota) = \mathrm{Wh}_H(\iota)$.*

In this case the more subtle invariant $\mathrm{Wh}_H(\iota)$ is determined by the cruder invariant $t_0(\iota)$ and is a PL invariant. We can apply the above results to classify certain types of actions of $G$ on the $n$-disk $D$.

3.11. DEFINITION. Let $G$ be a compact Lie group and let $M$ be a $G$-manifold. $M$ is called a *semilinear $G$-disk* if, for each closed subgroup $H \subset G$, $M^H$ is diffeomorphic to $D^k$ for some $k \geqq 0$. Observe that if $\dim M^H \geqq 5$, this is equivalent to the homotopic equivalence of the pair $(M^H, \partial M^H)$ to $(D^k, S^{k-1})$.

If $M$ is an $n$-dimensional semilinear $G$-disk, then $G$ acts linearly on $M_x$, where

$x \in M^G$. This representation is equivalent to a representation $\alpha = \alpha(n)\colon G \to O(n)$. Here $\alpha$ is uniquely determined up to linear equivalence independent of $x$, since $M^G$ is assumed connected. $\alpha$ determines an orthogonal action of $G$ on $D^n$ and there is an equivariant embedding $\hat{j}\colon D^n_\alpha \to M - \partial M$, where $D^n_\alpha$ denotes the $G$-manifold $D^n$ with action coming from $\alpha$. We then also have $j = \hat{j} \mid S^{n-1}_\alpha$ : $S^{n-1}_\alpha \hookrightarrow M_0 = M - \hat{j}(\mathring{D}^n_\alpha)$.

For any $G$-manifold $M$, we say $M$ satisfies the codimension $i$ condition if dim $M^H$ − dim $M^{H'} \neq i$, for $H \subset H' \subset G$ and dim $M^{H'} > 0$. The semilinear disk $M$ above satisfies the codimension $i$ condition if and only if $D^n_\alpha$ does. If it satisfies the codimension 2 condition then the map $j\colon S^{n-1}_\alpha \hookrightarrow M_0$ is a $G$ $h$-cobordism.

3.12. THEOREM. *If $M$ is a semilinear $G$-disk and satisfies the codimension 2 condition, and if* dim $M_H \geqq 6$ *for $M_H \neq \varnothing$ and $H \neq G$, then* $(\alpha(M), \mathrm{Wh}^i_H(j))$ *form a complete set of $G$-diffeomorphism invariants of $M$. That is, two semilinear $G$-disks satisfying the hypothesis have the same invariant if and only if they are $G$-diffeomorphic. Further, given $D^n_\alpha$ satisfying the hypothesis and a family* $(W^i_H)$, $W^i_H \in \mathrm{Wh}(\pi_1(S^{n-1}_{\alpha, H}{}^{i}))$, *there exists a semilinear $G$-disk $M$ with invariants* $(\alpha, W^i_H)$.

PROOF. This follows from 3.5 and 3.6.

Suppose now that $G$ is finite and in addition to the above hypothesis, $M$ satisfies the codimension 1 condition. Then $M^{(H)}$ is both connected and simply connected. By 3.10 and 3.12, $(\alpha(M), t_0(j))$ form a complete set of invariants. By 1.33, $t_0(j) = t_0(j')$, where $j'\colon p \to M^G \subset M$ is a map of a point into $M$. Now $t_0(j') = -t_0(\hat{j})$, where $\hat{j}\colon M \to p$. Thus we can conclude.

3.13. THEOREM. *Let $G$ be a finite group. Then the $G$-diffeomorphism classes of semilinear $G$-disks $M$ which satisfy the codimension* 1 *and* 2 *conditions and for which* dim $M^H \geqq 6$ *for $H \neq G$ are in* 1-1 *correspondence with pairs $(\alpha, a)$, where $a \in$* Wh$(G; Z)$ *and $\alpha$ is a conjugacy class of orthogonal representations of $G$ such that $D^n_\alpha$ satisfy the above conditions. The correspondence is given by $M \to (\alpha(M), -t_0(\hat{j}))$, where $\alpha(M)$ is the action of $G$ on $M_x$, $x \in M^G$ and $\hat{j}\colon M \to p$, $p$ a point.*

Analogous to semilinear $G$-disks we can define semilinear $G$-Euclidean spaces and semilinear $G$-spheres. In the latter case, it is convenient to require not that $M^H$ be diffeomorphic to $S^k$ but that either $M^H$ be pl homeomorphic to $S^k$ or empty. The problem of finding a complete set of $G$-diffeomorphism invariants for semilinear $G$-spheres is as difficult as it is interesting. Using the engulfing techniques of **[CMY]** one can show:

3.14. THEOREM. *Let $G$ be a compact Lie group. Let $M$ be a semilinear $G$-Euclidean space satisfying the codimension* 2 *condition and with* dim $M_H \geqq 5$ *if $M_H \neq \varnothing$ and $H \neq G$. Then $M$ is $G$-diffeomorphic to the linear representation $\alpha(M)$.*

3.15. COROLLARY. (a) *Let $M$ be a semilinear $G$-sphere which satisfies the codimension* 2 *condition with* dim $M_H \geqq 5$, *for $M_H \neq \varnothing$ and $H \neq G$. Assume also that $M^G \neq \varnothing$. Then $M$ is $G$-homeomorphic to $S(1 \oplus \alpha(M - p))$, where $p \in M^G$ and where we have written $S(\rho)$ for $S_\rho$. The homeomorphism can be taken to be smooth on $M - (p)$.*

(b) *Let $M$ be a semilinear $G$-disk such that $\partial M$ satisfies the hypothesis of* (a). *Then $M$ is $G$-homeomorphic to $D_\alpha$, where $\alpha = \alpha(M)$.*

PROOF. (a) follows from 3.14 by removing the point $p$ from $M^G$. (b) follows from (a) and the fact that $M - (q)$ is $G$-diffeomorphic to $\partial M \times [0, 1)$ by $G$-engulfing, where $q$ is a fixed point in the interior of $M$.

REMARK. 3.12 and 3.15(b) show that our torsion invariants are not topological invariants, nor is any nontrivial function of them a topological invariant.

We now apply our results on semilinear $G$-disks to get results on semilinear $G$-spheres. For the remainder of this section, $G$ will be finite. We utilize some notions from **[R1]**, and refer there for further elaboration.

Let $\alpha: G \to O_n$ be a fixed orthogonal representation of $G$. We let $O_\alpha$ be the centralizer of $\alpha(G)$ on $O_n$. The unit disk and unit sphere in $R^n$ with the induced action of $G$ will be denoted as above by $D_\alpha$ and $S_\alpha$. A smooth $G$-manifold $M$ is called a $(G, \alpha)$-manifold if $M^G \neq \emptyset$ and if, for each $x \in M^G$, the induced action of $G$ on $M_x$ is equivalent to $\alpha$. For example, $D_\alpha$ is a $(G, \alpha)$-manifold as is $S_{\alpha \oplus 1}$. If $M$ is a $(G, \alpha)$-manifold then there is a canonical reduction of $t(M) \mid M^G$ to $O_\alpha$, where $t(M)$ is the tangent bundle of $M$. A further reduction of $t(M) \mid M^G$ to the component of the identity of $O_\alpha$ is called a $G$-orientation of $M$. Such a reduction always exists if each component of $M$ is simply connected. On $(G, \alpha)$-oriented manifolds with connected fixed point sets of positive dimension one can define equivariant connected sums and also boundary connected sums if the fixed point sets of the boundaries are connected and of positive dimension.

3.16. DEFINITION. Let $D_\alpha(G)$ be the set of $G$-oriented $G$-diffeomorphism classes of semilinear $(G, \alpha)$-disks. Let $S_\alpha(G)$ be the corresponding set of classes of semilinear $(G, \alpha)$-spheres. If $\dim D_\alpha^G \geqq 2$, then $D_\alpha(G)$, under boundary connected sums, is an Abelian monoid with linear objects as identities: if $\dim D_\alpha^G \geqq 1$, then $S_\alpha(G)$, under connected sum, is an Abelian monoid with the linear objects as identities.

Passing to boundaries we have the map $j_{\alpha+1}: D_{\alpha+1}(G) \to S_\alpha(G)$. If $\dim D_\alpha^G \geqq 1$ we can remove the interior of an embedded $D_\alpha$ in a semilinear $(G, \alpha)$-sphere and get a map $k_\alpha: S_\alpha(G) \to D_\alpha(G)$. The maps $j_{\alpha+1}$, $k_\alpha$ are morphisms when the ranges and domains are monoids. Suppose in $S_\alpha(G)$ we divide out by the equivalence relation of oriented $G$ $h$-cobordisms. Call the resulting set $C_\alpha(G)$. When $S_\alpha(G)$ is a monoid, $C_\alpha(G)$ inherits this structure and we have an exact sequence of monoids $D_{\alpha+1}(G) \to^{j_{\alpha+1}} S_\alpha(G) \to C_\alpha(G) \to 0$ when $\dim D_\alpha^G \geqq 1$. If $X$ is a semilinear $(G, \alpha)$-sphere, and $X'$ is $X$ with the opposite orientation, the connected sum of $X$ and $X'$ is 0 in $C_\alpha(G)$ since it bounds $I \times k_\alpha(X)$. Hence if $\dim D_\alpha^G \geqq 1$, $C_\alpha(G)$ is an Abelian group. For $M \in D_\alpha(G)$ and $j: M \to p$, we have $-t_0(j) \in \mathrm{Wh}(G; Z)$ is well defined and induces $\hat{t}_0: D_\alpha(G) \to \mathrm{Wh}(G; Z)$. If the domain of $\hat{t}_0$ is a monoid, $\hat{t}_0$ is a morphism. If $D_\alpha$ satisfies the hypothesis of 3.13, then, by 3.13, $\hat{t}_0$ is a bijection and thus $D_\alpha(G)$ is an Abelian group. Thus if $D_{\alpha+1}$ satisfies the hypothesis of 3.13 and $\dim D_\alpha^G \geqq 1$, we have both $D_{\alpha+1}(G)$ and $C_\alpha(G)$ Abelian groups and thus by the above exact sequence $S_\alpha(G)$ is an Abelian group. The basis for getting results about $S_\alpha(G)$ is the following theorem:

3.17. THEOREM. *There exists an automorphism $\psi_\alpha$:* $\mathrm{Wh}(G; Z) \to \mathrm{Wh}(G; Z)$, *with $\psi_\alpha^2 = \mathrm{Id}$, depending only on $\alpha$, such that the following diagram commutes*:

$$\begin{array}{ccccc} \mathrm{Wh}(G;Z) & & \xrightarrow{1-\psi_{\alpha\oplus 1}} & & \mathrm{Wh}(G;Z) \\ \uparrow \hat{t}_0 & & & & \uparrow \hat{t}_0 \\ D_{\alpha\oplus 1}(G) & \xrightarrow{j_{\alpha+1}} & S_\alpha(G) & \xrightarrow{k_\alpha} & D_\alpha(G) \end{array}$$

*Further,* $\psi_{\alpha\oplus 1} = -\psi_\alpha$.

PROOF. Let $M$ represent an element of $D_{\alpha+1}(G)$, $x \in (\mathring{M})^G$, $\mathring{M} = M - \partial M$. Then $x \in D^G_{\alpha\oplus 1} \subset \mathring{M}^G$, where $D_{\alpha\oplus 1}$ is a small linear disk neighborhood of $x$ in $\mathring{M}$. Let $W = M - \mathring{D}_{\alpha\oplus 1}$. Then $W$ is a $G$-cobordism between $\partial_0 W = S_{\alpha\oplus 1}$ and $\partial_1 W = \partial M$, with $H_*(\partial_i W^H) = H_*(W^H)$ for all $H \subset G$. Thus each inclusion $\iota_i\colon \partial_i W \to W$ is a $t_0$-equivalence. By definition, $\hat{t}_0(M) = t_0(\iota_0)$. We have a $G$-embedding $(D_\alpha \times I, D_\alpha \times (0), D_\alpha \times (1)) \to^{\rho} (W, \partial_0 W, \partial_1 W)$. Let $\partial_i W^+ = \partial_i(W) - \rho(\mathring{D}_\alpha \times (i))$ and $W^+ = W - \rho(\mathring{D}_\alpha \times I)$. Then $t_0(\iota_i \mid \partial_i W^+\colon \partial_i W^+ \to W^+) = t_0(\iota_i)$. Now $\partial_1 W^+ = k_\alpha j_{\alpha+1} M$. Let $\lambda\colon W^+ \to p$. Then we have $t_0(\lambda) + t_0(\iota_0 \mid \partial_0 W^+) = t_0(\lambda) + t_0(\iota_0) = 0$ since $\lambda_0 \mid \partial_0 W_0^+ \to p$ is the map $D_\alpha \to p$ and has zero torsion since $D_\alpha$ is $G$-collapsible. Thus $t_0(\lambda) = -t_0(\iota_0)$. We have $\hat{t}_0(k_\alpha j_{\alpha+1}(M)) = -t_0(\hat{\lambda})$, where $\hat{\lambda}\colon \partial_1 W^+ \to p$ and $t_0(\hat{\lambda}) = t_0(\iota_1) + t_0(\rho) = t_0(\iota_1) - t_0(\iota_0)$. Thus $\hat{t}_0(k_\alpha j_{\alpha+1}(M)) = t_0(M) - \hat{t}_0(\iota_1)$. Thus to complete the proof we must find an automorphism $\psi_{\alpha+1}$ of $\mathrm{Wh}(G; Z)$, with $\psi^2_{\alpha+1} = \mathrm{Id}$, and such that $t_0(\iota_1) = \psi_{\alpha+1} t_0(\iota_0)$. We now proceed to derive such $\psi_{\alpha+1}$ and thus complete the proof of 3.17.

Let $M$ be a compact $G$-manifold such that, for each $H \subset G$, if $M^H \neq \varnothing$ then $M^H$ is orientable and connected. If $\dim M^H = m_H$ then $N(H)$ acts on $H_{m_H}(M^H, \partial M^H) = Z$ and thus yields a homomorphism $\lambda^M_H\colon N(H)/H \to Z/2Z = (0, 1)$. If $M^H = \varnothing$ we take $\lambda^M_H = 0$. We use this to define an antiautomorphism $\gamma^M_H$ of $Z(N(H)/H)$ given by

$$\gamma^M_H\Big(\sum_i c_i h_i\Big) = \sum_i (-1)^{\lambda^M_H(h_i)} c_i h_i^{-1}.$$

This in turn induces an automorphism $\gamma^M_H$ of order 2 on $\mathrm{Wh}(N(H)/H)$, the classical Whitehead group. Since

$$\mathrm{Wh}(G; Z) = \sum_{[H] \in \mathrm{Conj}(G)} \mathrm{Wh}(N(H)/H)$$

we can fit the $\gamma^M_H$ together to define an automorphism $\gamma^M$ of order 2 on $\mathrm{Wh}(G; Z)$ by the formula $\gamma^M = \sum(-1)^{m_H}\gamma^M_H$. This is independent of the coset representation $H$ we have chosen.

3.18. THEOREM. *Let* $\iota_0\colon M \to N$ *be a* $G$ *h-cobordism with* $M^H$ *connected and simply connected for each* $H \subset G$. *Let* $\iota_1\colon \overline{\partial N - M} \to N$. *Then* $t_0(\iota_1) = \gamma^m t_0(\iota_0)$.

PROOF. (a) Assume $M$ satisfies the codimension 1 and 2 conditions. We prove the theorem by induction on the number of $[H]$ such that $M^H \neq \varnothing$. Let $H$ be a maximal isotopy subgroup of $M$. We can then assume there exist two $G$ $h$-cobordisms $\iota_0\colon M \to N'$ and $\iota_0'\colon M' \to N''$, where $M' = \overline{\partial N' - M}$, such that $M = N' \cup_{M'} N''$ and such that $(N')^{[H]} = I \times M^{[H]}$ as a $G$-manifold and further such that $N'' = M' \times \bigcup_{E \mid (M' \times (1))^H} E$, where $E$ is a $G$-disk bundle over a $G$-manifold $V$ and the map $(M' \times (1))^{[H]} \hookrightarrow V$ is a $G$ $h$-cobordism and $E \mid (M' \times (1))^{[H]}$ is a $G$-normal tube of $(M' \times (1))^{[H]}$ in $M' \times (1)$. It will suffice to prove the formula for $\iota_0: M \hookrightarrow N'$ and $\iota_0': M' \hookrightarrow N''$ with respect to $\iota_1: \overline{\partial N' - M} \hookrightarrow N'$ and $\iota_1': \partial N'' - M' \hookrightarrow N''$.

If $\bar{E}$ is a normal $G$-disk bundle for $N'^{[H]}$ in $N'$ such that $E \mid M^{[H]}$ is a normal $G$-disk bundle in $M$, then since $(N')^{[H]} = I \times M^{[H]}$, $\bar{E} = I \times \bar{E} \mid M^{[H]}$. Let $\bar{E}_0$ be the associated open disk bundle. Let $M_0 = M - \bar{E}_0 \mid M^{[H]}$ and $N_0' = N' - \bar{E}_0$. Then our codimension hypothesis implies that $\bar{\iota}_0: M_0 \hookrightarrow N_0'$ is also a $G$ $h$-cobordism satisfying the hypothesis of the theorem. Let $\bar{\iota}_1: \overline{\partial N_0' - M_0} \to N_0'$. Then $t_0(\iota_i) = t_0(\bar{\iota}_1) = \gamma^{m_0} t_0(\bar{\iota}_0)$ by the induction hypothesis. Clearly $\gamma^{m_0} t_0(\bar{\iota}_0) = \gamma^m t_0(\iota_0)$ since $(N')^{[H]} = I \times M^{[H]}$. Hence the formula is true for $\iota_0$ and $\iota_1$. Now $t_0(\iota_0') = t_0(j)$, where $j: (M')^{[H]} \to V$ and $t_0(\iota_1') = t_0(k)$, where $k: \overline{\partial V - (M')^{[H]}} \to V$. Here the formula holds by the classical duality formula [**M**, S 10].

(b) Suppose $M$ does not satisfy the codimension 1 and 2 conditions. Let $M_1 = \overline{\partial N - M}$. Then there exists a complex linear representation $\mu: G \to U_k$ with $M \times D_\mu$ satisfying the codimension 1 and 2 conditions. Consider $\iota_0 \times \mathrm{Id}: M \times D_\mu \to N \times D_\mu$. Clearly $t_0(\iota_0 \times \mathrm{Id}) = t_0(\iota_0)$. Then $\partial(N \times D_\mu) - M \times D_\mu = M_1 \times D_\mu \cup_{M_1 \times S_\mu} N \times S_\mu = W$. We have the composite $M_1 \times D_\mu \to^j W \to^k N \times D_\mu$ and $t_0(\iota_1) = t_0(j) + t_0(k)$. Now $t_0(j) = t_0(j')$, where $j': M_1 \times S_\mu \to N \times S_\mu$. $j' = \iota_1 \times \mathrm{Id}$. Since $S_\mu^H$ is an odd sphere, $\varepsilon(S_\mu) = 0$ in $K_0(B(G))$ and the Product Theorem 1.27 says that $t_0(j') = 0$. Hence $t_0(\iota_1) = t_0(k) = \gamma^{M \times D_\mu} t_0(\iota_0)$ by part (a). But $\gamma^{M \times D_\mu} = \gamma^M$. Q.E.D.

We would like to apply 3.18 to prove 3.17 by taking $\psi_{\alpha\oplus 1} = \gamma^{S\alpha\oplus 1}$. The only difficulty is that $(S_{\alpha\oplus 1})^H$ may not be connnected or simply connected and so $\iota_0: S_{\alpha\oplus 1} \to W$ may not be a $G$ $h$-cobordism. However, we can remedy this difficulty by reapplying the argument of 3.17 above. That is, we take a complex representation $\mu$ of $G$ such that $(S_{\alpha\oplus\mu\oplus 1})^H$ is connected and simply connected for every $H \subset G$ and such that $S_{\alpha\oplus\mu\oplus 1}$ satisfies the codimension 1 and 2 condition. We replace $M$ by $M \times D_\mu$ and we have

$$t_0(\iota_0) = t_0(M) = \hat{t}_0(M \times D_\mu) = t_0(\bar{\iota}_0),$$

where $\bar{\iota}_0: \partial_0 \tilde{W} \to \tilde{W}$, and $\tilde{W} = M \times D_\mu - D^0_{\alpha\oplus 1} \times \tilde{D}^0_\mu$, where $\tilde{D}_\mu$ is a small disk around the origin in $D_\mu$. Then $\bar{\iota}_0$ satisfies the hypothesis of 3.18. We have $\bar{\iota}_1: \partial_1 \tilde{W} \to \tilde{W}$, $\partial_1 \tilde{W} = \partial M \times D_\mu \cup_{\partial M \times S_\mu} M \times S_\mu$. By Theorem 3.18,

$$t_0(\bar{\iota}_1) = \gamma^{S_{\alpha\oplus\mu\oplus 1}} t_0(\iota_0).$$

Now $\partial_1 \tilde{W} = A_1 \cup A_2$, where $A_1 = \partial M \times D_\mu$, $A_2 = M \times S_\mu$ and $A_1 \cap A_2 = \partial M \times S_\mu$ and $\tilde{W} = B_1 \cup B_2$, where $B_1 = W \times D_\mu$, $B_2 = M \times S_\mu \times I$ and $B_1 \cap B_2 = W \times S_\mu \times I$. The map $\bar{\iota}_1$ is a map of triples $(A_1, A_2, A_1 \cap A_2) \to (B_1, B_2, B_1 \cap B_2)$. Thus

$$t_0(\bar{\iota}_1) = t_0(\bar{\iota}_1 \mid A_1) + t_0(\bar{\iota}_1 \mid A_2) - t_0(\bar{\iota}_1 \mid A_1 \cap A_2).$$

By the product theorem, $t_0(\bar{\iota}_1 \mid A_1) = t_0(\iota_1)$, $t_0(\bar{\iota}_1 \mid A_2) = 0$, and $t_0(\bar{\iota}_1 \mid A_1 \cap A_2) = 0$. Therefore

$$t_0(\iota_1) = t_0(\bar{\iota}_1) = \gamma^{A_{\alpha\oplus\mu\oplus 1}} t_0(\iota_0) = \gamma^{S_{\alpha\oplus 1}} t_0(\iota_0).$$

This completes the proof of 3.17.

REMARK. Since $\psi_{\alpha\oplus 1} = -\psi_\alpha$, $2a = (1-\psi_\alpha)a + (1-\psi_{\alpha\oplus 1})a$, for any $a \in \mathrm{Wh}(G; Z)$. The special case of 3.17 for relatively free actions was first proved in **[RS]**.

We would like now to know what $\psi_\alpha$ is. In general this question is quite difficult. However, the following is an easy consequence of 6.10 of **[M]** and our definitions.

3.19. THEOREM. *If $\alpha$ comes from a complex representation, then* $\psi_\alpha \otimes \mathrm{Id}$: $\mathrm{Wh}(G; Z) \otimes Q \to \mathrm{Wh}(G; Z) \otimes Q$ *is* $-\mathrm{Id}$.

3.20. COROLLARY. *Let $\alpha \oplus 1$ be complex. If $D_{\alpha\oplus 1}$ satisfies the hypothesis of* 3.13 *and if* $\dim D_\alpha^G \geqq 1$, *so that $D_\alpha(G)$ and $S_\alpha(G)$ are Abelian groups and $\hat{\imath}_0$ an isomorphism, then, as vector spaces over $Q$,* $\dim \mathrm{Wh}(G; Z) \otimes Q \leqq \dim S_\alpha(G) \otimes Q$.

REMARKS. (a) It follows from Corollary 6.3 of **[M]** that if $G$ is an Abelian group with an element not of order 1, 2, 3, 4, or 6 or if $G$ is a solvable group with a normal subgroup $H$ such that $o(G/H)$ is relatively prime to 2 and 3, then $\dim \mathrm{Wh}(G; Z) \otimes Q > 0$. Hence 3.20 allows us to construct infinite families of pairwise distinct semilinear $G$-spheres of odd dimension. These spheres are even distinct PL, but by 3.15, they are all equivariantly topologically equivalent. This shows (see 3.22) that the oriented Reidemeister torsion is not a topological invariant.

(b) In order to define $k_\alpha$: $S_\alpha(G) \to D_\alpha(G)$ uniquely, we have assumed that $\dim D_\alpha^G \geqq 1$. If $D_\alpha^G = (0)$, then $k_\alpha$ is not uniquely defined. However, let us replace the sets $D_{\alpha\oplus 1}$ and $S_\alpha$ by $\mathring{D}_{\alpha\oplus 1}$ and $\mathring{S}_\alpha$, respectively. $\mathring{D}_{\alpha\oplus 1}(G) =$ group of $G$-oriented basepoint preserving diffeomorphism classes of basepointed $(G, \alpha\oplus 1)$ semilinear disks; $\mathring{S}_\alpha(G)$ is the group of basepointed $(G, \alpha)$ semilinear spheres. In the case of $D_{\alpha\oplus 1}(G)$, the basepoint lies in the boundary. We now have that $k_\alpha$ is well defined and everything goes through as above. Corollary 3.20 is vacuous in this case.

(c) The sequence $S_{\alpha\oplus 1}(G) \to^{k_{\alpha\oplus 1}} D_{\alpha\oplus 1}(G) \to^{j_{\alpha\oplus 1}} S_\alpha(G)$ is always exact.

We now investigate the Reidemeister torsion of semilinear $G$-spheres. $\tau = \tau(\alpha)$ will denote $\alpha$-torsion when it is defined. Note that if $X \in S_\alpha(G)$, then $\tau(\alpha)$-torsion is defined for $X$. In particular, when $\alpha$ or $\alpha \oplus 1$ is complex, $\tau(\alpha) =$ oriented Reidemeister torsion. The assignment $N \to \tau(N)$ does *not* induce a morphism of monoids $S_\alpha(G) \to \tilde{K}_1(B(G); \tau)$. Let $\alpha$ be a representation of $G$. For any $\tau$-acyclic manifold $N$, i.e., any $G$-manifold whose complex is $\tau$-acyclic, we define $\tau_\alpha(N) = \tau(N) - \tau(S_{\alpha\oplus 1})$. Then it is easily checked that $\tau_\alpha$ does define a morphism of monoids $S_\alpha(G) \to \tilde{K}_1(B(G); \tau)$.

If $N = \partial M$, where $M \in D_\alpha(G)$, we can use 3.18 to relate $\hat{\imath}_0(M)$ and $\tau_{\alpha-1}(N)$. Using the same notation as in 3.17, we set $W = M - \mathring{D}_\alpha$. Then $\iota_1$: $N \to W$, $\iota_0$: $S_0 \to W$ are $t_0$-equivalences and thus are $\tau$-equivalences. We have $\tau(N) = \tau(W) - \tau(\iota_1)$ and $\tau(W) = \tau(S_\alpha) + \tau(\iota_0)$. $\tau(\iota_0)$ and $\tau(\iota_1)$ are the images under $\gamma$: $\mathrm{Wh}(G; Z) \to K_1((BG); \tau)$ of $t_0(\iota_0)$ and $t_0(\iota_1)$. Therefore $\tau(\iota_1) = \gamma(\psi_\alpha(t_0(\iota_0))) = \gamma(\psi_\alpha(\hat{\imath}_0(M)))$. We thus have

3.21. THEOREM. *If $N = \partial M$, where $M \in D_\alpha(G)$, then* $\tau_{\alpha-1}(N) = \gamma(1-\psi_\alpha)(\hat{\imath}_0(M))$.

When $k_{\alpha-1}$ is defined, this last term is $\gamma(\hat{\imath}_0(k_{\alpha-1}(N)))$ by 3.17. In this case, we do not get anything new. However, $\tau_{\alpha-1}(N)$ is defined even when $k_{\alpha-1}(N)$ is not, i.e., when $N^G = \emptyset$. Hence 3.21 provides us with a method of constructing distinct semilinear $G$-spheres without fixed points.

For example, we have

3.22. THEOREM. *Let $a \in \mathrm{Wh}(G; Z) = \sum_{[H] \in \mathrm{Conj}(G)} \mathrm{Wh}(N(H)/H)$. This permits us to write $a = \sum_{[H]} a_H$. Suppose $\alpha$ is a complex representation of $G$ with* $\dim S^H_\alpha \geqq 5$ *for $a_H \neq 0$. Let $\tau$ be oriented torsion. For any $\tau$-acyclic manifold $V$, let $\hat{\tau}(V)$ be the image of $\tau(V)$ in $\tilde{K}_1(B(G); \tau) \otimes Q$. Then there exists an $M \in D_\alpha(G)$ with $\hat{\tau}(\partial M) = \hat{\tau}(S_\alpha) + 2\hat{\gamma}(a)$, where $\hat{\gamma}\colon \mathrm{Wh}(G; Z) \to \tilde{K}_1(B(G); \tau) \otimes Q$.*

PROOF. This follows from the Realization Theorem 3.5 and from 3.21 and 3.19.

Thus for $G = Z/mZ$ and $m \neq 1, 2, 3, 4, 6$ we can generate infinite distinct actions of $G$ on $S^{2k+1}$, for $k \geqq 2$, According to [**M**, 12.5], for $G$ finite Abelian, $\ker \hat{\gamma}\colon \mathrm{Wh}(G; Z) \to \tilde{K}_1(B(G); \tau) \otimes Q$ consists of the elements of finite order in $\mathrm{Wh}(G; Z)$. Thus if we let the element $a$ of 3.22 run over elements of infinite order, we generate infinite families of distinct semilinear $G$-spheres. Suppose $D_\alpha$ satisfies the hypothesis of 3.13. Then the $M$ in 3.22 is determined by $a$. Since $\mathrm{Wh}(G; Z)$ is finitely generated, the number of semilinear $(G, \alpha - 1)$-spheres bounding semilinear $(G, \alpha)$-disks and having the same torsion invariant is finite.

Let $M$ be a $G$-manifold. The previous theorem provides us with a method for constructing infinite families of differentiably distinct $G$-actions of $G$ on $M$, under very general conditions.

Suppose $H$ is a normal subgroup of $G$, and suppose $M$, as an $H$-manifold, is $\tau$-acyclic, where $\tau$ is oriented torsion. Let $x \in M^H$. Suppose that the representations of $H$ at $M_{gx}$, $g \in G$, are all equivalent to a representation $\alpha$, with $\alpha \oplus 1$ complex. Let $a \in \mathrm{Wh}(H; Z)$, with $a = \sum_{H' \in \mathrm{Conj}(H)} a_{H'}$. We assume as in 3.22 that $\dim S^{H'}_{\alpha\oplus 1} \geqq 5$ if $a_{H'} \neq 0$. Then we can construct a semilinear $(H, \alpha)$-sphere $N$ with $\tau_\alpha(N) = 2\gamma(a)$ by 3.22. By assumption, $G \times_H D_\alpha \subset G \times_H N$ and $G \times_H D_\alpha \subset M$ are $G$-submanifolds. We can then form $M' = G \times_H N \cup_{G \times_H D_\alpha} M$. $M'$ is a $G$-manifold diffeomorphic to $M$ and $\tau$-acyclic as an $H$-manifold. Further, $\tau_\alpha(M') = \tau_\alpha(M) + 2(o(G/H))\ \hat{\gamma}(a)$. Hence for $\hat{\gamma}(a) \neq 0$, $M'$ is not $H$- and hence not $G$-diffeomorphic to $M$. Thus an infinite family of $a$, such that the infinite family $\hat{\gamma}(a)$ are all distinct, yields an infinite family of distinct actions of $G$ on $M$. In general these actions will be topologically equivalent, but not PL equivalent.

From this construction we get the following theorem.

3.23. THEOREM. *Let $\alpha$ be a complex respresentation of $G$; let $H$ be a normal subgroup of $G$. Suppose there exists an element of infinite order $a$, $a = \sum_{[H'] \in \mathrm{Conj}(H)} a_{H'} \in \mathrm{Wh}(H; Z)$. Assume further that* $\dim S^{H'}_{\alpha|H} \geqq 5$ *if $a_{H'} \neq 0$. Then there exists an infinite family of pairwise differentiably distinct smooth actions of $G$ on $S_\alpha$. In fact, these actions are distinct as $H$-actions. In particular, if $\alpha = \beta \oplus 1$, the image of $S_\beta(G) \to S_{\beta|H}(H)$ is infinite.*

**4. de Rham's theorem and a homotopy classification of representations**. In this section we prove the theorem of de Rham [**D**] that if $\alpha$ and $\alpha'$ are two representations of a compact Lie group such that $S_\alpha$ is $G$-diffeomorphic to $S_{\alpha'}$ then $\alpha \cong \alpha'$. This result is a consequence of the fact that for finite Abelian groups the torsion invariants determine the representation. In fact our proof is rather close to de Rham's argument, and shows in a striking manner the power of these invariants.

Another result of this section is that, for $G$ finite Abelian, we give necessary and sufficient conditions, in terms of the torsion of $S_\alpha$ and $S_{\alpha'}$, for the homotopy equivalence of $S_\alpha$ and $S_{\alpha'}$.

We will adopt the following notation in this section. $\alpha: G \to O(n)$ will denote a representation, and $S_\alpha$ will denote, as before, the $(n-1)$-sphere with the induced action of $G$. # will denote join. We remind the reader that $S_{\alpha\oplus\beta} = S_\alpha \# S_\beta$. We write $\alpha = \alpha'$ when the two representations are linearly, hence orthogonally, equivalent. $I: G \to SO(1)$ will denote the trivial action. Then $S_{\alpha\oplus I} = S_\alpha \# S_I = S_\alpha \# S^0$ is the usual suspension. We say $\alpha$ is *G-orientable*, or just *orientable*, if for each $H \subset G$ and each $x \in N(H)/H$, $x$ acts trivially on $H_*(S_\alpha^H)$, i.e., if $\tau(\alpha) = \tau$ = oriented torsion (Definition 1.26(b)).

If $\alpha$ is an orientable representation, then $\tau(S_\alpha) \in \tilde{K}_1(B(G;\tau))$ is defined, where in this section $\tau$ will always mean oriented torsion.

4.1. THEOREM. *If $\alpha$ and $\alpha'$ are orientable representations of a finite group $G$ and if $\tau(S_\alpha) = \tau(S_{\alpha'})$ and if* $\dim \alpha \leqq \dim \alpha'$, *then $\alpha + 2kI = \alpha'$.*

PROOF. For $\psi$ a representation of $G$, $x \in G$, let $\psi_x$ be $\psi$ restricted to the cyclic subgroup generated by $x$. If $\alpha_x + 2kI = \alpha'_x$ and $\alpha_y + 2k'I = \alpha'_y$, then by dimensionality $k = k'$. By the theory of characters the linear equivalence class of $\psi$ is determined by that of $\psi_x$ as $x$ runs over $G$. Since $\tau$ is functorial with respect to subgroups, it will suffice to prove the theorem for $G$ cyclic.

Assuming $G$ is cyclic, we can write $\tau(S_\alpha) = \sum_{H\subset G} \tau_H(S_\alpha)$ and $\tau(S_{\alpha'}) = \sum_{H\subset G}\tau_H(S_{\alpha'})$. Write $\varepsilon(X)$ for the Euler characteristic of $X$. We have $\varepsilon(S_\alpha^G) = \tau_G(S_\alpha) = \tau_G(S_{\alpha'}) = \varepsilon(S_{\alpha'}^G)$. Thus $\dim(S_\alpha^G) - \dim(S_{\alpha'}^G) = 0 \pmod 2$. We have $S_\alpha = S_{\alpha_0} \# S_\alpha^G$ and $S_{\alpha'} = S_{\alpha'_0} \# S_{\alpha'}^G$, where $S_{\alpha_0}^G = S_{\alpha'_0}^G = \varnothing$. Thus it will suffice to show $\alpha_0 = \alpha'_0$. We will do this by induction on $\dim \alpha_0 + \dim \alpha'_0$. Let $H \neq G$ be a maximal isotopy group which actually occurs in $S_{\alpha_0}$. Then $G/H$ acts freely on $S_{\alpha_0}^H$ and thus is cyclic. By [M], $\tau_H(S_{\alpha_0}) \neq 0$. Thus $\tau_H(S_{\alpha'_0}) = 0$ and $H$ occurs as an isotopy subgroup on $S_{\alpha'_0}$. If $H$ were not a maximal isotopy subgroup of $S_{\alpha'_0}$, then $H \subset H'$, where $H'$ is a maximal isotopy subgroup of $S_{\alpha'_0}$. But reversing the above argument shows that $H'$ is then an isotopy subgroup of $S_{\alpha_0}$, contradicting the choice of $H$. Hence $H$ is also a maximal isotopy subgroup of $S_{\alpha'_0}$. Thus $G/H$ acts freely on $S_{\alpha'_0}^H$. Since $\tau_H(S_{\alpha_0}) = \tau_H(S_{\alpha'_0})$, 12.7 of [M] shows that $S_{\alpha_0}^H$ and $S_{\alpha'_0}^H$ are linearly equivalent as $G/H$-, and hence as $G$-manifolds. Thus we have $S_{\alpha_0} = S_{\alpha_{00}} \# S_{\alpha_0}^H$ and $S_{\alpha'_0} = S_{\alpha'_{00}} \# S_{\alpha'_0}^H$, with $S_{\alpha'_0}^H = S_{\alpha'_0}^H$.

For $G$-complexes $X$ and $Y$, we have a relative $G$-cellular isomorphism $(X \# Y, X + Y) \to (S(X \times Y), p_0 + p_1)$ and $\tau(S(X \times Y), p_0 + p_1) = -\tau(X \times Y)$; hence $\tau(X \# Y) = \tau(X) + \tau(Y) - \tau(X \times Y)$. Applying 1.28(b), we get $\tau(S_{\alpha_0}) = \tau(S_{\alpha_{00}}) + \tau(S_{\alpha_0}^H)$ and $\tau(S_{\alpha'_0}) = \tau(S_{\alpha'_{00}}) + \tau(S_{\alpha'_0}^H)$. Since $\tau(S_{\alpha_0}^H) = \tau(S_{\alpha'_0}^H)$ and $\tau(S_{\alpha_0}) = \tau(S_{\alpha'_0})$, we have $\tau(S_{\alpha_{00}}) = \tau(S_{\alpha'_{00}})$. We can now apply the induction hypothesis to show $\alpha_0 = \alpha'_0$. Q.E.D.

For any finite group $G$ and any representation $\alpha$, $\alpha \oplus \alpha$ is always orientable, and thus we can define $\tau_2(\alpha) = \tau(S_{\alpha\oplus\alpha})$. If $\tau_2(\alpha) = \tau_2(\alpha')$, then by the previous theorem, $\alpha \oplus \alpha \oplus 2kI = \alpha' \oplus \alpha'$. But then $\alpha \oplus kI = \alpha'$. Since $\tau_2(\alpha)$ is a piecewise linear invariant and a smooth invariant of $S_\alpha$, we have

4.2. COROLLARY. *If $\alpha$ and $\alpha'$ are representations of a finite group G with $S_\alpha$ smoothly or* pl *G-equivalent to $S_{\alpha'}$, then $\alpha = \alpha'$.*

Hence, we have the theorem of de Rham.

4.3. THEOREM (DE RHAM). *If G is a compact Lie group and $\alpha$ and $\alpha'$ are two representations of G such that $S_\alpha$ is G-diffeomorphic to $S_{\alpha'}$, then $\alpha = \alpha'$.*

PROOF. By character theory we need only show that $\alpha_x = \alpha'_x$ for elements $x \in G$ of finite order. This follows from 4.2.

We now study the homotopy classification of representations. For $G$-complexes $X$ and $Y$ we write $X \sim_G Y$ if there is a $G$-homotopy equivalence from $X$ to $Y$. This is equivalent to having a $G$-map $f: X \to Y$ such that for each $H \subset G, f^H: X^H \to Y^H$ is a homotopy equivalence. We say two representations $\alpha$ and $\alpha'$ are homotopically equivalent if $S_\alpha \sim_G S_{\alpha'}$.

We consider first orientable representations $\alpha$ and $\alpha'$ of $G$. If $f$: $S_\alpha \to S_{\alpha'}$ is a $G$-homotopy equivalence, $t_0(f)$ is defined as well as is $\tau(S_\alpha)$ and $\tau(S_{\alpha'})$. We then have the relation $\tau(S_{\alpha'}) - \tau(S_\alpha) = \gamma(t_0(f))$, where $\gamma$: $\mathrm{Wh}(G; Z) \to \tilde{K}_1(B(G); \tau)$ is the natural map. Hence this relationship gives a necessary condition for two orientable representations to be $G$-homotopically equivalent. We can prove a converse when $G$ is Abelian.

For $X$ a semilinear $G$-sphere, $\varepsilon(X^G) = 0$ or 2. Thus $\eta(X) = (-1)^{\varepsilon(X^G)/2}$ is well defined. We can use this to construct an additive morphism on representations.

4.4. THEOREM. *For G Abelian and for orientable representations, let $\lambda(\alpha) = (\eta(S_\alpha)\ \tau(S_\alpha), \eta(S_\alpha)) \in \bar{K}_1(B(G); \tau) + Z/2Z$. Then $\lambda$ is additive on representations and thus induces a homomorphism $\hat{\lambda}$: $RSO(G) \to \bar{K}_1(B(G); \tau) + Z/2Z$, where $RSO(G)$ is the subring of the real representation ring generated by orientable representations. Further* Ker $\hat{\lambda} = Z$*, generated by 2I.*

PROOF. Once we show $\lambda$ is additive it will follow from 4.1 that Ker $\hat{\lambda} = \{2I\}$. As was shown in the proof of 4.1, $\tau(S_\alpha \# S_{\alpha'}) = \tau(S_\alpha) + \tau(S_{\alpha'}) - \tau(S_\alpha \times S_{\alpha'})$. The product formula 1.28 (a) and an easy calculation yield that $\lambda(\alpha \oplus \alpha') = \lambda(\alpha) + \lambda(\alpha')$. Q.E.D.

Note that $\gamma$: $\mathrm{Wh}(G; Z) \to \bar{K}_1(B(G); \tau)$ induces an action of $\mathrm{Wh}(G; Z)$ on $\bar{K}_1(B(G); \tau) + Z/2Z$ given by $a(c, j) = (c + \gamma(a), j)$.

4.5. THEOREM. *If G is an Abelian group and $\alpha$ and $\alpha'$ are oriented representations of G, then $S_\alpha$ and $S_{\alpha'}$ are G-homotopically equivalent if and only if*

(1) dim $\alpha$ = dim $\alpha'$,

(2) *there exists an $a \in \mathrm{Wh}(G; Z)$ with $a\lambda(\alpha) = \lambda(\alpha')$.*

PROOF. The necessity of the conditions are clear from the earlier discussion. To show sufficiency we first replace condition (1) by

(1′) dim $S_\alpha^G$ = dim $S_{\alpha'}^G$.

To show conditions (1′) and (2) are sufficient it will suffice to consider the case $S_\alpha^G = S_{\alpha'}^G = 0$.

Let $H \not\subset G$ be a maximal isotopy subgroup of $S_\alpha$. Then $S_\alpha^H \neq \emptyset$ and $G/H$ acts freely on $S_\alpha^H$. Thus $G/H$ is cyclic. Then $\lambda(\alpha)_H \neq 0$ and $\lambda(\alpha)_H \in \tau(G/H; Q)$, where the latter group is a quotient of the group of units of $Q[\varepsilon]/(\varepsilon^m = 1)$ (see 1.25). In fact

$$\lambda(\alpha)_H = (\varepsilon^{r_1} - 1)(\varepsilon^{r_2} - 1)\cdots(\varepsilon^{r_k} - 1),$$

where $(r_i, m) = 1$ (see [**M**, 12.5]).

$a_H$ is an element of $\mathrm{Wh}(G/H)$, where since $G/H$ is cyclic, $\mathrm{Wh}(G/H)$ is a quotient of the group of units of $Z[\varepsilon]/(\varepsilon^m = 1)$. (See [**B**, 11, 7.3].) Now $(a_H)\,\lambda(\alpha)_H \neq 0$ in $\tau(G/H; Q)$, for if it were

$$(a_H)(\varepsilon^{r_1} - 1)(\varepsilon^{r_2} - 1)\cdots(\varepsilon^{r_k} - 1) = \mp\, \varepsilon^l + k\Big(\sum_{j=1}^{m} \varepsilon^j\Big)$$

in $Z[\varepsilon]/(\varepsilon^m = 1)$. But this is impossible since under the augmentation map into $Z$, the left-hand side is 0 and the right-hand side is $\mp 1 + km \neq 0$ if $m > 1$. Thus $\lambda'(\alpha)_H \neq 0$, and thus $S^H_{\alpha'} \neq \varnothing$. If $H$ were not a maximal isotopy subgroup of $S_{\alpha'}$, one could find $H'$, $H \not\subset H'$, with $H'$ a maximal isotopy subgroup of $S_{\alpha'}$. Running this same argument in the other direction shows $S^{H'}_\alpha \neq \varnothing$ which contradicts the choice of $H$. Thus $H$ is a maximal isotopy subgroup of $S_{\alpha'}$. Now $S_\alpha = S^H_\alpha \# S_\rho$ and $S_{\alpha'} = S^H_{\alpha'} \# S_{\rho'}$, and as $\lambda(\rho) = \lambda(\alpha) - \lambda(\alpha)_H$ and $\lambda(\rho') = \lambda(\alpha') - \lambda(\alpha')_H$ and $(a_H)\,\lambda(\alpha)_H = \lambda(\alpha')_H$ we have $(a - a_H)\,\lambda(\rho) = \lambda(\rho')$. Thus by induction on the dimension of $\alpha$, we can assume $S_\rho \sim_G S_{\rho'}$. We need only show that $S^H_\alpha \sim_G S^H_{\alpha'}$. This is the same as $S^H_\alpha \sim_{G/H} S^H_{\alpha'}$ and this is implied by $(a_H)\,\lambda(\alpha)_H = \lambda(\alpha')_H$, by [**M**, 12.1 and 12.10]. Hence we have shown that conditions (1′) and (2) suffice to show $\alpha$ and $\alpha'$ are $G$-homotopic.

It remains to show conditions (1) and (2) are sufficient. Suppose that $\alpha$ and $\alpha'$ satisfy only condition (2). Observe that $a_G\,\lambda(\alpha)_G = \lambda(\alpha')_G$ implies $\lambda(\alpha)_G = \lambda(\alpha')_G$ since $a_G = 0$. Hence $\varepsilon(S^G_\alpha) = \lambda(\alpha)_G = \lambda(\alpha')_G = \varepsilon(S^G_{\alpha'})$. Thus $\dim(S^G_\alpha) = \dim(S^G_{\alpha'})$ (mod 2). Thus replacing one of the two representations, say $\alpha$, by $\alpha + 2kI$, we can assume condition (1′) and condition (2) are satisfied for $\alpha + 2kI$ and $\alpha'$. But then $S_{\alpha \oplus 2kI} \sim_G S_{\alpha'}$ and this is only possible if $\dim \alpha + 2kI = \dim \alpha'$. If $\dim \alpha = \dim \alpha'$ to begin with, then $k = 0$. Q.E.D.

ADDENDUM. The argument above shows we could take the $G$-homotopy equivalence between $S_\alpha$ and $S_{\alpha'}$ as isovariant. Hence two orientable representations of $G$ are $G$-homotopically equivalent if and only if they are $G$-homotopically equivalent via an equivalence that preserves orbit type.

We now turn to arbitrary representations $\alpha$ of $G$. Here the formulation is not quite as neat since we no longer have a simple product theorem. However, there still is a satisfactory theorem.

A natural approach to generalizing the previous theorem to nonorientable presentations would be to replace oriented torsion by $\alpha$-torsion in the previous theorem. If $S_\alpha \sim_G S_{\alpha'}$ then necessarily ${}^\alpha\tau = {}^{\alpha'}\tau$, where ${}^\alpha\tau$ is $\alpha$-torsion (see Definition 1.26(b)). Then as before, if $S_\alpha \sim_G S_{\alpha'}$, there exists an $a \in \mathrm{Wh}(G; Z)$ with $\gamma(a) + {}^\alpha\tau(S_\alpha) = {}^\alpha\tau(S_{\alpha'})$. Unfortunately, this condition is not quite sufficient as can be seen by looking at free linear actions of $Z/2Z$ on even-dimension spheres of different dimension. Therefore we need a stronger invariant. To define it we need a couple of simple propositions:

4.6. PROPOSITION. *Let $\alpha$ be a representation of $G$ with $S^G_\alpha = 0$, and let $G$ be Abelian.*

*If $H \subset G$ with $G/H$ acting nontrivially on $H_*(S_\alpha^H)$ then there exists $H'' \supset H$ with $G/H''$ acting nontrivially on $H_*(S_\alpha^{H''})$ and $G/H'' = Z/2Z$.*

PROOF. We can assume $G$ acts effectively on $S_\alpha$, for otherwise we could replace $G$ by $G/\ker(\alpha)$. Let $H'$ be a maximal isotopy subgroup of $S_\alpha$ with $H \subset H'$. Then either

(1) $H' = (e)$ and $G$ acts freely on $S_\alpha$, or

(2) $H \neq (e)$ and there exists $\rho \neq 0$ with $S_\alpha = S_\alpha^{H'} \# S_\rho$. In case (1), $H = (e)$. Hence $G$ acts freely and nonorientably on $S_\alpha$. Thus $G = Z/2Z$. In case (2), we have $S_\alpha^H = S_\alpha^{H'} \# S_\rho^H$, and $G/H'$ acts freely on $S_\alpha^{H'}$. If $G/H'$ acts nontrivially on $H_*(S_\alpha^{H'})$, we can take $H' = H''$. Otherwise, $G/H$ acts nontrivially on $H_*(S_\rho^H)$. Hence the proposition follows by induction on the dimension of $\alpha$.

4.7. PROPOSITION. *Let $(H_1, H_2, \cdots, H_k)$ be those subgroups of $G$ with $G/H = Z/2Z$. Let $\alpha = \alpha_0 + kI$, where $S_{\alpha_0}^G = \varnothing$. Then*

$$S_\alpha = S_\alpha^G \# S_{\alpha_0}^{H_1} \# S_{\alpha_0}^{H_2} \# \cdots \# S_{\alpha_0}^H \# S_{\alpha_2},$$

*where $\alpha_2$ is an orientable representation if $G$ is Abelian.*

PROOF. We may as well assume $S_\alpha^G = \varnothing$ and $\alpha = \alpha_0$. $H_1, H_2, \cdots, H_k$ are maximal normal subgroups of $G$. It follows that $(S_\alpha^{H_i})^{H_j} = \varnothing$, $H_1 \neq H_j$, and hence $S_\alpha = S_\alpha^{H_1} \# S_\alpha^{H_2} \# \cdots \# S_\alpha^{H_k} \# S_{\alpha_2}$ for some unique $\alpha_2$.

It remains to show that $\alpha_2$ is orientable. Suppose $H \subset G$ with $G/H$ acting nontrivially on $H_*(S_{\alpha_2}^H)$. Then $H \subset H'$, with $G/H' = Z/2Z$ and $S_{\alpha_2}^{H'} \neq \varnothing$. Then $H' = H_i$ but $S_{\alpha_2}^{H_i} = \varnothing$. Thus no such $H$ can exist. Q.E.D.

There exists a unique representation $\alpha_1$ of $G$ with $S_{\alpha_1} = S_\alpha^G \# S_{\alpha_0}^{H_1} \# \cdots \# S_{\alpha_0}^{H_k}$. Then $\alpha = \alpha_1 \oplus \alpha_2$.

4.8. PROPOSITION. *Let $G$ be Abelian. Then $\alpha$ is $G$-homotopically equivalent to $\alpha'$ if and only if $\alpha_1 = \alpha_1'$ and $\alpha_2$ is $G$-homotopically equivalent to $\alpha_2'$.*

PROOF. Clearly, the second condition implies the first. Assume $S_\alpha \sim_G S_{\alpha'}$. Then $\dim S_\alpha^G(S_\alpha^{H_i}) = \dim S_{\alpha'}^G(S_{\alpha'}^{H_i})$. Thus

$$\dim S_{\alpha_0}^{H_i} = \dim S_\alpha^{H_i} - \dim S_\alpha^G - 1 = \dim S_{\alpha'}^{H_i} - \dim S_{\alpha'}^G - 1 = \dim S_{\alpha_0'}^{H_i}.$$

Since $S_{\alpha_0}^{H_i}$ is the space of a $G/H = Z/2Z$ free linear action, it is determined as a $G/H$-space and hence as a $G$-space by its dimension. Thus $S_{\alpha_0}^{H_i} = S_{\alpha_0'}^{H_i}$ for all $i$. Thus $\alpha_1 = \alpha_1'$. Since $S_\alpha \sim_G S_{\alpha'}$, $S_{\alpha \oplus \alpha_1} \sim_G S_{\alpha' \oplus \alpha_1}$. Both $\alpha \oplus \alpha_1$ and $\alpha' \oplus \alpha_1$ are orientable representations since $\alpha$, $\alpha_2'$ and $2\alpha_1 = 2\alpha_1'$ are. Thus oriented torsion is defined and $\tau(S_{\alpha \oplus \alpha_1}) - \tau(S_{\alpha' \oplus \alpha_1}) = \gamma(a)$, $a \in \mathrm{Wh}(G; Z)$. Now

$$\tau(S_{\alpha \oplus \alpha_1}) = \tau(S_{\alpha_2 \oplus 2\alpha_1}) = \tau(S_{2\alpha_1}) \pm \tau(S_{\alpha_2})$$

and

$$\tau(S_{\alpha' \oplus \alpha_1}) = \tau(S_{\alpha_2' \oplus \alpha_1' \oplus \alpha_1}) = \tau(S_{\alpha_2' \oplus 2\alpha_1}) = \tau(S_{2\alpha_1}) \pm \tau(S_{\alpha_2'}).$$

Thus $\tau(S_{\alpha_2}) - \tau(S_{\alpha_2'}) = \gamma(\pm a)$. Since $S_{\alpha_2}^G = S_{\alpha_2'}^G = \varnothing$, we conclude from the proof of 4.5 that $S_{\alpha_2} \sim_G S_{\alpha_2'}$. Q.E.D.

Since $\alpha_2$ is orientable, 4.8 and 4.5 characterize when two representations are homotopically equivalent. We can formulate this explicitly as follows.

4.9. THEOREM. *Let $G$ be an Abelian group. Let $\alpha$ be a representation of $G$. Let $\alpha = \alpha_1 \oplus \alpha_2$ be the decomposition of $\alpha$ determined by* 4.7. *Let $\lambda(\alpha) \in \bar{K}_1(B(G);\tau) + Z^0 + Z^1 + \cdots + Z^k$ be given by*

$$\lambda(\alpha) = (\tau(S_{\alpha_2}), \dim S_\alpha^G + 1, \dim S_\alpha^{H_1} - \dim S_\alpha^G, \cdots, \dim S_\alpha^{H_k} - \dim S_\alpha^G),$$

*where $\dim \varnothing = -1$, and $H_1, H_2, \cdots, H_k$ are the subgroups of $G$ of index 2. Then $\lambda(\alpha \oplus \alpha') = \lambda(\alpha) \oplus \lambda(\alpha')$ and hence $\lambda$ induces a homomorphism, still called $\lambda$, from the real representation ring $RO(G)$ of $G$: $\lambda\colon RO(G) \to \bar{K}_1(B(G);\tau) + Z^0 + Z^1 + \cdots + Z^k$. Then $\lambda$ is a monomorphism. Further $\alpha$ is $G$-homotopic to $\alpha'$ if and only if $\lambda(\alpha) - \lambda(\alpha') = \gamma(a)$, for some $a \in \mathrm{Wh}(G; Z)$, where $\gamma\colon \mathrm{Wh}(G; Z) \to \tilde{K}_1(B(G); \tau)$ is the natural map.*

PROOF. This follows easily from 4.1 and 4.8.

ADDENDUM. The arguments of 4.1 and 4.8 show that homotopical equivalence and isovariant homotopical equivalence are identical relationships for representations of finite Abelian groups.

4.10. COROLLARY. *If $\alpha, \alpha', \beta, \beta'$ are representations of $G$, then $S_{\alpha\oplus\beta} \sim_G S_{\alpha'\oplus\beta'}$ and $S_\alpha \sim_G S_{\alpha'}$ implies $S_\beta \sim_G S_{\beta'}$. Thus stable $G$-homotopy equivalence implies $G$-homotopy equivalence for representations of a finite Abelian group.*

4.11. COROLLARY. *Let $J(G) = RO(G)/(\alpha - \alpha')$, $\alpha$ and $\alpha'$ $G$-homotopically equivalent. Then the map $\lambda$ of* 4.9 *induces a monomorphism, $\tilde{\lambda}\colon J(G) \to \tilde{K}_1(B(G);\tau)/\gamma(\mathrm{Wh}(G;Z)) + Z^0 + Z^1 + \cdots + Z^k$.*

REMARK. Theorem 4.5 admits a generalization to oriented semilinear actions of finite Abelian groups.

**5. *G* pl manifolds.** For finite groups, it makes sense to drop the assumption of smoothness and consider the broader category of pl-manifolds and pl actions. This category lacks certain fundamental features of the smooth category, though a certain subcategory does have nice features which allow us to develop the theorems of §3 in this context. Further, the category does have certain advantages over the smooth category when it comes to studying actions on $n$-spheres, and will in fact allow us to derive certain information about smooth actions.

In discussing the pl category we follow the definitions and terminology of Hudson **[H]**. A $G$-simplicial complex $K$ is a simplicial complex with an action of $G$ on it through simplicial isomorphisms. Such a $K$ is always a $G$-$CW$ complex. A $G$ pl space $X$ is a pl space with an action of $G$ on it such that, for some triangulation $\psi\colon K \to X$, the induced $G$-action of $K$ turns $K$ into a $G$-simplicial complex. Such a $\psi$ will be called a $G$-triangulation of $X$. Observe that if $K$ is a $G$-simplicial complex and $K'$ any subdivision of $K$, there exists a further subdivision $K''$ of $K'$ such that $K''$ is a $G$-simplicial complex. Such a subdivision will be called a $G$-subdivision. It follows that if $\psi$ and $\psi'$ are $G$-triangulations of $X$, a $G$ pl space, there exists $G$-subdivisions such that with respect to those subdivisions $(\psi')^{-1} \circ \psi$ is a $G$-equivariant simplicial isomorphism.

It follows that for compact $G$ pl spaces $X$, $Y$ and a nice functor $T$ on $B(G)$, the notion of a $T$-equivalence $f\colon X \to Y$ is well defined and depends only on the homotopy class of $f$, where $f$ is a continuous map. Further, the invariant $T(f) \in \tilde{K}_1(B(G); T)$ is well defined for such an $f$. However, the theory of §3 for manifolds in this category fails, as the following examples shows.

5.1. EXAMPLE. Let $X \in S_\alpha(G)$ with $X$ not pl equivalent to $S_{\alpha \oplus 1}$. (See 3.20 and 3.23 for many examples.) Let $Y$ be the cone on $X$. $Y$ collapses $G$ equivariantly to $Y^G$ = cone over $X^G$, which is a disk and collapses to any $y \in Y^G$. Hence $\hat{t}_0(Y) = t(j), j\colon Y \to p$, is 0. For $x \in X^G$ not the cone point, has a collapsible neighborhood which is $G$ pl equivalent to $D_{\alpha \oplus 1}$. But $Y$ is not $G$ pl equivalent to $D_{\alpha \oplus 1}$ since $\partial Y = X$ is not $G$ pl equivalent to $S_{\alpha \oplus 1}$. Thus Theorem 3.13 fails for this $Y$. As a consequence there is no simple analogue of the $s$-cobordism theorem in this category. An even more serious disadvantage is shown by this example: There is no reasonable theory for $G$-regular neighborhoods in this category. Also this example shows that the $G$-covering isotopy theorem fails in this category.

We circumvent these difficulties by adopting a restricted definition of $G$ pl manifolds:

5.2. DEFINITION. A $G$ pl space is called a linear $G$-disk, ($G$-sphere) if for some $G$-triangulation $K$ of $X$, there is a piecewise differentiable $G$-homeomorphism $f\colon K \to D_\alpha$ ($f\colon K \to S_\alpha$) for some representation $\alpha$ of $G$. A $G$ pl space $X$ is called a $G$ pl manifold if each $x \in X$ has a $G_x$-invariant neighborhood which is a linear $G_x$-disk.

*Note.* According to this definition, the $Y$ of Example 5.1, while a manifold on which $G$ acts pl, is *not* a $G$ pl manifold.

One can now generalize Chapters 1, 2 and 6 of Hudson **[H]** to our category, modifying his arguments in a relatively obvious way. In particular, we have a well-defined notion of $G$-regular neighborhood and the analogue of Theorem 2.16 of Hudson.

5.3. GENERALIZED REGULAR NEIGHBORHOOD THEOREM. *If $N_1$ and $N_2$ are two $G$-regular neighborhoods of the compact $G$ pl space $X$ in the $G$ pl manifold $M$ which meet $\partial M$ regularly, there exists a $G$-ambient isotopy of $M$ fixed on $X$ and throwing $N_1$ onto $N_2$.*

Replacing the notion of tubular neighborhood with regular neighborhood, 3.1—3.24 go through in the $G$ pl manifold category. As a consequence we have the following theorem.

5.4. THEOREM. *Let $X$ be a pl semilinear $G$-disk satisfying the codimension 1 and 2 conditions, and let $\dim M^H \geqq 5$ for $H \neq G$. Then $X$ is $G$ pl isomorphic to the triangulation of a smooth semilinear $G$-disk $\mathring{X}$. Further the $G$-diffeomorphism type of $\mathring{X}$ is determined by $X$.*

PROOF. This follows from 3.13 applied to both categories along with the de Rham theorem.

REMARK. One can prove a sharper form of 5.4 using $G$-smoothing theory and a $G$ Cairns-Hirsch theorem. Specifically one can remove the codimension 1 and 2 restrictions.

Let us denote the sets of semilinear($G$-$\alpha$)-disks and spheres in the $G$ pl category by

$S_\alpha^{\mathrm{pl}}(G)$ and $D_\alpha^{\mathrm{pl}}(G)$. The advantage of the $G$ pl category is that the map $k_\alpha: S_\alpha^{\mathrm{pl}}(G) \to D_\alpha^{\mathrm{pl}}(G)$ is injective and that the following sequence is exact:

$$0 \to S_{\alpha\oplus 1}^{\mathrm{pl}}(G) \xrightarrow{k_{\alpha+1}} D_{\alpha\oplus 1}^{\mathrm{pl}}(G) \xrightarrow{j_{\alpha+1}} S_\alpha^{\mathrm{pl}}(G).$$

It follows that $S_{\alpha\oplus 1}^{\mathrm{pl}}(G)$ can be identified with the kernel of the composite

$$D_{\alpha\oplus 1}^{\mathrm{pl}}(G) \xrightarrow{j_{\alpha+1}} S_\alpha^{\mathrm{pl}}(G) \xrightarrow{k_\alpha} D_\alpha^{\mathrm{pl}}(G).$$

Thus:

5.5. THEOREM. *Suppose $\alpha$ satisfies the codimension* 1 *and* 2 *conditions and* $\dim D_\alpha^H \geqq 6$ *for $H \neq G$. Then $S_{\alpha\oplus 1}^{\mathrm{pl}}(G)$ is isomorphic to the subgroup of* Wh$(G; Z)$ *of all elements fixed under $\psi_{\alpha\oplus 1}$. The isomorphism is given by $N \to \hat{\imath}_0 k_{\alpha+1}(N)$, $N \in S_{\alpha\oplus 1}^{\mathrm{pl}}(G)$.*

PROOF. This follows from the pl version of 3.16.

Consider the commutative diagram:

$$\begin{array}{ccccc} \mathrm{Wh}(G;Z) & \xrightarrow{1-\psi_{\alpha+2}} & \mathrm{Wh}(G;Z) & \xrightarrow{1-\psi_{\alpha+1}} & \mathrm{Wh}(G;Z) \\ \uparrow{\scriptstyle \hat{\imath}_0} & & \uparrow{\scriptstyle \hat{\imath}_0} & & \uparrow{\scriptstyle \hat{\imath}_0} \\ D_{\alpha\oplus 2}^{\mathrm{pl}}(G) & \xrightarrow{k_{\alpha+1}j_{\alpha+2}} & D_{\alpha\oplus 1}^{\mathrm{pl}}(G) & \xrightarrow{k_\alpha j_{\alpha+1}} & D_\alpha^{\mathrm{pl}}(G) \end{array}$$

If $\hat{\imath}_0$ is an isomorphism and the top row is exact, so is the bottom row. Then $S_{\alpha+1}^{\mathrm{pl}}(G) = \mathrm{kernel}\ k_\alpha j_{\alpha+1} = \mathrm{image}\ j_{\alpha+2}k_{\alpha+1} = \mathrm{image}\ j_{\alpha+2}$. If we tensor with $Q$, the top row is always exact since $\psi_{\alpha+1} = -\psi_{\alpha+2}$ and $\psi_{\alpha+2}^2 = \mathrm{Id}$. Thus under the above hypotheses,

$$S_{\alpha\oplus 1}^{\mathrm{pl}} \otimes Q = j_{\alpha+2}(D_{\alpha\oplus 2}^{\mathrm{pl}}(G) \otimes Q).$$

Since under appropriate hypotheses $D_{\alpha\oplus 2}^{\mathrm{pl}}(G) = D_{\alpha\oplus 2}(G)$, we have:

5.6. THEOREM. *Let $\alpha$ satisfy the codimension* 1 *and* 2 *conditions and assume $D_\alpha^H$ is of dimension $\geqq 5$ if $H \neq G$. Then $S_{\alpha\oplus 1}(G) \otimes Q \to S_{\alpha\oplus 1}^{\mathrm{pl}}(G) \otimes Q$ is onto. Hence some nonzero multiple of every* pl $(G, \alpha\oplus 1)$*-semilinear sphere is a smooth triangulation of a smooth semilinear $(G, \alpha \oplus 1)$-sphere. If in addition,* Wh$(G; Z)$ *has no torsion and $\alpha \oplus 1$ is complex, $S_{\alpha\oplus 1}^{\mathrm{pl}}(G) = (e)$ and $S_{\alpha\oplus 2}(G) \otimes Z[1/2] \to S_{\alpha\oplus 2}^{\mathrm{pl}}(G) \otimes Z[1/2]$ is onto. Thus twice any* pl $(G, \alpha \oplus 2)$*-sphere is a smooth triangulation of a smooth one.*

PROOF. This follows from the above discussion and 3.19.

**6. Stable theory.** The results so far on semilinear spheres and disks involve codimension assumptions along with assumptions on lower bounds for the dimensions of the fixed point sets. In this section, we will stabilize these results, getting broader and more elegant formulations at the cost of some sharpness. Using the product theorems of §1, the present results are simply reformulations of the earlier ones.

Suppose $N \in D_\alpha(G)$ and $N' \in D_{\alpha'}(G)$. Then $N \times D_{\alpha'}$ and $N' \times D_\alpha$ are elements of $D_{\alpha\oplus\alpha'}(G)$ with $\hat{\imath}_0(N) = \hat{\imath}_0(N \times D_{\alpha'})$ and $\hat{\imath}_0(N') = \hat{\imath}_0(N' \times D_\alpha)$. For some representation $\mu$ of $G$, $N \times D_{\alpha'} \times D_\mu$ and $N' \times D_\alpha \times D_\mu$ (elements of $D_{\alpha\oplus\alpha'\oplus\mu}(G)$) satisfy the codimension 1 and 2 conditions as well as the lower bound restrictions on the fixed point sets. We thus have the first of our stability theorems:

6.1. THEOREM. *Let $N$ and $M$ be semilinear $G$-disks. Then $\hat{\imath}_0(N) = \hat{\imath}_0(M)$ if and only if for some linear representations $\psi$ and $\psi'$ of $G$, $N \times D_\psi$ and $M \times D_{\psi'}$ are $G$-diffeomorphic. This is also true in the $G$* pl *category.*

We can formulate a type of stability theorem for spheres in the pl category as follows. Given $N \in S_\alpha(G)$ and $N' \in S_{\alpha'}(G)$, we can form $N \wedge N' = N \times N'/N \vee N'$, where this is independent of the choice of basepoint if $\dim D_\alpha^G$ and $\dim D_{\alpha'}^G > 0$. $N \wedge N'$ is a pl sphere with a pl $G$-action but not a $G$ pl manifold since the action near the basepoint is not in general linear. We can also write $N \wedge N' = C(V) \cup_V (k_\alpha(N) \times k_{\alpha'}(N'))$, where $V = \partial(k_\alpha(N) \times k_{\alpha'}(N'))$ and $C(V)$ is the cone over $V$.

6.2. THEOREM. *Let $N \in S_\alpha(G)$ and $N' \in S_{\alpha'}(G)$. Then $\hat{\imath}_0(k_\alpha(N)) = \hat{\imath}_0(k_{\alpha'}(N'))$ if and only if there exist linear representations $\psi$ and $\psi'$ of $G$ such that $N \wedge S_\psi$ and $N' \wedge S_{\psi'}$ are $G$-equivariantly* pl *isomorphic.*

PROOF. This follows from 6.1 and the fact that $k$ is injective in the pl *category*.

*Note*. By 3.22, if $\alpha, \alpha'$ are complex, $G$ Abelian and $\mathrm{Wh}(G; Z)$ has no torsion, $\hat{\imath}_0(k_\alpha(N))$ and $\hat{\imath}_0(k_{\alpha'}(N'))$ are determined by the torsion invariants $\tau_\alpha(N)$ and the $\tau_{\alpha'}(N')$, respectively.

The Stability Theorem 6.1 can be reinterpreted by introducing certain functors of $G$.

6.3. THEOREM. *Let $D(G)$ be the Grothendieck group on $G$ diffeomorphism classes of semilinear $G$-disks under the operation $\times$, Cartesian product. Then the map $N \to \hat{\imath}_0(N)$ induces a homomorphism $t_0^\times\colon D(G) \to \mathrm{Wh}(G; Z)$ whose kernel is the real representation ring $RO(G)$ of $G$. $t_0^\times$ splits and this yields an isomorphism*

$$D(G) \cong \mathrm{Wh}(G; Z) + RO(G).$$

PROOF. This follows easily from 6.1 and Product Theorem 1.26. Since $D^{\mathrm{pl}}(G) = D(G)$ the same results hold in the $G$ pl category.

To formulate an analogue for spheres we must go outside the category of $G$ pl manifolds. If $X$ is a $G$ pl space with $X^H$ pl isomorphic to a sphere or the empty set for each $H \subset G$, $X$ will be called a *weak* semilinear $G$-sphere. The notion of $G$-orientable weak semilinear $G$-sphere is defined exactly as in the more restricted case. We let $S(G)$ ($S^+(G)$) be the Grothendieck group of equivariant pl isomorphism classes of (orientable) weak semilinear $G$-spheres under the operation of join, $\#$. For orientable $G$-spheres, the oriented torsion $\tau$ is defined and this induces as in 4.4 a homomorphism $\hat{\lambda}\colon S^+(G) \to \tilde{K}_1(B(G); \tau) + Z/2Z$. We also have $e\colon S^+(G) \to Z$ given by $e(N) = \dim N + 1$. We can now reformulate and refine de Rham's theorem for finite groups $G$.

6.4. THEOREM. *For any finite group $G$ the natural map $j\colon RO(G) \to S(G)$ is injective.*

PROOF. The subgroup $RSO(G)$ is of index 2 in the torsion free group $RO(G)$ and $j|\colon RSO(G) \to S^+(G)$ is a subgroup of index 2 in $S(G)$. Hence it suffices to show $j|$ is injective. As in the proof of de Rham's theorem it suffices to prove this for $G$ abelian. But then the composite

$$RSO(G) \to S^+(G) \xrightarrow{\hat{\lambda}\times e} \tilde{K}_1(B(G);\tau) \times Z/2Z \times Z$$

is injective by 4.4. Q.E.D.

**7. Some generalizations and intriguing questions.** We have concentrated on semilinear cells and spheres in order to derive the sharpest possible classification results. However, our invariants are defined much more broadly. In passing to broader classes of objects one does not expect to get the classification theorems we have derived earlier, but at the same time one may still find these invariants useful in distinguishing various actions.

Given $G$ we can successively broaden the family of smooth semilinear $G$-disks to smooth homotopy semilinear $G$-disks, and further to smooth $R$-homology semilinear $G$-disks, where $R$ is a commutative ring with identity, by replacing the conditions on the fixed point sets in our original definition by the corresponding homotopy and homology conditions. Given a representation $\alpha$ of $G$ we can define the analogous set $D_\alpha(G;\pi)$ and $D_\alpha(G;R)$, where we also require for a manifold $X$ to define an element of $D_\alpha(G;R)$ that $X^H$ be an orientable homology $R$-disk. This last condition is of course automatically satisfied for $R$ of odd order or $R$ a field of characteristic 0. We have natural maps $D_\alpha(G) \to D_\alpha(G;\pi) \to D_\alpha(G;Z) \to D_\alpha(G;R)$. We can form the analogous sets in the $G$ pl category. All the sets are monoids when $D_\alpha(G)$ is. We can also form $S_\alpha(G;\pi)$ and $S_\alpha(G;R)$ and again we have the maps $S_\alpha(G) \to S_\alpha(G;\pi) \to S_\alpha(G;Z) \to S_\alpha(G;R)$. Observe that $S_\alpha(G) = S_\alpha(G;\pi)$ provided $\dim S^H_{\alpha\oplus 1} \neq 3, 4$ for all $H \subset G$. In general these sets can differ only if the Poincaré conjecture fails.

There is a Whitehead torsion functor $D_\alpha(G;R) \to \mathrm{Wh}(G;R)$ defined. It seems more fruitful to consider a slightly stronger functor. We have the natural map $\lambda\colon Z \otimes B(G) \to R \otimes B(G)$ induced by the morphism $Z \to R$. Let $\iota\colon B(G) \to Z \otimes B(G)$ and $\iota_r\colon B(G) \to R \otimes B(G)$ be the inclusion functors. Let $C(\lambda)$ be the full subcategory of $C(\iota)$ consisting of those $X$ such that $\lambda(X)$ is universally acyclic in $C(\iota_r)$. In other words $C(\lambda) = \lambda^{-1}C_0(\iota_r)$. We have $C_0(\iota) \subset C(\lambda)$ and $C(\lambda)$ inherits an $ES$ structure from $C(\iota)$.

7.1. DEFINITION. We define $\mathrm{Wh}(G; Z \to R) = K_0(C(\lambda))/\bigcup_i i(\pm f)$, where $f$ is an isomorphism in $B(G)$ and $i(f)$ is the associated acyclic complex concentrated in dimensions $i$ and $i-1$. Clearly we have natural homomorphisms $\mathrm{Wh}(G;Z) \to \mathrm{Wh}(G;Z\to R) \to \mathrm{Wh}(G;R)$ and the $R$ Whitehead torsion of a $R$-homology semilinear $G$-disk factors back to $\hat{\iota}_0 \otimes R\colon D_\alpha(G;R) \to \mathrm{Wh}(G;Z\to R)$.

In this more general context we state explicitly the following analogue of 3.16.

7.2. THEOREM. *There exists an automorphism* $\psi_\alpha\colon \mathrm{Wh}(G;Z\to R) \to \mathrm{Wh}(G;Z\to R)$ *with* $\psi_\alpha^2 = \mathrm{Id}$ *and depending only on* $\alpha$ *such that*

$$\begin{array}{ccccc}
 & & \mathrm{Wh}(G;Z\to R) & \xrightarrow{1-\psi_{\alpha\oplus 1}} & \mathrm{Wh}(G;Z\to R) \\
 & & \uparrow{\scriptstyle \hat{\iota}_0\otimes R} & & \uparrow{\scriptstyle \hat{\iota}_0\otimes R} \\
D_{\alpha+1}(G;R) & \xrightarrow{j_{\alpha\oplus 1}} & S_\alpha(G;R) & \xrightarrow{k_\alpha} & D_\alpha(G;R)
\end{array}$$

*commutes. Further* $\psi_{\alpha\oplus 1} = -\psi_\alpha$.

We also have the morphism $\tau_\alpha \otimes R\colon S_\alpha(G;R) \to \tilde{K}_1(B(G);\tau;Z\to R)$, where

$\tau = {}^{\alpha}\tau$ is $\alpha$-torsion and $\tilde{K}_1(B(G); \tau; Z \to R)$ is defined analogously to Wh($G$; $Z \to R$) replacing the phrase "universally acyclic" by "$\tau$-acyclic".

The interest of the above is the following. Let $p$ be a prime and $G$ be a $p$-group, i.e., $o(G) = p^n$. Let $Z_{(p)}$ be the integers localized at $p$. Then by Smith theory any disk on which $G$ acts smoothly or pl is a $Z_{(p)}$-homology semilinear $G$-disk, and any sphere on which $G$ acts is a $Z_{(p)}$-homology semilinear $G$-sphere. For $\hat{\iota}_0 \otimes Z_{(p)}$ one can also prove the Realization Theorem 3.5. Hence for such $G$ the invariants $\hat{\iota}_0 \otimes Z_{(p)}$ and $\tau \otimes Z_{(p)}$ are always defined and there are many examples which are not integral semilinear $G$-disks or spheres. In fact, L. Jones [J] has shown that there are many actions of $G$ on disks whose fixed point sets are not integrally acyclic.

The invariant $\hat{\iota}_0 \otimes Z_{(p)}: D_\alpha(G; Z_{(p)}) \to \mathrm{Wh}(G; Z \to Z_{(p)})$ yields information on reduced integral homology of the fixed point set, $\bar{H}_*(X^G)$, where $X \in D_\alpha(G; Z_{(p)})$. Precisely, let $A_p$ be the Abelian category of all finite Abelian groups of order prime to $p$. Then $\bar{H}_i(X^G) \in A_p$ for all $i$. Thus we have an element $\varepsilon(\bar{H}_*(X^G)) = \sum(-1)^i \bar{H}(X^G)$ of $K_0(A_p)$ and this defines a morphism $\varepsilon: D_\alpha(G; Z_{(p)}) \to K_0(A_p)$. An easy argument shows that $\varepsilon$ factors through $\varepsilon'$: $\mathrm{Wh}(G; Z \to Z_{(p)}) \to K_0(A_p)$. Thus we have a commutative diagram:

$$\begin{array}{ccc} D_\alpha(G; Z_{(p)}) & \xrightarrow{\hat{\iota}_0 \otimes Z_{(p)}} & \mathrm{Wh}(G; Z \to Z_{(p)}) \\ & {\scriptstyle\varepsilon}\searrow \quad \swarrow{\scriptstyle\varepsilon'} & \\ & K_0(A_p) & \end{array}$$

Hence:

7.3. PROPOSITION. *Let $G$ be a $p$-group. Suppose $X$, $Y$ are two $Z_{(p)}$-homology disks on which $G$ acts such that $\overline{H_*(X^G)}$ and $\overline{H_*(Y^G)}$ are concentrated in dimensions $i$ and $j$ respectively. If*

$$\hat{\iota}_0 \otimes Z_{(p)}(X) = \hat{\iota}_0 \otimes Z_{(p)}(Y), \quad \textit{then} \quad o(\overline{H_i(X^G)}) = o(\overline{H_j(Y^G)}).$$

REMARK. When $D_\alpha$ and $\dot{D}_\alpha^G$ are even dimensional, the morphism $\varepsilon$ is a mod $p$ $G$ $h$-cobordism invariant and thus represents an obstruction for an element $X$ in $D_\alpha(G; Z_{(p)})$ to be mod $p$ $G$ $h$-cobordant to an element $X'$ in $D_\alpha(G; Z)$. Further $K_0(A_p)$ is torsion free and thus $\varepsilon'$ annihilates the torsion of $\mathrm{Wh}(G; Z \to Z_{(p)})$.

To study the invariant $\varepsilon$, we consider $p$-pointlike manifolds $W$. That is, $W$ is an orientable $n$-manifold with boundary such that $H_i(W) \in A_p$ for $i > 0$ and $H_j(W, \partial W) \in A_p$ for $j \neq n$. Note that this implies $H_i(\partial W) \in A_p$ for $i \neq 0, n-1$. For any Abelian group $F$, let $\bar{F} = F$ if $F \in A_p$ and $\bar{F} = 0$ if $F \notin A_p$. Then for any pair of spaces $(X, Y)$ of finite homological rank over $Z$ we let $\varepsilon(X, Y) = \sum_i (-1)^i \overline{H_i(X, Y)}$ which is an element of $K_0(A_p)$. For any $p$-pointlike manifold $W$ the exact homology sequence yields $\varepsilon(W) = \varepsilon(\partial W) + \varepsilon(W, \partial W)$. However, Poincaré duality says $H_i(W, \partial W) = \overline{H^{n-i}(W)} = \overline{H_{n-i-1}(W)}$, the last equality being given by the universal coefficient theorem. Thus if $n = 1$ (mod 2), $\varepsilon(W, \partial W) = \varepsilon(W)$ and if $n = 0$ (mod 2), $\varepsilon(W, \partial W) = -\varepsilon(W)$. Thus if $n = 1$ (mod 2), $\varepsilon(\partial W) = 0$ and if $n = 0$ (mod 2), $\varepsilon(\partial W) = 2\,\varepsilon(W)$. This motivates the following which is an easy calculation.

7.4. PROPOSITION. *The following diagram is commutative.*

$$\begin{array}{ccc} K_0(A_p) & \xrightarrow{1-\psi'\alpha+1} & K_0(A_p) \\ \uparrow{\scriptstyle \varepsilon'} & & \uparrow{\scriptstyle \varepsilon'} \\ \mathrm{Wh}(G; Z \to Z_{(p)}) & \xrightarrow{1-\psi_{\alpha+1}} & \mathrm{Wh}(G; Z \to Z_{(p)}) \\ \uparrow{\scriptstyle \hat{\imath}_0 \otimes Z_{(p)}} & & \uparrow{\scriptstyle \hat{\imath}_0 \otimes Z_{(p)}} \\ D_{\alpha\oplus 1}(G; Z_{(p)}) & \xrightarrow{k_\alpha j_{\alpha\oplus 1}} & D_\alpha(G; Z_{(p)}) \end{array}$$

*where* $\psi'_{\alpha+1} = (-1)^{\dim D^G_\alpha}$.

It is probably true that

7.5. CONJECTURE. If $\alpha \oplus 1$ is a complex representation then $\psi_{\alpha\oplus 1} = -\,\mathrm{Id}$.

REMARK. Suppose $X \in S_\alpha(G; Z_{(p)})$ with $\dim(D^G_\alpha) = 0 \pmod 2$. Then $2\varepsilon(k_\alpha(X)) = \varepsilon(\partial k_\alpha(X)) = 0$. Since $K_0(A_p)$ is torsion free, $\varepsilon(k_\alpha(X)) = 0$. Thus $\varepsilon(X) = \varepsilon(k_\alpha(X)) = 0$.

The following is a very optimistic conjecture which is considerably more delicate than the previous one. For $X \in S_\alpha(G; Z_{(p)})$ there is a natural $G$-map of degree 1, $\rho: X \to S_{\alpha\oplus 1\alpha}$, obtained by collapsing the complement of a $D_\alpha \subset X$ to a point. Let $S^F_\alpha(G; Z_{(p)}) \subset S_\alpha(G; Z_{(p)})$ be those elements $X$ such that the stable $G$-tangent bundle of $X$ pulls back from some $G$-bundle over $S_{\alpha\oplus 1}$ under $\rho$. Define $D^F_\alpha(G; Z_{(p)})$ similarly.

7.6. CONJECTURE. *Let* $\alpha$ *satisfy the conditions of* 3.13. *Then the image* $D^F_{\alpha\oplus 1}(G; Z_{(p)}) + S_\alpha(G) \to^{j_{\alpha\oplus 1}+\iota} S^F_\alpha(G; Z_{(p)})$ *contains the kernel of*

$$(\hat{\imath}_0 \otimes Z_{(p)}) \circ k_\alpha | : S^F_\alpha(G; Z_{(p)}) \to \mathrm{Wh}(G; Z \to Z_{(p)}).$$

The proof of this conjecture would involve showing that certain $G$ surgery obstructions are captured by $\hat{\imath}_0 \otimes Z_{(p)}$. A consequence of these two conjectures is that for $\alpha$ complex

$$D^F_{\alpha\oplus 1}(G; Z_{(p)}) + S_\alpha(G) \xrightarrow{j_\alpha \oplus_{1+\iota}} S^F_\alpha(G; Z_{(p)})$$

is onto. This result would be very interesting even for the group $G = Z/pZ$. It is true but nontrivial for $G = (e)$.

Let $G$ be a finite group and $X$ a finite $G$-$CW$ complex which is pointlike in integral homology. If $G_p$ is a $p$-Sylow subgroup of $G$ then the induced action of $G_p$ on $X$ defines an invariant $\hat{\imath}_0 \otimes Z_{(p)}(X) \in \mathrm{Wh}(G_p; Z \to Z_{(p)})$. If $N(G_p)$ is the normalizer of $G_p$ in $G$, then $N(G_p)$ acts on $G_p$ and hence on $\mathrm{Wh}(G_p; Z \to Z_{(p)})$. Then $t_0 \otimes Z_{(p)}(X) \in \mathrm{Wh}^{N(G_p)}(G_{(p)}; Z \to Z_{(p)})$, the latter being the subgroup fixed under the action of $N(G_p)$. If $G'_p$ is another Sylow subgroup, then $G_p$ and $G'_p$ are conjugate in $G$ and this induces a canonical identification of $\mathrm{Wh}^{N(G_p)}(G_{(p)}; Z \to Z_{(p)}) = \mathrm{Wh}^{N(G'_p)}(G'_p; Z \to Z_{(p)})$. We can denote this common group by $\mathrm{Wh}(G; Z; p)$. Let $\lambda = \times_{p|o(G)}\, t_0 \otimes Z_{(p)}$. Then $\lambda(X) \in \times_{p|o(G)} \mathrm{Wh}(G; Z; p)$ yields an invariant of the pointlike $G$-complex. Unfortunately, this invariant is rather weak. Suppose for example, $X$ is a semilinear $G$-disk. Then $\lambda(X)$ does not determine $t_0(X)$ when $G$ is not of prime power order.

To devise and motivate a stronger invariant let us reformulate Smith theory in our context. For $W \in B(G)$, let $C(W)$ be the singular 0 chains of $W$ modulo the basepoint $*$. $C(W)$ is a module over $Z(G)$. For any complex $X = (X_n, d_n)$ of $R \otimes B(G)$, we have the corresponding chain complex $C(X) = (R \otimes C(X_n), d_n')$ which is a chain complex of modules over $R(G)$. We define $H_i(X) = H_i(C(X))$. If $X$ is a complex of $Z \otimes B(G)$, then for any commutative ring $R$ with identity, $X$ projects to a unique complex in $R \otimes B(G)$ which we denote by $R \otimes X$. For such an $X$ we define $H_i(X; R) = H_i(R \otimes X) = H_i(R \otimes C(X))$. Smith theory works in our categories [**B2**], and we have

7.6. THEOREM (SMITH). *Let $G$ be a $p$-group and $X$ a complex of $Z/pZ \otimes B(G)$. Then $X^G$ is a complex of $Z/pZ \otimes B(G)$ and for each $n$,*

$$\dim\Big(\sum_{i \geqq n} H_i(X^G)\Big) \leqq \dim\Big(\sum_{i \geqq n} H_i(X)\Big)$$

*and*

$$\sum (-1)^i \dim H_i(X^G) = \sum (-1)^i \dim H_i(X) \mod p$$

*where dimension means dimension as a vector space over $Z/pZ$. In particular, if $X$ is acyclic, i.e. $H_i(X) = 0$ for all $i$, then $X \in C_0(\iota)$, i.e. $X$ is universally acyclic (where $\iota: B(G) \to Z/pZ \otimes B/(G)$).*

Now let $G$ be an arbitrary finite group. It follows from 7.6 and the universal coefficient theorems that if $X$ is a complex in $Z \otimes B(G)$, which is acyclic, then for any $p$-subgroup $H$ of $G$, $H_i(X^H; Z_{(p)}) = 0$. ($X^H$ is a complex in $Z \otimes B(N(H))$.)

Now let $\iota: B(G) \to Z \otimes B(G)$ be the natural inclusion and $\hat{C} \subset C(\iota)$ be the full subcategory of all acyclic $X$. Define $\mathrm{Wh}\hat{\ }(G) = K_0(\hat{C})/(i(\pm f))$, where $f$ is an isomorphism in $B(G)$. Then the discussion in the previous paragraph implies that for any $p$-subgroup $H$ of $G$ we have a morphism

$$\mu(H): \mathrm{Wh}\hat{\ }(G) \to \mathrm{Wh}^{N(H)}(H; Z \to Z_{(p)}).$$

Now if $X$ is any finite $G$-$CW$ complex which has the integral homology of a point, there is a torsion $\hat{\tau}_0(X) \in \mathrm{Wh}\hat{\ }(G)$ defined and we can factor the $\lambda$ defined above as $\lambda(X) = \times_{p|o(G)} \mu(G_p)(\hat{\tau}_0(X))$. The group $\mathrm{Wh}\hat{\ }(G)$ is rather mysterious. It would be interesting to have computations for a $G$ not of prime power order. Clearly there is a natural homomorphism $j$: $\mathrm{Wh}(G; Z) \to \mathrm{Wh}\hat{\ }(G)$.

7.7. CONJECTURE. *$j$ is injective.*

REMARK. For $G$ a $p$-group, we have a factorization

$$\mathrm{Wh}(G; Z) \to \mathrm{Wh}\hat{\ }(G) \to \mathrm{Wh}(G; Z \to Z_{(p)}) \to \mathrm{Wh}(G; Z_{(p)}),$$

and the composition is injective for $G = Z/p^nZ$. Thus the conjecture is reasonable.

## BIBLIOGRAPHY

The following is a selective rather than a full bibliography. A relatively complete bibliography of material prior to 1966 can be found in [**M**].

**B1** H. Bass, *Algebraic K-theory*, Benjamin, New York, 1968. MR **40** #2736.

**B2** G. Bredon, *Equivariant cohomology theories*, Lecture Notes in Math., vol. 34, Springer-Verlag, Berlin and New York, 1967. MR **35** #4914.

**B3** ———, *Introduction to compact transformation groups*, Academic Press, New York, 1972.

**C** M.M. Cohen, *A course in simple homotopy theory*, Springer-Verlag, Berlin and New York, 1970.

**CMY** E. H. Connell, D. Montgomery and C. T. Yang, *Compact groups in $E^n$*, Ann. of Math. (2) **80** (1964), 94–103; correction, ibid. (2) **81** (1965), 194. MR **29** #189.

**D** G. de Rham, *Reidemeister's torsion invariant and rotations of $S^n$*, Differential Analysis (published for the Tata Institute of Fundamental Research, Bombay), Oxford Univ. Press, London, 1964. pp. 27–36. MR **32** #8355.

**DMK** G. de Rham, S. Maumary and M. A. Kervaire *Torsion et type simple d'homotopie*, Lecture Notes in Math., vol. 48, Springer-Verlag, Berlin and New York, 1967. MR **36** #5943.

**H** J. F. P. Hudson, *Piecewise linear topology*, Benjamin, New York, 1969. MR **40** #2094.

**I1** S. Illman, *Whitehead torsion and group actions*, Ann. Acad. Sci. Fenn. Ser. AI Math. No. 588, 1974. MR **51** #14075.

**I2** ———, *Smooth equivariant triangulations of G-manifolds for G a finite group*, preprint, 1977.

**J** L. Jones, *The converse to the fixed point theorem of P. A. Smith*, Ann. of Math. (2) **94** (1971), 52–68. MR **45** #4427.

**LR** R. Lashof and M. Rothenberg, *G- smoothing theory*, these Proceedings, Part I, pp. 211–266.

**KS** K. W. Kwun, and R. H. Szczarba, *Product and sum theorems for Whitehead torsion*, Ann. of Math. (2) **82** (1965), 183–190. MR **32** #454.

**M** J. W. Milnor, *Whitehead torsion*, Bull. Amer. Math. Soc. **72** (1966), 358–426. MR **33** #4922.

**Q** D. Quillen, *Higher algebraic K-theory*—I, Lecture Notes in Math., vol. 341, Springer-Verlag, Berlin and New York, 1973, pp. 85–147. MR **49** #2895.

**R1** M. Rothenberg, *Differential group actions on spheres*, Proc. Advanced Study Inst. on Algebraic Topology, Aarhus Univ., Aarhus, 1970.

**RS** M. Rothenberg, and J. Sondow, *Nonlinear smooth representations of compact Lie groups* (preprint).

**W1** J. H. C. Whitehead, *On $C^1$ complexes*, Ann. of Math. (2) **41** (1940), 809–824. MR **2**, 73.

**W2** ———, *Combinatorial homotopy*. I, Bull. Amer. Math. Soc. **55** (1949), 213–245. MR **11**, 48.

**W3** ———, *Simple homotopy types*, Amer. J. Math. **72** (1952), 1–57. MR **11**, 735.

**BQ** W. Browder and F. Quinn, *A surgery theory for G manifolds and stratified sets*, Manifolds, Univ. of Tokyo Press, Tokyo, 1973.

University of Chicago

Proceedings of Symposia in Pure Mathematics
Volume 32, 1978

# DIFFERENTIAL ACTIONS OF COMPACT SIMPLE LIE GROUPS ON HOMOTOPY SPHERES AND EUCLIDEAN SPACES*

M. DAVIS, W. C. HSIANG AND W. Y. HSIANG

**Introduction.** In the past decade, there have been many advances in the theory of differentiable actions on manifolds. Most topologists were concerned with actions of finite groups or of $S^1$. Most of the results stem out of surgery theory and equivariant handlebody theory. (See [3], [19], [27] for samples of this type of result.) Inspired by a simple result of [16], we thought that it might be interesting to consider 'large' group actions; more specifically, what can we say about differentiable actions of compact simple Lie groups on homotopy spheres or Euclidean spaces when the dimension of the group is relatively large compared to the dimension of the ambient space? At the first glance, it looked as if it would be an impossible task, since we had already run into so much trouble trying to understand $S^1$ or $Z_p$ actions and since a general equivariant embedding theorem [24], [25] says that the *orbit structure* of an *orthogonal* action on a Euclidean space can be as complicated as one wishes. So, how can we even begin to ask a sensible question? After some thought, we set forth the following problem.

*Let $M^n$ be an n-*dim *homotopy sphere or a Euclidean space and let G be a compact simple Lie group of dimension* (*as space*) *greater than n, i.e.*

$$\dim G > \dim M = n. \tag{1}$$

*Classify all the differentiable actions of G on $M^n$.*

*AMS* (*MOS*) *subject classifications* (1970). Primary 57E15, 57E25, 57D60.

*Key words and phrases.* Differential actions of compact Lie groups, homotopy spheres, Euclidean spaces.

*These lectures were given by W. C. Hsiang. All three authors were partially supported by NSF grants during the preparation of this manuscript.

We are now close to a satisfactory final answer to this problem and this is a progress report to date.

Let us first explain why it is a reasonable problem. When $G$ is 'small' (e.g., $Z_p$), we do not have much choice of the orbit types and we are forced to study the global classification problem immediately. This is the reason why surgery theory and equivariant handlebody theory are natural tools for these actions. On the other hand, if $G$ is a compact simple Lie group and $\dim G > \dim M$, then no isotropy group is the trivial group. Only certain well-understood orthogonal actions have this property (that no isotropy group is trivial). For a general differentiable action, we may ask if we can always find an orthogonal model to compare it with, and if so how can we measure its deviation from its orthogonal model. If $M^n$ is a homotopy sphere, we can ask another even more interesting question: How is the differential structure of $M^n$ determined by the action? For example, can a homotopy sphere $M^n \notin bP_{n+1}$ allow a differentiable action of a compact simple Lie group $G$ with $\dim G > \dim M$? If the answer of the last question is 'no' then the degree of symmetry of a homotopy sphere $M^n \notin bP_{n+1}$ as defined in **[17]** is at most $n$. (Cf. **[12]**, **[21]**, **[26]**.) In this paper, we shall summarize the results on these questions up to now. Of course, one can ask the same questions in general without the restriction $\dim G > \dim M$, but we do not even know the orbit structure of an arbitrary orthogonal action. It is generally very complicated, although the action is theoretically determined by its weights. So, it is a good idea to stick to the original program.

We next translate (1) into a statement about isotropy subgroups. Recall a result of Montgomery-Samelson-Yang **[23]**. Let $\Psi: G \times M \to M$ be an action with $M/G$ connected. Order the conjugacy classes of isotropy subgroups by inclusion. They have a *unique absolute minimum* $(H_\Psi)$ and the union of the orbits of the type $G/H_\Psi$ is an open dense subset $M_{(H_\Psi)}$ of $M$. Call $H_\Psi$ the *principal isotorpy subgroup* of $\Psi$ and $G/H_\Psi$ the principal orbit type. Clearly, (1) implies

$$(H_\Psi) \neq (\mathrm{id}), \tag{2}$$

i.e., the principal isotropy subgroup $H_\Psi$ of $\Psi$ is nontrivial. So, our problem can be altered to the classification problem of differentiable actions of compact Lie groups on homotopy spheres or Euclidean space with nontrivial principal isotropy subgroups.

Let us now fix a differentiable action $\Psi: G \times M^n \to M^n$ satisfying (2) on a homotopy sphere or a Euclidean space $M^n$. In order to focus our attention, we examine the following three aspects of $\Psi$.

(A) *Orbit structure of $\Psi$ and picking up an orthogonal model for $\Psi$.*

Does there exist a (unique) orthogonal representation $\phi$ on the unit sphere or the Euclidean space (of the same dimension) such that $\Psi$ and $\phi$ have the same orbit types? It turns out that the answer is always 'yes'. The precise result is stated in §I.

(B) *Is $\Psi$ a pullback? Is $\Psi$ homologically modeled?*

$(M, \Psi)$ and its orthogonal model $(N, \phi)$ are naturally stratified spaces **[4]**, **[5]**, **[6]** where a 'stratum' is the union of orbits of a given type. The stratifications of $(M, \Psi)$ and $(N, \phi)$ induce stratifications of the orbit spaces $M/\Psi$ and $N/\phi$. Let $\pi_M: M \to M/\Psi$, $\pi_N: N \to N/\phi$ be the projections. We say that $(M, \Psi)$ is a pullback from $(N, \phi)$ via $f$ if there is a commutative diagram of maps **[1]**, **[5]**, **[6]**

(3)
$$\begin{array}{ccc} (M, \Psi) & \xrightarrow{F} & (N, \phi) \\ \Big\downarrow{\scriptstyle \pi_M} & & \Big\downarrow{\scriptstyle \pi_N} \\ M/\Psi & \xrightarrow{f} & N/\phi \end{array}$$

such that $F$ is a stratified (= isovariant and normally transverse) map and $f$ is the induced stratified map of orbit spaces. We say that $(M, \Psi)$ is *homologically modeled via* $F$ if $F$ and $f$ induce homological isomorphisms (with appropriate coefficients) on the corresponding strata. (We do not want to be too precise here. Cf. §II.)

Since the aswer to (A) is affirmative, we can ask whether $(M, \Psi)$ is a pullback from its orthogonal model $(N, \phi)$ via some map and whether the map actually makes it homologically modeled. So far, we can prove that if the rank of $G$ is greater than 4, then $(M, \Psi)$ is always a pullback and is always homologically modeled. Some more details will be given in §II.

Let us remark that it is absolutely necessary for us to stay at the homological level for $F$ and $f$. In other words, it is impossible to ask for $F$ and $f$ to induce homeomorphism (or even homotopy equivalence) on the corresponding strata because we can easily alter the fundamental group of some stratum of $M/\Psi$ or $(M, \Psi)$ and obtain a pullback diagram

(4)
$$\begin{array}{ccc} (M, \Psi') & \xrightarrow{F'} & (N, \phi) \\ \Big\downarrow{\scriptstyle \pi'_M} & & \Big\downarrow{\scriptstyle \pi_N} \\ M/\Psi' & \xrightarrow{f} & N/\Psi \end{array}$$

such that $F'_*$, $F_*$ and $f_*, f'_*$ are homologically indistinguishable.

(C) *Concordance.* We say that two differentiable actions $\Psi_i = G \times M_i \to M_i$ $(i = 0, 1)$ are *concordant* if there is a differentiable action of $G$ on $M \times I$ such that the restriction to $M \times i$ is equivalent to $\Psi_i$ $(i = 0, 1)$. (For regular $O(m)$ actions, we have to take the orientations of some strata into account too.) Clearly, *concordance* is an equivalence relation. After what we said about homological modeling and the possibility of changing $\pi_1$ of some strata of the orbit space of $M$, it is natural to consider the classification problem for these actions at the concordance level rather than at the honest equivariant diffeomorphism level (where there are too many). The point is that concordance relationship translates into a homology $h$-cobordism relation on each stratum at the orbit space level. In §III, we shall summarize the concordance classification results. Roughly speaking, the result can be stated as follows: *If the rank of* $G > 7$ *and* $G \neq O(m)$, $(M, \Psi)$ *is concordant to either the corresponding orthogonal action or a known action on a Brieskorn manifold* [7], [8], [9]. In fact, the majority of the actions are concordant to the orthogonal ones. But we do have some new concordance class of $O(m)$ actions [9].

One consequence of the results of this paper is the following theorem.

THEOREM. *Let* $M^n$ *be a homotopy sphere and assume that* $M^n \notin bP_{n+1}$. *Then the degree of symmetry of* $M^n$ *is* $\leqq 2n$.

About half of the results summarized in this paper are new and the details will appear elsewhere.

**I. Orbit structure of $\Psi$ and picking up an orthogonal model for $\Psi$.** Let $\Psi : G \times M^n \to M^n$ be a differentiable action of a compact Lie group $G$ on a manifold $M^n$. If we have an orthogonal action $\psi$ (on the Euclidean space or the unit sphere) such that $\Psi$ and $\psi$ have the same orbit types and the slice representations of the corresponding orbits are also the same, then we say that $\psi$ is the *orthogonal model* of $\Psi$.

THEOREM 1.1. *Let $M^n$ be an n-*dim *acyclic manifold or a homology sphere and let $\Psi$: $G \times M^n \to M^n$ be a differentiable action of a (connected) simple Lie group $G$ with $(H_\Psi) \neq$* (id). *Then, there is a unique orthogonal action $\psi$ on the corresponding n-*dim *Euclidean space or on the unit sphere such that $\psi$ is the orthogonal model of $\Psi$.*

This theorem is due to the second two authors. It actually took us more than ten years to prove this theorem. Let us indicate the history of this theorem. In **[11**, I], we proved that if $G$ is a connected classical group (*i.e.*, $G = SO(m)$, $SU(m)$ or $Sp(m)$) and $n \leqq ((m-1)^2/2)$, $11 \leqq m$ for $SO(m)$, $n \leqq (m-1)^2/2$, $8 \leqq m$ for $SU(m)$ and $n \leqq (m-1)^2$, $8 \leqq m$ for $Sp(m)$ respectively, then the orbits are all (real, complex or quaternionic) Stiefel manifolds. The corresponding $\psi$ for $\Psi$ can be picked up as follows. Let $\rho_m$ be the standard representation of $G = SO(m)$, $SU(m)$ or $Sp(m)$ into $O(m)$, $O(2m)$ or $O(2m)$ respectively. Then, $\psi$ must be $k\rho_m + l1$ for suitably chosen $k$ and $l$, where 1 denotes the trivial representation of $G$ on $R^1$. In the course of determining the orbits of $\Psi$, we developed a formula to compute the Pontriagin classes of the normal bundle of an orbit and used it to show that, under present dimension assumption, every orbit has to be covered by a Stiefel manifold. So, we determined the identity components of the isotropy subgroups. Then, we restricted the action to a maximal torus or a maximal $Z_2$-torus and applied the classical Smith theory and a formula of Borel on the dimensions of the fixed points of actions of elementary abelian groups on homology sphere (or acyclic manifold) to the restricted action to prove Theorem 1.1 for $G = SO(m)$, $SU(m)$ or $Sp(m)$ (under the dimension restriction). The method of the proof is even of independent interest today, but the restriction of the dimension of $G$ is too severe to give us the full theorem.

In the meantime, we tried to analyze what are the essential tools in representation theory if we want to determine the orbit structure of an orthogonal action. Two things come out loud and clear: The weights of the representation and the group generated by reflections (with respect to the planes perpendicular to some roots). Then, we naturally asked ourselves whether we can formulate some topological substitutes for these important tools so that they are strong enough to detect the orbit structure for a general differentiable action $\Psi$ if $(H_\Psi) \neq$ (id). If $M^n$ is an acyclic manifold, one of the authors introduced the '*geometric weight system of* $\Psi$' in **[13]**, as follows. Fix a maximal torus $T \subset G$. By Smith theory, the fixed point set $F$ of $\Psi | T$ is acyclic (and hence nonempty). It is easy to see that the local representation of $T$ at a point $p \in F$ is independent of $p$. We call this local representation the 'geometric weight system of $\psi$'. In **[11**, II], we used this simple notion to prove that for acyclic manifolds we can always find an orthogonal representation $\psi$ such that the identity component of $H_\Psi$ is the same as $H_\psi$.

After this, it became clear to us that we need a good structure theorem for a group generated by *differentiable reflections*[1] on acyclic manifolds and homology

[1]A diffeomorphism $r$: $M \to M$ of a connected oriented manifold $M$ is a (differentiable) reflection if (i) $r$ reverses the orientation and $r^2 = \text{id}$, (ii) the set $\{x \in M \mid r(x) \neq x\}$ is disconnected.

spheres [2], [11, III], [20]. We publish a proof for differentiable reflections on acyclic manifolds and claimed (perhaps, extravagantly) that the proof would also go through for the homology sphere case in [11, III]. Bredon recently has given a simpler proof and corrected some of our gaps [2]. In any case, there is the following theorem.

THEOREM 1.2. *Let $W$ be a Coxeter group and $M^n$ be an* $n$-dim *acyclic manifold or a homology sphere. Let $\Psi: W \times M^n \to M^n$ be a group generated by differentiable reflections on $M^n$. Then, we have a group generated by orthogonal reflections $\psi$: $W \times N^n \to N^n$ where $N^n$ the* $n$-dim *Euclidean space or the unit sphere on a stratified map $F: M^n \to N^n$ which induces the commutative diagram*

$$\begin{array}{ccc} M^n & \xrightarrow{F} & N^n \\ \downarrow{\scriptstyle \pi_M} & & \downarrow{\scriptstyle \pi_N} \\ M^n/W & \xrightarrow{f} & N^n/W \end{array}$$

*where $F_*, f_*$ induce homology isomorphism on each stratum.*

In other words, a group generated by differential reflections on an acyclic manifold or a homology sphere is homologically modeled on the usual orthogonal group generated by reflections.

With geometric weight systems and groups generated by differentiable reflections as our topological substitutes for weights and orthogonal reflections, we proved Theorem 1.1 for the case that $M^n$ is an acyclic manifold[2].

But, we still have more technical difficulties if $M^n$ is a homology sphere. The action restricted to a maximal torus $T$, $\Psi|T$, may not have any fixed point if $M^n$ is a homology sphere, and the 'geometric weight system' defined as above may not be definable. In [14], [15], *rational weight system* and *$Z_p$-weight system* for topological actions on homology sphere are introduced, and it was proved that there are some relations between the rational weights and $Z_p$-weights due to the fact that when we suitably restrict to the action to some codim 1 subtori (or $Z_p$-subtori) the fixed point sets exist. Using these facts as the substitute for geometric weights, W. Y. Hsiang finally proved that there is an orthogonal representation $\psi$ on the unit sphere such that $(H_\Psi) = (H_\psi)$ for $M^n$ a homology sphere. After that, we can prove Theorem 1.1.

The list of such orthogonal representations appeared in [11, II]. This list is easier to describe if we assume that $\text{rk}(G) > 7$.

Let $S = A/K$ be an irreducible compact symmetric space. Consider the isotropy representation $\lambda$ of $K$ on the tangent space of $S$ at the base point. In general, $K$ (locally) splits up as a product of one, two or four factors such that the restrictions $\psi_1$ of $\lambda$ to one (simple) factor $G$ have nontrivial principal isotropy subgroup. One can show that if $G$ is a compact simple Lie group and $\psi$ is an orthogonal representation of $G$ with $(H_\psi) \neq (\text{id})$, then $\psi = \psi_1 + l1$ for $l \geqq 0$. These representations are the orthogonal models for actions with nontrivial principal isotropy subgroups. Let us divide them into three classes of actions according to the nature of their models.

[2]In [11], we thought that $F_4$ is an exceptional case for Theorem 1.1, but after we checked through the argument, we see that Theorem 1.1 actually is also valid for $F_4$.

(a) *Regular actions.* $G = \mathrm{O}(m)$, $\mathrm{U}(m)$ or $\mathrm{Sp}(m)$ *and the orthogonal model is* $k\rho_m + l1$ *for* $k < m$. (Strictly speaking, we should consider the simple group $\mathrm{SU}(m)$ instead of $\mathrm{U}(m)$, but the modification is trivial.) Under the present circumstances, $S$ is a real, complex or quaternionic Grassmannian manifold and $\phi$ is the sum of the trivial representation $l1$ on $R^l$ with the representation $\phi_1 = k\rho_m$: $G \times M_{m,k} \to M_{m,k}$ of $G$ on the $(m \times k)$-matrices with real, complex or quaternionic entries. A differentiable action is *regular* if its orthogonal model is $k\rho_m + l1$.

(b) *Actions of the adjoint type.* In this case, $K$ has only one factor. So, $\phi = \lambda + l1$ where $\lambda$ denotes the isotropy representation of $S = A/K$ at the base point. See **[28]** for the list of orthogonal representations of the adjoint type. A differentiable action is of *adjoint type* if it is modeled on an orthogonal representation of adjoint type. Note that $\phi = \mathrm{Ad}_G$ is of the adjoint type.

(c) *Actions of near adjoint type.* Suppose that $K$ has two simple factors but $S$ is not a Grassmannian manifold. Let $G$ be a simple factor of $K$ such that the restriction $\phi_1$ of the isotropy representation $\lambda$ of $K$ to $G$ has nontrivial principal isotropy subgroup $H_{\phi_1}$. If a differentiable action $\phi$ is modeled on $\phi_1 + l1$, then it is called an *action of near adjoint type.* (Of course, regular actions are of near adjoint type. But because they were studied first by us and their nature is different from the other actions which are called of near adjoint type, we separate them out.)

A typical example of near adjoint type is as follows. Let $G = \mathrm{SU}(m)$ and let $\phi_1 = \Lambda^2\rho_m$ where $\rho_m$ is the standard representation of $\mathrm{SU}(m)$. Then, $\phi = \phi_1 + l1$ is of near adjoint type. Note that

$$\begin{aligned}(H_\phi) &= (\mathrm{SU}(2) \times \cdots \times \mathrm{SU}(2)) \qquad \left(\left[\frac{m}{2}\right] \text{copies}\right) \text{if } m \text{ is odd,}\\ &= \left(\frac{\mathrm{SU}(2) \times \cdots \times \mathrm{SU}(2)}{Z_2}\right) \qquad \left(\frac{m}{2} \text{ copies}\right) \text{if } m \text{ is even.}\end{aligned}$$

(This model is the most difficult one of the actions of near adjoint type.)

**II. $\Psi$ is a pullback from $\phi$.** It follows from Theorem 1.1 that a differentiable action $\Psi: G \times M^n \to M^n$ of a compact simple Lie group on an acyclic manifold or a homology sphere with $(H_\Psi) \neq (\mathrm{id})$ always has an orthogonal model $\phi$. We next ask ourselves whether $\Psi$ is a pullback from $\phi$ and whether $\Psi$ is homologically modeled on $\phi$.

From now on, we shall only state the results for the more interesting case that $M^n$ is a homology sphere. The corresponding acyclic space case is similar and easier and we shall leave it to the reader to formulate the results.

THEOREM 2.1. (A) *Let $M^n$ be a homology sphere and let $\Psi: G \times M^n \to M^n$ be an action with $\phi$ as its orthogonal model* (*where $\phi$ is as in* §I). *Then, there is a commutative diagram of stratified maps*

$$\begin{array}{ccc} M^n & \xrightarrow{F} & N^n \\ \Big\downarrow{\scriptstyle \pi_M} & & \Big\downarrow{\scriptstyle \pi_N} \\ M^n/\Psi & \xrightarrow{f} & N^n/\phi \end{array}$$

*such that* $(M^n, \Psi)$ *is a pullback via* $f$.

(B) *$F$ can be chosen to be of degree* 1, *except in the case that* $G = \mathrm{O}(m)$ *and* $\phi = k\rho_m$ *for $k$ even and $m$ odd* (*in which case, $F$ may be of odd degree*).

For $G = \mathrm{O}(m)$, $\mathrm{U}(m)$ or $\mathrm{Sp}(m)$ and $\phi = 2\rho_m + l1$, this theorem was proved in **[1]**. (See **[10]**, **[12]**, **[18]** also.) For the other regular case, it was proved in **[6]**. For locally smooth actions of adjoint type and some of near adjoint type, the corresponding theorem appeared in **[29]**. The full proof for the smooth actions of adjoint and near adjoint type will be published elsewhere in the future.

The next two theorems tells us when a stratified map of orbit space is induced by a homology isomorphism of total spaces.

THEOREM 2.2. *Suppose that $M$, $M'$ are $G$-manifolds having a fixed orthogonal model $\phi$ as in* §I. *Also suppose that* $G \neq \mathrm{O}(m)$ *and* $\phi \neq k\rho_m + l1$. *Let* $F$: $M \to M'$ *be a stratified map and let* $f$: $B = M/G \to B' = M'/G$ *be the induced map. Finally, let* $f_\alpha = f \mid B_\alpha$ *be the restriction of $f$ to the $\alpha$-stratum of $B$* ($\{\alpha\}$ *index the strata*). *Then,* $F_* = H_*(M; Z) \to H_*(M'; Z)$ *is an isomorphism if and only if* $(f_\alpha)_*$: $H_*(B_\alpha;Z) \to H_*(B'_\alpha; Z)$ *is an isomorphism for each* $\alpha$.

For the regular case, this theorem was proved in **[6]**. For the actions of adjoint type or near adjoint type, it essentially follows from Theorem 1.2. We shall publish the proof of this theorem with Theorem 2.1 elsewhere.

For $G = \mathrm{O}(m)$ and $\phi = k\rho_m + l1$, the analogous theorem is a little more difficult to state. First let us recall some notation of **[6]**, **[9]**. Let $(M, \Psi)$ be an $\mathrm{O}(m)$-action and let $B$ be the orbit space. The strata can be indexed by $\{0, 1, \cdots, k\}$ so that $M_i$ denotes the orbits of type $\mathrm{O}(m)/\mathrm{O}(m-i)$. Set $F = M^{\mathrm{O}(m-i)}$. Then, $F$ is an $\mathrm{O}(i)$-manifold modeled on $k\rho_i + l1$. We have that

$$B_i = M_i/\mathrm{O}(m) \cong F_i/\mathrm{O}(i).$$

Define

$$E_i = (F_i \cup F_{i-1})/\mathrm{SO}(i).$$

Let $E_0 = B_0$, $E_{k+1} = B_k$. The smooth manifold $E_i$ is the doubled branched cover of $B_i \cup B_{i-1}$ along $B_{i-1}$.

THEOREM 2.3. *Suppose that* $G = \mathrm{O}(m)$, $\phi = k\rho_m + l1$ *and $M$, $M'$ are $G$-manifolds having orthogonal model $\phi$. Let* $F$: $M \to M'$ *be a stratified map and, for each $i$, let* $f_i$: $E_i \to E'_i$ *be the induced map. Then,* $F_*$: $H_*(M; Z) \to H_*(M'; Z)$ *is an isomorphism if and only if for each $i$* $(0 \leqq i \leqq k+1)$ $(f_i)_*: H_*(E_i; Z) \to H_*(E'_i; Z)$ *is an isomorphism.*

This theorem is due to Davis. The proof will appear in **[6]**.

By Theorem 2.1 we have a degree one stratified map $F$: $M \to N$ from our differentiable $G$-manifold $M$ to its orthogonal model. Since $M$ is acyclic or a homology sphere, $F$ is a homology equivalence. Theorems 2.2 and 2.3 show that the induced map $f$: $M/G \to N/G$ must be a 'homology equivalence on each stratum' (in a modified sense when $G = \mathrm{O}(m)$, $\phi = k\rho_m + l1$). Thus, Theorems 2.2 and 2.3 make precise the notion of '$M$ being homologically modeled on $N$ via $F$'.

**III. Concordance.** Recall that two differentiable actions $\Psi_i: G \times M_i \to M_i$ $(i = 0, 1)$ on oriented manifolds are concordant if there is a differentiable action on $M \times I$ (oriented) so that the restriction to $M \times i$ is equivalent to $\Psi_i$. (In the case of O($m$) actions modeled on $\phi = k\rho_m + l1$ $(k \leq m)$, where $k$ is even, we also require an orientation for $M$ and $M^{O(1)}$ as part of the structure.) In this section, we discuss the classification of differentiable actions on homotopy spheres modeled on $\phi$ up to concordance (where $\phi$ is as in §I).

Both the arguments and the results for regular actions are different from those concerning actions of adjoint or near adjoint type. We, therefore, divide our discussion into two cases.

(A) *Regular actions.* ($G$ = O($m$), U($m$) or Sp($m$) and $\phi = k\rho_m + l1$.) We will use the notation $G(m)$ to denote either O($m$), U($m$) or Sp($m$). There are two typical examples of these actions.

(1) The unit sphere of the representation space of $\phi$.

(2) Let $\Sigma^{2s+1}$ be the Brieskorn sphere defined as the intersection of the unit sphere in $C^{2+s}$ with the hypersurface $f^{-1}(\varepsilon)$ where

$$f(u, v, z_1, \cdots, z_s) = u^p + v^q + z_1^2 + \cdots + z_s^2$$

and where $\varepsilon$ is a sufficiently small number. For suitably chosen $p$ and $q$, $\Sigma^{2s+1}$ will be a homotopy sphere (for $s > 1$). If $V^{2s+2}$ is the intersection of $f^{-1}(\varepsilon)$ with the unit disc in $C^{2+s}$, then $V^{2s+2}$ is a parallelizable manifold with boundary $\Sigma^{2s+1}$. O($s$) acts by operating on the last $s$ coordinates and the invariant submanifold $V^{2s+2}$ becomes an O($s$) manifold modeled on $2\rho_s + 21$. Write $s = dtm + (l - 2)$ where $d$ = 1, 2 or 4 as $G(m)$ = O($m$), U($m$) or Sp($m$) and considered the embedding $t\rho_m + (l - 2)1: G(m) \to$ O($s$). The restriction of the O($s$)-action to $G(m)$ gives $\Sigma^{2s+1}$ the structure of a regular $G(m)$-sphere (homologically) modeled on $\phi = k\rho_m + l1$ where $k = 2t$. These actions on Brieskorn spheres can often be distinguished from one another by the index or Kervaire invariant of the fixed point set of $V^{2s+2}$.

The fact that the regular $G(m)$ action on the Brieskorn sphere $\Sigma^{2s+1}$ extends to an action on a parallelizable manifold $V^{2s+2}$ is not accidental. In fact, we have the following theorem.

THEOREM 3.1. *Let $\Psi: G(m) \times M \to M$ be an action on a homotopy sphere $M$ which is modeled on $k\rho_m + l1$. Then, $M$ equivariantly bounds a parallelizable manifold $V$ and the map $F: M \to N$ of Theorem 2.1 extends to a stratified map $H$: $(V,M) \to (W, N)$ where $W$ is the unit disc in the representation space of $k\rho_m + l1$. and where $N = \partial W$ is the unit sphere.*

*Thus, we get a commutative diagram*

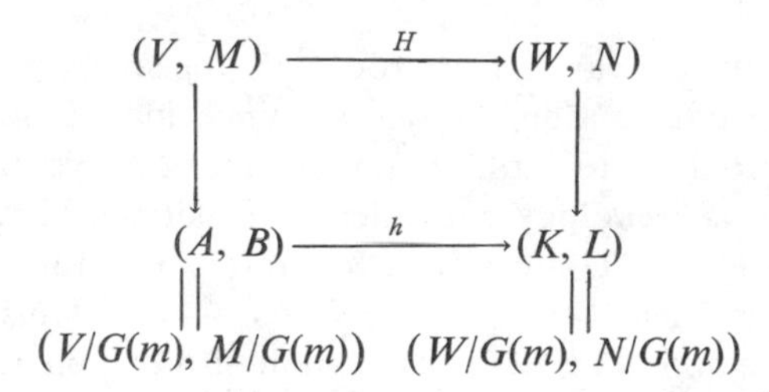


*as in Theorem 2.1.*

The problem of constructing a concordance of $M$ to the orthogonal action is equivalent to the problem of doing equivariant surgery on $V$ to a contractible manifold. For U($m$) or Sp($m$), in view of Theorem 2.2, this amounts to trying to do surgery rel $B$ on $h$ to a homology equivalence on each stratum. In the O($m$) case, this program must be modified as in Theorem 2.3.

Let $\Theta^d(k, m, l)$ denote the set of concordance classes of regular $dkm + (l - 1)$ modeled on $k\rho_m + l1$ (where $d = 1$ for O($m$), 2 for U($m$) and 4 for Sp($m$)). For $l > 0$, $\Theta^d(k, m, l)$ has an abelian group structure induced by taking equivariant connected sum at fixed points. Let $P_l = L_l(1)$. For $d = 2$ or 4, we have a homomorphism $\sigma_0 \colon \Theta^d(k, m, l) \to P_l$ by taking the surgery obstruction at the fixed point set of $V$ (i.e., the surgery obstruction of $h_0 \colon (A_0, B_0) \to (K_0, L_0)$). It turns out that for the U($m$) and Sp($m$) cases, $\sigma_0$ is the only obstruction to completing this surgery. Thus, we have the following theorem, which was proved in [7].

THEOREM 3.2. *Let $d = 2$ or 4. Assume $l \neq 4$ and if $l \leqq 3$ also assume that $k \geqq 3$ for* U($m$) *and $k \geqq 2$ for* Sp($m$). *If $k$ is odd or if $l = 0$, then $\Theta^d(k, m, l)$ is the trivial group. If $k$ is even and $l > 0$, then $\sigma_0$ is an isomorphism.*

In particular, this shows that for regular U($m$) or Sp($m$) actions (with the mild assumption of the above theorem), the above typical examples are the only possibilities which can occur up to concordance. Let us briefly indicate the idea of the proof. To prove this theorem we must compute the obstruction to doing surgery on $h \colon (A, B) \to (K, L)$ to a homology equivalence on each stratum (cf. [4]). Thus we must solve a sequence of surgery problems indexed by $\{0, 1, \cdots, k\}$. The definition of the problem at the $i$th stage depends on the solution at the stages $0, 1, \cdots, i - 1$. Suppose $r_i = Y_i \to K_i$ is the problem at stage $i$. The point is that part of $Y_i$ is a fiber bundle over the cobordism at stage $i - 1$. The fiber is $CP^{k-i}$ or $QP^{k-i}$. If $k - i$ is even, we can alter the cobordism at stage $i - 1$ (by adding a Milnor manifold or a Kervaire manifold) so that the surgery obstruction of $r_i$ is zero. On the other hand, if $k - i$ is odd, we note that there is a $CP^{k-i+1}$ bundle over $Y_i$ which bounds the $(i + 1)$-stratum. Since in this case $(k - i + 1)$ is even, it follows that the obstruction to doing surgery is automatically zero. The only time one of these arguments does not work is when $i = 0$ and $k$ is even in which case we meet a legitimate obstruction $\sigma_0$.

The corresponding theorem for regular O($m$) actions is much more complicated (though it is more interesting). This result is due to the first two authors and John Morgan. We refer the interested reader to [9] for details.

(B) *Actions of adjoint type and near adjoint type.* Corresponding to Theorem 3.2, we have the following theorem for actions of adjoint type and near adjoint type.

THEOREM 3.3. *Let $\Psi \colon G \times \Sigma^n \to \Sigma^n$ be an action of a compact simple Lie group $G$ (rk $G > 7$) of adjoint type or near adjoint type on a homotopy sphere $\Sigma^n$ and let $\phi$ be its orthogonal model. Then, $\Psi$ is concordant to $\phi$. In particular, $\Sigma^n = S^n$.*

This theorem is closely related to Theorem 1.2. It turns out that the stratification of the orbit space is 'more or less' homologically modeled on the join of a sphere with a simplex. From this, we can directly produce a concordance between $\Psi$ and $\phi$. The details of this theorem will appear with Theorems 2.1 and 2.2 elsewhere.

## References


**1.** G. Bredon, *Biaxial actions* (mimeographed notes).

**2.** ———, *Group generated by differentiable reflections* (mimeographed notes).

**3.** W. Browder, *Surgery and the theory of differentiable transformation groups* (Proc. Conf. on Transformation Groups, New Orleans, La., 1967), Springer, New York, 1968, pp. 1–46. MR **41** #6242.

**4.** W. Browder and F. Quinn, *A surgery theory for G-manifolds and stratified sets*, Manifolds—Tokyo 1973, pp. 27–36, Univ. of Tokyo Press, Tokyo, 1975. MR **51** #11543.

**5.** M. Davis, *Smooth G-manifolds as collections of fiber bundles* (to appear).

**6.** ———, *Regular* $O_n$, $U_n$, $Sp_n$ *manifolds* (to appear).

**7.** M. Davis and W. C. Hsiang, *Concordance classes of regular* $U_n$ *and* $Sp_n$ *actions on homotopy spheres*, Ann. of Math. (to appear).

**8.** M. Davis, W. C. Hsiang and W. Y. Hsiang, *Differentiable actions of adjoint type and near adjoint type* (to appear).

**9.** M. Davis, W. C. Hsiang and J. Morgan, *Concordance classes of regular* $O_n$ *actions on homotopy sphere* (to appear).

**10.** D. Erle and W. C. Hsiang, *On certain unitary and sympletic actions with three orbit types*, Amer. J. Math. **94** (1972), 289–308. MR **46** #4558.

**11.** W. C. Hsiang and W. Y. Hsiang, *Differentiable actions of compact connected Lie groups*. I, II, III, Amer. J. Math. **89** (1967), 705–786; Ann. of Math. **92** (1970), 189–223; ibid. **99** (1974), 220–256. MR **36** #304; **42** #420; **49** #11550.

**12.** ———, *The degree of symmetry of homotopy spheres*, Ann. of Math. (2) **89** (1969), 52–67. MR **39** #978.

**13.** W. Y. Hsiang, *On the geometric weight system of differentiable compact transformation groups on acyclic manifolds*, Invent. Math. **12** (1971), 35–47. MR **46** #916.

**14.** ———, *On the splitting principal and the geometric weight system of topological transformation groups*. I (Proc. 2nd Conf. on Compact Transformation Groups, Amherst, Mass.), Lecture Notes in Math., vol. 298, Springer-Verlag, Berlin and New York, 1972, pp. 334–402.

**15.** ———, *Cohomology theory of topological transformation groups*, Ergebnisse der Math. und ihrer Grenzgebiete, Band 85, Springer-Verlag, Berlin and New York, 1975.

**16.** ———, *On classification of differentiable* SO($n$) *actions on simply connected $\pi$-manifolds*, Amer. J. Math. **88** (1966), 137–153. MR **33** #1856.

**17.** ———, *On the bound of the dimensions of isometry groups of all possible riemannian metrics on an exotic sphere*, Ann. of Math. (2) **85** (1967), 351–358. MR **35** #4935.

**18.** K. Jänich, *Differenzierbare Mannigfatigkeiten mit Rand als Orbiträume differenzierbarer G-Mannigfaltigkeiten ohne Rand*, Topology **5** (1966), 301–320. MR **34** #2030.

**19.** L. Jones, *The converse to the fixed point theorem of P. A. Smith*. I, Ann. of Math. (2) **94** (1971), 52–68. MR **45** #4427.

**20.** J. Koszul, *Lecture on groups of transformations*, Tata Institute of Fundemental Research, Bombay, 1965. MR **36** #1571.

**21.** B. Lawson and S. T. Yau, *Scalar curvature, nonabelian group actions, and the degree of symmetry of exotic spheres*, Comment. Math. Helv. **49** (1974), 232–244. MR **50** #11300.

**22.** J. Levine, *Knot cobordism groups in codimension two*, Comment. Math. Helv. **44** (1969), 229–244. MR **39** #7618.

**23.** D. Montgomery, H. Samelson and C. T. Yang, *Exceptional orbits of highest dimension*, Ann. of Math. (2) **64** (1956), 131–141. MR **17**, 1224.

**24.** G. D. Mostow, *Equivariant embedding in Euclidean space*, Ann. of Math. (2) **65** (1957), 432–446. MR **19**, 291.

**25.** R. Palaìs, *Embedding of compact differentiable transformation groups in orthogonal representations*, J. Math. Mech. **6** (1957), 673–678. MR **19**, 1181.

**26.** R. Schultz, *Differentiable group actions on homotopy spheres*. I, Invent. Math. **31** (1975), 105–128.

**27.** C. T. C. Wall, *Surgery on compact manifolds*, Academic Press, New York, 1970.

**28.** J. Wolf, *Space of constant curvature*, McGraw-Hill, New York, 1967.

**29.** P. Yang, *Adjoint type representations of classical groups on homology spheres*, Thesis, Princeton Univ., Princeton, N.J., 1974.

INSTITUTE FOR ADVANCED STUDY

PRINCETON UNIVERSITY

UNIVERSITY OF CALIFORNIA, BERKELEY

Proceedings of Symposia in Pure Mathematics
Volume 32, 1978

# ON THE RATIONAL HOMOTOPY GROUPS OF THE DIFFEOMORPHISM GROUPS OF DISCS, SPHERES AND ASPHERICAL MANIFOLDS

F. T. FARRELL* AND W. C. HSIANG**

**0. Introduction.** Recently, topologists became very interested in computing $\pi_i \mathrm{Diff}(M^n)$ [26], [27], [24], [10], [9]. In this note, we shall compute

$$\pi_i \mathrm{Diff}(M^n) \otimes Q \qquad (i \ll n)$$

when $M^n$ is $(D^n, \partial)$, a sphere or an aspherical manifold of certain type. (Here, $D^n$ denotes the $n$-disc.)

Roughly speaking, we have the following result where $i \ll n$

$$\text{(A)} \qquad \pi_i \mathrm{Diff}(D^n, \partial) \otimes Q = \begin{cases} Q & \text{for } i = 4k-1 \text{ and } n \text{ odd}, \\ 0 & \text{otherwise.} \end{cases}$$

$$\text{(B)} \qquad \pi_i \mathrm{Diff}(S^n) \otimes Q = \begin{cases} 0 & \text{for } i \neq 4k-1, \\ Q & \text{for } i = 4k-1 \text{ and } n \text{ even}, \\ Q \oplus Q & \text{for } i = 4k-1 \text{ and } n \text{ odd.} \end{cases}$$

(C) *If $M^n$ is an orientable closed aspherical manifold satisfying Conjectures* 1 *and* 2 *stated below, then*

$$\pi_i \mathrm{Diff}(M^n) \otimes Q = \begin{cases} \mathit{center}(\pi_1 M^n) \otimes Q & \mathit{for}\ i = 1, \\ \bigoplus_{j=1}^{\infty} H_{(i+1)-4j}(M^n, Q) & \mathit{for}\ i > 1 \ \mathit{and}\ n\ \mathit{odd}, \\ 0 & \mathit{for}\ i > 1\ \mathit{and}\ n\ \mathit{even}. \end{cases}$$

*In particular,* (C) *holds for solvmanifolds and flat Riemannian manifolds* (*closed orientable*).

---

*AMS* (*MOS*) *subject classifications* (1970). Primary 55A05, 57D50.

*Supported by grants from the National Science Foundation.
**Partially supported by National Science Foundation grant GP34324X1.

CONJECTURE 1. *The surgery map*

$$\theta: [M^n \times D^i, \partial; G/\mathrm{Top}, *] \otimes \boldsymbol{Q} \to L_{n+i}(\pi_1 M^n) \otimes \boldsymbol{Q}$$

*is an isomorphism.*

This is a stronger version of Novikov's conjecture. We also need its algebraic $K$-theoretic analogue. Let $\underline{\underline{K}}_{\boldsymbol{Z}}$ be the algebraic $K$-theory spectrum for the ring $\boldsymbol{Z}$ and $h_*(\ ;\underline{\underline{K}}_{\boldsymbol{Z}})$ the generalized homology theory with respect to $\underline{\underline{K}}_{\boldsymbol{Z}}$.

CONJECTURE 2. *The map defined in* **[14]**

$$\lambda_* \otimes \mathrm{id}: h_*(M_u; \underline{\underline{K}}_{\boldsymbol{Z}}) \otimes \boldsymbol{Q} \to K_*(\boldsymbol{Z}[\pi_1 M^n]) \otimes \boldsymbol{Q}$$

*is an isomorphism.*

(In both conjectures, $M^n$ is a closed orientable aspherical manifold.)

These statements are surprisingly strong, although their proof is embarrassingly easy. Essentially, they are consequences of **[24]**, **[1]**, **[10]**, **[9]**. The main thrust comes from the recent work of Waldhausen **[24]**. It relates algebraic $K$-theory to pseudo-isotopy theory very nicely. In particular, it translates the rational homotopy groups of pseudoisotopies of a disc to a problem of computing the homology of the arithmetic group $\mathrm{GL}_n(\boldsymbol{Z})$ with coefficients in $M_n(\boldsymbol{Q})$ with respect to the adjoint representation. It follows directly from **[20]**, **[1]**, **[2]**, **[6]** that $H_i(\mathrm{GL}_n(\boldsymbol{Z}); M_n(\boldsymbol{Q})) \to^{\mathrm{tr}} H_i(\mathrm{GL}_n(\boldsymbol{Z}), \boldsymbol{Q})$ is an isomorphism for $i \ll n$ under the trace map $\mathrm{tr}: M_n(\boldsymbol{Q}) \to \boldsymbol{Q}$. In §I, we almost paraphrase the background material from **[20]** for the reader's convenience and quote the necessary results from **[1]**, **[2]**, **[6]** to prove the above isomorphism. From this, we derive (A), (B) in §II from **[10]**, **[9]** and (C) in §IV assuming the conjectures are verified. In §III, we verify Conjecture 2 for flat Riemannian manifolds following the method of **[5]** where Conjecture 1 was verified for these manifolds. Conjectures 1 and 2 for solvmanifolds were verified in **[25]** and **[19]** respectively.

**I. A vanishing theorem for** $H^*(\mathrm{SL}_n(\boldsymbol{Z}); \rho)$. In this section, we shall prove the following vanishing theorem which is a consequence of Y. Matsushima and S. Murakami **[16]**, M. S. Raghunathan **[20]**, A. Borel **[1]**, **[2]** and H. Garland and W. C. Hsiang **[6]**. This theorem will be used in the next section.

THEOREM 1.1. *Let* $\rho$ *be a nontrivial irreducible representation of* $\mathrm{SL}_n(\boldsymbol{R})$; *then* $H^p(\mathrm{SL}_n(\boldsymbol{Z}); \rho)$ *vanishes for* $p \leqq (n-1)/4$, $5 \leqq n$.

*In particular,* $H^p(\mathrm{SL}_n(\boldsymbol{Z}); \mathrm{Ad}) = 0$ *for* $p \leqq (n-1)/4$, $5 \leqq n$.

First, we recall the results of **[16]**, **[20]**. Consider the symmetric space $X = K\backslash G$ where $K = SO_n(\boldsymbol{R})$ and $G = \mathrm{SL}_n(\boldsymbol{R})$. Let $\rho$ be a finite-dimensional representation of $G$ on a complex vector space $F$, and $\Gamma$ be a congruence subgroup of $\mathrm{SL}_n(\boldsymbol{Z})$ containing no element of finite order except the identity. For example, take $\Gamma$ to be the congruence subgroup of level 3, i.e.,

$$\Gamma = \mathrm{Ker}(\mathrm{SL}_n(\boldsymbol{Z}) \to \mathrm{SL}_n(\boldsymbol{Z}_3)).$$

So, $\Gamma$ acts freely on $X$ (on the right). The representation $\rho$ defines a locally constant sheaf $L(\rho)$ and a vector bundle $E(\rho)$ on $X/\Gamma$. Since $X$ is homeomorphic to a Euclidean space,

$$H^p(\Gamma; \rho) = H^p(X/\Gamma; L(\rho)). \tag{1.1}$$

Now, let $A(\Gamma, X, \rho)$ denote the complex whose $p$th graded component $A^p(\Gamma, X, \rho)$ is the space of all $F$-valued $C^\infty$ $p$-forms $\eta$ on $X$ satisfying

$$\eta \circ R_\gamma = \rho(\gamma^{-1})\eta \quad \text{for } \gamma \in \Gamma \tag{1.2}$$

(where $R_\gamma$ denotes the action of $\gamma \in \Gamma$ on $X$), and whose coboundary operator $d$ is exterior differentiation. Let $H^p(\Gamma, X, \rho)$ denote the cohomology group of this complex. It follows from the de Rham theorem that

$$H^p(X/\Gamma; L(\rho)) \simeq H^p(\Gamma, X, \rho). \tag{1.3}$$

We now introduce a Hermitian metric along the fibers of $E(\rho)$. Let $\mathfrak{g}$ be the Lie algebra of right invariant vector fields on $G$ and $\underline{k}$ the subalgebra corresponding to $K$. Let $\mathfrak{g} = \underline{k} \oplus \underline{p}$ be the Cartan decomposition. Now, let $\langle\ ,\ \rangle$ denote a Hermitian positive definite form on $F$ such that, with respect to this form $\rho(v)$ is unitary for all $v \in K$, and $\rho(Y)$ is Hermitian symmetric for all $Y \in \underline{p}$. (We also denote by $\rho$ the induced representation of $\mathfrak{g}$.) Recall $E(\rho)$ is obtained as the quotient space of $X \times F$ by the diagonal action of $\Gamma$

$$(x, f)\gamma = (x\gamma, \rho(\gamma^{-1})f). \tag{1.4}$$

For $x$, $y = x\gamma \in X$ ($\gamma \in \Gamma$) and $\tilde{x}$, $\tilde{y} = \tilde{x}\gamma \in G$ the elements lying over $x$ and $y$, we define

$$\langle(x, f), (y, g)\rangle = \langle\rho(\tilde{x})f, \rho(\tilde{y})g\rangle \tag{1.5}$$

for $f, g \in F$. It defines a metric along the fibers of $E(\rho)$. On the other hand, we have a natural metric on $X$ given by the Killing form with the obvious change. Choose a scalar multiple such that, if $\{X_\alpha\}_{N+1 \leq \alpha \leq m}$ and $\{X_i\}_{1 \leq i \leq N}$ are orthonormal bases of $\underline{k}$ and $\underline{p}$ respectively with respect to the negative of the restriction, and the restriction of this form to $\underline{k}$ and $\underline{p}$ respectively, then the volume form on $\underline{k}$ defined by $X_{N+1} \wedge \cdots \wedge X_m$ is the normalized Haar measure. (The projections of the $\{X_i\}_{1 \leq i \leq N}$ at identity of course are taken to be the orthonormal basis of the metric on $X$.)

The metric on $X/\Gamma$ deduced from that on $X$, and the metric along the fibers of $E(\rho)$, together enable one to define a scalar product on the subspace of $A^p(\Gamma, X, \rho)$ of $E(\rho)$-valued forms on $X/\Gamma$ with compact support. A simple explicit formula is given as follows.

Let $p\colon G \to X$ denote the natural projection, and $X_1, \cdots, X_N$ a basis of $\underline{p}$ (considered as right invariant vector fields) orthonormal with respect to the above chosen scalar multiple of the Killing form. Then, for $\xi, \eta \in A^p(\Gamma, X, \rho)$ with compact support, the scalar product $(\xi, \eta)$ is given by

$$(\xi, \eta) = \frac{1}{p!}\int_{G/\Gamma} \sum_{i_1, \cdots, i_p=1}^{N} \langle\rho(g)\xi_{i_1\cdots i_p}(g), \rho(g)\eta_{i_1\cdots i_p}(g)\rangle \tag{1.6}$$

where $\xi_{i_1\cdots i_p}$ and $\eta_{i_1\cdots i_p}$ are the functions on $G$ defined by $\xi_{i_1\cdots i_p}(x) = p^*\xi(X_{i_1}(x), \cdots, X_{i_p}(x))$ and $\eta_{i_1\cdots i_p}(x) = p^*\eta(X_{i_1}(x), \cdots, X_{i_p}(x))$, $x \in G$.

One can define an adjoint differential operator

$$\delta\colon A^p(\Gamma, X, \rho) \to A^{p-1}(\Gamma, X, \rho) \tag{1.7}$$

with respect to the above scalar product, i.e., $\delta$ satisfies

$$(1.8) \qquad (\delta\xi, \eta) = (\xi, d\eta),$$

where $\xi \in A^{p+1}(\Gamma, X, \rho)$ and $\eta \in A^p(\Gamma, X, \rho)$ have compact support. The Laplacian

$$(1.9) \qquad \Delta = d\delta + \delta d\colon A^p(\Gamma, X, \rho) \to A^p(\Gamma, X, \rho)$$

is an elliptic operator of order 2.

Lifting to $G/\Gamma$, we have an isomorphism $\Phi$ of $A^p(\Gamma, X, \rho)$ to the complex $A_p^0(\Gamma, X, \rho)$ of $F$-valued $C^\infty$ $p$-forms on $G/\Gamma$ satisfying

$$(1.10) \qquad \begin{aligned} \eta \circ L_n &= \rho(v)\eta \quad \text{for } v \in K, \\ i(Y)(\eta) &= 0 \qquad\ \text{for } Y \in \underline{K}, \end{aligned}$$

provided with the differential operator via the isomorphism $\Phi$. (We again identify right invariant vector fields on $G$ with their projections on $G/\Gamma$.) Note that elements of $A^p(\Gamma, X, \rho)$ with compact support correspond again to forms with compact support, and for two such forms ( , ) denotes again their scalar product obtained from that on $A^p(\Gamma, X, \rho)$ via $\Phi$. With this notation, we have

$$(1.11) \qquad (\xi, \eta) = \frac{1}{p!}\int_{G/\Gamma} \sum_{i_1, \cdots, i_p=1}^{N} \langle \xi_{i_1\cdots i_p}, \eta_{i_1\cdots i_p} \rangle$$

for $\xi, \eta \in A_0^p(\Gamma, X, \rho)$ with compact support.

If we denote again by $\Delta$ the operator on $A_0^p(\Gamma, X, \rho)$ obtained through $\Phi$ from $\Delta$ on $A^p(\Gamma, X, \rho)$ then $\Delta = \Delta_D + \Delta_\rho$ where

$$(1.12) \qquad \Delta_D\colon A_0^p(\Gamma, X, \rho) \to A_0^p(\Gamma, X, \rho), \qquad \Delta_\rho\colon A_0^p(\Gamma, X, \rho) \to A_0^p(\Gamma, X, \rho)$$

are operators defined by the following formulae. For $\eta \in A_0^p(\Gamma, X, \rho)$, let $\eta_{i_1\cdots i_p} = \eta(X_{i_1}, \cdots, X_{i_p})$. (Recall that $X_1, \cdots, X_N$ is a basis of $\underline{p}$.) Then,

$$(1.13) \qquad \begin{aligned} (\Delta_D\eta)_{i_1\cdots i_p} = &- \sum_{k=1}^{N} X_k^2 \eta_{i_1\cdots i_p} \\ &+ \sum_{k=1}^{N}\sum_{u=1}^{p} (-1)^{u-1}(-[X_{i_u}, X_k]\, \eta_{ki_u\cdots \hat{i}_1\cdots i_p}) \end{aligned}$$

and

$$(1.14) \qquad \begin{aligned} (\Delta_\rho\eta)_{i_1\cdots i_p} = &\sum_{k=1}^{N} \rho(X_k)^2 \eta_{i_1\cdots i_p} \\ &+ \sum_{k=1}^{N}\sum_{u=1}^{p} (-1)^{u-1} \rho([X_{i_u}, X_k])\eta_{ki_1\cdots \hat{i}_u\cdots i_p}. \end{aligned}$$

Moreover, if $\eta$ has compact support,

$$(1.15) \qquad (\Delta_D\eta_{i_1\cdots i_p}, \eta_{i_1\cdots i_p}) \geqq 0.$$

We say that $E(\rho)$ is $W^p$-elliptic if there is a constant $c > 0$ such that for any $\eta \in A_0^p(\Gamma, X, \rho)$ with compact support, we have

$$(1.16) \qquad (\Delta\eta, \eta) \geqq c(\eta, \eta).$$

For the orthonormal basis elements $X_{i_1}, \cdots, X_{i_p}$ of $\underline{p}$ and $\beta \in \mathrm{Hom}_{\boldsymbol{R}}(\Lambda^p\underline{p}, F)$, if the operator

$$(1.17) \qquad \Delta_\rho\colon \mathrm{Hom}_{\boldsymbol{R}}(\Lambda^p\underline{p}, F) \to \mathrm{Hom}_{\boldsymbol{R}}(\Lambda^p\underline{p}, F)$$

defined by

$$(1.18)\quad \begin{aligned}(\Delta_\rho\beta)(X_{i_1}, \cdots, X_{i_p}) &= \sum_{k=1}^{N} \rho(X_k)^2\beta(X_{i_1}, \cdots, X_{i_p}) \\ &+ \sum_{k=1}^{N}\sum_{u=1}^{p}(-1)^{u-1}\rho([X_{i_u}, X_k])\,\beta(X_k, X_{i_1}, \cdots, \hat{X}_{i_u}, \cdots, X_{i_p})\end{aligned}$$

is such that the Hermitian quadratic form

$$(1.19)\quad \frac{1}{p!}\sum_{i_1, \cdots, i_p=1}^{N} \langle \Delta_\rho\beta(X_{i_1} \cdots X_{i_p}), \beta(X_{i_1} \cdots X_{i_p})\rangle$$

is positive definite, then $E(\rho)$ is $W^p$-elliptic by (1.12), (1.13), (1.14) and (1.15).

LEMMA 1.2. *If $\rho$ is a nontrivial irreducible (complex) representation of* $\mathrm{SL}_n(\boldsymbol{R})$, *then $E(\rho)$ is $W^p$-elliptic for $p \leqq n - 1$.*

This is a special case of Proposition 2.2 of [**2**].

We say that $\xi \in A_0^p(\Gamma, X, \rho)$ is square-integrable if

$$(1.20)\quad \frac{1}{p!}\int_{G/\Gamma}\sum_{i_1, \cdots, i_p=1}^{N}\langle \xi_{i_1\cdots i_p}, \xi_{i_1\cdots i_p}\rangle < \infty.$$

LEMMA 1.3. *If $E(\rho)$ is $W^p$-elliptic, then any closed form in $A_0^p(\Gamma, X, \rho)$ which is square-integrable is a coboundary.*

This is Lemma 5 of [**20**].

LEMMA 1.4. *Let $\rho$ be a (complex or real) representation of* $\mathrm{SL}_n(\boldsymbol{R})$. *Then every cocycle of $A_0^p(\Gamma, X, \rho)$ is cohomologous to a square-integrable form in $A_0^p(\Gamma, X, \rho)$ for $p \leqq (n-1)/4$.*

This is a special case of Theorem 7.5 of [**1**] and Theorem 3.1 of [**6**].

It is clear that Theorem 1.1 follows immediately from the above three lemmas.

**II. Diffeomorphisms of discs and spheres.** In this section, we compute some rational homotopy groups of the diffeomorphism groups of discs and spheres.

THEOREM 2.1. *Let $D^n$ be the n-disc, $\Sigma^n$ an arbitrary n-dimensional homotopy sphere and $0 < i < n/6 - 7$; then*

$$\text{(A)}\quad \pi_i\mathrm{Diff}(D^n, \partial) \otimes \boldsymbol{Q} = \begin{cases} \boldsymbol{Q}, & n \text{ odd and } i = 4k-1, \\ 0, & \text{otherwise.}\end{cases}$$

$$\text{(B)}\quad \pi_i\mathrm{Diff}(\Sigma^n) \otimes \boldsymbol{Q} = \begin{cases} 0, & i \neq 4k-1, \\ \boldsymbol{Q}, & n \text{ even and } i = 4k-1, \\ \boldsymbol{Q}\oplus\boldsymbol{Q}, & n \text{ odd and } i = 4k-1.\end{cases}$$

The main thrust of the proof of the above theorem is Waldhausen's remarkable work on concordance theory [**24**]. He reduces the computation of the rational homotopy groups of the differentiable concordance space $\mathscr{P}(D^n)$ of $D^n$ to a question on the group cohomology of $\mathrm{GL}_n(\boldsymbol{Z})$ with coefficients in $M_n(\boldsymbol{Q})$ under the adjoint action. Let us first recall his result. Let $M_n(\boldsymbol{Q})$ be the vector space over $\boldsymbol{Q}$ of $(n \times n)$-matrices and let $\mathrm{GL}_n(\boldsymbol{Z})$ act on $M_n(\boldsymbol{Q})$ by the adjoint action, i.e.,

$$(2.1)\quad \mathrm{Ad}(g)(x) = gxg^{-1}$$

for $g \in \mathrm{GL}_n(\boldsymbol{Z})$ and $x \in M_n(\boldsymbol{Q})$. Let $\mathrm{GL}_n(\boldsymbol{Z})$ act trivially on $\boldsymbol{Q}$. There is an equivariant map

$$\text{(2.2)} \qquad \mathrm{tr}\colon M_n(\boldsymbol{Q}) \to \boldsymbol{Q}$$

by sending $x$ to its trace. So, we have a homomorphism

$$\text{(2.3)} \qquad \mathrm{tr}_*\colon H_*(\mathrm{GL}_n(\boldsymbol{Z}); M_n(\boldsymbol{Q})) \to H_*(\mathrm{GL}_n(\boldsymbol{Z}); \boldsymbol{Q}).$$

Let us stabilize $M_n(\boldsymbol{Q})$ by sending $x \in M_n(\boldsymbol{Q})$ to

$$\begin{pmatrix} x & 0 \\ 0 & 0 \end{pmatrix} \in M_{n+1}(\boldsymbol{Q}).$$

So, we have a limit homomorphism

$$\text{(2.4)} \qquad \mathrm{tr}_*\colon \lim_{n\to\infty} H_*(\mathrm{GL}_n(\boldsymbol{Z}); M_n(\boldsymbol{Q})) \to \lim_{n\to\infty} H_*(\mathrm{GL}_n(\boldsymbol{Z}); \boldsymbol{Q}).$$

THEOREM 2.2 [24]. *Provided $i < n/6 - 7$,*

$$\pi_i \mathscr{P}(D^n) \otimes \boldsymbol{Q} \simeq K_{i+2}(\boldsymbol{Z}) \otimes \boldsymbol{Q}$$

*if and only if the homomorphism* $\mathrm{tr}_*$ *of* (2.4) *is an isomorphism.*

LEMMA 2.3. *The homomorphism* $\mathrm{tr}_*$ *of* (2.4) *is an isomorphism. Hence,*

$$\pi_i \mathscr{P}(D^n) \otimes \boldsymbol{Q} \simeq K_{i+2}(\boldsymbol{Z}) \otimes \boldsymbol{Q} \qquad (i < n/6 - 7).$$

PROOF. (A) First observe $\mathrm{tr}_*$ of (2.4) is an isomorphism if and only if the corresponding homomorphism on cohomology

$$\text{(2.5)} \qquad \mathrm{tr}^*\colon \lim_{n\to\infty} H^*(\mathrm{GL}_n(\boldsymbol{Z}); M_n(\boldsymbol{Q})) \to \lim_{n\to\infty} H^*(\mathrm{GL}_n(\boldsymbol{Z}); \boldsymbol{Q})$$

is an isomorphism. Moreover, extending the action to $M_n(\boldsymbol{R})$ in the obvious way, it suffices to prove that

$$\text{(2.6)} \qquad \mathrm{tr}^*\colon \lim_{n\to\infty} H^*(\mathrm{GL}_n(\boldsymbol{Z}); M_n(\boldsymbol{R})) \to \lim_{n\to\infty} H^*(\mathrm{GL}_n(\boldsymbol{Z}); \boldsymbol{R})$$

is an isomorphism. These facts are exercises in homological algebra together with the observation that $\mathrm{Hom}_{\boldsymbol{Q}}(M_n(\boldsymbol{Q}), \boldsymbol{Q})$ and $M_n(\boldsymbol{Q})$ are isomorphic $\mathrm{GL}_n(\boldsymbol{Z})$-modules.

(B) Consider the short exact sequence of $\mathrm{GL}_n(\boldsymbol{Z})$-modules

$$\text{(2.7)} \qquad 0 \longrightarrow \underline{\mathfrak{g}} \longrightarrow M_n(\boldsymbol{R}) \xrightarrow{\mathrm{tr}} \boldsymbol{R} \longrightarrow 0$$

where $\underline{\mathfrak{g}}$ are the matrices of trace zero. Since (2.7) splits, Lemma 2.3 reduces to showing $H^p(\mathrm{SL}_n(\boldsymbol{Z})\colon \underline{\mathfrak{g}})$ vanishes for all $n$ sufficiently large (depending on $p$). This follows from Theorem 1.1 since $\underline{\mathfrak{g}}$ is the adjoint representation of $\mathrm{SL}_n(\boldsymbol{R})$ restricted to $\mathrm{SL}_n(\boldsymbol{Z})$.

Let $x \mapsto \bar{x}$ denote conjugation on $K_i\boldsymbol{Z}$ and also the induced maps on $K_i(\boldsymbol{Z}) \otimes \boldsymbol{Q}$ and $K_i(\boldsymbol{Z}) \otimes \boldsymbol{R}$.

LEMMA 2.4. *For all $x \in K_i(\boldsymbol{Z}) \otimes \boldsymbol{Q}$, we have*

$$\bar{x} = -x \quad \textit{when } i > 0,$$
$$\bar{x} = x \quad \textit{when } i = 0.$$

PROOF. For $i = 0$, the argument is easy and hence omitted. For $i > 0$, it suffices to demonstrate the corresponding statement for all $x \in K_i(\mathbf{Z}) \otimes \mathbf{R}$. But [1], $K_i(\mathbf{Z}) \otimes \mathbf{R}$ can be identified with a subspace of $H_i(\mathrm{SL}_n(\mathbf{Z}), \mathbf{R})$ for $n$ sufficiently large relative to $i$. Also by Borel [1], $H_i(\mathrm{SL}_n(\mathbf{Z}), \mathbf{R})$ can be equated with the relative Lie algebra cohomology $H_i(\underline{\mathfrak{g}}, \underline{k}; \mathbf{R})$ where $\underline{\mathfrak{g}}$ and $\underline{k}$ are the Lie algebras of $\mathrm{SL}_n(\mathbf{R})$ and $SO_n$, respectively. (Again $n$ is large relative to $i$.) Recall $\underline{\mathfrak{g}}$ consists of the matrices in $M_n(\mathbf{R})$ with trace zero and $\underline{k}$ those which are skew-symmetric. Under these identifications, $x \mapsto \bar{x}$ is induced by the Lie algebra automorphism of $\underline{\mathfrak{g}}$: $A \mapsto -A^*$ where $A^*$ denotes $A$ transposed. But, the effect of this automorphism on $H_i(\underline{\mathfrak{g}}, \underline{k}; \mathbf{R})$ is easily calculated; it induces multiplication by $(-1)^i$. Since $K_i(\mathbf{Z}) \otimes \mathbf{R}$ vanishes when $i$ is even ($i > 0$), this proves Lemma 2.4.

We next study the conjugation $x \mapsto \bar{x}$ induced on $K_i(\Lambda)$ for an arbitrary ring $\Lambda$ equipped with an anti-involution —. First define an involution on $\mathrm{GL}_n(\Lambda)$ by sending $A \in \mathrm{GL}_n(\Lambda)$ to $(\bar{A}^*)^{-1}$ where $\bar{A}^*$ is the conjugate transpose of $A$, i.e., $(\bar{A}^*)_{ij} = \bar{A}_{ji}$. Let $\underline{K}_\Lambda$ be the algebraic $K$-theory spectrum for the ring $\Lambda$; then $A \mapsto (\bar{A}^*)^{-1}$ induces an involution

$$—: \underline{K}_\Lambda \to \underline{K}_\Lambda \tag{2.8}$$

at the spectrum level and hence the desired conjugation on $K_i(\Lambda)$. When $\Lambda = \mathbf{Z}[\pi]$, we put the anti-involution on $\Lambda$ given by $(\sum n(g)g)^- = \sum w(g)n(g)g^{-1}$ where $w: \pi \to \{\pm 1\}$ is a homomorphism. If $\pi = \pi_1 M^n$, $w$ is taken to be the first Stiefel-Whitney class of $M^n$.

For any spectrum $\underline{E}$, let $h_*(\ ;\underline{E})$ be the generalized homology theory with respect to $\underline{E}$. In particular, Loday [14] constructs a homomorphism

$$\lambda_*: h_*(B\Gamma; \underline{K}_{\mathbf{Z}}) \to K_*(\mathbf{Z}\Gamma) \tag{2.9}$$

for any group $\Gamma$. Let $\underline{S}$ denote the sphere spectrum; since $K_0(\mathbf{Z}) = \mathbf{Z}$, we can define a map of spectrums $\varphi: \underline{S} \to \underline{K}_{\mathbf{Z}}$ so that the induced map $h_0(*; \underline{S}) \to h_0(*; \underline{K}_{\mathbf{Z}})$ is an isomorphism. Since $h_*(\ ; \underline{S}) \otimes \mathbf{Q}$ is ordinary homology theory $H_*(\ ; \mathbf{Q})$, we can define a map

$$\varphi_*: H_*(\ ; \mathbf{Q}) \to h_*(\ ; \underline{K}_{\mathbf{Z}}) \otimes \mathbf{Q}. \tag{2.9.1}$$

For any group $\pi$, we use $\bar{K}_n(\pi)$ to denote the cokernel of the composite

$$H_n(B\pi, \mathbf{Q}) \xrightarrow{\varphi_n} h_n(B\pi; \underline{K}_{\mathbf{Z}}) \otimes \mathbf{Q} \xrightarrow{\lambda_n \otimes \mathrm{id}} K_n(\mathbf{Z}\pi) \otimes \mathbf{Q}. \tag{2.9.2}$$

(Note $\bar{K}$ is *not* ordinary reduced $K$-theory.) The conjugation on $K_i(\mathbf{Z}\pi)$ induces one on $\bar{K}_i(\pi)$. Let $\bar{K}_i^{\pm}(\pi) = \{x \mid \bar{x} = \pm x \text{ for } x \in \bar{K}_i(\pi)\}$; then $\bar{K}_i(\pi) = \bar{K}_i^+(\pi) \oplus \bar{K}_i^-(\pi)$.

Let $\mathscr{B}(M)$ be the space introduced in [10]. The following lemma is from [9].

LEMMA 2.5. *Let $M^n$ be a compact orientable manifold such that $\pi_i(M^n) = 0$ for $1 < i < n$. Let $0 < j < n/6 - 7$; then*

$$\pi_j \mathscr{B}(M^n) \otimes \mathbf{Q} = \begin{cases} \bar{K}_{j+2}^-(\pi) & \text{for } n \text{ odd}, \\ \bar{K}_{j+2}^+(\pi) & \text{for } n \text{ even}. \end{cases}$$

(The proof of this lemma depends on Lemma 2.3.)

PROOF OF THEOREM 2.1. Let $M^n$ be either $(D^n, \partial)$ or $\Sigma^n$ and apply Lemma 2.5 together with Lemma 2.4. Since $\pi$ is trivial, we obtain, for $0 < i < n/6 - 7$,

$$\pi_i\mathscr{B}(M^n) \otimes Q = \begin{cases} Q & \text{for } n \text{ odd and } i = 4k - 1, \\ 0 & \text{otherwise.} \end{cases} \tag{2.10}$$

Next, consider the fibration (1) from [10, p. 401]

$$\mathfrak{g}(M) \to \mathrm{Diff}(M) \to \mathrm{Aut}(M). \tag{2.11}$$

Since $\pi_i$ $\mathrm{Aut}(M^n) \otimes Q = 0$ for $i < n$, we have

$$\pi_i \mathrm{Diff}(M^n) \otimes Q = \pi_i\mathfrak{g}(M^n) \otimes Q \tag{2.12}$$

for $i < n$.

Let us first discuss case (A), i.e., assume $M^n = (D^n, \partial)$. Consider the braid diagram (7) from [10, p. 402]; in particular, look at the exact sequence

$$\to \pi_{i+1}\mathscr{S}(M \times (S^1, 1)) \to \pi_i\mathscr{B}(M) \to \pi_i\mathfrak{g}(M) \to \pi_i\mathscr{S}(M \times (S^1, 1)) \to . \tag{2.13}$$

In this case, $\pi_i\mathscr{S}(M \times (S^1, 1)) \otimes Q = 0$ for $i \geqq 0$; therefore,

$$\pi_i\mathscr{B}(M) \otimes Q \simeq \pi_i\mathfrak{g}(M) \otimes Q. \tag{2.14}$$

Concatenating formulas (2.10), (2.12) and (2.14) proves case (A).

Now consider case (B), i.e., $M^n = \Sigma^n$. Looking at another exact sequence in the braid diagram (7) [10, p. 402], we obtain

$$0 \to \pi_i\mathscr{S}(M^n \times (S^1, 1)) \otimes Q \to [\Sigma^{i+1}M^+, G/0] \otimes Q \to L_{n+i+1}(1) \otimes Q \to 0. \tag{2.15}$$

Since $\Sigma^{i+1}M^+$ is homotopically equivalent to $S^{i+1} \vee S^{n+i+1}$, (2.15) yields

$$\pi_i\mathscr{S}(M^n \times (S^1, 1)) \otimes Q = \pi_{i+1}G/0 \otimes Q = \begin{cases} Q, & i = 4k - 1, \\ 0, & \text{otherwise.} \end{cases} \tag{2.16}$$

Since (2.13) is still valid, the proof of case (B) is now an easy calculation using formulas (2.10), (2.12), (2.13) and (2.16).

**III. Rational $K$-theory of Bieberbach groups.** In [5], we calculated $L_n(\Gamma) \otimes Q$ for Bieberbach groups $\Gamma$, i.e., fundamental groups of flat Riemannian manifolds. In this section, we calculate $K_n(Z\Gamma) \otimes Q$ for the same class of groups. We only sketch the necessary arguments since they are analogous to those given in detail for $L_n(\Gamma) \otimes Q$ in [5] and easier. As in §II, let $\underline{\underline{K}}_Z$ be the algebraic $K$-theory spectrum for the ring $Z$, $h_*(\ ;\underline{\underline{K}}_Z)$ the generalized homology theory with respect to $\underline{\underline{K}}_Z$, and recall Loday's homomorphism

$$\lambda_n\colon h_n(B\Gamma; \underline{\underline{K}}_Z) \to K_n(Z\Gamma). \tag{3.1}$$

THEOREM 3.1. *If $\Gamma$ is a Bieberbach group, then $\lambda_n$ is rationally an isomorphism, i.e.,*

$$\lambda_n \otimes \mathrm{id}\colon h_n(B\Gamma; \underline{\underline{K}}_Z) \otimes Q \to K_n(Z\Gamma) \otimes Q$$

*is an isomorphism.*

(An application of this result will be given in §IV.)

To prove this theorem, we need an action of Swan's ring on $K$-theory. Let $F$ be a finite group; recall Swan's ring $G_0(ZF)$ is obtained via a Grothendieck construction applied to the category Rep $F$. An object of Rep $F$, called an $F$-representation, is a left $F$-module which is finitely generated and free as an abelian group. (See Swan [22] for a detailed discussion of $G_0(\mathbf{Z}F)$.) When $F$ is the quotient of a group $\Gamma$, we have a pairing of categories.

$$\text{Rep } F \times \mathscr{P}(\Gamma) \to \mathscr{P}(\Gamma) \tag{3.2}$$

where $\mathscr{P}(\Gamma)$ is the category of finitely generated projective $\mathbf{Z}\Gamma$-modules. This pairing associates to an $F$-representation $V$ and a $\Gamma$-projective module $P$ the $\Gamma$-projective module $V \otimes P$. (Tensor product is over $\mathbf{Z}$.) The $\Gamma$-structure on $V \otimes P$ is defined using the diagonal action, i.e.,

$$\gamma(v \otimes x) = \gamma' v \otimes \gamma x \tag{3.3}$$

where $\gamma \in \Gamma$, $v \in V$, $x \in P$ and $\gamma'$ denotes the image of $\gamma$ in $F$. This pairing induces via Quillen's $Q$-construction [19] an action of $G_0(\mathbf{Z}F)$ on $K_n(\mathbf{Z}\Gamma)$ ($n \geqq 0$). Under this action, $K_n(\mathbf{Z}\Gamma)$ becomes a $G_0(\mathbf{Z}F)$-module. In fact, letting $p\colon \Gamma \to F$ denote the quotient map, $K_n(\mathbf{Z}p^{-1}(S))$ is a Frobenius module over the Frobenius functor $G_0(\mathbf{Z}S)$ as $S$ varies over the subgroups of $F$. In particular, the following formula is valid:

$$i_*(ri^*(x)) = i_*(r)x \tag{3.4}$$

where $x \in K_n(\mathbf{Z}\Gamma)$ and $r \in G_0(\mathbf{Z}S)$. Here, $i$ denotes both the inclusion map of $S$ in $F$ and $p^{-1}(S)$ in $\Gamma$ and $i_*$, $i^*$ represent the induction and transfer (restriction) maps, respectively. (See Quillen [14] for the constructions of $i_*$ and $i^*$. Beware Quillen's notation is the opposite of ours; namely, his $i^*$ denotes the induction and $i_*$ the transfer map.)

Now, let $\Gamma$ be a Bieberbach group and $F$ its holonomy; in particular, $\Gamma$ is an extension of a free abelian group by the finite group $F$. By standard induction techniques (formula (3.4) is crucial), the general calculation of $K_n(\mathbf{Z}\Gamma) \otimes \mathbf{Q}$ reduces to the more restricted case when $F$ is cyclic. Then, $\Gamma$ is a poly-$\mathbf{Z}$ group (cf. [5, Lemma 4.1]). But, for poly-$\mathbf{Z}$ groups, Theorem 1 is verified by using a result of Quillen [19, p. 122, exercise]. This completes our outline of the proof of Theorem 1.

**IV. Novikov's conjecture and diffeomorphisms of compact aspherical manifolds.** Let $M^n$ be an oriented manifold and $x \in H^i(\pi, 1; \mathbf{Q})$ a rational cohomology class of $K(\pi, 1)$. Given a homomorphism $\pi_1 M \to \pi$, we have a natural map

$$f\colon M^n \to K(\pi, 1). \tag{4.1}$$

Let $L_*(M^n) \in H^{4*}(M^n; \mathbf{Q})$ be the total $L$-genus of $M^n$. Consider the value

$$L(x)(M^n) = \langle L_*(M^n) \cup f^*(x), [M^n]\rangle \in \mathbf{Q}. \tag{4.2}$$

It is called the higher signature of $M^n$ associated to $x$.

Several years ago, Novikov made the following conjecture.

CONJECTURE 4.1. *If $M^n$ is a closed manifold, then $L(x)$ is a homotopy invariant.*

Since then, the conjecture has been verified for various $\pi$: Novikov [18]; Rohlin [21]; Farrell and Hsiang [8], [4]; Kasparov [11], [12]; Lusztig [15]; Miščenko [17] and Cappell [3]. The most interesting case for this conjecture is when $M^n$ itself is a compact $K(\pi, 1)$. In this special case, Conjecture 4.1 has the following form.

CONJECTURE 4.1′. *Let $M^n$ be a closed manifold which is a $K(\pi, 1)$. Then, the surgery map of* **[25]**, **[13]**

$$\theta: [M^n, G/\mathrm{Top}] \otimes \boldsymbol{Q} \to L_n(\pi, w_1(M^n)) \otimes \boldsymbol{Q}$$

*is a monomorphism.*

Actually, there is no counterexample to the following much stronger conjecture [7, problem list].

CONJECTURE 4.2. *Let $M^n$ be a closed manifold which is a $K(\pi, 1)$; then*

$$\theta: [M^n \times D^k, \partial; G/\mathrm{Top}, *] \otimes \boldsymbol{Q} \to L_{n+k}(\pi, w_1(M^n)) \otimes \boldsymbol{Q}$$

*is an isomorphism of groups.*

(Note $\theta$ is a group homomorphism even when $k = 0$ [**13**, Essay V, Appendix C].)

In fact, there is no counterexample *even* when we *do not* tensor with $\boldsymbol{Q}$; this strongest conjecture is equivalent to $\mathscr{S}^{\mathrm{Top}}(M^n \times D^k) = 0$ $(n > 4)$ **[25]**, **[13]**. Since $(G/\mathrm{Top}) \times Z$ may be identified with Quinn's spectrum $\underline{\underline{L}}_{\boldsymbol{Z}}$, $[M^n \times D^k, \partial; G/\mathrm{Top}, *]$ may be viewed as the generalized cohomology groups with coefficients in $\underline{\underline{L}}_{\boldsymbol{Z}}$. Hence, an algebraic $K$-theory analogue of Conjecture 4.2 can be formulated using Loday's map (3.1).

CONJECTURE 4.3. *If $M^n$ is a closed manifold which is a $K(\pi, 1)$, then*

$$\lambda_* \otimes \mathrm{id}: h_*(M^n; \underline{\underline{K}}_{\boldsymbol{Z}}) \otimes \boldsymbol{Q} \to K_*(\boldsymbol{Z}[\pi]) \otimes \boldsymbol{Q}$$

*is an isomorphism.*

This conjecture has been verified for various $\pi$: Quillen **[19]**, Waldenhausen **[23]**, and this paper, §III. Conjecture 4.1 has a geometric meaning in terms of stable homeomorphisms. So far, no additional geometric significance has been attached to Conjectures 4.2 and 4.3, though topologists consider them as very important problems. We shall indicate some interesting applications of Conjectures 4.2 and 4.3 to diffeomorphism groups. First, we need a lemma.

LEMMA 4.4. *Suppose $M^n$ is an orientable closed manifold which is a $K(\pi, 1)$. If Conjecture 4.3 is true for $M^n$, then*

$$\bar{K}_i^+(\pi) = 0 \quad \text{and} \quad \bar{K}_i^-(\pi) = \bigoplus_{j=1}^{\infty} H_{(i-1)-4j}(M^n, \boldsymbol{Q}).$$

(Recall $\bar{K}_i^{\pm}(\pi)$ was defined in §II.)

PROOF. Since $M^n$ is orientable, $w_1(M^n) = 1$ and the involution on $\boldsymbol{Z}[\pi]$ sends $\sum n(g)g$ to $\sum n(g)g^{-1}$. It follows from the construction of $\lambda_*$ that the compatible involution on $h_*(M^n; \underline{\underline{K}}_{\boldsymbol{Z}})$ is induced by the involution on $\underline{\underline{K}}_{\boldsymbol{Z}}$. By Lemma 2.4, that involution induces multiplication by $-1$ on $K_i(\boldsymbol{Z}) \otimes \boldsymbol{Q}$ for $i > 0$. So, the lemma follows.

Now we are ready to state the main result of this section.

THEOREM 4.5. *Let $M^n$ be an orientable closed manifold which is a $K(\pi, 1)$.*

(A) *If Conjecture 4.2 is verified for $M^n$ and* $\mathrm{Wh}_2(\pi)$ *is a torsion group, then for $0 < i < n/6 - 7$ we have*

$$\pi_i \operatorname{Diff}(M^n) \otimes \mathbf{Q} = \begin{cases} \bar{K}_3^-(\pi) \oplus (\mathit{center}(\pi) \otimes \mathbf{Q}) & \text{for } i = 1, n \text{ odd;} \\ \bar{K}_3^+(\pi) \oplus (\mathit{center}(\pi) \otimes \mathbf{Q}) & \text{for } i = 1, n \text{ even;} \\ \bar{K}_{i+2}^-(\pi) & \text{for } i > 1, n \text{ odd;} \\ \bar{K}_{i+2}^+(\pi) & \text{for } i > 1, n \text{ even.} \end{cases}$$

(B) *If both Conjecture* 4.2 *and Conjecture* 4.3 *are verified for* $M^n$, *then for* $0 < i < n/6 - 7$ *we have*

$$\pi_i \operatorname{Diff}(M^n) \otimes \mathbf{Q} = \begin{cases} \mathit{center}(\pi) \otimes \mathbf{Q} & \text{for } i = 1; \\ \bigoplus_{j=1}^{\infty} H_{(i+1)-4j}(M^n, \mathbf{Q}) & \text{for } i > 1 \text{ and } n \text{ odd;} \\ 0 & \text{for } i > 1 \text{ and } n \text{ even.} \end{cases}$$

PROOF. Since $M^n$ is a $K(\pi, 1)$, we have

$$\pi_i \operatorname{Aut}(M^n) = \begin{cases} \text{center } \pi & \text{for } i = 1, \\ 0 & \text{for } i > 1. \end{cases} \tag{4.3}$$

Fibration (2.11) is still valid; hence we obtain from it and (4.3)

$$\pi_i \mathfrak{g}(M) \otimes \mathbf{Q} \simeq \pi_i \operatorname{Diff}(M) \otimes \mathbf{Q} \quad \text{for } i > 1. \tag{4.4}$$

Also, we get the following exact sequence:

$$0 \to \pi_i \mathfrak{g}(M) \to \pi_1 \operatorname{Diff}(M) \to \operatorname{center}(\pi) \to \pi_0 \mathfrak{g}(M). \tag{4.5}$$

One of the exact sequences in the braid diagram (7) [**10**, p. 402] is

$$\begin{aligned} &\to \pi_i \mathscr{S}(M^n \times (S^1, 1)) \to [\Sigma^{i+1} M^+, G/0] \\ &\xrightarrow{\theta} L_{n+i+1}(\pi) \to \pi_{i-1} \mathscr{S}(M^n \times (S^1, 1)) \to . \end{aligned} \tag{4.6}$$

Since we assume Conjecture 4.2 is verified for $M^n$, we see from (4.6) that

$$\pi_i \mathscr{S}(M^n \times (S^1, 1)) \otimes \mathbf{Q} = 0 \quad \text{for } i \geqq 0. \tag{4.7}$$

Inserting (4.7) into sequence (2.13) (it is still exact), we obtain

$$\pi_i \mathscr{B}(M) \otimes \mathbf{Q} \simeq \pi_i \mathfrak{g}(M) \otimes \mathbf{Q} \quad \text{for } i \geqq 0. \tag{4.8}$$

Since we assumed $\mathrm{Wh}_2(\pi) \otimes \mathbf{Q} = 0$, Theorem 2.1 of [**10**] together with (4.8) yields $\pi_0 \mathfrak{g}(M) \otimes \mathbf{Q} = 0$. This combined with (4.5) gives

$$\pi_1 \operatorname{Diff}(M) \otimes \mathbf{Q} \simeq (\pi_1 \mathfrak{g}(M) \otimes \mathbf{Q}) \oplus (\operatorname{center}(\pi) \otimes \mathbf{Q}). \tag{4.9}$$

Concatenating formulas (4.4) and (4.9) with (4.8) and Lemma 2.5 proves part (A) of Theorem 4.5; part (B) follows from (A) and Lemma 4.4.

COROLLARY 4.6. *If* $M^n$ *is a closed orientable aspherical manifold such that either* $\pi_1 M$ *is a poly-*$\mathbf{Z}$ *group or* $\pi_1 M$ *contains an abelian subgroup with finite index, then for* $0 < i < n/6 - 7$ *we have*

$$\pi_i \operatorname{Diff}(M^n) \otimes \mathbf{Q} = \begin{cases} \mathit{center}(\pi) \otimes \mathbf{Q} & \text{for } i = 1; \\ \bigoplus_{j=1}^{\infty} H_{(i+1)-4j}(M^n, \mathbf{Q}) & \text{for } i > 1 \text{ and } n \text{ odd;} \\ 0 & \text{for } i > 1 \text{ and } n \text{ even.} \end{cases}$$

REMARK 4.7. *In particular, Corollary* 4.6 *applies to solvmanifolds and flat Riemannian manifolds* (*closed orientable*).

PROOF. When $\pi_1 M^n$ is a poly-$\boldsymbol{Z}$ group, Conjecture 4.2 was verified by Wall [25] and Conjecture 4.3 by Quillen [19]. When $\pi_1 M^n$ is a Bieberbach group (i.e., contains an abelian subgroup with finite index), Conjecture 4.2 was verified by Farrell and Hsiang [5] and Conjecture 4.3 is §III above.

REMARK 4.8. Cappell [3] has verified Conjecture 4.2 for a large class of aspherical manifolds $M^n$ (those with $\pi_1 M^n$ constructible in a finite number of steps from the trivial group via amalgamated free products and *HNN* constructions). Hence, the conclusion of Theorem 4.5(A) for $i > 1$ is valid for these $M^n$. Waldhausen [23] has calculated $K_i(\boldsymbol{Z}\pi_1 M)$ for this same class of manifolds. In his calculation certain exotic "Nil-groups" occur; if these are always torsion groups, then Conjecture 4.3 would be verified for this class of manifolds and hence the conclusion of Theorem 4.5(B) would apply to them.

## REFERENCES


**1.** A. Borel, *Stable real cohomology of arithmetic groups*, Ann. Sci. École Norm. Sup. **7** (1974), 235–272.

**2.** A. Borel and N. R. Wallach, *Seminar on the cohomology of discrete subgroups of semisimple groups*, Inst. for Advanced Study, 1976 (preprint).

**3.** S. E. Cappell, *On homotopy invariance of higher signatures*, Invent. Math. **33** (1976), 171–179.

**4.** F. T. Farrell and W. C. Hsiang, *Manifolds with* $\pi_1 = \mathrm{Z} \times_\alpha G$, Amer. J. Math. **95** (1973), 813–848.

**5.** ———, *Rational L-groups of Bieberbach groups*, Comment. Math. Helv. (to appear).

**6.** H. Garland and W. C. Hsiang, *A square integrability criterion for the cohomology of arithmetic groups*, Proc. Nat. Acad. Sci. U.S.A. **59** (1968), 354–360. MR **37** # 4084.

**7.** A. Hattori, editor, *Manifolds*, Univ. of Tokyo Press, Tokyo, 1975.

**8.** W. C. Hsiang, *A splitting theorem and the Künneth formula in algebraic K-theory*, Algebraic *K*-theory and its geometric applications, Lecture Notes in Math., vol. 108, Springer-Verlag, Berlin and New York, 1969, pp. 72–77. MR **40** # 6560.

**9.** W. C. Hsiang and B. Jahren, (to appear).

**10.** W. C. Hsiang and R. W. Sharpe, *Parametrized surgery and isotopy*, Pacific J. Math. **67** (1976), 401–459.

**11.** G. G. Kasparov, *The homotopy invariance of the rational Pontrjagin numbers*, Dokl. Akad. Nauk SSSR **190** (1970), 1022–1025. (Russian) MR **44** # 4769.

**12.** ———, *Topological invariance of elliptic operators*. I, *K-homology*, Izv. Akad. Nauk SSSR, Ser. Mat. **39** (1975), 796–838. (Russian)

**13.** R. C. Kirby and L. C. Siebenmann, *Foundational essays on topological manifolds, smoothings, and triangulations*, Ann. of Math. Studies, Princeton Univ. Press, Princeton, N. J., 1977.

**14.** J.-L. Loday, *K-théorie algébrique et représentations de groupes*, Ann. Sci. École Norm. Sup. **9** (1976), 309–377.

**15.** G. Lusztig, *Novikov's higher signature and families of elliptic operators*, J. Differential Geometry **7** (1972), 229–256. MR **48** # 1250.

**16.** Y. Matsushima and S. Murakami, *On vector bundle valued harmonic forms and automorphic forms on symmetric Riemannian manifolds*, Ann. of Math. (2) **78** (1963), 365–416. MR **27** # 2997.

**17.** A. S. Miščenko, *Infinite-dimensional representations of discrete groups, and higher signatures*, Izv. Akad. Nauk SSSR Ser. Mat. **38** (1974), 81–106. (Russian) MR **50** # 14848.

**18.** S. P. Novikov, *Homotopic and topological invariance of certain rational classes of Pontrjagin*, Dokl. Akad. Nauk SSSR **162** (1965), 1248–1251. (Russian) MR **31** # 5219.

**19.** D. Quillen, *Higher algebraic K-theory*. I, Algebraic *K*-Theory I, Lecture Notes in Math., vol. 341, Springer-Verlag, Berlin and New York, 1973, pp. 85–147. MR **49** # 2895.

**20.** M. S. Raghunathan, *Cohomology of arithmetic subgroups of algebraic groups*. II, Ann. of Math. (2) **87** (1968), 279–304.

**21.** V. A. Rohlin, *Pontrjagin-Hirzebruch class of codimension* 2, Izv. Akad. Nauk SSSR Ser. Mat. **30** (1966), 705–718. (Russian) MR **35** # 2295.

**22.** R. G. Swan, *K-theory of finite groups and orders*, Lecture Notes in Math., vol. 149, Springer-Verlag, Berlin and New York, 1970.

**23.** F. Waldhausen, *Algebraic K-theory of generalized free products*, Ann. of Math. (to appear).

**24.** ———, *Algebraic K-theory of topological spaces*. I, these PROCEEDINGS, Part I, pp. 35–60.

**25.** C. T. C. Wall, *Surgery on compact manifolds*, Academic Press, New York, 1970.

**26.** J. Cerf, *La stratification naturelle des espaces de fonctions différentiables réelles et le théorème de la pseudo-isotopie*, Inst. Hautes Études Sci. Publ. Math. **39** (1970).

**27.** A. Hatcher and J. Wagoner, *Pseudo-isotopies of compact manifolds*, Asterisque, No. 6, Société Mathématique de France, Paris, 1973. MR **50** # 5821.

PENNSYLVANIA STATE UNIVERSITY
INSTITUTE FOR ADVANCED STUDY

PRINCETON UNIVERSITY

Proceedings of Symposia in Pure Mathematics
Volume 32, 1978

# GROUP ACTIONS ON DISKS, INTEGRAL PERMUTION REPRESENTATIONS, AND THE BURNSIDE RING

ROBERT OLIVER

The Burnside ring $\Omega(G)$ of a finite group $G$ was originally defined to be the Grothendieck group on all finite $G$-sets (finite sets with $G$-action); addition is induced by disjoint union and multiplication by Cartesian product. tom Dieck **[1]** has given an alternative geometric definition for $\Omega(G)$: it is the ring of all equivalence classes of compact manifolds with smooth $G$-action, where two manifolds $M$ and $N$ are called equivalent if $\chi(M^H) = \chi(N^H)$ for all subgroups $H \subseteq G$. Again, addition and multiplication are induced by disjoint union and Cartesian product, respectively. The correspondence between the two definitions is given by considering finite $G$-sets as zero-dimensional manifolds.

With tom Dieck's definition, one can consider the problem of identifying which elements of the Burnside ring can be represented by actions of $G$ on (sufficiently high dimensional) disks. More specifically, an ideal $\Delta(G) \subseteq \Omega(G)$ is defined:

$$\Delta(G) = \{x \in \Omega(G) \mid 1 + x = [D],\ D \text{ some disk with } G\text{-action}\}.$$

Calculating $\Delta(G)$ turns out to be the key problem in identifying up to homotopy type which compact manifolds can occur as fixed-point sets of smooth $G$-actions on disks.

Another problem, posed by Dress, is to identify which elements $x \in \Omega(G)$ can be expressed as a difference $x = [S_1] - [S_2]$ of two $G$-sets, whose permutation modules $\boldsymbol{Z}(S_1)$ and $\boldsymbol{Z}(S_2)$ are isomorphic as $\boldsymbol{Z}G$ modules. It turns out that an element $x$ can be expressed in that fashion if and only if $x \in \Delta(G)$.

This paper is a summary of work related to characterizations and computations of $\Delta(G)$. §1 gives the background for the result just described, characterizing algebraically the geometrically defined $\Delta(G)$. §2 describes computations of $\Delta(G)$ made for certain classes of groups $G$, and §3 deals with applications to the prob-

*AMS (MOS) subject classifications* (1970). Primary 57E25; Secondary 20C10.

lem of identifying which spaces can occur as fixed-point sets of smooth $G$-actions on disks. The details of these results will appear later.

**1.** To study smooth $G$-actions on disks, the problem is first reduced to one of studying finite contractible "$G$-complexes": finite CW complexes upon which $G$ acts, permuting the cells, and such that any cell mapped to itself under the action of some $g \in G$ is mapped via the identity. Illman [7] showed that any manifold with smooth $G$-action has the structure of a $G$-complex. Conversely, it was shown in [8] that any finite $G$-complex is $G$-homotopy equivalent to some smooth action of $G$ on a manifold (on a disk, if the complex is contractible). Thus, up to $G$-homotopy equivalence, the category of finite contractible $G$-complexes is equivalent to that of disk actions. Also, $\Omega(G)$ can now be regarded as consisting of equivalence classes of finite $G$-complexes.

A *G-resolution* is defined to be an $n$-dimensional $(n - 1)$-connected (for some $n$) finite $G$-complex $X$ such that $H_n(X)$ is a projective $\mathbf{Z}G$-module. If $H_n(X)$ is a free $\mathbf{Z}G$-module, then it can be killed by adding $(n + 1)$-cells to $X$; thus producing a contractible complex. Swan [11] has shown that any projective $\mathbf{Z}G$-module is locally isomorphic to a free module (i.e., isomorphic after localizing at any prime $p$); thus $G$-resolutions may be regarded as "local approximations" to finite contractible $G$-complexes.

For any $G$-resolution $X$, invariants are defined as follows. Let $n = \dim(X)$, so that $H_n(X)$ is the one nonzero homology group, and let $S$ be the free $G$-set such that $|S| = \mathrm{rk}_{\mathbf{Z}}(H_n(X))$ (thus, $\mathbf{Z}(S)$ is the free $\mathbf{Z}G$-module to which $H_n(X)$ is locally isomorphic). Now, invariants for $X$ can be defined:

$$\varphi(X) = [X] + (-1)^{n+1}[S] - 1 \in \Omega(G),$$
$$\gamma_G(X) = (-1)^n([H_n(X)] - [\mathbf{Z}(S)]) \in \tilde{K}_0(\mathbf{Z}G).$$

These are defined in a way so that adding free $G$-orbits of $(n + 1)$-cells to $X$ to cancel out part or all of $H_n(X)$ does not change $\varphi(X)$ and $\gamma_G(X)$. Furthermore, if $X$ is contractible, $\varphi(X) = [X] - 1$, and so (referring back to the original definition)

$$\Delta(G) = \{\varphi(X) \mid X \text{ a finite contractible } G\text{-complex}\}.$$

The advantage of $G$-resolutions is that the group of elements in $\Omega(G)$ which occur as $\varphi(X)$, for $G$-resolutions $X$, is much easier to identify, and gives a first approximation to $\Delta(G)$. Let $\mathscr{G}^1$ denote the class of all finite groups $G$ which have normal subgroups $P \lhd G$ of prime power order, such that $G/P$ is cyclic. (These are the groups which Dress calls "cyclic mod $p$" for some prime $p$.) Define an ideal $\Phi(G) \subseteq \Omega(G)$ by

$$\Phi(G) = \{x \in \Omega(G) \mid \mathrm{Res}^G_H(x) = 0 \text{ for all } H \subseteq G, H \in \mathscr{G}^1\}.$$

($\mathrm{Res}^G_H$ denotes the map from $\Omega(G)$ to $\Omega(H)$ induced by restriction of group action.) It was shown in Theorem 2 of [8] that $x = \varphi(X)$ for some $G$-resolution $X$ if and only if $x \in \Phi(G)$.

Consider now the map

$$\varphi \times \gamma_G\colon \{G\text{-resolutions}\} \to \Omega(G) \times \tilde{K}_0(\mathbf{Z}G).$$

By simple geometric manipulations, its image is seen to be a subgroup. Define

$$B_0(G) = \{\gamma_G(X) | X \text{ a } G\text{-resolution}, \varphi(X) = 0\}.$$

One then gets induced a well-defined map $\Gamma_G: \Phi(G) \to \tilde{K}_0(ZG)/B_0(G)$, defined by setting $\Gamma_G(\varphi(X)) = \gamma_G(X) \pmod{B_0(G)}$ for any $G$-resolution $X$.

Now, $\Gamma_G(x) = 0$, for $x \in \Phi(G)$, if and only if $x = \varphi(X)$ for some $G$-resolution $X$ with $\gamma_G(X) = 0$, i.e., whose nonzero homology group is stably free. That group can be killed by adding cells in free orbits, thus producing a contractible complex $X' \supseteq X$ with $\varphi(X') = \varphi(X)$. It follows that $\Gamma_G$ has kernel precisely $\Delta(G)$.

One can now study isomorphisms between permutation modules in an analogous manner. "Local approximation" of such isomorphisms are given by 4-tuples $(S_1, S_2, P_1, P_2)$, where $S_1$ and $S_2$ are finite $G$-sets, $P_1$ and $P_2$ are projective $ZG$-modules of equal rank, and $Z(S_1) \oplus P_1 \cong Z(S_2) \oplus P_2$. Invariants for such 4-tuples are defined:

$$\varphi'(S_1, S_2, P_1, P_2) = [S_1] - [S_2] \in \Omega(G),$$
$$\gamma'_G(S_1, S_2, P_1, P_2) = [P_2] - [P_1] \in \tilde{K}_0(ZG).$$

For any two $G$-sets $S_1$ and $S_2$, there exist projective modules $P_1$ and $P_2$ of equal rank, such that $Z(S_1) \oplus P_1 \cong Z(S_2) \oplus P_2$, if and only if $Z(S_1)$ and $Z(S_2)$ are locally isomorphic. Dress [**2**, Proposition 9.6] showed that this is the case if and only if $[S_1] - [S_2] \in \Phi(G)$.

Endo and Miyata [**4**] defined subgroups of $\tilde{K}_0(ZG)$:

$$C^q(ZG) = \{[P_1] - [P_2] \in \tilde{K}_0(ZG) | Z(S_1) \oplus P_1 \cong Z(S_2) \oplus P_2, \text{ some } S_1, S_2\},$$
$$\tilde{C}^q(ZG) = \{[P_1] - [P_2] \in \tilde{K}_0(ZG) | Z(S) \otimes P_1 \cong Z(S) \oplus P_2, \text{ some } S\}.$$

Similarly to $\Gamma_G$, one now uses $\varphi'$ and $\gamma'_G$ to define a map $\Gamma'_G: \Phi(G) \to \tilde{K}_0(ZG)/\tilde{C}^q(ZG)$, whose image is $C^q(ZG)/\tilde{C}^q(ZG)$, and whose kernel is the group $\{x \in \Omega(G) | x = [S_1] - [S_2] \text{ with } Z(S_1) \cong Z(S_2)\}$.

For any finite group $G$, a maximal order of $QG$ is defined to be a subgroup of $QG$ which is maximal among all $Z$-lattices of $QG$ which are subrings. If $\mathfrak{M}$ is a maximal order of $QG$ containing $ZG$, then an induction map

$$\mathrm{Ind}_{ZG}^{\mathfrak{M}}: \tilde{K}_0(ZG) \to \tilde{K}_0(\mathfrak{M})$$

is defined: $\mathrm{Ind}_{ZG}^{\mathfrak{M}}([P]) = [\mathfrak{M} \otimes_{ZG} P]$. This map is always a surjection, and its kernel, denoted by $D(ZG)$, is independent of the choice of maximal order.

THEOREM 1. (1) $B_0(G) = \tilde{C}^q(ZG) = D(ZG)$ *for any* $G$.

(2) *Let* $\mathfrak{M} \supseteq ZG$ *be a maximal order, and identify* $\tilde{K}_0(ZG)/D(ZG)$ *with* $\tilde{K}_0(\mathfrak{M})$. *Then for any* $x = [S_1] - [S_2] \in \Phi(G)$,

$$\Gamma_G(x) = \Gamma'_G(x) = \left[\frac{\mathfrak{M} \otimes_{ZG} Z(S_1)}{torsion}\right] - \left[\frac{\mathfrak{M} \otimes_{ZG} Z(S_2)}{torsion}\right] \in \tilde{K}_0(\mathfrak{M}).$$

(By Corollary 21.5 of [**9**], all torsion-free $\mathfrak{M}$-modules are projective.)

In other words, $\Gamma_G = \Gamma'_G$, and the image of the map $\varphi \times \gamma_G$: {$G$-resolutions} $\to \Omega(G) \times \tilde{K}_0(ZG)$ is the same as the set $\{([S_1] - [S_2], [P_2] - [P_1]) | Z(S_1) \oplus P_1 \cong Z(S_2) \oplus P_2\}$.

COROLLARY 1. *For any* $x \in \Omega(G)$, *one can express* $x = [S_1] - [S_2]$ *such that* $Z(S_1) \cong Z(S_2)$, *if and only if* $x \in \Delta(G)$.

COROLLARY 2. $\Phi(G)/\Delta(G) \cong \mathrm{Im}(\Gamma_G) = C^q(ZG)/\tilde{C}^q(ZG)$.

One also gets, as a consequence of the result $\tilde{C}^q(\mathbf{Z}G) = D(\mathbf{Z}G)$:

COROLLARY 3. *If $M$ and $N$ are two finitely generated stably isomorphic torsion-free $\mathbf{Z}G$-modules, then $M \oplus \mathbf{Z}(S) \cong N \oplus \mathbf{Z}(S)$ for some finite $G$-set $S$.*

Let $K(G, \mathbf{Z})$ denote the Grothendieck group of all finitely generated torsion-free $\mathbf{Z}G$-modules, under direct sum. Then Corollaries 1 and 3 give the following additional characterization of $\Delta(G)$:

COROLLARY 4. *$\Delta(G)$ is the kernel of the map $\Omega(G) \to K(G, \mathbf{Z})$ defined by sending $[S]$ to $[\mathbf{Z}(S)]$.*

Swan [**11**] has shown that $\tilde{K}_0(\mathbf{Z}G)$ is finite for any finite group $G$; thus $\Delta(G)$ has finite index in $\Phi(G)$. It follows from the definition of $\Phi(G)$ that these two groups have rank equal to the number of conjugacy classes of subgroups $H \subseteq G$ with $H \notin \mathscr{G}^1$. In particular, $\Delta(G)$ is nonzero if and only if $G \notin \mathscr{G}^1$. The smallest example of such a group is $D_6$, the dihedral group of order 12, and one can explicitly exhibit two distinct finite $D_6$-sets whose permutation representations are isomorphic. (This is made easier by the fact that $\tilde{K}_0(\mathbf{Z}D_6) = 0$.)

PROPOSITION 1. *Let $a, b \in G = D_6$ be generators, with $a^6 = b^2 = 1$, and $bab^{-1} = a^{-1}$. Define $G$-sets:*

$$S_1 = 2(G/G) \cup (G/\langle a^3\rangle) \cup (G/\langle a^2\rangle) \cup (G/\langle b\rangle) \cup (G/\langle a^3b\rangle) \cup (G),$$
$$S_2 = (G/\langle a\rangle) \cup 2(G/\langle a^3, b\rangle) \cup (G/\langle a^2, b\rangle) \cup (G/\langle a^2, a^3b\rangle) \cup 2(G).$$

*Then $\mathbf{Z}(S_1) \cong \mathbf{Z}(S_2)$ as $\mathbf{Z}G$-modules.*

One can attempt to clarify the relationship between these characterizations of $\Delta(G)$ by considering chain complexes. Assume $X$ is a finite contractible $G$-complex of dimension $n$, and let $S_i$ $(0 \leq i \leq n)$ denote the set of $i$-cells of $X$, with $G$-action induced by the action on $X$. Let $S_{-1}$ be a point. The augmented chain complex for $X$ is then an exact sequence:

$$0 \to \mathbf{Z}(S_n) \to \mathbf{Z}(S_{n-1}) \to \cdots \to \mathbf{Z}(S_0) \to \mathbf{Z}(S_{-1}) \to 0.$$

Furthermore, one easily checks that $\varphi(X) = \sum_{i=-1}^{n} (-1)^i[S_i]$.

Thus, $\Delta(G)$ is by definition the set of all $x \in \Omega(G)$ which can be expressed as $x = \sum_{i=-1}^{n} (-1)^i[S_i]$, for some exact sequence $0 \to \mathbf{Z}(S_n) \to \cdots \to \mathbf{Z}(S_{-1}) \to 0$ which can be realized as the chain complex of a $G$-complex. Corollary 1 says that this is the same as the group of all $x$ which can be written $x = \sum_{i=0}^{1}(-1)^i[S_i]$ for some length two exact sequence $0 \to \mathbf{Z}(S_1) \to \mathbf{Z}(S_0) \to 0$ (not necessarily realizable as a $G$-complex).

A natural group to consider now would be the group of all $x$ which can be expressed $x = \sum_{i=0}^{n}(-1)^i[S_i]$ for some *arbitrary* (finite) exact sequence

$$0 \to \mathbf{Z}(S_n) \to \mathbf{Z}(S_{n-1}) \to \cdots \to \mathbf{Z}(S_0) \to 0.$$

If it turned out that only elements of $\Delta(G)$ could be expressed in this fashion, then this would provide a "link" for explaining Corollary 1.

To make matters more interesting, however, this is not the case. The simplest example occurs for $G \cong \mathbf{Z}_2^2$ (where $\Delta(G) = 0$). Let $a$, $b$, $c$ denote the nontrivial elements of $G$; then there is an exact sequence:

$$0 \longrightarrow Z \xrightarrow{(1+b,\,-1-b)} Z(G/\langle c\rangle) \oplus Z(G/\langle a\rangle) \xrightarrow{(1+c,\,1+a)} Z(G)$$
$$\xrightarrow{1-a} Z(G/\langle b\rangle) \xrightarrow{1} Z \longrightarrow 0.$$

Let $X(G, Z)$ denote the Grothendieck group on all finitely generated torsion-free $ZG$-modules, modulo short exact sequences. A reasonable conjecture, supported by the above example, would be that an element $x \in \Omega(G)$ can be expressed as the alternating sum of $G$-sets in an exact sequence if and only if $x$ lies in the kernel of the map $\Omega(G) \to X(G, Z)$ defined by sending $[S]$ to $[Z(S)]$. Dress [3] has studied $X(G, Z)$, and shown that modulo torsion, it is isomorphic to the rational representation ring. Thus, the kernel of the above map has rank equal to the number of conjugacy classes of noncyclic subgroups of $G$, and is larger than $\Delta(G)$ whenever $G$ is not cyclic.

**2.** In order to develop a procedure for computing $\Gamma_G$ and its kernel, certain special simple summands of the group ring $QG$ are studied. Assume first that $G$ has a normal, cyclic subgroup $Z_n \lhd G$ which is its own centralizer. Let $Q\zeta_n$ denote the extension of $Q$ generated by the $n$th roots of unity; this can be regarded as a quotient ring of $QZ_n$, and one defines

$$Q^{\#}G = Q\zeta_n \underset{QZ_n}{\otimes} QG.$$

This is a quotient ring of $QG$, and has the form of a twisted group ring $Q\zeta_n[G/Z_n]$. The action of $G/Z_n$ on $Q\zeta_n$, induced by the conjugation action on $Z_n$, is effective: $Z_n$ was assumed self-centralizing. Thus, $Q^{\#}G$ is actually a simple ring; moreover, it can be shown to be the unique simple summand of $QG$ upon which $G$ acts effectively. Thus, when it exists, $Q^{\#}G$ is a well-defined summand of $QG$, independent of the choice of the normal subgroup $Z_n \lhd G$.

Assume that $G$ is a group for which $Q^{\#}G$ exists, and write $QG = Q^{\#}G \oplus A$. If $\mathfrak{M} \supseteq ZG$ is a maximal order in $QG$, then $\mathfrak{M}$ splits as a sum $\mathfrak{M} = \mathfrak{M}^{\#} \oplus \mathfrak{M}'$, where $\mathfrak{M}^{\#} \subseteq Q^{\#}G$ and $\mathfrak{M}' \subseteq A$ are maximal orders. Let $p\colon \mathfrak{M} \to \mathfrak{M}^{\#}$ be the projection; one may then define a map $\Gamma^{\#}_G\colon \Phi(G) \to \tilde{K}_0(\mathfrak{M}^{\#})$ to be the composite

$$\Phi(G) \xrightarrow{\Gamma_G} \tilde{K}_0(\mathfrak{M}) \xrightarrow{p_*} \tilde{K}_0(\mathfrak{M}^{\#}).$$

One first reduces computation of $\Gamma_G$ to computations of $\Gamma^{\#}_H$ for certain subquotients $H$ of $G$.

THEOREM 2. *Let $G$ be a finite group. For any $H \lhd K \subseteq G$, let*

$$p_{K/H}\colon \Omega(G) \to \Omega(K/H)$$

*denote the map: $p_{K/H}([S]) = [S/H]$. Then, for any $x \in \Phi(G)$, $\Gamma_G(x) = 0$ if and only if*

$$(\Gamma^{\#}_{K/H} \circ p_{K/H})(x) = 0$$

*for all $H \lhd K \subseteq G$ such that $K$ is hyperelementary, and such that $Q^{\#}[K/H]$ is defined.*

It remains to describe the computation of $\Gamma^{\#}_G$. Let $G$ be any group for which $Q^{\#}G$ is defined, and let $Z_n \lhd G$ be a self-centralizing normal subgroup. Identify $Z_n$ with the group $\langle\zeta_n\rangle$ of $n$th roots of unity.

DEFINITION. Let $S$ be any finite $G$-set, and let $\nu = \{e_1, \cdots, e_m\}$ be a set of elements of $S$, with one representative from each free orbit of $\boldsymbol{Z}_n$. For any $g \in G$, elements $a_1^g, \cdots, a_m^g \in \boldsymbol{Z}_n$ and $\sigma_g \in \Sigma_m$ are defined, such that $g \cdot e_i = a_i^g \cdot e_{\sigma_g i}$ for all $i$. Then a cocycle $c'(S, \nu) \in Z^1(G; \langle \pm \zeta_n \rangle)$ is defined ($\langle \pm \zeta_n \rangle$ denoting the group of all roots of unity in $\boldsymbol{Z}\zeta_n$) by

$$c'(S, \nu)(g) = \operatorname{sign}(\sigma_g) \cdot a_1^g \cdots a_m^g.$$

One now checks by computations:

LEMMA 1. (1) *$c'(S, \nu)$ is a cocycle.*

(2) *If $S'$ is another $G$-set with orbit representatives $\nu'$, then*

$$c'(S, \nu) + c'(S', \nu') = c'(S \cup S', \nu \cup \nu')$$

*(writing $Z^1(G; \langle \pm \zeta_n \rangle)$ additively).*

(3) *If $\nu_1$ and $\nu_2$ are two sets of orbit representatives for $S$, then*

$$c'(S, \nu_1) - c'(S, \nu_2) \in B^1(G; \langle \pm \zeta_n \rangle).$$

*Thus, for any finite $G$-set $S$, one gets induced a well-defined cohomology class $c'(S) \in H^1(G; \langle \pm \zeta_n \rangle)$; and this yields a homomorphism $c' : \Omega(G) \to H^1(G; \langle \pm \zeta_n \rangle)$.*

By Proposition 4 of Chapter VII in **[10]**, one has an exact sequence

$$0 \longrightarrow H^1(G/\boldsymbol{Z}_n; \langle \pm \zeta_n \rangle) \xrightarrow{\pi^*} H^1(G; \langle \pm \zeta_n \rangle) \longrightarrow H^1(\boldsymbol{Z}_n; \langle \pm \zeta_n \rangle)$$

($\pi^*$ induced by the projection $\pi: G \to G/\boldsymbol{Z}_n$). The second map is given by restriction; since $\Phi(G)$ is contained in the kernel of $\operatorname{Res}^G_{\boldsymbol{Z}_n}: \Omega(G) \to \Omega(\boldsymbol{Z}_n)$, it follows that $c'(\Phi(G))$ lies in the image of $\pi^*$. So $c'|\Phi(G)$ lifts to a map

$$c_G: \Phi(G) \to H^1(G/\boldsymbol{Z}_n; \langle \pm \zeta_n \rangle).$$

THEOREM 3. *For any $x \in \Phi(G)$, $\Gamma_G^\sharp(x) = 0$ if and only if $x$ lies in the kernel of the composite*

$$\Phi(G) \xrightarrow{c_G} H^1(G/\boldsymbol{Z}_n; \langle \pm \zeta_n \rangle) \longrightarrow H^1(G/\boldsymbol{Z}_n; (\boldsymbol{Z}\zeta_n)^*)$$

*(the second map being induced by inclusion of coefficients).*

The computation procedures behind Theorems 2 and 3 seem somewhat unwieldy, but they can actually be applied directly to the computation of $\Gamma_G$ and $\Delta(G)$ for many families of groups. The following examples extend results of Endo and Miyata **[5]** and **[6]**; they studied the group $C^q(\boldsymbol{Z}G)/\tilde{C}^q(\boldsymbol{Z}G) \cong \operatorname{Im}(\Gamma_G)$ in relation to a problem on rationality of function fields of algebraic tori.

PROPOSITION 2. *$\Gamma_G = 0$ (and thus $C^q(\boldsymbol{Z}G) = \tilde{C}^q(\boldsymbol{Z}G)$) if $G$ meets any of the following conditions:*

(1) *$G \in \mathscr{G}^1$.*

(2) *$G$ is nilpotent.*

(3) *$G$ is dihedral or quaternionic.*

(4) *$G \cong \mathrm{SL}(2, q)$ or $\mathrm{PSL}(2, q)$ for any prime power $q$.*

(*Note.* The case of dihedral or quaternionic groups is the only one where Theorem 3 need be directly applied.)

PROPOSITION 3. *$\operatorname{Im}(\Gamma_G)$ has exponent 2 if $G$ is metacyclic.*

In fact, the rank of $\mathrm{Im}(\Gamma_G)$ can be computed explicitly for such $G$; in particular, for any $k \geqq 0$, there are metacyclic groups $G$ with $\mathrm{Im}(\Gamma_G) \cong (\boldsymbol{Z}_2)^k$.

In addition to this, explicit examples can be constructed to prove:

PROPOSITION 4. *For any $n$ there exists a group $G$ such that* $\mathrm{Im}(\Gamma_G)$ *has exponent $n$.*

**3.** A last application of these results is to the problem of classifying spaces which can occur as fixed point-sets of actions of a given finite group $G$ on disks. The following result was proven in **[8]**:

THEOREM 4. *For any finite group $G$ not of prime power order, there is an integer $n_G$ such that for any finite complex $F$, $F$ is homotopy equivalent to the fixed point set of some smooth action of $G$ on a disk, if and only if* $\chi(F) \equiv 1 \pmod{n_G}$.

Define a map $\chi_G\colon \Omega(G) \to \boldsymbol{Z}$ by setting $\chi_G([S]) = |S^G|$. From Theorem 4, it is then clear that $n_G$ is a generator of the image under $\chi_G$ of $\Delta(G)$. In **[8]**, it was shown that $n_G = 0$ if and only if $G \in \mathscr{G}^1$, and that $q | n_G$, for any prime $q$, if and only if $G/P$ is $q$-hyperelementary for some subgroup $P \lhd G$ of prime power order. Furthermore, it was seen that $n_G$ is a product of distinct primes (or 0, or 1) when $G$ is not hyperelementary.

Theorems 2 and 3 now allow the computation of $n_G$ in the remaining cases. One gets the following result:

THEOREM 5. *Let $G$ be a finite group. Then $n_G = 4$ if and only if the following conditions all hold*:

(1) *$G$ is 2-hyperelementary with cyclic 2-Sylow subgroup.*

(2) *$G \notin \mathscr{G}^1$, but the subgroup of index 2 is in $\mathscr{G}^1$.*

(3) *Let $\boldsymbol{Z}_n \lhd G$ be the largest odd order subgroup, and*

$$\alpha \in \mathrm{Aut}(\boldsymbol{Z}_n) \cong \mathrm{Gal}(\boldsymbol{Q}\zeta_n/\boldsymbol{Q})$$

*the element induced by the conjugation action of a generator of the 2-Sylow subgroup. Then there is no unit $u \in (\boldsymbol{Z}\zeta_n)^*$ such that $\alpha(u) = -u$.*

*Otherwise $n_G = 0$, 1, or a product of distinct primes.*

If $G$ has a normal subgroup $P$ of prime power order such that $G/P$ is $q$-hyperelementary, then one can show by elementary Smith theory considerations (see §1 of **[8]**) that $\chi(D^G) \equiv 1 \pmod{q}$ for any smooth action of $G$ on a disk $D$. Thus, Theorems 4 and 5 show that groups meeting conditions (1) to (3) above are the *only* ones for which there are nonelementary restrictions on what the fixed-point set of a disk action can be. The smallest example of such a group is the one generated by elements $x$ and $y$ with relations $x^{15} = y^4 = 1, = yxy^{-1}x^2$.

## REFERENCES


**1.** T. tom Dieck, *The Burnside ring of a compact Lie group*. I, Math. Ann. **215** (1975), 235–250. MR **52** #15510.

**2.** A. Dress, *Contributions to the theory of induced representations*, Algebraic $K$-Theory. II, Springer-Verlag, 1973, 183–240. MR **52** #5787.

**3.** ———, *Über ganzzahlige Permutationsdarstellungen einiger endlicher Gruppen*, Arch. Math. (Basel) **25** (1974), 231–240. MR **50** #13219.

**4.** S. Endo and T. Miyata, *Quasi-permutation modules over finite groups*, J. Math. Soc. Japan **25** (1973), 397–421. MR **47** #6823.

**5.** S. Endo and T. Miyata, *Quasi-permutation modules over finite groups*. II, Math. J. Soc. Japan **26** (1974), 698–713. MR **50** #13226.

**6.** ———, *On the projective class group of finite groups*, Osaka J. Math. **13** (1976), 109–122.

**7.** S. Illman, *Equivariant algebraic topology*, Thesis, Princeton Univ., 1972.

**8.** R. Oliver, *Fixed point sets of group actions on finite acyclic complexes*, Comment. Math. Helv. **50** (1975), 155–177. MR **51** #11556.

**9.** I. Reiner, *Maximal orders*, Academic Press, New York, 1975. MR **52** #13910.

**10.** J.-P. Serre, *Corps locaux*, Hermann, Paris, 1968. MR **50** #7096.

**11.** R. Swan, *Induced representations and projective modules*, Ann. of Math. (2) **71** (1960), 552–578. MR **25** #2131.

STANFORD UNIVERSITY

Proceedings of Symposia in Pure Mathematics
Volume 32, 1978

# AUTOMORPHISMS OF MANIFOLDS

DAN BURGHELEA

TABLE OF CONTENTS



**0. Introduction.** The geometric topology, in the present stage of its development, is mainly concerned with manifolds: differentiable, piecewise linear and topological. As it is well known the first problem of the geometric topology is to understand the structure of manifolds and their classification; the next important problem is to find out the structure of the automorphisms of manifolds, and along these lines the understanding of the *homotopy type of the space* (topological group) *of automorphisms* seems to be very important. Besides its general mathematical

*AMS* (*MOS*) *subject classifications* (1970). Primary 55D50, 57A99, 57C99, 55B20; Secondary 57C10, 57C35, 57D40, 57A35, 55D99, 18F25.

*Key words and phrases*. Diffeomorphisms, homeomorphisms, concordances, homotopical localization, algebraic Whitehead theory, algebraic $K$-theory, nilpotency.

interest which is certainly related with topology, geometry and mechanics, the researches on the homotopy type of the group of automorphisms has suggested and stimulated beautiful mathematical problems and progresses in the field of homotopy theory and algebra.

The present "article" is a sort of survey about the progresses towards the understanding of the homotopy type of the groups of automorphisms, which have grown up from the ideas of Cerf, Morlet, Quinn, Antonelli, Kahn, Hatcher, Wagoner, Lashof, Rothenberg and Burghelea. It is certainly incomplete in references and technical results but we hope not in driving ideas.

The categories where the objects of the geometric topology live in are $\mathscr{D}iff$, $\mathscr{P}\ell$ $\mathscr{T}op$, which we will denote from now on by $\mathscr{A}, \mathscr{B}, \cdots$. We will write $\mathscr{A} \leqq \mathscr{B}$ to indicate that $\mathscr{A}$ is finer than $\mathscr{B}$, that is any object, respectively morphism, of $\mathscr{A}$ has a well-defined structure of object, respectively morphism, of $\mathscr{B}$. The objects of a geometric category are compact manifolds with boundary (possibly empty). For such manifolds we consider the topological group of $C^\infty$-diffeomorphisms, $\mathscr{D}iff(M)$ endowed with the $C^\infty$-topology if $M$ is a differentiable manifold, $\mathscr{H}omeo\,(M)$ the topological group of homeomorphisms endowed with the compact open topology if $M$ is a topological manifold and the simplicial group of $p\ell$-homeomorphisms $\mathscr{P}\ell\,(M)$ if $M$ is a $p\ell$-manifold (which is apparently less natural as in fact the entire $\mathscr{P}\ell$-category but very useful for the understanding of $\mathscr{D}iff\,M$ and $\mathscr{T}op\,M$).

Since our objective is the homotopy type of these topological groups, it will be convenient to use the semisimplicial description of these spaces (see **[B.L.R.]**).

For a geometric category $\mathscr{A}$ and a manifold $M \in \text{ob}\,\mathscr{A}$ with $\partial M = \partial_+ M \cup \partial_- M$, $\partial_+ M \cap \partial_- M$ a codimension one submanifold of $\partial M$, we define $\mathscr{A}(M; \partial_- M)$ the s.s.-group of $\mathscr{A}$-automorphisms whose $k$-simplexes are $\mathscr{A}$-automorphisms

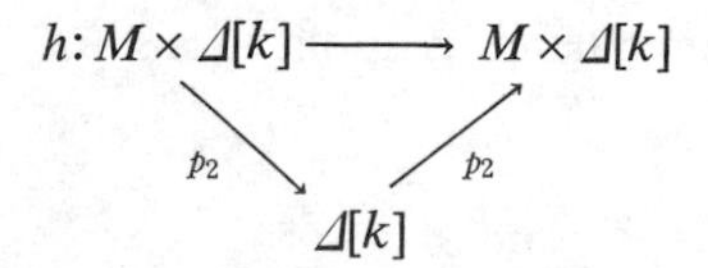

commuting with the projections on $\Delta[k]$ and with $h|\partial_+ M \times \Delta[k] = \text{id}$. In particular, if we take $\partial_- M = \phi$ we denote $\mathscr{A}(M; \phi)$ by $\mathscr{A}(M)$ the s.s.-group of $\mathscr{A}$-automorphisms which restrict to identity on $\partial M$. We also use the notation $\mathscr{H}(X)$ for the s.s.-associative monoid of the polyhedron homotopy equivalences of $X$, $X$ being a finite complex.

The problem we are interested in can now be easily formulated.

*Problem* A. *Describe the homotopy type of $\mathscr{A}(M)$, i.e., homotopy groups, Postnikov invariants, etc.*

*Problem* B. *Describe, from the homotopy point of view, the difference between $\mathscr{A}(M)$ and $\mathscr{B}(M)$ for $\mathscr{B} \geqq \mathscr{A}$ and $M \in$ ob $\mathscr{A}$, i.e., the homotopy type of the s.s.-complex $\mathscr{B}(M)/\mathscr{A}(M)$.*

Until 1968–1969 very little was known about the homotopy type of $\mathscr{A}(M)$ for manifolds of dim $M \geqq 3$. One knew only:

(1) $\mathscr{A}(M)$ have the homotopy type of a countable $CW$-complex, a general but very weak result;

(2) the group of connected components (in fact how to compute this group) of $\mathscr{A}(M)$, a particular but deep and stimulative result [C] (for $\pi_1(M) = 0$);

(3) some nontrivial homotopy groups of $\mathscr{D}iff\, S^n$ and $\mathscr{D}iff\, D^n$ computations due to Novikov, Milnor, Munkres (see **[N]** and **[Mu]**), and, of course,

(4) the contractibility of $\mathscr{P}\ell(D^n)$, and $\mathscr{T}op(D^n)$ obtained by the "so-called" *conning construction* or the Alexander trick;

(5) the homotopy type of $\mathscr{A}(M^n)$ for $n \leqq 2$; for manifolds of dimension $n \leqq 2$ there is no difference between the classification of differentiable, piecewise linear and topological manifolds and automorphisms, and these facts are reflected in the homotopy equivalence of $\mathscr{D}iff(M)$, $\mathscr{P}\ell(M)$, $\mathscr{T}op(M)$. All these spaces have the homotopy type described in **[G]**, their connected component of the identity being a compact Lie group ($SO(2)$, $SO(3)$, $S^1 \times S^1$, $S^1$); this is a consequence of the pioneering work of Kneser, M. H. Hamstrong, S. Smale, J. Eells and C. Earle, etc.;

(6) even for the simplest possible manifolds, namely $M = T^n$, $S^n$, $D^n$ if $n \geqq 5$, $\mathscr{D}iff(M^n)$ has the homotopy groups rich enough and does not have the homotopy type of a finite $CW$-complex **[N]**, **[A.B.K.]$_2$**.

The new developments (after 1968–1969), more precisely those which we intend to discuss in this article, have been greatly influenced by:

(1) Smoothing theory for $p\ell$-manifolds as it has been developed by Hirsch-Mazur, Lashof-Rothenberg, and the smoothing theory for topological manifolds, respectively the triangulation of topological manifolds, as it has been developed by Lashof-Rothenberg, Kirby-Siebenmann;

(2) Browder-Novikov-Sullivan-Wall celebrated work about surgery;

(3) Cerf's ideas about concordances.

We will explain these new developments as follows:

Problem B in Part I. The homotopy type of $\mathscr{A}(M)$, at least in stable ranges and localized to odd primes (the meaning of stable ranges will be explained in Part II, §5), is reconstructed from the homotopy type of the s.s.-group of block automorphisms $\tilde{\mathscr{A}}(M)$ (which will be described in Part I), and ${}^{\mathscr{A}}\mathscr{S}^{+}(M)$ or ${}^{\mathscr{A}}\mathscr{S}^{-}(M)$, $\infty$-loop spaces associated to $M$, whose homotopy type are tangential homotopy invariants (there is much evidence to believe they are actually homotopy invariants).

These $\infty$-loop spaces are direct factors of the $\infty$-loop space $\mathscr{S}^{\mathscr{A}}(M)$ defined using concordances and which is a homotopy invariant; $\mathscr{S}^{\mathscr{A}}$, ${}^{\mathscr{A}}\mathscr{S}^{+}$, ${}^{\mathscr{A}}\mathscr{S}^{-}$ are new homotopy functors with values in the weak homotopy category of $\infty$-loop spaces[1], intimately connected to the algebraic $K$-theory and Whitehead theory of higher order, and they provide the homotopy theory with new objects to study. The discussion on concordances and of the functors $\mathscr{S}^{\mathscr{A}}$, ${}^{\mathscr{A}}\mathscr{S}^{+}$, ${}^{\mathscr{A}}\mathscr{S}^{-}$ is contained in Part II.

Initially we intended to collect some important applications in Part III about:

(1) Homotopy type of $Top(n)$, $Pl(n)$.

(2) When $f$: $M^{n+k} \to P^n$, a continuous map between two compact manifolds, is homotopic to a locally trivial bundle.

(3) What kind of differentiable homeomorphisms from $\Sigma^n$, a homotopy sphere, to $S^n$ do exist ?

Since at the last moment we heard about new results which can complete sub-

[1]The objects of this category are $\infty$-loop spaces and the morphisms weak homotopy classes of $\infty$-loop space maps; a homotopy $f_t$ between two $\infty$-loop space maps $f_0$ and $f_1$ requires $f_t$ to be an $\infty$-loop space map for any $t$ and we say that $f_0$ and $f_1$ are weak homotopic if $[f_0]_k$ and $[f_1]_k$, the Postnikov $k$-terms of $f_0$ and $f_1$, are homotopic for any $k$.

stantially what we know about, we believe it premature to discuss these aspects indicating only as references **[B.L.]**$_1$, **[B.L.]**$_2$, **[B.L.R.]**, **[L]**.

Before starting our survey we mention some definitions and notations. According to the usual conventions in homotopy theory we say that:

(1) Two spaces (s.s.-complexes) $X$ and $Y$ have the same $k$-homotopy type (or are $k$-equivalent) if their $k$th Postnikov terms are homotopy equivalent.

(2) $G$ is a weak group if it is a group in the homotopy category of base pointed s.s.-complexes or spaces.

(3) $f$ is a $k$-homotopy equivalence, resp. homotopy injection, resp. a homotopy surjection, etc., if $f_k: X_k \to Y_k$ (where $X_k, Y_k, f_k$, are the Postnikov terms of $X$, $Y$, $f$) is a homotopy equivalence, resp. a homotopy injection, resp. a homotopy surjection, $\cdots$.

(4) $f$ and $g$ are weak homotopic if $f_k$ and $g_k$, the $k$th Postnikov terms of $f$ and $g$, are homotopic for any $k$.

One oberves that if $f: X_h \to Y_k$ is a $k$-homotopy equivalence from $X$ to $Y$ there exist a $(k+1)$-dimensional $CW$-complex $K$ and the maps $f: K \to X$, $g: K \to Y$ so that $\pi_i(f) = \pi_i(g)$ if $i \leqq k+1$; moreover one can assume that $g_* \circ f_*^{-1} = \sigma_*^{-1} \circ \varphi_* \circ \rho_*$ on $\pi_i(X)$ for $i \leqq k$ where $\sigma$ and $\rho$ are the canonical projections $\rho: X \to X_r$, $\sigma: Y \to Y_r$.

One says that two $n$-manifolds $V$ and $V'$ have the same tangential $k$-homotopy type if there exists a $k$-homotopy equivalence $f: V_k \to V'_k$ as above and for $f$ and $g$ constructed as above $f^*(\tau(V)) = g^*(\tau(V'))$. (More details about these considerations can be found in **[B.L.R.]**.)

We will denote by $\mathscr{P}$ the category of finite polyhedra and continuous maps and by $\mathscr{P}^{\mathscr{A}}$ the category whose objects are pairs $(X, \xi)$ consisting of a finite polyhedron and a stable $\mathscr{A}$-euclidean bundle, and morphisms $f: (X, \xi) \to (Y, \eta)$ are continuous maps with $f^*(\eta)$ stable isomorphic to $\xi$.

## PART I

**1. Classical groups of geometric topology and Problem B.** Let us denote by $\mathscr{A}(n)$ the s.s.-group of $\mathscr{A}$-automorphisms of $R^n$ fixing the origin. For $\mathscr{A} = \mathscr{D}iff$, $\mathscr{A}(n)$ is homotopy equivalent to $O(n)$—the orthogonal group. For $\mathscr{A} = \mathscr{P}\ell$ or $\mathscr{T}op$ we keep the standard notation, $Pl(n)$, and $Top(n)$.

It is also convenient to denote by $G(n)$ the s.s.-associative monoid (weak group) of proper homotopy equivalences of $R^n$ which fix the origin.

We have the obvious commutative diagram

$$\begin{array}{ccccccc}
O(n) & \hookrightarrow & O(n+1) & \hookrightarrow \cdots \rightarrow & O(n+i) & \hookrightarrow & O(\infty)=O \\
\downarrow & & \downarrow & & \downarrow & & \downarrow \\
Pl(n) & \hookrightarrow & Pl(n+1) & \hookrightarrow \cdots \rightarrow & Pl(n+i) & \hookrightarrow \cdots \rightarrow & Pl(\infty)=Pl \\
\downarrow & & \downarrow & & \downarrow & & \downarrow \\
Top(n) & \hookrightarrow & Top(n+1) & \hookrightarrow \cdots \rightarrow & Top(n+i) & \hookrightarrow \cdots \rightarrow & Top(\infty)=Top \\
\downarrow & & \downarrow & & \downarrow & & \downarrow \\
G(n) & \hookrightarrow & G(n+1) & \hookrightarrow \cdots \rightarrow & G(n+i) & \hookrightarrow \cdots \rightarrow & G(\infty)=G
\end{array}$$

where $\mathscr{A}(\infty)$, respectively $G(\infty)$, are $\lim_{\to} \mathscr{A}(n)$ respectively $\lim_{\to} G(n)$. The homotopy type of $O$, $Pl$, $Top$ are partly understood (and we assume for the purpose of our considerations they are known). For instance one knows $O$, $G/Top$, $G/Pl$, $Top/Pl \sim Top(n)/Pl(n) \sim K(Z_2, 3)$ (for $n \geqq 5$); $G$ is the "so-called" *spectrum of spheres*, its homotopy groups being the stable homotopy groups of spheres.

The results which will be discussed in this paper give valuable information about the homotopy types of the groups $\mathscr{A}(n)$ (see **[B.L.]**$_1$ and **[B.L.]**$_2$). The groups $\mathscr{A}(n)$, $\mathscr{A}(\infty)$, are usually called the classical groups of the geometric topology.

There exists a natural map $\Psi_n^{\mathscr{A}}: \mathscr{A}(n)/\mathscr{A}(n-1) \to \Omega\mathscr{A}(n+1)/\mathscr{A}(n)$ *called suspension* and defined as follows: For $h$ an $\mathscr{A}$-automorphism of $R^n$ fixing the origin and $0 \leqq t \leqq 1$, consider $\Psi_n^{\mathscr{A}}: I \times \mathscr{A}(n) \to \mathscr{A}(n+1)$ defined by the formula

$$\Psi_n^{\mathscr{A}}(t, h) = R_t \circ \left(\begin{array}{c|c} 1 & 0 \\ \hline 0 & h \end{array}\right) \circ R_{-t}$$

where $R_t$ is the rotation of angle $t\pi$ in the plane of the first two coordinates of $R^{n+1}$. Clearly $\Psi_n^{\mathscr{A}}$ defines $\Psi_n^{\mathscr{A}}: \mathscr{A}(n)/\mathscr{A}(n-1) \to \Omega\mathscr{A}(n+1)/\mathscr{A}(n)$ (see **[B.L.]**$_2$) and the following diagram is commutative:

$$\begin{array}{ccc}
O(n)/O(n-1) & \xrightarrow{\Psi_n^{\mathscr{O}}} & \Omega O(n+1)/O(n) \\
\downarrow & & \downarrow \\
Pl(n)/Pl(n-1) & \xrightarrow{\Psi_n^{\mathscr{P}}} & \Omega Pl(n+1)/Pl(n) \\
\downarrow & & \downarrow \\
Top(n)/Top(n-1) & \xrightarrow{\Psi_n^{\mathscr{T}}} & \Omega Top(n+1)/Top(n) \\
\downarrow & & \downarrow \\
G(n)/G(n-1) & \xrightarrow{\Psi_n^{G}} & \Omega G(n+1)/G(n)
\end{array}$$

If we denote by $F_n^{\mathscr{A},\mathscr{B}}$ the homotopy theoretic fibre of $\mathscr{A}(n)/\mathscr{A}(n-1) \to \mathscr{B}(n)/\mathscr{B}(n-1)$, $\mathscr{B} \geqq \mathscr{A}$, $\Psi_n$ induces $\Psi_n^{\mathscr{B},\mathscr{A}}: F_n^{\mathscr{B},\mathscr{A}} \to \Omega F_{n+1}^{\mathscr{B},\mathscr{A}}$. For any $\mathscr{A}$-manifold $M^n$, let $\mathscr{P}^{\mathscr{A}}(M^n)$ be the principal $\mathscr{A}(n)$-bundle associated to the tangent bundle; clearly $\mathscr{P}^{\mathscr{A}}(M^n)/\partial M^n$ contains $\mathscr{P}^{\mathscr{A}}(\partial M)$ as principal subbundle. If $\mathscr{B} \geqq \mathscr{A}$, hence $\mathscr{A}(n) \subseteq \mathscr{B}(n)$, to the principal bundle $\mathscr{P}^{\mathscr{A}}(M^n)$ we associate the bundle

$$(*) \qquad \mathscr{B}(n)/\mathscr{A}(n) \cdots \mathscr{E}_{\mathscr{B},\mathscr{A}} \to M$$

whose fibre is $\mathscr{B}(n)/\mathscr{A}(n)$; this bundle $\mathscr{E}_{\mathscr{B},\mathscr{A}}$ has a canonical cross-section $s$. $\mathscr{E}_{\mathscr{B},\mathscr{A}}(M)/\partial M$ contains $\mathscr{E}_{\mathscr{B},\mathscr{A}}(\partial M)$ as subbundle, and the canonical cross-section of $\mathscr{E}_{\mathscr{B},\mathscr{A}}(\partial M)$ over $\partial M$ is the canonical cross-section of $\mathscr{E}_{\mathscr{B},\mathscr{A}}(\partial M)$. With the bundle $\mathscr{P}^{\mathscr{A}}(M)$ we can also associate the bundle ${}^n\mathscr{F}_{\mathscr{B},\mathscr{A}}(M^n)$ with fibre $F_n^{\mathscr{B},\mathscr{A}}$ and the bundle ${}^{n+1}\mathscr{F}_{\mathscr{B},\mathscr{A}}(M^n)$, resp. $\Omega({}^{n+1}\mathscr{F}_{\mathscr{B},\mathscr{A}}(M^n)$; with fibre $F_{n+1}^{\mathscr{B},\mathscr{A}}$, resp. $\Omega F_{n+1}^{\mathscr{B},\mathscr{A}}$. Then the suspension $\Psi_n^{\mathscr{B},\mathscr{A}}$ extends to:

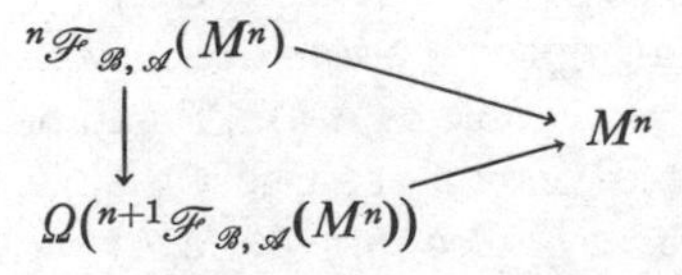


Let us denote by $\Gamma(\mathscr{E}_{\mathscr{B},\mathscr{A}})$ the s.s.-complex of cross-sections of $(*)$ which agree to $s$ on $\partial M$, and by $\Gamma(\mathscr{E}_{\mathscr{B},\mathscr{A}}; \partial_- M)$ the s.s.-complex of cross-sections of $(*)$ which agree to $s$ on $\partial_+ M$ and lie in $\mathscr{E}_{\mathscr{B},\mathscr{A}}(\partial M)$ if restricted to $\partial_- M$. The differential of a $\mathscr{B}$-automorphism defines the map

$$d_{\mathscr{B},\mathscr{A}}\colon \mathscr{B}(M;\partial_- M)/\mathscr{A}(M;\partial_- M) \to \Gamma(\mathscr{E}_{\mathscr{B},\mathscr{A}};\partial_- M).$$

The following result has its origin in the pioneering work of Cerf; it has been stated and sketchily proved by Morlet [**Mo**] and proved in [**B.L.**]$_1$. (We call $f\colon X \to Y$ an I.H.E.-map if it induces an injective correspondence for connected components and a homotopy equivalence on any connected component.)

THEOREM I. 1.1. *Assume $M^n$ an $\mathscr{A}$-manifold. Then*

$$d_{\mathscr{B},\mathscr{A}} : \mathscr{B}(M^n;\partial_- M)/\mathscr{A}(M^n;\partial_- M) \to \Gamma(\mathscr{E}_{\mathscr{B},\mathscr{A}};\partial_- M)$$

*is an I.H.E.-map provided $\mathscr{A} = \mathscr{D}\mathit{iff}$, $\mathscr{B} = \mathscr{P}\ell$ or $\mathscr{A} = \mathscr{D}\mathit{iff}, \mathscr{P}\ell$, $\mathscr{B} = \mathscr{T}\mathit{op}$ and $n \neq 4$, $\partial_- M = \emptyset$ or $\mathscr{A} = \mathscr{D}\mathit{iff}, \mathscr{P}\ell$, $\mathscr{B} = \mathscr{T}\mathit{op}$ and $n \neq 4, 5$.*

*If $M^n$ is parallelizable and $\partial_- M = \emptyset$ then $\Gamma(\mathscr{E}_{\mathscr{B},\mathscr{A}}) \sim$* Map $(M, \partial M; \mathscr{B}(n)/\mathscr{A}(n))$. ■

COROLLARY I. 1.2. (a) $\mathscr{D}\mathit{iff}(D^n) \sim \Omega^{n+1}(Pl(n)/O(n))$,

(b) $\mathscr{D}\mathit{iff}(D^n) \sim \Omega^{n+1}(Top(n)/O(n))$. ■

COROLLARY I. 1.3. *For $M = N \times I$, $\partial_- M = N \times \{1\}$ the differential $d_{\mathscr{B},\mathscr{A}}$: $\mathscr{B}(N \times I;\partial_- M = N \times \{1\})/\mathscr{A}(N \times I;\partial_- M) \to \Gamma(\mathscr{E}_{\mathscr{B},\mathscr{A}};\partial_- M) = \Gamma(\mathscr{F}^n_{\mathscr{B},\mathscr{A}})$ is an I.H.E.-map provided $\mathscr{B} = \mathscr{P}\ell$ or, $\mathscr{B} = \mathscr{T}\mathit{op}$ and $n \neq 4, 5$.* ■

COROLLARY I. 1.4. (a) $\mathscr{D}\mathit{iff}(D^n; D^{n-1}_-) \sim \Omega^n F_n^{\mathscr{P}\ell, \mathscr{D}\mathit{iff}}$,

(b) $\mathscr{D}\mathit{iff}(D^n; D^{n-1}_-) \sim \Omega^n F_n^{\mathscr{T}\mathit{op}, \mathscr{D}\mathit{iff}}$, $n \neq 4, 5$. ■

**2. Block automorphisms.** The surgery methods, among many other things, give homotopy criteria to decide when a homotopy equivalence is homotopic to an $\mathscr{A}$-automorphism. They are also able to say how many such automorphisms (up to an equivalence) correspond (i.e., are homotopic) to a homotopy equivalence. The right equivalence of $\mathscr{A}$-automorphism which follows from the surgery methods turns out to be not "isotopy" but "concordance". To understand the difference it is very useful to recall their definitions simultaneously:

"$h_1, h_2\colon M \to M$ are isotopic iff there exists an $\mathscr{A}$-automoprhism $H\colon M \times I \to M \times I$ *commuting with the projection on I*, with $H/M \times \{0\} = h_1$ and $H/M \times \{1\} = h_2$; they are concordant iff there exists an $\mathscr{A}$-automorphism $H\colon M \times I \to M \times I$ with $H|M \times \{0\} = h_1$, $H|M \times \{1\} = h_2$".

This definition suggests the introduction of another s.s.-group $\tilde{\mathscr{A}}(M)$ (which contains $\mathscr{A}(M)$ as a subgroup) the group of block-automorphisms; this group is naturally related to the surgery methods and smoothing theory methods. It has been defined simultaneously by many authors: Morlet, Quinn, Rourke-Sanderson-Casson, Antonelli-Burghelea-Kahn (see [**B.L.R.**]).

A $k$-simplex of $\tilde{\mathscr{A}}(M)$ is an $\mathscr{A}$-automorphism $h\colon M \times \Delta[k] \to M \times \Delta[k]$ with $h|M \times d_i(\Delta[k]) \subset M \times d_i(\Delta[k])$ and $h|\partial M \times \Delta[k] = \mathrm{id}$.[2] Obviously $\mathscr{A}(M) \subset \tilde{\mathscr{A}}(M)$ can be viewed as an approximation of $\mathscr{A}(M)$. The interest of $\tilde{\mathscr{A}}(M)$ comes first from the fact that both Problems A and B can be, at least theoretically, solved via the *right* parametrized version of surgery theory and smoothing theory. We explain below what we mean by at *least theoretically solved* and give the results of [**A.B.K.**]$_1$ (which have been also proved in part by Quinn in his unpublished thesis). For this purpose let us denote by $\mathscr{H}(M)$ the space of continuous maps

[2]We also assume some technical requirements namely to be "product-like" near corners (see [B. L. R.]) which allows us to define *the degeneracies*.

which are simple homotopy equivalences which restrict to identity on $\partial M$. For technical reasons we will describe $\mathscr{H}(M)$ as an s.s.-associative monoid (or weak-group) whose $k$-simplexes are homotopy equivalences $h: M \times \Delta[k] \to M \times \Delta[k]$ with $h(M \times d_i\Delta[k]) \subset M \times d_i\Delta[k]$. $h|\partial M \times \Delta[k] = \mathrm{id}$. The results of **[A.B.K.]**$_1$ allow us to reconstruct $\tilde{\mathscr{A}}(M)$ from:

(1) $\mathscr{H}(M)$, (2) Maps$(M, G/Top)$ or Maps$(M, G/Pl)$ or Maps$(M, G/O)$, and (3) the algebraic $L$-theory of Wall which is a functor $L$ from the category of groups with orientation;[3] $\mathscr{G}_0$ to the homotopy category of $\infty$-loop spaces $\Omega^h$ so that $\pi_i(L(G, \omega)) = L_i(G, \omega)$.

By Alexander trick $\widetilde{\mathscr{P}\ell}(D^n)$ and $\widetilde{\mathscr{T}op}(D^n)$ are contractible and by smoothing theory $\widetilde{\mathscr{D}iff}(D^n) \sim \Omega^{n+1}(Pl/O)$.

THEOREM I. 2.1 **[A. B. K.]**$_1$. (a) $\widetilde{\mathscr{P}\ell}(M^n)/\widetilde{\mathscr{D}iff}(M^n) \to$ Maps$(M^n, \partial M^n; Pl/O, *)$ *is a homotopy equivalence*,

(b) $\widetilde{\mathscr{T}op}(M^n)/\widetilde{\mathscr{D}iff}(M^n) \to$ Maps$(M^n, \partial M^n; Top/O, *)$ *is a homotopy equivalence for* $n \geqq 5$,

(c) $\widetilde{\mathscr{T}op}(M^n)/\widetilde{\mathscr{D}iff}(M^n) \to$ Maps$(M^n, \partial M^n; Top/Pl, *)$ *is a homotopy equivalence for* $n \geqq 5$. ■

THEOREM I. 2.2 **[A. B. K.]**$_1$. *For* $n \geqq 5$ *the natural map* $\tilde{d}: \mathscr{H}(M)/\tilde{\mathscr{A}}(M) \to$ Maps$(M, \partial M; G/\mathscr{A}, *)$ *has as fibre* $\Omega^{n+1}(L(\pi_1(M), \omega))$ *where the orientation* $\omega$: $\pi_1(M) \to \mathbf{Z}_2$ *is the first Stiefel-Whitney class.* ■

Theorem I. 2.2 makes clear the importance of the algebraic $L$-theory for our geometric problem. Most often by an algebraic theory we understand a system of functors $T_n$, defined on an "algebraic category" like the category of groups, groups with orientations rings, etc., with values in the category of groups, satisfying some naturality properties which resemble homology, $K$-theory, etc. All the nice algebraic theories we know can be obtained from functors defined on an "algebraic category" with values in the weak homotopy category of $\infty$-loop spaces, composed with the homotopy-groups functors. It was Quinn **[Q]** who first observed that the algebraic $L$-theory can be obtained in this way.

Another algebraic theory which relates the two sorts of algebraic $L$-theories is the algebraic $A$-theory invented by Rothenberg; it was proved in **[B]**$_1$ that this theory can be obtained from a functor $A: \mathscr{G} \to \Omega^h$ with $\mathscr{G}$ the category of groups. The geometric significance of the algebraic $A$-theory becomes deeper when one compares $\tilde{\mathscr{A}}(M)$ and $\tilde{\mathscr{A}}(M \times S^1)$.

THEOREM I. 2.3 **[B]**$_1$. *For any manifold* $M^n \in$ ob $\mathscr{A}$ *there exists an obvious map* ${}^{\mathscr{A}}\omega: \tilde{\mathscr{A}}(M) \times \Omega\tilde{\mathscr{A}}(M) \to \tilde{\mathscr{A}}(M \times S^1)$ *if* $\partial M \neq \phi$ *and* ${}^{\mathscr{A}}\omega: \tilde{\mathscr{A}}(M) \times \Omega\tilde{\mathscr{A}}(M) \times S^1 \to \tilde{\mathscr{A}}(M \times S^1)$ *if* $\partial M = \emptyset$ *so that its homotopy theoretical fibre* ${}^{\mathscr{A}}F(M)$ *is homotopy equivalent to* $\Omega^n A(\pi_1(M))$ *if* $n \geqq 5$. *Moreover* $\Omega^4 A(G) \sim A(G)$ *and if* $Wh_1(G) = 0$ *then* $A(G)$ *is contractibie. (The homotopy groups of* $A(G)$ *are the so-called Rothenberg groups* $A_i(G)$.) ■

[3]A group with orientation means a group $G$, together with a group-homomorphism $\omega: G \to Z_2$ and a morphism of groups with orientations $l: (G_1, \omega_1) \to (G_2\ \omega_2)$ is a group-homomorphism $l$: $G_1 \to G_2$ making commutative tee diagram

$$\begin{array}{ccc} G_1 & \xrightarrow{l} & G_2 \\ & {}_{\omega_1}\searrow \quad \swarrow_{\omega_2} & \\ & Z_2 & \end{array}$$

*Observation.* From Theorem I. 2.3 it is clear that $A(G)_{\text{odd}}$ is contractible since $A_i(G)$ are all 2-primary groups; hence $\tilde{\mathscr{A}}(M \times S^1)_{\text{odd}}$ is homotopy equivalent to $(\tilde{\mathscr{A}}(M) \times \Omega\tilde{\mathscr{A}}(M))_{\text{odd}}$ if $\partial M \neq \emptyset$ and to $(\tilde{\mathscr{A}}(M) \times \Omega\tilde{\mathscr{A}}(M) \times S^1)_{\text{odd}}$ if $\partial M = \emptyset$, $n \geqq 4$.

We also know that the "localization to odd primes" (which loses the 2-primary information) substantially simplifies the description of the structure of the algebraic $L$-theory as it was noticed by Novikov, Karoubi, etc. We will see in Part II how much we get outside the "prime 2" (i.e., localizing to odd primes) about the structure of $\mathscr{A}(M)$.

I should mention that since the appearance of **[A.B.K.]**$_1$ a lot of work has been done towards the computation of $\pi_i(\tilde{\mathscr{A}}(M))$ for $\mathscr{A} = \mathscr{D}iff$ but it is not the intention of this article to report about it.

## PART II

**3. Some history about concordances.** The introduction of $\mathscr{C}^{\mathscr{A}}(M)$ for the purpose of the understanding of $\pi_0(\mathscr{D}iff(M))$ $(\mathscr{C}^{\mathscr{A}}(M) = \mathscr{A}(M \times I; \partial_-(M \times I) = M \times \{1\})$ is due to Cerf. He noticed first the isomorphism between the homotopy groups of $B\mathscr{C}^{\mathscr{D}}(M)$ and the relative homotopy groups of $(\mathscr{F}, \mathscr{F}_0)$ where $\mathscr{F}$ denotes the space of all Morse functions on $M \times I$ and $\mathscr{F}_0$ the space of Morse functions with no critical points (both endowed with the $C^\infty$-topology), and he began the investigation of these relative homotopy groups filtrating $\mathscr{F}$ by codimension. In order to describe the low dimension strata which are necessary for the computation of $\pi_1(\mathscr{F}, \mathscr{F}_0)$ one uses the Thom theory of universal unfoldings. Cerf worked out the case of simply-connected manifolds and some of the geometry involved in the nonsimply-connected case was developed by Chenciger and Laudenbach. The full computation of $\pi_1(\mathscr{F}, \mathscr{F}_0)$ has been done in the nonsimply-connected case by Hatcher and Wagoner **[H.W.]** and Volodin. In our considerations the importance of $\mathscr{C}^{\mathscr{A}}(M)$ comes from the following reasons:

(1) "Concordances" allow us to formulate the disjunction lemma as noticed first by Morlet, and this has remarkable consequences as one can see from Theorems II. 4.2, II. 4.3, II. 4.4, II. 4.5 (§4) below.

(2) "Concordances" have good stability properties, as it was noticed first by Hatcher **[H]** for $\mathscr{A} = \mathscr{P}\ell$.

Remarkable connections between "concordances" and the classical groups of geometric topology were noticed long ago by Kuiper and Lashof **[K.L.]**, and they are some basic ingredients in the proof of stability for $\mathscr{A} = \mathscr{D}iff$.

Recall that for a manifold $M \in \text{ob}\, \mathscr{A}$ we denote by $\mathscr{C}^{\mathscr{A}}(M)$ the s.s.-group $\mathscr{A}(M \times I; M \times \{1\})$. Analogously we define $\tilde{\mathscr{C}}^{\mathscr{A}}(M)$ as the s.s.-complex whose $k$-simplexes are $\mathscr{A}$-automorphisms $h\colon \Delta[k] \times I \times M \to \Delta[k] \times I \times M$ with

$$h(d_i\, \Delta[k] \times I \times M) \subset d_i\Delta[k] \times I \times M, \quad h \mid \Delta[k] \times (I \times \partial M \cup \{0\} \times M) = \text{id}$$

plus some extra properties (near corners) which allow us to define the degeneracies. As one easily notices, $\tilde{\mathscr{C}}^{\mathscr{A}}(M)$ is contractible but this s.s.-group is still interesting since it permits a simple geometric description of $B\mathscr{C}^{\mathscr{A}}(M)$, the classifying space of $\mathscr{C}^{\mathscr{A}}(M)$, as the quotient $\tilde{\mathscr{C}}^{\mathscr{A}}(M)/\mathscr{C}^{\mathscr{A}}(M)$. Having defined $\mathscr{C}^{\mathscr{A}}(M)$ and $B\mathscr{C}^{\mathscr{A}}(M)$ we can formulate for them Problem A and Problem B. The homotopy reduction of Problem B has been already discussed in Part I since $\mathscr{C}^{\mathscr{A}}(M) = \mathscr{A}(M \times I; M \times \{1\})$ but we will come back to it in connection with "stability".

**4. Morlet's disjunction lemma and its applications.** As we have mentioned above, one of the main reasons of our interest for concordances comes from the possibility to formulate the "disjunction lemma". This lemma, a beautiful piece of geometric topology, has been stated and sketchily proved by Morlet **[Mo]**. A complete proof following Morlet's ideas has been given in **[B.L.R.]** and a different proof has been also produced by K. Millet **[Mi]**.

Let $(V^n, \partial V^n)$ be a compact manifold with boundary and $f: (D^p, \partial D^p) \hookrightarrow (V^n, \partial V^n)$, $g: (D^q, \partial D^q) \hookrightarrow (V^n, \partial V^n)$ be two proper embeddings, i.e., they intersect transversally $\partial V$ and $f^{-1}(\partial V)$, resp. $g^{-1}(\partial V)$, are $\partial D^p$, resp. $\partial D^q$. Assume that $f(D^p) \cap g(D^q) = \emptyset$. For an embedding $f$ as above we define $\mathscr{C}^{\mathscr{A}}_{\text{emb}}(D^p, V; f)$ the s.s.-complex whose $k$-simplexes are embeddings $h: \Delta[k] \times I \times D^p \to \Delta[k] \times I \times V^n$ commuting with the projection on $\Delta[k]$ and satisfying the following supplementary properties:

(1) $h_t: (I \times D^p) \to (I \times V^n)$ intersects transversally $\partial(I \times V^n)$ with $h_t^{-1}(\partial(I \times V^n)) = \partial(I \times D^p)$.

(2) $h_t \mid I \times \partial D^p = \text{id}_I \times f$ and $h_t \mid \{0\} \times D^p = f$.

THEOREM II. 4.1 (MORLET'S DISJUNCTION LEMMA). *If $n - q \geqq 3$ and $n - p \geqq 3$ then $\pi_i(\mathscr{C}^{\mathscr{A}}_{\text{emb}}(D^p, V; f), \mathscr{C}^{\mathscr{A}}_{\text{emb}}(D^p, V \backslash g(D^q); f) = 0$ for $i \leqq 2n - p - q - 5$. (If $\mathscr{A} = \mathscr{T}op$ one requires $n \geqq 5$.)* ■

We might say that Theorem II. 4.1 has been stated by Morlet only for $\mathscr{A} = \mathscr{D}iff$ and $\mathscr{P}\ell$. For $\mathscr{A} = \mathscr{T}op$ this theorem is due to Erick Pedersen (see Appendix in **[B.L.R.]**).

The handlebody structure of a manifold, the obvious relation of spaces of "concordances of embeddings" associated to a handlebody decomposition and the relations between spaces of embeddings and concordances of embeddings (as for instance the fibration $\mathscr{E}m\ell^{\mathscr{A}}(D^{p+1}, M^n \times I; f \times \text{id}) \to \mathscr{C}^{\mathscr{A}}_{\text{emb}}(D^p, M^n; f) \to \mathscr{E}m\ell^{\mathscr{A}}(D^p, M; f)$ for $n - p \geqq 3$ permits the proof of the following theorems **[B.L.R.]**:

THEOREM II. 4.2. *Let $V^n$ be a compact k-connected manifold, $n \geqq 5$ and $D^n \subset$* Int $V^n$. *Then*

(a) $\mathscr{A} = \mathscr{D}iff$; $\pi_j(\mathscr{A}(V^n)/\mathscr{A}(D^n)) \to \pi_j(\tilde{\mathscr{A}}(V^n)/\tilde{\mathscr{A}}(D^n))$ *is an isomorphism if* (1) $j \leqq 2k - 3$ ($k \leqq n - 4$ *if* $\partial V$ *not* 1-*connected or* $n = 5$), (2) $j \leqq 2k - 2$ *if* $(k + 1) < n/2$ *and either* $k \equiv 2, 4, 5, 6 \pmod 8$ *or* $\tau V$ *is trivial over the* $(k + 1)$-*skeleton.*

(b) $\mathscr{A} = \mathscr{P}\ell$ *or* $\mathscr{T}op$; $\pi_j(\mathscr{A}(V)) \to \pi_j(\tilde{\mathscr{A}}(V))$ *is an isomorphism if*

(1) $j \leqq \inf(2k - 3, k + 2)$ ($k \leqq n - 4$ *if* $\partial V$ *not* 1-*connected or* $n = 5$),

(2) $j \leqq 2k - 2$ *if* $(k + 1) < n/2$ *and either* $k = 2, 4$ *or* $k = 3$ *and* $\tau V$ *is trivial over the* 4-*skeleton.* ■

THEOREM II. 4.3. *Let $V^n$ be a compact k-connected manifold $n \geqq 5$ and $D^n \subset$* Int $V^n$. *Then*:

(a) $\mathscr{A} = \mathscr{D}iff$; $\pi_j(\mathscr{C}^{\mathscr{A}}(D^n)) \to \pi_j(\mathscr{C}^{\mathscr{A}}(V))$ *is an isomorphism for*

(1) $j \leqq 2k - 4$ ($k \leqq n - 4$ *if* $\partial V$ *not* 1-*connected or* $n = 5$),

(2) $j \leqq 2k - 2$ *if* $k \leqq n - 4$ (($k + 1) < n/2$ *and either* $k = 2, 4, 5, 6 \pmod 8$ *or* $\tau V$ *is trivial over the* $(k + 1)$-*skeleton*;

(b) $\mathscr{A} = \mathscr{P}\ell$ *or* $\mathscr{T}op$; $\pi_j(\mathscr{C}^{\mathscr{A}}(V)) = 0$ *if*

(1) $j \leqq \inf(2k - 3, k + 2)$, $k \leqq n - 4$ *if* $\partial V$ *not* 1-*connected or*

(2) $j \leqq 2k - 2$ *if* $(k + 1) < n/2$ *and either* $k = 2, 4$ *or* $k = 3$ *and* $\tau V$ *is trivial over the 4-skeleton of* $V$. ■

THEOREM II. 4.4. *Suppose* $V^n$, $V'^n$ *are k-connected compact manifolds of the same r-tangential homotopy type* $n/2 > r + 1 \geqq k$. *Then* $\tilde{\mathscr{A}}(V)/\mathscr{A}(V)$ *and* $\tilde{\mathscr{A}}(V')/\mathscr{A}(V')$ *have the same j-homotopy type for*

(a) $\mathscr{A} = \mathscr{D}iff$ *if* $j \leqq \inf(2r - 1, r + k - 1)$,

(b) $\mathscr{A} = \mathscr{P}\ell$ *or* $\mathscr{T}op$ *if* $j \leqq \inf(2r - 1, r + k - 1, r + 3)$. ■

THEOREM II. 4.5. *Suppose* $V^n$ *and* $V'^n$ *are k-connected compact manifolds of* $\dim n \geqq 5$ *and* $V$ *and* $V'$ *of the same r-tangential homotopy type* $n/2 > r + 1 \geqq k$. *Then* $\mathscr{C}^{\mathscr{A}}(V)$ *and* $\mathscr{C}^{\mathscr{A}}(V')$ *have the same j-homotopy type for*

(a) $\mathscr{A} = \mathscr{D}iff$; $j \leqq \inf(2r - 2, r + k - 2)$,

(b) $\mathscr{A} = \mathscr{P}\ell, \mathscr{T}op$; $j \leqq \inf(2r - 2, r + k - 2, r + 2)$. ■

These theorems show that for both $\mathscr{C}^{\mathscr{A}}(M)$ and $\tilde{\mathscr{A}}(M)/\mathscr{A}(M)$ the Postnikov $\lambda^{\mathscr{A}}(n)$th term depends only on the tangential homotopy type of $M$, where $\lambda^{\mathscr{A}}: Z_+ \to Z_+$ is an increasing function with $\lambda^{\mathscr{A}}(n) \geqq n/2$ for $n \geqq 12$.

**5. Stability.** The next remarkable fact about concordances are their stability properties. The "stability" has been discovered for $\mathscr{A} = \mathscr{P}\ell$ by A. Hatcher **[H]** and then proved for $\mathscr{A} = \mathscr{D}iff$ by Burghelea and Lashof **[B.L.]**$_2$. The proof for $\mathscr{A} = \mathscr{T}op$ is a trivial consequence of the relationship between $\mathscr{P}\ell$-concordances and $\mathscr{T}op$-concordances for a $p\ell$-manifold. Unfortunately for a moment the proof of the "stability property" of concordances for $\mathscr{A} = \mathscr{D}iff$, $\mathscr{T}op$ depends on the proof for $\mathscr{A} = \mathscr{P}\ell$; we do not know a proof simultaneously true in all geometric categories. To explain the stability property we describe the "transfer map" for concordances as we will briefly sketch below.

Let $E \to^{\xi} B$ be a locally trivial $\mathscr{A}$-bundle, $E, B \in \text{ob}\, \mathscr{A}$, called for short an $\mathscr{A}$-bundle. There exists a well-defined homotopy class ${}^{\mathscr{A}}l^{\xi}$: $\tilde{\mathscr{C}}^{\mathscr{A}}(B)/\mathscr{C}^{\mathscr{A}}(B) \to \tilde{\mathscr{C}}^{\mathscr{A}}(E)/\mathscr{C}^{\mathscr{A}}(E)$ called the "transfer map" satisfying the following properties:

(1) If $E_1 \to^{\xi_1} E_2 \to^{\xi_2} E_3$ are two $\mathscr{A}$-bundles then $l^{\xi_2 \cdot \xi_1} = l^{\xi_1} . l^{\xi_2}$.

(2) If $N^n \subset B^n$, $\xi$: $E \to B$ is an $\mathscr{A}$-bundle, $\xi' = \xi$: $\xi^{-1}(N) \to N$ is the restriction of $\xi$ to $N$ and

$$i_{N,B}: \tilde{\mathscr{C}}^{\mathscr{A}}(N)/\mathscr{C}^{\mathscr{A}}(N) \to \tilde{\mathscr{C}}^{\mathscr{A}}(B)/\mathscr{C}^{\mathscr{A}}(B)$$
$$i_{E',E}: \tilde{\mathscr{C}}^{\mathscr{A}}(E')/\mathscr{C}^{\mathscr{A}}(E') \to \tilde{\mathscr{C}}^{\mathscr{A}}(E)/\mathscr{C}^{\mathscr{A}}(E)$$

$E' = \xi^{-1}(N)$, are the natural s.s.-maps induced by the inclusions $N^n \subset B^n$ and $E' \subset E$, then $l^{\xi} \cdot i_{N,B} \sim i_{E',E} \cdot l^{\xi'}$.

(3) If $\mathscr{B} \geqq \mathscr{A}$ and $\gamma^{B}_{\mathscr{B},\mathscr{A}}$: $\tilde{\mathscr{C}}^{\mathscr{A}}(B)/\mathscr{C}^{\mathscr{A}}(B) \to \tilde{\mathscr{C}}^{\mathscr{B}}(B)/\mathscr{C}^{\mathscr{B}}(B)$ are the s.s.-maps defined by regarding $\mathscr{A}$-concordances as $\mathscr{B}$-concordances, then ${}^{\mathscr{B}}l^{\xi} \cdot \gamma^{B}_{\mathscr{B},\mathscr{A}} \sim \gamma^{E}_{\mathscr{B},\mathscr{A}} \cdot l^{\xi}$.

If $\xi$ is the trivial bundle and $K$ its fibre we will write $l^K$ instead of $l^{\xi}$.

The reason we call $l^{\xi}$ "transfer map" comes from the fact that, for $\mathscr{B} \geqq \mathscr{A}$, $l^{\xi}$ induces

$$l^{\xi}_{\mathscr{B},\mathscr{A}}: \mathscr{C}^{\mathscr{B}}(M)/\mathscr{C}^{\mathscr{A}}(M) \to \mathscr{C}^{\mathscr{B}}(E)/\mathscr{C}^{\mathscr{A}}(E),$$

and $\mathscr{C}^{\mathscr{B}}(\cdots)/\mathscr{C}^{\mathscr{A}}(\cdots)$ in "stable ranges" behaves like a generalized homology theory while $l^{\xi}_{\mathscr{A},\mathscr{B}}$ behaves like the transfer morphism associated to the fibration $E \to M$.

Since we are able to prove the expected algebraic properties of the "transfer map" only for the trivial bundle, we will define $l^\xi$ only in this particular case. If $K$ is an $\mathscr{A}$-manifold with empty boundary define $\Omega l^K\colon \mathscr{C}^{\mathscr{A}}(M) \to \mathscr{C}^{\mathscr{A}}(M \times K)$ by $\Omega l^K(h) = h \times \mathrm{id}_K$. A similar definition works for $\tilde{\mathscr{C}}^{\mathscr{A}}(\cdots)$ and therefore it induces $l^K$: $\tilde{\mathscr{C}}^{\mathscr{A}}(M)/\mathscr{C}^{\mathscr{A}}(M) \to \tilde{\mathscr{C}}^{\mathscr{A}}(M \times K)/\mathscr{C}^{\mathscr{A}}(M \times K)$. Assume now $\partial K \neq \emptyset$ and in this case regard $K$ as $K_0 \cup \partial K \times [0, 1]$, i.e., we specify a particular collar of the neighborhood which we identify with $\partial K \times [0, 1]$ and denote by $K_0$ the closure of the complement of this collar. We define again the group homomorphism $\Omega l^K$: $\mathscr{C}^{\mathscr{A}}(M) \to \mathscr{C}^{\mathscr{A}}(M \times K)$ as follows:

For a concordance $h \in \mathscr{C}(M)$, i.e., $h\colon M \times I \to M \times I$, $I = [0, 1]$, we consider the $\mathscr{A}$-automorphism $h \times \mathrm{id}_K\colon M \times K_0 \times I \to M \times K_0 \times I$; $h \times \mathrm{id}_K$ is not yet a concordance since $h \times \mathrm{id}_{K_0}$ is not identity if restricted to $M \times \partial K_0 \times I$, but $h \times \mathrm{id}_{\partial K_0}$ is a concordance. We construct the concordance $\Omega l^K(h)$ taking $\Omega l^K(h) = h \times \mathrm{id}_{K_0}$ on $M \times K_0 \times I$ and rotating the concordance $h \times \mathrm{id}_{\partial K_0}$ around $M \times \partial K_0$ inside $M \times \partial K_0 \times [0, 1] \times I$ as indicated in Figure 1.

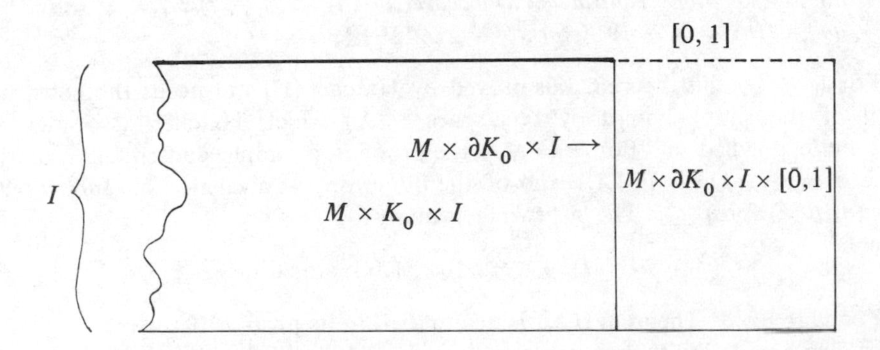


FIGURE 1

To be precise represent a point $u$ in $[0, 1] \times I$ (Figure 2) by its polar coordinates $(r, \theta)$, $r$ being the distance from $u$ to $A$ and $\theta = \measuredangle BAu \in [0, \pi/2]$:

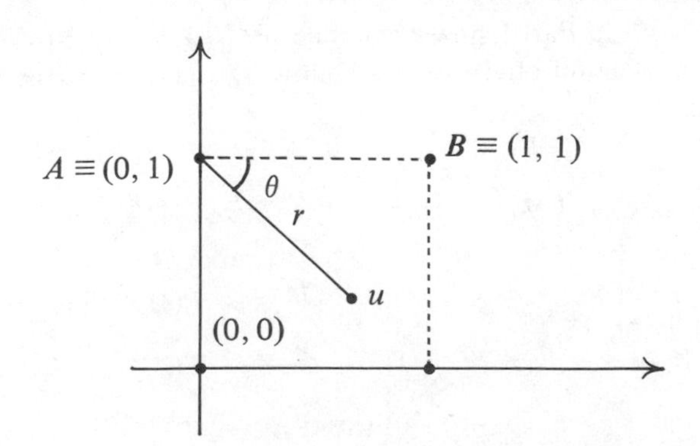


FIGURE 2

write $h(m, t) = (h_1(m, t), h_2(m, t)) \in M \times I$ and define $\Omega l^K(h)$ by the formula:

$$\Omega l^K(h)(m, k, t) = (h_1(m, t), k, h_2(m, t)) \quad \text{for } k \in K_0,$$
$$\Omega l^K(h)(m, k, t) = \begin{cases} h_1(m, 1 - r), k', 1 - h_2(m, 1 - r), \theta & \text{for } 0 \leqq r \leqq 1, \\ (m, k', r, \theta) & \text{if } r \geqq 1, m \in M, k_l \in \partial K_0, \quad (r, \theta) \in [0, 1] \times I. \end{cases}$$

$\Omega l^K$ defined in this way is an s.s.-group homomorphism and it extends (by a similar formula) to a group homomorphism $\Omega l^K: \tilde{\mathscr{C}}^{\mathscr{A}}(M) \to \tilde{\mathscr{C}}^{\mathscr{A}}(M \times K)$; hence it induces the s.s.-map $l^K: \tilde{\mathscr{C}}^{\mathscr{A}}(M)/\mathscr{C}^{\mathscr{A}}(M) \to \tilde{\mathscr{C}}^{\mathscr{A}}(M \times K)/\mathscr{C}^{\mathscr{A}}(M \times K)$. It is pretty obvious that $l^K$ (up to a homotopy) does not depend on the chosen collar.

A particular case of this construction is the case when $K = I$ and $K = D^r$. From now on we denote $l^I$ by $\Sigma$ and $l^{D^r}$ by $\Sigma_r$; it is easy to observe that $(\Sigma)^r \sim \Sigma_r$ because of property (a) of the "transfer map".

We are ready now to state the stability theorems.

THEOREM II. 5.1. *For any geometric category $\mathscr{A}$, there exists an increasing function $\omega^{\mathscr{A}}: Z_+ \to Z_+$ with $\lim_{n\to\infty} \omega^{\mathscr{A}}(n) \to \infty$ so that the transfer map $\Sigma$, called also "suspension", $\Sigma: \tilde{\mathscr{C}}^{\mathscr{A}}(M)/\mathscr{C}^{\mathscr{A}}(M) \to \tilde{\mathscr{C}}^{\mathscr{A}}(M \times I)/\mathscr{C}^{\mathscr{A}}(M \times I)$ induces an $\omega^{\mathscr{A}}$(dim $M$)-homotopy equivalence; moreover $\omega^{\mathscr{A}}(n) \geqq (n - 7)/3$ for $\mathscr{A} = \mathscr{P}\ell$ and $\mathscr{T}op$ and $\omega^{\mathscr{A}}(n) \geqq (n - 19)/6$ for $\mathscr{A} = \mathscr{D}iff$.* ■

For $\mathscr{A} = \mathscr{P}\ell$ this theorem was proved by Hatcher **[H]** and needs the construction of the simple-homotopy type-space $\mathscr{S}(X)$ which Hatcher associates to any finite polyhedron. For $\mathscr{A} = \mathscr{T}op$ the proof is a simple consequence of the case $\mathscr{A} = \mathscr{P}\ell$, Theorem I.1.1 and of the homotopy equivalence of *Top*/*Pl* with *Top(n)*/*Pl(n)* for $n \geqq 5$. The last two facts imply

$$\tilde{\mathscr{C}}^{\mathscr{P}\ell}(M)/\mathscr{C}^{\mathscr{P}\ell}(M) \sim \tilde{\mathscr{C}}^{\mathscr{T}}(M)/\mathscr{C}^{\mathscr{T}}(M) \quad \text{for dim } M \geqq 5.$$

For $\mathscr{A} = \mathscr{D}iff$ Theorem II.5.1 is due to **[B.L.]**$_2$; its proof is based on:

(i) The compatibility between the "suspension"-map for classical groups and the suspension "$\Sigma$",

(ii) the Kuiper-Lashof results **[K.L.]** which connect the topological concordances (resp. *pl*-concordances) to *Top*$(n + 1)$/*Top*$(n)$ (resp. *Pl*$(n + 1)$/*Pl*$(n)$),

(iii) the truth of Theorem II. 5.1 for $\mathscr{A} = \mathscr{P}\ell$.

The proof goes as follows.

With the notations of Part I, §1 we can identify $\Gamma(\mathscr{F}^n_{\mathscr{B},\mathscr{A}}(M))$ to $\Gamma(\mathscr{E}_{\mathscr{B},\mathscr{A}}(M \times I))$ and (i) means the commutativity of the following diagram whose horizontal lines are I.H.E.-maps.

$$\begin{array}{ccc} \mathscr{C}^{\mathscr{B}}(M)/\mathscr{C}^{\mathscr{A}}(M) & \xrightarrow{\quad d \quad} & \Gamma(\mathscr{F}^n_{\mathscr{B},\mathscr{A}}(M), s) \\ & & \downarrow \Gamma(\phi^{\mathscr{B},\mathscr{A}}_n) \\ \downarrow \Sigma & & \Gamma(\Omega\mathscr{F}^{n+1}_{\mathscr{B},\mathscr{A}}(M)) \\ & & \wr\wr \\ \mathscr{C}^{\mathscr{B}}(M \times I)/\mathscr{C}^{\mathscr{A}}(M \times I) & \longrightarrow & \Gamma(\mathscr{F}^{n+1}_{\mathscr{B},\mathscr{A}}(M \times I)) \end{array}$$

FIGURE 3

If $M$ has a trivial tangent bundle then the diagram above becomes

$$\begin{array}{ccc} \mathscr{C}^{\mathscr{B}}(M)/\mathscr{C}^{\mathscr{A}}(M) & \longrightarrow & \text{Maps}(M, \partial M; F^n_{\mathscr{B},\mathscr{A}}, *) \\ \downarrow \Sigma & & \downarrow \\ \mathscr{C}^{\mathscr{B}}(M \times I)/\mathscr{C}^{\mathscr{A}}(M \times I) & \longrightarrow & \text{Maps}(M, \partial M; \Omega F^{n+1}_{\mathscr{B},\mathscr{A}}, *) \\ & & \wr\wr \\ & & \text{Maps}(M \times I, \partial(M \times I); F^{n+1}_{\mathscr{B},\mathscr{A}}, *) \end{array}$$

FIGURE 4

A careful analysis (for $M = S^n \times D^k$) combined with Theorem II.4.5 converts the diagram above (Figure 4) into the following diagram

$$\begin{array}{ccc} \pi_i(\mathscr{C}^{\mathscr{T}_{\!\!\mathscr{P}}}(S^n \times D^k)) & \xrightarrow{\pi_i(\Sigma)} & \pi_i(\mathscr{C}^{\mathscr{T}_{\!\!\mathscr{P}}}(S^n \times D^{k+1})) \\ \downarrow \pi_i(d) & & \downarrow \pi_i(d) \\ \pi_i(\Omega^k F^{n+k}_{\mathscr{T}_{\!\!\mathscr{P}},\mathscr{D}\mathscr{I}}) & \xrightarrow{\pi_i(\Omega^k\psi)} & \pi_i(\Omega^{k+1} F^{n+k+1}_{\mathscr{T}_{\!\!\mathscr{P}},\mathscr{D}\mathscr{I}}) \end{array}$$

with $\pi_i(d)$ an isomorphism for $i \leqq 2n - 4$ and an epimorphism for $i \leqq 2n - 3$. Since by Kuiper-Lashof's results $\mathscr{C}^{\mathscr{T}_{\!\!\mathscr{P}}}(S^n)$ and $F^n_{\mathscr{T}_{\!\!\mathscr{P}},\mathscr{D}\mathscr{I}}$ are homotopy equivalent by a homotopy equivalence $d$ which induces for homotopy groups the homomorphism $\pi_i(d)$, we conclude that $\psi^r_{\mathscr{T}_{\!\!\mathscr{P}},\mathscr{D}\mathscr{I}}\colon F^r_{\mathscr{T}_{\!\!\mathscr{P}},\mathscr{D}\mathscr{I}} \to F^{r+1}_{\mathscr{T}_{\!\!\mathscr{P}},\mathscr{D}\mathscr{I}}$ is $r + s - 1$ connected for $2s + 1 \leqq \omega^{\mathscr{T}\mathscr{o}\mathscr{p}}(r)$. Using this fact together with the truth of Theorem II. 5.1 for $\mathscr{A} = \mathscr{T}\!\mathscr{o}\!\mathscr{p}$ one concludes Theorem II. 5.1 for $\mathscr{A} = \mathscr{D}\!\mathscr{i}\!\mathscr{f}\!\mathscr{f}$.

COROLLARY II. 5.2. *If $\xi\colon E \to M$ is an $\mathscr{A}$-bundle with fibre a disc, then $l^\xi$: $\tilde{\mathscr{C}}^{\mathscr{A}}(M)/\mathscr{C}^{\mathscr{A}}(M) \to \tilde{\mathscr{C}}^{\mathscr{A}}(E)/\mathscr{C}^{\mathscr{A}}(E)$ induces an $\omega^{\mathscr{A}}(\dim M)$-homotopy equivalence.* ■

Let $D^r \subset K^r$, $K^r \in \text{ob}\,\mathscr{A}$ and let us consider:

(1) $\tilde{\mathscr{C}}^{\mathscr{A}}(M)/\mathscr{C}^{\mathscr{A}}(M) \to^{\Sigma^r} \tilde{\mathscr{C}}^{\mathscr{A}}(M \times D^r)/\mathscr{C}^{\mathscr{A}}(M \times D^r) \to^{i} \tilde{\mathscr{C}}^{\mathscr{A}}(M \times K)/\mathscr{C}^{\mathscr{A}}(M \times K)$,
(2) $\tilde{\mathscr{C}}^{\mathscr{A}}(M)/\mathscr{C}^{\mathscr{A}}(M) \to^{l^K} \tilde{\mathscr{C}}^{\mathscr{A}}(M \times K)/\mathscr{C}^{\mathscr{A}}(M \times K)$,
(3) $\varphi\colon \tilde{\mathscr{C}}^{\mathscr{A}}(M)/\mathscr{C}^{\mathscr{A}}(M) \to G$,

with $G$ a weak commutative weak group and $i$ induced by the inclusion $M \times D^r \subset M \times K^r$.

THEOREM II. 5.3. *$\Sigma^2 \circ l^{K'} \sim x(K)\,(\varphi \circ i \circ \Sigma^r)$ if $K' = K \times I$.* ■

This last theorem was proved for $\mathscr{A} = \mathscr{P}\!\ell$ by Hatcher (see **[H]**); a proof simultaneously valid for all geometric categories is given in **[B.L.]**$_3$.

**6. Topological nilpotencies.** For an $\mathscr{A}$-manifold $M^n$ and a base pointed $\mathscr{A}$-manifold $(V, v_0)$ (for instance $V = R$, $v_0 = 0$ or $V = S^1$, $v_0 = e$) we define $\mathscr{E}\mathit{mb}^{\mathscr{A}}(M, M \times V)$, respectively $\mathscr{E}\mathit{mb}^{\mathscr{A}}_{m_0}(M, M \times V)$, the s.s.-complex of $\mathscr{A}$-embeddings which restrict on $\partial M$ respectively $\partial M \cup m_0$ to the canonical embedding $i$, $i(m) = (m, v_0)$; we denote by $\mathscr{E}^v\mathit{mb}^{\mathscr{A}}(M, M \times V)$, respectively $\mathscr{E}^v\mathit{mb}^{\mathscr{A}}_{m_0}(M, M \times V)$, the union of those connected components of $\mathscr{E}\mathit{mb}^{\mathscr{A}}(M; M \times V)$, respectively $\mathscr{E}\mathit{mb}^{\mathscr{A}}_{m_0}(M, M \times V)$, whose embeddings composed by the canonical projection on $V$ are homotopic to the constant map.

Let $j^{\mathscr{A}}(M)\colon \mathscr{E}^v\mathit{mb}^{\mathscr{A}}(M, M \times S^1) \to \mathscr{E}\mathit{mb}^{\mathscr{A}}(M, M \times R)$ if $\partial M \neq \emptyset$ respectively $j^{\mathscr{A}}_{m_0}(M)\colon \mathscr{E}^v\mathit{mb}^{\mathscr{A}}_{m_0}(M, M \times S^1) \to \mathscr{E}\mathit{mb}^{\mathscr{A}}_{m_0}(M, M \times R)$ if $\partial M = \emptyset$, be the obvious s.s.-maps induced by the lifting in the covering $M \times R \to M \times S^1$, and $\mathscr{N}^{\mathscr{A}}(M)$

the homotopy theoretic fibre of $j^{\mathscr{A}}(M)$, respectively $j^{\mathscr{A}}_{m_0}(M)$ (since the homotopy type of the homotopy theoretic fibre of $j^{\mathscr{A}}_{m_0}(M)$ if $\partial M = \emptyset$ does not depend on $m_0$, we delete $m_0$ from our notation). Analogously we have $\mathscr{C}j^{\mathscr{A}}(M)\colon \mathscr{C}^{\mathscr{A}}_{\text{emb}}(M, M \times S^1) \to \mathscr{C}^{\mathscr{A}}_{\text{emb}}(M, M \times R)$ and $\mathscr{C}\mathscr{N}^{\mathscr{A}}(M)$ the homotopy theoretic fibre of $\mathscr{C}j^{\mathscr{A}}(M)$. Theorem II. 6.1 describes the properties of $\mathscr{N}(M)$ and $\mathscr{C}\mathscr{N}(M)$, and justifies the name of topological nilpotency, respectively topological concordance nilpotency, as a topological analogy for the algebraic nilpotency. The algebraic nilpotency defined for a group $G$ is related to the algebraic $K$-theory of $G$ by the following formula: $K_i(G \times Z) \cong K_i(G) \oplus K_{i-1}(G) \oplus \mathscr{N}\!\mathit{il}^+_i(G) \oplus \mathscr{N}\!\mathit{il}^-_i(G)$.

THEOREM II. 6.1. (1) *There exists a homotopy equivalence* $\mathscr{C}\gamma^{\mathscr{A}}(M)\colon \mathscr{C}^{\mathscr{A}}(M \times S^1) \to \mathscr{C}^{\mathscr{A}}(M \times I) \times B\mathscr{C}^{\mathscr{A}}(M \times I) \times \mathscr{C}\mathscr{N}^{\mathscr{A}}(M)$ *which is natural with respect to inclusions* $M^n \subset N^n$ *and the category* $\mathscr{A}$.

(2) *There exists a homotopy equivalence* $\gamma^{\mathscr{A}}(M)\colon \mathscr{A}(M \times S^1) \to \mathscr{A}(M \times I)^0 \times B\mathscr{A}(M \times I) \times \mathscr{N}^{\mathscr{A}}(M)^0 \times \boldsymbol{T}$ *with* $\boldsymbol{T} = S^1$ *if* $\partial M = \emptyset$ *and* $\boldsymbol{T} = \text{pt}$ *if* $\partial M \neq \emptyset$; $G^0$ *denotes the base point connected component of* $G$. $\gamma^{\mathscr{A}}(M)$ *is natural with respect to inclusions* $M^n \subset N^n$ *and the category* $\mathscr{A}$.

(3) $\mathscr{N}^{\mathscr{A}}(M) \to \mathscr{N}^{\mathscr{B}}(M)$ *and* $\mathscr{C}\mathscr{N}^{\mathscr{A}}(M) \to \mathscr{C}\mathscr{N}^{\mathscr{B}}(M)$ *are homotopy equivalences for* $M$ *an* $\mathscr{A}$*-manifold* $\mathscr{B} \geqq \mathscr{A}$, *and* $\dim M \geqq 4$ *if* $\mathscr{B} = \mathscr{T}\!\mathit{op}$.

(4) *If* $M^n$ *and* $N^n$ *are* $k$*-tangential homotopy equivalent then* $\mathscr{N}(M^n)$ *and* $\mathscr{N}(N^n)$, *respectively* $\mathscr{C}\mathscr{N}(M^n)$ *and* $\mathscr{C}\mathscr{N}(N^n)$, *are* $\inf(k-2, n-1/2)$*-homotopy equivalent, respectively* $\{\inf(k-2, n-1/2)-1\}$*-homotopy equivalent.* ∎

The proof of (1), (2) and (3) can be found in **[B.L.R.**, Chapter VI], or $[\mathbf{B}]_2$ and of (4) in $[\mathbf{B}]_1$. For $M = M' \times I$, (1) and (2) follow immediately since we can find a homotopy inverse $\kappa\colon \mathscr{A}(M \times S^1) \to \mathscr{A}(M \times I)$ for the inclusion $\mathscr{A}(M \times I) \subseteq \mathscr{A}(M \times S^1)$. The proof of (3) follows from Theorem I. 1.1 while of (4) from Theorems II. 4.4. and II. 4.5. and Theorem I. 2.3.

One should notice that one can define a transfer map $\mathscr{N}l^{\xi}\colon \mathscr{C}\mathscr{N}(M) \to \mathscr{C}\mathscr{N}(E)$ which satisfies the properties (1) and (2) of the transfer map $l^{\xi}$ and $l^{\xi \times \text{id}_{S^1}} \sim \Omega l^{\xi} \times l^{\xi} \times \mathscr{N}l^{\xi}$.

**7. The homotopy functors $\mathscr{S}^{\mathscr{A}}$ and $\mathscr{S}\mathscr{N}$.** The considerations of §5 imply Theorem II. 7.1 which beside its interest in the study of the homotopy type of automorphisms directs the attention of the homotopy theory towards new kinds of functors. Before stating Theorem II. 7.1, recall that we have denoted by $\Omega^{wh}$ the category whose objects are $\infty$-loop spaces and morphisms weak homotopy classes of $\infty$-loop space maps; we say $f, g\colon X \to Y$ are weak homotopic if their $k$th Postnikov terms $[f]_k$ and $[g]_k$ are homotopic for any $k$.

THEOREM II. 7.1. *For any geometric category* $\mathscr{A}$ *there exists a functor* $\mathscr{S}^{\mathscr{A}}\colon \mathscr{P} \to \Omega^h$ *with the following properties*:

(a) $\mathscr{S}^{\mathscr{A}}$ *are homotopy functors, i.e. if* $f, g\colon X \to Y$ *are homotopic then* $\mathscr{S}^{\mathscr{A}}(f), \mathscr{S}^{\mathscr{A}}(g)$ *are weak homotopic* $\infty$*-loop space maps.*

(b) *If* $f\colon X \to Y$ *is* $k$*-connected then* $\mathscr{S}^{\mathscr{A}}(f)$ *is* $k$*-connected.*

(c) *If* $E \to^{\xi} B$ *is a locally trivial* $\mathscr{A}$*-bundle* $E$, $B$ *compact* $\mathscr{A}$*-manifolds there exists a "transfer map"* $l^{\xi}\colon \mathscr{S}^{\mathscr{A}}(B) \to \mathscr{S}^{\mathscr{A}}(E)$, *a morphism in* $\Omega^{wh}$ *so that*:

(i) $E_1 \to^{\xi_1} E_2 \to^{\xi_2} E_3$ *are two bundles as above then* $l^{\xi_1 \circ \xi_1} \sim l^{\xi_1} \circ l^{\xi_2}$.

(ii) *If*

$$\begin{array}{ccc} E_1 & \xrightarrow{\xi_1} & B_1 \\ \Big\downarrow f_E & & \Big\downarrow f_B \\ E_2 & \xrightarrow{\xi_2} & B_2 \end{array}$$

*is a cartesian diagram with bundles as above then* $l^{\xi_2} \circ \mathcal{S}^{\mathcal{A}}(f_B) \sim \mathcal{S}^{\mathcal{A}}(f_E) \circ l^{\xi_1}$.

(d) *If* $\xi$ *is the trivial bundle, i.e.,* $\xi \colon B \times K \to B$ *is the projection on* $B$ *with* $K$ *a compact connected* $\mathcal{A}$*-manifold and* $i \colon B \to B \times K$ *a cross-section* $i(b) = b \times K_0$ *then* $l^K$ *is homotopic to* $\chi(K)$ $(\mathcal{S}^{\mathcal{A}}(i))$.

(e) *If* $\mathcal{B} \geqq \mathcal{A}$, *then for any* $X$ *there exists a* $\infty$*-loop space map* $\gamma^{(X)}_{\mathcal{B},\mathcal{A}} \colon \mathcal{S}^{\mathcal{A}}(X) \to \mathcal{S}^{\mathcal{B}}(X)$, *and* $\gamma_{\mathcal{B},\mathcal{A}}(X)$ *defines a natural transformation of functors whose homotopy theoretic fibre* $\mathcal{F}_{\mathcal{B},\mathcal{A}}(X)$ *is a* $\infty$*-loop space whose homotopy groups* $\pi_i(\mathcal{F}_{\mathcal{B},\mathcal{A}}(X)) = H_i(X; \mathcal{W}(\mathcal{B}, \mathcal{A}))$.

(f) *for any* $X \in \text{ob}\ \mathcal{A}$ *there exists a well-defined homotopy class* ${}^{\mathcal{A}}\beta(X) \colon \underline{\mathcal{H}}(X) \to \mathcal{S}^{\mathcal{A}}(X)$ *natural with respect to* $\mathcal{A}$ *and homotopy equivalences in the sense that if* $f \colon X \to Y$ *is a homotopy simple equivalence which induces* $f_\# \colon \underline{\mathcal{H}}(X) \to \underline{\mathcal{H}}(Y)$ *the diagram*

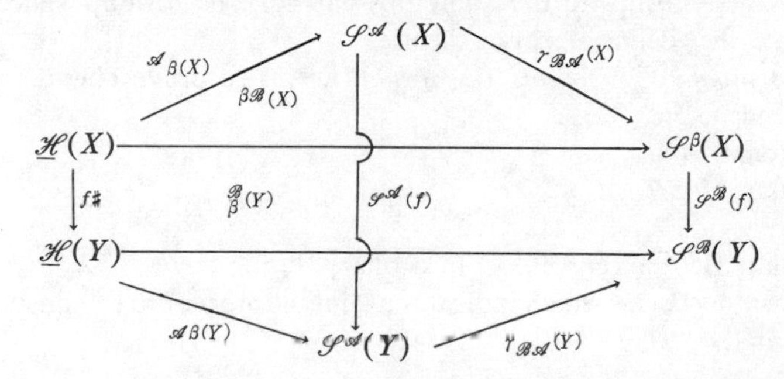


*is commutative; moreover* $\beta$ *is compatible with the transfer map, i.e., the diagram*

$$\begin{array}{ccc} \underline{\mathcal{H}}(X) & \xrightarrow{\beta(X)} & \mathcal{S}^{\mathcal{A}}(X) \\ \Big\downarrow \times \text{id}_k & & \Big\downarrow l^k \\ \underline{\mathcal{H}}(X \times K) & \xrightarrow{\rho(X \times K)} & \mathcal{S}^{\mathcal{A}}(X \times K) \end{array}$$

*is commutative.*

(g) *There exist an increasing function* $\omega^{\mathcal{A}} \colon Z_+ \to Z_+$ *called "the stable range"*

$$\omega^{\mathcal{A}}(n) \geqq \begin{cases} (n - 7)/3 & \text{if } \mathcal{A} = \mathcal{P}l, \mathcal{T}op, \\ (n - 19)/6 & \text{if } \mathcal{A} = \mathcal{D}iff, \end{cases}$$

*and a natural map (with respect to inclusions* $N^n \subset M^n$ *and* $\mathcal{B} \geqq \mathcal{A}$) ${}^{\mathcal{A}}i(M) \colon \tilde{\mathcal{C}}^{\mathcal{A}}(M)/\mathcal{C}^{\mathcal{A}}(M) \approx B\mathcal{C}^{\mathcal{A}}(M) \to \mathcal{S}^{\mathcal{A}}(M)$ *which is an* $\omega^{\mathcal{A}}(\dim M)$*-homotopy equivalence.*

(h) $\gamma_{\mathcal{T}op, \mathcal{P}l}$ *is an isomorphism,* $\mathcal{S}^{\mathcal{P}l}(\text{pt}) \sim *$, $\mathcal{S}^{\mathcal{D}iff}(\text{pt})$ *is the* $(-1)$*-component of the* $\Omega$*-spectrum associated to* $\mathcal{W}_{\mathcal{P}l, \mathcal{D}iff}$. ■

THEOREM II. 7.2. *There exists a functor* $\mathcal{S}\mathcal{N} \colon \mathcal{P} \to \Omega^{wh}$ *which satisfies*:

(a), (b), (c) *and* (d) *from Theorem* II. 7.1 *with* $\mathcal{S}^{\mathcal{A}}$ *replaced by* $\mathcal{S}\mathcal{N}$.

(e) $\mathcal{S}^{\mathcal{A}}(X \times S^1)$ *is naturally homotopy equivalent to* $\mathcal{S}^{\mathcal{A}}(X) \times B\mathcal{S}^{\mathcal{A}}(X) \times B\mathcal{S}\mathcal{N}(X)$.

(f) *For any* $\mathcal{A}$*-manifold* $M^n$ *there exists a map (natural with respect to inclu-*

*sions* $N^n \subset M^n$ *and the category* $\mathscr{A}$), $i^{\mathscr{A}}(M)$: $\mathscr{C}\mathscr{N}^{\mathscr{A}}(M) \to \mathscr{S}\mathscr{N}(M)$ *which is an* $\omega^{\mathscr{A}}(n)$*-homotopy equivalence.* ■

*Comments on Theorems* II. 7.1 *and* II. 7. 2. (i) (g) of Theorem II. 7.1 and (f) of Theorem II. 7.2 imply the unicity of $\mathscr{S}^{\mathscr{A}}$ and $\mathscr{S}\mathscr{N}$ up to an isomorphism of functors.

(ii) Hatcher and Wagoner, Hatcher and Volodin have computed $\pi_0(\mathscr{S}^{\mathscr{A}}(X))$, $\pi_1(\mathscr{S}^{\mathscr{A}}(X))$ and partly $\pi_2(\mathscr{S}^{\mathscr{A}}(X))$ in terms of an algebraic higher order Whitehead theory (see Hatcher—these PROCEEDINGS).

(iii) Hatcher [H] discovered a way to describe $\mathscr{S}^{\mathscr{D}\mathscr{I}}$ in terms of $\mathscr{S}^{\mathscr{B}}$.

(iv) The map ${}^{\mathscr{A}}\mathscr{B}(X)$: $\underline{\mathscr{H}}(X) \to \mathscr{S}^{\mathscr{A}}(X)$ induces the principal fibration $\Omega\mathscr{S}^{\mathscr{A}}(X) \to \mathscr{E}^{\mathscr{A}}(X) \to \underline{\mathscr{H}}(X)$ (denoted by $* \,^{\mathscr{A}}(X)$) which is a homotopy invariant of $X$, and will play an important role in what follows.

(v) The functors $\mathscr{S}^{\mathscr{A}}$ and $\mathscr{S}\mathscr{N}$ can be extended to the category of countable $CW$-complexes whose skeleta are finite $CW$-complexes; in particular if restricted to $K(G, 1)$ for $G$ a finitely presented group, $\mathscr{S}^{\mathscr{A}}$ and $\mathscr{S}\mathscr{N}$ define functors from the category of these groups to the homotopy category of $\infty$-loop spaces and composing with $\pi_i$ they define algebraic theories.

*About the proof of Theorems* II. 7.1 *and* II. 7.2. To prove Theorems II. 7.1 and II. 7.2 we need:

(1) Theorem II. 4.4,

(2) Corollary II. 5.2,

(3) Theorem II. 5.3,

(4) Corollary II. 7.3, a corollary of Corollary II. 5.2,

(5) Proposition II. 7.4 which indicates a general procedure to construct functors with values in $\Omega^{wh}$.

COROLLARY II. 7.3. *The natural inclusion*

$$\mathscr{A}_{k-1}: \mathscr{C}^{\mathscr{A}}(M \times D^{k-1})/\mathscr{C}^{\mathscr{A}}(M \times D^{k-1}) \to \underline{\tilde{\mathscr{A}}}(M \times D^k)/\underline{\mathscr{A}}(M \times D^k)^4$$

*is an* inf $(k - 4, (n + k - 6)/2)$*-homotopy equivalence.*

This corollary follows from the commutative diagram

$$\begin{array}{ccccc}
\tilde{\mathscr{A}}(M\times D^k)/\mathscr{A}(M\times D^k) & \to & \underline{\tilde{\mathscr{A}}}(M\times D^k)/\underline{\mathscr{A}}(M\times D^k) & \to & \tilde{\mathscr{A}}(\partial(M\times D^k))/\mathscr{A}(\partial(M\times D^k)) \\
\uparrow & & \uparrow{\scriptstyle \theta_{k-1}} & & \uparrow \\
\tilde{\mathscr{A}}(M\times D^k)/\mathscr{A}(M\times D^k) & \to & \mathscr{C}^{\mathscr{A}}(M\times D^{k-1})/\mathscr{C}^{\mathscr{A}}(M\times D^{k-1}) & \to & \tilde{\mathscr{A}}(M\times D^{k-1})/\mathscr{A}(M\times D^{k-1})
\end{array}$$

where horizontal lines are fibrations, and from Theorem II. 4.4 applied to the inclusion $M \times D^{k-1} \subset \partial(M \times D^k)$ which is $(k - 2)$-connected.

There exists an alternative way to describe the category $\Omega^{wh}$ up to an isomorphism, which will be very adequate for our presentation. To explain this way let us introduce the notion of Postnikov tower in $\Omega^h$ as a sequence $\{X_n, \bar{p}^X_{n+1} | \bar{p}^X_{n+1}: X_{n+1} \to X_n$, morphism in $\Omega^h\}$ which satisfies:

(i) $\pi_i(X_n) = 0$, $i \geqq n + 1$,

(ii) $\pi_i(\bar{p}^X_{n+1})$ is an isomorphism for $i \leqq n$.

[4] We have shortened the notation of Part I $\tilde{\mathscr{A}}(M \times D^n; \partial(M \times D^n))$, respectively $\mathscr{A}(M \times D^n; \partial(M \times D^n))$, to $\underline{\tilde{\mathscr{A}}}(M \times D^n)$ respectively $\underline{\mathscr{A}}(M \times D^n)$.

A morphism of Postnikov towers $\{\bar{f}_n\}: \{X_n, \bar{p}^X_{n+1}\} \to \{Y_n, \bar{p}^Y_{n+1}\}$ is a sequence $\bar{f}_n: X_n \to Y_n$ of morphisms in $\Omega^h$ so that $\bar{p}^Y_n \cdot \bar{f}_n = \bar{f}_{n-1} \cdot \bar{p}^X_{n-1}$. If $\mathscr{P}\Omega^h$ denotes the category of Postnikov towers in $\Omega^h$, it is not difficult to see that:

PROPOSITION II. 7.4. *There exists an isomorphism of categories* $\Omega^{wh} \rightsquigarrow \mathscr{P}\Omega^h$. ■

Consequently, in order to construct a functor $F$ from a category $\mathscr{C}$ to $\Omega^{wh}$ it suffices to construct a functor from $\mathscr{C}$ to $\mathscr{P}\Omega^h$.

PROOF OF II. 7.1 (IDEAS). Observe first that it suffices to construct $\mathscr{S}^{\mathscr{A}}$ on the category $\mathscr{A}$ of $\mathscr{A}$-manifolds and continuous maps, since the homotopy category of $\mathscr{A}$ and of $\mathscr{P}$ are isomorphic. The reader will guess that $\mathscr{S}^{\mathscr{A}}(M)$ is defined as $\lim_{\to k} \tilde{\mathscr{C}}^{\mathscr{A}}(M \times D^k)/\mathscr{C}^{\mathscr{A}}(M \times D^k)$ for $\Sigma$: $\tilde{\mathscr{C}}^{\mathscr{A}}(M \times D^k)/\mathscr{C}^{\mathscr{A}}(M \times D^k) \to \tilde{\mathscr{C}}^{\mathscr{A}}(M \times D^{k+1})/\mathscr{C}^{\mathscr{A}}(M \times D^{k+1})$ and $\mathscr{S}^{\mathscr{A}}(f)$ will be constructed using the transfer map. The details of the proof appeal to (1)—(4). In order to prove (f) we notice that because of the homotopy commutativity of the diagram

$$\begin{array}{ccccc}
\tilde{\mathscr{C}}\mathscr{A}(M)/\mathscr{C}\mathscr{A}(M) \longrightarrow \cdots \longrightarrow & \tilde{\mathscr{C}}\mathscr{A}(M\times I^k)/\mathscr{C}\mathscr{A}(M\times I^k) & \xrightarrow{\Sigma} & \tilde{\mathscr{C}}\mathscr{A}(M\times I^{k+1})/\mathscr{C}\mathscr{A}(M\times I^{k+1}) & \longrightarrow \cdots \\
\downarrow & \downarrow \theta_k & & \downarrow \theta_{k+1} & \\
\tilde{\underline{\mathscr{A}}}(M\times I)/\underline{\mathscr{A}}(M\times I) \longrightarrow \cdots \longrightarrow & \underline{\mathscr{A}}(M\times I^{k+1})/\underline{\mathscr{A}}(M\times I^{k+1}) & \xrightarrow{\times \mathrm{id}_I} & \underline{\mathscr{A}}(M\times I^{k+2})/\underline{\mathscr{A}}(M\times I^{k+2}) & \longrightarrow \cdots
\end{array}$$

there exists a well-defined weak homotopy equivalence (isomorphism in $\Omega^{wh}$) $\theta$: $\mathscr{S}^{\mathscr{A}}(M) \to \lim_{\to k} \tilde{\underline{\mathscr{A}}}(M \times D^k)/\underline{\mathscr{A}}(M \times D^k)$; consequently it suffices to construct ${}^{\mathscr{A}}\beta(M) : \underline{\mathscr{H}}(M) \to \lim_{\to k} \tilde{\underline{\mathscr{A}}}(M \times D^k)/\underline{\mathscr{A}}(M \times D^k)$ for any $\mathscr{A}$-manifold $M$.

Let us consider $\underline{\mathscr{A}}^R_0(M \times I^k) \subseteq \underline{\mathscr{A}}(M \times I^k)$ the s.s.-group of $\mathscr{A}$-isomorphisms of the trivial disc bundle $M \times D^k \to M$ and the commutative diagram

$$\begin{array}{ccc}
\tilde{\underline{\mathscr{A}}}(M\times I^k)/\underline{\mathscr{A}}^R_0(M\times I^k) & \xrightarrow{\beta_k} & \tilde{\underline{\mathscr{A}}}(M\times I^k)/\underline{\mathscr{A}}(M\times I^k) \\
\downarrow \cdots\times \mathrm{id}_I & & \downarrow \cdots\times \mathrm{id}_I \\
\tilde{\underline{\mathscr{A}}}(M \times I^{k+1})/\underline{\mathscr{A}}^R_0(M \times I^{k+1}) & \xrightarrow{\beta_{k+1}} & \tilde{\underline{\mathscr{A}}}(M \times I^{k+1})/\underline{\mathscr{A}}(M \times I^{k+1})
\end{array}$$

which by passing to $\lim_{\to k}$ induce ${}^{\mathscr{A}}\beta(M)$: $\lim_{\to k} \tilde{\underline{\mathscr{A}}}(M \times I^k)/\underline{\mathscr{A}}^R_0(M \times I^k) \to \lim_{\to} \tilde{\underline{\mathscr{A}}}(M \times I^k)/\underline{\mathscr{A}}(M \times I^k)$. The first limit identifies (naturally) up to homotopy to $\underline{\mathscr{H}}(M)$ while the second to $\mathscr{S}^{\mathscr{A}}(M)$ via $\theta$.

The proof of Theorem II. 7.2 goes on the same lines. One defines the transfer map for $\mathscr{C}\mathscr{N}$, and using the decomposition stated by Theorem II. 6.1 one checks that all the properties of the transfer map proved for $\tilde{\mathscr{C}} \cdots / \mathscr{C} \cdots$ hold for $\mathscr{C}\mathscr{N} \cdots$ and also that (1), $\cdots$, (4) have analogs for $\mathscr{C}\mathscr{N} \cdots$.

**8. A natural involution and its applications**. We begin this section with some algebraic and homotopy theoretic considerations.

*Properties of $Z(\frac{1}{2})$-modules*. Let $(M, \tau)$ be a commutative $Z(\frac{1}{2})$-module $M$ with involution $\tau: M \to M$, i.e., $\tau^2 = \mathrm{id}$ and $\tau$ is a $Z(\frac{1}{2})$-morphism. $f: (M_1, \tau_1) \to (M_2, \tau_2)$ is called a morphism of modules with involutions if $f$ is a $Z(\frac{1}{2})$-morphism and $\tau_2 \circ f = f \circ \tau_1$. We have

(a) $M = M^s \oplus M^a$.
(b) $f(M^s_1) \subset M^s_2$, $f(M^a_1) \subset M^a_2$ and $f = f^s \oplus f^a$ with $f^s = f|_{M^s_1}$ and $f^a = f|_{M^a_1}$.
(c) If $\cdots \to (M_i, \tau_i) \to^{f_i} (M_{i+1}, \tau_{i+1}) \to^{f_{i+1}} (M_{i+2}, \tau_{i+2}) \to$ is an exact sequence

of modules with involutions then

$$\cdots \longrightarrow M_i^s \xrightarrow{f_i^s} M_{i+1}^s \xrightarrow{f_{i+1}^s} M_{i+2}^s \xrightarrow{f_{i+2}^s} M_{i+3}^s \longrightarrow \cdots$$

and

$$\cdots \longrightarrow M_i^a \xrightarrow{f_i^a} M_{i+1}^a \xrightarrow{f_{i+1}^a} M_{i+2}^a \xrightarrow{f_{i+2}} M_{i+3}^a \longrightarrow \cdots$$

are exact sequences.

The algebraic splitting we just described corresponds to a geometric splitting we will describe below. A weak commutative group with involution $(X, \tau)$ is an $H$-space $X$, whose multiplication satisfies up to homotopy the axioms of "commutative groups" and $\tau$ is an $H$-map with $\tau^2$ homotopic to identity. $f: (X_1, \tau_1) \to (X_2, \tau_2)$ is a morphism of a weak commutative group with involutions if $f$ is an $H$-map with $\tau_2 \circ f$ homotopic to $f \circ \tau_1$. Observe also that for any weak commutative group we have two canonical involutions $\mathrm{id}_X$ and $\nu: X \to X$; $\nu$ represents the "inverse" with respect to the multiplication on $X$.

*Properties of weak commutative groups with involutions.* A based pointed space (s.s.-complex) $(X, x)$ is called an odd-space or a $Z(\frac{1}{2})$-space if $\pi_1(X, x)$ is abelian and all homotopy groups $\pi_i(X, x)$ are $Z(\frac{1}{2})$-modules. An odd-weak commutative group with involution $(X, \tau)$ is a $Z(\frac{1}{2})$-weak commutative group with involution if the space $(X, x)$ is a $Z(\frac{1}{2})$ -space. By localization to odd primes we pass functorially from weak commutative groups with involution to $Z(\frac{1}{2})$-weak commutative groups with involution. Using "general homotopy theory" one can prove that:

(a) For any $(X, \tau)$, a $Z(\frac{1}{2})$-weak commutative group with involution, there exist two $Z(\frac{1}{2})$-weak commutative groups $X^s$ and $X^a$ together with a homotopy equivalence of weak commutative groups with involution $h(X, \tau): (X, \tau) \to (X^s \times X^a, \mathrm{id}_{X^s} \times \nu_{X^a})$.

(b) If $f: (X_1, \tau_1) \to (X_2, \tau_2)$ is a morphism of $Z(\frac{1}{2})$-weak commutative groups with involution, then there exist the morphisms $f^s: (X_1^s, \mathrm{id}_{X_1^s}) \to (X_2^s, \mathrm{id}_{X_2^s})$ and $f^a: (X_1^a, \nu_{X_1^a}) \to (X_2^a, \nu_{X_2^a})$ of $Z(\frac{1}{2})$-weak commutative groups with involution so that the following diagram is (homotopy) commutative.

$$\begin{array}{ccc} X_1 & \xrightarrow{f} & X_2 \\ \downarrow{\scriptstyle h(X_1, \tau_1)} & & \downarrow{\scriptstyle h(X_2, \tau_2)} \\ X_1^s \times X_1^a & \xrightarrow{f^s \times f^a} & X_2^s \times X_2^a \end{array}$$

(c) Let $(X_1, \tau_1) \to^f (X_2, \tau_2) \to^g (X_3, \tau_3)$ be a fibration with $(X_i, \tau_i)$ $Z(\frac{1}{2})$-weak commutative groups with involutions and $f$, $g$ morphisms. Then $X_1^s \to^{f^s} X_2^s \to^{g^s} X_3^s$ and $X_1^a \to^{f^a} X_2^a \to^{g^a} X_3^a$ are fibrations. This geometric splitting corresponds entirely to the algebraic splitting in the following sense; if $K$ is a space (s.s.-complex) then the set $[K, X]$ of the homotopy classes of maps has a natural structure of $Z(\frac{1}{2})$-module with involution and $[K, X]^s = [K, X^s]$, respectively $[K, X]^a = [K, X^a]$.

Let $\tau: [0, 1] \to [0, 1]$ be the involution given by $\tau(\alpha) = 1 - \alpha$. The conjugation with $\mathrm{id}_M \times \tau$ induces on $\mathscr{A}(M \times I)$, $\tilde{\mathscr{A}}(M \times I)$, $\Omega\tilde{\mathscr{A}}(M \times I)/\mathscr{A}(M \times I)$, $\mathscr{C}^{\mathscr{A}}(M \times I)$, $\tilde{\mathscr{C}}^{\mathscr{A}}(M \times I)/\mathscr{C}^{\mathscr{A}}(M \times I)$ an involution $\tau$ so that all natural s.s.-maps and group-homomorphisms which will be involved in our considerations are equivariant with respect to $\tau$. Moreover, all the s.s.-complexes except the last one are weak commutative groups with involution while the last one is a weak group with involution which becomes a weak commutative group with involution if we assume $M^n \sim N^{n-1} \times I$. All our knowledge about the structure of $\mathscr{A}(M)$ comes from a partial understanding of the fibrations

$$(1)_r \qquad \Omega\tilde{\mathscr{A}}(M \times I^r)/\mathscr{A}(M \times I^r) \longrightarrow \mathscr{A}(M \times I^r) \longrightarrow \tilde{\mathscr{A}}(M \times I^r)$$

and

$$(2)_{r-1} \qquad \tilde{\mathscr{A}}(M \times I^r)/\mathscr{A}(M \times I^r) \xrightarrow{i_r} \tilde{\mathscr{C}}^{\mathscr{A}}(M \times I^{r-1})/\mathscr{C}^{\mathscr{A}}(M \times I^{r-1}) \xrightarrow{\pi_{r-1}} \tilde{\mathscr{A}}(M \times I^{r-1})/\mathscr{A}(M \times I^{r-1})$$

which will be done using the algebraic and the geometric decomposition we have discussed above and Theorem II.5.1; one obtains (see **[B.L.]**$_3$, §4) the following:

PROPOSITION II. 8.1. (*All s.s.-complexes and maps in this statement are assumed to be localized to odd primes.*)

(a) $\tilde{\mathscr{A}}(M \times I) \to \tilde{\mathscr{A}}^a(M \times I)$ *is a homotopy equivalence.*

(b) $\{\pi_i(\tilde{\mathscr{C}}^{\mathscr{A}}(M \times I)/\mathscr{C}^{\mathscr{A}}(M \times I))\}^a = 0$ *if* $i \leqq \omega^{\mathscr{A}}(\dim M + 1)$.

(c) $\pi_1 \cdot \Sigma \cdot i_1$ *induces the isomorphism* $2 \cdot \mathrm{id}$ *for* $\pi_r^s(\tilde{\mathscr{A}}(M \times I)/\mathscr{A}(M \times I))$.

(d) $\mathscr{A}^a(M \times I) \to \tilde{\mathscr{A}}(M \times I)$ *induces for homotopy groups an isomorphism in dimension* $i \leqq \omega^{\mathscr{A}}(\dim M + 1)$ *and an epimorphism in dimension* $i \leqq \omega^{\mathscr{A}}(\dim M + 1) + 1$.

(e) *The fibration* $\Omega\tilde{\mathscr{A}}(M \times I)/\mathscr{A}(M \times I) \to \mathscr{A}(M \times I) \to \tilde{\mathscr{A}}(M \times I)$ *which is classified by* $\tilde{\mathscr{A}}(M \times I) \to \tilde{\mathscr{A}}(M \times I)/\mathscr{A}^{\mathscr{A}}(M \times I)$ *is* $\omega^{\mathscr{A}}(\dim M + 1)$*-trivial in the sense that the classifying map is* $\omega^{\mathscr{A}}(\dim M + 1)$*-trival.*

(c′) $\pi_1 \circ \Sigma \circ i_1$ *and* $2\,\mathrm{id}$ *are* $\omega^{\mathscr{A}}(\dim M+1)$*-homotopic, and* $\pi_2 \circ \Sigma^2 \circ i_1$ *is* $\omega^{\mathscr{A}}(\dim M+1)$*-trivial.*

(c″) $\pi_{2s-1} \circ \Sigma^{2s-1} \circ i_1$ *is an* $\omega^{\mathscr{A}}(\dim M+1)$*-homotopy equivalence and* $\pi_{2s} \circ \Sigma^{2s} \circ i_1$ *is* $\omega^{\mathscr{A}}(\dim M + 1)$*-trivial.* ∎

For any $G$, a $Z(\frac{1}{2})$-weak group, and $K$ an arbitrary s.s.-complex, the set of homotopy classes $[K, G]$ is a group whose elements are all uniquely divisible by 2, hence a $Z(\frac{1}{2})$-module; consequently for any $f \in [K, G]$ and $\alpha \in Z(\frac{1}{2})$, $\alpha f$ is a well-defined element of $[K,G]$. As for $k \geqq 1$, $\tilde{\mathscr{A}}(M \times I^k)/\mathscr{A}(M \times I^k), \tilde{\mathscr{C}}^{\mathscr{A}}(M \times I^k)/\mathscr{C}^{\mathscr{A}}(M \times I^k)$ are $Z(\frac{1}{2})$-weak groups (since we have assumed all these spaces localized to odd primes) we define the homotopy classes $p_{k+1}, j_k$:

$$p_{k+1}\colon \tilde{\mathscr{C}}^{\mathscr{A}}(M \times I^k)/\mathscr{C}^{\mathscr{A}}(M \times I^k) \longrightarrow \tilde{\mathscr{A}}(M \times I^{k+1})/\mathscr{A}(M \times I^{k+1}),$$
$$j_k\colon \tilde{\mathscr{A}}(M \times I^k)/\mathscr{A}(M \times I^k) \longrightarrow \tilde{\mathscr{C}}^{\mathscr{A}}(M \times I^k)/\mathscr{C}^{\mathscr{A}}(M \times I^k)$$

by $p_{k+1} = \frac{1}{2}\pi_{k+1} \circ \Sigma$, $j_k = \frac{1}{2}\Sigma \circ i_k$ for $k \geqq 1$ and the homotopy class

$$\omega_k^u\colon \tilde{\mathscr{A}}(M \times I^k)/\mathscr{A}(M \times I^k) \longrightarrow \tilde{\mathscr{A}}(M \times I^{k+2u})/\mathscr{A}(M \times I^{k+2u})$$

by $\omega_k^u = \frac{1}{2}\pi_{k+2u} \circ \Sigma^{2u+1} \circ i_k$ ($j_k$ can be defined even for $k = 0$ at the $\omega^{\mathscr{A}}(\dim M)$th Postnikov term level), with $i_{k+1}$ and $\pi_k$ the maps in the fibration

$$(k) \qquad \tilde{\mathscr{A}}(M \times I^{k+1})/\mathscr{A}(M \times I^{k+1}) \xrightarrow{i_{k+1}} \tilde{\mathscr{C}}^{\mathscr{A}}(M \times I^k)/\mathscr{C}^{\mathscr{A}}(M \times I^k) \xrightarrow{\pi_k} \tilde{\mathscr{A}}(M \times I^k)/\mathscr{A}(M \times I^k).$$

THEOREM II. 8.2. *If $l \leqq \omega^{\mathscr{A}}(\dim M + k)$ then the Postnikov lth term of the sequence (k) is a fibration which satisfies:*

(1) $p_{k+1}\, i_{k+1} \sim \mathrm{id}$, $\pi_k\, j_k \sim \mathrm{id}$.

(2) $\pi_k\, i_{k+1} \sim 0$, $p_{k+1}\, j_k \sim 0$.

(3) $j_k\, \pi_k + i_{k+1}\, p_{k+1} \sim \mathrm{id}$ *(the lth Postnikov term of $\tilde{\mathscr{C}}^{\mathscr{A}}(M \times I^k)/\mathscr{C}^{\mathscr{A}}(M \times I^k)$ is a weak commutative group since it is homotopy equivalent to $[\mathscr{S}^{\mathscr{A}}(M \times I^k)]_l$).*

(4) *The diagram (Figure 5) is (homotopy) commutative*

$$\begin{array}{ccccc}
\left[\tilde{\mathscr{A}}(M \times I^{k+1})/\mathscr{A}(M \times I^{k+1})\right]_l & \underset{p_{k+1}}{\overset{i_{k+1}}{\rightleftarrows}} & \left[\tilde{\mathscr{C}}^{\mathscr{A}}(M \times I^k)/\mathscr{C}^{\mathscr{A}}(M \times I^k)\right]_l & \underset{j_k}{\overset{\pi_k}{\rightleftarrows}} & \left[\tilde{\mathscr{A}}(M \times I^k)/\mathscr{A}(M \times I^k)\right]_l \\
\downarrow 2\mathrm{id} & & \downarrow \Sigma & & \downarrow \tfrac{1}{2}\omega_k^1 \\
\left[\tilde{\mathscr{A}}(M \times I^{k+1})/\mathscr{A}(M \times I^{k+1})\right]_l & \underset{j_{k+1}}{\overset{\pi_{k+1}}{\leftrightarrows}} & \left[\tilde{\mathscr{C}}^{\mathscr{A}}(M \times I^{k+1})/\mathscr{C}^{\mathscr{A}}(M \times I^{k+1})\right]_l & \underset{p_{k+2}}{\overset{i_{k+2}}{\leftrightarrows}} & \left[\tilde{\mathscr{A}}(M \times I^{k+2})/\mathscr{A}(M \times I^{k+2})\right]_l
\end{array}$$

FIGURE 5

(5) $\omega_k^u = \omega_{k+2u-2}^1 \circ \omega_{k+2u-4}^1 \circ \cdots \circ \omega_{k+2}^1 \circ \omega_k^1$.

(6) *The following diagram (Figure 6) is (homotopy) commutative with $\omega_{k+1}^u$ homotopy equivalence, and natural (up to homotopy) with respect to inclusion $M^n \subset V^n$.*

$$\begin{array}{ccccc}
\left[\tilde{\mathscr{A}}(M \times I^{k+1})/\mathscr{A}(M \times I^{k+1})\right]_l & \underset{p_{k+1}}{\overset{i_{k+1}}{\rightleftarrows}} & \left[\tilde{\mathscr{C}}^{\mathscr{A}}(M \times I_k)/\mathscr{C}^{\mathscr{A}}(M \times I^k)\right]_l & \underset{j_k}{\overset{\pi_k}{\rightleftarrows}} & \left[\tilde{\mathscr{A}}(M \times I^k)/\mathscr{A}(M \times I^k)\right]_l \\
\downarrow \omega_{k+1}^s & & \downarrow \Sigma^{2s} & & \downarrow \omega_k^s \\
\left[\tilde{\mathscr{A}}(M \times I^{k+2s+1})/\mathscr{A}(M \times I^{k+2s+1})\right]_l & \underset{p_{k+2s+1}}{\overset{i_{k+2s+1}}{\leftrightarrows}} & \left[\tilde{\mathscr{C}}^{\mathscr{A}}(M \times I^{k+2s})/\mathscr{C}^{\mathscr{A}}(M \times I^{k+2s})\right]_l & \underset{j_{k+2s}}{\overset{\pi_{k+2s}}{\rightleftarrows}} & \left[\tilde{\mathscr{A}}(M \times I^{k+2s})/\mathscr{A}(M \times I^{k+2s})\right]_l
\end{array} \quad \blacksquare$$

FIGURE 6

## 9. The functors ${}^{\mathscr{A}}\mathscr{S}^{\pm}$ and the structure theorem.

THEOREM II. 9.1. *For any geometric category $\mathscr{A}$, there exist two functors ${}^{\mathscr{A}}\mathscr{S}^{\pm}$: $\mathscr{P}^{\mathscr{A}} \to \Omega^{wh}$ (see §0 for the definition of $\mathscr{P}^{\mathscr{A}}$) so that:*

(a) *${}^{\mathscr{A}}\mathscr{S}^{\pm}$ are homotopy functors.*

(b) *If $f\colon (X, \xi) \to (Y, \eta)$ is a morphism with $f$ k-connected, ${}^{\mathscr{A}}\mathscr{S}^{\pm}(f)$ are k-connected.*

(c) *There exists a functorial isomorphism between the functors $\mathscr{S}_{\mathrm{odd}}^{\mathscr{A}} \circ \pi^{\mathscr{A}}$ ($\mathscr{S}_{\mathrm{odd}}^{\mathscr{A}}(X) = (\mathscr{S}^{\mathscr{A}}(X))_{\mathrm{odd}}$) and ${}^{\mathscr{A}}\mathscr{S}^{+} \times {}^{\mathscr{A}}\mathscr{S}^{-}$ defined by ${}^{\mathscr{A}}p^{\pm}(X, \xi)\colon {}^{\mathscr{A}}\mathscr{S}_{\mathrm{odd}}(X) \to {}^{\mathscr{A}}\mathscr{S}^{\pm}(X, \xi)$ with $\pi^{\mathscr{A}} \circ \mathscr{P}^{\mathscr{A}} \to \mathscr{P}$ the forgetful functor.*

(d) *For any $(K, \eta) \in \mathrm{ob}\, \mathscr{P}^{\mathscr{A}}$ there exists a "transfer map" ${}^{\pm}l^{(K,\eta)}\colon {}^{\mathscr{A}}\mathscr{S}^{\pm}(X, \xi) \to {}^{\mathscr{A}}\mathscr{S}^{\pm}(X \times K;\ \xi \times \eta)$ which is a natural transformation of functors from ${}^{\mathscr{A}}\mathscr{S}^{\pm}$ to ${}^{\mathscr{A}}\mathscr{S}^{\pm}0(\cdots \times (K, \eta))$ so that:*

(i) ${}^{\pm}l^{(K_1 \times K_2,\, \eta_1 \times \eta_2)} \sim {}^{\pm}l^{(K_2, \eta_2)} \circ {}^{\pm}l^{(K_1, \eta_1)} \sim {}^{\pm}l^{(K_1, \eta_1)} \circ {}^{\pm}l^{(K_2, \eta_2)}$,

(ii) ${}^{\pm}l^{(K,\eta)} \sim \chi(K) \cdot {}^{\mathscr{A}}\mathscr{S}^{\pm}(i)$ *with i given by $i(x) = (x, k_0) \in X \times K$,*

(iii) *via the identification of $\mathscr{S}_{\mathrm{odd}}^{\mathscr{A}}(X)$ with ${}^{\mathscr{A}}\mathscr{S}^{+}(X, \xi) \times {}^{\mathscr{A}}\mathscr{S}^{-}(X, \xi)$, $(l^K)_{\mathrm{odd}} \sim {}^{+}l^{(K,\xi)} \times {}^{-}l^{(K,\xi)}$.*

(e) *If $\mathscr{B} \geqq \mathscr{A}$ for any $(X, \xi) \in \mathrm{ob}\, \mathscr{P}^{\mathscr{A}}$ there exists a morphism $\gamma_{\mathscr{B},\mathscr{A}}^{\pm}(X, \xi)\colon {}^{\mathscr{A}}\mathscr{S}^{\pm}(X, \xi) \to {}^{\mathscr{B}}\mathscr{S}^{\pm}(X, \xi)$, so that $\gamma_{\mathscr{B},\mathscr{A}}^{\pm}$ defines a natural transformation between ${}^{\mathscr{A}}\mathscr{S}^{\pm}$ and ${}^{\mathscr{B}}\mathscr{S}^{\pm} \circ \pi^{\mathscr{B},\mathscr{A}}$ with $\pi^{\mathscr{B},\mathscr{A}}\colon \mathscr{P}^{\mathscr{B}} \to \mathscr{P}^{\mathscr{A}}$ the forgetful functor.*

(f) *For any $(X, \xi) \in \mathrm{ob}\, \mathscr{P}^{\mathscr{A}}$ there exists a map ${}^{\mathscr{A}}\beta^{\pm}(X, \xi)\colon \mathscr{H}(K) \to {}^{\mathscr{A}}\mathscr{S}^{\pm}(X, \xi)$ so that ${}^{\mathscr{A}}\beta^{\pm}(X, \xi) = {}^{\mathscr{A}}P^{\pm}(X, \xi) \circ {}^{\mathscr{A}}\beta(X)$.*

(g) *For any manifold $M^k$ and $n$ with $2n \geqq k$ there exists the homotopy class*

$$i^+_{2n}(M)\colon (\tilde{\mathscr{A}}(M^k \times I^{2n-k})/\mathscr{A}(M^k \times I^{2n-k}))_{\text{odd}} \to {}^{\mathscr{A}}\mathscr{S}^+(M;\tau(M))$$

*respectively*

$$i^-_{2n}(M)\colon (\tilde{\mathscr{A}}(M^k \times I^{2n+1-k})/\mathscr{A}(M^k \times I^{2n+1-k}))_{\text{odd}} \to {}^{\mathscr{A}}\mathscr{S}^-(M;\tau(M))$$

*natural with respect to inclusions $N^k \subset M^k$ and the category $\mathscr{A}$ which is $\omega^{\mathscr{A}}_{(2n)}$ respectively $\omega^{\mathscr{A}}(2n+1)$-homotopy equivalence.* ■

THEOREM II. 9.2. *There exist the functors ${}^{\mathscr{A}}\mathscr{N}^{\pm}\colon \mathscr{P}^{\mathscr{A}} \to \Omega^{wh}$ which satisfy* (a), (b), (c) *and* (d) *of Theorem* II. 9.1 *with ${}^{\mathscr{A}}\mathscr{S}^{\pm}$ replaced by ${}^{\mathscr{A}}\mathscr{N}^{\pm}$ and $\mathscr{S}^{\mathscr{A}}$ replaced by $\mathscr{S}\mathscr{N}$ as well as:*

(e) *${}^{\mathscr{A}}\mathscr{N}^{\pm}$ and ${}^{\mathscr{B}}\mathscr{N}^{\pm} \circ \pi^{\mathscr{B},\mathscr{A}}$ are naturally isomorphic.*

(f) *For any $M^k \in$ ob $\mathscr{A}$ and $n$ so that $2n \geqq k$ there exists*

$${}^{\mathscr{A}}i_{2n}(M^k)\colon \mathscr{N}(M^k \times I^{2n-k})_{\text{odd}} \to {}^{\mathscr{A}}\mathscr{N}^+(M, \tau(M))$$

*respectively*

$${}^{\mathscr{A}}i_{2n+1}(M^k)\colon \mathscr{N}(M^k \times I^{2n+1-k})_{\text{odd}} \to {}^{\mathscr{A}}\mathscr{N}^-(M, \tau(M))$$

*which is natural with respect to inclusions $N^k \subset M^k$ and the category $\mathscr{A}$, commutes with $\omega^s_r$, and is an $\omega^{\mathscr{A}}(2n)$-, resp. $\omega^{\mathscr{A}}(2n+1)$-homotopy equivalence.* ■

*Comments about Theorems* II. 9.1 *and* II. 9.2. (i) (g) of Theorem II. 9.1 and (f) of Theorem II. 9.2 imply the unicity of ${}^{\mathscr{A}}\mathscr{S}^{\pm}$ and ${}^{\mathscr{A}}\mathscr{N}^{\pm}$ up to an isomorphism of functors.

(ii) The maps ${}^{\mathscr{A}}\beta^{\pm}(X)\colon \underline{\mathscr{H}}(X) \to {}^{\mathscr{A}}\mathscr{S}^{\pm}(X, \xi)$ induce the principal fibrations $\Omega^{\mathscr{A}}\mathscr{S}^{\pm}(X, \xi) \to {}^{\mathscr{A}}\mathscr{E}^{\pm}(X, \xi) \to \underline{\mathscr{H}}(X)$ denoted by ${}^{\mathscr{A}}*^{\pm}(M)$, which are homotopy invariants of $(X, \xi)$ and consequently for any manifold $M^n$ the fibrations ${}^{\mathscr{A}}*^{\pm}(M)$: $\Omega^{\mathscr{A}}\mathscr{S}^{\pm}(M, \tau(M)) \to {}^{\mathscr{A}}\mathscr{E}^{\pm}(M, \tau(M)) \to \underline{\mathscr{H}}(M)$ are tangential homotopy invariants of $M$.

(iii) Clearly (f) implies that $*^{\mathscr{A}}(M)$ fibrewise localized to odd primes is the Whitney sum of ${}^{\mathscr{A}}*^+(M)$ and ${}^{\mathscr{A}}*^-(M)$ for any $\xi$; hence we have the commutative diagram

$$\begin{array}{ccccc}
{}^{\mathscr{A}}\mathscr{S}^{\pm}(X, \xi) & \longrightarrow & {}^{\mathscr{A}}\mathscr{E}^{\pm}(X, \xi) & \longrightarrow & \underline{\mathscr{H}}(X) \\
\uparrow {\scriptstyle {}^{\mathscr{A}}p^{\pm}(X,\xi)} & & \uparrow & & \uparrow {\scriptstyle \approx} \\
\mathscr{S}^{\mathscr{A}}(X) & \longrightarrow & \mathscr{E}^{\mathscr{A}}(X) & \longrightarrow & \underline{\mathscr{H}}(X)
\end{array}$$

hence the triviality of $*^{\mathscr{A}}(X)$ implies the triviality of ${}^{\mathscr{A}}*^{\pm}(X, \xi)$ for any $\xi$.

THEOREM II. 9.3 (THE MAIN THEOREM). *Let $M^n$ be an $\mathscr{A}$-manifold and let $**(M)$ be the pull-back of ${}^{\mathscr{A}}*^{\varepsilon(n)}(M)$ by the natural map $\tilde{\mathscr{A}}(M) \to \underline{\mathscr{H}}(M)$*

$$**(M)\colon \Omega^{\mathscr{A}}\mathscr{S}^{\varepsilon(n)}(M, \tau(M)) \longrightarrow \boldsymbol{E}^{\mathscr{A}}(M) \longrightarrow \tilde{\mathscr{A}}(M)$$

*where*

$$\varepsilon(n) = \begin{cases} + & \text{if } n \text{ is even,} \\ - & \text{if } n \text{ is odd.} \end{cases}$$

*Then the fibrewise odd localization of* $\Omega\tilde{\mathscr{A}}(M^n)/\mathscr{A}(M^n) \to \mathscr{A}(M^n) \to \tilde{\mathscr{A}}(M^n)$ *and* $**(M)$ *are* $\omega^{\mathscr{A}}(n)$*-isomorphic, in particular* $[\boldsymbol{E}^{\mathscr{A}}(M^n)_{\mathrm{odd}}]\omega^{\mathscr{A}}(n)$ *and* $[\mathscr{A}(M^n)_{\mathrm{odd}}]\omega^{\mathscr{A}}(n)$ *are homotopy equivalent.* ■

COROLLARY II. 9.4. (1) *Let* $M^n$ *be an* $\mathscr{A}$*-manifold with nonempty boundary and* $p\colon M \to \partial M$ *so that* $i \circ p \to \mathrm{id}$. *Then* $**(M)$ *is trivial and consequently* $[\mathscr{A}(M^n)_{\mathrm{odd}}]_l$ *and* $[\Omega^{\mathscr{A}}\mathscr{S}^{\varepsilon(n)}(M, \tau(M^n)) \times \tilde{\mathscr{A}}(M)_{\mathrm{odd}}]_l$, $l \leqq \omega^{\mathscr{A}}(n)$, *are homotopy equivalent.*

(2) *If* $M^n$ *is* $\mathscr{A}$*-isomorphic to* $N^{n-1} \times S^1$, *then the restriction of* $**(M)$ *to the connected component of the identity of* $\tilde{\mathscr{A}}(M_n)$ *is trivial; hence* $(\mathscr{A}(M^n))_{\mathrm{odd}} \to \tilde{\mathscr{A}}(M)_{\mathrm{odd}}$ *is split surjective.*

(3) *Let* $K$ *be an* $\mathscr{A}$*-manifold with* $\partial k = \varnothing$ *and* $\chi(K) = 0$ *and let us consider the diagram*:

$$\begin{array}{ccc} {}^{\mathscr{A}}(M \times K)_{\mathrm{odd}} & \xrightarrow{e} & \tilde{\mathscr{A}}(M \times K)_{\mathrm{odd}} \\ \nearrow \cdots \times \mathrm{id}_K & & \nearrow \cdots \times \mathrm{id}_K \\ \mathscr{A}(M)_{\mathrm{odd}} & \xrightarrow{e} & \tilde{\mathscr{A}}(M)_{\mathrm{odd}} \end{array}$$

*Then for any* $L$ *which has the homotopy type of a CW-complex of dimension* $\leqq \omega^{\mathscr{A}}(n)$ *and any* $f\colon L \to \tilde{\mathscr{A}}(M)_{\mathrm{odd}}$ *the composition* $\underline{f}\colon L \to^{f} \tilde{\mathscr{A}}(M)_{\mathrm{odd}} \to^{\cdots \times \mathrm{id}_k} \tilde{\mathscr{A}}(M \times K)_{\mathrm{odd}}$ *has a lifting* $\underline{\underline{f}}\colon L \to \mathscr{A}(M \times K)_{\mathrm{odd}}$ *with* $e \cdot \underline{\underline{f}} \sim \underline{f}$. ■

The proof of Theorems II. 9.1–II. 9.3 and Corollary II. 9.4 can be found in **[B.L.]**$_3$, §5; however we find it instructive to point out some steps.

About the proof of Theorem II. 9.1: One constructs the functors ${}^{\mathscr{A}}\mathscr{S}^{\pm}$ as functors defined on the category $\mathscr{A}_0$ of compact $\mathscr{A}$-manifolds and continuous maps $f\colon M^n \to N^n$ with $f_*(\tau(N)) \sim^s \tau(M)$ with values in the category of Postnikov towers of $\Omega^h$ according to Proposition II. 7.4. For any $n$ one chooses an increasing sequence $r_1^n < r_2^n < r_3^n < \cdots r_{i+1}^n \cdots$ with $2r_1^n \geqq n$ and $\omega^{\mathscr{A}}(r_i^n) \geqq i$ for any geometric category $\mathscr{A}$.

For a manifold $M \in \mathrm{ob}\,\mathscr{A}$ one defines

$${}^{\mathscr{A}}\mathscr{S}^{\varepsilon}(M, \tau(M))_i = [\tilde{\mathscr{A}}(M \times D^{\varepsilon_i^n - n})/\mathscr{A}(M \times D^{\varepsilon_i^n - n})]_i$$

where $\varepsilon_i^n = 2r_i^n$ if $\varepsilon = +$ and $\varepsilon_i^n = 2r_i^n + 1$ if $\varepsilon = -$. As one follows from Theorem II. 7.2 one can construct the homotopy equivalence

$${}^{\varepsilon}\Sigma\colon {}^{\mathscr{A}}\mathscr{S}^{\pm}(M^n, \tau(M^n)) \to {}^{\mathscr{A}}\mathscr{S}^{\pm}(M^n \times I, \tau(M^n) \times \tau(I)),$$
$${}^{\varepsilon}\Sigma_i = \left[\omega^{r_i^n}_{\varepsilon_i^{n+1} - n}\right]_i^{-1} \cdot \left[\omega^{r_i^{n+1}}_{\varepsilon_i^n - n}\right]_i$$

where $[f]_i$ denotes the $i$th Postnikov term of $f$. If $f\colon M^n \subset N^n$ is a 0-codimensional embedding ${}^{\mathscr{A}}\mathscr{S}^{\pm}(f)$ is obviously defined; if $f$ is a continuous map $f\colon M^n \to N^k$ so that $f_*(\tau(N^k)) \sim^s \tau(M)$, choose an embedding $\tilde{f}\colon M^n \times D^{2s-n} \to N^k \times D^{2s-k}$ (for a very big $s$) so that $\tilde{f} \mid M^n \times \{0\}$ is homotopic to $i \circ f$ when $i(x) = (x, 0)$, and define ${}^{\mathscr{A}}\mathscr{S}^{\varepsilon}(f) = {}^{\varepsilon}\Sigma^{-1} \circ {}^{\mathscr{A}}\mathscr{S}^{\varepsilon}(\tilde{f}) \circ {}^{\varepsilon}\Sigma^{2s-n}$. Proposition II. 7.1 and Theorem II. 7.2 allow us to verify all the statements of Theorem II. 9.1 except (f).(f) will follow from inspecting the diagram (Figure 7)

$$\begin{array}{ccccc}
\mathscr{H}(M) & \xleftarrow{j_1} & \tilde{\mathscr{A}}(M \times D^{2N+1})/\mathscr{A}_0^R(M \times D^{2N+1}) & \xrightarrow{d} & \tilde{\mathscr{A}}(\partial(M \times D)^{2N+1})/\mathscr{A}(\partial(M \times D^{2N+1})) \\
\uparrow d & \nwarrow \bar{d}_2, \bar{d}_3 & \uparrow i \quad \searrow \pi & \nearrow d_1' & \uparrow i' \\
 & & \underline{\underline{\tilde{\mathscr{A}}}}(M\times D^{2N+1})/\underline{\underline{\mathscr{A}}}{}_0^R(M\times D^{2N+1}) & & \underline{\underline{\tilde{\mathscr{A}}}}(M\times D^{2N+1})/\underline{\underline{\mathscr{A}}}(M \times D^{2N+1}) \\
 & & \downarrow d & & \\
\tilde{\mathscr{A}}(M\times S^{2N}) & \xrightarrow{\pi'} & \tilde{\mathscr{A}}(M \times S^{2N})/\mathscr{A}_0(M \times S^{2N}) & \xrightarrow{\pi''} & \tilde{\mathscr{A}}(M\times S^{2N})/\mathscr{A}(M \times S^{2N})
\end{array}$$

FIGURE 7

where $\underline{\underline{\tilde{\mathscr{A}}}}(M \times D^{2N+1})$ is the subgroup of $\tilde{\mathscr{A}}(M \times D^{2N+1})$ consisting of those block automorphisms which are identity on $\partial M \times D^{2N+1}$,

$$\underline{\underline{\mathscr{A}}}(M \times D^{2N+1}) = \underline{\underline{\tilde{\mathscr{A}}}}(M \times D^{2N+1}) \cap \mathscr{A}(M \times D^{2N+1}),$$
$$\underline{\underline{\mathscr{A}}}{}_0^R(M \times D^{2N+1}) = \underline{\underline{\tilde{\mathscr{A}}}}(M \times D^{2N+1}) \cap \mathscr{A}_0^R(M \times D^{2N+1})$$

(for the definition of $\mathscr{A}_0^R\cdots$see §7)

where the arrows are defined as follows: $j, j_1, j_2, j_3$ associate to any block automorphism of $M \times D^{2N+1}$ or $M \times S^{2N}$ the homotopy equivalence which one obtains composing on the left by the canonical inclusion $M \to M \times D^{2N+1}$ or $M \to M \times S^{2N}$ and on the right by the projection $M \times D^{2N+1} \to M$ or $M \times S^{2N} \to M$ ($N$ very big). $\pi$, $\pi'$, $\pi''$ are induced by the obvious group-factorizations, $i$ is given by the inclusion $\underline{\underline{\tilde{\mathscr{A}}}}\cdots \subset \tilde{\mathscr{A}}\cdots$ and $i'$ is induced by the inclusion of $M \times S^{2N}$ in $\partial(M \times D^{2N+1})$; $d$ and $d'$ are induced by restriction to the boundary $\partial(M \times D^{2N+1})$.

For the proof of Theorem II. 9.2 one repeats the arguments in the proof of Theorem II. 9.1 with $\tilde{\mathscr{A}}(M \times D^k)/\mathscr{A}(M \times D^k)$ replaced by $\mathscr{N}(M \times D^k)$.

PROOF OF THEOREM II. 9.3. In order to prove Theorem II. 9.3 we consider the diagram:

$$\begin{array}{ccccccc}
\Omega^{\mathscr{A}}\mathscr{S}^{\varepsilon(n)}(M, \tau(M)) & \longrightarrow & {}^{\mathscr{A}}\mathscr{E}^{\varepsilon(n)}(M, \tau(M)) & \longrightarrow & \mathscr{H}(M) & \longrightarrow & {}^{\mathscr{A}}\mathscr{S}^{\varepsilon(n)}(M, \tau(M)) \\
 & & & & \uparrow & & \\
\Omega\tilde{\mathscr{A}}(M \times S^{2N})/\mathscr{A}(M \times S^{2N}) & \to & \mathscr{A}(M\times S^{2N}) & \to & \tilde{\mathscr{A}}(M \times S^{2N}) & & \\
\uparrow \Omega I^{S^{2N}} & & \uparrow I^{S^{2N}} & & \uparrow I^{S^{2N}} & & \\
\Omega\tilde{\mathscr{A}}(M)/\mathscr{A}(M) & \longrightarrow & \mathscr{A}(M) & \longrightarrow & \tilde{\mathscr{A}}(M) & &
\end{array}$$

FIGURE 8

We first observe that $\Omega I^{S^{2N}}$ is an $\omega^{\mathscr{A}}$ (dim $M$) + 1-homotopy equivalence since, considering the diagram

$$\begin{array}{lccccc}
(1) & \tilde{\mathscr{A}}(M)/\mathscr{A}(M) & \longleftarrow & \mathscr{C}^{\mathscr{A}}(M)/\mathscr{C}^{\mathscr{A}}(M) & \longleftarrow & \tilde{\mathscr{A}}(M\times I)/\mathscr{A}(M\times I) \\
 & \downarrow & & \downarrow & & \downarrow \\
(2) & \tilde{\mathscr{A}}(M\times K)/\mathscr{A}(M\times K) & \leftarrow & \tilde{\mathscr{C}}^{\mathscr{A}}(M\times K)/\mathscr{C}^{\mathscr{A}}(M\times K) & \leftarrow & \tilde{\mathscr{A}}(M\times K\times I)/\mathscr{A}(M\times K\times I)
\end{array}$$

the fibrations (1) and (2) are $\omega^{\mathscr{A}}(\dim M)$-trivial by Theorem II. 4.2 and because $K = S^{2N}$ and $\chi(S^{2N}) = 2$, $l^{S^{2N}}$ is an $\omega^{\mathscr{A}}(\dim M)$-homotopy equivalence; hence $\Omega l^{S^{2N}}$ is an $\omega^{\mathscr{A}}(\dim M)$-homotopy equivalence.

From Figure 7 we observe that there exists a $(\omega^{\mathscr{A}}(\dim M) + 2)$-homotopy equivalence $\pi: [\tilde{\mathscr{A}}(M \times S^{2N})/\mathscr{A}(M \times S^{2N})]_l \to [{}^{\mathscr{A}}\mathscr{S}^{\varepsilon(n)}(M, \tau(M))]_l$, $l \leqq \omega^{\mathscr{A}}(\dim M) + 2$, making commutative the diagram:

$$\begin{array}{ccc} [\underline{\mathscr{H}}(M)]_l & \longrightarrow & [{}^{\mathscr{A}}\mathscr{S}^{\varepsilon(o)}(M^n)]_l \\ \uparrow {\scriptstyle [j]_l} & & \vdots\uparrow {\scriptstyle \pi} \\ [\tilde{\mathscr{A}}(M \times S^{2N})]_l & \longrightarrow & [\tilde{\mathscr{A}}(M \times S^{2N})/\mathscr{A}(M \times S^{2N})]_l \end{array}$$

Since any 3 consecutive terms of the first 2 lines in Figure 8 form a fibration, $j$ and $\pi$ induce $\pi'$ so that the following diagram is commutative:

$$\begin{array}{ccccc} [\Omega^{\mathscr{A}}\mathscr{S}^{\varepsilon(n)}(M, \tau(M))]_u & \longrightarrow & [{}^{\mathscr{A}}\mathscr{C}^{\varepsilon(n)}(M, \tau(M))]_u & \longrightarrow & [\underline{\mathscr{H}}(M)]_u \\ \uparrow {\scriptstyle [\Omega\pi]_u} & & \vdots\uparrow {\scriptstyle \pi'} & & \uparrow {\scriptstyle [j]_u} \\ [\Omega\tilde{\mathscr{A}}(M \times S^{2N})/\mathscr{A}(M \times S^{2N})]_u & \longrightarrow & [\mathscr{A}(M \times S)^{2N})]_u & \longrightarrow & [\tilde{\mathscr{A}}(M \times S^{2N})]_u \\ \uparrow {\scriptstyle [\Omega l^{S^{2N}}]_u} & & \uparrow & & \uparrow \\ [\Omega\tilde{\mathscr{A}}(M)/\mathscr{A}(M)]_u & \longrightarrow & [\mathscr{A}(M)]_u & \longrightarrow & [\tilde{\mathscr{A}}(M)]_u \end{array}$$

Since $\Omega\pi$ is an $u$-homotopy equivalence for $u \leqq \omega^{\mathscr{A}}(\dim M) + 2$ the assertion follows.

About Corollary II. 9.4: (1) follows immediately since $\tilde{\mathscr{A}}(M) \to \underline{\mathscr{H}}(M)$ factors through $\mathscr{H}(M)$ and because of the hypothesis, $\mathscr{H}(M) \to \underline{\mathscr{H}}(M)$ is homotopic to the constant map.

(2) follows from (1) and from the decomposition given by Theorem I. 2.3.

(3) follows from the fact that $\mathscr{S}^{\mathscr{A}}(M)_{\text{odd}} \to \mathscr{S}^{\mathscr{A}}(M \times K)_{\text{odd}}$ is homotopic to the constant map since $\chi(K) = 0$ (by Theorem II. 7.1).

**10. Problems.** We end up this survey with a few problems whose solution will give a deeper understanding of the functors $\mathscr{S}^{\mathscr{A}}$, ${}^{\mathscr{A}}\mathscr{S}^{\pm}$, $\mathscr{S}\mathscr{N}$ and $\mathscr{N}^{\pm}$ and at the same time will push a little further our knowledge about the structure of $\mathscr{A}(M)$.

(P.1) The estimate of $\omega^{\mathscr{G}}(n)$ is definitely susceptible to improvements; find the right values of $\omega^{\mathscr{A}}(n)$.

(P.2) Does (d) remain true for $\mathscr{A}$-locally trivial bundle with a cross-section? (Important consequences will follow even though this happens only for sphere bundles associated with vector bundles; for instance this will imply that ${}^{\mathscr{A}}\mathscr{S}^{\pm}$ are actually defined on $\mathscr{P}$.)

(P.3) What can be said about $l^{\xi}$ when $\xi$ is a finite cover?

(P.4) Study the fibration $*^{\mathscr{A}}(M)$.

(P.5) Compute $\pi_i(\mathscr{S}^{\mathscr{A}}(X))$, $\pi_i(\mathscr{W}_{\mathscr{A}, \mathscr{G}})$, $\pi_i(\mathscr{S}\mathscr{N}(X))$.

(P.6) What is the relationship between $\pi_i(\mathscr{S}^{\mathscr{A}}(K(G, 1)))$ and the higher order $K$-theory, respectively Whitehead theory (with various possible definitions)?

(P.7) What is the relationship between $\pi_i(\mathscr{S}\mathscr{N}(K(G, 1)))$ and the algebraic nilpotency (eventually higher order algebraic nilpotency)?

(P.8) Is $\pi_i(\mathscr{S}\mathscr{N}(K(G, 1))) \otimes z(\frac{1}{2}) = 0$ if $G$ is a finitely generated free abelian group?

(P.9) Does ${}^{\mathscr{A}}\beta(x) : \underline{\mathscr{H}}(X) \to \mathscr{S}^{\mathscr{A}}(X)$ deloop to a map $B({}^{\mathscr{A}}_{\beta}(X)): B\underline{\mathscr{H}}(X) \to B\mathscr{S}^{\mathscr{A}}(X)$ which satisfies the same properties as ${}^{\mathscr{A}}\beta(X)$?

(P.10) Prove that ${}^{\mathscr{A}}\mathscr{S}^{\pm}$ and $\mathscr{N}^{\pm}$ factor through $\mathscr{P}$.

ADDED IN PROOF. (i) Apparently, in Corollary II. 9.4 (2) we are able to prove only the triviality of the fibration $\Omega^{2\mathscr{A}}\mathscr{S}^{\varepsilon(n)}(M, \tau(M)) \to \Omega E^{\mathscr{A}}(M) \to \Omega\tilde{\mathscr{A}}(M)$.

(ii) Problem (6) is settled by Waldhausen's paper, these PROCEEDINGS.

(iii) Problem (7) was settled by the author and will be published in a forthcoming paper as well as problem (10) which was affirmatively answered.

## BIBLIOGRAPHY


**[A.B. K.]$_1$** P. Antonelli, D. Burghelea and P. Kahn, *Concordance homotopy groups of geometric automorphism groups*, Lecture Notes in Math., vol. 215, Springer-Verlag, Berlin and New York, 1971.

**[A. B. K.]$_2$** ———, *The nonfinite homotopy type of some diffeomorphism groups*, Topology **11** (1972), 1–49. MR **45** #1193.

**[B]$_1$** D. Burghelea, *The structure of block automorphisms of*, Topology (to appear).

**[B]$_2$** _____, *On the decomposition of the automorphisms groups of*, Rev. Roumaine Math. Pures Appl. (1977).

**[B. L.]$_1$** D. Burghelea and R. Lashof, *The homotopy type of the space of diffeomorphisms*. I, II, Trans. Amer. Math. Soc. **196** (1975), 1–36; ibid. **196** (1975), 37–50.

**[B. L.]$_2$** ———, *Stability of concordances and suspension homeomorphism*, Ann. of Math. (to appear).

**[B.L.]$_3$** ———, *The homotopy structure of the group of automorphisms in stable range and new homotopy functors* (to appear).

**[B. L. R.]** D. Burghelea, R. Lashof and M. Rothenberg, *Groups of automorphisms of manifolds*, Lecture Notes in Math., vol. 473, Springer-Verlag, Berlin and New York, 1975. MR **52** #1738.

**[C]** J. Cerf, *La stratification naturelle des espaces de fonctions differentiables réeles et le théorème de la pseudoisotopie*, Inst. Hautes Étude Sci. Publ. Math. No. **39** (1970), 5–173. MR **45** # 1176.

**[G]** A. Gramain, *Groupe des diffeomorphismes et espace de Teichmüller d'une surface*, Seminaire Bourbaki, 1973, Exp. No. 424. MR **50** #17.

**[H. W.]** A. Hatcher and J. Wagoner, *Pseudoisotopies of compact manifolds*, Astérisque, No. 6, Société Mathématique de France, Paris, 1973. MR **50** #5821.

**[H]** A. Hatcher, *Higher simple homotopy type*, Ann. of Math. (2) **102** (1975), 101–137. MR **52** #4305.

**[K. L.]** N. Kuiper and R. Lashof, *Microbundles and bundles*. I, II, Invent. Math. **1** (1966), 1–17; ibid. Invent. Math. **1** (1966), 243–259. MR **35** #7339; **35** #7340.

**[K. S.]** R. Kirby and L. Siebenmann, *Papers on hauptvermutung*,

**[L]** R. Lashof, *The immersion approach to triangulation and smoothing*, Proc. Advanced Study Inst. on Algebraic Topology (Aarhus, 1970), Vol. II, pp. 282–322. Math. Inst., Aarhus Univ., Aarhus, 1970. MR **43** #2715a.

**[Mi]** K. Millet, *Piecewise linear concordances and isotopies*, Mem. Amer. Math. Soc. **153** (1974).

**[Mo]** C. Morlet, *Plongements et automorphismes des vanétès*-Cours Pecot-Collège de France 1969.

**[Mu]** J. Munkres, *Concordance inertia groups*, Advances in Math. **4** (1970), 224–235. MR **41** #7390.

**[N]** S. P. Novikov, *Differentiable sphere bundles*, Izv. Akad. Nauk SSSR Ser. Mat. **29** (1965), 71–96. (Russian) MR **30** #4366.

**[Q]** F. Quinn, Thesis, Princeton Univ., Princeton, N. J., 1969.


INSTITUTE OF ATOMIC PHYSICS, BUCAREST, RUMANIA

Proceedings of Symposia in Pure Mathematics
Volume 32, 1978

# BASED FREE COMPACT LIE GROUP ACTIONS ON HILBERT CUBES

I. BERSTEIN AND J. E. WEST*

**1. Introduction.** The Hilbert cube is a surprisingly ubiquitous and versatile object. Its compactness, convexity, and multifaceted substructure endow it richly from the geometric point of view, yet its infinite-dimensionality has a soothing effect, lending plasticity and universality, and thus linking problems of topological classification particularly closely with the analogous homotopy-theoretic and simple-homotopy-theoretic questions. This interplay between the geometric and the homotopy-theoretic is particularly intriguing, at least to us, in the setting of transformation groups.

In this context, we mean by transformation groups the various actions of familiar, well-understood groups on the spaces in question rather than the study of the topological type of the full group of homeomorphisms, which has a completely different character.[1]

There is as yet virtually no literature on the subject of transformation groups on Hilbert cube manifolds although the second author has been proselytizing vigorously ever since the preliminary version of **[40]** began to circulate.[2] The beauty of that

*AMS* (*MOS*) *subject classifications* (1970). Primary 57E10, 57E30, 57A20.

*Research partially supported by the NSF. This is an expanded version of the second half of the second author's talk at the A.M.S. Summer Institute on Topology at Stanford, 1976.

[1]It must be said that the study of the full groups of homeomorphisms of Hilbert cube manifolds is one of the most dramatic and important branches of infinite-dimensional topology, including such standout results as R. Wong's proof **[36]** of the contractibility of the homeomorphism group of the Hilbert cube, the Chapman and Fathi-Visetti proofs **[10]**, **[18]** of local contractibility for the homeomorphism group of any compact Hilbert cube manifold, and the very recent proofs by S. Ferry and H. Torunczyk that these groups are, indeed, Hilbert manifolds **[19]**, **[28]**.

[2]The only other papers specifically relevant to this discussion seem to be **[35]**, which is highly specialized, **[13]**, **[31]**, **[36]**–**[39]**, **[42]**, **[43]** and Chapter 7 of **[15]**, with which there is considerable overlap here. (There has been a free sharing of thoughts among D. A. Edwards, H. Hastings, and the present authors.)

paper is its simplicity and the ease with which one may construct examples covered by its hypotheses. In it, Wong considered $Z_p$-actions on the Hilbert cube with unique fixed point, for $p$ prime, and took as his standard model of the Hilbert cube the countably infinite product of the unit disc in the complex plane with itself. He then defined the standard action of $Z_p$ on this model to be the simultaneous multiplication by the principal $p$th root of unity and gave the following simple characterization of those $Z_p$-actions on Hilbert cubes which are topologically equivalent (i.e., conjugate) to his standard:

THEOREM (R. WONG [**40**]). *A $Z_p$-action on the Hilbert cube with unique fixed point is topologically equivalent to the standard if and only if the fixed point has a basis of contractible, equivariant neighborhoods.*

With this theorem and the theorems of R. D. Anderson, the second author, and R. D. Edwards which collectively imply that any countably infinite product of nondegenerate AR's is a Hilbert cube [**1**], [**32**], [**11**], one can obtain immediately many examples of equivalent, but not obviously so, examples of $Z_p$-actions on the Hilbert cube with unique fixed point. In particular, if we have a countable collection of compact ANR's, each equipped with a free $Z_p$-action and if we extend each action to the cone on the ANR in the natural way so that the interval parameter is invariant, then we obtain an equivariantly contractible action on an AR with unique fixed point. The diagonal (simultaneous) action on the product of these cones is then a $Z_p$-action on a Hilbert cube with unique fixed point—the infinite vertex—satisfying Wong's hypotheses and is thus standard. Hence, if we let $Z_p$ act on itself by translation or choose any collection of free $Z_p$-actions on $S^3$ giving lens spaces $L(p, q)$, we obtain the "same" action on the Hilbert cube.

In this paper we define in a natural way semifree $G$-actions on Hilbert cubes with single fixed point whenever $G$ is a Lie group, and we show that the classification of such actions up to conjugacy reduces in large measure to the classification of the orbit spaces of the free parts of the actions up to proper homotopy, entirely so for finite $G$ all automorphisms of which are inner, e.g., finite complete groups $G$. (This recovers Wong's theorem cited above, once one investigates the standard orbit spaces.) These orbit spaces are locally compact classifying spaces. This development provides a geometric motivation for the study of the proper homotopy theory of locally compact classifying spaces in general, which we do not discuss here.

Restricting ourselves to compact $G$ and taking our construction as the standard, we relate our version of $B_G$ to Milnor's infinite join construction and show that our spaces are proper-homotopy equivalent to the infinite mapping cylinder of the inclusions $G/G \hookrightarrow (G*G)/G \hookrightarrow (G*G*G)/G \hookrightarrow \cdots$, which we feel should be taken as a standard because proper homotopy theory reduces to ordinary homotopy theory in them. Since there is not a satisfactory "Whitehead" theorem in the "proper" category, at least for the infinite-dimensional complexes we are usually dealing with, we are reduced to ad hoc methods of studying these orbit spaces, so much so, in fact, that we cannot assert the existence of nonstandard actions for a single finite group, although we believe they exist. (However, we note that T. Tucker has produced wild translations of $R^3$ by modifying Whitehead's contractible 3-manifold construction [**29**] and that this produces nonstandard $Z$-actions on the Hilbert cube, too.)

Since these standard spaces are relatively simple, we are able to pass from proper homotopy theory to homotopy theory of the end of the nonsingular orbit space and to show that the orbit spaces are standard if their ends are movable in the sense of Borsuk. Using our movability criterion, we are able to show that, for $G$ finite, any based-free $G$-action on the Hilbert cube which splits as a diagonal action on a product of finite-dimensional spaces must be standard.

Finally, we establish another homotopy-theoretic criterion for the existence of nonstandard $G$-actions, i.e., actions the orbit spaces of the free parts of which are not proper-homotopy equivalent to those of the standard actions. We have not as yet been able to use this criterion in any substantive way, however, and by way of illustration of the delicacy of the situation, we provide an extremely simple example which shows how easily it may be satisfied if a relative finiteness condition (forced by the compactness of the Hilbert cube) is relieved.

**2. Notational conventions.** We use $Q$ to denote any Hilbert cube; as a model for $Q$, we ordinarily think of the infinite product $J^\infty = \prod_{i=1}^\infty J_i$, where $J = [-1, 1]$, but we use the symbol generally to denote any Hilbert cube, and we introduce a host of other models between which we do not discriminate. Because we are always dealing with an action on (a) $Q$ with a single fixed point, we need a symbol for the complement of the fixed point, and we use $Q_0$. When we are dealing with homotopies, we use the upper case, e.g., $F: X \times I \to Y$, for the entire one-parameter family of mappings and indexed lower case type for the members of that family e.g., $f_0, f_t: X \to Y$.

**3. The standard actions.** Any $Z_2$-action on $Q$ with unique fixed point worthy of the name "standard" *must be* multiplication by "$-1$". Proceeding from this toward a natural definition which will hold for all compact Lie groups, we note that $J = [-1, 1]$ is the cone on $Z_2$ and that for $G$ any compact Lie group, the underlying space, being a polyhedron, has the property that its cone $C(G) = G \times I/(G \times \{0\})$ is a compact, contractible polyhedron and thus **[32]** its countably infinite product $C(G)^\infty$ is a Hilbert cube. Therefore, left translation of $G$ on itself extends naturally to the cone (by the rule $g(h, t) = (gh, t)$) and thence to $C(G)^\infty$. The resulting action has only the infinite vertex as fixed point, being free on the complement.[3] We take this to be *standard based-free G-action on a Hilbert cube.* Any semifree $G$-action on a Hilbert cube will be called "standard" if it is equivariantly homeomorphic with this.

This definition works perfectly well for noncompact Lie groups with one adjustment (to give compactness), so although we do not address this situation, we suggest the definition: An action of the group $G$ on a space is based-principal if there is a unique fixed point and on the complement of that point the action is principal. If $G$ is a noncompact Lie group, let $G_+ = G \cup \{\infty\}$ be its one-point compactification and extend the left translation of $G$ on $G$ to $G_+$ so that the point "at infinity" is fixed. Let $C_+(G_+) = G_+ \times I/(G_+ \times \{0\} \cup \{\infty\} \times I)$ be the reduced cone of $G_+$ (reduced at "infinity"). Now, $C_+(G_+)$ is a compact AR, so by Ed-

[3]This construction is obviously functorial on the category of compact ANR's with $G$-actions and equivariant maps; if we restrict ourselves to compact AR's, we may omit the coning operation, keeping in mind that if we begin with a single point, we shall not construct a Hilbert cube.

wards' theorem that compact metric AR's are $Q$-factors **[11]** and the result **[1]**, **[32]** that infinite products of $Q$-factors are $Q$'s, $C_+(G_+)^\infty$ is a Hilbert cube; it is a simple matter *to check that the G-action* $\alpha_G$ *induced* on $C_+(G_+)$ is based-principal.[4] *A based-principal G-action on Q is standard if it is equivariantly homeomorphic to* $\alpha_G$.

**4. A reduction to proper homotopy theory.** Let $G$ be a compact Lie group and $\alpha: G \times Q \to Q$ a based-free $G$-action. If $\beta$ is another based-free $G$-action, we say that $\alpha$ and $\beta$ are *equivalent* (conjugate) if there is a homeomorphism $f: Q \to Q$ such that $f(\alpha(g, q)) = \beta(g, f(q))$, for each $g \in G$ and all $q \in Q$, and we say that $\alpha$ and $\beta$ are *weakly equivalent* if for some automorphism $\tau$ of $G$, there is a homeomorphism $f: Q \to Q$ for which $f(\alpha(g, q)) = \beta(\tau(g), f(q))$ for each $g \in G$ and all $q \in Q$.

Our standing hypothesis is that space, $Q_0/\alpha$, of nonsingular orbits be a Hilbert cube manifold. (This is of course true if $G$ is finite, as the orbit map $p_\alpha$ is then a covering map; in general, $p_\alpha$ is the projection of a principal $G$-bundle by Gleason's theorem **[20]**.) Our program is to study $Q_0/\alpha$ from the viewpoint of $Q$-manifold theory. To begin our investigation we note the following crucial fact.

(4.1). PROPOSITION. *The complement of any point in the Hilbert cube is contractible.*

PROOF. As $Q$ is homogeneous **[22]**, **[2]**, the point in question may be taken to be the "corner" point $(1, 1, 1, \cdots)$, the complement of which is even convex.

(4.2). COROLLARY. *The principal G-bundle* $Q_0 \to Q_0/\alpha$ *is a universal bundle.*

(4.3). PROPOSITION. *There is an equivariant cell structure on* $Q_0$, *i.e., a commutative diagram*

$$\begin{array}{ccc} L \times Q & \xrightarrow{h'} & Q_0 \\ {\scriptstyle p\times \mathrm{id}}\downarrow & & \downarrow{\scriptstyle p_\alpha} \\ K \times Q & \xrightarrow{h} & Q_0/\alpha \end{array}$$

*where* $L \to^{p} K$ *is a principal G-bundle of CW-complexes, h and h′ are homeomorphisms, h′ equivariant.*

PROOF. The orbit space $Q_0/\alpha$ is a $Q$-manifold by hypothesis, so the bottom line of the diagram exists by Chapman's triangulation theorem [9], **[11]**. Let $f: K \to K \times \{0\} \to K \times Q \to Q_0/\alpha$ be the inclusion, and let $L = f^*(Q_0 \to^{p_\alpha} Q_0/\alpha)$ be the induced $G$-bundle over $K$. Now $L$ is a CW-complex because $K$ and $G$ are and the map $L \to K$ is a locally trivial bundle. The equivariant homeomorphism $h'$ is now easily constructed using the local triviality of the bundle $Q_0 \to Q_0/\alpha$ over $g(\sigma \times Q)$ for each simplex $\sigma$ of $K$ and of the bundle $L \times Q \to K \times Q$ over $\sigma \times Q$. (We have equivariant diagrams

$$\begin{array}{ccccccccc} L \times Q & \xleftarrow{\varphi} & G \times \sigma \times Q & \to & G \times f(\sigma \times Q) & \xrightarrow{\psi} & p_\alpha^{-1}(f(\sigma \times Q)) & \to & Q_0 \\ \downarrow & & \downarrow & & \downarrow & & \downarrow & & \downarrow \\ K \times Q & \longleftarrow & \sigma \times Q & \longrightarrow & f(\sigma \times Q) & \longrightarrow & f(\sigma \times Q) & \longrightarrow & Q_0/\alpha \end{array}$$

[4] This construction is also clearly functorial on the locally compact ANR $G$-spaces.

where $\varphi$ and $\psi$ are given by the local triviality. Taken together, these diagrams define $h'$.)

(4.4). DEFINITION. The End, $E(X)$, of a locally compact space is the inverse system of those subsets of $X$ the complements of which have compact closures and inclusions. Each member (i.e., set) of the system is a neighborhood of $E(X)$.

(4.5). PROPOSITION. *There is a cofinal sequence $N_1 \supset N_2 \supset N_3 \supset \cdots$ of neighborhoods of $E(K)$ with the property that if $N_i' = p^{-1}(N_i) \subset L$, then for each $i$, $N_i'$ is null-homotopic in $N_{i-1}'$.*

PROOF. Consider $Q_0$ again as being the complement of the "corner" point (1, 1, 1, $\cdots$), and note that (1, 1, 1, $\cdots$) has a nested basis of convex neighborhoods which remain convex when it is deleted. If we select a sequence $V_1 \supset V_2 \supset V_3 \supset \cdots$ of these with null intersection and a sequence $N_1 \supset N_2 \supset N_3 \supset \cdots$ of neighborhoods of $E(K)$ such that for each $i$, $h'(N_{i-1}' \times Q) \supset V_i \supset h'(N_i') \supset V_{i+1}$, then we are done.

We are now in a position to prove the main result of this section.

(4.6). THEOREM. *If $\alpha$ and $\beta$ are based-free actions on $Q$ of the compact Lie group $G$ such that $Q_0/\alpha$ and $Q_0/\beta$ are $Q$-manifolds, then each proper-homotopy equivalence from $Q_0/\alpha$ to $Q_0/\beta$ is properly homotopic to a homeomorphism.*

PROOF. Let $K_\alpha$ and $K_\beta$ be CW-complexes such that $K_\alpha \times Q$ and $K_\beta \times Q$ are homeomorphic to $Q_0/\alpha$ and $Q_0/\beta$, respectively **[9]**, and note that in light of (4.5) there is a cofinal sequence $N_1 \supseteq N_2 \supseteq \cdots$ of neighborhoods of the end of $K_\alpha$ and of $K_\beta$ in each case with the property that, with due attention to change of base point isomorphisms, the inclusion-induced homomorphisms on the fundamental groups factor through $\pi_0(G)$ as follows:

$$\begin{array}{ccccccccc} \cdots \longrightarrow & \pi_1(N_{n+1}) & \longrightarrow & \pi_1(N_n) & \longrightarrow & \pi_1(N_{n-1}) & \longrightarrow \cdots \\ & \searrow & \nearrow & \searrow & \nearrow & \searrow & \\ \cdots \longrightarrow & \pi_0(G) & \xrightarrow{\text{id}} & \pi_0(G) & \xrightarrow{\text{id}} & \pi_0(G) & \longrightarrow \cdots \end{array}$$

(That is, the fundamental group is essentially constant at infinity.) This ensures the coalescence of the simple-homotopy structures (see **[27]**) so that all proper-homotopy equivalences are simple. By the argument given in **[32]** (which extends automatically to the case of CW-complexes through **[33]** (see also **[10]**, **[11]**) ), if $f: K_\alpha \to K_\beta$ is a simple homotopy equivalence, then $f \times \text{id}: K_\alpha \times Q \to K_\beta \times Q$ is properly homotopic to a homeomorphism.

(4.7). COROLLARY. *Let $\alpha$ be a based-free $G$-action on $Q$ with the property that every homotopy equivalence of $Q_0/\alpha$ with itself is homotopic to a proper homotopy equivalence. Then a based-free $G$-action $\beta$ on $Q$ is conjugate to $\alpha$ if and only if $Q_0/\beta$ is proper homotopy equivalent to $Q_0/\alpha$.*

PROOF. As conjugacy implies the proper homotopy equivalence of the spaces of nonsingular orbits, we need only prove the converse. Let $\beta$ be a based-free $G$-action on $Q$ with $Q_0/\beta$ proper homotopy equivalent to $Q_0/\alpha$. As $Q_0 \to^{p_\alpha} Q_0/\alpha$ and $Q_0 \to^{p_\beta} Q_0/\beta$ are principal $G$-bundles, universal by (4.2), the orbit spaces $Q_0/\alpha$, etc., are classifying spaces for principal $G$-bundles or, equivalently, principal (= free, since $G$ is compact, Lie) $G$-spaces. (See **[6**, Chapter II**]**.) By Theorem (4.6),

$Q_0/\alpha$ is homeomorphic to $Q_0/\beta$. Let $f$: $Q_0/\alpha \to Q_0/\beta$ be a homeomorphism, and let $f^*(\beta) \to^p Q_0/\alpha$ be the induced bundle, with (equivariant) bundle map $f_*: f^*(\beta) \to Q_0$ covering $f$. As $f$ is a homotopy equivalence, $f^*(\beta)$ is universal; therefore there is a classifying map $g$: $Q_0/\alpha \to Q_0/\alpha$ inducing a bundle $g^*f^*(\beta) \to^{p'} Q_0/\alpha$ isomorphic to $Q_0 \to^{p_\alpha} Q_0/\alpha$ over the identity. Let $h$: $Q_0 \to g^*f^*(\beta)$ be an isomorphism. Then $h$ is equivariant, and if $g$ can be chosen to be a homeomorphism, then $f_* \circ g_* \circ h$: $Q_0 \to Q_0$ is an equivariant homeomorphism from $\alpha$ to $\beta$, and its extension to $Q$, the one-point compactification of $Q_0$, is the homeomorphism by which to conjugate. But $g$ must be a homotopy equivalence in order to retrieve a universal bundle, by functoriality, so it may be replaced up to homotopy by a proper homotopy equivalence which, by Theorem (4.6), may in turn be replaced up to homotopy by a homeomorphism. Homotopic maps induce isomorphic (equivariantly homeomorphic) bundles, so we may assume that $g$ is a homeomorphism.

(4.8). COROLLARY. *Based-free involutions of the Hilbert cube are conjugate if and only if their spaces of nonsingular orbits are proper-homotopy equivalent.*

PROOF. In this case, $G = Z_2$ and for any based-free involution $\alpha$, the orbit map $P_\alpha$: $Q_0 \to Q_0/\alpha$ is a covering projection and $Q_0/\alpha$ is a locally compact $Q$-manifold $K(Z_2, 1)$. The homotopy classes of homotopy equivalences of a $K(G, 1)$ with itself are in one-to-one correspondence with the automorphisms of $G$ ($G$ discrete), so the hypothesis of Corollary (4.7) is always satisfied as $Z_2$ has only the trivial automorphism.

More generally, we have

(4.9). COROLLARY. *Based-free actions of a finite group on the Hilbert cube are weakly equivalent if and only if their spaces of nonsingular orbits are proper-homotopy equivalent.*

PROOF. We redo (4.7) in the context of covering spaces. Let $G$ be finite, and let $\alpha$ and $\beta$ be two based-free $G$-actions on $Q$ with proper-homotopy equivalent nonsingular orbit spaces $Q_0/\alpha$ and $Q_0/\beta$. As $G$ is finite, $p_\alpha$: $Q_0 \to Q_0/\alpha$ and $p_\beta$: $Q_0 \to Q_0/\beta$ are universal coverings, and $Q_0/\alpha$ and $Q_0/\beta$ are $Q$-manifold $K(G, 1)$'s. By Theorem (4.6) they are homeomorphic. Let $f$: $Q_0/\alpha \to Q_0/\beta$ be a homeomorphism. By picking base points $\tilde{x}_0 \in Q_0$, $x_0 = p_\alpha(\tilde{x}_0)$, $y_0 = f(x_0)$, and $\tilde{y}_0 \in p_\beta^{-1}(y_0)$, we may construct a homeomorphism $\tilde{f}$: $Q_0 \to Q_0$ covering $f$ by selecting, for each $\tilde{x} \in Q_0$, a path $\tilde{w}$: $(I, 0, 1) \to (Q_0, \tilde{x}_0, \tilde{x})$ and setting $f(\tilde{x}) = (f \circ p \circ \tilde{w}(1))^\sim$, where $(f \circ p \circ \tilde{w})^\sim$ is the lifting of $f \circ p \circ \tilde{w}$ which begins at $\tilde{y}_0$.

Now let $h_\alpha$: $G \to \pi_1(Q_0/\alpha, x_0)$ and $h_\beta$: $G \to \pi_1(Q_0/\beta, y_0)$ be the isomorphisms of $G$ with the fundamental group induced by the choice of base points, i.e., $h_\alpha(g) = p_\alpha \circ \tilde{w}_g$, where $\tilde{w}_g$ is any path in $Q_0$ from $\tilde{x}_0$ to $\alpha(g, \tilde{x}_0)$, and let $f_*$ be the isomorphism induced on the fundamental group by $f$. Then $\tilde{f} \circ \alpha(g, \tilde{x}) = \beta(h_\beta^{-1} \circ f_\# \circ h_\alpha(g), \tilde{f}(\tilde{x}))$.

(4.10). COROLLARY. *If $G$ is finite and all its automorphisms are inner, based-free $G$-actions on $Q$ are conjugate if and only if their spaces of nonsingular orbits are proper-homotopy equivalent.*

PROOF. Proceed as in (4.9), and then let $h = f_\#^{-1} \circ h_\beta \circ h_\alpha^{-1}$: $\pi_1(Q_0/\alpha, y_0) \to \pi_1(Q_0/\alpha, y_0)$. Choose a loop $k$ at $x_0$ such that $h$ is conjugation by $[k]$. Changing the base point from $\tilde{x}_0$ to $\tilde{k}(1)$ in the above results in an isomorphism $h'_\alpha$ of $G$ with $\pi_1(Q_0/\alpha, x_0)$ differing from $h_\alpha$ by conjugation by $k$; thus $h_\beta^{-1} \circ f_\# \circ h'_\alpha = \mathrm{id}_G$, and the above construction with the new base point provides an equivariant homeomorphism which extends to $Q$.

(4.11). REMARK. Finite groups all automorphisms of which are inner include the complete groups (which in additon have no center). Examples are $S_n$ (the symmetric group on $n$ symbols), $n \neq 6$, and the automorphism groups of simple groups of composite order **[26]**.

**5. The standard orbit spaces.** At this point, we return to our standard actions and investigate their orbit spaces, which are easily seen to be the simplest possible objects, wherein proper homotopy theory and homotopy theory coincide. Thus, for these actions the hypothesis of Corollary (4.7) holds and, in fact, is a characterization of them.

Recall Milnor's infinite join construction of universal $G$-bundles: $E_G$ = inj lim $\{G \to G*G \to G*G*G \to \cdots\}$, $G$ acts diagonally on the join factors, and $B_G = E_G/G$ **[24]**. This is virtually what we have as our standard orbit spaces, except that ours are locally compact. Let $TB_G$, for "telescope" of $B_G$, be the mapping cylinder of the direct system $G \to G*G \to G*G*G \to \cdots$, i.e., $TB_G$ is the union of the mapping cylinders of the maps of the sequence with the target and of the $n$th mapping cylinder identified with the domain end of the $(n + 1)$st. Note that $TB_G$ is naturally homeomorphic to $\bigcup_{n=1}^{\infty} (*_{i=1}^{n} G_i) \times [n, \infty)$, which is the model we shall use.

(5.1). PROPOSITION. *If $\alpha$ is the standard based-free G-action on Q, then $Q_0/\alpha$ is proper-homotopy equivalent to $TB_G$.*

PROOF. We do explicitly the case that $G$ is compact; it should be clear to the reader that with minor modifications, the proof presented below works for noncompact groups as well. Let $X_G = \{y = ((x_1, t_1), (x_2, t_2), \cdots) \in C(G)^\infty | t_n \neq 0 \Rightarrow t_{n-k} \leqq 1/(k + 1)\}$. We construct an equivariant strong deformation retraction of $C(G)^\infty$ to $X_G$ as follows: Let $X_G(n) = \{y \in C(G)^\infty | m \leqq n \text{ and } t_m > 0 \Rightarrow t_{m-k} \leqq 1/(k + 1)\}$ be intermediate stages, and let $X_G(n, m) = \{y \in X_G(n) | \ t_{n+1} > 0 \Rightarrow \text{if } i = (n + 1) - k \leqq m, \text{ then } t_i \leqq 1/(k + 1)\}$ be intermediate stages between *them*. Let $f_{n,m}: X_G(n, m) \to X_G(n, m + 1)$ be the equivariant retraction which "projects in the plane of the $(n + 1)$st and $(m + 1)$st cone coordinates at a slope of $2^{2(n+1)}$", that is, in formulae,

$$f_{n,m}((x_1, t_1), (x_2, t_2), \cdots) = ((x_1, s_1), (x_2, s_2), \cdots),$$

where $s_i = t_i$ unless $i = m + 1$ or $n + 1$, $s_{m+1} = \max\{1/(n - m), s_{m+1} - t_{n+1}/2^{2(n+1)}\}$, and $s_{n+1} = t_{n+1} - 2^{2(n+1)}(t_{m+1} - s_{m+1})$. Setting $f_n = f_{n,n} \circ \cdots \circ f_{n,1}: X_G(n) \to X_G(n+1)$, we note that the slope condition guarantees that the infinite composition $f = \cdots \circ f_n \circ f_{n-1} \circ \cdots \circ f_1$ is a well-defined mapping (equivariant) and that no point is sent by $f$ to the infinite vertex $v$ except $v$ itself. It is immediate that we can produce equivariant homotopies sliding along the projection lines to obtain an equivariant strong deformation retraction $F$: $C(G)^\infty \times I \to C(G)^\infty$ such that $F^{-1}(v) = \{v\} \times I$.

Now using the fact that for any spaces $Y$ and $Z$ we have a homeomorphism $g$: $C(Y) \times C(Z) \to C(Y*Z)$ by the formula

$$((y, s), (z, t)) \to \begin{cases} ((y, 1 - s/2, z), t), & \text{if } t \geqq s, \\ ((y, 1 - ((s - t)/s \cdot (1 - s/2) + s/2), z), s), & \text{if } t \leqq s, \end{cases}$$

where $C(Y) = (Y \times I)/(Y \times \{0\})$ and $Y*Z = Y \times I \times Z/\equiv$, with $(y, 0, z) \equiv (y, 0, z')$ and $(y, 1, z) \equiv (y', 1, z)$, which is clearly equivariant, we can immediately construct an equivariant homeomorphism $h\colon X_G - \{v\} \to E_G$ inductively as another infinite composition: Let $Y_n = (1/n)C(G)_1 \times (1/(n-1))C(G)_2 \times \cdots \times C(G)_n$ and, identifying $Y_n$ with its natural image in $C(G)^\infty$ as $Y_n \times \{v\} \times \{v\} \times \cdots$, note that $X_G = \bigcup_{n=1}^\infty Y_n$. Let $Y_n' = (1/n)C(G)_1 \times (1/n)C(G)_2 \times \cdots \times (1/n)C(G)_n$, where $tC(G) = G \times [0, t]/G \times \{0\}$, and let $h_0\colon X_G \to X_G' = \bigcup_{n=1}^\infty Y_n'$ be an equivariant homeomorphism, e.g., $h_0((x_1, t_1), (x_2, t_2), \cdots) = ((x_1, t_1'), (x_2, t_2'), \cdots)$, where $t_n' = c_n(t_n)$ and $c_n\colon I \to I$ is a homeomorphism which sends each $1/k \to 1/(k + n - 1)$. Now, letting

$$Z_n = C(\underbrace{G*G* \cdots *G}_{n}) \times C(G)_{n+1} \times C(G)_{n+2} \times \cdots,$$

let $g_n = g \times \mathrm{id}\colon Z_n \to Z_{n+1}$. Then the infinite composition $h_1 = \cdots \circ g_n \circ g_{n-1} \circ \cdots \circ g_1$ maps $C(G)^\infty$ into $C(\bigcup_{n=1}^\infty(\ast_{i=1}^n G_i))$ because of the following considerations:

(1) $g((1/n)C(A) \times (1/n)C(B)) = (1/n)C(A * B)$, so $g_n \circ \cdots \circ g_1(Y_n') = (1/n)C(\ast_{i=1}^n G_i)$,

(2) $g((x_1, s), (x_2, 0)) = ((x_1, 0, x_2), s)$, so $h_1(Y_n' - Y_{n+1}') = C((\ast_{i=1}^{n+1} G_i) \times \{0\} \times G \times \{0\} \times G \times \{0\} \times \cdots) - (1/(n+1))C(\ast_{n=1}^\infty G_n)$ and the coordinate string is eventually constant ($G_1 \times \{0\} \times G_2 = G_1$ in $G_1 * G_2$, and $(x_n, 0) = v$ in $C(G_n)$),

(3) the two preceding considerations give both well-definedness and continuity of $h_1$, allowing us to conclude in addition that its image is precisely $\bigcup_{n=1}^\infty (1/n)C(\ast_{i=1}^n G_i)$, which is also the image of $TE_G$ under the embedding $h_2$ which merely inverts the real coordinate, together with the vertex, to which no other point has been sent by $h_1$. The mapping $h = h_2^{-1} \circ h_1 \circ h_0\colon X_G \to TE_G$ is thus a homeomorphism and completes the proof.

**6. Some rudimentary proper homotopy theory.** We have now shown that the proper homotopy theory of infinite-dimensional CW-complexes plays a major role in the study and classification of based-free $G$-actions on the Hilbert cube when $G$ is a compact Lie group, through the study of the spaces of nonsingular orbits, which are locally compact, one-ended, often $Q$-manifolds, classifying spaces, and, inevitably, of infinite homotopy dimension. This last, as it turns out, immeasurably complicates our problem, which, basically, is to ascertain when a given homotopy equivalence is or is not homotopic to a proper homotopy equivalence. The difficulty of the situation is starkly illuminated by the fact that there is no known sufficient "Whitehead theorem" in the proper homotopy theory of infinite-dimensional complexes—and most straightforward generalizations of the theorem "If $f\colon X \to Y$ induces isomorphisms on all homotopy groups, then it is a homotopy equivalence" are demonstrably false without some rather strong tameness hypothesis at infinity and/or finite-dimensionality condition on either the domain or the target. (A review of this problem with several instructive examples is given in **[15]** which also contains cross-references to the analogous problem in shape theory, which is exceptionally closely related to the present problems. Indeed, there is a "forgetful" functor from proper homotopy theory to shape theory via Chapman's "Complement Theorem" **[8]**. See also **[5]**, **[14]** and **[16]**). For the proper homotopy theory of finite-dimensinoal complexes, where the situation is well in hand, see **[27]**, **[7]**, **[17]**, **[25]**, **[21]**, and **[23]**; but there are oversights in both **[7]** and **[17]**, where the infinite-dimensional case is treated, as is indicated by the examples in §5.5 of **[15]**.

We give here an extremely restricted introduction which addresses only the situation at hand and is for that reason able to dispense with the terminology of pro-groups. We begin by showing that our standard orbit spaces satisfy the hypothesis of (4.7). A locally compact space $X$ is said to admit a *proper deformation to infinity* if there is a proper mapping $F: X \times [0, 1) \to X$ such that $F|X \times \{0\}$ is the "identity". This condition effectively destroys the distinction between proper homotopy theory and homotopy theory, as has been observed by several authors, e.g., **[12]**, **[41]**.

(6.1). PROPOSITION. *If $X$ admits a proper deformation to infinity, then every mapping $f: Y \to X$ of a connected, locally compact, metric space is homotopic to a proper mapping; if $f$ is proper on the closed subset $A$ of $Y$, then $f$ is homotopic relative to $A$ to a proper map.*

PROOF. Continuity dictates that there is a closed neighborhood $N$ of $A$ in $Y$ on which $f$ is proper. Let $g: (Y, A, Y - \mathring{N}) \to ([0, 1], 0, 1)$ be a map and let $h: Y \to [0, 1)$ be a proper map. Then set $H: Y \times I \to X$ to be $H(y, t) = F(f(y), tg(y)h(y))$, where $F$ is a proper deformation of $X$ to infinity. This is the desired homotopy.

(6.2). COROLLARY. *If $G$ is a compact Lie group, then a based-free $G$-action $\alpha$ on $Q$ is standard if and only if $Q_0/\alpha$ admits a proper deformation to infinity.*

PROOF. Being a classifying space for $G$, $Q_0/\alpha$ is homotopy equivalent to $G_0/\sigma$, where $\sigma$ is the standard. As $Q_0/\sigma$ admits a proper deformation to infinity, (6.1) shows that every self-homotopy equivalence of $Q_0/\sigma$ is homotopic to a self-proper-homotopy equivalence. Then (4.7) yields that $\alpha$ is conjugate to $\sigma$ if and only if $Q_0/\alpha$ is proper-homotopy equivalent to $Q_0/\sigma$, which is, by (6.1), true if and only if $Q_0/\alpha$ admits a proper deformation to infinity.

Corollary (6.2) recovers Wong's theorem **[40]**. These spaces have a property similar to (6.1) when they are the domains of maps which are "proper on skeleta".

(6.3). PROPOSITION. *Let $X$ be a locally finite $CW$-complex which admits a proper deformation to infinity. If $f: X \to Y$ is a mapping which is proper on each $n$-skeleton of $X$, then there is a proper deformation $H: X \times I \to X$ of $X$ such that $h_0 =$ id and $f \circ h_1: X \to Y$ is proper; if $f$ and $g$ are two proper maps and $G: X \times I \to Y$ is a homotopy between them which is proper on each skeleton of $X \times I$, then there is a proper homotopy $H: X \times I \to Y$, stationary on $X \times \{0, 1\}$, such that $h_0 =$ id and $H \circ (f_1 \times \text{id}): X \times I \to Y$ is a proper homotopy between $f$ and $g$.*

PROOF. Let $F: X \times [0, 1) \to X$ be a proper deformation to infinity, and let $X = \bigcup_{n=1}^{\infty} X_n$ and $Y = \bigcup_{n=1}^{\infty} Y_n$ be expressions of $X$ and $Y$ as increasing towers of compacta, subcomplexes in the case of $X$. Assume that $F(X \times [n, \infty)) \cap X_n = \emptyset$, and let $i(j, k) = 1 + \max\{i | f^{-1}(Y_k) \cap X^{(j)} \not\subset X_i\}$, where $X^{(j)}$ is the $j$-skeleton of $X$.

Now choose a function $k: X \to R$ (= reals) such that, for each $n$-simplex $\alpha^n$, $k(\alpha^n) \subset [i(n+1, n), \infty)$. Now, let $h = f \circ F \circ (\mathrm{id} \times k)$. Then $h(\alpha^n) \cap Y_n = \varnothing$, if we assume, as we may, that $F$ is cellular and hence sends $X^{(n)} \times [0, 1)$ into $X^{(n+1)}$; moreover, $h$ is proper on skeleta, being a composition of maps with that property. We observe that $h$ is proper by noting that if it were not, then it would send an infinite-dimensional set into some $Y_n$ by virtue of being proper on skeleta, and this it cannot do by the construction, since $h^{-1}(Y_n) \subset X^{(n-1)}$. It is obvious that the homotopy $H: X \times I \to X$ given by $H(x, t) = F(x, tk(x))$ is proper and that $h = f \circ h_1$.

If, now, we have proper maps $f, g: X \to Y$ and a homotopy $G: X \times I \to Y$ with $g_0 = f$, $g_1 = g$ which is proper on skeleta, we may do the same thing as before, using a proper deformation to infinity of $X \times I$ but keeping the parameter $k = 0$ on $X \times \{0, 1\}$, since $H$ will by continuity be proper on a neighborhood of $X \times \{0, 1\}$.

Proposition (6.7) shows that although there may be no "proper" Whitehead theorem, there is in our case a "proper on skeleta" theorem. [This is in fact true in general; to be precise, if $f: X \to Y$ is a map inducing isomorphisms on all homotopy groups and on the pro-groups pro-$\pi_n(E(X))$, $n \geqq 0$, then $f$ is homotopic to a map which is proper on skeleta, has a homotopy inverse $g$ which is proper on skeleta, and the compositions $g \circ f$ and $f \circ g$ are homotopic to the identity by homotopies which are proper on skeleta. (See **[3]**, **[7]**, **[17]**.)] The peculiarities of the structures with which we are dealing allow us to dispense with the pro-notation.

(6.4). LEMMA. *If $G$ is a compact Lie group, $\alpha$ is a based-free $G$-action on $Q$.*

$$\begin{array}{ccc} L \times Q & \xrightarrow{h'} & Q_0 \\ {\scriptstyle p \times \mathrm{id}} \downarrow & & \downarrow {\scriptstyle p_\alpha} \\ K \times Q & \xrightarrow{h} & Q_0/\alpha \end{array}$$

*is an equivariant cell structure on $\alpha$ as in* (4.3), *and $N_1 \supseteq N_2 \supseteq N_3 \supseteq \cdots$ is a cofinal sequence of neighborhoods of the end $E(K)$ of $K$ as in* (4.5) *such that $N_i' = p^{-1}(N_i)$ is null-homotopic in $N_{i-1}'$ for each $i > 1$, then for each $n$, $i$, and $k < i$, we have*

$$\mathrm{im}(\pi_n(N_{i-1}, N_i) \to \pi_n(N_{i-k-1}, N_i)) = \mathrm{im}(\pi_n(N_{i-k}, N_i) \to \pi_n(N_{i-k-1}, N_i)).$$

PROOF. As we are dealing with a fibration, we may replace the pairs $(N_j, N_{j+k})$ by $(N_j', N_{j+k}')$ for each $j$, if $n \geqq 1$, and the result is trivial.

(6.5). LEMMA. *With the hypotheses of* (6.4) *and the assumption of connectivity for each $N_i$, there is a homotopy $F: K \times I \to K$ with $f_0 = \mathrm{id}$, $f_1(N_i) \subset N_{i+1}$, and $F(N_i^{(j)} \times I) \subset N_{i-j}$, for each $i$ and each $j < i$.*

PROOF. This follows by induction on the skeleta of $K$, using (6.4): For each vertex, $v$, of $K$, let $i(v) = \max\{i \mid v \in N_i\}$, and let $d(v) = \max\{\dim\{\beta \mid v \in \beta$ and $\beta$ is a simplex of $K\}\}$. Pick $f_1(v) \in N_{i(v)+d(v)+1}^{(0)}$ and define $F$ on $K^{(0)} \times I$ so that $F(\{v\} \times I) \subset N_{i(v)}$. Now, inductively, if $F$ is defined on $K \times \{0\} \cup K^{(n)} \times I$ so that, with $i(\beta)$ and $d(\beta)$ the extension to all simplices of these functions, $f_1(\beta^n) \subset N_{i(\beta)+d(\beta)+1-n}$ and $F(\beta^n \times I) \subset N_{i(\beta)-n}$, for each $n$-simplex $\beta^n$ of $K$, apply

Lemma (6.4) to the element of $\pi_n(N_{i(\gamma)-n}, N_{i(\gamma)+d(\gamma)+1-n})$ represented by $F|(\partial\gamma \times I \cup \gamma \times \{0\})$ for each $\gamma^{n+1}$ to extend $F$ to $K^{(n+1)} \times I$.

(6.6). PROPOSITION. *With the hypotheses of* (6.4), *there is a deformation* $F$: $K \times [0, 1) \to K$ *with* $f_0 = \mathrm{id}$ *which is proper on each skeleton of* $K$.

PROOF. Apply successively an infinite number of homotopies satisfying Lemma (6.5).

(6.7). PROPOSITION. *If* $\alpha$ *is a based-free action on* $Q$ *of a compact Lie group and* $f: X \to Q_0/\alpha$ *is a mapping of a connected, locally finite CW-complex, then* $f$ *is homotopic to a mapping which is proper on skeleta; if* $f$ *is proper on a subcomplex* $A$, *the homotopy may be taken relative to* $A$.

PROOF. This is the same as that of (6.1), using (6.6).

(6.8). COROLLARY. *If* $\alpha$ *is the standard based-free G-action on* $Q$, *then any map* $f$: $Q_0/\alpha \to Q_0/\beta$, *for any based-free H-action* $\beta$ *on* $Q$, *is homotopic to a proper mapping* ($G$, $H$ *compact, Lie*).

**7. A homotopy-theoretic criterion for standardness.** In this section, we establish a criterion, movability of the end of $Q_0/\alpha$, for standardness which is homotopy-theoretic, more properly, pro-homotopy-theoretic or shape-theoretic, rather than proper-homotopy-theoretic.

(7.1). DEFINITION. A based-free $G$-action $\alpha$ on $Q$ is *movable* if $E(Q_0/\alpha)$ is movable, meaning that for each neighborhood $N$ of $E(Q_0/\alpha)$, there is another neighborhood $U_N$ of $E(Q_0/\alpha)$ contained in $N$ with the property that for every neighborhood $W$ of $E(Q_0/\alpha)$ there is a map $f$: $U_N \to W$ which is homotopic in $N$ to the inclusion $i$: $U_N \to N$.

(7.2). REMARK. This very strong condition was invented by K. Borsuk to rule out pathologies such as the solenoids of van Danzig (inverse limits of circles). It provides a setting in which the inverse limits of the homotopy groups of neighborhoods of finite-dimensional compacta in ANR's carry sufficient information to give a Whitehead theorem in shape theory [**25**], [**23**], [**21**].

(7.3). THEOREM. *A based-free action on* $Q$ *of a compact Lie group is standard if and only if it is movable.*

PROOF. The standard actions are clearly movable by construction. (Indeed, it is immediate that an action is movable if its space of nonsingular orbits admits a proper deformation to infinity.) We must show that if $E(Q_0/\alpha)$ is movable, then $Q_0/\alpha$ admits a proper deformation to infinity.

Let $K$ be a simplicial complex triangulating $Q_0/\alpha$, i.e., such that there is a homeomorphism from $K \times Q$ to $Q_0/\alpha$ [**9**], and, using the movability, select a cofinal sequence $N_1 \supseteq N_2 \supseteq \cdots$ of neighborhoods of the end $E(K)$ which are subcomplexes, are connected, and satisfy the conclusion of (4.5) as well as the condition that $N_i$ deform into a subset of $N_j$ through maps into $N_{i-1}$ for each $j \geqq i$ and each $i > 1$. We produce a proper homotopy of $K$ from the identity to a mapping which carries each $N_i$ into $N_{i+1}$. Infinite iteration of this homotopy provides a proper deformation of $K$ to infinity.

Let $A_i = N_i - \mathring{N}_{i+1}$, $B_i$ = boundary of $N_i$, and $d(i)$ = dimension of $A_{i-1} \cup A_i \cup A_{i+1}$.

Choose deformations $F^i: N_i \times I \to N_{i-1}$ such that $f_1^i(N_i) \subset N_{i+d(i)}$. Now, $f_1^i$ and $f_1^{i-1}$ restrict to mappings of $B_i$ into $N_{i-1+d(i-1)}$ which are homotopic in $N_{i-2}$. We may employ Lemma (6.4) inductively to produce a homotopy $G^i: B_i \times I \to N_{i+1}$ with $g_0^i = f_1^{i-1}$, $g_1^i = f_1^i$ and the property that the mapping $g^i: B_i \times S^1 \to N_{i-2}$ given by

$$g(x, e^{2\pi it/3}) = \begin{cases} f_{3t}^{i-1}(x), & \text{if } 0 \leqq t \leqq 1/3, \\ g_{3t-1}^i(x), & \text{if } 1/3 \leqq t \leqq 2/3, \\ f_{3-3t}^i(x), & \text{if } 2/3 \leqq t \leqq 1, \end{cases}$$

extends to $B_i \times D^2$ in $N_{i-3-d(i)}$.

Now, applying $F^{i-2}$ yields a deformation $H^i: B_i \times S^1 \times I \to N_{i-3}$ from $g^i = h_0^i$ to a mapping $h_1^i: B_i \times S^1 \to N_{i-2+d(i-2)}$ which extends to a map of $B_i \times D^2$ into $N_{i-3-d(i-1)}$. Another inductive application of Lemma (6.4) produces an extension of $h_1^i$ to a map $j^i: B_i \times D^2 \to N_{i-3}$, so we see that in fact there is a homotopy $J^i: (B_i \times I) \times I \to N_{i-3}$ which is the inclusion on $B_i \times \{0\} \times I$, is $F^i$ on $B_i \times I \times \{0\}$, is $F^{i-1}$ on $B_i \times I \times \{1\}$, and is $G^i$ on $B_i \times \{1\} \times I$. Using a regular neighborhood of $B_i$ in $A_i$, we can, by deforming to $B_i$ in this regular neighborhood and then partially across the second parameter of $J^i$, make a continuous transition from $F^i$ to $F^{i-1}$ while preserving the property of mapping into $N_{i+1}$ when the first parameter has value 1 and staying in $N_{i-3}$. From this it is easy to see how to interpolate between these homotopies to obtain a proper-homotopy (which for each $i > 3$ carries $N_i \times I$ into $N_{i-3}$) from the identity to a mapping carrying each $N_i$ into $N_{i+1}$ as desired.

**8. Products of finite-dimensional actions are standard.** In this section, we apply our movability criterion to show that a based-free $G$-action on $Q$ ($G$ compact, Lie) is standard if it splits into the diagonal action on some product of based-free $G$-actions on finite-dimensional compact absolute retracts.

(8.1). LEMMA. *Let $\alpha$ be a based-free $G$-action on $Q$ with base point $q_0$, where $G$ is a compact Lie group. There is a homotopy $F: Q \times I \to Q$ such that $f_0$ = identity, $F^{-1}(q_0) = \{q_0\} \times \{0\}$, each $f_t$ is the identity off the $t$-neighborhood of $q_0$, and for each $n > 0$ there is a strictly increasing, continuous function $g_n$: $(0, 1] \to (0, 1]$ with $\lim_{t \to 0} g_n(t) = 0$ such that for $t_1, \cdots, t_n \in [0, 1]$, $(p^{-1} \circ p) \circ f_{t_n} \circ (p^{-1} \circ p) \circ f_{t_{n-1}} \circ (p^{-1} \circ p) \circ \cdots \circ (p^{-1} \circ p) \circ f_{t_1}(Q)$ misses the $g_n(\max\{t_1, \cdots, t_n\})$-neighborhood of $q_0$, where $p: Q \to Q/\alpha$ is the orbit map.*

PROOF. The validity of this proposition is independent of the choice of metric and follows from the fact that $q_0$ may be assumed to be a "corner point" of $Q$ and the compactness of $Q$ and $G$. A convenient model for $(Q, q_0)$ is the following: In $l_2$, the Hilbert space of all square-summable sequences of real numbers, let

$$\bar{Q} = \{\bar{q} = (t_1, t_2, \cdots) | t_1 = 0 \text{ and for each } i > 1, |t_i| \leqq 1/2^i\}.$$

Let $q_0 = (1, 0, 0, \cdots)$, and let $Q = \{t\bar{q} + (1 - t)q_0 | 0 \leqq t \leqq 1, \bar{q} \in \bar{Q}\}$. This is a Hilbert cube by [22], as it is an infinite-dimensional, convex compactum in $l_2$. Now set $f_s(t\bar{q} + (1 - t)q_0) = (t\bar{q} + rq_0)$, where $r = \min\{1 - t, 1 - s/2\}$. Then $f_s$ satisfies all the desired conditions using the $l_2$ norm. To construct the functions $g_n$, we have only to consider the spaces $Q \times G^n \times I^n = X_n$ and maps $h_n\colon X_n \to X_n$ given by

$$h_n(q, (g_1, \cdots, g_n), (t_1, \cdots, t_n))$$
$$= (\alpha(g_n, f_{t_n} \circ \alpha(g_{n-1}, f_{t_{n-1}} \circ \cdots \circ \alpha(g_1, f_{t_1}(q)) \cdots), (g_1, g_2, \cdots, g_n), (t_1, \cdots, t_n)).$$

The image of $h_n$ meets $\{q_0\} \times G^n \times I^n$ in the set $\{q_0\} \times G^n \times \{(0, 0, \cdots, 0)\}$. As $X_n$ is compact, if we let $C_t = \{(t_1, \cdots, t_n) \in I^n |$ at least one $t_i \geqq t\}$ and $d_n(t) =$ minimum distance between $h_n(X_n) \cap Q \times G_n \times C_t$ and $\{q_0\} \times G^n \times C_t$, we obtain a nondecreasing function $d_n\colon I \to I$. Taking a strictly increasing, continuous function $g_n$ such that $g_n(t) \leqq d_n(t)$ for each $t > 0$ finishes the argument.

Before proceeding, it is convenient to set up some notation. (See Figure 1.) Let, for each $i$, $\alpha_i$ be a based-free $G$-action on a finite-dimensional, compact absolute retract $X_i$ with fixed point $x_i$. The product of the $X_i$'s is a Hilbert cube, which we shall denote by $Q$. The diagonal action on $Q$ will be $\alpha$, and its fixed point, $q_0$. Let $p_j$ denote the projection $p_j\colon Q_i \to X_j$ as well as the induced map $p_j\colon Q/\alpha \to X_j/\alpha_j$ and the restrictions $p_j\colon Q - \{x_j\} \times \prod_{i=1; i\neq j}^{\infty} X_i \to X_j - \{x_j\}$ and $p_j\colon E_j = Q/\alpha - p_j^{-1}(x_j) \to (X_j - \{x_j\})/\alpha = B_j$. Note that this last is a locally trivial fiber bundle with Hilbert cube fiber $Q'_j = \prod_{i=1; i\neq j}^{\infty} X_i$ and structure group $G$ acting based-freely with fixed point $x'_j = (x_1, \cdots, \hat{x}_j, \cdots)$, where, as usual, the symbol "$\hat{x}_j$" means that $x_j$ has been deleted. We denote the action of $G$ on $Q'_j$, which is the diagonal action, by $\alpha'_j$, We also need notation for the complementary projections and bundles. Let $p'_j$ denote the projection $p'_j\colon Q \to Q'_j$ and the other maps derived from it as in the case of $p_j$, above. In particular, we shall use $E'_j = Q/\alpha - p_j'^{-1}(x'_j)$ and $B'_j = (Q'_j - x'_j)/\alpha'_j$. (Note that we denote by $q_0$, $x_j$, and $x'_j$ both these points as defined above and their images under the various maps.) In the bundles, the natural cross-sections (to the base points in the fibers) will be called the "zero-sections", as will their images. This notation will be standard throughout this section. Also, we assume all subspaces and quotient spaces have metrics induced from some $G$-invariant metric on $Q$.

(8.2). LEMMA. *Let $U$ be any neighborhood of the zero-section in $E'_j$, and let $N$ be any neighborhood of $x'_j$ in $Q'_j$. There is a neighborhood $L \leqq N$ and a deformation $H\colon Q_0/\alpha \times I \to Q_0/\alpha$ which restricts to a fiber-preserving homotopy on $E_j$ such that*

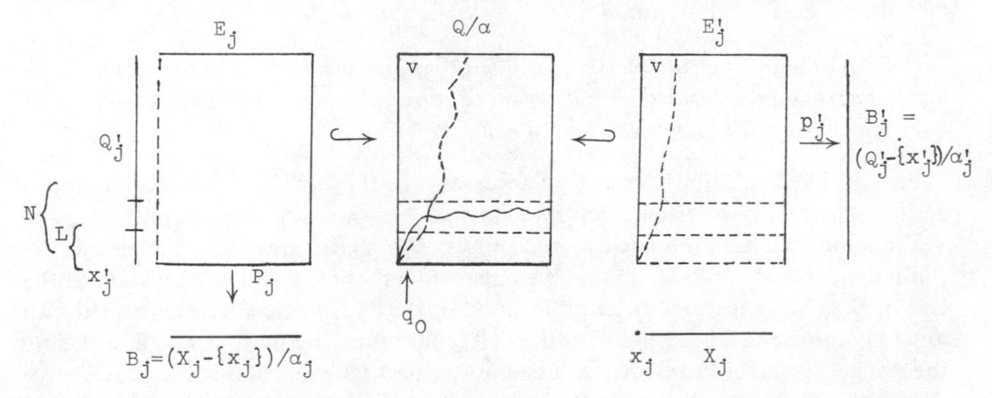


FIGURE 1

(a) $h_1(Q_0/\alpha) \cap (L \times X_j)/\alpha \subset U$,
(b) $h_0 = identity$,
(c) $h_t$ *is the identity on the zero-section of* $E'_j$ *for each* $t$,
(d) $h_t$ *is the identity off* $(N \times X_j)/\alpha$.

PROOF. We use Lemma (8.1). Let us assume that $N$ is invariant; otherwise, replace it by one which is. Let $n$ = dimension of $X_j$, and let $\mathscr{U}$ be an open, locally finite trivializing cover for the bundle $p_j: E_j \to B_j$ which is of order $(n+1)$ or less. Let $t_0 > 0$ be small enough that the $t_0$-neighborhood of $x'_j$ in $Q'_j$ lies in $N$. By Lemma (8.1), there is a deformation $F: Q'_j \times I \to Q'_j$ such that, for each $t$, $f_t$ is the identity off the $t$-neighborhood of $x'_j$ and all compositions of $n + 1$ functions of the form $\alpha'_j(g_i, f_{t_i}(\cdot))$ carry $Q'_j$ off the $g(\max\{t_1, t_2, \cdots, t_{n+1}\})$-neighborhood of $x'_j$. Let $L$ be the $g(t_0)$-neighborhood of $x'_j$ in $Q'_j$. Now let $\{C_U\}_{U \in \mathscr{U}}$ be a closed cover of $B_j$ such that each $C_U \subset U$ and let, for each $U$, $\lambda_U: (B_j, B_j - U, C_U) \to (I, 0, 1)$ be a Urysohn function. If $\varphi_U: Q'_J \times U \to p_j^{-1}(U) \to E_j$ is a coordinate chart for $E_j$ over $U$, define $k_U: E_j \times I \to E_j$ by

$$k_U(x, t) = \begin{cases} \varphi_U(f_{\lambda(u)\cdot t}(q), u), & \text{if } (q, u) = \varphi_U^{-1}(x), \\ x, & \text{if } p_j(x) \notin U. \end{cases}$$

Letting $K: E_j \times I \to E_j$ be defined by composing the $k_U$'s in some definite order produces a deformation of $E_j$ with the property that $k_0$ = identity, each $k_t$ is the identity off $(N_t \times (X_j - \{x_j\}))/\alpha$, where $N_t$ is the $t$-neighborhod of $x'_j$ in $Q'_j$, and $k_t(E_j) \cap (N_{g(t)} \times (X_j - \{x_j\}))/\alpha = \varnothing$.

To finish, we have only to set $h_t(x) = k_{\mu(p_j(x))\cdot t}(x)$, for some mapping $\mu: B_j \to (0, t_0]$ such that $\mu$ extends to $X_j/\alpha$ with $\mu(x_j) = 0$ and has the property that $U \supset ((N_{t_0} - N_{g(\mu(u))}) \times X_j)/\alpha \cap p_j^{-1}(u)$ for each $u$ in some neighborhood $W$ of $x_j$ in $B_j$. Such a mapping is easy to construct, using compactness to observe that there is a nondecreasing mapping $\nu: I \to I$ with the property that if $x \in B'_j$ and $d(x, x'_j) \geqq t$, then $U \supset (N_{\nu(t)} \times Q'_j)/\alpha \cap p_j'^{-1}(x)$. If $\nu$ is chosen to be strictly increasing with $\nu(0) = 0$, then we may set $\mu(u) = g^{-1} \circ \nu^{-1}(d(u,x_j))$ or $t_0$, whichever is smaller, when $d(u, x_j)$ is in the image of $\nu$, and simply $t_0$, when $d(u, x_j)$ is not in the image of $\nu$.

We shall say that a based-free $G$-action $\alpha$ on $Q$ *splits into finite-dimensional actions* if there is an equivariant homeomorphism converting it into the diagonal action on a product of finite-dimensional, compact AR's, each of which is equipped with a based-free $G$-action.

(8.3). THEOREM. *Let $G$ be a compact Lie group. A based-free $G$-action $\alpha$ on a Hilbert cube is standard if and only if its space of nonsingular orbits is a Hilbert cube manifold and $\alpha$ splits into finite-dimensional actions.*

PROOF. The conditions are clearly necessary, as they are built into the standard action. Therefore, we assume that $\alpha$ splits and that $Q_0/\alpha$ is a $Q$-manifold.

We adopt the notation fixed at the outset. We shall show that, under the hypotheses, $\alpha$ is movable. To do this, we must show that for any neighborhood of the end of $Q_0/\alpha$ there is a second neighborhood of the end which can be deformed into any neighborhood of the end with the deformation taking place in the first. Using the twisted product structure, we assume the first neighborhood is a basic "product" neighborhood of $q_0$ and refactor the product so that the neighborhood is of the form $(U \times \prod_{n=2}^{\infty} X_i)/\alpha$. We perform our deformations one "direction" at a time, employing the contractibility of the fibers of the bundles $E'_j \to B'_j$ and using our deformation Lemma (8.2) to deform $Q_0/\alpha$ into $E'_j$, which at once allows us to use the setting of bundles and ensures that no point is inadvertently dragged off the end of $Q_0/\alpha$, i.e., through or onto the point $q_0$, base point of the action $\alpha$.

For each $i$, let $\{U_j^i\}_{j=1}^{\infty}$ be a nested basis of equivariant neighborhoods of $x_i$ in $X_i$. Let $\{V_i = \prod_{j=1}^{i} U_i^j \times \prod_{j>i} X_j\}_{i=1}^{\infty}$ be the indicated basis of $q_0$ in $Q$, denoting by $V_i^*$ the deleted neighborhood $V_i - \{q_0\}$ and setting $(W_i, W_i^*) = (V_i, V_i^*)/\alpha$.

To select the neighborhood in $W_i^*$ which will deform into any $W_j^*$ in $W_i^*$, refactor $Q$ as $\prod_{j=1}^{\infty} Y_j$, where $Y_1 = \prod_{k=1}^{i} X_k$ and, for $j > 1$, $Y_j = X_{(j-1)+i}$. Now, as $Y_1$ is an AR, the bundle $E'_1 \to B'_1$ with fiber $Y_1$ deforms to the zero-section by a fiber-preserving homotopy $F$. Let $A$ be any neighborhood of the zero-section of $E'_1$ such that $F(A \times I) \subset W_i^*$. Now apply Lemma (8.2) to find that, for some $j > i$, there is a deformation $H$: $W_j^* \times I \to Q_0/\alpha$ restricting to a fiber-preserving homotopy on $E_1$ and carrying $W_j^*$ into $A$. The composition of the two deformations, first $H$ and then $F$, deforms $W_j^*$ into the zero-section of $E'_1$ with the deformation occurring in $W_i^*$. If, now, we wish to deform $W_j^*$ further into any $W_k^*$, we repeat this procedure in $B'_1 = ((\prod_{j=i+1}^{k} x_j \times \prod_{j>k} X_j) - \{y'_1\})/\alpha$ rather than in $Q_0/\alpha$, with no restriction in the way of a neighborhood in which the deformation must be contained.

**9. Is all of the above vacuous?** We cannot yet prove the existence of a single nonstandard based-free compact Lie group action on $Q$. In particular, by Theorem (7.1), we cannot construct an "immobile" action. (We have only tried to do so for the case $G = Z_2$.) However, we are passionately convinced of their existence for the following reasons: (1) We have tried and, especially the second author, failed to prove their nonexistence, with all our false proofs foundering on approximately the same point at which the proof given of Theorem (8.3) fails to cover the general case—namely that without the finite-dimensionality of $B_1$, we cannot prove Lemma (8.2) in the strength we need, having to settle for a deformation carrying $E_1$ off the zero-section without any control on how far away from the zero-section we can keep it, which, coupled with the fact that the infinite-dimensionality of the base $B'_1$ prevents us from constructing a uniformly continuous fiber-preserving contraction of $E'_1$ to the zero-section, causes us to lose control of the deformation into $B'_1$ constructed in the proof ot the theorem. (2) If we lift any of the conditions in Theorem (9.4) below, we have examples. (3) If we lift the restriction to compact groups, then the example of Tucker referred to in the introduction provides nonstandard examples of $Z$-actions derived directly from the classical mainstream of manifold topology.

In this section, we reduce the question of the existence of nonstandard based-free $G$-actions on $Q$ from the movability criterion of Theorem (7.1) to a homotopy-theoretic question involving the essentiality of indefinitely long compositions of maps between relatively finite pairs of the form $(K, B_G)$. This result was obtained simultaneously and independently by H. Hastings.

Let $\alpha$, $\sigma$ be based-free actions on $Q$ of the compact Lie group $G$ with $\sigma$ standard.

(9.1). LEMMA. *There is an embedding of $\sigma$ in $\alpha$ as an equivariant retract.*

PROOF. Let $f$: $Q_0/\sigma \to Q_0/\alpha$ be a classifying map for $\sigma$. By Corollary (6.8), we may assume $f$ to be proper. Triangulating $Q_0/\alpha$ as $K_\alpha \times Q$ [9] ,we see that $f$ may be deformed to a closed embedding, $g$. Now, by the covering homotopy theorem, the equivariant homeomorphism

$$\xrightarrow[Q_0 \to g^*(\alpha) \to Q_0]{\tilde{h}}$$

covering $g$ is the desired embedding, as $g$ must be a homotopy equivalence and a deformation retraction of $Q_0/\alpha$ to $g(Q_0/\sigma)$ must classify $\alpha$.

(9.2). LEMMA. *The embedding $h$ of* (9.1) *may be taken so that the pair* $(Q_0, h(Q_0))$ *may be given an equivariant cell structure as* $(E, TE_G) \times Q$ *for some contractible complex $E$, where $TE_G$ is as in* §5.

PROOF. Triangulate [9] $Q_0/\alpha$ as $K_\alpha \times Q_1 \times Q_2$, where $Q_i = Q, i = 1, 2$. Deform $f$ to a closed embedding $g$ in $K_\alpha \times Q_1 \times \{q_0\}$, for some $q_0$ in $Q$. The pair $(Q_0/\alpha, g(Q_0/\sigma))$ now satisfies a relative triangulation theorem of Chapman [9], so that a triangulation of $g(Q_0/\sigma)$ as $TB_G \times Q$ extends to one of $Q_0/\alpha$. Lifting this to a cell structure on $Q_0$ as in the proof of (4.3) gives the desired cell structure.

We now denote by $E(\alpha, \sigma)$ a cofinite sequence of nested neighborhoods $(K_i, L_i)$ of the end $E(Q_0/\alpha)$, where $L_i = K_i \cap TB_G$ and is a strong deformation retract of $L_{i-1}$, e.g., $L_i = \bigcup_{j=i}^{\infty}((\ast_{k=1}^{i} G_k)/G \times [i, \infty))$. In view of (4.5), which forces essential constancy of $\pi_n$ at the end for each $n$, and the extra "room" afforded by the factor of $Q$ in the "cell structure", which allows us to assume that $K_\alpha = K'_\alpha \times I^n$ and that $TB_G$ is contained in $K'_\alpha \times \{0\}$, we may alter the cell structure by attaching discs to $K_i$ in $K_{i-1}$ to assure that the pair $(K_i, L_i)$ is relatively 1-connected.

(9.3). LEMMA. *The pair $(K_i, L_i)$ is dominated, relative to $L_i$, by a relatively finite pair $(X_i, L_i)$.*

PROOF. It is a CW pair and relatively 1-connected. By an easily proven relative version of Wall's conditions for finite domination [30], it suffices to show that (1) $\pi_1(K_i)$ is finitely presented, and for some $n$, (2) $H_j(K_i, L_i)^\sim$ is finitely generated over the group ring of $\pi_1(K_i)$, where "$\sim$" denotes the universal cover, (3) $H_j(K_i, L_i)^\sim = 0$ for $i > n$, and (4) $H^{n+1}(K_i, L_i, \mathscr{B}) = 0$, for all local coefficients $\mathscr{B}$. These are easily verified: (1) is true because $\pi_1(K_i) \cong \pi_0(G)$, which is finite. (2) is true from the fact that in the inclusion of pairs, $(K_i, L_i) \to (K, L_i) \to (K, K_i)$ we have $L_i \to K$ a homotopy equivalence, hence $\tilde{L}_i \to \tilde{K}_i$ is, also, and by employing the exact sequence of the triple $(\tilde{K}, \tilde{K}_i, \tilde{L}_j)$, we find $H_j(K_i, L_i)^\sim \cong H_{j+1}(K, K_i)^\sim$ for each $j$. Since $K_i$ is cofinite in $K$ and we are dealing with a finite covering, $\tilde{K}_i$ is cofinite in $\tilde{K}$ and $H_j(K, K_i)^\sim$ is finitely generated over $Z$, hence $Z\pi_1(K_i)$. (3) follows from the considerations of (2), since $\dim(\tilde{K} - \tilde{K}_i)$ is finite. (4) also follows from the fact that we can use the exact sequence of the triple $(K, K_i, L_i)$ and the cofiniteness of $K_i$ in $K$, together with the fact that, since all fundamental groups are isomorphic, we can extend any local system to $K$.

We remark, before stating our theorem, that a map $f$: $(X, A) \to (Y, B)$ of pairs is essential if it cannot be deformed into $B$ rel $A$.

(9.4). THEOREM (BERSTEIN, WEST AND HASTINGS). *Let $G$ be a compact Lie group. There is a nonstandard, based-free $G$-action on $Q$ if and only if there exists a sequence*

$$\cdots \xrightarrow{\bar{f}_i} (E_i, E_G) \xrightarrow{\bar{f}_{i-1}} (E_{i-1}, E_G) \longrightarrow \cdots \xrightarrow{f_i} (E_1, E_G)$$

*of principal G-bundles of CW-complexes and bundle maps, each the identity on* $E_G$, *satisfying the following with* $f_{i-1}: (B_i, B_G) \to (B_{i-1}, B_G)$ *denoting the induced mappings of orbit spaces:*

(i) *each* $(E_i, E_G)$ *is relatively finite,*
(ii) *each* $(B_i, B_G)$ *is relatively* 1*-connected,*
(iii) *each* $\bar{f}_i$ *is null-homotopic, and*
(iv) *each finite composition* $f_1 \circ \cdots \circ f_i: (B_{i+1}, B_G) \to (B_1, B_G)$ *is essential.*

PROOF. If $\alpha$ is a nonstandard based-free $G$-action on $Q$, then by Lemmas (9.1) and (9.2) we may equivariantly embed $\sigma$, the standard action, in $\alpha$ and give $Q_0$ an equivariant cell structure as $(E, TE_G) \times Q \to^{p_\alpha} (B, TB_G) \times Q$. Now consider $E(\alpha, \sigma)$, where we assume, from Proposition (4.5), that $p_\alpha^{-1}(K_i)$ is null-homotopic in $p_\alpha^{-1}(K_{i-1})$, for each $i > 1$. Using Lemma (9.3), we choose relatively finite dominations $(K_i, L_i) \to^{u_i} (X_i, L_i) \to^{d_i} (K_i, L_i)$. Let $E_i = d_{i*}(p_\alpha^{-1}(K_i))$. Then as $d_i \circ u_i \simeq \mathrm{id}_{(K_i, L_i)}$, we can cover $u_i$ by a bundle map $u_{i*}: (p_\alpha^{-1}(K_i, L_i) \to d_{i*}(p_\alpha^{-1}(K_i, L_i))$ which is an equivariant left homotopy inverse to $d_{i*}$. Now, the sequence

$$\cdots \to (E_i, p_\alpha^{-1}(L_i)) \xrightarrow[u_{i-1*} \circ i_{i-1} \circ d_{i*}]{} (E_{i-1}, p_\alpha^{-1}(L_{i-1})),$$

where $i_{i-1}$ is the inclusion $p_\alpha^{-1}(K_i, L_i) \to p_\alpha^{-1}(K_{i-1}, L_{i-1})$, satisfies (i), (ii), and (iii) by construction and (iv), if we neglect some initial terms of the sequence, because $\alpha$ is nonstandard and therefore, by Theorem (7.1), $E(B)$ is immobile; hence for some $i$ there exists no $j$ such that $K_j$ will deform into each $K_m$ inside $K_i$. This, however, given the structure of $L_i$, which is just $TB_G$ and therefore deforms to infinity, implies that no $K_j$ deforms into $L_i$ in $K_i$. Now, the construction of the $X_i$'s from the $K_i$'s shows that (iv) holds. In order to conclude this portion of the proof, we have only to replace the $L_i$'s with a single copy of $B_G$ and adjust matters so that the maps are the identity on it.

To establish the other half of the theorem, we need only replace $E_G$ by $TE_G$ in each pair, take the (infinite) mapping cylinder $M$ of the resultant sequence $\cdots \to (E_i, TE_G) \xrightarrow{\bar{f}_{i-1}} (E_{i-1}, TE_G) \to \cdots$, look at $M \times Q$, which is a $Q$-manifold, by **[33]**, and apply **[13]** (in a trivial case) to conclude that its one-point compactification $M'$ is a $Q$-manifold. (One could avoid **[13]** by noting that $M'$ is locally contractible at the compactification point, hence an AR, hence a Hilbert cube factor by Edwards' theorem **[11]**, and by applying Theorem 1 of **[34]**.) Now, $M'$ inherits a based-free $G$-action from $M$, with base point at infinity, but it is not contractible. This may be remedied by extending the sequence with the addition of any "$\to^{\bar{f}_0} (TE_G, TE_G)$".

(9.5). EXAMPLE. This simple example shows how easy it is to satisfy all hypotheses except (i) in Theorem (9.4) for $G = \boldsymbol{Z}_2$. Let $S^1 \subset S^2 \subset \cdots \subset S^n \subset \cdots \subset S^\infty$ be the standard sequence of unit spheres on which $\boldsymbol{Z}_2$ acts by the antipodal map. The quotients are the real projective spaces $\boldsymbol{RP}^1 \subset \boldsymbol{RP}^2 \subset \cdots \subset \boldsymbol{RP}^n \subset \cdots \subset \boldsymbol{RP}^\infty$. Let $S_k^n$ and $\boldsymbol{RP}_k^n$, $k = 1, 2, \cdots$, denote copies of $S^n$ and $\boldsymbol{RP}^n$, respectively. Form the disjoint union $\coprod_{n=1}^\infty S_n^{2n} \coprod S^\infty$ and let $\tilde{X} = (\coprod_{n=1}^\infty S_n^{2n}) \coprod S^\infty)/\equiv$, where $S_n^{2n} \equiv S^n$ for each $n = 1, 2, \cdots$. We have an induced antipodal $\boldsymbol{Z}_2$-action on $\tilde{X}$ with orbit space $X$. We also have a natural shift $\sigma: \coprod_{n=1}^\infty S_n^{2n} \to \coprod_{n=1}^\infty S_n^{2n}$ given by the inclusions $S_n^{2n} \to S_{n+1}^{2n+2}$. This map respects the identifications and is equivariant, yielding an equivariant shift, $\tilde{\sigma}: \tilde{X} \to \tilde{X}$ which is the identity on $S^\infty$ and null-homotopic. The induced shift $\sigma: X \to X$ on the orbit space is essential, as is $\sigma^k$ for any $k$, since for any $n > k$, the induced homomorphism $\sigma_*^k: H_{2n}(\boldsymbol{RP}_n^{2n}; \boldsymbol{Z}_2) \to H_{2n}(\boldsymbol{RP}_{n+k}^{2(n+k)}; \boldsymbol{Z}_2)$ is an isomorphism and in the diagram

$$\begin{array}{ccc} H_{2n}(\boldsymbol{RP}_n^{2n}; \boldsymbol{Z}_2) & \xrightarrow{\sigma_*^k} & H_{2n}(\boldsymbol{RP}_{n+k}^{2(n+k)}; \boldsymbol{Z}_2) \\ \downarrow j_* & & \downarrow j_* \\ H_{2n}(X, \boldsymbol{RP}^\infty; \boldsymbol{Z}_2) & \xrightarrow[\sigma_*^k]{} & H_{2n}(X, \boldsymbol{RP}^\infty; \boldsymbol{Z}_2) \end{array}$$

$j_*$ (induced by inclusion) is not zero because $\boldsymbol{RP}_{n+k}^{2(n+k)} \cap \boldsymbol{RP}^\infty = \boldsymbol{RP}^{n+k}$ and $2n > n + k$.

## References


1. R. D. Anderson, *The Hilbert cube as a product of dendrons*, Notices Amer. Math. Soc. **11** (1964), 572. Abstract #614–149

2. ———, *Topological properties of the Hilbert cube and the infinite product of open intervals*, Trans. Amer. Math. Soc. **126** (1967), 200–216. MR **34** #5045.

3. M. Artin and B. Mazur, *Etale homotopy*, Lecture Notes in Math., vol. 100, Springer-Verlag, Berlin and New York, 1969. MR **39** #6883.

4. K. Borsuk, *On movable compacta*, Fund. Math. **66** (1969/70), 137–146. MR **40** #4925.

5. ———, *Theory of shape*, Monografie Mat., Tom 59, PWN, Warsaw, 1975.

6. G. E. Bredon, *Introduction to compact transformation groups*. Pure and Appl. Math., Vol. 46, Academic Press, New York, 1972.

7. E. M. Brown, *Proper homotopy theory in simplifical complexes*, Topology Conf. (Virginia Polytechnic Inst. and Univ.), Lecture Notes in Math., vol. 375, Springer-Verlag, Berlin and New York, 1974. MR **50** #8513.

8. T. A. Chapman, *On some applications of infinite-dimensional manifolds to the theory of shape*, Fund. Math. **76** (1972), 181–193. MR **47** #9530.

9. ———, *All Hilbert cube manifolds are triangulable* (mimeographed notes).

10. ———, *Homeomorphisms of Hilbert cube manifolds*, Trans. Amer. Math. Soc. **182** (1973), 227–239. MR **51** #9067.

11. ———, *Lectures on Hilbert cube manifolds*, C. B. M. S. Regional Conference Series in Math., no. 28, Amer. Math. Soc., Providence, R. I., 1976.

12. T. A. Chapman and S. Ferry, *Obstruction to finiteness in the proper category* (mimeographed).

13. T. A. Chapman and L. C. Siebenmann, *Finding a boundary for a Hilbert cube manifold* (mimeographed).

14. J. Dydak, *The Whitehead and Smale theorems in shape theory.*, Inst. of Math., Polish Academy of Sciences, preprint 87.

15. D. A. Edwards and H. M. Hastings, *Čech and Steenrod homotopy theories with applications to geometric topology*, Lecture Notes in Math., vol. 542, Springer-Verlag, Berlin and New York, 1976.

16. D. A. Edwards and R. Geoghegan, *Infinite-dimensional Whitehead and Vietoris theorems in shape and pro-homotopy*, Trans. Amer. Math. Soc. **219** (1976), 351–360.

17. F. T. Farrell, L. R. Taylor, and J. B. Wagoner, *The Whitehead theorem in the proper category*, Compositio Math. **27** (1973), 1–23. MR **48** #12545.

18. A. Fathi and Y. M. Visetti, *Deformation of open embeddings of Q-manifolds*. Trans. Amer. Math. Soc. **224** (1976), 427–435.

19. S. Ferry, *The homeomorphism group of a compact Q-manifold is an ANR* (mimeographed).

20. A. M. Gleason, *Spaces with a compact Lie group of transformations*, Proc. Amer. Math. Soc. **1** (1950), 35–43. MR **11**, 497.

**21.** J. E. Keesling, *On the Whitehead theorem in shape theory*, Fund. Math. **92** (1976), 247–253.

**22.** O. H. Keller, *Die Homoiomorphie der kompakten, konvexen Mengen im Hilbertschen Raum*, Math. Ann. **105** (1931), 748–758.

**23.** S. Mardesic, *On the Whitehead theorem in shape theory*. I, Fund. Math. **91** (1976), 51–64.

**24.** J. W. Milnor, *Construction of universal bundles*. I, Ann. of Math. (2) **63** (1956), 272–284. MR **17**, 994.

**25.** M. Moszynska, *The Whitehead theorem in the theory of shapes*, Fund. Math. **80** (1973), 221–263. MR **49** #3922.

**26.** J. J. Rotman, *The theory of groups*: *an introduction*, 2nd ed., Allyn and Bacon, Boston, Mass., 1973.

**27.** L. C. Siebenmann, *Infinite simple homotopy types*, Indag. Math. **32** (1970), 479–495. MR **44** #4746.

**28.** H. Torunczyk, *Homeomorphism groups of compact Hilbert cube manifolds which are manifolds* (mimeographed).

**29.** T. W. Tucker, *Some non-compact* 3-*manifold examples giving wild translations of* $R^3$ (mimeographed).

**30.** C. T. C. Wall, *Finiteness conditions for CW-complexes*, Ann. of Math. (2) **81** (1965), 55–69. MR **30** #1515.

**31.** J. E. West, *Extending certain transformation group actions in separable, infinite-dimensional Fréchet spaces and the Hilbert cube*, Bull. Amer. Math. Soc. **74** (1968), 1015–1019. MR **38** #1510.

**32.** ———, *Infinite products which are Hilbert cubes*, Trans. Amer. Math. Soc. **150** (1970), 1–25. MR **42** #1055.

**33.** ———, *Mapping cylinders of Hilbert cube factors*, General Topology and Appl. **1** (1971), no. 2, 111–125. MR **44** #5984.

**34.** ———, *The subcontinua of a dendron form a Hilbert cube factor*, Proc. Amer. Math. Soc. **36** (1972), 603–608. MR **47** #1006.

**35.** ———, *Induced involutions on Hilbert cube hyperspaces*, Proc. of the Auburn Topology Conf., 1976, Topology Proceedings **1**, Auburn Univ., pp. 281–293.

**36.** R.Y. -T. Wong, *On homeomorphisms of certain infinite-dimensional spaces*, Trans. Amer. Math. Soc. **128** (1967), 148–154. MR **35** #4892.

**37.** ———, *Extending homeomorphisms by means of collaring*, Proc. Amer. Math. Soc. **19** (1968), 1443–1447. MR **38** #2747.

**38.** ———, *Stationary isotopies of infinite-dimensional spaces*, Trans. Amer. Math. Soc. **156** (1971), 131–136. MR **43** #1230.

**39.** ———, *Extending homeomorphisms in compactifications of Fréchet spaces*, Proc. Amer. Math. Soc. **25** (1970), 548–550. MR **42** #1071.

**40.** ———, *Periodic actions on the Hilbert cube*, Fund. Math. **85** (1974), 203–210. MR **50** #11303.

**41.** ———, *Noncompact Hilbert cube manifolds* (mimeographed).

**42.** V.-T. Liem, *Factorizations of free finite group actions on compact Q-manifolds* (mimeographed).

**43.** L. S. Newman, Jr., *Applications of group actions on finite complexes to Hilbert cube manifolds*, Proc. Amer. Math. Soc. **62** (1977), 183–187.

CORNELL UNIVERSITY

Proceedings of Symposia in Pure Mathematics
Volume 32, 1978

# REALIZING AUTOMORPHISMS OF $\pi_1$ BY HOMEOMORPHISMS

P. E. CONNER AND FRANK RAYMOND

**1. Introduction.** Let $X$ be a reasonable type of space. Let $\mathscr{H}(X)$ be the group of self-homeomorphisms of $X$ and $\mathscr{E}(X)$ the $H$-space of self-homotopy equivalences. The inclusion $i\colon \mathscr{H}(X) \subsetneq \mathscr{E}(X)$ is an $H$-space homomorphism. By passing to homotopy $i$ induces homomorphisms $i_*\colon \pi_j(\mathscr{H}(X), 1_X) \to \pi_j(\mathscr{E}(X), 1_X)$, $j \geqq 0$. Saying that $i_*$ is onto, for $j = 0$, is equivalent to saying that every self-homotopy equivalence can be deformed to a homeomorphism.

We wish to describe some geometrically very interesting spaces for which $i_*$ is *onto for all j*. The spaces are known as *injective Seifert fiber spaces* and were defined by Conner and Raymond in 1969 and 1971 [**C-R-2, 3,** and **4**]. There are extensive generalizations of the classical Seifert manifolds. The "classical" Seifert manifolds $X$ are 3-dimensional boundaryless manifolds $M^3 = X$ which *singularly* fiber onto boundaryless surfaces $Y$, $\mu\colon M^3 \to Y$, so that $\mu^{-1}(y)$ is homeomorphic to a circle $S^1$, for each $y \in Y$. The mapping is not necessarily a fiber-bundle map for some fibers may wrap around nearby singular fibers. Explicitly, each point $y \in Y$ has a neighborhood $D^*$ so that $\mu\colon \mu^{-1}(D^*) \to D^*$ is the orbit mapping

$$\mu\colon (S^1, S^1 \times_{(\mathbf{Z}/p\mathbf{Z})} D^2) \xrightarrow{/S^1} D^2/(\mathbf{Z}/p\mathbf{Z}) = D^*$$

of the natural circle action of $S^1$ on the solid torus $S^1 \times_{\mathbf{Z}/p\mathbf{Z}} D^2$. Here $p \in \mathbf{Z}^+$ and the "*singular*" *fiber* (if $p > 1$) is the orbit through the center of the disk $D$. All other orbits wrap around this orbit $p$ times. This local action does not necessarily extend to all of $M^3$.

A rather straightforward kind of generalization of Seifert's 3-manifolds is just to consider any space with an (effective) action of a $k$-torus. However, we realized that almost all of the time the classical Seifert manifolds had special coverings so

*AMS (MOS) subject classifications* (1970). Primary 55F55, 55A10, 57A65, 57E05; Secondary 20Exx, 20E40, 20F55, 20J05, 57E30, 57E10.

that the singular fibering lifted to a nonsingular fibering. Moreover, in most of these cases the lifted nonsingular fibering was actually a trivial product bundle of $S^1 \times \boldsymbol{R}^2$. This sort of *uniformization* could actually explain and unify many of the results concerning the classical Seifert fiberings. This note is a resume of some of the results in **[C-R-1]**.

1.1. *Injective Seifert fiber spaces.* Let $Y$ be an arbitrary space. We wish to describe all spaces $X$ and mappings $\mu: X \to Y$ onto $Y$ so that $\mu^{-1}(y) = F_y$, where $F_y$ is a $k$-torus, $T^k$, or the quotient of a $k$-torus by the action of a finite group of isometries acting freely on $T^k$. Thus, $F_y$ will be a flat manifold of dimension $k$. The fiber $F_y$ will depend upon $y$. We also want this singular fibering to be "uniformizable". We shall construct all such spaces $X$ directly and characterize the singular fiberings $\mu$.

Let $W$ be a simply connected space and $N$ a discrete group acting properly discontinuously from the right on $W$ so that the orbit space $W/N = Y$. The orbit mapping $\nu: (W, N) \to^{/N} W/N = Y$ is a regular branched covering mapping. The stabilizer, $N_w$, of $w \in W$ is a *finite* subgroup of $N$ and the branching index at $w \in W$ is the order of $N_w$, $|N_w|$. We form $X' = T^k \times W$ and try to impose actions of $T^k$ and $N$ on $X'$. We define the action of $T^k$ by just left translation along the first factor. We let $\mu': X' \to W$ be the projection onto the 2nd coordinate. We now wish to impose a right $N$-action on $X'$ so that the map $\mu'$ is equivariant.

Choose $\phi: N \to \mathrm{GL}(k, \boldsymbol{Z})$, a homomorphism. Since $\mathrm{GL}(k, \boldsymbol{Z}) \subset \mathrm{GL}(k, \boldsymbol{R})$, $\phi$ induces a representation of $N$ into the automorphisms of $T^k$. We want a right $N$-action so that

$$(tx')\,\alpha = \phi\,(\alpha^{-1})\,(t)\,(x'\,\alpha),$$

for all $t \in T^k$, $\alpha \in N$. This essentially means that we wish to impose a right action of the semidirect product $T^k \circ N$ on $X'$.

Now suppose we have succeeded in imposing such an $N$-action. We will then have the following commutative diagram:

$$\begin{array}{ccc} (T^k, T^k \times W, N) & \xrightarrow[\mu']{/T^k} & (W, N) \\ \exists\nu' \Big\downarrow /N & & \Big\downarrow \nu \\ (X = (T^k \times W)/N) & \xrightarrow[\exists\mu]{} & W/N = Y \end{array}$$

To try to understand $\mu^{-1}(y)$ we choose $w \in W$, $x' \in X'$ so that $\mu'(x') = w$, $\nu(w) = y$. Then $\mu^{-1}(y) = T^k/N_w$, the quotient of the $k$-torus by the action of the finite group $N_w$. The action of $N$ on $X'$ will be properly discontinuous and, in general, the orbit mapping $\nu': X' \to X$ will be a regular branched covering mapping. Clearly, this covering $\nu'$ will be *unbranched* if and only if the action of $N_w$ on $T^k = T^k(x')$ is free, for each $w \in W$. From the equivariance it is not hard to see that the action of $N_w$ on $T^k$ is equivalent to an action of a finite group of isometries on a flat $T^k$. The characterization is as follows **[C-R-4]**:

Let $a \in H^2_\phi(N; \boldsymbol{Z}^k)$. Corresponding to each class $a$ there is a *group extension* $a: 1 \to \boldsymbol{Z}^k \to \pi \to N \to 1$, with operators $\phi: N \to \mathrm{GL}(k, \boldsymbol{Z})$. For each $w \in W$, there is the inclusion $i_w: N_w \to N$, and the *induced extension*

$$i_w^*(a): 1 \to \boldsymbol{Z}^k \to \pi_w \to N_w \to 1.$$

If for *each* $w \in W$, the group $\pi_w$ is *torsion free*, we shall call $(a)$ *a Bieberbach class.* (Each $\pi_w$ is then a classical Bieberbach group, i.e., the fundamental group of a flat closed $k$-dimensional manifold with holonomy group $N_w$.) We denote the set of Bieberbach classes by $B_\phi$. This subset of $H^2_\phi(N; \mathbf{Z}^k)$ is usually not a subgroup.

THEOREM 1. *The equivariant equivalence class of left $T^k$-right $N$ actions on $X'$ is in natural one-one correspondence with the elements of $H^2_\phi(N; \mathbf{Z}^k)$.*

*The covering $\nu'$ is unbranched, if and only if, $a \in B_\phi \subset H^2_\phi(N; \mathbf{Z}^k)$. If $a$ is a Bieberbach class ($a \in B_\phi$) then the homotopy exact sequence of the covering projection*

$$1 \to \pi_1(X', x') \to \pi_1(X, x) \to N \to 1$$

*coincides with the group extension determined by a.*

If $a \in B_\phi \subset H^2_\phi(N; \mathbf{Z}^k)$ we denote the resulting *injective Seifert fiber space* $X$ by $(T^k, X, \phi, N)$ and the resulting *injective Seifert fibering* by $\mu: X \to Y$. Note that $Y$ parametrizes a whole class of singularity-free immersions of the $k$-torus and that $i^*_w(a)$ describes both the action of $(T^k, N_w)$ and the fundamental group of $\pi_1(\mu^{-1}(y)) = \pi_1(T^k/N_w)$.

1.2. EXAMPLES. A. Let $(T^k, X)$ be an action. Suppose the evaluation mapping $\mathrm{ev}^x: (T^k, 1) \to (X, x)$, defined by $\mathrm{ev}^x(t) = tx$, induces a *monomorphism*

$$\mathrm{ev}^x_*: \pi_1(T^k, 1) \to \pi_1(X, x);$$

then $(T^k, X)$ is called an *injective action.* It can be seen that $(T^k, X)$ *is an injective Seifert fibre space* and the *orbit map* $\mu: X \to X/T^k$ *is the associated Seifert fibering.* Of course, $N = \pi_1(X, x)/\mathrm{image}(\mathrm{ev}^x_*)$, $X'$ is the covering space of $X$ corresponding to image$(\mathrm{ev}^x_*)$ and $\phi: N \to \mathrm{GL}(k, \mathbf{Z})$ is trivial. (The image of the evaluation homomorphism $\mathrm{ev}^x_*$ for any path connected group is always a *central* subgroup of $\pi_1(X, x)$.)

B. Let $W$ = a point, $N$ any finite group. Then the set of $k$-dimensional compact flat manifolds with holonomy group $N$ are given by the $(T^k, X, \phi, N)$ for all $a \in B_\phi \subset H^2_\phi(N; \mathbf{Z}^k)$.

C. The *classical*, closed, orientable Seifert manifolds $X$ with an orientable decomposition space $Y$ (with the restriction that $Y$ is allowed to be the 2-sphere $S^2$ only if the number of singular fibers is greater than 3, or if exactly 3, then their multiplicities $\alpha_1, \alpha_2, \alpha_3$ satisfy $1/\alpha_1 + 1/\alpha_2 + 1/\alpha_3 \leqq 1$) correspond to Bieberbach classes $a \in B \subset H^2(N; \mathbf{Z})$. Here $N$ is a uniform discrete subgroup of $\mathrm{PSL}(2, \mathbf{R})$ or the Euclidean group $E(2)$. These fiberings are the orbit mappings of an injective circle action and the manifolds $X$ are then identified with the closed, orientable 3-manifolds which are $K(\pi, 1)$'s and admit effective circle actions.

On the other hand, the *classical* closed orientable Seifert manifolds $X$ with a nonorientable decomposition space $Y$ (with the restriction that $Y$ is allowed to be $\mathbf{RP}_2$ only if the number of singular fibers exceeds 1) correspond to Bieberbach classes $a \in B_\phi \subset H^2_\phi(N; \mathbf{Z})$. Now $N$ is a uniform discrete subgroup of *all* the isometries of the hyperbolic plane or Euclidean plane *excluding* reflections. In this case, $\phi: N \to \mathrm{GL}(1, \mathbf{Z}) \cong \mathbf{Z}/2\mathbf{Z}$ factors through $\pi_1(Y)$ so that $\phi$ restricted to each standard generator is nontrivial. These manifolds are also $K(\pi, 1)$'s. They admit no circle action except for a single exception.

D. The whole theory can be done in different ways. A particularly interesting situation arises when $N$ is a *holomorphic* group of automorphisms of a complex manifold $W$ (or an analytic space $W$) and $\phi: N \to \mathrm{GL}(2k, \boldsymbol{Z}) \subset \mathrm{GL}(k, \boldsymbol{C})$. There is then a corresponding cohomological classification of *holomorphic* Seifert fiberings ($\mu: X \to W/N$ is holomorphic). The (holomorphic) Bieberbach classes are now elements of $H^1_\phi(N; \mathscr{T})$ with $\mathscr{T}$ a suitable (and computable) *sheaf*. There is a coboundary $\delta: H^1_\phi(N; \mathscr{T}) \to H^2_\phi(N; \boldsymbol{Z}^{2k})$ which assigns to a holomorphic Seifert fiber structure its underlying (differential) Seifert fiber structure. For example, by choosing $N$ uniform discrete in $\mathrm{PSL}(2, \boldsymbol{R})$ or in the isometries of the complex line, and $\phi: N \to \mathrm{GL}(2, \boldsymbol{Z}) \subseteqq \mathrm{GL}(1, \boldsymbol{C})$, the Bieberbach classes $\tau \in H^1_\phi(N; \mathscr{T})$ determine (holomorphically) the *elliptic fiber spaces*, in the sense of *Kodaira*, which are $K(\pi, 1)$'s. The coboundary, $\delta: H^1_\phi(N; \mathscr{T}) \to H^2_\phi(N; \boldsymbol{Z}^2)$, assigns to the elliptic surface its underlying differential Seifert fiber space structure. Two elements $\tau$ and $\tau'$ have the same image, $\delta\tau = \delta\tau'$, if and only if $\tau$ can be (holomorphically) *deformed* through holomorphic fiberings to $\tau'$. A consequence of this classification is that two elliptic surfaces of the above type are *diffeomorphic*, if and only if they have *isomorphic fundamental groups*. Another consequence: if $\delta(\tau)$ has *finite* order it can be deformed through holomorphic Seifert fiberings to an elliptic surface which is *algebraic*. In fact, all algebraic elliptic surfaces which are $K(\pi, 1)$'s arise this way.

E. It should be pointed out that the injective Seifert fibering is *not* the orbit map of $k$-toral action on $X$ unless $\phi: N \to \mathrm{GL}(k, \boldsymbol{Z})$ is *trivial*.

F. A good source of examples of $(W, N)$ arises from $W = K\backslash G$, where $G$ is a Lie group with at most a finite number of connected components and $K$ is a maximal compact group. $W$ is diffeomorphic to some Euclidean space $\boldsymbol{R}^n$. Suppose $N = \Gamma \subset G$ is uniform and discrete. Then

$$\nu: (K\backslash G, N) \xrightarrow{/N} Y = W/N = K\backslash G/N$$

is a branched covering map with $Y$ compact. In particular, if $(T^k, X, \phi, N) \to K\backslash G/N$ is an injective Seifert fibering, then $X$ is a $K(\pi, 1)$-manifold. In fact, more generally

THEOREM 2. *If $a \in H^2_\phi(N; \boldsymbol{Z}^k)$ and $W$ is a contractible manifold with $W/N$ compact, then a necessary and sufficient condition for $a: 1 \to \boldsymbol{Z}^k \to \pi \to N \to 1$ to be the fundamental group of a closed $K(\pi, 1)$-manifold is that $\pi$ be torsion free. (This is equivalent to $a \in B_\phi$ in this case.)*

Of course $(T^k, X, \phi, N)$ will be our closed $K(\pi, 1)$-manifold.

**2.** Out$(\pi; \boldsymbol{Z}^k)$. In §1 we have described our nice class of spaces: injective Seifert fiber spaces. In this section we describe how one computes, for $a: 1 \to \boldsymbol{Z}^k \to \pi \to N \to 1$, $\mathrm{Out}(\pi; \boldsymbol{Z}^k)$, the outer automorphisms of $\pi$ that are invariant on $\boldsymbol{Z}^k$ (that is, the automorphisms of $\pi$ that leave $\boldsymbol{Z}^k$ invariant modulo all inner automorphisms).

Let Aut($a$) denote the group of automorphisms $\Phi$ of $N$ so that there exists a $g \in \mathrm{GL}(k, \boldsymbol{Z})$ with $\Phi^*(a) = g_*(a)$. $\mathrm{Inn}(N) \subset \mathrm{Aut(a)}$, and define $\mathrm{Out(a)} = \mathrm{Aut(a)}/\mathrm{Inn}(N)$. Let $K_1$ denote those automorphisms of $\pi$ which induce the identity on $N$ and are invariant on $\boldsymbol{Z}^k$. Let $K = K_1/(K_1 \cap \mathrm{Inn}(\pi))$.

THEOREM 3. *We have the 2 exact sequences:*

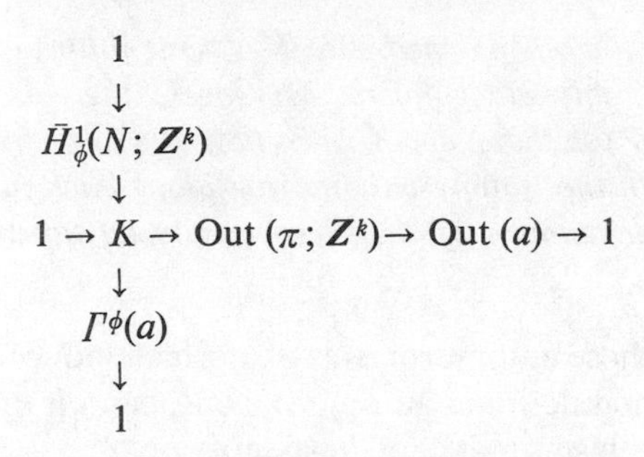


Here, $\bar{H}^1_\phi(N; \boldsymbol{Z}^k)$ is a well-defined quotient of $H^1_\phi(N; \boldsymbol{Z}^k)$ and $\Gamma^\phi(a)$ is just those elements $g$ of the centralizer of $\phi(N)$ in $\mathrm{GL}(k, \boldsymbol{Z})$ modulo the image of the center of $N$ so that $g_*(a) = a$. The proof is computational [**C-R-1**, §6].

2.2. $\mathrm{Out}(\pi; \boldsymbol{Z}^k)$ is the full outer automorphism group $\mathrm{Out}(\pi)$, whenever $\boldsymbol{Z}^k$ is characteristic. For example, $\boldsymbol{Z}^k$ could be the center of $\pi$ or $N$ has no abelian normal subgroups.

2.3. EXAMPLE. Let $N = \boldsymbol{Z}$ and $\phi = \phi(1) \in \mathrm{GL}(k, \boldsymbol{Z})$. Let $f(x)$ be the characteristic polynomial of $\phi$. Suppose

(1) all roots are distinct and not complex roots of unity,

(2) $f(x)$ is irreducible over $Q$,

(3) $f(x) \not\equiv (x^n f(1/x))/f(0)$.

Then,

$$\mathrm{Out}(\pi) \cong H^1_\phi(\boldsymbol{Z}; \boldsymbol{Z}^k) \circ \Gamma^\phi(a).$$

Here $\bar{H}^1_\phi(\boldsymbol{Z}; \boldsymbol{Z}^k) = H^1_\phi(\boldsymbol{Z}; \boldsymbol{Z}^k) = \boldsymbol{Z}^k/(I - \Phi)\boldsymbol{Z}^k$, a finite abelian group whose order is $|f(1)|$, $\Gamma^\phi(a)$ is a finitely generated abelian group isomorphic to the centralizer of $\phi$ in $\mathrm{GL}(k, \boldsymbol{Z})$ modulo the cyclic group generated by $\phi$, $\mathrm{Out}(a) = 1$, and $\mathrm{Out}(\pi; \boldsymbol{Z}^k) = \mathrm{Out}(\pi)$. The semidirect product structure is the obvious one. In this example one can even show that $\mathrm{Out}(\pi)$ lifts back into $\mathscr{H}(X)$.

Here are some explicit cases. Let $k = 2$, $\Phi = \left(\begin{smallmatrix} 0 & 1 \\ 1 & 1 \end{smallmatrix}\right)$; then $\mathrm{Out}(\pi) \cong \boldsymbol{Z}/2\boldsymbol{Z}$.

Let $k = 2$, $\Phi = \left(\begin{smallmatrix} 1 & 2s \\ 0 & 1 \end{smallmatrix}\right)$, then $\mathrm{Out}(\pi) \cong (\boldsymbol{Z}/s\boldsymbol{Z} \oplus \boldsymbol{Z}/s\boldsymbol{Z}) \circ \mathrm{GL}(2, \boldsymbol{Z})$. In this case $\Gamma^\phi(a) = 0$, and $\mathrm{Out}(a) = \mathrm{GL}(2, \boldsymbol{Z})$. (Eigenvalues are 1.)

**3. The group of fiber preserving homeomorphisms of $(T^k, X, \phi, N)$.**

3.1. Let $a \in B_\phi$ and on $X'$ define $G'(T^k, X', \phi, N)$ by the group of all homeomorphisms $h: X' \to X'$ so that there exists $g \in \mathrm{GL}(k, \boldsymbol{Z})$ such that

(1) $h(tx') = g(t)\, h(x')$,

(2) $h(x' \alpha) = h(x') \cdot H_*(\alpha)$,

for some automorphism $H_*$ of $B$. Condition (2) implies that $h$ induces a homeomorphism $H$ on $X$. Denote this homomorphism $h \to H$ also by $\nu'$ and its image by $G(T^k, X, \phi, N)$. The kernel of $\nu'$ is $N$. $G(T^k, X, \phi, N)$ is called the *group of fiber preserving homeomorphisms of the injective Seifert fiber space* $(T^k, X, \phi, N)$.

Notice that for any $f \in \mathscr{E}(X)$, there is an isomorphism $f_*: \pi_1(X, x) \to \pi_1(X, f(x))$. By choosing a path from $x$ to $f(x)$, $f_*$ induces an automorphism which is unique up to an inner automorphism. Therefore we get a map $\Psi: \mathscr{E}(X) \to \mathrm{Out}\pi_1(X)$ which factors through $\pi_0(\mathscr{E}(X), 1_X)$.

THEOREM 4. *Let $(T^k, X, \phi, N)$ be an injective Seifert fiber space corresponding to a Bieberbach class $a \in B \subseteq H^2_\phi(N; \boldsymbol{Z}^k)$. There exists a subgroup $G_1$ of the fiber preserv-*

*ing homeomorphisms which is mapped onto* $K \subseteq \operatorname{Out} \pi$ *by* $\Psi \circ i$. *Each element* $h$ *of* $G_1$ *moves points in* $X$ *only along fibers. Moreover, there is a subgroup* $G_0 \subset G_1$ *mapping onto* $\bar{H}^1_\varphi(N; \boldsymbol{Z}^k)$ *so that* $h$ *and* $h_1$ *in* $G_0$ *have the same image if and only if they are isotopic in* $G_0$ *(with the isotopy only moving along each fiber). If* $(W, N)$ *is differentiable, then all ingredients of the theorem can be taken in the differentiable category.*

Note $K$, which are those automorphisms of $\pi$ which induce the identity on $N$ and are invariant on $\boldsymbol{Z}^k$ modulo inner automorphisms, can *always* be realized by fiber preserving homeomorphisms inducing the identity on $Y$.

3.2. To realize the full outer automorphism group by $G(T^k, X, \phi, N)$ one must know something about $(W, N)$. If $\theta\colon N \to N$ is an automorphism of $N$, then we may define a new action of $N$ on $W$, $(W, N; \theta)$, by $w * \alpha = w \cdot \theta(\alpha)$, where $\cdot$ denotes the old action.

THEOREM 5. *Let* $\bar{\theta} \in \operatorname{Out}(a) \subset \operatorname{Out}(N)$, *then* $\bar{\theta}$ *lies in the image of the composite* $G(T^k, X, \phi, N) \to \operatorname{Out}(\pi) \to \operatorname{Out}(a)$, *if and only if, for some representative automorphism* $\theta$, *the action* $(W, N)$ *is equivalent to the action* $(W, N; \theta)$.

Of course, if $\boldsymbol{Z}^k$ is characteristic in $\pi$, $\operatorname{Out}(\pi; \boldsymbol{Z}^k) = \operatorname{Out} \pi$. It is also easy to check that if $(W, N; \theta)$ is equivalent to $(W, N)$ for some one representative $\theta$ of $\bar{\theta}$, it is equivalent for each representative of $\bar{\theta}$.

3.3. DEFINITION. An action (resp. smooth action) $(W, N)$ is *rigid* (resp. *smoothly rigid*) if $(W, N; \theta)$ is equivariantly homeomorphic (resp. equivariantly diffeomorphic) to $(W, N)$ for each automorphism $\theta$ of $N$.

COROLLARY 1. *If* $(W, N)$ *is rigid (resp. smoothly rigid) then the homomorphism of fiber preserving automorphisms* $\Psi \circ i\colon G(T^k, X, \phi, M) \to \operatorname{Out}(\pi; \boldsymbol{Z}^k)$ *is surjective.*

3.4. EXAMPLES. (1) Let $N$ be a uniform discrete subgroup of the Euclidean group $E(n) = \boldsymbol{R}^n \circ O(n)$. That is, $N$ is a cyrstallographic group. $(\boldsymbol{R}^n, N)$ is smoothly rigid.

(2) Let $N$ be discrete in $G$, semisimple, connected, centerless Lie group without compact and 3-dimensional factors so that $G/N$ has finite volume. Then $(W, N) = (K\backslash G, N) = (\boldsymbol{R}^n, N)$ is smoothly rigid, where $K$ is a maximal compact subgroup.

(3) If $N$ is uniform discrete in a simply connected solvable Lie group $G$, then $(G, N) = (W, N) \approx (\boldsymbol{R}^n, N)$ is smoothly rigid. These last two results are direct consequences of the rigidity theorems of G. D. Mostow.

(4) Let $(W, N) = (\boldsymbol{R}^2, N)$, where $N$ is any properly discontinuous action of orientation preserving homeomorphisms with compact quotient. Then $(\boldsymbol{R}^2, N)$ is topologically equivalent to an action of a uniform discrete subgroup of $E(2)$ or $\mathrm{PSL}(2, \boldsymbol{R})$. These latter actions are known to be smoothly rigid. It is this case when $k = 1$ that yields the fact: *A homotopy equivalence between two classical Seifert fiberings can be deformed to a fiber preserving diffeomorphism* provided that the fundamental groups are *not finite nor cyclic.* Where $k = 2$, we are dealing essentially with elliptic surfaces. There are also interesting generalizations to the nonorientable, noncompact, and/or boundary cases.

(5) Let $N$ be any finite group and $W$ be a point. Then $a \in B_\phi \subset H^2_\phi(N; \boldsymbol{Z}^k)$ represents a flat, closed $k$-dimensional manifold. Trivially, (point, $N$) is smoothly rigid. Since $\boldsymbol{Z}^k$ is characteristic in $\pi_1(X)$, for $(a)$ representing $(T^k, X, \phi, N)$, every

automorphism of $\pi_1(X)$ can be realized by an element of $G(T^k, X, \phi, N)$. It is also not hard to see that $G(T^k, X, \phi, N)$ consists of the *affine diffeomorphisms* of $X$ with respect to the usual Riemannian connection. This yields an alternative proof of a theorem of Charlap and Vasquez.

3.5. The proofs of Theorems 3, 4, 5 and Corollary 1 are the main results of **[C-R-1]**. They depend very essentially upon the cohomological formulation of the Seifert fiberings. It turns out that the geometric problem of realizing the elements of $\mathrm{Out}(\pi; \boldsymbol{Z}^k)$ by the elements of $G(T^k, X, \phi, N)$ is faithfully represented by the purely algebraic and computational proof of Theorem 3.

(Corollary 1 represents an overkill. There are many instances of the condition of Theorem 5 holding without the full action $(W, N)$ being rigid.)

In examples (1), (2), (4) and (5), $\mathrm{Out}(\pi) = \mathrm{Out}(\pi; \boldsymbol{Z}^k)$.

**4. Aspherical manifolds.**

4.1. A *closed* connected manifold which is also a $K(\pi, 1)$ is called *aspherical.* If we look at the examples of the last section (make $N$ uniform in example (2)) all of these Seifert fiber spaces $(T^k, X, \phi, N)$ are aspherical manifolds. They represent the only possible homotopy types when $\pi$ contains $\boldsymbol{Z}^k$ as a normal subgroup and has $N$ as a quotient group. It is unknown whether there exist nonhomeomorphic aspherical manifolds with isomorphic fundamental groups.

Since $\Psi_*: \pi_0(\mathscr{E}(X), 1_X) \to \mathrm{Out}(\pi_1(X))$ is an isomorphism for any $K(\pi, 1)$-space, Corollary 1 says that every homotopy equivalence can be deformed to fiber preserving diffeomorphism for our particular examples.

We promised earlier some theorems about domination. Since $\phi$: $N \to \mathrm{GL}(k, \boldsymbol{Z}) \equiv \mathrm{Aut}(T^k)$ is a homomorphism, put $G = H^0_\phi(N; T^k)$. This is a closed subgroup of $T^k$. The connected component of the identity $G_0$ of $G$ is isomorphic to a subtorus $T^s$ and we can regard $G = G_0 \times G/G_0 \subset T^s \times T^{k-s}$. For convenience, assume that (kernel $\phi$) $\cap$ center$(N) = e$, and let $(T^k, X, \phi, N)$ be an injective Seifert fiber space. With these assumptions, $H^1_\phi(N; \boldsymbol{Z}^k) = \bar{H}^1_\phi(N; \boldsymbol{Z}^k)$ and the torsion subgroup is isomorphic to $G/G_0$.

THEOREM 6. *$G \subset G(T^k, X, \phi, N)$ and $G_0 = T^s$ acts effectively as an injective $T^s$ action on $X$ so that image* $(\mathrm{ev}^X_* \ \pi_1(T^s, 1))$ *is the entire center of $\pi$. Moreover, $\Psi \circ i(G) = G/G_0 =$ torsion subgroup of $H^1_\phi\ (N; \boldsymbol{Z}^k)$.*

Notice how the theorem enables one to realize certain finite subgroups of the outer automorphism group by finite groups of homeomorphisms of $X$, and that the center of $\pi$ is hit by the largest possible injective toral action. (For aspherical manifolds the only possible connected compact Lie subgroups of $\mathscr{H}(X)$ are injective toral actions and we see that in our case the largest possible toral action does actually occur.)

4.2. Let $\mathscr{E}_0(X)$ and $\mathscr{H}_0(X)$ denote the corresponding connected components of the identity $1_X$ in $\mathscr{E}(X)$ and $\mathscr{H}(X)$ respectively. Let $X$, having the homotopy type of a finite complex, be a $K(\pi, 1)$-space and an injective Seifert fiber space with (kernel $\phi$) $\cap$ center$(N) = e$.

COROLLARY 2. *$i: \mathscr{H}_0(X) \to \mathscr{E}_0(X)$ is a domination. Moreover, if $\Psi \circ i: \mathscr{H}(X) \to \mathrm{Out}(\pi)$ is surjective, then $i: \mathscr{H}(X) \to \mathscr{E}(X)$ is a domination.*

The point there is that $\mathscr{E}(X)$ has the homotopy type of $T^s \times \mathrm{Out}(\pi_1(X))$ where

$s$ = rank of the center of $\pi_1$ by virtue of D. Gottlieb's theorem. Now, apply Theorem 6.

## References

**C-R-1.** P. E. Conner and Frank Raymond, *Deforming homotopy equivalences to homeomorphisms in aspherical manifolds*, Bull. Amer. Math. Soc. **83** (1977), 36–85.

**C-R-2.** ———, *Actions of compact Lie groups on aspherical manifolds*, Topology of Manifolds, edited by J. C. Cantrell and C. H. Edwards, Jr., Markham, Chicago, Ill., 1969, pp. 227–264. MR **42** #6839.

**C-R-3.** ———, *Injective actions of the toral groups*, Topology **10** (1971), 283–296. MR **43** #6937.

**C-R-4.** ———, *Holomorphic Seifert fiberings* (Proc. of the 2nd Conf. on Compact Transformation Groups. Part II), Lecture Notes in Math., vol. 299, Springer-Verlag, Berlin and New York, 1972, pp. 124–204.

LOUISIANA STATE UNIVERSITY

THE UNIVERSITY OF TEXAS

THE UNIVERSITY OF MICHIGAN

Proceedings of Symposia in Pure Mathematics
Volume 32, 1978

# GROUP ACTIONS, VECTOR FIELDS, AND CHARACTERISTIC NUMBERS

FRANCIS X. CONNOLLY

In a well-known set of papers, Bott [**5**], [**6**] and then Atiyah and Singer [**1**, Chapter 8] and also Baum and Cheeger [**2**] obtain formulas for the characteristic numbers of the tangent bundle of a compact Riemannian or complex manifold in terms of the zeros of some given Killing or holomorphic vector field.

Here we provide a very brief proof, in a more general setting, of such formulas, which uses no analysis and allows for any bundle; thus, it applies to $S^1$ actions on topological manifolds.

We get the above-mentioned formulae at once in the Riemannian and Hermitian case. (Our approach does not yield the full holomorphic case, however.) This method is generally more flexible than [**5**]; it can be used to get obstructions to extending $Z/k$ actions to circle actions, and we use it in §III to study $S^1$ actions on $CP^n$. The general setting will be a compact manifold $M^n$ with an $S^1$ action, together with some equivariant bundle $\xi$ on $M$. Given a characteristic class, $C(\xi)$ in $H^n(M; Q)$, and given any invariant submanifold $N$ containing the fixed set, we show how to compute $\langle C(\xi), [M] \rangle$ in terms of the behavior of the action near $N$. This result, Theorem 1.4, is our main theorem.

We then show how the result of Bott follows when $\xi = \tau(M)$ and $N =$ the fixed set.

As an application of these ideas we turn to $S^1$ actions on $CP^n$. Here Ted Petrie has given formulas relating the weights of the fixed point representations and certain more "globally" defined integers [**8**]. Our main result allows us to reprove and improve these relations. See 3.2.

All homology shall be with $Q$ coefficients until we mention otherwise.

**I.** Given a compact Lie group $G$ with universal bundle $EG \to^{\pi} BG$, we employ the functor $X \mapsto X_G = X \times_G EG$ from $G$ spaces and $G$ maps to spaces and maps.

---

*AMS (MOS) subject classifications* (1970). Primary 57D20, 57E15.

See [3] or [7]. We write $i_X\colon X \to X_G$ for the fiber inclusion. Given a $G$ equivariant bundle on $X$ (complex, real, or a microbundle) say $p\colon E \to X$, then $p_G\colon E_G \to X_G$ is a bundle with the same fiber (also complex, real, or micro). Note $i_X^* E_G = E$.

If $G = S^1$ and $G$ acts trivially on $F$, then $F_G = F \times BS^1$ so that $H^*(F_G) = H^*(F) \otimes Q[t] = H^*(F)[t]$ where $t \in H^2(BS^1)$ is the first Chern class of $\zeta$, the canonical line bundle on $BS^1$. Note that $i_F^*\colon H^*(F)[t] \to H^*(F)$ is given by:

$$i_F^*(\phi(t)) = \phi(0). \tag{1.1}$$

(1.2). Let $G = S^1$. Assume $G$ acts on the compact oriented manifold $M^n$ with fixed set $F$. Assume $N \subset M$ is $G$-invariant, contains $F$, and is a compact oriented $Q$-homology manifold. Thus we could take $N = F$ (see [3, Chapter 5, Theorem 3.2]). Assume further that $N$ is a $G$-deformation retract of some $G$-invariant neighborhood $U$. The following facts, well known in the $C^\infty$ case, are proved in the Appendix:

(1.3). (a) There are "Thom isomorphisms" $\Phi\colon H^*(N) \to H^*(M, M - N)$ and $\Phi_G\colon H^*(N_G) \to H^*(M_G, (M - N)_G)$ such that $i_M^* \Phi_G = \Phi i_F^*$. Similarly in homology. (One uses here the excision $(U, U - N) \approx (M, M - N)$.)

(b) In homology, $\Phi[M] = [N]$.

(c) If $S\colon N \to (M, M - N)$ is inclusion, define the Euler class $\chi_G = S_G^* \Phi_G(1) \in H^*(N_G)$. Then $S_G^* \Phi_G(a) = a \cup \chi_G$ for all $a \in H^*(F_G)$.

(d) There is a Gysin exact sequence

$$H^i((U - N)_G) \leftarrow H^i(N_G) \xleftarrow{\cup \chi_G} H^{i-q}(N_G) \leftarrow H^{i-1}((U-N)_G)$$

where $q$ = codimension of $N$ in $M$; $q$ may take different values on different components of $N$.

In the above and in what follows a bundle shall be understood to be a microbundle, a real bundle or a complex bundle; a characteristic class for it shall thus mean a class in $H^*(B\,\mathrm{Top}(m))$, $H^*(BSO(m))$ or $H^*(BU(m))$, etc.

Here is the main theorem:

(1.4). THEOREM. *Let $G = S^1$ act on the compact, oriented manifold $M^n$. Let $N \supset F$ be an invariant $Q$-homology submanifold satisfying* (1.2). *Let $\xi^m$ be an $S^1$-equivariant bundle on $M$ and $C(\xi) \in H^n(M)$ some characteristic class for $\xi$. Let $\chi_G$ be as in* (1.3)(c). *Then*

$$\langle C(\xi), [M] \rangle = \langle i_N^*\{C((\xi|_N)_G)/\chi_G\}, [N] \rangle. \tag{1.5}$$

PROOF. We first claim that $H^i((M - N)_G) = H^{i-1}((U - N)_G) = 0$ for $i \geqq n$. To see this set $X = M - N$ or $U - N$. The natural map $p\colon X_G \to X/G$ satisfies: $p^{-1}(Gx) = BG_x$ for $x \in X$, and $G_x \neq S^1$. So $G_x$ is finite. Therefore $BG_x$ is acyclic over $Q$. So $p$ is a homology equivalence by the Vietoris-Begle theorem. Now $X/G$ is $n - 1$ dimensional so $H^i(X_G) = 0$ if $i \geqq n$. In case $X = U - N$ we note that no component of $U - N$ is compact so $H^n(U - N) = 0$. The principal $S^1$ bundle: $G \to (U - N) \times E \to (U - N)_G$ then has a Gysin sequence

$$H^{n+1}((U-N)_G) \leftarrow H^{n-1}((U-N)_G) \leftarrow H^n((U-N) \times EG)$$

from which we see at once that $H^{n-1}((U - N)_G) = 0$. This proves the claim.

From the claim we see at once, by exactness, that, if $j$ denotes inclusion

$H^n(M_G) \xleftarrow{j_G} H^n(M_G, (M-N)_G)$ is onto, and $H^n(N_G) \xleftarrow{\cup\chi_G} H^{n-q}(N_G)$ is an isomorphism. This gives precise meaning to division by $\chi_G$ in the right side of (1.4).

Now consider the following commuting diagram:

$$\begin{array}{ccccc} H^n(M) & \xleftarrow{i_M^*} & H^n(M_G) & \xleftarrow{j_G^*} & H^n(M_G,(M-N)_G) \\ & & \downarrow S_0^* \quad S_G^* \swarrow & & \uparrow \Phi_G \\ & & H^n(N_G) \xleftarrow{\cup\chi_G} H^{n-q}(N_G) & \xrightarrow{i_N^*} & H^{n-q}(N) \end{array} \tag{1.6}$$

Here $S_0$ denotes inclusion. Since $j_G^*$ is onto we write $j_G^{*^{-1}}(x)$ for any $y$ such that $j_G^* y = x$. We have:

$$\begin{aligned} \langle C(\xi),[M]\rangle &= \langle C(i_M^*\xi_G), [M]\rangle = \langle C(\xi_G), i_{M*}[M]\rangle \\ &= \langle \Phi_G\{\Phi_G^{-1} j_G^{*^{-1}} C(\xi_G)\}, (j_G i_M)_*[M]\rangle \\ &= \langle \Phi_G\{S_0^* C(\xi_G)/\chi_G\}, (j_G i_M)_*[M]\rangle \quad \text{by (1.6)} \\ &= \langle \{C(\xi_G|N_G)/\chi_G\}, \Phi_G(j_G i_M)_*[M]\rangle \quad \text{by (1.6)} \\ &= \langle \{C((\xi|_N)_G)/\chi_G\}, i_{N*}[N]\rangle \quad \text{by 1.3(a) and (b)} \\ &= \langle i_N^*\{C((\xi|_N)_G)/\chi_G\}, [N]\rangle. \quad \text{Done} \end{aligned}$$

**II.** We now show how (1.4) leads immediately to the formula of Bott. In this section we assume $S^1$ acts in a smooth (resp. almost complex) fashion on the smooth (resp. almost complex) manifold $M^{2n}$, with $N = F$ and $\xi = TM$. Put an invariant metric on $TM$. The characteristic class $C$ can be specified by a symmetric polynomial $f(X_1, \cdots, X_n)$. Hence if $\xi_i$ are oriented 2-plane bundles (= complex line bundles since $U(1) = SO(2)$), $C(\xi_1 \oplus \cdots \oplus \xi_n) = f(y_1, y_2, \cdots, y_n)$ where $y_i$ is the first Chern class of $\xi_i$. Cf. [**4**].

We use the splitting principle to rewrite (1.5) in terms of symmetric polynomials.

Let $\nu$ denote the normal bundle of $F$ in $M$. $\nu$ has a complex structure relative to which the $S^1$ action on each fiber of $\nu$ splits into one dimensional representations specified by weights $\alpha_1, \alpha_2, \cdots, \alpha_q$, $q = \dim_C \nu$. These $\alpha_i$ are only well defined up to sign in the real case.

Let $P \to^{\pi} F$ be the "flag bundle" associated to $TM|_F = TF \oplus \nu$, whose fiber over each $x \in F$ consists of all sequences $(\eta_1, \eta_2, \cdots, \eta_{n-q}, \xi_1, \cdots, \xi_q)$ of mutually orthogonal oriented 2-planes (resp. complex lines) such that $\eta_i \subset TF_x$ and $\xi_i \subset \nu_x$, $\xi_i$ is $S^1$ invariant with weight $\alpha_i$. We therefore obtain a canonical splitting:

$$\pi^*(TM|_F) = \eta_1 \oplus \cdots \oplus \eta_{n-q} \oplus \xi_1 \oplus \cdots \oplus \xi_q.$$

Now $P$ is a trivial $G$ space, $\pi^*(TM|_F)$ is a $G$-bundle, so we form $\pi_G: P_G \to F_G$ and get:

$$\pi_G^*((TM|_F)_G) = (\eta_1)_G \oplus \cdots \oplus (\eta_{n-q})_G \oplus (\xi_1)_G \oplus \cdots \oplus (\xi_q)_G.$$

By the splitting principle $\pi_G^*: H^*(F_G) \to H^*(P_G)$ is injective [**4**, pp. 485–487]. Each $\eta_i, \xi_i$ is a complex line bundle, so set $x_i = C_1(\eta_i)$, $y_i = C_1(\xi_i)$.

Now $P_G = P \times BG$, $E(\eta_i)_G = E(\eta_i) \times BG$ so in $H^*(P_G) = H^*(P)[t]$ we compute $C_1((\eta_i)_G) = x_i$.

However, $E(\xi_i)_G = E(\xi_i) \times_{S^1} ES^1 = (E(\xi_i) \otimes_C C) \times_{S^1} ES^1$ where $S^1$ acts on the first factor by $(v \otimes z)t = v \otimes zt^{\alpha_i}$. So

$$E(\xi_i)_G = E(\xi_i) \hat{\otimes} (C \times_{S^1} ES^1) \qquad \text{(exterior tensor products)}$$
$$= E(\xi_i) \hat{\otimes} E(\xi^{\alpha_i}), \qquad \zeta = \text{canonical bundle.}$$

So we compute:

$$C_1((\xi_i)_G) = C_1(\xi_i \hat{\otimes} \zeta^{\alpha_i}) = C_1(\xi_i) + \alpha_i C_1(\zeta) = y_i + \alpha_i t \in H^*(P)[t]. \tag{2.1}$$

Hence we get:

$$C(\pi_G^*(TM|_F)_G) = f(x_1, \cdots, x_{n-q}, y_1 + \alpha_1 t, \cdots, y_q + \alpha_q t). \tag{2.2}$$

The Euler class $\chi_G = \chi(\nu_G)$ has symmetric polynomial $Y_1 \cdot Y_2 \cdots Y_q$, so if we view $\pi_G^*: H^*(F)[t] \to H^*(P)[t]$ as an inclusion map we are entitled to write:

$$C(TM|_F)_G / \chi_G = f(x_1, \cdots, x_{n-q}, y_1 + \alpha_1 t, \cdots, y_q + \alpha_q t) \cdot \prod_j (y_j + \alpha_j T)^{-1} \tag{2.3}$$

in $H^*(F)[t]$.

Let $u \in H^0(F)$ be any element. Let $\hat{\imath}_u: H^*(F)[t] \to H^*(F)$ be $\hat{\imath}_u(\varphi(t)) = \varphi(u)$. So $\hat{\imath}_0 = i_F^*$. But if $\varphi(t) \in \sum_{i=0}^{2(n-q)} H^i(F_G)$ we note that $\langle \hat{\imath}_u \varphi(t), [F] \rangle = \langle i_F^* \varphi(t), [F] \rangle$ (independent of $u$!). Moreover, $\hat{\imath}_u \chi_G = \prod_{j=1}^q (y_j + \alpha_j u)$ is invertible if $u$ is a unit. So, putting (2.3) and (1.5) together we get:

$$\langle C(TM), [M] \rangle = \langle f(x_1, \cdots, x_{n-q}, y_1 + \alpha_1 u, \cdots, y_q + \alpha_q u) \prod (y_j + \alpha_j u)^{-1}, [F] \rangle \tag{2.4}$$

for any unit $u \in H^0(F)$. Here $q$ and the $\alpha_j$ vary on the different components of $F$. This is the formula which appears in [**2**] as 2.3 if $u = 1$ and in [**1**, 8.11, 8.13], if $u = \sqrt{-1}$. However, we have only proved it, at the moment, for compact 1-parameter groups. The transition to arbitrary Killing vector fields is easily managed as follows:

(2.5). Let $Y$ be any Killing vector field on $M^{2n}$ where $M$ is a compact oriented Riemannian (resp. Hermitian) manifold. If $M$ is almost complex, assume also that $Y$ preserves this almost complex structure. Let $T$ denote the closure of the one parameter group of $Y$ in $I(M)$, the isometry group of $M$. $I(M)$ is a compact Lie group [**8**] so $T$ is a torus, and its fixed set $F$ is the set of zeroes of $Y$. One can choose a complex structure for the normal bundle $\nu$ of $F$ in $M$, relative to which the representation of $T$ on $\nu$ is a sum of one dimensional representations. The weights of these are homomorphisms $\alpha_j: t \to R$, where $t$ is the Lie algebra of $T$. Let $t_0 = \{X \in t: \forall j, \alpha_j(X) \neq 0\}$. Note $Y \in t_0$. Define a continuous function $\varphi_u: t_0 \to \boldsymbol{R}$ for each unit $u \in H^0(F)$ by:

$$\varphi_u(X) = \Big\langle f(x_1, \cdots, x_{n-q}, y_1 + u\alpha_1(X), \ldots, y_q + u\alpha_q(X)) \cdot \prod_{j=1}^{q} (y_j + u\alpha_j(X))^{-1}, [F] \Big\rangle \tag{2.6}$$

where the $x$'s and $y$'s are as before.

As explained above, this is independent of the unit $u$. But clearly $\varphi_u(tX) = \varphi_{tu}(X) = \varphi_u(X)$ if $t \neq 0$. By (1.5) and (2.4) we conclude, therefore, that $\langle C(TM), [M] \rangle = \varphi_u(X)$ if $X$ is any vector of $t_0$ tangent to a circle subgroup $G$ of $T$ for which $F$ is the fixed set of $G$. But such vectors $X$ are dense in $t_0$. So $\varphi_u(X)$ is constant in $u$ and $X$. We conclude:

(2.7). COROLLARY. *Let $Y$, $M^{2n}$, $F$, $C$, $f$ be as in* (2.5). *Then* $\langle C(TM), [M]\rangle = \varphi_u(Y)$ *where* $u \in H^0(F)$ *is any unit and* $\varphi_u$ *is defined by* (2.6).

**III. Applications to $S^1$ actions on $CP^n$.** Let $M^{2n}$ be a smooth manifold with a homotopy equivalence $f: M \to CP^n$. Assume $S^1$ acts on $M$ with isolated fixed points. There are always $n+1$ of these: $p_0 \cdots p_n$. In **[8]** Petrie observes that if $\eta = f^*\zeta$ ($\zeta$ = canonical line bundle on $CP^n$) then the action of $S^1$ on $M$ lifts to an $S^1$ action on $\eta$ making $\eta$ an equivariant complex bundle. So $\eta|_{p_i}$ is a 1-dimensional representation of $S^1$ with weight $a_i$, say. An alteration of the lifting to a new $S^1$ action on $\eta$ changes these $a_i$ to $a_i - \theta$, for some $\theta \in Z$, and each $\theta \in Z$ can be so obtained. Here Petrie relies on work of Stewart **[10]**.

Now assume $S^1$ acts smoothly. $S^1$ acts on the tangent space of $M$ at $p_i$ with weights $x_{ij}$, $j = 1, \cdots, n$. For linear actions of $S^1$ on $CP^n$ it turns out that $\{x_{ij}: j = 1, \cdots, n\} = \{a_j - a_i : j \neq i\}$. For exotic actions, this is false. However, it is still possible to obtain relations between the $x_{ij}$ and the $a_j - a_i$. Petrie proves that

$$\prod_{j=1}^{n} x_{ij} = \prod_{j\neq i} (a_i - a_j) \tag{3.1}$$

(see **[8**, Part II, §2, pp. 137–139]).

We first of all show how our theorem provides a simple proof of (3.1). We may as well assume $i = 0$. In (1.4) we let $\xi = n\eta = \sum_{j=1}^{n} \eta_j$ where $\eta_j$ is given the $S^1$ action such that at $p_i$ its weight is $a_i - a_j$. Then we get by (1.5):

$$\begin{aligned} 1 &= \langle C_n(\xi), [M]\rangle = \langle C_n(\xi_G)/\chi_G, [F]\rangle \\ &= \sum_{i=1}^{n} \left\langle \prod_{j=1}^{n} (C_1(\eta_j)_G) \Big/ \chi_G, [p_i] \right\rangle = \sum_{i=0}^{n} \prod_{j=1}^{n} (a_i - a_j)t \Big/ \prod_{j=1}^{n} x_{0j}\, t \\ &= \prod_{j=1}^{n} (a_0 - a_j) \Big/ \prod_{j=1}^{n} x_{0j}. \end{aligned}$$

This proves (3.1). Now we show how to strengthen this to the following:

(3.2). PROPOSITION. *Suppose $M$ as above is tangentially equivalent to $CP^n$. Then*

$$\sum_{i=0}^{n} f(x_{i1}, \cdots, x_{in}) = \sum_{i=0}^{n} f(a_i - a_0, \cdots, a_i - a_n) \tag{3.3}$$

*for any function $f$ such that $f(X_1, X_2, \cdots, X_n) \cdot X_1 \cdots X_n$ is a symmetric polynomial of degree $n$ in the $X_i$.*

PROOF. Let $g(X_1, \cdots, X_n)$ be a symmetric polynomial of degree $n$. Let $C$ be the characteristic class corresponding to $g$. Now:

$$\langle C(\tau M), [M]\rangle = \langle C((n+1)\eta), [M]\rangle \tag{3.4}$$

since $M \approx CP^n$. Let $\eta_j$ denote $\eta$ with the $S^1$ structure whose weight at $p_i$ is $a_i - a_j$. Proceeding as above we note:

$$\langle C(\tau M), [M]\rangle \sum_{i=1}^{n} \langle C(\tau M_G|(p_i)_G)/\chi_G, [p_i]\rangle = \sum_{i=0}^{n} g(x_{i1}, \cdots, x_{in}) \Big/ \prod_{j=1}^{n} x_{ij}.$$

On the other hand, in the same way,

$$\langle C((n+1)\eta), [M] \rangle = \sum_{i=0}^{n} g(a_i - a_0, \cdots, a_i - a_n) \Big/ \prod_{j=1}^{n} x_{ij}$$

and by (3.1), this is $= \sum_{i=0}^{n} g(a_i - a_0, \cdots, a_i - a_n)/\prod_{j \neq i} (a_i - a_j)$. Since $f$ has the form $g/X_1 \cdots X_n$ for suitable $g$, the result now follows from (3.4).

**Appendix.** We must prove (1.3)(a), (b), (c), (d). Choose a fundamental class $[N]$ for $N$, and define $\Phi: H^*(N) \to H^*(M, M - N)$ to be the composite:

$$H^*(N) \xrightarrow{\cap [N]} H_*(N) \xrightarrow{i_*} H_*(U) \xrightarrow{\cap (U)^{-1}} H^*(U, U - N) \xrightarrow{e^{*-1}} H^*(M, M - N)$$

where $e$ and $i$ are inclusion maps. $\Phi$ is an isomorphism clearly. The corresponding map in homology is defined in the same way, so it follows that $\Phi[M] = [N]$.

Consider $\Phi(1)$ in $H^*(M, M - N)$. The Serre spectral sequence converging to $H^*(M_G, (M - N)_G)$ has $E_2^{**} = H^*(BG; H^*(M, M - N))$ and $\Phi(1)$ is a permanent cycle for dimensional reasons. $\Phi(1)$ thus determines a unique class $\Phi_G(1) \in H^*(M_G (M - N)_G)$ in lowest filtration. Let $\Phi_G: H^*(N_G) \to H^*(M_G, (M - N)_G)$ be the composite

$$H^*(N_G) \xrightarrow{r_G^*} H^*(U_G) \xrightarrow{\cup e^*\Phi_G(1)} H^*(U_G(U - N)_G) \xrightarrow{e_G^{*-1}} H^*(M_G, (M - N)_G)$$

where $r: U \to N$ is retraction and $e$ is excision. Note that $\Phi_G$ passes to a map from the Serre spectral sequence of $N_G$ over $B_G$ to the Serre spectral sequence of $(M_G, (M - N)_G)$ over $BG$. This map is an isomorphism at the $E_2$ level since it is the map of coefficients in $H^*(BG; —)$ induced by $\Phi$. Thus $\Phi_G$ is an isomorphism and by now (a), (b), and (c) are clear. The Gysin sequence is obtained as usual from the exact sequence of the pair $(U_G, (U - N)_G)$ by replacing $H^*(U_G)$ with $H^*(N_G)$ and $H^*(U_G, (U - N)_G)$ by $H^*(N_G)$.

## BIBLIOGRAPHY


1. M. Atiyah and I. Singer, *The index of elliptic operators*. III, Ann. of Math. (2) **87** (1968), 546–604. MR **38** #5245.

2. P. Baum and J. Cheeger, *Infinitesimal isometries and Pontrjagin numbers*, Topology **8** (1968), 173–194. MR **38** #6627.

3. A. Borel, *Seminar on transformation groups*, Princeton Univ. Press, Princeton, N. J., 1960. MR **22** #7129.

4. A. Borel and F. Hirzebruch, *Characteristic classes and homogeneous spaces*. I, Amer. J. Math. **80** (1958), 458–538. MR **21** #1586.

5. R. Bott, *Vector fields and characteristic numbers*, Michigan Math. J. **14** (1967), 231–244.

6. ———, *A residue formula for holomorphic vector fields*, J. Differential Geometry **1** (1967), 311–330.

7. Wu Yi Hsiang, *Cohomology of topological transformation groups*, Springer-Verlag, Berlin and New York, 1976.

8. T. Petrie, *Smooth $S^1$ actions on homotopy complex projective spaces and related topics*, Bull. Amer. Math. Soc. **78** (1972), 105–153. MR **45** #6029.

9. N. Steenrod and S. Myers, *The group of isometries of a Riemannian manifold*, Ann. of Math. (2) **40** (1939), 400–416.

10. T. Stewart, *Lifting the action of a group in a fiber bundle*, Bull. Amer. Math. Soc. **66** (1960), 129–132. MR **22** #2994.


UNIVERSITY OF NOTRE DAME

Proceedings of Symposia in Pure Mathematics
Volume 32, 1978

# INFINITE REGULAR COVERINGS AND KLEINIAN GROUPS

RAVINDRA S. KULKARNI

The following is a report on the lecture delivered at the conference. The proofs of the results stated herein are contained in the papers [**K**$_1$], [**K**$_2$], [**K**$_3$] which are in the process of publication.

We reconsider a problem of H. Hopf (cf. [**H**]) namely, find conditions on a noncompact manifold to be a covering space of a closed manifold. For retaining generality it is useful to consider the problem in a wider context: find conditions on a noncompact "space" to admit an *infinite* properly discontinuous group of homeomorphisms.

The latter problem arises in a natural way in the classical theory of "Kleinian" groups which is one type of generalization of the theory of "Fuchsian" groups. (Recall that a "Fuchsian" group is a discrete subgroup of the group of orientation-preserving isometries of the hyperbolic plane. Such a group can be represented by Möbius transformations with real coefficients. A "Kleinian" group is a discrete subgroup of Möbius transformations (with real or complex coefficients) which acts discontinuously on some nonempty open subset of the Riemann sphere.) The "set of discontinuity" of a Kleinian group—which is constructed in a special way, cf. §1 below—is an important class of 2-dimensional infinite regular covering spaces. An outstanding problem in the theory of Kleinian groups is to find its higher dimensional analogues, say, e.g., in discrete subgroups of holomorphic transformations of hermitian symmetric spaces or in discrete subgroups of conformal transformations of the standard $n$-sphere, etc., cf. [**A**$_1$], [**A**$_2$], [**B**, §V]. It turns out that the theory of ends of spaces and ends of groups explains the topological genesis of the classical theory and provides a first step towards its higher dimensional generalizations. This is discussed in §§1, 2 and 3.

There is another significant class of infinite regular covering spaces which arises

*AMS* (*MOS*) *subject classifications* (1970). Primary 30A58, 57E30.

in differential geometry and which satisfies certain "finitistic" restrictions, e.g., a noncompact connected homogeneous space of a Lie group—which is a good candidate for being an infinite regular covering space—has *finitely generated* homology; a discrete, torsionfree subgroup of $GL_n(\boldsymbol{R})$ has *finite* cohomological dimension, etc. It turns out that these types of restrictions lead to surprisingly strong conclusions. A typical example of such previously known results is the duality theorem of Milnor, cf. [**M**, §4] or infinite cyclic coverings. This is discussed in §§4, 5 and 6.

**1. Construction of the limit set and the set of discontinuity.** Let $X$ be a locally compact Hausdorff space and $\Gamma$ a group acting effectively on $X$. Let

$L_0 = L_0(\Gamma) =$ the closure of the set of points in $X$ with infinite isotropy group.

$L_1 = L_1(\Gamma) =$ the closure of the set of cluster points of $\Gamma$-orbits of points in $X - L_0(\Gamma)$.

$L_2 = L_2(\Gamma) =$ the closure of the set of cluster points of $\Gamma$-orbits of compact subsets of $X - \{L_0(\Gamma) \cup L_1(\Gamma)\}$.

The set $\Lambda = \Lambda(\Gamma) = L_0(\Gamma) \cup L_1(\Gamma) \cup L_2(\Gamma)$ is called the *limit set* of $\Gamma$. Its complement $\Omega = \Omega(\Gamma) = X - \Lambda(\Gamma)$ is the *set of discontinuity* of $\Gamma$. The group $\Gamma$ is said to have the *Kleinian property* if $\Omega \neq \varnothing$. It is not difficult to see that $\Gamma$ acts properly discontinuously on $\Omega$.

REMARK. In the theory of Kleinian groups (where $X$ is the Riemann sphere and $\Gamma$ is a group of Möbius transformations) the definition of limit set is taken to be $L_0(\Gamma)$. This is easily seen to be inadequate in the general case. Indeed it is a special feature of the conformal geometry that $L_0(\Gamma)$ happens to be $\Lambda(\Gamma)$.

From now on for the sake of simplicity and since it is adequate in our applications we shall assume further that $X$ is connected, locally connected, compact, Hausdorff and with a countable base for topology. Suppose that $\Gamma$ is a group of homeomorphisms of $X$ with the Kleinian property and let $\Omega$, $\Lambda$ be as above. Then $\bar{\Omega}$ is compact. Let $\partial\Omega = \bar{\Omega} - \Omega$ be the set-theoretic boundary of $\Omega$. We shall say that a point $p \in \partial\Omega$ is *$\Gamma$-accessible* if it is a cluster point of a $\Gamma$-orbit of a point in $\Omega$. The following is a crucial property of a group with the Kleinian property.

THEOREM 1. *Suppose that $\bar{\Omega} = X$ and $\Gamma\backslash\Omega$ is connected. Then each component of $\partial\Omega$ contains a $\Gamma$-accessible point.*

The structure of $\partial\Omega$ is described in

THEOREM 2. *Suppose that $\bar{\Omega} = X$ and $\Omega$ is connected. Then $\partial\Omega$ has either $\leqq 2$ components or else it has uncountably many components and it is a perfect set.*

**2. An extension of a theorem of Hopf.** Let $\Omega$ be an arcwise connected, locally connected, locally compact, Hausdorff space with a countable base for topology. Let $\Omega_E$ denote the end-compactification of $\Omega$ (cf. Freudenthal [**F**$_1$]) and $\varepsilon(\Omega) = \Omega_E - \Omega$ the space of ends of $\Omega$. Let $\Gamma$ be a group acting properly discontinuously on $\Omega$. It is well known that $\Gamma$ has a canonical extension to a group of homeomorphisms of $\Omega_E$. Similarly there is a notion of the set of ends of $\Gamma$ (cf. [**H**], [**F**$_2$], [**S**]). Hopf's main results in [**H**] may be reformulated as follows

THEOREM 3 (HOPF). *Suppose that $\Gamma\backslash\Omega$ is compact. Then $\#\varepsilon(\Omega) = \#\varepsilon(\Gamma)$ and $\#\varepsilon(\Omega)$ is either $\leqq 2$ or else $\varepsilon(\Omega)$ is a perfect set.*

It is worth remarking that for $\Omega$ and $X$ as in §1, Theorem 2, the space $X$ is closely related to $\Omega_E$. Indeed let $X_1$ be the quotient space of $X$ under the equivalence relation: if $x, y \in X$ then $x \sim y$ iff $x = y$ or else $x, y$ belong to the same component of $\partial\Omega$. Then there is a canonical map $\phi: \Omega_E \to X_1$ such that $\phi|\Omega$ is a homeomorphism. Indeed $\phi$ itself is often a homeomorphism, e.g., if $X$ is a compact simply connected manifold.

Hopf's main concern was the structure of universal covering spaces of compact manifolds which explains the hypothesis about compactness of $\Gamma\backslash\Omega$. As remarked earlier this hypothesis should be removed as far as possible. This is done in the following technical result. As in §1 we say that an end is $\Gamma$-accessible if it is a cluster point of a $\Gamma$-orbit of a point in $\Omega$.

THEOREM 4. *Suppose that every end of $\Omega$ is $\Gamma$-accessible.*[1] *Then*
(i) $\#\varepsilon(\Omega) \leqq 2$ *or else* $\varepsilon(\Omega)$ *is a perfect set*;
(ii) (*the end-inequality*) *if $\Gamma$ is finitely generated then* $\#\,\varepsilon(\Omega) \leqq \#\,\varepsilon(\Gamma)$.

**3. Applications to Kleinian groups.** The end-inequality coupled with Stallings' theorem on groups with infinitely many ends leads to several results of which the following two are typical. The first gives a group-theoretical structure of a Kleinian group. The second is a result in discrete groups of conformal transformations of the standard $n$-sphere. Concerning the first result it is worth noticing that its statement as well as its proof is "elementary"—one does not have to invoke any special features of the theory of Riemann surfaces!

THEOREM 5. *Let $\Omega$ be a connected surface of genus* 0. *Let $\Gamma$ be a finitely generated torsion free group acting properly discontinuously on $\Omega$. Then* $\Gamma \approx \Gamma_0 * \Gamma_1 * \cdots * \Gamma_r$, $r < \infty$ (*free product*) *where $\Gamma_0$ is a finitely generated free group and for $i > 0$ each* $\Gamma_i \approx \pi_1$ (*a compact surface*).

THEOREM 6. *Let $\Gamma$ be a finitely generated torsion free group of conformal transformations of the standard $n$-sphere such that $\Lambda(\Gamma)$ is totally disconnected. Then* $\Gamma \approx \Gamma_0 * \Gamma_1 * \cdots * \Gamma_r$, $r < \infty$ (*free product*) *where $\Gamma_0$ is a finitely generated free group and for $i > 0$ each $\Gamma_i$ is a Bieberbach group in dimension $n$.*

(Recall that Bieberbach groups in dimension $n$ are certain finite extensions of a free abelian group of rank $\leqq n$, cf. [**W** Chapter 3]. The groups $\Gamma$ of Theorem 6 may be considered as $n$-dimensional analogues of "Schottky groups" which is a well-studied class of Kleinian groups.)

**4. The case of a single $\pi_1$-stable end.** Let $\Omega$ be an arcwise connected, locally connected, locally simply connected, locally compact, Hausdorff space with a countable base for its topology and with a single end $\varepsilon$. We say that $\Omega$ is *simply connected at infinity* (notation: $\pi_1(\varepsilon) = 1$) if given a compact set $C$ there exists a compact set $D$ containing $C$ such that $\Omega - D$ is connected and simply connected. More generally one may formulate a more or less obvious but rather technical notion of the

[1] More generally we may consider the set-up in §1, Theorem 2, and replace $\#\varepsilon(\Omega)$ by $\#\{$components of $\partial\Omega\}$. Then Theorem 4 is still valid.

"fundamental group at infinity" (notation: $\pi_1(\varepsilon)$) under conditions of "$\pi_1$-stability at infinity". The procedure is analogous to that followed by Siebenmann **[Si]** so we omit the details. There is a noncanonical homomorphism $\phi: \pi_1(\varepsilon) \to \pi_1(\Omega, *)$. However the conjugacy class, normal closure and characteristic closure of $\phi(\pi_1(\varepsilon))$ in $\pi_1(\Omega, *)$ are canonically defined. (Recall that the normal resp. characteristic closure of a subset $S$ of a group is the intersection of all normal resp. characteristic subgroups containing $S$.) We denote the characteristic closure of $\phi(\pi_1(\varepsilon))$ in $\pi_1(\Omega, *)$ by $\hat{\pi}_1(\varepsilon)$.

THEOREM 7. *Let $\Omega$ be as above and suppose that $\Omega$ admits an infinite properly discontinuous group of homeomorphisms. Then $\hat{\pi}_1(\varepsilon) = \pi_1(\Omega, *)$. In particular if $\Omega$ is simply connected at infinity it is simply connected.*

EXAMPLE. An (open) Möbius band $\Omega$ does not admit an infinite properly discontinuous group of homeomorphisms—for, in this case $\hat{\pi}_1(\varepsilon)$ is of index 2 in $\pi_1(\Omega, *)$.

**5. The ring $H_c^*(\Omega, F)$.** Let $\Omega$ be a locally compact Hausdorff space, $F$ a field and $H_c^*(\Omega, F)$ the cohomology ring with compact support of $\Omega$ with coefficients in $F$. The following result is more or less well known.

THEOREM 8. *Let $\Omega$ and $F$ be as above and suppose that $\Omega$ admits a properly discontinuous infinite group $\Gamma$ of homeomorphisms.*

*Suppose moreover that either*

(1) *The set $\Gamma_0 = \{\gamma \in \Gamma \mid \gamma$ is homotopic to identity by a proper homotopy$\}$ is infinite*;

*or*

(2) *$H_c^*(\Omega, F)$ is finitely generated. Then the multiplication in $H_c^*(\Omega, F)$ is trivial.*

Here is an application of these considerations to the space form problem in differential geometry. Recall that the space form problem, in its wider sense, asks: given a Lie group $G$ and a closed subgroup $H$ classify discrete subgroups $\Gamma$ of $G$ which act freely and properly discontinuously on $G/H$. $\Gamma\backslash G/H$ is called a "space form" of $G/H$. A homogeneous space $G/H$ when $G$ and $H$ have finitely many components and $K \supset L$ resp. are maximal compact subgroups of $G \supset H$ resp. has the structure of a vector bundle on the compact manifold $K/L$, cf. Mostow **[Mo]**. This fact should explain the hypothesis of the following.

THEOREM 9. *Let $F$ be the total space of an oriented vector bundle $\xi$ on an oriented compact manifold. Then $E$ admits an infinite properly discontinuous group iff $\xi$ has a nowhere vanishing section.*

It follows in particular that certain noncompact homogeneous spaces (e.g., an open Möbius band) do not admit even "topological" space forms with infinite deck-transformation groups.

Here is a notable application which uses Theorems 7 and 8. Before stating it we recall the notion of the "cohomology at infinity" of $\Omega$ which is simply the cohomology of the complex $C^*(\Omega)/C_c^*(\Omega)$ where $C^*(\Omega)$ resp. $C_c^*(\Omega)$ is the singular cochain complex resp. singular cochain complex with compact support of $\Omega$. We denote it by $H_\varepsilon^*(\Omega)$.

THEOREM 10. *Let $\Omega$ be a connected, paracompact n-dimensional manifold with a single end $\varepsilon$. Suppose that $\Omega$ is simply connected at infinity. $H_*(\Omega, \boldsymbol{Z})$ is finitely generated and $H^*_\varepsilon(\Omega, \boldsymbol{Z}) \approx H^*(S^{n-1}, \boldsymbol{Z})$. Suppose moreover that $\Omega$ admits an infinite properly discontinuous group of homeomorphisms. Then $\Omega$ is contractible. If $n \geqq 5$ then $\Omega \approx \boldsymbol{R}^n$.*

COROLLARY. *Let $M^n$ be a compact manifold such that $M^n - \{a \text{ point}\}$ admits an infinite, properly discontinuous group of homeomorphisms. Then $M^n$ is a homotopy n-sphere. If $n \geqq 5$ it is homeomorphic to the n-sphere.*

**6. Covering spaces of finite type.** It is convenient to introduce the following.

DEFINITION. An oriented $n$-dimensional manifold $\Omega$ is said to be a covering space of finite type if (i) $\Omega$ has finitely generated homology/$\boldsymbol{Z}$. (ii) $\Omega$ is a regular covering space of a compact oriented manifold with covering group $\Gamma$. (iii) There exists an integer $d$ such that for every subgroup $\Gamma'$ of finite index in $\Gamma$ we have $H^d(\Gamma', \boldsymbol{Z}) \neq 0$ and $H^l(\Gamma', \boldsymbol{Z}) = 0$ for $l > d$.

THEOREM 11. *Let $\Omega^n$ be a covering space of finite type with $\Gamma$ and $d$ as above. Then $d \leqq n$, $H^{n-d}(\Omega, \boldsymbol{Z}) \approx \boldsymbol{Z}$ and $H^k(\Omega, \boldsymbol{Z}) = 0$ for $k > n-d$.*

Taking $\Gamma = \boldsymbol{Z}$ in Theorem 1.1 it is easy to deduce the following duality theorem due to Milnor [**M**, §4] who proved it in a different way.

THEOREM 12 (MILNOR). *Let $\Omega^n$ be an infinite cyclic covering of an oriented compact manifold such that $H^*(\Omega, \boldsymbol{Z})$ is finitely generated. Then $\Omega$ satisfies Poincaré duality in dimension $n - 1$.*

EXAMPLE. Let $\Omega$ be a tubular neighborhood of the figure 8 in $\boldsymbol{R}^n$, $n \geqq 2$. Then $\Omega$ is *not* a covering space of finite type.

It is also easy to deduce from Theorem 1.1 the following

THEOREM 13. *Let $\Omega^n$, $n \leqq 4$, be a simply connected covering space of finite type. Then $\Omega$ is either contractible or has the homotopy type of $S^r$, $2 \leqq r \leqq n - 1$.*

REMARK. There are restrictions on a group $\Gamma$ if it is to appear as a deck-transformation group in a covering space of finite type, e.g., $\Gamma$ cannot be $F$ = a non-abelian free group or $F \times \boldsymbol{Z}$.

Finally I should also like to mention some curious phenomena already hinted at by Hopf, which show a significant contrast between regular and nonregular coverings, e.g., let $\Omega^2$ be a paracompact manifold admitting an infinite, properly discontinuous group $\Gamma$ of homeomorphisms. Then $\Omega$ has genus either 0 or infinity. If moreover $H_*(\Omega, \boldsymbol{Z})$ is finitely generated then $\Omega \approx \boldsymbol{R}^2$ or $\boldsymbol{R} \times S^1$. Of these $\boldsymbol{R} \times S^1$ regularly covers *only* itself, Möbius band, torus or the Klein bottle, but nonregularly it covers *every* surface except $\boldsymbol{R}^2$, $S^2$ and the projective plane! Here is another phenomenon which follows by the methods of this section.

THEOREM 14. *Let $\Omega$ be a regular covering of a finite CW complex $B$ with a finitely presented deck-transformation group $\Gamma$. Let $A$ be a finite abelian group. Then there are only finitely many $A$-fold coverings $\tilde{\Omega}$ of $\Omega$ (up to isomorphism) which regularly cover $B$.*

(Note that by contrast not necessarily regular or connected $A$-fold coverings of $\Omega$ are classified by $H^1(\Omega, A)$ which can very well be uncountable!)

## References

[$A_1$] L. Ahlfors, *Kleinsche Gruppen in der Ebene und im Raum*, Festband 70, Geburtstag R. Nevanlinna, pp. 7–18, Springer, Berlin, 1966. MR **35** #1777.

[$A_2$] ———, *Hyperbolic motions*, Nagoya Math. J. **29** (1967), 163–166. MR **38** #2297.

[B] A. Borel, *Lex fonctions automorphes de plusiers variables complexes*, Bull. Soc. Math. France **80** (1952), 167–182. MR **14**, 1077.

[$F_1$] H. Freudenthal, *Über die Enden topologischer Räume und Gruppen*, Math. Z. **33** (1931), 692–713.

[$F_2$] ———, *Über die Enden discreter Räume und Gruppen*, Comment. Math. Helv. **17** (1945), 1–38. MR **6**, 277.

[H] H. Hopf, *Enden offener Räume und unendliche discontinuierliche Gruppen*, Comment. Math. Helv. **16** (1944), 81–100. MR **5**, 272.

[$K_1$] R. S. Kulkarni, *Some topological aspects of Kleinian groups*, Amer. J. Math. (to appear).

[$K_2$] ———, *Groups with domains of discontinuity* (to appear).

[$K_3$] ———, *Infinite regular coverings* (to appear).

[M] J. Milnor, *Infinite cyclic coverings*, Conf. on Topology of Manifolds (1967), 115–133, Prindle, Weber and Schmidt, Boston, Mass., 1968. MR **39** #3497.

[Mo] G. D. Mostow, *On covariant fiberings of Klein spaces*, Amer. J. Math. **77** (1955), 247–278. MR **16**, 795.

[Si] L. Siebenmann, Thesis, Princeton University, 1965.

[S] J. Stallings, *Group theory and 3-dimensional manifolds*, Yale Univ. Press, New Haven, Conn. 1971.

[W] J. Wolf, *Spaces of constant curvature*, third ed., Publish or Perish, Boston, Mass., 1974. MR **49** #7958.

Indiana University

BCDEFGHIJ–CM–89876543210